Lecture Notes in Networks and Systems

Volume 557

The series "Lecture Notes in Networks and Systems" publishes the latest developments in Networks and Systems—quickly, informally and with high quality. Original research reported in proceedings and post-proceedings represents the core of LNNS.

Volumes published in LNNS embrace all aspects and subfields of, as well as new challenges in, Networks and Systems.

The series contains proceedings and edited volumes in systems and networks, spanning the areas of Cyber-Physical Systems, Autonomous Systems, Sensor Networks, Control Systems, Energy Systems, Automotive Systems, Biological Systems, Vehicular Networking and Connected Vehicles, Aerospace Systems, Automation, Manufacturing, Smart Grids, Nonlinear Systems, Power Systems, Robotics, Social Systems, Economic Systems and other. Of particular value to both the contributors and the readership are the short publication timeframe and the world-wide distribution and exposure which enable both a wide and rapid dissemination of research output.

The series covers the theory, applications, and perspectives on the state of the art and future developments relevant to systems and networks, decision making, control, complex processes and related areas, as embedded in the fields of interdisciplinary and applied sciences, engineering, computer science, physics, economics, social, and life sciences, as well as the paradigms and methodologies behind them.

Indexed by SCOPUS, INSPEC, WTI Frankfurt eG, zbMATH, SCImago.

All books published in the series are submitted for consideration in Web of Science.

For proposals from Asia please contact Aninda Bose (aninda.bose@springer.com).

Abdalmuttaleb M. A. Musleh Al-Sartawi ·
Anjum Razzaque · Muhammad Mustafa Kamal
Editors

From the Internet of Things to the Internet of Ideas: The Role of Artificial Intelligence

Proceedings of EAMMIS 2022

Springer

Editors
Abdalmuttaleb M. A. Musleh Al-Sartawi
Accounting Finance and Banking Department
Ahlia University
Manama, Bahrain

Anjum Razzaque
College of Business and Finance
Ahlia University
Manama, Bahrain

Muhammad Mustafa Kamal
School of Strategy and Leadership
Coventry University
Coventry, UK

ISSN 2367-3370 ISSN 2367-3389 (electronic)
Lecture Notes in Networks and Systems
ISBN 978-3-031-17745-3 ISBN 978-3-031-17746-0 (eBook)
https://doi.org/10.1007/978-3-031-17746-0

This Springer imprint is published by the registered company Springer Nature Switzerland AG
The registered company address is: Gewerbestrasse 11, 6330 Cham, Switzerland

Foreword

Since the popularization of the Internet in the 1990s, the cyberspace has continued to expand and evolve. From the metaverse, non-fungible tokens (NFTs) to cryptocurrencies, the Internet of Things (IoT) is transforming and reshaping the economy and society at large. The applications and usage of the Internet are multifaceted and expanding on a daily basis. The Internet of Things (IoT) and Internet of Ideas (IoI) are novel approaches that have ingrained the Internet into various facets of our lives including personal, professional, and societal aspects. At the heart of the metaverse is the vision of an immersive virtual experience. The metaverse is no longer science fiction. It is catalyzed by emerging technologies such as 5G, extended reality, and artificial intelligence, making the digital cyberspace a reality. The IoT primarily embodies electrical devices that are connected to the Internet and which have helped in creating virtual ecosystems, augmented realities, and content creation.

The question here is whether IoT is sufficient to support the impending and continued growth of the digital society. Based on scholars and other professionals, the emergence of Industry 5.0 is human-centric. Therefore, it is believed that IoT needs to be more comprehensive and includes a 'bank' of social ideas, also known as Internet of Ideas (IoI), thus enabling capacity building across concepts, designs, ideas, fields, people, resources, networks, and know-how. To support interoperability and to transcend the spatial and social boundaries of the digital era, IoI is needed to interact across distances, multi-disciplines, languages, cultures, and even generations.

This conference proceedings covers contemporary digitalization concepts such innovation in the digital age and Internet of Ideas with topics related to augmented press and robot journalists, knowledge management capabilities, meta-analysis of AI research, and big data systems architecture and data security fundamentals. The book also raises important issues regarding cybersecurity, artificial intelligence, and Internet of Things addressing special topics on the use of artificial intelligence in combating crimes, and the effect of information sensitivity and usage on consumer privacy. Other topics are related to the COVID-19 pandemic and how it played a role in accelerating digitalization leading to the emergence of Industry 5.0. This book, therefore, aims to offer novel knowledge not only to researchers, but also practitioners, professionals, and policymakers.

Abdalmuttaleb M. A. Musleh Al-Sartawi
Ahlia University
Manama, Bahrain

Anjum Razzaque
Ahlia University
Manama, Bahrain

Muhammad M. Kamal
Coventry University
Coventry, UK

Preface

European, Asian, Middle Eastern, North African Conference on Management and Information Systems (EAMMIS) gets organized by Bridges Foundation in cooperation with Coventry University. EAMMIS 2022 probes a deeper academic and scholarly interpretation through its theme: From the Internet of Things to the Internet of Ideas: The role of Artificial Intelligence. EAMMIS is an international platform for researchers, academicians, practitioners, and industry professionals, to present research findings offering deeper insight into novel theories that embrace theoretical, managerial, and practical implications within the fields of management and information systems. In extensive consultation with the EAMMIS executive committee and the faculty of the organizing universities, it gets decided that it is best not to hold only a physical EAMMIS 2021 conference, given the uncertainty of current circumstances. Our organizing committee volunteers have invested ample effort in assuring the quality assurance governance of the EAMMIS 2022 conference. Many of such stakeholders are already dealing with additional workloads because of COVID-19. Such is an event for both scientific and social interactions. We believe we can achieve such a conference through a hybrid conference platform. The conference's call for papers has asked for submissions in full research papers. As a result, it attracted many submissions from 150 universities and 51 countries around the world. Submissions were reviewed in a double-blind process by at least two specialized academics in the relevant fields. The EAMMIS 2022 conference received a total of 306 submissions, and overall, 60 papers got accepted with an acceptance rate of 20%. The final set of proceedings have gotten grouped according to track, each outlined below.

The EAMMIS 2022 proceedings book gets organized into 60 chapters. Such chapters cover a strategical alignment of a wide variety of topics that get categorized within the following tracks: Track 1: Innovation in the digital era, Track 2: Sustainable technologies, artificial intelligence, and Internet of Ideas, Track 3: Cybersecurity, artificial intelligence, and Internet of Things, and Track 4: Internet of Ideas, artificial intelligence, and the business systems of the future.

The Track 1 papers address the key innovations in the current digital era, Industrial Revolution 4.0. The works here stem from innovative technologies, including artificial intelligence and IoT, which offer solutions in the form of reengineered business processes serving implications within various sectors. Furthermore, different types of clustering algorithms and software-defined networking algorithms get examined. Additionally, the papers tackle various topics in social media usage during the COVID-19 pandemic.

The papers in Track 2 explore sustainable technologies, artificial intelligence, and the Internet of Ideas. The works here deal with machine learning for sustainability and a review on Industry 4.0 and 5.0 management for sustainable technologies. Moreover, the papers investigate sustainable innovations such as the diffusion of IT and knowledge management competence and capabilities for empowering and business analytics strategies and tools for empowering sustainable corporate development.

The intellectual contributions within Track 3 (Cybersecurity, artificial intelligence, and Internet of Things) address a critical topic: the utilization of blockchain, cryptocurrencies, and fintech for AI and IoT in the digital eras. Papers investigate knowledge management in the digital era about the accounting, oil and gas sectors, and other global private and public sectors. Another paper explores a digital sustainability reporting framework leaping organizations towards the Industry 5.0 from the lens of an accounting discipline with implications managerial and theoretical implications. Ultimately, a couple of intellectual contributions discuss blockchain technologies for sustainable development. This track also addresses big data systems governance, cybersecurity, and privacy. Papers examined, in detail, financial statement and fintech fraud and preventative measures and unethical practices through a detailed literature review. Additionally, such a track explores corporate and environmental governance and sustainability of the financial industry as well as intellectual contributions on the role of gender in telecommuting amidst the COVID-19 period for assuring long-term economic sustainability.

Finally, Track 4 addresses the Internet of Ideas, artificial intelligence, and the future business systems. Papers within this track discuss novel ideas such as smart algorithms and blockchain technology for enhancing the competitiveness of financial systems.

The EAMMIS 2022 conference, theme: from the Internet of Things to the Internet of Ideas: the role of artificial intelligence, got held on 10 and 11 June 2022. The hybrid venue got hosted by Coventry University, UK. The EAMMIS 2022 success rests on the shoulders of the synergic efforts of various individuals, committees, and organizations, mainly the Bridges Foundation and Coventry University.

Abdalmuttaleb M. A. Musleh Al-Sartawi
Ahlia University
Manama, Bahrain

Anjum Razzaque
Ahlia University
Manama, Bahrain

Muhammad M. Kamal
Coventry University
Coventry, UK

Acknowledgements

This book offers a discussion platform for researchers, scholars, students, and academics to provide insights and explore the opportunities and challenges of sustainable technologies, including artificial intelligence (AI), the Internet of Things (IoT) and Internet of Ideas (IOI) in the digital era. This book has been inspired by Prof. Abdulla Yusuf Al Hawaj, the founding president of the Ahlia University in the Kingdom of Bahrain. His immense support and efforts in envisioning the true sense of sustainable development and incorporating the right blend of technology has motivated us to honour him with this EAMMIS 2022 proceedings.

We want to express our gratitude to the contributing authors. Their valuable submissions helped formulate a holistic comprehension of how technologically we can transform from the Internet of Things to the Internet of Ideas and how the role of artificial intelligence facilitates such a transition amidst the fourth industrial revolution. Moreover, the implications of the contributions of such chapters bare a significant impact on education, information systems, the economy, and the financial industry. Moreover, we thank the reviewers who voluntarily reviewed the chapters. Finally, we thank the authors who also served as reviewers.

Special thanks go to the efforts and support of various individuals, committees, and organizations who contributed to the success of the European, Asian, Middle Eastern, North African Conference on Management and Information Systems (EAMMIS) 2022. EAMMIS 2022 got held on 10 and 11 June 2022. We want to thank Bridges Foundation, the strategic partner, and Coventry University, UK, the host institution.

Finally, we thank the book series editor Prof. Dr. Janusz Kacprzyk and the Springer editorial team, including the editorial director Dr. Thomas Ditzinger. We also thank the senior editorial assistant Mr. Holger Schaepe for the immense support and acceptance of this valuable offering to students, practitioners, and professionals in the form of state-of-the-art knowledge.

Abdalmuttaleb M. A. Musleh Al-Sartawi
Anjum Razzaque
Muhammad Mustafa Kamal

Contents

Innovation in the Digital Era

Innovation in Digital Era: Transforming Trucking Industry for Operational Efficiency Through Digital Transformation

Mahendra Parihar and Nagaraju Dasari

Abstract The purpose of current study is to understand trucking industry and challenges associated with its operation and transform trucking industry through Digital Transformation taking help from Digital India Program started by Government of India to enhance digital transformation given its impact on operation efficiency of various businesses in general and trucking industry in particular with special reference to its operational efficiency. Having look at current scenario of trucking industry with reference to nature of operation, ownership profile, areas of operation, regulatory and legislative measures relating to trucking industry, enroute operation challenges, higher operating cost and delays in operation, and many more are not allowing the trucking industry to enjoy economies of scale with special reference to small operators. Thus, in other words, the study found that trucking operation has numerous challenges which hampers the growth of the sector as observed during conduct of the study. However, through penetration of technology via digital transformation in trucking segment especially with reference to its operation in particular we can provide some good relief and can transform the trucking sector into most appropriate segment from viability of operation i.e. operational efficiency. The major implication of the study is that the result is based on limited data and information due to Covid-19 pandemic (Second Wave). Nevertheless, this can form basis to understand the process of digital transformation and its application to transform trucking operation to solve its operational issue and many more. The utility of the study can be derived from the point that this paper is one of the few studies conducted recently that provides insight into process of transforming trucking industry through digital transformation and possible economic benefits derived from digital transformation.

Keywords Transportation · Innovation · Technology · Digital transformation · Economic growth

---

M. Parihar (✉)
Mukesh Patel School of Technology Management and Engineering, SVKM's NMIMS University, Mumbai, India
e-mail: mahendra_parihar@rediffmail.com

N. Dasari
Department of Mathematics and Statistics, Manipal University Jaipur, Jaipur, India

A. M. A. Musleh Al-Sartawi et al. (eds.), *From the Internet of Things to the Internet of Ideas: The Role of Artificial Intelligence*, Lecture Notes in Networks and Systems 557,
https://doi.org/10.1007/978-3-031-17746-0_1

1 Introduction

Road Goods Transport Industry (RGTI) is one of the key components in the field of transportation in any economy. Road goods transport industry or trucking industry especially in Indian context has its own significance given its various inherent characteristics which makes it more convenient as compare to other modes of transportation in some respect. Having look at trucking industry in India with special reference to its growth over period of time, it is quite uneven. This is because of negligence of policy makers and thus didn't receive required support from regulatory and legislative front despite the fact that trucking industry in particular considered as back bone of our economy.

Further, given the ownership profile nature of operation, market structure, etc. the trucking industry in India facing numerous challenges from socio-economic point of view in general and operation viability in particular. The situation of small operators in particular is not good especially in terms of load carry, distance covered and so on. However, given the various components of trucking operation, it is presumed that if technical support in the form of digitalization via digital transformation and its penetration in various components of trucking operation, can bring certain specific solutions. Thus, in order to transform trucking industry, digital transformation is must especially in India. This is because having look at nature of trucking industry in some of the other economies of world, we find good level of digital transformation already taken place, whereas in India it is just a beginning in the form of electronic toll collection via Fast-tag, e-Challen, etc. for trucking operation. Again, the current study doesn't deny the presence of digital transformation in trucking industry, but it was mostly limited to manufacturing activities and some Government procedures only.

Now, given the role played by trucking industry especially in past few decades which is being indicated by its share in GDP and its popularity compared to other modes of transportation especially railway in terms of passenger movement and freight handles, any kind of inadequacy in road freight transportation will have huge cost to our nation directly as well as indirectly. As observed, in case of freight movement, it has increased from less than 20% in 1951 to 80% now and expected to grow up and may established around 85% [34], literature suggests that in past sometime road freight transportation has grown at CAGR of around 12% whereas rail freight transportation grew at about 1.5% [5]. Furthermore, given the expected annual growth of Indian economy by around 7 to 7.5% implies a doubling of freight transport output every 10–13 years [38].

Moreover, although trucking industry has emerged as dominant mode, it has not been able to come out of its traditional unorganized framework with large number of small operators i.e. more than 77% of road fleet is under small operators which was around 95% in 1993 [37] and 80% in 1998 [2]. Again, given the growth in trucking population over period of time i.e., currently more than 1 crore goods vehicles which were only 87,000 in 1951 [3], 5 lakhs in 1980-81 [10] and about 20 lakhs in 1996 [4].

Further, the speeding up of the process of economic development exert pressure on trucking industry for mobility and distribution will result into more demand for trucking services. However, productivity of truck operators depends on fleet size, its utilization, adequate infrastructure, trained manpower, freight structure, information availability, etc. but being dominated by small operators with reference to ownership profile and scattered nature of the industry, the truck operators especially small operators does not enjoy economies of scale in their operation. This results into challenges emerges for viability of operation. Again, the existing structure along with existing challenges to trucking industry also make it difficult to introduce new ideas and achieve various objectives of trucking industry such as operational efficiency, road safety, technology upgradation, professional management, etc. The industry also has vast opportunity and potential to generate employment [3].

Moreover, taking help from Government of India's "Digital India Program" and try to penetrate more technology in operation of trucking industry with especial reference to enroute operation, loading and unloading, freight information (rates and availability), finance requirement, legislative and regulatory measures, etc. this is because transforming trucking industry through "Digital Transformation" although would be a challenge but surely beneficial for industry itself with special reference to small operators along with other stake holders and to the economy as whole too. However, one need to be careful about very important thing i.e. implementing new technologies are not enough for Digital Transformation, but in order to have true Digital Transformation, we need to make sure that trucking industry should take advantage of the possibilities that new technology provides through the transformation.

Thus, the problem states that given the lack of appropriate data base relating to trucking industry with regard to Government Administration, enroute barriers, ownership profile, viability of operation, etc. along with lack of information flow i.e. asymmetric information, it is necessity to have digital transformation with reference to trucking industry for its better future.

2 Literature

Trucking industry as a back bone of our economy need to sustain at any cost. But in reality it is not happening especially having look at socio-economic condition of small operators in particular. On manufacturing side, on technology front hardly some marginal development or in fact modification has been done but mostly we are using same old vintage technology of 1960's and that's why if we look at any truck, we may find modification in cabin, body and some other accessories but if you look at mileage from truck engine point of view, there are hardly any changes i.e. even in past around 6 decades back or 2–3 decades back, the mileage of truck is not so much different from today's mileage despite of changes in quality and quantity of road infrastructure. However, if we look at mileage of car segment, it is almost double or more than double due to technological improvement. Thus, the low penetration of technology in trucking industry and its operation posing major challenges to its operational efficiency as well as sustainability. Therefore, it is important now to re-look very differently at trucking industry and its operation.

Further, having look at some of the recent developments on policy front from Government of India and some of the major developmental programs initiated such as Digital India, JAM Trinity, Make in India, Atmanirbhar Bharat, etc. we may find some impacts of it on trucking operations too i.e. very recently through policy implementation Government of India made Fast-tag mandatory on highways, and this is the beginning of technological transformation in trucking industry in India.

Moreover, it is important to note that the operational efficiency of trucking industry mostly get affected by delay due to various reasons, earlier it was mainly due to octroi [7] along with other reasons, and now we have it due to tolls and other regulatory or legislative measures [27]. The loss of time is harmful for trucking industry as well as entire economy because delay in operation leading to underutilization of trucking capacity [5]. This also resulted into increase in corruption in trucking industry and relevant literature reviewed gives the impression that major part of corruption lies to concerned authorities i.e. traffic police and local authorities-police [19].

Further, as we are aware that viability is one of the major challenges for small operator in particular, if the movement of the trucks are low on one hand and high fuel cost along with wastage of fuel due to congestion, corruption and graft on the other hand resulting into financial trouble. This is because literature suggests that the movement of truck was around 5000 km per month only [8] and increased to round 100,000 km per annum [38] but still less than a quarter of those in developed countries. However, one of the recent study indicated that movement of trucks has improved further marginally due to improved road infrastructure [28], but trucking industry still not able to enjoy the full benefits of improved road infrastructure due to various regulatory and legislative measures. As observed earlier too that delay in operation resulting into 20–30% travel time loss which has direct impact on operational efficiency [23].

However, transforming trucking industry for operational efficiency only be possible through "Digital Transformation" and for this the competitiveness of trucking sector be improved on various front so it can tackle all possible challenges, primarily through technology penetration. In this connection one of the recent literature [34] examined competition issues in trucking industry and analyzed freight rates, operator's costs, profitability along with limitations on movement due to enroute delays and revealed that unless these problems will not be solved, the industry will continue facing same challenges because of its ownership profile i.e. majority of them are small operators. The study also revealed that with the increase in freight handled share by road transport, the truck population will increase but challenges will continue. Therefore, undertaken current study is of the opinion that with the digital transformation of trucking industry, we can address the operation related issues in much better way to some good extent.

Since road transport has linkages with economic development and social integration [9], the truck population has increased over time due to increase in population and other economic activities, but the dominant feature of trucking industry even remains same i.e. it is highly unorganized and under small operators [37, 33] despite the fact that road transport has accounted its highest share in GDP as compare to other modes in India [11]. On the other hand, the trucking industry also has vast opportunities for employment directly and indirectly [3, 29], and therefore it is important that concerned Government should take some quick steps and allow the technology to penetrate in trucking operations via digital transformation, the only way to improve operational efficiency and achieve economies of scale. This is because it was presumed that in case of trucking industry small operators would be handling 80% of freight and only 20% by transport companies, primarily large operators.

Further, in one of the recent move Government has made Fast-tag mandatory because huge operating time was getting lost at toll plazas due to long queues, congestion on one hand and many other challenges related to trucking industry on other hand [1]. Having look at delay costs and extra fuel consumption due to traffic jams, congestion at toll plazas, Indian economy was bearing unnecessary additional cost of thousands of crores of rupees directly and indirectly [36]. Hence, learning from literature, we presumed that digital transformation be one of the problem solving mechanism for trucking operation and therefore GoI has taken right step in this direction by making Fast-tag mandatory. This is because electronic toll collection is benefiting the trucking industry and Indian economy too in many respects [15]. Although, there are many other segments (which are discussed later in this paper) of trucking industry where digital transformation equally required for operational efficiency.

However, the action taken in itself too delayed as compared to some of the other countries of the world i.e. with reference to digital transformation initiatives in trucking operation Government of India made it mandatory in 2021 only where as in country like Japan, China has done well back more than a decade or so [31]. Further, given the fact that logistics cost in India is almost 1.5 times of developed nations [20] and India's 35th rank in logistics performance index, which is quite low considered the size of Indian economy [39], the Government of India realized that for socio-economic empowerment, digitalization of economy and its various segments is need for hour and thus, for digital transformation process started "DIGITL INDIA" Program in 2015 [32]. Hence, the Fast-tag implementation and making it mandatory is a part of digitalization process although at small level but good beginning for trucking industry and operational efficiency in particular.

But, on the contrary, the success of digital transformation in any economic system in general and trucking industry in particular depends on creation of an eco-system to take maximum advantage of digitalization process [22], otherwise if economic advantages are not available in particular, the transformation is of no use. This, is because in case of trucking industry especially during trucking operation after making Fast-tag mandatory, no doubt the trucks are not wasting time at toll plazas, no additional fuel wastage, less pollution, etc. but due to rent seeking activities i.e. once the truck crosses the toll-gate, just next after some distance local police along with traffic police (in some cases local goons too) forcibly stop the vehicle and ask for money which is illegal primarily. If truck operator refuse to pay, then these people don't allow them to move and harassed truck operators by way of asking for unnecessary documents and booked the operators somehow under some charges/fault. Given this, the truck operators don't take advantages of improved road infrastructure and their operational efficiency suffers. However, it is learnt from literature that with the digital transformation, we can help the trucking industry for its operational efficiency. Some of the glimpse we have already experienced during Covid-19 Pandemic i.e. how digital transformation helped various sectors during this critical situation [26]. Hence, through current study, attempt being made to look at possibility of digital transformation in trucking industry to encountered challenges faced by the industry which ultimately impacting operational efficiency.

3 Objectives

1. To examine challenges faced by trucking industry during operation in particular.
2. To examine and evaluate the possibility of Digital Transformation in trucking operation.
3. To analyze the advantages received by trucking industry due to Digital Transformation.
4. To examine the impediments in the process of Digital Transformation to transform trucking industry and find out possible solution to it.
5. To recommend possible guidelines on the basis of the study.

4 Methodology

Since independence whatever has been done mostly on the front of road transport development, majorly focused on road infrastructure only. Again, in the field of manufacturing units connected to vehicles (which uses the road infrastructure) has been taken care-of under various policy measures through industrial development. However, almost nothing has been done to the operation and operator part of road transportation. Given no appropriate presentation of trucking sector towards regulatory and legislative circle as well as Government circle, the trucking industry has been neglected so far and this resulted into emergence of various challenges. Moreover, the emerged challenges creating problem for sustainability of trucking industry in itself with especial reference to small operators.

However, current Study is of an opinion that provided the use of technology through technological transformation in various segments of trucking operation so that the trucking sector can take advantages of technological transformation and overcome many of the existing challenges to some good extent and also avoid upcoming unforeseen challenges which may emerges due to diseconomies in trucking operation. Thus, the methodology adopted for the current study includes examining the various challenges faced by trucking industry and small operators in particular during their operation. The study also examines current scenario with reference to operational efficiency.

The methodology also includes evaluating the possibility of technological transformation through appropriate technology so the operators get maximum advantages. This will be done by following structured approach during conduct of the study by way of relevant literature review, use of secondary data and information, collection of primary information through survey/observations etc. Further, keeping in mind the relevance of the study and role of trucking industry, it is important to examine the impact of GDP growth on trucking industry with reference to truck population as well as future truck demand i.e. impact of freight handled on truck population, and therefore, data analysis technique will be employed through log–log regression model to time series data for around 20 years. Thus, given the mindfulness with reference to road freight transportation in our economy, and keeping in mind the importance of technology uses through technological transformation in the form of digital transformation to tackle various challenges in order to transform trucking industry and make it future ready with economies of scale, the significance of current study be considered from future policy framework.

The Model is:

$$Y_i = \beta_i + \beta_2 X_i + \beta_3 X_2 + U_i$$

Y_i = Truck population,
β_i------β_3 = parameters to be estimated,
X_i = GDP,
X_2 = Freight Handled by Road Transport (Trucking Industry)

5 Digital Transformation and Operational Efficiency: Result and Findings

Efficient and appropriate transport system is key to economic success in modern world. Historically, the invention of wheel was landmark in mankind journey which made them to move fast and explore new things. Currently, application of digital technology in the form of Artificial Intelligence, Machine Learning, Data Analytics, Deep Learning, Algorithms, etc. with certain data analytical techniques proving great support with reference to movement. However, having look at trucking industry somehow we are missing this spirit in Indian context. As road transport mode serves as one of the key factors in the development process of any economy in particular given size of our country and geographical dispersal of natural resources, trucking industry in particular has played a pivotal role in trade and commerce in past few decades.

Today, the trucking sector is facing many challenges across various dimensions which ultimately have an adverse effect on its operational efficiency. However, appropriate strategy for use of adequate technology via digital transformation can be an appropriate answer to the various challenges of trucking industry in India. Further, GoI can expand the limits of DIGITAL INDIA Program and bring trucking industry under its umbrella via various policy measures (as what they have done very recently by making Fast-tag mandatory) and enhance the operational efficiency. Hence, result can be discussed more appropriately by having look at certain challenges faced by trucking industry and to see how best the implementation of technology via digital transformation can solve or provide better solution so that survival and sustainability of small operators in particular can be taken care of. Mentioned below certain challenges discussed with technology implementation via digital transformation to transform trucking industry for its operational efficiency:

1. **Viability of Operation**: The basic problem of trucking operation is viability issue. This is primarily due to operational inefficiency as a result of asymmetric information, nature of trucking ownership i.e. large number of small operators and so on. Given this, trucking industry is not in position to take its operation to the level of economies of scale. In case of asymmetric information with reference to freight rates, freight availability and so on, the middlemen i.e. intermediaries take the advantage of this and make money on one hand, and on other hand given lack of information about load, many times the truck moves empty for up to few hundred kilometers. Although, some initiatives have been taken by few entrepreneurs but given the size of trucking industry and kind of people involved in it with inadequate resources, lot more is to be done. One possible way is to set-up of Transport Exchange at national level by developing appropriate model and solve the problem of information asymmetry via Digital Transformation in the form of Transport Exchange. Although it is not an easy task but not impossible too.
2. **Ownership Profile**: Given the ownership profile i.e. large number of truck operators are small operators and highly unorganized, and therefore unable to take market advantage due to inadequate resources available with them. The only solution is to bring them together and operate jointly so that we can share common costs and also develop a mechanism jointly to have check on truck operation. Thus, if trucking cooperatives be formed and with the help of Digital Transformation trucking operation be monitored, it may quite possible that trucking operation of small operators in particular will get operational efficiency.
3. **Enroute Barriers**: As discussed earlier enroute barriers resulting into delay in operation and hence operational inefficiency is the result. Enroute barriers includes Tolls, legislative and regulatory barriers, corruption by RTO and local police, etc. Although government has taken good step by making Fast-tag mandatory very recently, more technological penetration is required for other dimensions so that trucking industry can take maximum advantage of improved road infrastructure and improved its operational efficiency.

Table 1 Sample techniques analysis

	Truck population growth	
	GDP	Freight handled
Intercept	−6.49	1.07
Coefficient	0.84	0.88
R-Square	96.54	99.64
Std. error	0.03	0.01
t-stats	21.78	69.03
P-value	7.36929E−14	2.89801E−22

Source Our analysis

4. **Government Administration**: Implementation of JAM Trinity, online license, and registration of vehicles, etc. are the moves already taken in the form of digital transformation for transport sector in general and trucking industry in particular, but more technological penetration in the form of linking various Government's Schemes to the data of stakeholders and small operators in particular so they can take maximum benefits of welfare schemes. Further, the task related to taxation, overloading of vehicles, highway safety, finance, check on corruption, etc. be done by developing huge data base related to trucking operation using the concept like Industry 4.0 and digital transformation in particular.

Further, as per adopted methodology, the study also tried to find out relationship between truck population growth and economic growth as well as freight handled. As truck population growth indicates demand pattern of trucking services, and thus it is important to know its future trends based on changes in economic activities. Thus, as result shows that there is significant relationship between Dependent variable i.e. truck Population growth and Independent variables i.e. GDP Growth and Freight handled as mentioned in Table 1 below, it is important now to make sure that transformation of trucking industry is must to eliminate its certain bottleneck, and this can be done smoothly and efficiently via Digital Transformation only and hence we can achieve operational efficiency.

6 Discussion

Given the various inherent characteristics, the trucking sector has certain advantages over other modes in Indian context and therefore it is important that the operation must be efficient. However, given nature of operation, ownership profile, its presentation to concerned authorities, operational viability issue, enroute barriers, regulatory and legislative regimes, and most important the lack of coordination and lack of appropriate as well as timely information in trucking sector not allowing it to get operational efficiency.

Further, it is noticed that through digital transformation i.e. technological penetration in various dimension of trucking operation, we can collectively transform the trucking industry. However, given the lack of resources for its modernization and restructuring, the pace of technological transformation is too slow, unless recently Fast-tag made mandatory by government. How technological penetration will take place? Who will get benefited and How much? Given the various restrictions to the trucking industry is still a big question in front of stakeholders and policy makers too. Thus, it is suggested that one can take this journey forward from technological perspectives and develop various appropriate models in the form of Apps, Systems, Devices, Connections and so on and experiment it for benefiting purpose, because current study is an attempt only to provide basic framework from theoretical perspective based on literature and observations.

7 Recommendations

1. Along with transforming toll collection via GPS based after making Fast-tag mandatory, government should also develop and extend GPS services to tackle rent seeking activities.
2. Government should promote modernization of trucking industry so the new trucks come in market with all required technology for digital transformation.
3. Government can implement digital transformation mechanism in their own departments at fast pace and trained the man power adequately.
4. Government should extend financial, legislative and regulatory support to small operators in particular to form trucking cooperatives, which will further help in digital transformation process of trucking industry.
5. With the set-up of trucking cooperatives, if Government also helps in set-up of Transport Exchange to reduce information asymmetry via Digital Transformation, operational efficiency can be achieved in trucking industry.

8 Limitation

Limited availability of literature and relevant data on digital transformation in trucking industry with special reference to trucking operation are basic but major limitations of the study. Further, the study was conducted during second wave of Covid-19 pandemic and therefore limited interaction with concerned stakeholders was the possibility. Another limitation is that the study has considered only operational efficiency whereas there are another segments of trucking operation too where it can be taken into consideration. However, the study can be expanded in future with more aspects related to trucking industry.

References

1. Amrin S (2019) An introduction to Fast-tag: a game changer in automatic toll collection system in India. Int J Res Anal Rev 6(2)
2. AITD (1999) Trucking operations in India. Asian Institute of Transport Development, Ministry of Surface Transport, Government of India, New Delhi
3. CIRT (1994) Road goods transport in India - a study of its structure and organization. Central Institute of Road Transport, Pune
4. Dalvi MQ (1996) Does India need a new national transport policy. In: National conference, Department of Economics, University of Mumbai, Mumbai, 7–9 December 1996
5. Deloitte (2003) Modernisation of the trucking industry, Report submitted to the Advisory Group on Trucking Industry in India, All India Motor Transport Congress and All India confederation of Goods Vehicles Owners' Association
6. Deloitte (2016) Digital India- unlocking the trillion dollar opportunity, Deloitte Touche Limited, United Kingdom
7. GOI (1950) Report of the motor vehicle taxation enquiry committee, Ministry of Transport, Government of India, New Delhi
8. GOI (1967) Final report of the road transport taxation enquiry committee, Ministry of Transport, Government of India, New Delhi
9. GOI (1980) Report of the national transport policy committee, Planning Commission, Government of India, New Delhi
10. GOI (1987) Report of the planning group on road transport for the 20th century, Planning Commission, New Delhi
11. GOI (2012) Working group report of twelfth five year plan, Planning Commission, Government of India, New Delhi
12. GoI (2019) India's trillion-dollar digital opportunity, Ministry of Electronics and Information Technology, Government of India, New Delhi
13. IAMAI (2020) Digital India- present, 2025 and 2035, Internet and Mobile Association of India, Mumbai
14. ICFAI (2020) Digital transformation for socio-economic development of rural India. In: Proceedings of national conference, ICFAI Ranchi, Jharkhand 2020
15. Joshi et al (2017) A Comparative study of toll collection systems in India. Int J Eng Res Dev 13(11):68–71
16. Latta A, Singh T (2021) Digitalization in India and its impact on the economy. Int J Adv Res Sci Commun Technol 2(3)
17. Maiti D et al (2020) Digitalization and development: issues for India and beyond, digitalization and development (edited book). Springer, Singapore
18. Maiti M, Kayal P (2017) Digitalization: its impact on economic development and trade with special reference to services and MSME sector in India. Asian Econ Financ Rev 7(6):541–549
19. MDRA (2006) Corruption in trucking operation in India, marketing and development Research Associates, New Delhi
20. MGI (2019) Digital India-technology to transform a connected nation, McKinsey Global Institute, McKinsey and Company
21. MIT (2011) Digital transformation: a road map for billion-dollar organizations. MIT Centre for Digital Business and Capgemini Consulting
22. Mishra MK (2020) Digital transformation of public services and administration. ZBW-Leibniz Information Centre for Economics, Kiel-Hamberg
23. Musleh Al-Sartawi AMA (eds.) (2022) Artificial intelligence for sustainable finance and sustainable technology. In: ICGER 2021. Lecture notes in networks and systems, vol 423. Springer, Cham
24. Narula TK, Rana S (2017) (2017) Digitalization in India. Int J Emerg Technol Spec Issue-NCEIST 8(1):298–303
25. NHAI (2021) Toll Plaza, National Highway Authority of India, Government of India, New Delhi, 30th June 2021

26. OECD (2020) Digital transformation in the age of Covid-19: building resilience and bridging divides, Digital Economy Outlook 2020 Supplement, OECD-Paris
27. Parihar M (2013) Impact of tolls on viability of trucking operations in India, Supply Chain Pulse, ITM Centre for Supply Chain and Logistics Research, vol 4, no 2, September 2013
28. Parihar M (2014) Viability analysis of trucking operations in India, Supply chain pulse, ITM Centre for Supply Chain and Logistics Research, vol 5, no 2
29. Parihar M (2020) Strategy to achieve sustainable development goals through employment generation: a case of road freight transportation (trucking) industry in India. Supply chain pulse, ITM Centre for Supply Chain and Logistics Research, vol 11, no 2, July–December 2020
30. PWC (2017) Digital India-targeting inclusive growth. Price Waterhouse Cooper India
31. PWC (2018) Payment-newsletter-digitalization of toll payment. Price Waterhouse Cooper, PWC India
32. Shallu et al (2019) Digitalization in India- an innovative concept. Int J Eng Develop Res 7(1)
33. Sriraman et al (1998) Report of the sub-committee on financing, taxation, fares and freight and insurance submitted to the Steering Committee on Trucking Operations in India, Asian Institute of Transport Development, New Delhi
34. Sriraman et al (2006) Competition issues in the road goods transport industry in India with special reference to the Mumbai metropolitan region, competition Commission of India, New Delhi
35. Singh A, Parihar M (2018) Analysis of issues relating to work life balance in trucking operation: a human resource perspective. Int J Comput Sci Eng 5(8)
36. TCI (2015) Operational efficiency of freight transportation by road in India, Transport Corporation of India limited, 3rd edn.
37. Mission UN (1993) transport sector needs assessment in India, mission report. United Nations Development Programme, New York
38. World Bank (2005) Road transport service efficiency study, mimeograph. World Bank, Washington D.C.
39. World Bank (2018) World bank logistics report, Washington DC
40. World Economic Forum (2020) Digital transformation: powering the great reset. World Economic Forum, Geneva-Switzerland, July 2020

A Sensible Solution to an Unintended Consequence of Relative Grading

Nagaraju Dasari, Mohiddin Shaw Shaik, Mahendra Parihar, and Indu Priyanka Dasari

Abstract Existing literature discusses various issues associated with the teaching learning process, type of examinations to be conducted, methods of assessing students' performance, advantage and disadvantages of each method, psychological issues of teachers, students, parents etc. It has been observed that there is no mathematical tool to early predict the unexpected consequence of relative grading and to get away from this problem. Neither touching ethical issues of teaching learning process nor the tools to set proper question paper and evaluation, in this article some formulae were introduced through which, one can easily identify the courses which bound to produce unintended consequences with the relative grading. Set formulae were designed based on the vital parameters of normal distribution -that is the mean and standard deviation of the scores obtained by the students in a course examination.

Keywords Grades · Relative grading · Grading policy

1 Introduction

The sacredness of the education is largely dependent on the examination system. Education and exams are like the two sides of the same coin. They are complementary and have a symbiotic relationship. The Grading and reporting are the recent development in the examination system, because prior to 1850 this was not known to educational institutions across the globe. Since then, this paradigm shift has been troughing challenges to the educators to explore various methods of content delivery, set up certain parameters for question papers setting, adopt to suitable methods for

N. Dasari (✉) · I. P. Dasari
Manipal University Jaipur, Jaipur, India
e-mail: 1979dasari.nagaraju@gmail.com

M. S. Shaik
Narasaraopeta Engineering College, Narasaraopeta, AP, India

M. Parihar
Mukesh Patel School of Technology Management and Engineering, SVKM's NMIMS University, Mumbai, India

A. M. A. Musleh Al-Sartawi et al. (eds.), *From the Internet of Things to the Internet of Ideas: The Role of Artificial Intelligence*, Lecture Notes in Networks and Systems 557,
https://doi.org/10.1007/978-3-031-17746-0_2

assessing students' performance. Referring to existing literature we can classify grading methods broadly in two types (i) Traditional grading methods (ii) Innovative, non-traditional grading methods. Though the readers are familiar with the various grading methods to assess the performance, for the sack of completeness we are briefly describing each as under.

Traditional—more familiar Grading methods:

Normal Curve Method: is a method of assigning performance in which students' scores are scaled as per the frequency distribution. It assigns grades to students with respect to other students' performance in the class. Which is also called as relative grading method. The main drawback is method is it discourages collaboration, as competition becomes central.

Absolute Standards: gives the grades in accordance with the grade boundaries/marks range decided before the start of the course and which are known to the students well in advance. Through this method of grading students should always be performing well to score the marks in the desired range. But when the question paper or the assessment criteria for any of the course exams becomes tough, there may be possibility that no students getting top grade or majority may also fail to clear the course exam.

Pass/Fail Methods: In this method grade letter becomes binary and student receive a letter grade as either pass or fail. The main advantage with this method is that assessor can grade the students, purely based on their personal interest, zeal in exploring new topics and identifying new horizons to develop on their learning. One disadvantage with is method is students may require spending more time in learning course further may have to put in more efforts to pass the course.

Innovative, Non-traditional Grading Methods:

Self-paced Learning and Evaluation: In this method, leaners will have a choice to choose courses of their interest in learning. Therefore, students have self-control over the content that they wish to learn as per the time that they could devote for. Now-a-days more and more students are getting attracted to MOOCs (massive open online courses) as most of them provide this self-paced learning and evaluation.

Subjective Evaluation: In these, evaluations happen subjectively based on opinions, by influence, by the feelings of individuals. This evaluation is based on students understanding of the concepts, how it is presented. It is quite evident the answers written by the students would differ from one another. These kinds of evaluations, one can notice when a question paper is of case study type.

Objective Evaluation: In this evaluation students will have a choice to choose an option in each question paper which has to be completed in stipulated time. Most of the large-scale competitive examinations have been happening through this method because of answer scripts auto evaluation facility associated with these kinds of examinations. But there is a possibility of happening malpractice activities like sharing correct answers with one another with using advanced technology available at another end.

Blanket Grading: In this method all the students will be getting the same grade those who were part of the project team irrespective of level of individual contribution to the project all though it is expected an equal participation by the team in collaborative projects. To get fair evaluation out of this grading blind evaluations can be carried out by the team of another project. Two or more peer teams blind evaluations may bring more justification to the awarded grades.

Performance Checklists: Basically, the performance checklist is a super tool for the management to ensure that all the employees have been provided with the skills, tools, resources, and directions to do the work. But it may not be applicable to assess the students' performance.

Contract Grading: It demands cooperation between assessor and the assess. In fact, contract grading is exactly what these words suggest it works in contrast to other grading systems. In this, the course content to be studied, assignments to be submitted, number of quiz tests to be given and other assessment components and their weightage would be decided in contract with the instructor or the evaluator. Further the kind of assessment tools will also be decided on mutual understanding between students and instructors. In this method learning happen in cooperation between teacher and the students. The main advantage of this grading practice is that it reduces the grades competition among the larger group of students, but it remains with the smaller group of students who do not maintain good contract with the assessor. Another disadvantage with this system is that probably the course content to be studied, assessment tools to be used may not meet the defined expected outcomes of that course.

Of all these discussed methods, relative grading or normal curve method is widely accepted and implemented system in most of the educational institutions across the globe. Though it is popular among all other briefly discussed methods, if it is not implemented properly it is bound to produce unintended consequences, objections and hence a question on the method. From the existing literature it is evident to say that there is no mathematical tool to early predict the unexpected consequence of relative grading. Lack of mathematical tool or a programmed software to early predict the objectional grading on relative grounds has motivated us to develop such a tool and further to develop a coding to this algorithm to make it user friendly in our future work.

2 Literature Review

Sadler [17] examined the relationship between academic performance of the students with the awarded grades. It also suggested that there is a need to narrow down the existing considerable gap between actual academic performance of students and the awarded grades. Jeffrey Schinske et al. [10], an evidential study states that accuracy based grading would actually demonstrates students and hindrance in learning.

Because giving less credits for the huge amount of work done by the students in completing assignments, homework's would result in adverse implications towards accuracy-based grading. Even with this kind of blanket grading hardly would give encouraging results through grading system.

Alyson [3] analytically established that there is hardly any correlation between highly paid teachers' evaluation with the student's achievement. Therefore, the belief that students achievement is directly proportional to the evaluation methodology uses by high stake teachers is meaning less and non-evident. Susan M. Brookhart [19] study was a literature review on various grading systems evolved over 100 years across the educational institutions in the world. The study was carried out on the following five aspects (i) whether the grading system is reliable (ii) quantitative study on grade report cards of K-12 composition (iii) primary and secondary data on teacher's perception on grading (iv) what are the standard parameters for grading; and (v) how the grading happening in higher educational institutions. As an extension of the study carried out by Sadler [17], Lorin Anderson [13] has carried our his study on following aspects and his recommendations are as follows: (a) There should be direct or indirect relationship between the expected learning outcomes of the course with the type of assignments, questions that students were supposed to submit or answer (b) The teaching methodology and assessment tools used by the instructors should be the similar across all section students registered for the course (c) The minimum quality standards set by the teacher should be instructed to all the students at the beginning of the course, so that students get seriousness in their learning and further they can improve their own ability (d) It is very much important to get various stake holders feedback on the final draft of grading system before it actually get implemented in the system.

Meike Bonefeld [14] study suggests that teachers should feel more and more responsible and justifiable to the Noble profession by means of adopting accurate judgement on student's performance. Further there should be a defined mechanism for their judgement because errors on their part will have great impact on careers of students. Keri Schwab [11] study was a discussion on few alternate assessment strategies which enhances the students learning and performance. Dasari et al. [8, 9] were made an attempt to study the difficulty associated with the grading.

3 Motivation Behind the Study

In bringing more efficient and acceptable mechanism in assessing student's performance Normal distribution or relative grading system is widely implemented one across most of the institutions across the globe, among all other methods mentioned in the introduction. In other words, it is the system where the grading will be done with reference to whole class performance. Through this system the top performers will be getting highest grade and accordingly the remaining lower grades can be decided. Because in a class the level of learning, intellectual abilities of students vary from person to person and in a class these factors are expected to following normal distribution curve.

Any real-life phenomena either in material or non-material form which is associated with the time is found to follow normal curve. Though the teacher delivers the same content to all the students, the levels of learning from the lecturers will not be same for all students. Some students may get the things which the teacher delivers during the lecture itself and some may have to spend some time after the lecture for them to get the things clear. The same pattern can even be observed when students perform in a course exam. There were many interpersonal skills and psychological issues to be addressed at the individual student level in motivating them to step into next level of learning from the existing level.

Focusing on the pattern of student scores in a course exam, one may not find the score distribution of students following normal curve in every examination. Sometimes we may find positive or negative skewed distribution of the scores deviating from the expected pattern. In a normal distribution one could observe that the statistical parameters mean, median, mode would be agreeing on a single mark and the curve would be symmetrical about the mean scores. When peak of the normal curve shifts to the right of the mean scores, it is a negatively skewed distribution and is regarded as a positively skewed distribution when the peak of the curve shifts to the left of the mean scores.

In a process of identifying the cases where this unintended drift in the distribution occurs, after analysing secondary data we present here certain formulae which could help both the teachers and the management of the institutions to review the reasons for this short fall that is bound to happen at the later stage. The more positive or negative drifts would result into a situation to doubt the grading system which has evolved out of lot of refinements over past 100 years in assessing student's performance. But actual reason for these drifts is due to not maintain standards in course content, its delivery, assessment tools, question papers for the examinations etc. If everything goes fine the scores bound to fit into standard normal curve.

Our analysis and results are based on parameters of normal distribution, that is 'mean' and 'standard deviation' of the scores. It is known that for a data following normal distribution there would be 68% of the population spreads with in one standard deviation from the mean score and 95% of the population spreads with in 2 standard deviations from the mean and 99% of the population spreads within 3 standard deviations around the mean score. Considering above facts different institutions adopts different cut-off formulae for different letter grades. What eve the grade letter cut-offs for different grades the main pain point both for students and their parents to some extent to the teacher is that when a student not getting highest grade or even fails a course even after scoring good marks and in contrast if no student fails a course exam and gets highest grade on less scores due to the adopted relative grading based on parameters of normal distribution. The possibilities of happening of this is shown in Table 1. Therefore, this situation demands manual intervention by the management of the institutions before result made public. At the ground level these kinds of situations usually managements face at the last-minute during result declaration time. By that time management hardly finds time to investigate the cause.

Table 1 Mean vs standard deviation

s =	16		17		18		19		20		21	
	CLPG	CHPG	CLPG	CHPG	CLPG	CHPG	CLPG	CHPG	CLPG	CHPG	CLPG	CHPG
⋮	⋮	⋮	⋮	⋮	⋮	⋮	⋮	⋮	⋮	⋮	⋮	⋮
m = 40	8	72	6	74	4	76	2	78	0	80	-2	82
41	9	73	7	75	5	77	3	79	1	81	-1	83
42	10	74	8	76	6	78	4	80	2	82	0	84
43	11	75	9	77	7	79	5	81	3	83	1	85
⋮	⋮	⋮	⋮	⋮	⋮	⋮	⋮	⋮	⋮	⋮	⋮	⋮
57	25	89	23	91	21	93	19	95	17	97	15	99
58	26	90	24	92	22	94	20	96	18	98	16	100
59	27	91	25	93	23	95	21	97	19	99	17	101
60	28	92	26	94	24	96	22	98	20	100	18	102
61	29	93	27	95	25	97	23	99	21	101	19	103
62	30	94	28	96	26	98	24	100	22	102	20	104
63	31	95	29	97	27	99	25	101	23	103	21	105
64	32	96	30	98	28	100	26	102	24	104	22	106
65	33	97	31	99	29	101	27	103	25	105	23	107
66	34	98	32	100	30	102	28	104	26	106	24	108
67	35	99	33	101	31	103	29	105	27	107	25	109
68	36	100	34	102	32	104	30	106	28	108	26	110
⋮	⋮	⋮	⋮	⋮	⋮	⋮	⋮	⋮	⋮	⋮	⋮	⋮

This difficulty can better be addressed if the progress of the course is assessed periodically based on the initial one or two examinations student scores in a formative assessment suiting to relative grading. If the anticipation on grade letter cut-offs to fall under absolute range through the relative grading mechanism, the proposed tool would certainly be helpful for the intervention or the introspection in identifying suitable corrective measures.

4 Formulae to Address the Possible Consequences Due to Relative Grading

As it is known different institutions have been following their own set formulae to identify cutoff score for different letter grades, but all will definitely be playing around the mean and standard deviation of scores for grading students performance through the relative grading. Either of the cases where the expected set range falling within one standard deviation or two standard deviations or three standard deviations from the mean score, in order that the expected cutoffs for passing grade and highest grade to be within permissible range of marks. Following formulae are suggested case wise.

We use following notation in all the cases:

m—mean
σ—standard deviation
n—expected score to award passing grade
N—expected score to award highest grade
a, b—non-zero positive integers
$[n_1, n_2]$—score interval for passing grade
$[N_1, N_2]$—score interval for highest grade
CLPG—Cutoff marks for least passing grade
CHPG—Cutoff marks for highest passing grade

With an assumption that $m - (\frac{a}{b})\sigma$ and $m + (\frac{a}{b})\sigma$ to be the cut-off identification formulae set by an institution for awarding minimum passing grade and highest grade in a course respectively.

Case 1: If order that CLPG $\leq n$ and CHPG $\geq N$, the parameters should be satisfying the inequalities $m \leq \frac{n+N}{2}$ and $\sigma \geq \frac{b(N-m)}{a}$ or $m > \frac{n+N}{2}$ and $\sigma \geq \frac{b(m-n)}{a}$
Case 2: If the anticipation is CLPG $\geq n$ and CHPG $\geq N$, then $m > \frac{n+N}{2}$ and $\frac{b(N-m)}{a} \leq \sigma \leq \frac{b(m-n)}{a}$.
Case 3: If the anticipation is CLPG $\geq n$ and CHPG $\leq N$, then $m < \frac{n+N}{2}$ and $\sigma \leq \frac{b(m-n)}{a}$ or $m \geq \frac{n+N}{2}$ and $\sigma \leq \frac{b(N-m)}{a}$
Case 4: If the anticipation is CLPG $\leq n$ and CHPG $\leq N$, then $m < \frac{n+N}{2}$ and $\frac{b(m-n)}{a} \leq \sigma \leq \frac{b(N-m)}{a}$
Case 5: If the anticipation is $n_1 \leq$ CLPG $\leq n_2$, then $\frac{b(m-n_2)}{a} \leq \sigma \leq \frac{b(m-n_1)}{a}$
Case 6: If the anticipation is $N_1 \leq$ CHPG $\leq N_2$, then $\frac{b(N_1-m)}{a} \leq \sigma \leq \frac{b(N_2-m)}{a}$

Assuming that $m - 2\sigma$ and $m + 2\sigma$ are the cut-off marks for the least passing grade and least mark for awarding highest grade, we placed below a Table 2 explaining each case discussed above.

Table 2 Particular cases vs parameter values range

Cases	This happens only when
CLPG $\leq$ 40 and CHPG $\geq$ 75	m $\leq$ 57.5 and $\sigma \geq \frac{75-m}{2}$ or m > 57.5 and $\sigma \geq \frac{m-40}{2}$
CLPG $\geq$ 25 and CHPG $\geq$ 72	m > 48.5 and $\frac{72-m}{2} \leq \sigma \leq \frac{m-25}{2}$
CLPG $\geq$ 20 and CHPG $\geq$ 80	m < 50 and $\sigma \leq \frac{m-20}{2}$ or m $\geq$ 50 and $\sigma \leq \frac{80-m}{2}$
CLPG $\leq$ 35 and CHPG $\leq$ 85	m < 60 and $\frac{m-35}{2} \leq \sigma \leq \frac{85-m}{2}$
25 $\leq$ CLPG $\leq$ 35	$\frac{m-35}{2} \leq \sigma \leq \frac{m-25}{2}$
75 $\leq$ CHPG $\leq$ 90	$\frac{75-m}{2} \leq \sigma \leq \frac{90-m}{2}$

5 Conclusion

Through the tool introduced in this research paper, it can easily be identified the courses which definitely be producing adverse effect on the grades emerges out of relative grading policy. Using the mentioned formulae, a programming be developed in future so that authorities shall be alerted automatically for them to intervene and take necessary remedial measures before the result made public. More importantly, if this tool made available to the faculty members their teaching methodology and assessment pattern be introspected periodically during continuous evaluation process.

References

1. Abadie A, Athey S, Imbens GW, Wooldridge J (2017) When should you adjust standard errors for clustering? National Bureau of Economic Research Working Paper 24003
2. Gheondea-Eladi A, Gheondea A (2022) Bifurcation in the evolution of certainty in a small decision-making group by consensus. Br J Math Stat Psychol 75:88–115
3. Lavigne A (2014) Exploring the intended and unintended consequences of high-stakes teacher evaluation on schools, teachers, and students. Teachers College Rec 116(1)
4. Andreoni J, Brownback A (2017) All pay auctions and group size: grading on a curve and other applications. J Econ Behav Organ 137:361–373
5. Brownback A (2018) A classroom experiment on effort allocation under relative grading. Econ Edu Rev 62:113–128
6. Campos-Mercade P, Wengström E (2019) Threshold incentives and academic performance. Job market paper, pp 1–70
7. Çelik E, Çalık M (2022) Examining the relationship between sensation seeking, positive and negative experiences, emotional autonomy and coping strategies in adolescents. Int J Educ Psychol 11(1):68–91

8. Dasari N, Parihar M (2021) Application of computational technique to assess the performance of staff for sustainable business. In: 2021 2nd international conference on computation, automation and knowledge management (ICCAKM), Dubai, United Arab Emirates, pp 163–167
9. Dasari N, Parihar M, Shaw M (2021) Analysis of exertion with relative grading mechanism in academics using computational technique. IOP Conf Ser Mater Sci Eng 1099:012018
10. Schinske J, Tanner K (2014) Feature approaches to biology teaching and learning teaching more by grading less (or differently). CBE Life Sci Educ 13:159–166
11. Schwab K, Moseley B, Dustin D (2018) Grading grades as a measure of student learning. SCHOLE J Leis Stud Recreation Educ 33(2):87–95
12. Levitt SD, List JA, Neckermann S, Sadoff S (2016) The behavioralist goes to school: leveraging behavioral economics to improve educational performance. Am Econ J Econ Policy 8(4):183–219
13. Anderson LW (2018) A critique of grading: policies, practices, and technical matters. Educ Policy Anal Arch 26(49):1–31
14. Bonefeld M, Dickhauser O (2018) (Biased) grading of students' performance: students' names, performance level, and implicit attitudes. Front Psychol 9:1–13
15. Musleh Al-Sartawi AMA (2020) E-learning improves accounting education: case of the higher education sector of Bahrain. In: Themistocleous M, Papadaki M, Kamal MM (eds) Information systems. EMCIS 2020. Lecture notes in business information processing, vol 402. Springer, Cham
16. Prashanti E, Ramnarayan K (2019) Ten maxims of formative assessment. Adv Physiol Educ 43:99–102
17. Royce Sadler D (2009) Grade integrity and the representation of academic achievement. Stud High Educ 34(7):807–826
18. Cheung SKS, Lam FK, Kongkiti P, Harrison HY (2021) Shaping the future learning environments with smart elements: challenges and opportunities. Int J Educ Technol High Educ 18:16
19. Susan MB et al (2016) A century of grading research: meaning and value in the most common educational measure notes/citation information. Published in Rev Educ Res 86(4):803–848
20. Melkonyan T, Zeitoun H, Chater N (2022) The cognitive foundations of tacit commitments: a virtual bargaining model of dynamic interactions. J Math Psychol 108:102640

Modern Trends in the Visual Media Industry

Alaa Maki and Asal Al Shammari

Abstract We cannot discuss trends in media and video without mentioning images themselves and the technology that makes them possible. We simply cannot. The study addressed the modern trends in the visual media industry through an introduction and two types of research. To achieve the study's objectives, several questions were posed: What are the media means in terms of its origins? What are the effects of relying on modern media? What is the visual media industry? What is the contribution of modern television technologies to visual media? The study used descriptive, analytical, and historical approaches. The study concluded that recent trends in the visual media industry presented new concepts that allow a detailed presentation of the challenges facing visual media in modern society are represented in three factors that are political, technological, and economic; it explains the emergence of the media and the transformations that occurred in it. The visual media industry knew recent developments and trends during the twentieth century and at the beginning of the current century, thanks to the technological development in media and communication.

Keywords Modern trends · Visual media industry · Phenomenon of free media

1 Introduction

Visual media contribute to improving or destroying society's mental image at home and abroad. They may even reform or spoil the relations of individuals to each other within the same country, as well as the relationship of the nation with other countries of the world, which is contributed to the development of research development in the visual media industry as an extension of the theory of cultural industries, it is almost inseparable from it, since the end of the seventies, through many publications. Then, the research expanded within the framework of university research teams or

A. Maki · A. Al Shammari (✉)
University of Sharjah, Sharjah, United Arab Emirates
e-mail: aaakkof@sharjah.ac.ae

A. Maki
e-mail: aakkof@sharjah.ac.ae

A. M. A. Musleh Al-Sartawi et al. (eds.), *From the Internet of Things to the Internet of Ideas: The Role of Artificial Intelligence*, Lecture Notes in Networks and Systems 557,
https://doi.org/10.1007/978-3-031-17746-0_3

collective works, distinguished by combining the fields of culture, communication, and media from an economic perspective.

In recent years, there has been a disturbance in the dual technology of the media, which has led to low interest, the failure of advertising-based businesses and the poor role of the press, and political and social unrest has affected citizen confidence in the press and led to attacks on news media in many countries. Unless the internet is regulated, trust in the media will be rebuilt, and the relationship with the public strengthened, contributing to the emergence of modern trends on which the media, in general, will depend. So, modern trends through new and sophisticated technology will bring a wave of technological upheaval from the atom that will rely on artificial intelligence, big data, and new audio-visual interfaces. Despite all this, there is economic and political uncertainty that will pose further challenges to the survival of news media organizations [14].

Visual effects domain is based on new technologies. They very often offer a new way of telling an old story. The rapid development of digitalization in the mid-90 s brought about a new phenomenon. Computers provided a common platform which facilitated the process of spreading specialist software used in one artistic field into other activities.

2 Problem Statement

The problem of the study lies in the study of modern trends in the visual media industry, where the researcher found the scarcity of studies that have researched this topic due to the significant development in technology and subsequent technologies such as the Internet, telecommunications, IT industry has emerged visual media, and because of the availability of several elements contributed to the emergence of many trends in this area. Therefore, the problem of the study revolves around the main question: What are the modern trends in the media industry video?

3 Study Questions

1. Definition of the media in terms of its origin?
2. What are the effects of dependence on modern media?
3. What is the visual media industry?
4. How do modern television technologies contribute to visual media?

4 Importance of Study

The importance of the study lies in the importance of the information obtained from its sources, which dealt with modern trends in the visual media industry. Besides, it lies in the importance of visual media in the progress and development of all societies and the increase of cultural and intellectual awareness. Also, it is the first within the researcher's knowledge to address modern trends in the visual media industry, making it a new addition to the library and a nucleus for other similar studies.

5 Objectives

1. Reveal what the media is in terms of its origin.
2. Know the effects of dependence on modern media.
3. Determine what the visual media industry is.
4. Demonstrate the contribution of modern television technologies to visual media.

6 Media Richness Theory

It explains that the effectiveness of communication depends on the extent to which the medium is used, and it focuses more on the interactive forms of two-way communication between the communicator and the audience receiving the message.

Media richness theory has been adapted to include modern media communication such as advanced video and virtual video conferencing over the Internet.

The media richness theory states that all communication channels have certain characteristics that make them less or wealthier, and one of the most important goals of choosing a communication medium is to reduce the ambiguity of the message.

The richness of the media is a function of several characteristics, including the following: [2].

The ability to process multiple information signals simultaneously, the ability to facilitate instant reverberation, the ability to attract personal focus & the ability to use natural language.

7 Literature Review

[17] aimed to research the technical development and its impact on the performance of media institutions and aimed to identify the modern Sudanese television technologies used and indicate the extent of their use in the field of television programs and know the challenges facing the television. The study concluded that the TV cadre is not as

interested in modern trends as technologies exposing them to many difficulties and problems.

[11] study aimed to learn about the role of modern media in raising awareness and addressing security crises. The study presented the concept of security media and areas of security awareness and the role of visual media in addressing challenges. The study found an evident prevalence of modern means and systems such as television and that visual images are more effective in conveying the media message, especially based on understanding between the sender and the receiver.

[9] attempted to bridge this literary gap in studies that addressed the positive effects of visual media on social behavior. The study problem question focused on what effect visual media has on adolescents' social interactions, especially those with autism spectrum disorders? The study was applied to a sample of (20) participants, the questionnaire was distributed to the study sample as a tool for collecting study data, and the study found that teens who participate in visual media use video games more or less weekly through apps, computers and other forms of visual media.

[16] noted that in 2014 the media industry moved away from traditional media and audio-visual dominance to the digital age with smartphones, as it considers modern trends as the key to access in advanced and innovative technology and high-speed communication innovation, as well as intelligent phones, showed an increase in the deployment of digital applications in small towns and cities. The study found tremendous growth in the use of digital media despite the poor infrastructure in India, which is considered one of the obstacles facing the use of such media.

The study [6] dealt with the visual and audio media industry in modern society, where the study dealt with the origins of visual and audio media and their economic, organizational specificity, and its internal logic, which distinguishes it from publishing activities, journalistic activities, broadcasting activities, and broadcasting activity (visual and audio media) through economic, political and technology elements. The study also noted that the political element is the cause of the emergence of modern media and its continued presence in modern democracies, as it allows a space for discussion between the private interests and the interests of the authorities. The economic component includes the value of information products and markets. The industrial-technical component has undergone a significant shift in technical development and has had a profound impact on the organization of activities, evaluation of the product in the market, and impacted the relationship of the consumer audience with media products. The study found that what happens to the media in any modern society is the availability of political, economic technology.

The study [1] also examined visual media and the construction of political symbols from 2014 to 2018, and the problem of the study focused on the following questions: What are the most important tools used in political marketing? What effect does television media have on how the viewer sees political reality and its personality? What mechanisms do the media to create political symbols? The study used the field survey (survey) method by developing a questionnaire as a data collection tool. The study included all university students in Egypt through a random sample of (81) respondents. The study reached many results, the most important of which is that most university students have seen that satellite video channels have a crucial role in the political propaganda of the political leadership.

[15] concluded the extent to which visual media contribute to violence in young people. This study focused on measuring the impact of high school students in Nicosia with violent images in the visual media. The effects of realistic images on violent images have been discussed in visual media. The study found that television is the most influential tool in visual media.

[18] addressed the needs and satisfaction of Arab bloggers on YouTube. The study also aimed to understand the motivations of these bloggers in sharing and creating visual content. The study concluded that through video, bloggers seek to meet several needs: providing Arabic content related to non-media topics for the benefit of others, promoting a cough or brand for the benefit of the physical blogger, investing productive free time, sharing self-experiences, and finally making financial profit through YouTube by producing content that attracts many views.

First Topic—Media Sector

In the light of rapid scientific progress and development, modern means, and tools such as radio, newspapers, books, television, and the internet have emerged to reach a broad audience, the most famous and widely used media [5]. Mass Communication and mass media are necessary to maintain the partnership by maintaining the social presence and existence of the partnership. Those tools, which perform various functions such as visual media, public opinion raising, education, and entertainment, affect people's lifestyles, tastes, preferences, and habits concerning consumption, youth sensitivity, and behavior. Moreover, multiple visual media comes first television, internet, cinema, social media, etc.

First Requirement—The Origins of Modern Media

A) The political logic of the modern media emergence

The emergence of modern democratic bourgeois societies was conditional on creating a public space, where public and private interests converge, and the issue of state arbitrariness towards the emerging liberal economy was discussed at the time. In the authors' view, it is customary to accept the dominance of economic or technical logic unless they reassess political logic, i.e., liberalism. Political logic, in this aspect, is the origin of the demand for the media that stands behind the upsurge of the media's existence in its general frames in all societies, including its presence in the public space of each country and contributes to the development of public debates [17].

B) Economic logic of visual media

This logic of visual media is based on two pillars: the value of media products and the markets of media means. The value of media products includes the usage of the value of information (VOI) and the reciprocal value. Besides, the usage of the value varies from person to person and from group to group; no one information has the same value of usage for all people; it depends on the previous knowledge of mutual value, which can be referred to as the value of exchange as any other service. In addition, it is defined as the time of social action that is necessary to produce information [7].

The media markets consist of the final market, and it concerns the equipment (TV, radio, antennas... Your product is imaginary (radio and television programs, books and CDs, subscriptions to cable networks. This market is experiencing an involuntary growth starting with the purchase of a TV set, for example, or a tape drive, then the consumption of the products such as TV programs and tapes, then comes the market of advertisers, where part of the media product is sold again in the market of advertisers as advertising space in newspapers, magazines, television, and radio. However, the advertising market alone is not sufficient to ensure the emergence of new media. The role of advertising has twofold: It is a source of financing for the media if taken separately, on the one hand, and one of the reasons to focus on the industry and the fragmentation of the consumer market on the other [17].

In terms of the financial market, certain aspects of the media weigh heavily on the financial market, such as media activity often does not have a capital-intensive structure, except for the equipment industry, where media operators are usually very conservative in their relationship with capital for fear of taking over the media institution and distorting intellectual work. It is a traditional position in media and cultural rights. This attitude weakens the financial structure of media institutions [12].

C) Industrial logic for Visual Media

Researchers have distinguished since the eighties of the last century between two logics that govern the functioning of the media industries. Then they differentiated between these industries in terms of these two logics. The first logic refers to the structural conditions that determine the variables of the work of the actors in a sector in a specific historical period, namely, the rules that guide its structure and the functioning of a particular industrial sector, that define the peculiarities and detail the processes of creation, production, and consumption of cultural products. The logic here determines the rules of action between the various parties regardless of their will; that is, it describes the industry's major trends at a given moment [13].

The second logic is the forms that predominate the development and culture making at a given historical moment. It results from a particular connection between the functions of production and programming or publishing and broadcasting or distribution. The logic of broadcasting or streaming governs the presentation of products by mass broadcasting, instead of copying the original work model several times or distributing it by simultaneous broadcasting to a significant number of consumers.

Second Requirement: The Effects of Dependence on Modern Media

Studies [16] observed many effects as a result of less dependence on the media in three basic categories.

A) Cognitive effects: its distinct from those that affect behavior explicitly and evidently, but the two are clearly related and are embodied in many areas:

- **Ambiguity:** It is a problem either in the lack of information or the presence of conflicting and contradictory information for people to understand the meaning of a particular event or to find a correct interpretation of the event among many possible explanations. The research shows that when an unexpected event occurs, such as natural disasters or the assassination of a political leader, many public members are aware of these events through insufficient media information about what happened. However, they do not know what the event means and do not recognize how the public will identify that an event is being explained, and more information will be sought to resolve such ambiguity.

- In the outside world, they use the information of those means to form new trends about events such as environmental problems, energy crises, political corruption, crisis conditions, or wars, and thus new trends are constantly formed. The media means are not identical in their impacts on trends, where they follow selective processes—as confirmed by early studies, which plays a vital role in the process of forming trends, as well as opinion leaders in the local community select and direct people's attention to events and influence the content and intensity of the established trend.
- **Enlargement of people's belief systems:** This effect occurs in a media-dependent society. Charles-Cooley has used for a long time the term "enlargement" to indicate that people's knowledge and belief systems increase because they learn and know more about other people and things from the media.
- **Clarification of values:** Whatever the form of information provided by the media, it includes an explanation of the conflict between values, which motivates individuals to consider the positions of their values, and this clarification can cause pain in the goals and ways to achieve them and leads to multiple options. Also, it can cause a conflict between them to decide which is more important so that the priorities of values become apparent to the individuals.

B) Affective effects:

The effective processes generally indicate different feelings and emotions and the essential components of love and hate. In societies where reliance on the media is developed for information, emotional impact. The change in people can occur when certain information is presented through the media messages that affect individuals' feelings and responses.

Second Topic—Modern Television Technologies in the Visual Media

Since its inception and until today, the means of communication have been channels of human knowledge, and their importance is constantly increasing as a result of the development of communication technologies. The usage of modern television technologies in the coverage of news bulletins is an essential task in the industry, which gave it significant importance represented in the process of selecting television events and differentiating between the TV news, the news stuff, and its broadcast to draw the attention of viewers and those who are interested in this field. As well as to pursue, to study, and disclose the news stories and to designate the importance of each of them in relation to the environment of the political system from which it has emerged, and the possibility of its ability to meet its needs and its vision to the functions of news and its content in accordance with the standards adopted for news coverage and the function of the visual media. Modern media and television technologies are characterized by many important features, most notably the ability to transmit information from one system to another through communication systems, the ability to use a set of special devices to communicate information in several languages to influence the audiences receiving this special information [12].

First Requirement—Methods of Producing News Information Through Television Technology
Technology and its applications with semiotics helped identify advanced means in the production of news information via television images, using the works of the French semiotician "Roland Barths," which was included in the book "**Semantics**" published in 1967. This science is meant as the special significance of human behavior, where he identified several levels of image production and understanding by observing everything that exists. In this case, the reading and production of the image are based on a set of symbols and semantics that are considered a visual language using various reading mechanisms resulting from the sight of the eye that travels between the symbols to form the axis of vision, thereby breaching the limits of the image. Its apparent form of revealing the sub-content of its content is transformed to develop the process of dismantling its elements and its internal fabric to explain what these symbols and connotations conceal, both in-kind and implicit [13].

Second Requirement—Levels of Symbols, Visual Semantics, and Montage Technology
A) Levels of symbols and visual semantics:

The levels of symbols and visual semantics are represented by the sampling level, which is meant by the unification of the signifier with its meaning; the initial reading of the television visual material takes place as it constitutes evidence of significant meaning. It describes the shape in which the image appears and determines its structure, which contributes to understanding its structural characteristics and showing its main lines that are called lines of force that help in distributing its elements and forming its balance and unity of composition within it, whether it is experimental, diagnostic, engineering, spontaneous, or otherwise, given that each formation has its own peculiarity and significance that is discovered through the directions of the lines within the image and the determination of the focal point that is called the visual anchor [4].

1. **Reduction technique:** The reduction technique is practiced through the truncation or reduction of images, which is one of the basic design elements to be configured as color or lighting through technical devices with modern and advanced technologies.
2. **Truncation or deletion of a part of the image:** The unwanted part of the television images is truncated to shift the focus of the viewer's vision. Also, to limit it to the essential elements that serve the targeted subject in the image with the deletion of the below stimuli that may cause a kind of distraction.
3. **Changing the image size:** The process of changing the size of images is carried out by enlarging or decreasing one of the components of the subject of the image

to control the viewer's attention so that the stored information about the relative size of the elements in the image is compared with that information and the contrary perceptions suggested by the information material to form the desired meaning in the viewer's mind.

4. **The features of the digital camera:** With the introduction of digital systems and the emergence of advanced generations, which are endowed with modern technologies that have allowed the implementation of precise processes in the production of the television image of the visual news message, such as the diversity of shooting angles and the multiplicity of camera movements.

B) The change in the elements of television images: The change that affects television images through the usage of increasing the length or width, or modern technologies and sophisticated devices that allow exaggeration in the form of the image as desired to direct the viewer to the destination of a particular change adds new communicative dimensions that serve the purposes of the attitudes and behaviors [16].

C) Montage technique:
The technique of montage arises through installation techniques and methods that have become one of the basics of image production, especially with the use of modern technology means that allow combining various close, medium, and general shots, or a color image with another black and white..., etc. Also, that the most important thing that digital montage achieves is the suggestion through which it can change the meanings by changing the general background of the main subject.

Third Topic—Criteria and Elements of Spreading the News

The media highlights a particular type of news, namely the modern news that constitutes what was termed the scoop, and it is a value that the media seeks to achieve, and since news values are the focus in the media in general, so the television news is being a visual media in particular.

First Requirement—Modern Foundations in the Dissemination of News and Its Values

A) The Modern foundations in the dissemination of news

The first thing that should be noted is the necessity of distinguishing between the modern foundations of news dissemination. The features that must be available in the news are the qualities that must be available in the news and the conditions that must be met, such as accuracy, honesty, and objectivity, while reporting the news has become characterized by bias and imbalance due to the political and economic societal influence. Accordingly, those foundations referred to [1] represented in the following:

- **Standards for publishing the news:** The standard for publishing or broadcasting the news is the basis for which media outlets determine its content and the news's validity for broadcast or publication. Hence, many researchers have been interested in the issue of determining the foundations of the criteria that control the priority and preference of publication. Many call them elements, others call them adjectives, and others call them considerations. Others also worked hard to classify them according to the media and political system or to the nature of the media outlet that must influence in one way or another in determining or shaping those standards.

- **News value:** It is the element that must be all or part of the news in order to be considered news, and therefore news values are not the only basis for choosing news suitable for broadcast or publication, and with the difference in news values between one region and another, instead, within the same regions, some media researchers set criteria for measuring the level of news values, as the first criterion was that the news is a mirror that reflects events as they are, just as it reflects the image of the person standing in front of them. It is assumed that the news reflects current events and issues, regardless of who benefits or loses from the broadcast [10]. There is another trend that shows that the news has become less serious and more inclined to entertainment, this is due to the impact of economic controls and great competition to attract viewers, as the center of gravity has shifted from the standards of news work and the foundations of objectivity to standards aimed at creating an atmosphere characterized by excitement and exhibition. There is a set of ideas, beliefs, and ways of behavior called values in every society, which is a more comprehensive and broader concept than beliefs. However, beliefs mean a judgment, opinion, or direction related to a social reality that the individual embraces his correctness. This concept differs from the concept of values, which are often semi-codified systems. The individual uses it in his assessment of situations [13].
- **The characteristics of the news:** They are the qualities that must be present in the news, and they are the conditions that must be met in it, such as accuracy, honesty, and objectivity, while reporting the news has become characterized by bias and imbalance due to the social, political, and economic influence. This is the third criterion for measuring news values, as we find that economic considerations prevent the required justice in the news because news networks depend on the profits they earn [9].

News values are also considered to be the perceptual or semantic framework—explicit or implicit—that makes the communicator, an individual, a group of individuals, or an institution, able to perceive and select the news value that is most important to viewers among a large number of facts or information and can formulate and create its elements in a special news arrangement and to give it the opportunity to appear through one of the media means with degrees of prominence that derive from the nature of the medium used and its specific and public circumstance [17]

Based on the preceding, news values refer to everything that deserves an individual's attention and cares for economic, psychological, social, moral, or aesthetic considerations. This view excreted a general concept of values under which a set of considerations fall centered on interest, economic, psychological, social, moral, and aesthetic considerations [8].

Second Requirement—Visual Effects

In the past decade digital technology has become inherently involved in almost every artistic creation. Artists use computers at various stages of creation, employing them for diverse tasks [16]. Even if the final work is not directly connected with digital technology, various software can be used to communicate with other artists or collaborators, organize the creative process, gather resources, or publish the artistic outcomes [3]. Many professionals have had to reinvent the way they work or to convert to a completely different specialization. Their methods of seeking inspiration and developing a concept have also had to be adjusted so that they use the benefits of technological progress. Visual effects are one of the modern and advanced technologies used in television media, as they can be defined as a mixture of written media, including sound, image, text, and video clips. The text in the read news is represented in the form of words, which the viewer must perceive as news, while the image, which is more eloquent than the read news, where the moving image is a thing that makes it easier for the viewer to understand the information or what is mixed with sound [9]. Also, the visual effects are defined as the joint use of a group of media, such as films, music, and lighting, to transmit information visually on the television screen for the purposes of publishing and advertising towards the viewers using certain graphics (1, p11). [14] also defined visual effects as a group of technologies that allow the integration of many parts from different sources (texts, images, sounds), as they are combined through informatics.

For publishing and advertising purposes to viewers using certain graphics (1, p11). Moreover, [14] defined visual effects as a set of technologies that allow the integration of many parts from different sources (text, images, sounds), where it is combined through Informatics.

My research allows me to admit, looking back at my past professional career, that I have repeatedly undertaken the typical actions of a creative technologist quite unconsciously. However, this research is my first attempt to genuinely understand this new discipline and to comprehend the meaning of contemporary VFX tools in the palette of artistic possibilities.

Visual effects have several tools to refer to [1]:

- **Audio tools:** Represented in the headphones, the microphone, and the sound card, where the sound is one of the effects and if it is combined with the rest of the influences, it will give a distinctive application and great benefit. The sound is entered through the microphone, which transmits what it picks up and translates into machine language to the sound card that processes it, then stores and broadcasts it.
- **Visual tools:** These tools consist of images, videos, optical films, motion graphics, and other visual means on the television screen, aiming to convey information to the viewer. Texts can also be considered part of the visual tools because the viewer will read them and access their contents.

8 Conclusion and Results

In this research, modern trends in the visual media industry were addressed through an introduction and two chapters. The introduction included the study problem, importance, questions, objectives, methodology, and previous studies. The first topic dealt with the media sector, and it included two requirements. The first requirement dealt with the emergence of modern media, and the second requirement addressed the effects of relying on modern media. The second topic dealt with modern television technologies in visual media. It included two requirements that dealt with the first requirement: methods of producing news information through television technology and the second requirement: levels of symbols and visual semantics and montage technology.

At the End of the Study, the Following Results were Obtained

1. As 2021 has brought about even more new ways of doing things, the new normal should trend toward hybrid production to get the best of everything.
2. Over the last years, people's priorities have changed—so does the content they prefer to interact with. It must be adapted to a new reality and reflect the audience changing values and lifestyles.
3. Thanks to technological development in media and communications, the visual media industry has been aware of recent developments and trends during the twentieth century and at the beginning of the present century.
4. Recent trends in the visual media industry have introduced new concepts that allow a detailed presentation of the challenges facing visual media in modern society.
5. The visual media industry has distinguished between the types of logic in the media in general, represented by the logic of publishing, the logic of Written Journalism, and the logic of broadcasting.
6. The visual media industry has defined several criteria, including crucial general functional characteristics, economic cycle, creative professions, and market characteristics. In order to know everything that distinguishes the industry from these industries.
7. What determines the direction of any media in modern society is a combination of three levels: political, technological, and economic.
8. The study explained the emergence of the media and the occurred transformations.
9. The study came out with an explanation of the development and shortcomings of the media in industrial societies known as the democratic society.
10. Expressive and laconic footage that looks like a freeze-frame from an unseen movie is one of the key visual trends in 2021.
11. In a world crowded with content, it is crucial to stand out. New live footage is hard to come by, while the director's creativity level limits animated videos. Unusual and eye-catching—all these features make the animated content attractive.

12. New and creative content will continue to rule our media world. With the new season of every quality series comes new storylines, new heroes, and new twists. There will be loads more content every day across 2022.
13. Last year, we noticed content fatigue as a trend. How wrong can we be? Sign-up fatigue has also been considered, and one thing is for sure: the sheer number of shows and content leads to many transformations.
14. Let us not even mention the impressive ratings that linear TV had in 2021. We certainly do not want to live 2022 in lockdown. Moreover, even if ratings were good, the revenue was not. So let us not anticipate a full lockdown effect in 2022.
15. Another behavior that we saw in 2021 and is here to stay—despite what happens with Covid-19—is co-viewing. Co-viewing refers to people in the same household watching the same content simultaneously but not always in the same room or on the same device.
16. All 4 K is readily upscaled to 8 K using AI capabilities on the TV set, and there is now ample 4 K content via on-demand and broadcast platforms.
17. As soon as the industry sees a new opportunity, it heavily invests. The pandemic might have impacted that.
18. Covid-19 accelerated many technologies or habits already on the rise. eSports and cybercrime. 2021 came with more than one deadly virus.
19. According to pretty much everyone, Deepfakes will become more of a threat. It is simply becoming creepier.
20. Until 2021, we could have argued that, hey, other sectors were more significant pollution creators than us. Moreover, there is always a more significant polluter.
21. In 2022, and moving forward, it is time to realize that everybody has their part to play, and we all must pull together and do everything we can.
22. Streaming will be trending, and that the streaming war is not over.

References

1. Al-Hadidi AR (2019) Visual media and the building of political symbols: The Egyptian case study from 2014 to 2018, Arab Democratic Center. Published online https://democraticac.de/?p=60112 on April 18, 2019, check it out Wednesday, 24/11/2021. 2:07 PM
2. Kannan A (2019) "Media Theory", Yazouri Group for Publication and Distribution
3. Al-Sartawi A (2019) Assessing the relationship between information transparency through social media disclosure and firm value. Manage Account Rev 18(2):1–20
4. Arafa S (2021) Visual media and awareness formation in Egyptian society. J Misr Univ Stud 1(1):337–351
5. Ashour F (2013) The economy of the Audiovisual Media in Algeria. National Publishing House, Algeria
6. Ashour F (2015) The industry of visual and audio media. Algerian J Soc Sci Humanit 3(5):105–124
7. Clement YK, Louis S, Leung P, Lee N (2004) Impact and issues in new media toward intelligent societies. Hampton Press, Inc., U.S.A, p 8

8. Gupta N (2019) Influence of demographic variables on synchronisation between customer satisfaction and retail banking channels for customers of public sector banks of India. Int J Electron Bank 1(3):206–219
9. Hanley AJ (2014) Visual media and ASD: impact on social interactions of adolescents. Retrieved from Sophia, the St. Catherine University repository website: https://sophia.stkate.edu/msw_papers/324
10. Hannoon A, Abdalla Y, Al-Sartawi A, Khalid A (2021) Board of directors composition and social media financial disclosure: the case of the United Arab Emirates. In: Musleh Al-Sartawi AMA (eds) The big data-driven digital economy: artificial and computational intelligence. Studies in computational intelligence, vol 974. Springer, Cham
11. Helmi W (2013) The role of modern media in raising awareness, crisis, and security. Faculty of Social Sciences, Umm Al-Qura University. Research published in ResearchGate on the link: https://www.researchgate.net/publication/309195835_dwr_wsayl_alalam_alhdytht_fy_altwyt_wmwajht_alazmat_alamnyt
12. Ibrahim A (2017) integration between traditional and new media, T1, Oman Dar Al-Mu'taz Publishing and Distribution
13. Jameel I (2012) Media innovation in Arab satellite TV, 1st edn. Osama Publishing & Distribution House, Oman
14. Lule J (2012) Exploring media and culture, 1st edn. Flat World Knowledge Inc., Irvington
15. Lesinger FY, Oytun M, Yinal A The impact of violence of the visual media on high school students. Folklor/Edebiyat 25(97):676–702
16. Musleh Al-Sartawi AMA (eds) (2022) Artificial intelligence for sustainable finance and sustainable technology. In: ICGER 2021. Lecture notes in networks and systems, vol 423. Springer, Cham
17. Sheikh I (2006) Technical development and its impact on the performance of media institutions: a descriptive and analytical study in the television sector (unpublished master thesis). Omdurman Islamic University, Sudan
18. Yusuf N (2021) The needs and gratifications that YouTube offers to Arab video bloggers. Arab J Humanit 39 (159)

Electronic Advertising and Its Usage in Digital World

Sumaya Asgher Ali

Abstract The study aims to identify the definition of electronic advertising and its usage, advantages and disadvantages, and the types of electronic advertising (websites, social media and Mobile). The results of the study revealed that e-marketing has become one of the most important tools for advertising and promotion of services and products because it is characterized by ensuring access to the largest possible number of the target audience, predetermined in terms of age, gender, or even presence within a specific geographical area, which is reflected positively, significantly and effectively on increasing The volume of dealing on products and services, as well as obtaining the largest possible number of customers, and thus an increase in the rate of sales and profits, as a direct result of the e-marketing process that you have undertaken.

Keywords Electronic advertising · Social media advertising · Mobile advertising

1 Introduction

Electronic advertisements are a type of advertisements that have emerged as a result of the great technological development that the world has witnessed [5]. They are published on various websites and social media. Electronic advertisements are characterized by the speed of their spread, as they can reach a large number of people quickly, as the Internet has entered all global homes [12, 32]. And it has become everywhere in the world and electronic advertising design needs the presence of experts who are able to take advantage of global design programs in order to prepare electronic advertisements that reach the things that the company wants to deliver to the user in the shortest way, and electronic advertisements are characterized by the possibility of directing them, where the company can direct electronic advertisements are directed towards a specific segment, such as directing them to men or women, the elderly or children, or specific age groups [4]. Electronic advertisements

S. A. Ali (✉)
Cairo University, Giza, Egypt
e-mail: Sumayaaali3@gmail.com

A. M. A. Musleh Al-Sartawi et al. (eds.), *From the Internet of Things to the Internet of Ideas: The Role of Artificial Intelligence*, Lecture Notes in Networks and Systems 557,
https://doi.org/10.1007/978-3-031-17746-0_4

in our time have a great and distinctive role no less than the role played by television advertisements, so that electronic advertisements have surpassed them in some cases and for advertisements Electronics is of great importance, and its importance lies in a number of things, and through the lines of this article we will talk about the importance of electronic ads [29, 31].

Electronic advertising is considered the ad of the age, as it is characterized by flexibility and low costs compared to traditional methods, the speed of spread and the geographical expansion of advertising, the ability to accurately target segments, and the use of multimedia such as sounds, animations and videos, which makes advertising more attractive and influential than traditional methods, especially printed ones. of which [17]. With the great and rapid progress of technology and communications in our time and the increase in the number of Internet users tremendously, the Internet has become one of the largest commercial markets in the world, as it offers many advantages that are not found in traditional markets, and includes hundreds of millions of users who can Turning them into real customers for products, these reasons are only a small part of a wide range of reasons that push companies currently to work in the field of electronic advertising [22, 30].

2 Definition of Electronic Advertising

Electronic advertising is the art of using the Internet as a means of communicating marketing messages to a specific and targeted audience. It's useful for attracting visitors to website and displaying brand, but first and foremost, digital advertising is designed to convince target customer to engage in a particular action—such as making a purchase [13]. Electronic advertising is also called as online advertising or digital advertising, digital advertising is directed towards defining markets with unique and useful applications, and since the early 1990s, there has been a massive increase in the growth of online advertising, which has developed into a standard for small and large enterprises [1].

3 Types of Electronic Ads

Electronic advertisements evolve with the development of technology, so there are many different ways to display advertising messages over the Internet [2], the most prominent of which are:

Interstitial Banners

These are the billboards that are displayed between the pages of the website, meaning that the advertisement appears when you click to move from one page to another, with the ability to close the advertisement sometimes [9].

Pop-Ups and Pop-Unders
As its name shows, these ads pop up from the bottom of the screen or from the top to open a new and small window on the page. The pop-up ads can be noticed from the bottom right away, but the problem lies in the pop-up ads from the top, which are difficult to notice until after the browser window is closed [4]. This type of advertisement was very popular in the early days of the internet, but because of its inconvenience, most of the good internet browsers now have tools to block it, although it can cause a problem for sites that legitimately use pop-up technology to display user information [11].

Map Advertisement
Map ads are ads that are published in online map software and applications, such as Google Maps [3].

Floating Advertisement
Floating ads appear for several seconds as a layer above the page content but are a separate window that the user can usually close. DHTML or Flash is used to design floating ads, and an animated ad often ends up displaying a static billboard on the page [14].

Wallpaper Advertisement
This type of ad changes the background of the page you're viewing, but it's usually not clickable [19].

4 Payment Methods for Online Ads

Payment methods for electronic ads vary as follows.

Cost per Impression or Cost per Thousand Impressions
CPI means that an advertiser is paid for each time an ad appears on the publisher's page, and this method is commonly referred to as "cost per thousand impressions (CPM"), especially when the main objective of the campaign is brand promotion [6].

Cost per Click—CPC
This means that the advertiser is paid only when an interested customer clicks on the ad, and this method is usually associated with marketing through search results, also called "PPC". This method is also used in the pricing of banners when the goal of the campaign is to drive traffic, and sometimes in affiliate marketing when the goal is to get more traffic to a new website [8].

Cost per Acquisition—CPA
In this approach, the advertiser pays only when the ad leads to a specific earning, and the definition of earning varies from site to site, it may mean the user filling

out a form, uploading a file, or purchasing a product [18]. This is the best method for the advertiser [28], because he pays only when the advertisement achieves its specific goal, but it is the worst for the publisher, who does not get any compensation except when the advertisement succeeds, as the publisher in this case will depend on the conversion rate on the advertiser's site, which is out of his control. For this reason, this method may not be used frequently in billboards, and it is more related to commission marketing [26].

Flat Rate
Sometimes owners of low-traffic sites tend to sell ad space for a fixed monthly cost, regardless of the number of visits or impressions the ad generates, and this method may be suitable for advertisers who want to test campaigns targeting specific markets [28].

Cost per Engagement
This method is concerned with interactive ads, which are a new type of ads used in videos and applications, which require the mouse to be hovered over them in order to expand, and when they expand, they may contain a video, a game, or any other type of content [7]. However, it does not remove the user from the web page they are viewing, in which case advertisers pay only when the user completes the action specified in the advertisement [2]. 'Interactive ads' as its name implies requires user interaction, but if you're going to use this type of ads, how can you measure their success [5].

5 Importance of Electronic Advertisements

Electronic advertisements are of great importance, and their importance lies in a number of matters [24], and the most important and prominent of these are:

Electronic advertising is characterized by its rapid spread, as it is possible for the electronic advertisement to reach all parts of the world within a few seconds, and thus the company is able to target a large number of customers [1]. Through electronic advertisements, customers will be able to learn about all the characteristics of the commodity that is being advertised, and how to use this commodity [22]. Online advertising helps communicate offers that companies make to the products they make [27, 30]. Through electronic advertisements, the producing company is able to determine the category that it will target with this advertisement, as the companies that manufacture men's clothing determine the category of men and contribute to the delivery of the advertisement to men only, so that the advertisement does not reach groups that do not care about it [12]. Online advertising pushes many customers to visit the company's websites, and thus see other products it offers, which leads to an increase in sales [16]. Through electronic advertisements, companies reach a large number of markets, even reaching markets they would not have reached without electronic advertisements, due to the ability of electronic advertisements to transcend borders [13, 29]. Commercial advertisements allow companies to compete, through which companies learn about the products offered by other companies, and thus try to provide better offers and quality, which is beneficial to the consumer [1].Electronic advertisements help to make the company present in front of them continuously, due to the presence of an advertisement in a large number of sites that they frequent [4].

6 What Are the Different Types of Digital Advertising?

Search Ads
Search ads, also called search engine marketing (SEM), appear on search engine results pages (SERP). These are usually text ads that appear on top of or alongside organic search results [19].

Display Ads
Display ads are online ads that use text and visual elements, such as an image or animation, and can appear on websites, apps, and devices. They appear in or near the content of a website [4].

Online Video Ads
Online video ads are ads that use the video format. Out-of-stream video ads appear in places similar to display ads: on websites, apps, and devices. In-stream video ads appear before, during, or after video content [12].

Broadcast Media Ads
Also known as Over-the-Line Broadcasting (OTT), this is a specific type of video ad that appears in media broadcast content that is delivered over the Internet without a satellite or cable subscription [24].

Audio Advertising
In the context of digital ads, audio ads are those ads that run before, during, or after online audio content, such as streaming music or podcasts [17].

Social Media Ads
Social media ads appear on social media platforms, such as Twitter or LinkedIn [18].

7 The Best Social Networking Sites for Posting Electronic Ads

In order for electronic ads to succeed, they must be well marketed, and in sites where there are a large number of visitors, and social networking sites are among the most important of these places [12].

Facebook Site. Facebook is the best place to publish electronic ads, due to the presence of a large number of active users on the site [29], in addition to the presence of a large group of groups and pages that the advertiser can benefit from in order to deliver his advertisement to the largest possible segment [25].

Twitter: The Twitter platform is one of the most important social networking sites in which the advertiser can publish his electronic ads, as the platform witnesses the presence of a huge group of active users daily [13].

Instagram: Like Facebook and Twitter, Instagram provides a fertile place for advertisers to post electronic ads, which are in the form of images.

Snapchat: Snapchat is one of the most important social media platforms that advertisers can take advantage of in publishing electronic ads [3].

YouTube: YouTube is ideal for publishing electronic ads, as the company can publish its ads in the form of a video or through an advertisement that appears in the channels of celebrities [3].

8 The Most Important Advantages of Advertising on Social Networking Sites

1- Help with branding and acquaint the public with the services you provide [5].
2- Increasing the spread and reaching the largest number of potential and target customers [16].
3- Increase the number of visits to website [20].
4- It works to increase the interaction of comments and likes on the advertising content [3].
5- Converting potential clients to real clients interested in the service provided [17].
6- Ease of communication with customers via personal messages [11].

9 Electronic Advertising Objectives

Brand Awareness

Public awareness of the product or brand is one of the most important long-term goals that marketing workers seek to achieve. Introducing the consumer to the product or brand is the first step towards gaining his trust [27]. Electronic billboards (banners) mainly contain visual elements, which makes them the ideal channel to promote brand advertising [2].

Create Consumer Demand

Consumers cannot want to buy a product that they do not even know, and here lies the importance of advertisements, as they seek to convince consumers of products, and explain the reasons for their need, and modern electronic advertisements in this regard provide a great ability to promote sales offers, and thus stimulate consumer demand [18].

Meet the Demands of Consumers
Consumers are quick to search for ways to satisfy their desires as soon as they appear, and here comes the role of marketing workers to convince consumers that their products and brands are best able to satisfy those needs and desires [1].

Increase Sales and Engagement
All types of e-marketing seek to increase traffic to the online store, and thus increase sales in the long run. However, we find that electronic advertising has an immediate effect, that is, it increases visits and sales in the short and medium term. Unlike traditional advertising methods, electronic advertising is distinguished by its ability to convert a potential customer into an actual customer immediately, in addition to the ability to accurately measure its effectiveness [21].

The Main Difference
Electronic advertising is characterized by its ability to achieve sales immediately, and unlike traditional media, the consumer can move from advertising to the store with one click, and the Internet allows the ability to monitor and measure electronic activities accurately, which means the ability to customize advertisements, and measure the success of each one of them [3].

10 The Most Important Factors for the Success of Digital Advertising

Choose Keywords
The art of choosing keywords is one of the most important factors for the success of site or advertising campaign, as it is the first step to success. Keywords are those words that users search for through search engines, social media platforms, or Google Ads. You must put self in the user's place. What will you Looking or by writing a sentence? [2]

Attention Advertising Link
Pay attention to the design and form of the sponsored advertisement. ad must be characterized by credibility. If ad contains any promises and discounts, those promises must be real and not fake. There must be accuracy and attract attention, but with rationality and honesty [24].

Use of Analysis Tools
You must define a set of tools that will help you determine the target audience of gender, country, region or interests. Analytical tools to measure the success of the

advertising campaign, the reasons for its success, the number of upcoming visits and their source [4].

Choose Advertising Platform
You have to carefully choose the electronic platform on which you will do the advertising funded, whether it is social media platforms or search engine ads such as Bing and Google Ads "Google Ads" and these platforms are characterized by giving choices of age, geographic area and interests, and these are indispensable features [18].

11 Disadvantages of Social Media Marketing

No Intention of the Customer to Buy
One of the most important differences between advertising on social media and other means such as Google! On social media, they don't always have the intention of becoming customers for what you offer. This means that you can run a large ad campaign but run into the reality of users who have no intention of buying or requesting the service even if the ad click or impression rates are high [5].

Wrong Targeting is Costing You
If you do not select audience carefully and carefully, you may find that you are not getting the attention you want from target groups, social media audiences are large, and they are different from each other [23].

This could be a weakness for you if you cannot reach target groups or who are basically in the middle of this world full of users, and you may know them and know how to reach them, but you fail to formulate the advertisement that attracts their interest [24].

Crowded with Competitors
Social media platforms are always crowded with photos and videos. That's why you need to know how to make content stand out in this environment full of great stuff [16]. Thus, expressive images and distinctively written texts will certainly appear [29]. You must be innovative in the way name will float on the surface of social media ads. You can make special offers and prizes for followers or assign the task to a famous person on these sites and says Ad week [13].

88% of businesses are located on social networking sites, so if you are not able to direct business on these platforms, you should look for a specialized person to plan campaigns on these platforms [18].

12 Advantages of Electronic Advertising and Comparison with Traditional Advertising

Company Size Does Not Matter

One of the main advantages of electronic advertising is ability to delve into this field regardless of the size of the company you run. If company is emerging and you want to promote product via the Internet, you will not be worried about the size of the competitor, you can compete with the largest companies even if you are in the launch stage, in traditional advertising large companies are leading and taking the lead, the reason for this is the high cost of this type of advertising, which makes startups absolutely unable to compete, but in the field of electronic advertising, success is for the distinguished. The size of this competitor [7].

Low Cost

One of the most important advantages of electronic advertising is the low cost, as it is well known that advertising through traditional means is very high. One advertisement on TV may cost you tens of thousands of dollars, while electronic advertising is much less expensive than that, and this is what drives many to use it [9].

Ease of Measurement

In traditional advertising, there is a big problem with scaling. To illustrate this problem, suppose you place two ads on two different TV channels, and you want to stop one of the less effective ads, how will you choose? In other words, how will you measure and compare the effectiveness of both ads? This applies to all means of traditional advertising, not just television [10]. Newspapers, magazines, and yellow pages all have the same shortcomings in the measurement process [3].

As for electronic advertising, the matter is completely different. The measurement process can be done easily [5]. You can measure the number of visits to site that come from each advertisement you place, and thus you can easily compare between various ads and choose the best one. This comparison is made using many tools available on the Internet [20, 23], including free and paid ones, perhaps the most famous and most popular of these tools is Google Analytics, and this tool is only one of a very large number of tools that you can use in the measurements of electronic ads on the Internet [10].

Instant Reports and Results

In advertising on the Internet, you will start seeing the results of the advertisement as soon as you publish it on the Internet, the number of clicks on the advertisement will appear directly in front of you, and you can monitor the number of visits and the number of clicks in real time, and this is much better than traditional advertisements that do not enable you to know the extent of interaction with the advertisement The publication, and in the best cases, some traditional advertising companies provide you with results on the effectiveness of the advertisement, but these results are approximate and require a period of time to be ready, while the results in electronic advertising are immediate results [13].

The Possibility of Continuous Improvement of the Advertising Strategy
In e-marketing, you can track the customer during advertising campaign at every stage of his interaction with this campaign, starting from how he arrived at the advertisement to when he saw the advertisement, to clicking on the advertisement and then entering the target site with the advertisement, through this information you can improve marketing strategy [2]. Constantly, you can learn the weaknesses in this strategy, for example, you may notice that the ad site does not achieve a good viewership compared to another site, so you change the site. This method of gathering information enables you to continuously develop advertising strategy [8].

Improve Brand Reputation
You can improve the reputation of the brand and give visitors a good impression of the products through the site. The good content of the site and the distinctive and unique design plays a big role in improving the reputation of the products, for example, if the articles published on site are of a professional nature, this generates a feeling among visitors. company is an expert in its field of work [10].

In traditional advertising, advertising does not have the same role in improving the reputation of the product, as its impact is much less compared to electronic advertising [13].

Very Wide Market
In e-marketing, you can target a very large number of users, as the geographical boundaries will not stand in way, and if you manage to design the site in a manner commensurate with the search engines (this area is called "SEO" and it is a wide field and there are specialists in it and the goal is to increase ranking in search engine results) the world's markets will open up to you, on the other hand, if you think of targeting the same market size using traditional advertising methods, the cost is huge and even huge, and you will not be able to bear it unless you are one of the very large companies [22].

Propagation Speed
E-advertising is characterized by a very rapid spread. To make this point clear to you, let me give you the following example:

Suppose you place an advertisement on one of the well-known TV channels, what are the limits of the spread of this advertisement? In fact, the last point the advertisement will reach is the viewer, and it will not spread more than that. Suppose now that you place an advertisement on one of the sites for product [2]. A group of users liked this advertisement and in turn shared the advertisement on their accounts on social networks from Facebook, Twitter and others [10, 21], What happened is that these users turned from recipients of the advertisement to its publishers, and thus the rate of spread of advertisement increased, let's follow our example, after the process of sharing the advertisement, some people clicked the like button, this simple clicking process turned it from a recipient of the advertisement to a publisher for it because all the user's friends Whoever hit the like button will see what their friend liked. From this example, you can deduce the great speed of spread that can be achieved by electronic advertising compared to traditional advertising [7].

Controlled

In traditional e-marketing, you cannot control the person who sees the advertisement. When you place an advertisement on the TV, you cannot control the people who see this advertisement, which makes it annoying to some, or rather annoying to all people who are not interested in the subject of advertisement [4, 20], e-marketing is superior to traditional marketing in this regard, It depends on the concept of smart marketing, the goal of smart marketing is to display the advertisement to the right person, that is, to display the advertisement to the person who is interested in it. The popularity of product and company [12].

It Doesn't Matter When the Ad is Shown

In traditional advertising, the ad display time plays a role in determining the ad price, for example the price per second in television or radio advertising in high viewing hours (often in the evening) is much higher than advertising in the middle of the night when viewership rates are low, in electronic advertising Time is not important, as ad is placed 24 h a day, and customers are available 24 h a day, and this is what makes e-marketing more effective in targeting a larger number of customers [1, 19].

Electronic Advertising Problems

The study conducted by Advertising and Electronic Marketing touched on the comparison between traditional advertising media (radio, television and publications) and electronic advertising, and found that one of the most important problems faced by electronic advertising compared to others is the weak Internet infrastructure, weak communication and the lack of broadband users of the Internet compared to other countries [14]. The large number of advertisements on some websites may lead to a visitor's distraction and lack of interest in advertising, in addition to the unprofessional ways that some site owners do, such as automatic pop-ups or automatic transition to other sites that the visitor may not want to enter [16]. With the popularity of traditional advertising media such as television and radio, the study mentioned the pros and cons of each type of advertising separately and came to a logical conclusion based on international studies and on the real reality in both Syria and the world [17].

Mobile Advertising

It is a form of advertising conducted via mobile phones or other mobile devices. Advertisers buy mobile phone numbers data for their existing customer base as well as a list of potential customers' phone numbers [12, 18]. The message rarely goes unnoticed and thus qualifies as an effective marketing or advertising tool. Although it is quite new, it is undeniable that it dominates the advertising market. This is primarily due to the fact that there is an increase in the use of mobile devices such as smartphones and tablets, and mobile advertising can have a powerful impact when done in the right way and quickly become popular for this reason [11].

13 Conclusion

The study indicated that electronic advertising is a modern method that has opened a wide scope for companies and businessmen to promote and advertise easily via the Internet, as well as to keep pace with the remarkable development in the world of technology and the Internet. Online advertising can also reach the farthest limits and spread the product very quickly and for a wider group. Online advertising has been in a constant state of development since the opening of the Internet to commerce, and here are some of the different forms that advertising has taken so far. The results also revealed that online advertising has advantages, and there are many, and each of them positively affects the company's reputation. But, like anything else, this type has a downside that can hurt when it becomes intrusive and inappropriate and drives the consumer away to make a purchase. The results also revealed that companies specialized in e-marketing use online advertisements in marketing their various products and services. Through advertisements, especially through social networking sites, especially Facebook, Twitter and Instagram, these companies can find new targeted customers who are interested in what they offer, and they use companies specialized in e-marketing. Ads are an important way to bring targeted visitors to their websites, depending on the areas in which they work, as the large number of visitors to these sites helps and the visitor stays inside them for long periods. The results revealed that mobile advertising is that campaign whose main objective is to launch actions or programs for advertising content that allow users to communicate more with the target audience. One factor that distinguishes this advertising model is that it is eminently interactive formats that are developed through any technological device (laptops, mobile phones, tablets or other technology gadgets).

References

1. Pal AK, Shukla B (2020) Impact of digital marketing on consumer buying behaviour. Iconic Res Eng J 3:11 Page 209–220
2. Al-Sartawi A (2019) Assessing the relationship between information transparency through social media disclosure and firm value. Manage Account Rev 18(2):1–20
3. Al-Sartawi A (2020) Social media disclosure of intellectual capital and firm value. Int J Learn Intellect Cap 17(4):312–323
4. Ayedi A, Jerath K, Baghaie M (2018) Exclusive placement in online advertising. Market Sci 37(6):970–986
5. Beig FA, Khan MF (2018) Impact of social media marketing on brand experience: a study of select apparel brands on Facebook. Vision 22:264–275

6. Bhambra RK (2022) Adoption of online banking in Goa amidst the pandemic. Int J Electron Bank 3(2):163–175
7. Bruce NI, Murthi BPS, Rao RC (2017) A dynamic model for digital advertising: the effects of creative format, message content, and targeting on engagement. J Mark Res 54(2):202–218
8. Chen SC, Lin CP (2019) Understanding the effect of social media marketing activities: the mediation of social identification, perceived value, and satisfaction. Technol Forecast Soc Change 140:22–32
9. Chen X, Qasim H (2021) Does E-Brand experience matter in the consumer market? Explaining the impact of social media marketing activities on consumer-based brand equity and love. J Consumer Behav 20:1065–1077
10. Cheung ML, Pires GD, Rosenberger PJ III, Leung WKS, Ting H (2021) Investigating the role of social media marketing on value co-creation and engagement: an empirical study in China and Hong Kong. Austral Market J 29:118–131
11. Choi WJ, Sayedi A (2019) Learning in online advertising. Mark Sci 38(4):584–608
12. Di Minin E, Fink C, Hausmann A, Kremer J, Kulkarni R (2021) How to address data privacy concerns when using social media data in conservation science. Conserv Biol 35:437–446
13. Duff BRL, Segijn CM (2019) Advertising in a media multitasking era: considerations and future directions. J Advert 48(1):27–37
14. Dunnan L, Jamil K, Abrar U, Arain B, Guangyu Q, Awan FH (2020) Analyzing the green technology market focus on environmental performance in Pakistan. In: 2020 3rd international conference on computing, mathematics and engineering technologies (ICoMET). IEEE, Sukkur, pp 1–5
15. Dutot V (2020) A social identity perspective of social media's impact on satisfaction with life. Psychol. Market 37:759–772
16. Edelman B, Ostrovsky M, Schwarz M (2007) Internet advertising and the generalized second-price auction: selling billions of dollars worth of keywords. Am Econ Rev 97(1):242–259
17. Gul RF, Liu D, Jamil K, Baig SA, Awan FH, Liu M (2021) Linkages between market orientation and brand performance with positioning strategies of significant fashion apparels in Pakistan. Fashion Textiles 8:1–19
18. Hafez M (2021) The impact of social media marketing activities on brand equity in the banking sector in Bangladesh: the mediating role of brand love and brand trust. Int J Bank Market 39:1353–1376
19. Hummel P, McAfee RP (2017) Loss functions for predicted click-through rates in auctions for online advertising. J Appl Economet 32(7):1314–1328
20. Jarman HK, Marques MD, McLean SA, Slater A, Paxton SJ (2021) Social media, body satisfaction and well-being among adolescents: a mediation model of appearance-ideal internalization and comparison. Body Image 36:139–148
21. Karolak M, Razzaque A, Al-Sartawi A (2021) E-services and M-services using IoT: an assessment of the Kingdom of Bahrain. In: Musleh Al-Sartawi AM, Razzaque A, Kamal MM (eds) Artificial intelligence systems and the internet of things in the digital era, EAMMIS 2021. Lecture notes in networks and systems, vol 239. Springer, Cham
22. Lejeune MA, Turner J (2019) Planning online advertising using Gini indices. Oper Res 67(5):1222–1245
23. Liu-Thompkins Y (2019) A decade of online advertising research: what we learned and what we need to know. J Advert 48(1):1–13
24. Martínez-Martínez IJ, Aguado JM, Boeykens Y (2017) Ethical implications of digital advertising automation: the case of programmatic advertising in Spain. El profesional de la Información (EPI) 26(2):201–210
25. Musleh Al-Sartawi AMA (ed) (2022) Artificial intelligence for sustainable finance and sustainable technology. ICGER 2021. Lecture notes in networks and systems, vol 423. Springer, Cham
26. Kushwaha NK, Shukla B (2021) Effectiveness of online advertising in current scenario. Int J Res Cult Soc 5(7):16–20. ISSN 2456–6683

27. Razzaque A, Al-Sartawi AMM, Karolak M (2021) Rethinking tourism entrepreneurship amidst COVID-19 through the lens of artificial intelligence and big data analytics. In: Big data for entrepreneurship and sustainable development, pp 175–188. CRC Press
28. Schmeiser S (2018) Sharing audience data: strategic participation in behavioral advertising networks. Rev Ind Organ 52(3):429–450
29. Sakhnini Z, Elali W, Al-Sartawi A (2022) Behavioral and non-behavioral factors and the level of adapting and implementing Fintech and E-Banking in Bahrain: suggested model. In: Musleh Al-Sartawi AMA (ed) Artificial intelligence for sustainable finance and sustainable technology, ICGER 2021. Lecture notes in networks and systems, vol 423. Springer, Cham
30. Sanad Z, Shiwakoti R, Al-Sartawi A (2022) Gender diversity on corporate boards and earnings management: a review. In: Musleh Al-Sartawi AMA (eds) Artificial intelligence for sustainable finance and sustainable technology, ICGER 2021. Lecture notes in networks and systems, vol 423. Springer, Cham
31. Voorveld HAM, van Noort G, Muntinga DG, Bronner F (2018) Engagement with social media and social media advertising: the differentiating role of platform type. J Advert 47(1):38–54
32. West D, Koslow S, Kilgour M (2019) Future directions for advertising creativity research. J Advert 48(1):102–114

Technology-Enhanced Learning Acceptance in Pakistani Primary Education

Mohammed Habes, Saadia Anwar Pasha, Sana Ali, Mokhtar Elareshi, Abdulkrim Ziani, and Bahaeldin Ali Bashir

Abstract The integration of digital technology in pedagogy is a tremendous change in educational arenas and is perceived as revolutionizing the educational system. This paper examined the perceptions of primary school teachers in Pakistan regarding ICT incorporation used in pedagogy. Based on the theoretical framework of TAM and analyzed by PLS-SEM to assess the conceptual framework, the findings revealed that respondents considered ICT integration in pedagogy as a fruitful decision. In this regard, PEU and PU further strengthen the acceptance of ICT incorporation. However, participants indicated some fundamental barriers regarding ICT integration in primary education. Thus, the study concluded that teachers acknowledge ICT, being a fundamental component of the educational process, as a means of upgrading the pedagogical system in Pakistan.

Keywords Pedagogy · TAM · Pakistan · Educational system · ICT

M. Habes
Radio and TV Department, Yarmouk University, Zip Code 21163, Irbid, Jordan
e-mail: mohammad.habes@yu.edu.jo

S. A. Pasha · S. Ali
Allama Iqbal Open University Islamabad, Islamabad 44000, Pakistan
e-mail: saadia.pasha@aiou.edu.pk

S. Ali
e-mail: sana_leo1990@hotmail.com

M. Elareshi (✉)
College of Communication and Media, Al Ain University, P.O Box: 64141, Al Ain, UAE
e-mail: mokhtar.elareshi@aau.ac.ae

A. Ziani · B. A. Bashir
Mass Communication College, Umm Al Quwain University, P.O Box: 536, Umm Al Quwain, UAE
e-mail: drabdul.karim@uaqu.ac.ae

B. A. Bashir
e-mail: drbahaeldin.h@uaqu.ac.ae

A. M. A. Musleh Al-Sartawi et al. (eds.), *From the Internet of Things to the Internet of Ideas: The Role of Artificial Intelligence*, Lecture Notes in Networks and Systems 557,
https://doi.org/10.1007/978-3-031-17746-0_5

1 Introduction

Digital technology has enabled teachers to be more lenient and flexible in delivering their lessons than previously. This is because of increased exposure to new educational patterns as a result of the technological revolution and its integration in pedagogy. Technological changes in school education are aiding the advancement of every individual in society [1]. Pupils are comparatively more involved in their education through ICT as part of daily use. Despite several empirical investigations assessing ICT adoption in education, examination of technology adoption at the primary level from teachers' perceptions is still scarce. This research is an attempt to fill this gap through examining the perceptions of primary school teachers regarding ICT incorporation.

2 Literature Review

2.1 Technology and Education in Pakistan

According to Usman & Oyefolahan [2], web-based technologies and devices provide a dynamic communication environment for their users. These platforms are based on the principles of "create, share, receive, respond", providing users with an equal opportunity to access and use them. As a result, ICT demands an accessible digital environment where pupils can learn new skills/knowledge and ideas and share their experience with others. PCs are more personal, and they are also available domestically, with portable devices further increasing their potential contributions to education [23].

Notably, in Pakistan, where early education is going through several transitions, technology integration can be of greater significance [3, 4]. However, although Pakistani educational institutions have been able to keep pace with modern educational trends, technology acceptance and ICT integration is still an enormous challenge because of lack of infrastructure and strategic policies. In general, ICT is already integrated in most formal educational systems, although some do not have adequate policies for the practical execution of this [4, 24].

2.2 Knowledge Sharing

Knowledge sharing is seen as inevitably challenging and as an important concept [5]. Ghazali et al. [6] describe it two or more people communicating with each other through technology. When knowledge sharing becomes reciprocated, it is much more effective, having an in-depth and prolonged impact on one's perceptions. However, today's knowledge sharing is even more efficient and robust, with long-term impacts [7]. Consequently, there has been a growing appreciation of ICT integration in education during the past decade. This wider accessibility is one of the primary reasons behind higher Internet acceptance in most forms of education. Such knowledge sharing can predict users' acceptance of ICT due to its positive PEU and PU, constructive outcomes attributed to Internet technology adoption. As a result, the growth of ICT usage is a result of knowledge sharing and the receiving of information [8].

H1: Knowledge sharing has a positive impact on accessibility.
H2a: Knowledge sharing has a positive impact on PEU.
H2b: Knowledge sharing has a positive impact on PU.

2.3 Internet Accessibility

Accessibility refers to the availability of the Internet for everyone regardless of geographical or any other relevant boundaries. As the Internet has become an integral part of our daily activities, (non)-professional arenas have complete access to its facilities. For example, learners might prefer studying from home and using the Internet for their education in this way. They might also find that digital learning is easy and gives them greater hopes of meeting their educational goals. Zhao and Song [9] further broaden the idea of Internet-based learning being accessible as a result of simple usage patterns and sufficient compatibility to meet professional criteria.

H3a: Accessibility has a positive impact on PEU.
H3b: Accessibility has a positive impact on PU.

2.4 TAM and ICT Acceptance in Primary Education

Users' acceptance and adoption of technology has been an important topic for academics and technology developers. Several models have attempted to describe technology acceptance through different frameworks such as TAM, proposed by Davis [10]. Davis [10] primarily provides two independent factors, perceived ease of use (PEU) and perceived usefulness (PU), to examine the dynamic mechanisms behind technology acceptance. TAM provides a strong conceptual support for ICT incorporation and acceptance in education, especially for e-learning [11, 24]. For example, Aypay et al. [12] looked at TAM in terms of eLearning acceptance among pre-service teachers in Turkey and indicated that both PEU and PU significantly influenced behavioral intention to use the Internet in education.

H4: Perceived ease of use has a positive impact on perceived usefulness.
H5: Perceived ease of use has a positive impact on ICT acceptance.
H6: Perceived usefulness has a positive impact on ICT acceptance (Fig. 1).

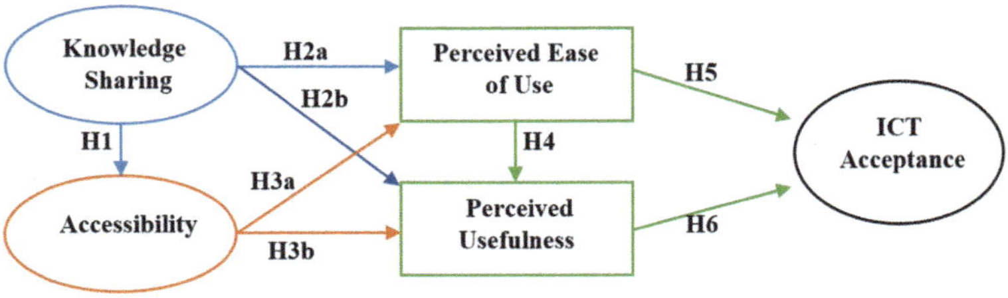

Fig. 1 The conceptual model

3 Research Methodology

3.1 Population and Sampling

This research examined primary teachers currently working in public educational institutions in Pakistan, based on a randomly selected of seven public primary schools in the city of Rawalpindi and in Islamabad, the capital of Pakistan. The study managed to conduct a sample size of n = 310 teachers, using a simple random sampling method.

3.2 Questionnaire

The study used a pre-structured, close-ended questionnaire that was adopted from the literature to measure statements regarding digital technology acceptance in primary education in Pakistan. These statements were measured on a five-point Likert scale (1 = strongly disagree and 5 = strongly agree). Finally, respondents were asked to state their demographic details.

3.3 Convergent Validity and Construct Reliability

The internal consistency of the research instrument was tested by conducting the convergent validity analysis. it helps in examining the trait well measured by the research indicators, to the extent to which the instrument can dig out generalizable and reliable results. Thus, the Cronbach Alpha (CA) values range from 0.667 to 0.826 and the Composite Reliability (CR) values range from 0.691 to 0.803. However, the CA value of PU (0.667) and the CR of the same variable (0.691) are less than the threshold value of 0.07, indicating the CR of the research model is established [13]. Moreover, the Factor Loading (FL) values of all the items are higher than the threshold value of 0.05. The Average Variance Extracted (AVE) values also surpass the threshold value of 0.05, concluding that the convergent validity is successfully established (Table 1).

Table 1 Convergent validity and construct reliability assessment

Construct	Items	FL	CA	AVE	CR
Knowledge sharing	KSG1	0.802	0.799	0.908	0.803
	KSG2	0.574			
	KSG3	0.720			
	KSG4	0.735			
Accessibility	ACS1	0.781	0.789	0.897	0.795
	ACS2	0.532			
	ACS3	0.722			
	ACS4	0.759			
Perceived ease of use	PEU1	0.804	0.732	0.881	0.733
	PEU2	0.718			
	PEU3	0.821			
Perceived usefulness	PU1	0.725	0.667	0.901	0.691
	PU2	0.690			
	PU3	0.711			
ICT acceptance	ITA1	0.832	0.826	0.857	0.830
	ITA2	0.859			
	ITA3	0.558			
	ITS4	0.696			

3.4 Discriminant Validity

The discriminant validity confirmation is a prerequisite in SEM for examining the relationship between latent variables [14]. The Fornell-Larcker criterion and the Heterotrait-Monotrait (HTMT) ratio criterion are essential for ensuring that the research model fits the discriminant validity criteria (Table 2). Thus, the study also involves discriminant validity analysis. The Fornell-Larcker criterion was run and found all the square values of AVE surpassing the correlation values. The HTMT is also fulfilled as the calculated HTMT value of 0.493 is smaller than the designated HTMT value of 0.85. Thus, the discriminant validity is successfully established.

Table 2 Fornell-Larcker criterion scale Heterotrait-Monotrait ratio scale

	KSG	ACS	PEU	PU	ITA	KSG	ACS	PEU	PU	ITA
KSG	**0.824**									
ACS	0.815	**0.804**				0.443				
PEU	0.089	0.729	**0.776**			0.042	0.442			
PU	0.087	0.746	0.606	**0.713**		0.028	0.411	0.004		
ITA	0.746	0.069	0.663	0.906	**0.734**	0.343	0.063	058	0.086	

Table 3 Coefficients of determination R^2

Variables	R^2	Strength
ACS	0.876	Strong
PEU	0.833	Strong
PU	0.769	Strong
ITA	0.791	Strong

4 Research Results

4.1 Descriptive Analysis

The descriptive analysis of the gathered data revealed that the majority of participants (61.9%) were male, 38.1% were female, and 3.5% were transgender. A total of 17.4% of respondents were 36–45 years old, 54.8% were aged 25–35, 10.3% 46–55, and 16.8% 56+. Moreover, 54.5% had 16 years of educational experience, 25.8% had university degrees or below, and 19.7% had reached postgraduate level. Finally, 50.1% of respondents came from Rawalpindi and 49.9% were from Islamabad.

4.2 Coefficients of Determination R^2

To determine the predictive potential of our research model, the coefficients of determination R^2 were calculated, as suggested by [15]. The coefficient of determination R^2 is the sum of squares due to the regression divided by the sum of total squares [16]. Thus, the coefficients of determination found the R^2 values ranging from 0.769 to 0.876, indicating that the predictive power of the proposed research model is strong (Table 3).

4.3 Hypotheses Testing

The results of testing the hypotheses indicated that KSG had a positive impact on ACS ($p \geq 0.000$, $t = 8.099$), PEU ($p \geq 0.000$, $t = 113.17$), and PU ($p \geq 0.000$, $t = 10.095$). Similarly, ACS had a positive impact on PEU ($p \geq 0.000$, $t = 10.039$) and PU ($p \geq 0.041$, $t = 9.566$). PEU had a positive impact on PU ($p \geq 0.010$, $t = 11.005$) and ITA ($p \geq 0.067$, $t = 4.518$). Finally, PU had a positive impact on ITA ($p \geq 0.022$, $t = 1.555$). Thus, the findings conclude that all the study hypotheses are significantly validated.

5 Discussions and Conclusion

Earlier research found that integrating ICT in education and teaching to be very complicated. Although, in early education, teachers' hesitancy used to present an obstacle to ICT integration, our study showed contradictory results. The respondents considered ICT integration in pedagogy as a fruitful decision. This result is supported by several studies [12] which found that ICT integration in education was mainly accelerated by its PEU and PU [17, 18]. However, the beneficial factors, such as PEU and PU, in integrating ICT in primary education are complex. Stakeholders, teachers, and pupils all acknowledge ICT as a means of upgrading the educational system in Pakistan. Similarly, our respondents also consider ICT integration in pedagogy as a step forward in keeping pace with modern educational criteria in Pakistan.

However, our respondents pointed to insufficient funds and lack of ICT awareness as two primary factors preventing ICT integration in primary education. This result is confirmed by Maqbool et al. [20], who found that the local Pakistani educational system did not have any standard training programmes to improve teacher education and professional skills [4]. Incapable and frequently varying educational policies are also significant factors hampering efforts to cope with modern educational challenges. Conventional methods of pedagogy are still the only methods of instruction and learning in Pakistan. On the other hand, several countries have stopped using conventional methods and are adopting modern learning and teaching methods, mainly assisted by ICT.

This research has provided us with an idea for implementing educational reforms in Pakistani primary education, but it also shows teachers' willingness to adopt contemporary educational patterns. This willingness is compatible with the proposition given by Hassan and Aziz [4] who highlighted the importance and dependence of ICT integration in education [21]. If the local Pakistani education system adopted ICT for primary education purposes, pupils and teachers would enjoy an updated and barrier-free education that also meets international standards [22].

5.1 Limitations and Future Research

As for the limitations and implications for future research, the sample size was limited to public schools and only teachers in two cities when other private institutions have similar concerns regarding ICT and therefore care should be considered when generalizing the results, perhaps by including large sample and pupils. Methodologically, a qualitative approach could be used to gain an in-depth understanding of the phenomenon.

References

1. Robles H, Guerrero J, LLinás H, Montero P (2019) Online teacher-students interactions using WhatsApp in a law course. J Inf Technol Educ 18:231–252. https://doi.org/10.28945/4321. Accessed 6 Nov 2021
2. Usman SH, Oyefolahan IO (2014) Encouraging knowledge sharing using web 2.0 technologies in higher education: a survey. Int J Manag Inf Technol 6(2):19–28
3. Mumtaz N, Saqulain G, Mumtaz N (2021) Online academics in Pakistan: COVID-19 and beyond. Pakistan J Med Sci 37(1):1–5
4. ul Hassan M, Aziz AA (2019) Investigating the use of computer technology for E-learning in Pakistani Maddaris: case of religious teachers. Int J Distance Educ E-Learning 5(1):44–56. http://irigs.iiu.edu.pk:64447/ojs/index.php/IJDEEL/article/view/792. Accessed 30 Mar 2022
5. Salloum SA, Al-Emran M, Habes MM, Alghizzawi M, Ghani MA, Shaalan K et al (2019) Understanding the impact of social media practices on e-learning systems acceptance. In: International conference on advanced intelligent systems and informatics. Springer, Cham, pp 360–369. https://doi.org/10.1007/978-3-030-31129-2_33. Accessed 31 Aug 2020
6. Al-Sartawi A (2020) Information technology governance and cybersecurity at the board level. Int J Crit Infrastruct 16(2):150–161
7. Adomi EE, Solomon-Uwakwe B (2019) Work related WhatsApp groups as knowledge sharing platforms among librarians in selected federal universities in Nigeria. J ICT Dev Appl Res 1(11):11–19
8. Danurdoro K, Wulandari D (2016) The impact of perceived usefulness, perceived ease of use, subjective norm, and experience toward student's intention to use internet banking. J Ekon dan Ekon Stud Pembang 8(1):17–22
9. Zhao S, Song J (2020) Students' perceptions of a learning support initiative for b-MOOCs. Int J Emerg Technol 15(21):179–194
10. Davis FD (1989) Perceived usefulness, perceived ease of use, and user acceptance of information technology. MIS Q 13(3):319–340
11. Granić A, Marangunić N (2019) Technology acceptance model in educational context: a systematic literature review. Br J Educ Technol 50(5):2572–2593
12. Aypay A, Çelik HC, Aypay A, Sever M (2012) Technology acceptance in education: a study of pre-service teachers in Turkey. Turkish Online J Educ Technol 11(4):264–272
13. Elareshi M, Habes M, Ali S, Ziani A (2021) Using online platforms for political communication in Bahrain election campaigns. Pertanika J Soc Sci Humanit 29(3):2013–2031
14. Henseler J, Ringle CM, Sarstedt M (2015) A new criterion for assessing discriminant validity in variance-based structural equation modeling. J Acad Mark Sci 43(1):115–135
15. Dufour J-M (1983) Coefficients of determination ∗. 2011
16. Figueiredo Filho DB, Silva Júnior JA, Rocha EC (2011) What is R2 all about? Leviathan (São Paulo) (3):60
17. Ma YJ, Gam HJ, Banning J (2017) Perceived ease of use and usefulness of sustainability labels on apparel products: application of the technology acceptance model. Fash Text 4(1):1–20
18. Zaman MS (2020) Impact of perceived ease of use and perceived usefulness of enterprise resource planning system adoption on end user acceptance [Unpublished Doctoral Dissertation, Capital University of Science and Technology, Islamabad]
19. AlSalem F (2016) Women and online political participation in Kuwait. In: Gunter B, Elareshi M, Al-Jaber K (eds) Social media in the Arab world: communication and public opinion in the Gulf States. I.B. Tauris, London and New York, pp 135–155
20. Maqbool S, Abdulmalik S, Maqbool S (2020) A correlational study between learning motivation and satisfaction in online courses. Humanit Soc Sci Rev 8(1):716–724. https://doi.org/10.18510/hssr.2020.8186716%7Cwww.hssr.in. Accessed 8 July 2021
21. Alhumaid K, Ali S, Waheed A, Zahid E (2020) COVID-19 & elearning: perceptions & attitudes of teachers towards E-learning acceptance in the developing countries. Multicult Educ 6:10–0

22. Shah U, Khan SH, Reynolds M (2020) Insights into variation in teachers' pedagogical relationship with ICT: a phenomenographic exploration in the Pakistani higher education context. Technol Pedagog Educ 541–555
23. Musleh Al-Sartawi AMA (eds) (2022) Artificial intelligence for sustainable finance and sustainable technology. In: ICGER 2021. Lecture notes in networks and systems, vol 423. Springer, Cham
24. Memdani L (2020) Demonetisation: a move towards cashless economy in India. Int J Electron Bank 2(3):205–211

Covid-19 Pandemic and Economic Growth in Jordan: Evidence from a Panel Cointegration Methodology

Mohamed Ibrahim Mugableh

Abstract The current study gains an insight into the determinants of economic growth (gross fixed capital formation, labor force, trade openness, economic freedom, and governance) taking into consideration the effect of the Covid-19 pandemic. It employs a panel co-integration methodology in Jordan over the period 1975–2020. The results show that gross fixed capital formation, labor force, trade openness, economic freedom, and governance indicators had major positive impacts on economic growth. The Covid-19 pandemic has had a substantial negative effect on economic growth. On the basis of the results, the government of Jordan ought to perform real institutional reforms and policies to promote human rights, and improve the rule of law, ensuring that resources are used efficiently and effectively in the pursuit of economic development. In addition, the promotion of trade such that new knowledge and leading technologies can be transferred, labor and capital productivity is increased, domestic and foreign competition is encouraged, and more international financial flows and investments are attracted. to the best of the author's knowledge and belief, the current study is the first to examine the determinants of economic growth in Jordan over the period, 1975–2020, taking into consideration the effect of the Covid-19 Pandemic.

Keywords Covid-19 · Co-integration · Economic growth · Jordan

1 Introduction

Economic growth has been considered a goal and obsession that all countries strive to achieve. It is one of the most important macroeconomic variables, which is reflected in the level of per capita income and welfare, and therefore the achievement of growth rates. Given the great importance of economic growth, economists across all schools and stages of thought have taken an interest and focused on the topic of growth, its explanation, and knowledge of its determinants. Economic

M. I. Mugableh (✉)
Faculty of Administrative and Financial Sciences, Department of Finance and Banking Science, Irbid National University, Irbid, Jordan
e-mail: dr.magableh@inu.edu.jo

A. M. A. Musleh Al-Sartawi et al. (eds.), *From the Internet of Things to the Internet of Ideas: The Role of Artificial Intelligence*, Lecture Notes in Networks and Systems 557,
https://doi.org/10.1007/978-3-031-17746-0_6

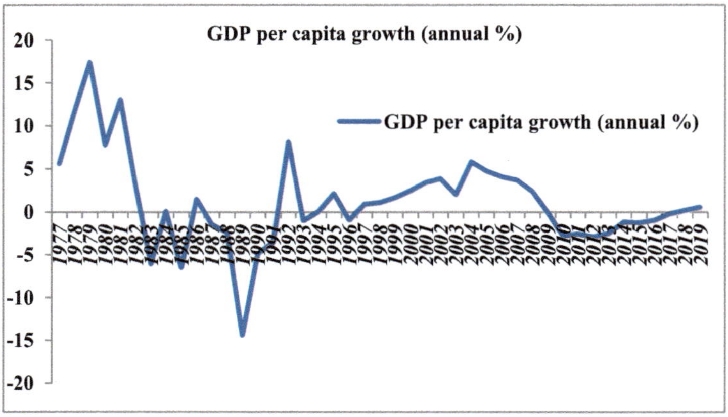

Fig. 1 GDP per capita growth (annual %) in Jordan over the period, 1977–2019 *Source* World Bank (2021)

growth has known different stages of development in Jordan that are consistent with the economic approach used. Recently, the government implemented the Jordan economic growth plan (JEGP) 2018–2022 to undertake private investments, public projects and required policy interventions. Successful implementation of this plan ensures a sustainable growth trajectory in the face of regional turmoil [31]. Figure 1 shows that gross domestic product (GDP) per capita registered an annual growth of 0.123 and 0.485% in 2018 and 2019, respectively. In light of this presentation, this study tries to find out the determinants of economic growth in Jordan over the period, 1975–2020 using quantitative methods that help to measure and predict the path of economic growth. These determinants are gross fixed capital formation, labor force, trade openness, economic freedom, Covid 19 pandemic, and governance indicators.

As a result of the outbreak of the Covid-19 pandemic, economic growth in Jordan is expected to stop [23]. Therefore, the Jordanian government imposed an economic shutdown that disrupted the economic activities of non-essential goods, in addition to limiting the movement of people in order to control the virus. Regarding the Jordanian economy during the lockdown period; some results reports indicated a decrease of 30% in production. On the other hand, statistics indicated that the food systems in Jordan witnessed a 40% decrease in production. Over 20% of jobs were lost during the lockdown, mostly in the service sector, followed by agriculture. The lockdown triggered a one-fifth decrease in household income, owing to a contraction in service sector operations, a decline in manufacturing production, and lower remittances from abroad [24]. Jordan's GDP growth rates registered negative records during 2020, ranging from −5.7% to −7.4%, depending on the speed at which the economy recovers. It is predicted that recovery will be slow. This period of economic growth presents prospects for long-term economic reform and systemic change. More economic diversification, greater resilience to economic shocks, and job growth should be the priorities of economic policies and incentives [13].

Numerous empirical studies attempted to examine the determinants of economic growth across developed and developing countries [3, 8, 10, 11, 17, 19–22, 25, 27, 29, 32]. However, the current study is the first to examine the determinants of economic growth in Jordan over the period, 1975–2020, taking into consideration the effect of Covid-19. The rest of this paper is structured as follows. Section 2 discusses the theoretical background of economic theories, while Sect. 3 provides a review of the literature. Section 4 defines data sources and variables. Section 5 deliberates methodology and results from the analysis, and finally, Sect. 6 concludes the paper and debates conclusions.

2 Theories of Economic Growth

The classical theory includes the views of [18, 28, 30]. Capital accumulation is the key to progress in economic growth. The higher the rate of profits, then, the higher the rate of capital accumulation and investment will be. Profits tend to decline due to increased competition. Once the profits start to decrease until the rate of profit reaches zero, capital accumulation stops, and the population stabilizes; then, the end result of growth is a recession. *The Keynesian theory* developed by [15] argued that aggregate demand could be influenced by economic policies by the government. The increases in government expenditures and lowering tax rates pull the economy out of the recession. Besides, monetary and fiscal policies are vital tools to manage the economy and control inflation and unemployment rates. The *new classical theory* developed by Solow (1970) argued that markets alone are generated maximum welfare. It rejected the Keynesian theory and limited the role of government. The new classical theory argued that trade liberalization enhances economic growth [9].

3 Literature Review

[1] examined the determinants of economic growth in Jordan including domestic investment, foreign direct investment, economic openness, and labor over the period, from 1985–2017. The results showed that economic growth is significantly influenced by the study's determinants. [24] examined the determinants of economic growth (i.e., demography, education, trade, and investment) using the Bayesian model averaging over the period, 1960–1990. [10] examined the determinants of economic growth including initial income, a fraction of Confucian, life expectancy, equipment investment, a fraction of Muslim, sub-Saharan dummy, rule of law, open economy, degree of capitalism, and a fraction of protestant over the period 1960–2016. [5] examined the determinants of economic growth in Kuwait (oil revenues, real exchange rate, real domestic output, inflation, global equity prices, short-term interest rates, foreign real output). The result showed that economic growth is significantly influenced by these determinants. [32] examined the determinants of economic growth in South

American countries over the period 1960–2008. The results showed that economic growth is driven most strongly by human and physical capital accumulation, foreign direct investment, and trade openness [4].

[8] examined the relationships between R&D intensity, waste generation, and economic growth. The results using the panel vector error correction model showed bidirectional causality between waste generation and economic growth. [29] examined the association between energy emissions, economic growth, and biomass energy. The results showed insignificant relationships among variables. [25] investigated the determinants of economic growth measured as its share in world GDP, in comparison with the per capita economic growth. These determinants are an investment, world population, exports, human capital, financial capital flows, and R&D investment. [19] investigated the impact of energy consumption on economic growth using a panel ARDL approach and a panel quantile regression model. The results documented that energy consumption increased the level of economic growth in 29 countries in the organization for economic cooperation and development.

4 Data and Variables Definitions

The current study examines the determinants of economic growth in Jordan over the period, 1980–2020. The definition of dependent and independent variables is given in Table 1.

5 Methodology and Results Analysis

The current study employs the Cobb–Douglas production function, which is a popular formula to estimate the relationship between real GDP and the factors of production as the following Eq. (1):

$$Y_t = A_t K_t^a L_t^B \tag{1}$$

Y_t: is the real GDP at time t, A: is the total factors of productivity or technological progress at the same time t, K and L are capital and labor factors of production during a specified period of time t, respectively. Equation (1) could be expanded as in Eq. (2):

$$Y_t = A_t K_t^a L_t^B TO_t^{\pi} E_t^{\mu} GI_t^{\lambda} \tag{2}$$

where, TO represents the trade openness at time t, E is economic freedom at time t, GI denotes governance indicators at time t, (α, β, π, μ, λ) denote the output elasticity of capital, labor, trade openness, economic freedom, and governance indicators respectively. Equation (2) could be rewritten as in Eq. (3) in logarithms for both sides:

Table 1 Definition of variables

Variable	Definition	Source	Website
EG—economic growth	the percentage of the annual change in the real gross domestic product (GDP). The GDP is the market value of all goods and services produced within a country's borders in a specific period of time and is usually calculated on an annual basis		
GFCF—gross fixed capital formation to GDP ratio	The value of net additions to the capital stock, such as tools, transportation assets, and equipment		
LF—labor force to GDP	Is the number of people that are employed or actively looking for work		
TO—Trade openness to GDP	Is a proxy of economic policies that either invite or prevent trade between countries. It is the sum of exports and imports to the country's GDP		
EF—economic freedom	Is the proxy where governments allow capital, goods, and labor to move freely and without constraints	2020 index of economic freedom	
Dummy variable—Covid 19	A dummy variable usually takes the value 1 to indicate the presence of the Covid 19 effect, while 0 to indicate the absence of the Covid 19 effect		
Governance	Is defined by the World Bank in which power is exercised in the management of a country's social and economic resources for development. The six indicators of governance are voice and accountability, political stability, government effectiveness, regulatory quality, rule of law, and control of corruption	World Bank Governance Indicators	

$$\text{Ln } Y_t = \text{Ln } A_t + \alpha \text{Ln } K_t + \beta \text{Ln } L_t + \pi \text{Ln } TO_t + \mu \text{Ln } E_t + \lambda \text{Ln } GI_{it} \quad (3)$$

For estimation purposes Eq. (3) is transformed into Eq. (4) as the following:

$$\text{Ln } Y_{it} = C + \alpha \text{Ln } K_{it} + \beta \text{Ln } L_{it} + \pi \text{Ln } TO_{it} + \mu \text{Ln } E_{it} + \lambda \text{Ln } GI_{it} + \delta DM_{it} + e_{it} \quad (4)$$

where C is the constant term, e_t represents the error term, and the total factor productivity has been excluded from Eq. (4) because it cannot be measured. K is measured by the gross fixed capital formation to GDP, L is measured by the growth rate of the labor force to GDP, TO is measured by the sum of exports and imports to GDP, E is measured by the 2020 index of economic freedom, GI represents the governance indicators (voice and accountability (VA), political stability (PS), government effectiveness (GE), regulatory quality (RQ), rule of law (RL), and control of corruption (CC)). DM is the dummy variable that represents the absence or occurrence of the global financial crisis (2007–2008) and Covid-19 (2019–present).

The first step in the analysis of the result is to use [12] stationary test for dynamic heterogeneous panels. Table 2 shows the stationary levels of dependent and independent variables included in Eq. (4). The null hypothesis (variables do include unit root) is rejected, and the alternative hypothesis (variables do not include unit root) is accepted. Thus, all variables are stationary at the first difference i.e., integrated at level I(1). After fulfilling the basic condition of co-integration, [14] panel co-integration test is applied. This test is a least square dummy variable estimator which is different from co-integration tests for the pure time-series data. The advantage of this test is that can be applied to panel data, especially for dummy variables. Table 3 provides the results of the [14] residual panel co-integration test. The null hypothesis (no co-integration exists among variables) is rejected and the alternative hypothesis is accepted (co-integration exists among variables). After confirming the existence of a panel co-integration among variables, the fully modified ordinary least square (FMOLS) test is applied to provide optimal unbiased estimates of co-integrating regressions.

The FMOLS test was originally designed by [26] and takes into account the problems of serial correlation effects and endogeneity. Table 4 provides the results of FMOLS, which are consistent with the economic theory and previous studies [6, 7, 16]. The estimated coefficients of independent variables except for the dummy variable (i.e., Covid-19) significantly and positively affect economic growth over the period of study. Covid-19 has a significant negative effect on economic growth in Jordan.

Table 2 Results analysis of stationarity test

Variables	Test in	By including	Statistics value	Probability	# of lags	Result
Ln Y_{it}	The first difference I(1)	Intercept and time trend	−2.250*	0.01	2	Stationary at the first difference I(1)
Ln K_{it}	The first difference I(1)	Intercept and time trend	−2.115*	0.01	2	Stationary at the first difference I(1)
Ln L_{it}	The first difference I(1)	Intercept and time trend	−3.135*	0.01	2	Stationary at the first difference I(1)
Ln TO_{it}	The first difference I(1)	Intercept and time trend	−4.123*	0.01	2	Stationary at the first difference I(1)
Ln E_{it}	The first difference I(1)	Intercept and time trend	−3.156*	0.01	2	Stationary at the first difference I(1)
Ln VA_{it}	The first difference I(1)	Intercept and time trend	−4.133*	0.01	2	Stationary at the first difference I(1)
Ln PS_{it}	The first difference I(1)	Intercept and time trend	−4.121*	0.01	2	Stationary at the first difference I(1)
Ln GE_{it}	The first difference I(1)	Intercept and time trend	−3.901*	0.01	2	Stationary at the first difference I(1)
Ln RQ_{it}	The first difference I(1)	Intercept and time trend	−3.101*	0.01	2	Stationary at the first difference I(1)
Ln RL_{it}	The first difference I(1)	Intercept and time trend	−2.522*	0.01	2	Stationary at the first difference I(1)
Ln CC_{it}	The first difference I(1)	Intercept and time trend	−3.156*	0.01	2	Stationary at the first difference I(1)

Notes (1) * denotes the significance at the 1% level. (2) *Source* Eviews software

Table 3 Results analysis of co-integration test

Equation	t-statistics	Probability	# of lags	Bandwidth selection	Result
(4)	−3.179*	0.01	2	Newey west automatic	Co-integrated

Notes (1) * denotes the significance at the 1% level. (2) *Source* Eviews software

Table 4 Results of the fully modified ordinary least square test

Variables	Coefficients	Probabilities	Significance (%)
C	10.0	0.01	1
Ln K_{it}	6.11	0.06	10
Ln L_{it}	7.10	0.07	10
Ln TO_{it}	9.12	0.04	5
Ln E_{it}	9.50	0.05	5
Ln VA_{it}	8.23	0.05	5
Ln PS_{it}	6.43	0.06	10
Ln GE_{it}	9.10	0.04	5
Ln RQ_{it}	9.12	0.04	5
Ln RL_{it}	10.1	0.03	5
Ln CC_{it}	7.16	0.07	10
DM_{it}	−6.1	0.06	10

Source Eviews software

6 Conclusions and Policy Implications

Over the period 1975–2020, this paper found that gross fixed capital formation, labor force, trade openness, economic freedom, and governance indicators had major positive impacts on economic growth. None of these variables had the greatest substantial positive impact on economic growth during the study period. The Covid-19 pandemic has had a substantial negative effect on economic growth. Indeed, governments that provide reliable and effective public services that are accountable and sensitive to their citizens' needs will ultimately create a democratic climate that is crucial for achieving desired economic development [2]. Furthermore, both foreign and domestic investors typically base their investment decisions on the application of good governance. In fact, the World Bank and the International Monetary Fund have stated that good governance is a requirement for receiving most external aid. Countries with strong public policy and governance can earn higher credit scores, which can lower capital costs and boost economic efficiency. As a result, Jordan's government ought to apply the following:

- Real institutional reforms and policies to promote human rights, and improve the rule of law, ensuring that resources are used efficiently and effectively in the pursuit of economic development.

- Improve educational quality and focus on developing the skills and expertise of their labor force.
- Develop specialization in the production of goods in which they have comparative advantages, as well as diversify their production and sources of national income so that they do not rely solely on exporting natural raw materials.
- The promotion of trade such that new knowledge and leading technologies can be transferred, labor and capital productivity is increased, domestic and foreign competition is encouraged, and more international financial flows and investments are attracted.
- Rising fixed and human capital investments should be accompanied by efforts to strengthen national cohesion and peaceful coexistence in order to maintain political and social stability.

References

1. Al-Smadi RW, Malkawi E (2020) Determinants of Jordanian economic growth: time series approach. J Crit Rev 7(2):534–541
2. Al-Sartawi A (2020) Does it pay to be socially responsible? Empirical evidence from the GCC countries. Int J Law Manage 62(5):381–394
3. Al-Qudah AA, Al-Okaily M, Alqudah G et al (2022) Mobile payment adoption in the time of the COVID-19 pandemic. Electron Commer Res. https://doi.org/10.1007/s10660-022-09577-1
4. Bhambra RK (2022) Adoption of online banking in Goa amidst the pandemic. Int J Electron Bank 3(2):163–175
5. Burney NA, Mohaddes K, Alawadhi A, Al-musallam M (2018) The dynamics and determinants of Kuwait's long-run economic growth. Econ Model 71:289–304
6. Bustos S, Yildirim MA (2020) Production ability and economic growth. Res Policy, in press
7. Emara N, El said A (2021) Financial inclusion and economic growth: the role of governance in selected MENA countries. Int Rev Econ Fin, in press
8. Gardiner R, Hajek P (2020) Municipal waste generation, R&D intensity, and economic growth nexus – a case of EU regions. Waste Manage 114:124–135
9. Gogia J, Chakraborty D (2022) Open banking: a revolution in the Tech-Fin industry. Int J Electron Bank 3(2):100–120
10. He Q, Xu B (2019) Determinants of economic growth: a varying-coefficient path identification approach. J Bus Res 101:811–818
11. Hossain MS, Mitra R (2013) The determinants of economic growth in Africa: a dynamic causality and panel co-integration analysis. Econ Anal Policy 43(2):217–226
12. Im IM, Pesaran MH, Shin Y (2001) Testing for unit roots in heterogeneous panels. J Econom 115(1):53–74
13. International Food Policy Research Institute Report (2020) Impact of COVID-19 on the Jordanian economy. http://ebrary.ifpri.org/utils/getfile/collection/p15738coll2/id/134132/filename/134343.pdf
14. Kao C (1999) Spurious regression and residual-based tests for cointegration in panel data. J Econom 90:1–44
15. Keynes JM (1936) The general theory of employment, interest, and money, 2007 edn. Palgrave Macmillan. ISBN 978-0-230-00476-4
16. Law SH, Ng CH, Kutan AM, Law ZK (2021) Public debt and economic growth in developing countries: nonlinearity and threshold analysis. Econ Model 98:26–40

17. Lee JW, Hong K (2012) Economic growth in Asia: determinants and prospects. Jpn World Econ 24:101–113
18. Mill JS (1965) Principles of political economy with some of their applications to social philosophy. Routledge and Kegan Paul, Toronto
19. Mugableh MI (2013) Analysing the CO2 emissions function in Malaysia: autoregressive distributed lag approach. Procedia Econ Financ 5:571–580
20. Mugableh MI (2015) Economic growth, CO2 emissions, and financial development in Jordan: equilibrium and dynamic causality analysis. Int J Econ Financ 7(7):98–105
21. Mugableh MI, Oudat MS (2018) Modelling the determinants of foreign portfolio investments: a bound testing and causality analysis for Jordan. Acad Account Financ Stud J 22(4):1–8
22. Mugableh MI, Oudat MS (2018) Economic growth and financial development nexus in Malaysia: dynamic simultaneous equations models. Asian J Fin Account 10(1):143–161
23. Musleh Al-Sartawi AMA (eds) (2022) Artificial intelligence for sustainable finance and sustainable technology. In: ICGER 2021. Lecture notes in networks and systems, vol 423. Springer, Cham
24. Musleh Al-Sartawi AMA (2020) E-learning improves accounting education: case of the higher education sector of Bahrain. In: Themistocleous M, Papadaki M, Kamal MM (eds) Information systems, EMCIS 2020. Lecture notes in business information processing, vol 402. Springer, Cham
25. Park JK, Ryu D, Lee K (2019) What determines the economic size of a nation in the world: determinants of a nation's share in world GDP vs. per capita GDP. Struct Chang Econ Dyn 51:203–214
26. Phillips PCB, Hansen BE (1990) Statistical inference in instrumental variables regression with I(1) process. Rev Econ Stud 57(1):99–125
27. Polykretis C, Alexakis DD (2021) Spatial stratified heterogeneity of fertility and its association with socio-economic determinants using geographic detector: the case study of Crete Island, Greece. Appl Geogr 127
28. Ricardo D (1951) 3rd edition 1821 On the principles of political economy and taxation. Cambridge University Press, Cambridge
29. Shah SAR, Naqvi SAA, Riaz S, Anwar S, ABbas N (2020) Nexus of biomass energy, key determinants of economic development and environment: a fresh evidence from Asia. Renew Sustain Energy Rev 133
30. Smith A (1776) The wealth of nations. https://www.amazon.com/Wealth-Nations-Adam-Smith/dp/1505577128
31. The Economic Policy Council Report (2018) Jordan Economic Growth Plan 2018 – 2022. http://extwprlegs1.fao.org/docs/pdf/jor170691.pdf
32. Vedia-Jerez, Chasco C (2016) Long-run determinants of economic growth in South America. J Appl Econ XIX(1):169–192

Artificial Intelligence in Educational Examinations

Abdulsadek Hassan, Sumaya Asgher Ali, and Nader Mohammed Sediq Abdulkhaleq

Abstract The study aims to identify the uses of artificial intelligence in exams with the aim of conducting exams in an effective manner and preventing cheating in these exams by using artificial intelligence applications. Errors that may occur and correction of the test electronically, and the results also showed that this technology is a faculty member of the university to carry out the process of monitoring and monitoring the performance of students during the performance of their tests with high accuracy, using modern digital means and artificial intelligence that can distinguish between natural and suspicious movements, so that every movement is monitored The student performs it during the test and analyzes and categorizes it.

Keywords Artificial intelligence · Examination · Cheating

1 Introduction

Artificial intelligence also provides opportunities for students who speak different languages or have visual or auditory problems. For example, presentation translator is an AI-based solution that does real-time translation. Using AI speech recognition, students can listen or read in their native language [15]. This will give possibilities to students who may not be able to attend school due to illness or who need to learn at a different level or in a particular subject that is not available at their school. AI can help break down the silos between traditional schools and grade levels [7]. The teacher spends an inordinate amount of time correcting assignments and tests. AI can step in and get these tasks done quickly while at the same time making recommendations on how to bridge learning gaps [19]. Artificial intelligence-based applications provide an opportunity to study in free time. Additionally, students can get feedback from teachers in real time [4]. AI-based platforms provide virtual mentors to track student progress. Of course, only human tutors can better understand students' needs, but it's good to get instant feedback and feedback from the virtual

A. Hassan (✉) · S. A. Ali · N. M. S. Abdulkhaleq
Ahlia University, Manama, Kingdom of Bahrain
e-mail: dr.sadek1964@yahoo.com

A. M. A. Musleh Al-Sartawi et al. (eds.), *From the Internet of Things to the Internet of Ideas: The Role of Artificial Intelligence*, Lecture Notes in Networks and Systems 557,
https://doi.org/10.1007/978-3-031-17746-0_7

tutor [3]. Modern technologies such as VR and gamification help to engage students in the learning process, making it more interactive [10]. The world's dependence on artificial intelligence is expected to increase in the next few years in many fields, including the field of education, so you will learn about the importance of artificial intelligence in education [17]. Machine learning (ML) and artificial intelligence (AI) are major drivers of growth and innovation across the board and the education sector is no different. According to studies, over 47% of educational management tools will be enabled with AI capabilities in the next three years [17]. While AI-powered solutions have been in the education technology field for some time, the education field has been slow to embrace them. However, the Corona pandemic has radically transformed the landscape, forcing teachers to rely on technology for virtual learning. Now, 86% of teachers say technology should be an essential part of education [16]. AI technologies have the potential to improve education, helping the education sector evolve to better benefit students and teachers [21].

2 Benefits of Artificial Intelligence for Students

A student's goal is very simple and that is to obtain a degree that proves their knowledge and artificial intelligence can help students achieve this goal and artificial intelligence can have a significant impact on students' educational journey by simplifying the education process, providing access to appropriate courses and improving communication with teachers and allowing more time to focus on other aspects of life some examples are:

Customization: Personalization is one of the biggest trends in education. With AI, students now have a personalized approach to learning programs based on their unique experiences and preferences [20]. AI can adapt to each student's level of knowledge, learning speed, and desired goals in order to get the most out of their education in addition, AI-powered solutions can analyze students' past learning history, identify weaknesses, and offer courses that are most suitable for improvement, providing many opportunities for a personalized learning experience [12].

Teaching: Although it is common for students to ask for extra help outside the classroom, many teachers do not have free time for students [11]. AI teachers and chatbots are ideal in these scenarios, but no chatbot can replace the teacher. However, AI tools can help students hone their skills and improve weaknesses outside the classroom. AI provides an individual learning experience without the need for having a teacher to answer questions at all hours of the day. The AI-powered chatbot can quickly answer students' questions [8].

Quick Responses: Nothing is more frustrating than asking a question and then waiting for an answer three days later. Teachers and faculty are often subjected to repetitive questions on a daily basis [9]. AI can help students find answers to their most common questions in seconds by supporting automation. This not only saves teachers a lot of time but also helps students spend less time tracking answers or waiting for their questions to be answered [10].

3 Global Access to 24/7 Learning

The tools available in artificial intelligence technology allow access to learning for all students at any time and from anywhere [7]. Each student learns at their own pace, and it is easy for them to explore what works for them without waiting for a teacher. In addition, students from all over the world can access high-quality education without incurring travel and living expenses.

4 Benefits of Artificial Intelligence for Teachers

Most teachers and faculty are not afraid to admit that they are short on time which is understandable given the number of daily tasks they have. Many teachers would like to spend more time teaching students individually and dive into research and further their education but do not have the ability to do so due to the lack of the time [10]. AI can help save teachers time by automating tasks, analyzing student performance, and closing the learning gap [13]. Here's how: AI can be used to automate most normal tasks including administrative work, classifying papers, assessing learning patterns in schools, answering general questions, and more [1]. According to some studies, teachers spend 31% of their time preparing for lessons, marking exams and doing administrative work. However, with automation tools teachers can automate manual processes leaving more time to focus on what matters most [14].

5 Benefits of Artificial Intelligence for Teachers

Most teachers and faculty are not afraid to admit that they are short on time which is understandable given the number of daily tasks they have [5]. Many teachers would like to spend more time teaching students individually and dive into research and further their education but do not have the ability to do so due to the lack of the time. AI can help save teachers time by automating tasks, analyzing student performance, and closing the learning gap [17]. Here's how:

Customization: Just as AI can customize courses for students, it can do the same for educators by analyzing students' learning capabilities and educational history. AI can give teachers a clear picture of topics and lessons to be re-evaluated. This analysis allows for the best educational program for students [14]. By analyzing the specific needs of each student, teachers and professors can adjust their courses to address the most common knowledge gaps or challenging areas before the student lags far behind their peers [2].

Answering Questions: With access to the school's database, AI-powered chatbots can answer a variety of general and recurring questions typically asked by students without involving a faculty member [5, 23]. AI allows teachers more time to focus on preparing lessons, researching curricula, or improving student engagement [18].

Task Automation: AI can be used to automate most normal tasks including administrative work, classifying papers, assessing learning patterns in schools, answering general questions, and more [7]. According to some studies, teachers spend 31% of their time preparing for lessons, marking exams and doing administrative work. However, with automation tools teachers can automate manual processes leaving more time to focus on what matters most [17]. From online textbooks to fully remote lectures, advances in educational technology are unprecedented. Today, artificial intelligence plays a role in helping students and teachers improve and automate learning and teaching tasks. As the field of artificial intelligence advances and innovates, we will see improved learning outcomes for all students and teachers [18]. From online textbooks to fully remote lectures, advances in educational technology are unprecedented. Today, artificial intelligence plays a role in helping students and teachers improve and automate learning and teaching tasks. As the field of artificial intelligence advances and innovates, we will see improved learning outcomes for all students and teachers [15].

6 How Does AI Drive Innovation to Produce an Enhanced Learning Experience?

AI simulates human intelligence processes in algorithmic services that improve processes, speed, and increase efficiency through learning, self-correction, and heuristics [21]. Artificial intelligence and machine learning are the main drivers of innovation in many professions, including education. AI promises to advance the education sector by improving learning and teaching for both students and teachers alike [16].

Among these are progressive education, which are initiatives that focus on experiential learning based on students' talents, needs, and desires to achieve the educational goals of both teacher and student [5]. "Artificial intelligence and machine learning are two tools that can play an instrumental role in helping educational institutions achieve progressive education goals [20]. These tools include adaptive learning systems, virtual reality, augmented reality, and serious game models. One can already see that AI systems are gaining ground and having a positive impact in the education sector [6]." In this regard, AI, along with delivering content at scale within a progressive learning model, is changing how people interact with educational information (and it has already changed in many respects), particularly in the context of lifelong learning. Among these areas is intelligent content creation, where artificial intelligence software automatically updates classroom content, such as digital textbooks, digital study guides, and digital content interfaces [2]. AI can also help increase universal access to education for global online learning communities, including providing content in different languages and skill levels while enabling more peer support among learners [21, 23]. Many AI and machine learning tools are already integrated into regular classroom programs, such as PowerPoint's presentation translator, which translates voice into synchronous text in several languages in real time [19]. These automatic translation tools help deliver content to a wider audience of students and professionals seeking to develop their skills and knowledge. There is a long list of other AI-powered content visualization and simulation efforts to enter the market. So, the question now is not about "whether" AI will affect education, but about "how" and "when" that will happen [10]. As with any of the previous educational methods, the focus in the use of artificial intelligence should be on what can contribute to enhancing the quality of teaching and learning [4].

7 The Importance of Artificial Intelligence Applications in Education

The applications of artificial intelligence in education are the latest method of learning, because of its great importance [7]. 1) The transfer of various human experiences, multi-machines and smart devices [14]. 2) Giving the opportunity for people with special needs and other people of all kinds to deal with machines, through the use of human language instead of programming language, so that it is not restricted to experts and specialists only [21]. 3) Smart devices are characterized by objectivity and accuracy that make their decisions completely far from falling into error, as well as making them neutral and independent even from external or personal interference, and far from racism [2]. 4) These smart systems will successfully take over the tasks of exploring places that are still unknown and will also contribute to rescue operations that occur during natural disasters, and will also perform many strenuous and dangerous work, and work in fields with different and complex details, which require sensitive and quick decisions that do not Include errors or delays [3].

8 The Most Important Initiatives Related to Artificial Intelligence in the Field of Education

There are many applications of artificial intelligence in education that are being developed, including artificial intelligence mentors for learners, further development of smart content, and a new way of personal development for teachers through virtual global conferences. Education may be a little slower in adopting artificial intelligence and machine learning, but changes have begun and will continue [4]. Thus, we have come to the end of this article today, and we have mentioned how artificial intelligence affects education, and it is expected that artificial intelligence systems will be employed in education more in the coming years [11].

(1) Thinkster Math: It is an application that combines the mathematics curriculum with a personal learning style. This application monitors the mental processing of each student, which gradually reveals on the IPad screen, and presents the user with different problems according to his ability, and once the user writes about the way he reached the answer The app analyzes its work and identifies why it went wrong or how it understood a specific part of the problem. The app improves all students' logical processing by providing instant, personalized feedback [9]. **(2) Brainly:** It is a social networking site for classroom questions. The program allows learners to ask homework questions and receive automatic, audited answers from their colleagues. This program helps to collaborate to reach correct answers. The program has a variety of experts in the subjects working to create classroom-like environment [15]. **(3) Content Technologies, Inc.** An artificial intelligence company for research, development, instructional design and production of solutions for content applications, uses "deep learning" to create specialized textbooks that fit course needs [18]. Educators enter curriculum descriptions into the content technology engine and then content technology machines use algorithms to produce personalized books and study materials... based on core concepts of the curriculum, and this initiative seeks to help publishers create effective textbooks that provide each learner with individualized, personalized learning [1]. **(4) Mika site:** This site at Carnegie Mellon University provides artificial intelligence-based teaching tools for busy learners, and students who lack personal interest, and Mika specializes in teaching to undergraduate students to bridge the gaps in crowded classrooms. The application guides the learning process of each student and makes the learner aware of their daily progress and adjusts lessons according to the needs of each student [3]. **(5) Netex Learning:** This site allows teachers to design the curriculum across multiple digital devices and the site helps even the most tech-savvy trainers integrate interactive elements such as audio, video and self-assessment into their digital lesson planning, all in a virtual platform for personalized learning [8]. On this site educators can create personalized student materials ready for publication on any digital platform while presenting video conferencing, digital discussions, personalized assignments, and educational analytics that demonstrate visual representations of each student's personal growth [13].

Electronic Exam Requirements: The successful preparation of electronic exams requires the availability of the infrastructure of computer laboratories, internet lines and specialized programs, in addition to the availability of manpower such as programmers, maintenance technicians and technical support [21]. Preparation also requires training of teachers in the skills of designing and building electronic tests. Training students on how to deal with the electronic exam environment is one of the most important requirements to be taken into consideration [3]. Awareness must also be spread in educational institutions about the importance and feasibility of using electronic tests in teaching and learning. As for the design of electronic tests, it is necessary to take into account some educational and technical standards to ensure their quality and achieve the desired goals [13].

The Use of Artificial Intelligence and Testing Systems. It is possible to use artificial intelligence techniques to change the regular test systems that uniformly evaluate students according to one test, which leads to a great injustice to the creative student, because these systems focus directly on typical answers in traditional tests [15], and even if some universities are trying currently, the methods of its tests have been changed to address the issue of innovation and reliance on abilities and creativity, but the traditional view still prevails, and the tests in their current method are seen as more like memory tests, in addition to the great burden that falls on the teacher, which is to correct a huge amount of tests annually—especially if these tests are in the form of essay questions—so artificial intelligence systems can correct a large part of these tests, by translating words and studying patterns very accurately [19]. Therefore, artificial intelligence systems can fill in the gaps in the individual differences between learners and can also "liberate" teachers from a large part of their responsibilities, which reflects positively on the educational and research process at the same time. In universities, for example, there will be greater opportunities for professors to focus on production [7]. At the same time, artificial intelligence systems will reduce a large part of the cost to universities—especially government ones—which will contribute to increasing the number of seats inside universities and institutes [2].

9 Advantages of Electronic Exams Using Artificial Intelligence Systems

1. Link the courses with the exam questions in a transparent and organized manner

In case of bridging the gap between syllabus content and exam questions, this feature is very useful. We will not talk here about the difference between electronic and traditional exams, the purpose of any exam is to ensure that the student has fully absorbed the content of the target course and therefore it is necessary that the exam elements reflect this content [13].Many people in learning and teaching are interested in the idea of linking course content to exam questions, and you find this the subject of many researches in the field of education, measurement, and evaluation. This linkage reassures us that we are on the right track and that what was actually meant to be taught was taught in the test questions [16].That is why the designers of the electronic test platform are interested in highlighting the cornerstone of any test, which is the content itself, for example, the hours of teaching the scheduled content and the weight of each part in it to create distinctive, clear and fair test questions for students and teachers alike [6].

2. The ability to measure students' cognitive levels
It is a hierarchy of six levels of cognition—remembering, understanding, applying, analyzing, evaluating, and creating—that move from remembering to constructing and hypothesizing [10].

3. Live monitoring capability
With electronic monitoring via camera and microphone, the test observer is able to see the student, control the video broadcast remotely, and intervene in real time if necessary. This gives the monitors an extra level of security [5].

Artificial Intelligence Tools to Effectively Manage Online Exams: With the rush to return to some of the normalcy and test students for the start of the new school year, many institutions scrambled to acquire electronic examination systems without paying attention to the most important [4].

10 Monitoring Tools Using Artificial Intelligence

Controlling the Student Screen: Once the online test starts, you can prevent the student from performing any action that you do not see in accordance with the rules of the test, such as opening any browser or electronic window next to the test as long as the time has not expired. This ensures that the exam is secured [4]. **Monitoring using artificial intelligence tools:** This feature can be used in conjunction with the availability of cheat detection reports provided by the electronic exam system. When the student clicks on anything displayed on the screen, during the exam, this action taken by the student is automatically saved, analyzed and added to the Instant access to the detailed test reports provided by the powerful analysis features of the electronic test system [2]. **Artificial intelligence tools to effectively manage online exams**: For something that only takes a few minutes to be able to design an electronic test and generate different question forms for the same test, and even get instant results. Work and restricting access to the question bank to specific people and setting work powers for each user, and others [5].

Electronic Exam Evaluation: Evaluation is one of the pillars of the educational system, and an important base for its development and the improvement of its outputs. The technical development, the cognitive explosion, and the tremendous acceleration in communication technologies have affected assessment methods and tools, and computer-based electronic tests have appeared in their design, construction, presentation, management, correction, and giving comprehensive reports of students' educational results and their scientific progress. In this article, we discuss electronic exams as one of the components of the e-learning environment [13]. Electronic Exams can be defined as that type of computer-based tests and its various technologies and multimedia, or via the Internet, to measure and evaluate the level of students' performance in a subject or course of study, so that they are automatically corrected and monitored, ensuring credibility and transparency in correction saves time and effort [13]. Electronic tests are divided into two main types: non-adaptive tests and adaptive tests. As for the non-adaptive ones, they are the electronic tests that match their counterparts from the traditional paper-based tests, in terms of the number of questions, their order, the degree of difficulty and the way they are presented [6]. As for the adaptive ones, they are electronic tests that adapt in terms of the number of questions, their arrangement and degree of difficulty with the student's level and progress, depending on his answer to the previous question, which affects the student's progress in the test, and thus provides the computer with sufficient information to judge the student's level without the need to display all test questions [7]. One of the most important goals achieved by electronic tests is to reveal the strengths and weaknesses of the student and contribute to addressing them, achieving equality between students, taking into account individual differences, speed and accuracy in obtaining results. It also facilitates the process of conducting statistical analyzes that will help in developing and improving teaching and learning [13].

11 Electronic Fraud and Artificial Intelligence

Now technology is being deployed by test-takers and mark-ups in order to reduce cheating—which is generally still a minor problem—and to create greater efficiency in test preparation and marking, and to help improve teaching and study. From traditional paper-and-pencil tests and textbook publishers, such as Pearson, to digital native companies, such as Coursera, electronic tools and artificial intelligence are being developed to reduce costs and enhance learning. For years, multiple-answer tests have allowed scanners to correct a test and monitor results without any human intervention. Now, technology is coming straight into the exam hall. Coursera has patented a system that includes taking pictures of students and verifying their identity through scanned documents [16]. These are plagiarism-detecting devices that can check answers to questions that come in the form of essays and search the web—or through other students' work—to determine if there has been a copying process. Surveillance cameras connected to the Internet can monitor test sites for fraud or any improper practice [21]. Even when students answer the questions, they provide clues that can be used to crack down on cheating. They leave electronic "fingerprints" such as keystrokes, speed or even typing style [5].

It's hard to prepare for people intent on cheating, but we try every possible method. [7] By assigning a randomly selected set of questions to each applicant, examiners can avoid the risks of papers being stolen, pre-circulated, or candidates copying others' answers while taking the test [8].

According to Chauvla, new batches of GMAT questions on each subject are quickly tested on candidates to ensure they are as difficult, and older questions are discarded once they have been used a certain number of times [14]. In some subjects, such as mathematics and finance, online publishers create unique questions by randomly changing the numerical variables used in test questions. "This method is very effective as long as the algorithms are programmed correctly [4]. The technology also helps the proofreaders who correct the more complex answers manually. When school students across the UK got their A-level test scores, the system for obtaining those scores was radically updated [3]. While applicants still routinely write answers by hand, each separate sheet is no longer mailed to a single corrective. Instead, they are scanned, separated, and sent separate questions to different examiners to be read electronically online [2].

This process ensures safe delivery, speed of correction, true traceability and quality assurance of markers and results. A number of testing systems are exploring ways to compare and evaluate the quality of marking for basic tests by cheaper and less experienced graders, allowing more experienced graders to focus on the process of judging more complex answers that require greater skills [17].

Educators also continue with machine learning and natural language processing to assess answers on more complex tests. "We're in the process of moving from assessing what the answer to a question was to how the answer was," says Tim Buzek, global head of product at Pearson [3]. The company is also testing an analysis of a visual survey of the steps involved in solving math problems, so that it can study and examine the methods students use and identify the stages in which students go wrong. It is also developing essay writing algorithms that analyze syntax to assess style, critical thinking, comprehension, and comprehension [6].

Much of the application of technology focuses less on final exams and detecting cheating, and more on intermediate exams designed to identify students' weaknesses and help improve them [5]. When you ask people to look at education in the past, they will talk about a teacher who was a source of motivation for them. The support provided through machine learning can be increased in specific tasks, allowing educators to spend more time focusing on the areas where they can make the most impact [18]. Tests are traditionally used to test students' abilities, while their results in tests when compiled are often used by parents and inspectors to judge the quality of schools. In Canada, with an education system that ranks very high in international assessments, school administrations have adopted a very different approach [13, 27]. Teachers and schools focus less on celebrating correct answers, and more on explaining how to respond to the most common wrong answers. This allows them to understand the students' weaknesses, so that they can enhance aspects of the curriculum [9].

12 Disadvantages of Artificial Intelligence Applications in Education

Artificial intelligence is superior to the human brain, which makes it powerful and therefore difficult to download [2]. Discharge of many workers and replace them with artificial intelligence systems in order to reduce costs directed to workers and invest them in smart devices [16]. Doing many remote attacks, because autonomous robots don't need to be controlled [10, 26]. Reproducing the tone of a person's voice, by extracting his voice signature, and constructing phrases he did not utter in the first place [12, 25]. Lethal weapons systems and the great burden they carry with them, as lethal weapons have been developed with artificial intelligence to the extent that they have the ability to make decisions regarding the killing or assassination of a person or group of people [14], Control of humans, as the capabilities of smart devices will exceed the capabilities of the human mind, so it is necessary to restrict machines and establish principles and ethics in them in order to avoid using them in matters that harm and harm humans [1, 22, 24].

13 Conclusion

The results revealed that AI applications make many aspects of our lives easier, but AI is also able to help teachers as well. In terms of artificial intelligence, artificial intelligence has helped teachers in the classroom by becoming able to take on administrative tasks. Teachers appreciate the time they can spend with their students, but much of that time can be occupied by tasks such as marking multiple-choice exams and assessing homework. The AI is already capable of correcting multiple-choice exams, and soon it will be able to categorize written claims as well. As artificial intelligence continues to be developed to complete these tasks, teachers will be given more time to spend with their students. The results also revealed that artificial intelligence is used with the free PowerPoint plug-in called "View Translator." This is useful because it is able to provide translations for the teacher speaking in real time. In this case, AI can eliminate boundaries and help ensure that all students understand the topic.

The results also revealed that AI can detect cases of plagiarism, examine answers to questions that come in the form of articles and search the web—or through other students' work—to determine if there is a copying process. Surveillance cameras connected to the Internet can monitor test sites for fraud or any improper practice.

References

1. Awang N, Hussin NS, Razali FA, Lyana S, Talib A (2020) Fraud triangle theory: calling for new factors. Editor. Board 7:54–64
2. Musleh Al-Sartawi AM, Hussainey K, Razzaque A (2022) The role of artificial intelligence in sustainable finance. J Sustain Finance Invest 1–6
3. Bozkurt A, Karadeniz A, Baneres D, Guerrero-Roldán AE, Rodríguez ME (2021) Artificial intelligence and reflections from educational landscape: a review of AI studies in half a century. Sustainability 13:800
4. Al-Sartawi A (2020) Social media disclosure of intellectual capital and firm value. Int J Learn Intellect Cap 17(4):312–323
5. Pawelczak D (2018) Benefits and drawbacks of source code plagiarism detection in engineering education. In: Proceedings of the IEEE global engineering education conference (EDUCON), Tenerife, Spain, pp. 1048–1056
6. Florea AM, Radu S (2019) Artificial intelligence and education. In: 2019 22nd international conference on control systems and computer science (CSCS). IEEE, Piscataway, pp 381–382
7. Greenspan H, van Ginneken B, Summers RM (2016) Guest editorial deep learning in medical imaging: overview and future promise of an exciting new technique. IEEE Trans Med Imaging 35:1153–1159
8. Hinojo-Lucena FJ, Aznar-Díaz I, Cáceres-Reche MP, Romero-Rodríguez JM (2019) Artificial intelligence in higher education: a bibliometric study on its impact in the scientific literature. Educ Sci 9:51
9. Jalali K, Noorbehbahani F (2017) An automatic method for cheating detection in online exams by processing the student's webcam images. In: Proceedings of the 3rd conference on electrical and computer engineering technology (E-Tech 2017), Tehran, Iran, pp 1–6
10. Musleh Al-Sartawi AMA (2020) E-learning improves accounting education: case of the higher education sector of Bahrain. In: Themistocleous M, Papadaki M, Kamal MM (eds) Information systems, EMCIS 2020. Lecture notes in business information processing, vol 402. Springer, Cham
11. Karolak M, Razzaque A, Al-Sartawi A (2021) E-services and M-services using IoT: an assessment of the Kingdom of Bahrain. In: Musleh Al-Sartawi AM, Razzaque A, Kamal MM (eds) Artificial intelligence systems and the internet of things in the digital era, EAMMIS 2021. Lecture notes in networks and systems, vol 239. Springer, Cham
12. Ghizlane M, Reda FH, Hicham B (2019) A smart card digital identity check model for university services access. In: Proceedings of the 2nd international conference on networking, information systems & security, New York, NY, USA, pp 1–4
13. Shukor NA, Tasir Z, Van der Meijden H (2020) An examination of online learning effectiveness using data mining. Procedia Soc Behav Sci 172:555–562
14. Al-Sartawi A (2020) Does it pay to be socially responsible? Empirical evidence from the GCC countries. Int J Law Manage 62(5):381–394
15. Roll I, Wylie R (2016) Evolution and revolution in artificial intelligence in education. Int J Artif Intell Educ 26:582–599
16. Sánchez-Aguayo M, Urquiza-Aguiar L, Estrada-Jiménez J (2021) Fraud detection using the fraud triangle theory and data mining techniques: a literature review. Computers 10:121
17. Al-Sartawi A (ed) (2021) The big data-driven digital economy: artificial and computational intelligence. In: Kacprzyk J (ed) Studies in computational intelligence, vol 974. Springer
18. Ullah, Xiao H, Barker T (2016) A classification of threats to remote online examinations. In: Proc. IEEE 7th annual information technology, electronics and mobile communication conference (IEMCON), Vancouver, BC, Canada, pp 1–7
19. Ullah, Xiao H, Barker T (2018) A dynamic profile questions approach to mitigate impersonation in online examinations. J Grid Comput 17(2):209–223

20. Sanad Z, Al-Sartawi A (2021) Financial statements fraud and data mining: a review. In: Musleh Al-Sartawi AM, Razzaque A, Kamal MM (eds) Artificial intelligence systems and the internet of things in the digital era, EAMMIS 2021. Lecture notes in networks and systems, vol 239. Springer, Cham
21. Chauhan N, Tekta P (2020) Fraud detection and verification system for online transactions: a brief overview. Int J Electron Bank 2(4):267–274
22. Shihadeh F (2020) Online payment services and individuals' behaviour: new evidence from the MENAP. Int J Electron Bank 2(4):275–282
23. Musleh Al-Sartawi AMA (eds) (2022) Artificial intelligence for sustainable finance and sustainable technology. In: ICGER 2021. Lecture notes in networks and systems, vol 423. Springer, Cham
24. Al-Qudah AA, Houcine A (2022) Stock markets' reaction to COVID-19: evidence from the six WHO regions. J Econ Stud 49(2):274–289. https://doi.org/10.1108/JES-09-2020-0477
25. Al-Qudah AA, Al-Okaily M, Alqudah G et al (2022) Mobile payment adoption in the time of the COVID-19 pandemic. Electron Commer Res. https://doi.org/10.1007/s10660-022-09577-1
26. Al-Qudah AA (2022) Artificial intelligence in practice: implications for information systems research, case study UAE companies. In: Musleh Al-Sartawi AMA (eds) Artificial intelligence for sustainable finance and sustainable technology, ICGER 2021. Lecture notes in networks and systems, vol 423. Springer, Cham. https://doi.org/10.1007/978-3-030-93464-4_23
27. Al-Okaily M, Alqudah HM, Al-Qudah AA, Alkhwaldi AF (2022) Examining the critical factors of computer-assisted audit tools and techniques adoption in the post-COVID-19 period: internal auditors perspective. VINE J Inf Knowl Manage Syst. Vol. ahead-of-print No. ahead-of-print. https://doi.org/10.1108/VJIKMS-12-2021-0311

The Future of e-learning in the post-Corona Era in Arab World

Abdulsadek Hassan, Enas Mahmoud Hamed Ahmed, Ahmed Mohamed Ahmed Sabek, Nehad Fawzy Elsaid Shalaby, and Abdulrahman Yaqoob Seyadi

Abstract The study aims to identify the future of e-learning in the post-Corona crisis, and study the most prominent future directions for the future of education during and after the Corona pandemic, which will chart the future of education and determine its identity, as the success of our educational systems in the future depends heavily on reading the future carefully and anticipating its features, The study reached many results, the most important of which are: The awakening of many educational systems in the world towards e-learning began after this pandemic, and even increased its conviction in it and took the direction towards it with full force during the crisis, and it is expected that interest in it and the expansion of its use even after the crisis, and in the near and far future, the results revealed that the outbreak of the Corona virus accelerated the movement of many education systems in the world to take many quick steps towards using and employing e-learning and spending on it, which will lead to its simplification and speed of use and the removal of many challenges and obstacles that stand in front of it.

Keywords e-learning · post-Corona era · Education

A. Hassan (✉)
Ahlia University, Manama, Kingdom of Bahrain
e-mail: dr.sadek1964@yahoo.com

E. M. H. Ahmed
Mass Communication, Department of Mass Communication and Children Culture, Faculty of Post Graduate Childhood Studies, Ain-Shams University, Cairo, Egypt

A. M. A. Sabek
Umm Al Qura University, Mecca, Saudi Arabia
e-mail: amsabek@uqu.edu.sa

N. F. E. Shalaby · A. Y. Seyadi
Qatar University, Doha, Qatar
e-mail: nshalaby@qu.edu.qa

A. M. A. Musleh Al-Sartawi et al. (eds.), *From the Internet of Things to the Internet of Ideas: The Role of Artificial Intelligence*, Lecture Notes in Networks and Systems 557,
https://doi.org/10.1007/978-3-031-17746-0_8

1 Introduction

Reading the future and anticipating its features is not easy, but it is not impossible, as there are many studies and research centers around the world that specialize in reading the future and there are many futurists who have emerged in their different specialties [28], and this is what can ease the trouble of researching the futures, especially in our Arab countries that It is rare that there are centers for future foresight or interest in this science, the science of futurism, and giving it what it deserves of the effort and attention it deserves [9]. As the study of future trends in various disciplines of education, industry, economy, finance, business, psychology, medicine and other multiple sciences is available to everyone with all its details and expectations, the field has become available to benefit from these studies in reading the future and employing them to serve humanity [15]. The outbreak of the Corona virus has accelerated the movement of many education systems in the world to take many quick steps towards using and employing e-learning and spending on it, which will lead to its simplification and speed of use and the removal of many challenges and obstacles that stand in front of it, many of which we discussed in the previous article [22], Where large sums of money will be allocated for scientific research and field work to solve these problems and overcome all the difficulties that exist in the face of e-learning and distance learning, in order to ensure its rapid spread and its effective use in the teaching and learning process [25].

The COVID-19 pandemic has created the largest disruption to education systems in history, affecting some 6.1 billion students on all continents. In more than 190 countries, closures of schools and other places of learning have affected 94% of students in the world, a proportion that rises to 99% in low- and lower-middle-income countries [26]. In fact, distance education is not new, and it has been known for decades in some developed countries, and it has become an urgent necessity, especially in times of crisis and the spread of epidemics that require social distancing; In 1892 the University of Chicago established the first independent Department of Correspondence Education. In 1956, Chicago Community Colleges introduced television service in teaching through educational channels, and NYSES University is the first open American university established in response to the desires of learners to make higher education available to them through non-traditional methods. Thus, the trend towards "digital education" was not something strange, it was expected, but the "Covid 19" pandemic hastened its emergence and pushed it to the fore [21].

2 Basic Requirements for the Development of "Distance Education" After the Corona Crisis

The COVID-19 pandemic has demonstrated our need for a new education system that is future-oriented and capable of facing challenges, including disasters and crises. This system is based on basic assumptions, the most important of which are the following: A—Orientation towards increasing investments in the development of infrastructure and technology for educational institutions, especially in rural and poor areas, which helps the flow of information between learning networks to an adequate extent. b—Redefining the role of the teacher [1]. C—Transcending the concept of providing education to the concept of quality associated with modifying and developing curricula and holding training courses for faculty members in schools and universities to manage curricula according to the formula that each country will adopt, whether it is distance education or hybrid education [11]. d—Develop scientific skills, enhance creativity, collaborative work, the ability to communicate and work in groups, and openness to the world and other cultures. e—The use of radio and television broadcasts in the so-called blended education. This can be done by reactivating these channels and developing their programs to become highly polarized multimedia. It is also possible to coordinate with the National Telecommunications Regulatory Authority to ensure the availability of discounted, subsidized packages for students, and to coordinate with donors and businessmen to provide personal computers in installments [18]. F—Concluding agreements and partnerships with some advanced universities to provide training and distance education, in order to exchange experiences and information, which contributes in developing and raising the scientific and practical level for members of the scientific body and graduates [4]. G—In order to avoid the psychological and social effects that e-learning leaves on students, teachers and parents, proper identification of expectations, creating motivation for the learner, developing a culture that supports e-learning, good planning for the school day, organizing the educational program, and paying attention to sound and visual effects should be taken into account [12].

3 Effects of the Pandemic on the Education Sector

Among the effects of the pandemic on education, according to the reports of international organizations such as UNESCO and the World Bank, are the following: [3]

1- **Millions of students are deprived of school education due to the closure of schools and universities.** The repercussions of the closure are exacerbated by students from the poor and disadvantaged groups, who have fewer educational opportunities outside school [9].

   There is much that can be done to reduce these effects, through distance learning strategies. The richer countries are the most willing to move to these strategies, although there is a great deal of effort and challenges facing teachers and parents [13].

2- **Unequal opportunities for students to benefit from digital learning platforms and distance learning opportunities.** The lack of access to technology is an obstacle to continuous learning, especially for students who come from disadvantaged families. Many of them do not have portable computers, and their

Internet connection is difficult; There are even those who do not have a study table, nor books, nor find any support from their parents as one would hope. On the other hand, there are others who have all of the above. Therefore, we must avoid widening these disparities in opportunities, and avoid increasing negative impacts on the learning of poor children [27].

Moreover, ensuring that children and young people return to schools when they reopen is a challenge, especially after they have been closed for a long time [9]. Therefore, there is a need to make use of television and radio channels to broadcast courses and curricula, as well as to coordinate with companies responsible for providing Internet service in order to reduce service prices [14].

3- **Deprivation of school feeding meals.** Many students rely on free or cheap meals provided by schools. If these meals cannot be delivered due to school closures, cash transfer programs should be expanded to compensate parents [19].
4- **Parents' unwillingness to teach their children remotely or at home.** When schools are closed, parents are often asked to facilitate the education of their children at home, and some find it difficult to do so, especially if they have limited education and resources [23].
5- **Disparity in childcare.** Working parents often leave their children alone as a result of school closures, which may lead to them falling prey to social networking sites and the Internet. This may lead to the spread of dangerous behavioral patterns, and to increase their exposure to Internet hackers, violence, bullying, blackmail and cybercrime by behavioral deviants [28].
6- **The effects of school closures on the family's economic situation.** The Corona pandemic has, in many cases, resulted in the inability of working parents to perform their work due to their time to take care of their children. This causes a decline in income in many cases, and negatively affects productivity [24].
7- **Pressure on the health care system.** Working women may be forced to miss work due to caring for their children during home quarantine and school closures. This means the absence of many health care providers from their workplaces, which are in dire need during health crises [16].
8- **Social isolation.** When schools close, many children and young people lose the social relationships that are essential to their learning and development [8]. The student not only learns science, mathematics and languages from the school, but also establishes social relations with his peers and develops his social skills [22]. Therefore, it is necessary to maintain contact with the school by any means [15].

If the role of parents has always been very important, it is even more important in this regard. Therefore, a great deal of help, which is provided by the ministries of education through the mass media, should also be directed to the parents [6]. In particular, messages through radio, television and SMS should be used in order to provide them with advice that will help them provide better support for their children [8].

4 The Use of Technology and Modern Methods of Teaching and Learning, and the Development of Academic Content in the post-Corona Period

1- Taking into account societal diversity and its variance across countries when applying distance learning in order to ensure that all learners have access to fair educational opportunities [2]. 2- Linking the process of transition to e-learning and distance learning with comprehensive plans for "digital transformation" at the level of countries, ministries, and relevant organizations [5]. 3- Giving the science of big data an increasing importance in the near future to facilitate the processes of analysis and decision-making, and to improve the results [14]. 4- Drawing up policies and developing executive programs to rehabilitate the human cadre (administrative and academic bodies and learners), and to inform them of all the necessary basic technologies that accompany distance education [10]. 5- Adopting policies that integrate educational means and technologies (mobile devices, television, synchronous and asynchronous learning, blended learning) to achieve greater interaction and improve learning motivation [7].

6- Recruiting experts and specialists to supervise the "flexible transition" to distance education and blended e-learning, and to develop perceptions and solutions to problems that may arise [25]. 7- Enacting legislation and regulations regulating distance education, and activating monitoring and follow-up tools, for the proper management of the educational process [4]. 8- Supporting, encouraging and training teachers to create educational content, organize student activities, set appropriate assessments, and formulate educational plans related to distance education [3]. 9- Taking into account young age groups (by relying on television broadcasts interspersed with special interactive activities and theatrical works, for example), and providing digital educational resources for learners with special needs [9]. 10- Going to the learners' parents as an integral part of the distance learning team and adopting dialogue and awareness as a means to support their children and support them morally and technically in organizing and following up on educational attainment [21]. 11- Preparing and developing the infrastructure of the telecommunications sector to keep pace with the distance education process and provide Internet services in areas not covered by the service, through satellites, depending on the appropriate technology in this field [15]. 12- Enhancing digital resources for practical fields of vocational and technical education by investing in the design of simulation programs, artificial intelligence, and virtual reality technologies [3].

5 Future Trends that Will Form the Interface of Education and Educational Systems

Among the most prominent of these trends are the following: The trend towards expanding the use of e-learning and distance learning, as the awakening of many educational systems in the world towards e-learning after this pandemic, and even increased their conviction in it and took by moving towards it with full force during the crisis, and it is expected that interest in it and the expansion of its use will continue even after the crisis, and in the near and long future [17]. The online education market in the Middle East is expected to expand at a compound annual rate of 9.8% during the period between 2017–2023 [23]. The market growth is likely to be driven by huge government investment and rapid adoption of online education and e-learning by educational and commercial institutions, especially after the Corona pandemic crisis [11].

This trend is reinforced by the fact that many prestigious universities in the world such as Harvard, University of London, Sydney, Johns Hopkins, University of Manchester and others have begun to invest in e-learning and transform many of their curricula and courses through e-learning [18]. The entry of universities and institutes with a global academic reputation into the full field of e-learning enhances expectations of revitalizing and developing the digital education industry, alongside global companies and partnerships, faster than experts estimated, when they expected this qualitative shift in the post 2030 stage [7]. The outbreak of the Corona virus has accelerated the movement of many education systems in the world to take many quick steps towards using and employing e-learning and spending on it, which will lead to its simplification and speed of use and the removal of many challenges and obstacles that stand in front of it, many of which we discussed in the previous article [15], Where large sums of money will be allocated for scientific research and field work to solve these problems and overcome all the difficulties that exist in the face of e-learning and distance learning, in order to ensure its rapid spread and its effective use in the teaching and learning process [18]. Perhaps the most prominent urgent challenges that need scientific treatment through sober studies and accurate scientific research to overcome them are the following: Developing and approving a missing strategic vision, policies and regulations to regulate the use of e-learning and distance education and to chart its features and future implications [8]. Many e-learning experts assert that what will help in accelerating its spread and increasing reliance on it and solving many of the problems before it is the emergence and spread of the fifth-generation technology (G5), which has now appeared in many developed, Arab and Gulf countries, which will greatly speed up the development and employment of tools, means and applications [27]. Modern educational technology for learning in the classroom and activating virtual reality in education in a large and fast way so that the e-learning process becomes acceptable and integrated as open classrooms that provide new ways of learning, which will help to reach a more effective and interactive digital learning environment in teaching and learning. This will contribute to the achievement of educational goals with ease and ease and encourage teachers and educators to increase its use. All of this confirms the promotion of the spread of e-learning in the future, even after the end of the Corona pandemic [3]. This does not mean that reliance will be only on e-learning and neglect of traditional education, but the two will go together through blended learning, as those in charge of Cisco, Zoom, Microsoft and other major technology companies that provided digital solutions for education during the crisis see that it will be inevitable Many distance education systems and institutions have to expand e-learning platforms and resort to the blended education system more in order to balance the educational needs that suit all educational systems and their various educational programs, and to achieve the various educational and learning goals, especially those that seek to develop skills that are difficult to achieve from through e-learning only [22]. The trend towards e-learning today, especially after this pandemic, is moving strongly towards enabling it, expanding its use, spending huge amounts of investment in it, installing it, publishing it, and solving all problems [15].

6 The Experiences of Arab Countries in Education During the Corona Period and Looking Forward to the Future

Corona forced the world to shift remotely into education, and although this is a new experience for the world, this had a great impact on learning, especially in the first basic classes, and university students, especially in the faculties of medicine, pharmacy, nursing, science and others [22]. There is no doubt that Arab universities need A tremendous technological renaissance, training for faculty members and staff, and the creation of a renewed educational environment that keeps pace with everything new and simulates any surprises that may be imposed on the world in the future [25]. The other face of the COVID-19 pandemic and its consequences for the educational process in Arab universities is embodied in providing an opportunity to activate the digital empowerment system for a university faculty member in order to achieve a quantitative leap from basic e-learning platforms to smart learning directly by employing the techniques of the Fourth Industrial Revolution and then avoiding having to control students and engineer assessments in ways that focus on their creativity [15], which has become an urgent requirement dictated by the movement of contemporary life on educational institutions to develop and improve the quality of their performance in order to interact and deal efficiently with crises and disasters and their consequences in the educational process [1]. The importance of integrating physical, digital and biological technologies and displacing the factors between them so that technology is a major part of the educational process, and then contributes effectively to treating the basic gap between the educational process on the one hand, and the challenges of the Fourth Industrial Revolution on the other hand [7], which represents a technological gap in the first place, and that from two perspectives: the first is the importance of the human resource in educational institutions as a first fact linked to agreement on a second fact, which is the necessity and importance of taking care of developing this element and developing its technical skills as a basis for developing the efficiency of educational institutions [8], and secondly the strategic importance of the applications of the Fourth Industrial Revolution as an indicator of added value and a source for achieving creativity digital in higher education institutions and a key indicator in facing the crisis of the COVID-19 pandemic and its repercussions on the educational system in Arab universities [13]. The importance of automating the education system in Arab universities, which is a practical process that points with the fingers of the thumb to the positive repercussions of the COVID-19 pandemic on education, with the aim of providing a mechanism to change the structure of education to change the value of the control unit in measuring the educational output to reach the desired value, while providing means to determine the compatibility of this output with the requirements of the market by working in the age of digitization [7].

The COVID-19 crisis has pushed the wheel of progress towards automating the education system in Arab universities, which contributes to providing information to those who perform the work and performs the calculations and recurring routine work about them through the digital transformation of courses and knowledge, and also facilitates communication with the parties to the educational process, which reduces time and speeds up work in collecting Information about the performance of each student individually, and it can highlight the strengths and weaknesses of each learner and then provide the necessary support to him in a timely manner; which made the educational process more quality and achieve the desired strategic goals [22]. The future of Arab universities will need a great deal of work to keep pace with the amazing development in information technology, as higher education before Corona is fundamentally different to after Corona, and strategies for self-structured learning must be introduced and focus on research and problem-solving approaches in order to bring out a generation that can keep pace with educational developments and the accompanying political transformations for those changes [3]. The future education of Arab universities after Corona requires benefiting from the Corona crisis for digital learning that fulfills the requirements of this type of education in light of the spread of crisis, disasters or other emergencies, and creates the need to keep pace with the rapid development of cyberspace and technology. Most importantly, the negatives of digital learning, and the missed positives of direct learning, are corrected by paying attention to the skills of laboratory experimentation, discovery skills, motor and sensory skills, and social negotiation [9]. The future of e-learning should not only be viewed in relation to distance learning, but it should be an auxiliary element for students and teachers in addition to in-person learning, which is indispensable, because it reflects human instinct and civic nature [15]. As most institutions of higher education in various countries of the world, especially in the Middle East, are generally optimistic about the results of the distance learning experience imposed on them by the Covid-19 pandemic, as distance education has become a primary method for providing knowledge in universities after the Corona virus pandemic, which forced skeptics [11]. It is useful to reassess and consider the future of higher education beyond the campus [24]. As a result of scientific and technological progress, the future of universities depends on making big and bold transformations, on top of which is the trend towards distance learning, so that electronic platforms are used for universities that include synchronous and asynchronous university education programs. This has many positives, including reducing study costs on the one hand, limiting the spread of the epidemic and making education. The most prestigious universities are within the reach of most people of different classes, so that the number of admissions doubles while maintaining the admission standards in those universities [15].

7 Foreseeing the Future of Distance Education and e-learning After Corona

The world is still suffering from the brunt of the spread of the Corona epidemic, and in light of the uncertainty about the decline or end of this epidemic, the option of distance education may not be considered an emergency shift in the educational process, and many questions are raised about the nature of the educational process after Corona, will things return to normal or is there an urgent need to conduct a comprehensive review of the educational process in the classroom? And the need to prepare to face any new emergency or education in light of coexistence with the Corona epidemic? Or, in short, can what happened to education as a result of the spread of the Corona epidemic be considered as an opportunity to reconsider the educational process? [16] Answering these questions and trying to anticipate the future of distance education and e-learning requires clarifying the problems that accompanied that experience, which can be summarized as follows: The weakness of the technological infrastructure necessary for the implementation of distance education and e-learning, so it is not possible to consider that the possibility of connecting to the Internet is available and possible for all students, in different countries of the world there are many remote or rural areas, or some social groups do not have access to the Internet [22].

The application of emergency distance education requires the presence of a special personal device dedicated to learning for each student independently, and this has become more difficult in light of the parents working from home, in addition to the presence of more than one student in the same house and the need for everyone to use computers or mobile devices for the purpose of learning or work [13]. The lack of prior preparation for teachers to use the Internet and technological tools and applications in the field of education, and the need for teachers to be trained professionally to accomplish all educational tasks online [2]. The widespread belief among students and parents that the emergency distance education came as a formal procedure to complete the educational process, and the lack of seriousness in pursuing distance education [19].

In an attempt to overcome these problems and determine the nature of the educational process after Corona, many proposals have emerged to develop the reality of distance education [5], including what was presented by the Dubai Future Foundation within its vision for the future of education as follows: Regulators in the education sector may begin to make several transformational changes in traditional education and develop innovative solutions such as parental training to teach, to change systems proactively rather than responding to unforeseen circumstances [3]. As learning outside the classroom becomes more prevalent, the involvement of students and their families in the design of learning tools and curricula will increase [8]. Fees for schools, software developers, and remote curriculum providers will change a lot, especially in education systems where families pay school fees, and schools may need less space when cohorts of students use school facilities on a rotating basis. A global debate will be launched to recognize the merit of distance education and the qualifications that can be acquired and learned using online education platforms. In addition to the above, some proposals can be made that contribute to the development of the educational process, enabling all parties to deal with any emergency with greater flexibility in the Kingdom of Saudi Arabia, as follows: Introducing disaster and crisis management in the Ministry of Education [3]. Attracting local experts from public and higher education [18]. Investing the national competencies of young men and women specialized in educational technology, teaching curricula, and educational administration [23]. Benefiting from successful international experiences and the need to harmonize and localize them according to the Saudi environment and the needs of the students of the country [9]. Converting courses to electronic, activating blended education, and training all education staff during normal days to employ it, such that direct education is 4 days, and the fifth day is electronic [4]. Using flipped learning or flipped classrooms, and training teachers on its pillars, such as active learning and technology [26]. Achieving equal opportunities and educational justice so that education reaches all regions, villages and deserts of the Kingdom in case of crises, using live and recorded television broadcasts, with extensive training for teachers, service providers, on communication skills, employing technology, and managing learning [23]. Providing mobile vehicles for the Internet and the fifth generation in cooperation with the Ministry of Communications and Information Technology in places that suffer from weak network or are outside coverage [14]. Establishing a department specialized in educational design, consisting of qualified cadres of educational technology specialists, working on formulating and designing curricula, studying reality and educational needs, designing scenarios, directing them, producing and evaluating them [5]. Urge teachers and students to bring their own personal devices into the classroom environment to perform their teaching assignments [12]. The adoption of blended education alongside traditional education, so that face-to-face education is combined in the classroom with the extensive use of handheld or portable tablets, applications and technological tools for teaching and learning, provided that this is done in a manner commensurate with the nature of students and their cognitive characteristics in different educational levels [16]. Allocating sufficient budgets for the distribution of hand-held devices to students in the different grades of study, and they are allocated to teaching and learning, and the production of educational content for all grades in digital form [25].

8 Conclusion

In conclusion, and looking ahead, it has become clear that digital education has become a strategic choice for every country that consciously appreciates the importance of the educational sector in the prosperity and development of society, and an indispensable alternative not in exceptional circumstances, as it is now to face the repercussions of the "Covid 19" pandemic, but also to build a new generation capable of benefiting from the data of the Fourth Industrial Revolution, and possessing the knowledge and skills that will enable it to engage in the jobs of tomorrow. In the end, we see that distance education can achieve its goals if the conditions are met for its success so that we can formulate appropriate scientific policies to deal with this future. Reading the future and anticipating its features is not easy, but it is not impossible, as there are many studies and research centers around the world that specialize in reading the future and there are many futurists who have emerged in their different specialties, and this is what can ease the trouble of researching the futures, especially in our Arab countries that It is rare that there are centers for future foresight or interest in this science, the science of futurism, and giving it what it deserves of the effort and attention it deserves. As the study of future trends in various disciplines of education, industry, economy, finance, business, psychology, medicine and other multiple sciences is available to everyone with all its details and expectations, the field has become available to benefit from these studies in reading the future and employing them to serve humanity.

References

1. Abbasi S, Ayoob T, Malik A, Memon SI (2020) Perceptions of students regarding E-learning during Covid-19 at a private medical college. Pak J Med Sci 36(Covid 19-S4):57–61
2. Al-Okaily M, Alqudah HM, Al-Qudah AA, Alkhwaldi AF (2022) Examining the critical factors of computer-assisted audit tools and techniques adoption in the post-COVID-19 period: internal auditors perspective. VINE J Inf Knowl Manage Syst. ahead-of-print No. ahead-of-print. https://doi.org/10.1108/VJIKMS-12-2021-0311
3. Alrefaie Z, Hassanien M, Al-Hayani A (2020) Monitoring online learning during COVID-19 pandemic; suggested online learning portfolio (COVID-19 OLP). MedEdPublish 9(1)
4. Al-Sartawi A (2019) Assessing the Relationship between information transparency through social media disclosure and firm value. Manage Account Rev 18(2):1–20
5. Alqahtani AY, Rajkhan AA (2020) E-learning critical success factors during the COVID-19 pandemic: a comprehensive analysis of e-learning managerial perspectives. Educ Sci 10:216. https://doi.org/10.3390/educsci10090216,1-16
6. Al-Qudah AA (2022) Artificial intelligence in practice: implications for information systems research, case study UAE companies. In: Musleh Al-Sartawi AMA (eds) Artificial intelligence for sustainable finance and sustainable technology, ICGER 2021. Lecture notes in networks and systems, vol 423. Springer, Cham. https://doi.org/10.1007/978-3-030-93464-4_23
7. Bovill C (2020) Co-creation in learning and teaching: the case for a whole-class approach in higher education. High Educ 79(1):1023–1037
8. Chayjan MR, Bagheri T, Kianian A, Someh NG (2020) Using data mining for prediction of retail banking customer's churn behaviour. Int J Electron Bank 2(4):303–320

9. Dhawan S (2020) Online Learning: a panacea in the time of COVID-19 crisis. J Educ Technol Syst 49(1):5–22. https://doi.org/10.1177/0047239520934018
10. Elumalai KV et al (2020) Factors affecting the quality of e-learning during the COVID-19 pandemic from the perspective of higher education students. J Inf Technol Educ Res 19:731–753
11. European Data Portal (2020) Education during COVID-19; Moving towards e-learning. https://www.europeandataportal.eu/en/impact-studies/covid-19/education-during-covid-19-moving-towards-e-learning
12. Hodges C, Moore S, Lockee B, Trust T, Bond A (2020) The difference between emergency remote teaching and online learning. Educause Rev 27(1):1–9. A.
13. Housawi A, Al Amoudi A, Alsaywid B, Lytras M, Bin Moreba YH, Abuznadah W et al (2020) Evaluation of key performance indicators (KPIs) for sustainable postgraduate medical training: an opportunity for implementing an innovative approach to advance the quality of training programs at the Saudi Commission for Health Specialties (SCFHS). Sustainability 12:8030
14. Hunukumbure AD, Horner PJ, Fox J et al (2021) An online discussion between students and teachers: a way forward for meaningful teacher feedback? BMC Med Educ 21:289
15. Ibrahim NK, Al Raddadi R, AlDarmasi M et al (2021) medical students' acceptance and perceptions of e-learning during the Covid-19 closure time in King Abdulaziz University, Jeddah. J Infect Public Health 14(1):17–23
16. Jebraeily M, Pirnejad H, Feizi A et al (2020) Evaluation of blended medical education from lecturers' and students' viewpoint: a qualitative study in a developing country. BMC Med Educ 20:482. https://doi.org/10.1186/s12909-020-02388-8
17. Karolak M, Razzaque A, Al-Sartawi A (2021) E-services and M-services using IoT: an assessment of the Kingdom of Bahrain. In: Musleh Al-Sartawi AM, Razzaque A, Kamal MM (eds) Artificial intelligence systems and the internet of things in the digital era, EAMMIS 2021. Lecture notes in networks and systems, vol 239. Springer, Cham
18. Liguori E, Winkler C (2020) From offline to online: challenges and opportunities for entrepreneurship education following the COVID-19 pandemic. Entrep Educ Pedagogy 3(4):346–351. https://doi.org/10.1177/2515127420916738
19. Mahyoob M (2020) Challenges of e-Learning during the COVID-19 pandemic experienced by EFL learners. Arab World Engl J 11(4):351–362. https://dx.doi.org/10.24093/awej/vol11no4.23
20. Musleh Al-Sartawi AMA (2020) E-learning improves accounting education: case of the higher education sector of Bahrain. In: Themistocleous M, Papadaki M, Kamal MM (eds) Information systems, EMCIS 2020. Lecture notes in business information processing, vol 402. Springer, Cham
21. Musleh Al-Sartawi AMA (eds) (2022) Artificial intelligence for sustainable finance and sustainable technology. In: ICGER 2021. Lecture notes in networks and systems, vol 423. Springer, Cham
22. Puljak L, Čivljak M, Haramina A, Mališa S, Čavić D, Klinec D et al (2020) Attitudes and concerns of undergraduate university health sciences students in Croatia regarding complete switch to e-learning during COVID-19 pandemic: a survey. BMC Med Educ 20(1):1–11. https://doi.org/10.1186/s12909-020-02343-7. PMID: 33167960
23. Ruthberg JS, Quereshy HA, Ahmadmehrabi S, Trudeau S, Chaudry E, Hair B et al (2020) A multimodal multi-institutional solution to remote medical student education for otolaryngology during covid-19. Otolaryngol Head Neck Surg 163:707–709. https://doi.org/10.1177/0194599820933599
24. Sahu P (2020) Closure of universities due to coronavirus disease 2019 (COVID-19): impact on education and mental health of students and academic staff. Cureus 12:e7541. https://doi.org/10.7759/cureus.7541
25. Widodo SFA, Wibowo YE, Wagiran W (2020). Online learning readiness during the Covid-19 pandemic. J Phys Conf Ser 1–4. ICOVEMAT 2020 1700 (2020) 012033 IOP Publishing. https://doi.org/10.1088/1742-6596/1700/1/012033

26. Yu J, Huang C, Han Z, He T, Li M (2020) Investigating the influence of interaction on learning persistence in online settings: moderation or mediation of academic emotions? Int J Environ Res Public Health 17(7):2320
27. Yu J, Huang C, Han Z, He T, Li M (2020) Investigating the influence of interaction on learning persistence in online settings: moderation or mediation of academic emotions? Int J Environ Res Public Health 17(1):1–21
28. Zalat MM, Hamed MS, Bolbol SA (2021) The experiences, challenges, and acceptance of e-learning as a tool for teaching during the COVID-19 pandemic among university medical staff. PLoS ONE 16(3):e0248758

Predicting the Pre-release Success of a Web Series' Season

Aeshna Jain, Sneha Aggarwal, Ruchika Swain, and Nidhi Arora

Abstract With the onset of COVID-19, OTT platforms have become popular. With this added popularity, many production companies tend to release their content on platforms like Netflix, Amazon Prime, Disney + Hotstar, etc. Through this research work, we tend to check the impact of different classical factors like genre, age certification, time of release, the platform of release, etc. as well as various social factors like the sentiment of the audience around the trailer, songs, and success of the previous season in predicting the success of the pre-release season of an English web series by creating our dataset. This will enhance the business strategies that production houses can use to improve their profits. We have trained different classification models like Decision Tree, Support Vector Machine, Multinomial Naive Bayes, and hyper tuned the parameters of Random Forest and K-Nearest Neighbours. We have also created a Multi-Layer Perceptron model and an ensemble classifier and trained them on our dataset. The best accuracy of 76.66% was achieved by the Hard Voting type ensemble classifier.

Keywords English web series · OTT · Multi-class classification

1 Introduction

Artificial Intelligence applies human intellectual activities through machines and has benefitted most domains, especially business predictive analysis and decision making. Predictive analysis helps us determine the future outcomes of a particular

A. Jain · S. Aggarwal (✉) · R. Swain · N. Arora
Indira Gandhi Delhi Technical University for Women, New Delhi, India
e-mail: sneha059btece18@igdtuw.ac.in

A. Jain
e-mail: aeshna040btece18@igdtuw.ac.in

R. Swain
e-mail: ruchika052btit18@igdtuw.ac.in

N. Arora
e-mail: nidhiarora@igdtuw.ac.in

A. M. A. Musleh Al-Sartawi et al. (eds.), *From the Internet of Things to the Internet of Ideas: The Role of Artificial Intelligence*, Lecture Notes in Networks and Systems 557,
https://doi.org/10.1007/978-3-031-17746-0_9

decision, based on the analysis and conclusions drawn by Machine Learning models on the available historical data. In the further sections, we have focused on business predictive analysis of one such emerging field-OTT and web series business.

1.1 Motivation

The OTT industry worldwide has gained tremendous popularity in the last two years with 88% of people using OTT platforms at least once a week. Online streaming platforms have made around Rs 3,100 crores in subscription revenue. 405 million dollars have been invested by Netflix for the development of authentic and licensed content over 2019 and 2020 [14]. This humongous investment in OTT makes it an interesting field to be researched. The currently available state-of-the-art predictive analysis models have limited their focus only to predicting the success of movies. Additionally, the current models consider various classical factors such as air date, genre, U/A certification, etc., even though with the advent of social media platforms like Twitter (77.75 million users), Youtube (>2 billion active users), etc., each review and opinion has gained immense predictive power. Through our research work, we aim to establish a link between critics' reviews and classical factors of a movie to predict the success of the pre-release season of an English web series.

2 Related Work

Though many factors play an integral role in the success of a particular season of an English web series, the way of their interaction is not crystal clear. This paper strives to determine such factors through various attributes.

In [1], the author aims to perform predictive analysis on various social and classical factors that might influence the success of a movie. The classical attributes considered include cast, director, producer, and genre. Various social attributes were considered as the response of users on social media. The paper concluded by claiming that it is both the classical features and user feedback that can be collected from various social media channels that can aid in predicting the gross success of an unreleased film accurately. The paper suggested that greater accuracy can be achieved with proper integration of more data and with the amalgamation of social and classical factors than considering the classical or social factors individually.

In [2], the author aimed to gauge the usefulness of blog data in the prediction of movie sales and ratings of users. The author deduced that there was a low correlation between the blog references and first-week movie sales reported by (Mishne, G. et. al. 2006) but discovered that succeeding weeks were way more predictive of blogs (with up to 0.86 correlation).

In [3], the authors highlighted the capability of data analytics in forming information systems that further brace business decision-making. Their regression analysis uncovered the impact of various crucial factors on movie profitability. The quantification of factors through their novel methods suggested by past theoretical studies (e.g., actor star powers, team expertise, and team diversity) also worked better in the prediction of profitability.

In [4], an ensemble classifier has been proposed by the authors, that combines the base learning classifier to build a single classifier, to improve the accuracy of sentiment classification techniques. The ensemble classifier showed better accuracy than stand-alone classifiers.

In [5], the following features were used by the author: star cast, director(s), producer(s), composer(s), cinematographer(s), MPAA rating, release month, and runtime to implement several prediction models to predict the success of multiple feature films. The paper claims that the performance of the predictors was moderately well on the test data.

In [6], Observing that the gross predictions of the traditional movie were based on numerical and categorical movie data from IMDb, the authors used the quantitative news data to help predict movie grosses. They found models using only news data were capable of achieving similar performance to those using IMDb data by analyzing regression and k-nearest neighbor models. Moreover, they found that better performance can be achieved by using an integration of IMDb data and news data with the improvement being statistically significant.

In [7], the authors have used the reviews of critics to predict the revenue of the opening weekend. Metadata for every movie was considered which consisted of the movie's name, its production house, the set of genres it belonged to, the scriptwriter(s), the director(s), the country of origin, star cast, the release date, its MPAA rating, and its run time. With respect to the text, seven review websites were chosen for movies and raw HTML was scraped to obtain the text of the reviews. The paper concluded by claiming that the features of the text that are extracted from pre-release reviews can replace and improve a powerful metadata-based prediction of revenue of movies over the first weekend.

In [8], two different data sources, YouTube and Box Office Mojo, were used by the authors to collect the required information to prepare their dataset. They have used the above data sources to form a dataset that contains various attributes like opening income, number of views, number of likes, number of dislikes, and number of comments. Linear regression was the most suitable method as concluded by the comparative analysis concerning the prediction of movies' gross income using these attributes.

In [9], the authors used multiple models and claimed that wide releases had a significantly positive coefficient and a similarly positive but less significant coefficient for limited releases. This suggested that most of the readers look at the reviews for the movies that they are interested in watching, mainly from marketing and advertisements. Hence, trailers and movie reviews both help the viewer to make a choice.

In [10], linear regression and logistic regression were the two different algorithms that were generalized for multiple labels were applied by the authors, and witnessed that logistic regression was the more accurate approach to predicting movie box office gross on data entries that included a movie id, the director, title, set of genres, rating, runtime, date of release, star cast, studio. They observed that the elimination of single actors and directors increased their accuracy to 50% accuracy. To make further improvements in the accuracy, they decided to apply k-means clustering on titles and complementary features and achieved an accuracy of 52% on its application [15].

In [11], authors found that using data that helps in analyzing the characteristics of how a movie's popularity propagates through social networks and its speed and extent over time can have direct correlation with gross revenue as the data is a representation of people's excitement and willingness to watch that movie.

[12] has rightly acknowledged the fact that the birth of TV and web series has contributed immensely to the growth of the entertainment industry. Prediction of their performance will aid streaming platforms/broadcasting networks to focus their investments to get maximum returns. They have limited their research to a single web series—The Big Bang theory—to extract all the factors that contributed to its success.

[13] has talked about various multiclass classification techniques that helped us in the application and improvement of accuracy of various models on the dataset.

3 Data Collection and Analysis

In this section, we aim to explain the collection and feature extraction. We have selected an amalgamation of social and classical factors to predict the value of the target variable. The classical factors are the defining factors of a series while social factors tend to capture the perspective of the audience. Herein, we have considered the following attributes in our dataset:

3.1 *Social Factors*

Official Trailer's Sentiment: The comments on the official trailer of the upcoming season were scrapped from Twitter, Youtube, and IMDb using the Selenium web driver. We have applied a pre-trained sentiment analysis model on an average of 10,000 comments per season to obtain the sentiment of the audience around the trailer, where positive is represented by 1, negative by -1, and neutral by 0.

Official Music Album's Sentiment: The comments on the official music album of the season were scrapped and analyzed similarly to Trailer's sentiment.

Previous Season's Rating: This attribute sets a movie and web series datasets apart. The success of the previous season is a measure of the anticipation of the audience for the upcoming season. IMDb ratings of all episodes of the previous season are scrapped and averaged out and given the following labels: 0–2: poor, 2–4: below average, 4–6: average, 6–8: good, 8–10: blockbuster.

3.2 Classical Factors

Genre: A one-hot label encoding has been done for the following genres respectively [Comedy, Romance, Action and Adventure, Crime and Mystery and Thriller, Sci-fi and Fantasy and SuperHero, Drama].

Certification: U/A labels: This attribute tells about the target audience of the particular season and can have only an 'A' (Adult: 18 and above 18 years of age) and 'U' (Universal: for all age groups) as its input values.

Languages: This attribute captures the number of languages a season is released in except English to determine how large of an audience it can cater to without causing any language barrier. It can have 0 (No other language except English), *1* (One more language except for English), *2* (Two more languages except for English) or *3* (Three or more than 3 languages except for English) as its input variables.

Air Date: It mentions the decade a particular season is released in as prevailing culture favours particular genres and hence leads to the success of specific category of web series.

Official Platform of Release: Owing to the subscription model, we have considered the most prominent streaming platforms in the world with diverse subscription models. We have used the following labels for the respective streaming sites: "Disney + Hotstar" (Disney + Hotstar), "Official Netflix" (Official Netflix), "Prime" (Amazon Prime Video), "CBS" (Television Broadcasting company), "Others" (Includes all the streaming platforms except the ones mentioned above).

3.3 Target Attribute

It is the label that helps to categorize a particular series based on how it is rated and perceived by the audience. The IMDb ratings of all episodes of the current season are scrapped and averaged out. The following labels are provided for the ratings lying in the respective range: 0–2: poor, 2–4: below average, 4–6: average, 6–8: good, 8–10: blockbuster. This is the attribute we aim to predict using the features mentioned above by applying various classification models to the dataset.

4 Machine Learning Models Used

4.1 Applying Pre-trained Sentiment Analysis Models

After we collected the CSV files with trailer and music album reviews, we implemented three most popular pre trained Sentiment Analysers-Text Blob, Vader and Flair. Flair was eventually selected as the obvious choice for it gave the best results, about 92.94% for sentiment analysis on a sample Youtube comment dataset on Kaggle and 69.7% for sentiment analysis on a sample Twitter comment dataset on Kaggle, and hence was the obvious choice for the analysis.

4.2 Applying Classification Models on the Dataset

After completing the preparation of the dataset, we applied various classification models to our dataset to predict the target variable:

Decision Tree: A Decision Tree classifier was selected because of it's speed and feature extraction approach. We have used the Decision Tree Classifier from sklearn.tree with a random state set tuned to 0 (Table 1).

Multinomial Naive Bayes: We trained the very scalable Multinomial Naive Bayes model on our dataset because it works well with discrete values We have used the MultinomialNB() from sklearn.naive_bayes with the default parameters (Table 2).

Support Vector Machine: Since Support Vector Machines (SVMs) are architectured for binary classification, we implemented the One Vs One approach by

Table 1 Confusion matrix of Decision Trees

Predicted label						
Actual Label		Average	Below average	Blockbuster	Good	Poor
	Average	5	1	0	0	0
	Below average	1	6	0	0	3
	Blockbuster	0	0	59	5	0
	Good	2	0	13	13	0
	Poor	2	1	0	1	8

Table 2 Confusion matrix of Multinomial Naive Bayes

Predicted label						
Actual Label		Average	Below average	Blockbuster	Good	Poor
	Average	2	0	2	1	1
	Below average	1	0	2	1	6
	Blockbuster	0	0	61	3	0
	Good	0	0	18	9	1
	Poor	0	0	3	0	9

Table 3 Confusion matrix of Support Vector Machine

Predicted label

Actual Label		Average	Below average	Blockbuster	Good	Poor
	Average	1	0	2	2	1
	Below average	0	0	2	1	7
	Blockbuster	0	0	57	7	0
	Good	0	0	15	12	1
	Poor	2	1	0	1	8

taking subsets of training data that helped in improving the Kernel performance. The parameter values of the SVM classifier were tuned to C = 1, kernel = linear (Table 3).

K-Nearest Neighbour: We have used KNeighborsClassifier from sklearn.neighbours and n_neighbours were set to 5. Later, on experimentation n was set to 4 as better results were seen by changing the parameter (Fig. 1 and Table 4).

Proposed Multi-Layer Perceptron Model: We built an MLP model by sequentially stacking multiple hidden layers and input and output layers. We later experimented on the number of hidden layers, nodes in the input layer, learning rate, momentum, loss function, batch size, and epoch to quote the best possible accuracy (Tables 5 and 6).

Random Forest: RandomForestClassifier was used from sklearn.ensemble library and further hyperparameter tuning was done to tune n_estimators to 4, random_state to 58 and criterion to 'Gini', and rest parameters to default to achieve the best accuracy (Tables 7 and 8).

Fig. 1 K value vs error rate

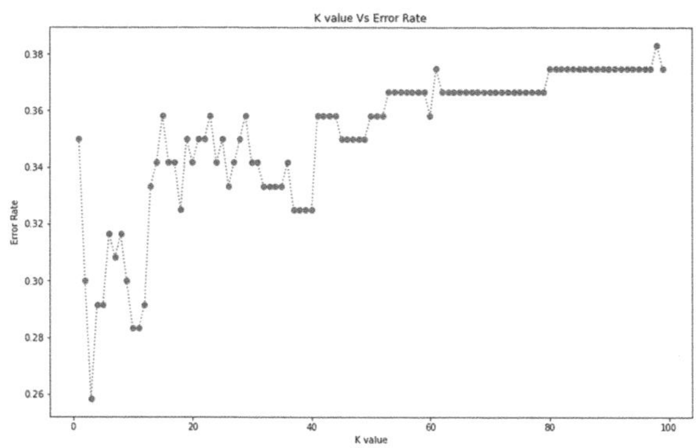

Table 4 Confusion matrix of K-Nearest Neighbours

Predicted label						
Actual Label		Average	Below average	Blockbuster	Good	Poor
	Average	5	1	0	0	0
	Below average	2	4	1	0	3
	Blockbuster	0	4	55	5	0
	Good	0	0	15	13	0
	Poor	1	1	1	1	8

Table 5 Hyper parameter tuning of Proposed MLP model

Parameters	Accuracy (%)
n = 50, hidden layers = "none", lr = 0.001, momentum = 0.1, loss = "sparse categorical cross entropy", batch size = 10, epochs = 150	45.33
n = 50, hidden layers = Dense (27, activation = 'relu'), lr = 0.001, momentum = 0.1, loss = "categorical cross entropy", batch size = 10, epochs = 200	47
n = 50, hidden layers = "none", lr = 0.001, momentum = 0.1, loss = "categorical cross entropy", batch size = 10, epochs = 200	52.14
n = 100, hidden layers = "none", lr = 0.001, momentum = 0.1, loss = "categorical cross entropy", batch size = 10, epochs = 200	53
n = 100, hidden layers = "none", lr = 0.001, momentum = 0.1, loss = "categorical cross entropy", batch size = 5, epochs = 150	54.65
n = 100, hidden layers = "none", lr = 0.001, momentum = 0.1, loss = "categorical cross entropy", batch size = 10, epochs = 120	52.9
n = 100, hidden layers = "none", lr = 0.001, momentum = 0.1, loss = "categorical cross entropy", batch size = 3, epochs = 150	49.5
n = 100, hidden layers = "none", lr = 0.01, momentum = 0.1, loss = "categorical cross entropy", batch size = 5, epochs = 150	41.17
n = 100, hidden layers = "none", lr = 0.0005, momentum = 0.1, loss = "categorical cross entropy", batch size = 5, epochs = 150	54.62
n = 100, hidden layers = "none", lr = 0.0005, momentum = 0.01, loss = "categorical cross entropy", batch size = 5, epochs = 150	57.98
n = 100, hidden layers = Dense (52, activation = 'linear'), lr = 0.0005, momentum = 0.01, loss = "categorical cross entropy", batch size = 5, epochs = 150	36.13

Ensemble Classifier: We have used a Voting Classifier from sklearn.ensemble comprising all the above models. Two types of voting have been implemented by us using the ensemble classifier namely hard voting and soft voting (Tables 9 and 10).

Table 6 Confusion matrix of MLP

Predicted label						
Actual Label		Average	Below average	Blockbuster	Good	Poor
	Average	0	0	2	2	2
	Below average	0	0	0	5	5
	Blockbuster	0	0	55	9	0
	Good	2	0	13	14	1
	Poor	0	0	3	1	8

Table 7 Hyper parameter tuning of Random Forest model

Parameters	Accuracy (%)
n_estimator = 100, max_features = "auto", max_depth = 8, min_sample_split = 5, min_sample_leaves = 2, random_state = 42, criterion = "gini", bootstrap = false	68.9
n_estimator = 100, max_features = "auto", max_depth = 8, min_sample_split = 5, min_sample_leaves = 2, random_state = 42, criterion = "entropy", bootstrap = false	65.52
n_estimator = 100, max_features = "sqrt", max_depth = 110, min_sample_split = 10, min_sample_leaves = 4, random_state = 42, criterion = "entropy", bootstrap = false	67.30
n_estimator = 50, max_features = "auto", max_depth = "none", min_sample_split = 2, min_sample_leaves = 1, random_state = 58, criterion = "gini", bootstrap = false	76.47
n_estimator = 10, max_features = "auto", max_depth = "none", min_sample_split = 2, min_sample_leaves = 1, random_state = "none", criterion = "gini", bootstrap = true	71.40

Table 8 Confusion matrix of Random Forest

Predicted label						
Actual Label		Average	Below average	Blockbuster	Good	Poor
	Average	5	1	0	0	0
	Below average	1	6	0	1	2
	Blockbuster	0	1	55	8	0
	Good	1	0	11	16	0
	Poor	0	3	0	1	8

Table 9 Confusion matrix of ensemble classifier: hard voting

Predicted label						
Actual Label		Average	Below average	Blockbuster	Good	Poor
	Average	6	0	0	0	0
	Below average	1	3	1	0	3
	Blockbuster	0	0	61	3	0
	Good	1	0	14	13	0
	Poor	0	1	1	1	9

Table 10 Confusion matrix of ensemble classifier: soft voting

Predicted label						
Actual Label		Average	Below average	Blockbuster	Good	Poor
	Average	5	1	0	0	0
	Below average	1	3	1	0	5
	Blockbuster	0	0	60	7	0
	Good	0	0	15	13	1
	Poor	0	1	1	1	9

5 Results and Observations

The dataset created in our paper is novel as it curates features that other datasets working with movie prediction systems do not include. In [1] the dataset used doesn't emphasize the social factor's influence in the pre-production stages, a scope we have covered by making use of the trailer's sentiment analysis, while other papers like [2] do the exact opposite by taking a dataset of 120 social factors and list out the relevant ones using correlation. The metric they use is KL divergence.

For our work, we have decided to stick with the standard classification metrics as our performance metrics. We have generated the classification matrix as a tabular ground for other metrics to be calculated. The null hypothesis assumed is that the ratings' label predicted is the same as the actual label. We have also inculcated accuracy for different models implemented, with an ensemble classifier giving us the best accuracy score.

We have used various multiclass classification models and hyper tuned the parameters of some to achieve the best possible accuracy for predicting the success of a pre-release season of an English web series (Table 11).

Table 11 Comparative performance of different models in terms of accuracy

Model name	Accuracy (%)
KNN classifier (where n = 4)	70.84
Decision Tree Classifier (Default parameters)	75.84
Multinomial Naive Bayes Classifier (Default Parameters)	67.50
Support Vector Machine (SVM) (kernel = 'linear', C = 1)	52.94
Proposed MLP model (no hidden layer),	57.983
Random Forest Classifier (n_estimators = 4, criterion = 'gini', random_state = 58)	76.47
Ensemble Classifier (using all the above mentioned models)	Soft: 75.00% Hard: 76.66%

6 Future Work and Conclusion

We have concluded that Hard-voting type Ensemble Classifier is the best model with an accuracy of 76.66%. Hard-voting classifier gives us the best result because of its non-bias nature and different models are considered. We take up the mode value of the combined outputs from individual models.

An additional observation was that Classical factors do not impact the ratings as strongly as the social features do. This indicated a shift in the box office collection paradigm, where the audiences' view of a series is more influential than any other standard feature.

References

1. Bhave A, Kulkarni H, Biramane V, Khamkar P (2015) Role of different factors in predicting movie success. In: 2015 international conference on pervasive computing (ICPC), Pune, Maharashtra, pp 541–545 (2015). https://doi.org/10.1109/PERVASIVE.2015.7087152
2. Sadikov E, Parameswaram A, Venetis P (2009) Blogs as predictors of movie success. In: Proceedings of the international AAAI conference on web and social media 2009, San Jose, California, USA, vol 3, no 1, pp 304–307
3. Lash MT, Zhao K (2016) Early predictions of movie success: the who, what, and when of profitability. J Manage Inf Syst 33(3):874–903. https://doi.org/10.1080/07421222.2016.1243969. The University of Iowa Lash
4. Saleena N (2018) An ensemble classification system for twitter sentiment analysis. Procedia Comput Sci 132:937–946. https://doi.org/10.1016/j.procs.2018.05.109
5. Cocuzzo D, Wu S Hit or flop: box office prediction for feature films. Stanford University
6. Zhang W, Skiena S (2009) Improving movie gross prediction through news analysis. In: Proceedings of the 2009 IEEE/WIC/ACM international joint conference on web intelligence and intelligent agent technology, NW Washington DC, United States, vol. 1. IEEE Computer Society, pp 301–304. https://doi.org/10.1109/WI-IAT.2009.53
7. Joshi M, Das D, Gimpel K, Smith NA (2010) Movie reviews and revenues: an experiment in text regression. In: Human language technologies: the 2010 annual conference of the North American chapter of the association for computational linguistics, Los Angeles, California, USA, pp 293–296

8. Rahim MS, Chowdhury AE, Islam MA, Islam MR (2017) Mining trailers data from youtube for predicting gross income of movies. In: 2017 IEEE Region 10 humanitarian technology conference (R10-HTC), Dhaka, Bangladesh. IEEE, pp 551–55 (2017). https://doi.org/10.1109/R10-HTC.2017.8289020
9. Kennedy A (2008) Predicting box office success: do critical reviews really matter? Berkeley projects
10. van der Merwe J, Eimon B (2013) Predicting movie box office gross. Stanford University
11. Yoo S, Kanter R, Cummings D (2011) Predicting movie revenue from IMDb data. Yoo2011PredictingMR. Stanford University
12. Fronzetti Colladon A, Naldi M (2019) Predicting the performance of TV series through textual and network analysis: the case of Big Bang Theory. PloS One 14(11):e0225306. https://doi.org/10.1371/journal.pone.0225306
13. Aly M (2005) Survey on multiclass classification methods: 2005. Neural Netw 19:1–9
14. Musleh Al-Sartawi AM, Razzaque A, Kamal MM (eds) (2021) Artificial intelligence systems and the internet of things in the digital era. In: EAMMIS 2021. Lecture notes in networks and systems, vol 239. Springer, Cham
15. Gupta M, Sikarwar TS (2020) Modelling credit risk management and bank's profitability. Int J Electron Bank 2(2):170–183

Impact of Financial Technology (FinTech) in the Field of Accounting

Vaishali Ojha, Ashish Mathur, G. B. S. Johri, Vandana Gupta, and Abhilasha Mathur

Abstract The prevalence of Technology has altered the way of life be it education, entertainment, finance and Accounting in not an exception to this. This digital transformation proved to be opportunity for many and threats to some. The Fin Tech also transformed the accounting activities and its management. FinTech like big data analysis, cloud accounting and other various advanced automated process are actually prepared the accountants or say elevated them to survive this new tech era in the best possible way. The accounting industry has undergone several blustery phases over the last decade. Several professional accounting organizations and bodies have also emphasized on imparting education on technology used for accounting, auditing and other associated business functions to make the accounting graduated industry ready. It is a major concern for all the educational institution across world to bridge this technological knowhow gap of what Industry demands from future accounting professional and what presently they know to keep up with advances in technologies. As a corollary, there is a rise in call for accounting professional who are well versed with Fintech. Through this paper researchers want to study the impact, opportunities and challenges faced accounting graduates and professors levied by FinTech.

Keywords Digital transformation · Fintech · Accounting education · Higher Education Institutions (HEIs)

V. Ojha (✉)
Department of Finance, KES Shroff College of Arts and Commerce, Mumbai, India
e-mail: vaishaliojha00@gmail.com

A. Mathur
Department of Commerce, IGNTU, Amarkantak, Madhya Pradesh, India
e-mail: ashish.mathur@igntu.ac.in

G. B. S. Johri
Department of Tourism, IGNTU, Amarkantak, Madhya Pradesh, India

V. Gupta · A. Mathur
Faculty of Commerce and Management, Lachoo Memorial College of Science and Technology, Jodhpur, India

A. M. A. Musleh Al-Sartawi et al. (eds.), *From the Internet of Things to the Internet of Ideas: The Role of Artificial Intelligence*, Lecture Notes in Networks and Systems 557,
https://doi.org/10.1007/978-3-031-17746-0_10

1 Introduction

According to prospect of the industry 4.0, in which the digital transformation will neither be an option nor a competitive advantage for the companies, rather it will part of the business model to face the ever-challenging demands of all the stakeholder i.e. employees, customers and other parties. The 4th age of Industry (Industry 4.0) includes several technologies such as IoT (Internet of Things), Artificial Intelligence, distributed ledge technologies, 3-D printing, mixed-reality [8]. It is very pertinent to know how this digital transformation and innovation is giving new models of business, workforce and workplace overall. Digital transformation is beacon of the new normal which will build companies of the future- companies who will adopt various technologies right from designing the product and services to the successful delivery and its accounting. The world economic forum's report published in 2020 on the future of Jobs in their study of 15 industries and 26 economies covered by the report stated that by 2025, many jobs can be displaced by machines from human and approx. 97 million new job profiles will emerge for machines, algorithms and humans [5]. A survey conducted in the year 2018 reveals that in the demand for Data Analysts and Scientists, Robotics Engineers, Application and Software developers, Specialist of AI and ML (Machine Learning) [2]. The transformation in accountancy sector brought about a change in the job profiles, with rise in job positions like IoTs specialists, PAS (Process Automation Expert) etc. At the opposite end of the scale, there some job roles which are probably displaced by new technologies to name some of them- Data Entry Clerks, Payroll Clerks, Administrative and Executive Secretaries, Accountant and Auditors etc. Gone are the days of traditional ways of accounting, these FinTech tool and its adaptation in accounting work entail a new kind of accountant who is tech savvy as well as capable of making best use of these technologies for value driven businesses [6].

1.1 Objectives of the Research

This research is aimed at achieving following objectives:

- To explore various Financial Technology (FinTech) which are useful for accounting work
- To know the various usage of Financial Technology (FinTech) in accounting
- To analyse the perception for Financial Technology (FinTech) among the respondents
- For study point of you above mentioned objectives are primarily concerned with various technologies, its uses and the perception with reference to accounting work.

2 Review of Literature

Digital Transformation: Digital transformation is absorption of technological competencies by building on the mindset change and process re-engineering so that it can contribute to identify and exploit new ways of reducing costs, quality improvement, value creation and enhancement of employee and customer experience [1].

Financial Technology: FinTech means technology-empowered innovation which enables the organisation for technology driven models, processes, products and applications. FinTech led companies focuses more on innovations and technological advancements [3].

FinTech Startups: FinTech are those start-ups who provide technological solution for banking, investing, accounting and other business functions [4].

Financial Technology in Accounting: Adopting technology for accounting process will help in improving operations and reducing costs. The significant to the digital transformation of accounting and financing is not replacing the human with technology rather it is uniting people and technology together for providing value driven services and products to the society. It was also stated that FinTech can be very useful in performing accounting work in following ways: [2, 7]

Accounts Payable and receivable processing	• AI-powered invoice management system • Using accounting codes for each invoices
Supplier onboarding	• using machines instead of humans for check the creditworthiness of the suppliers
Procurement	• Paperless purchanes and procurement process • One example is: Robo Advising
Audits	• Artificial intelligence- capable of auditing 100% of financial transaction of the company
Periodical closure	• Consolidation and reconcilation of data with FinTech (Big data analysis and cloud accounting etc)
Expense management	• Using FinTech to read receipts, audit expenses and later reviewing and approving them
AI Chatbots	• Using Artificial Intelligence for customer griviance handling effectively and efficiently

Source: The Digital Transformation Of Accounting And Finance - Artificial Intelligence, Robots And Chatbots (forbes.com)

This research is primarily based on primary data collected through an online questionnaire. The scope of this study is to analyses the role of FinTech in managing finances, particularly decisions pertaining to investment and to provide suggestions to readers for further study, to concerned association and authorities for policy framework. The convenience sampling method has been used.

Hypothesis
For this research following are the hypothesis that are being taken into consideration:

Ho: The usefulness of Financial Technology (FinTech) in accounting is not same across all the level of education.
H1: The usefulness of Financial Technology (FinTech) in accounting is same across all the level of education.
Ho: The usefulness of Financial Technology (FinTech) in accounting among males is not significantly different from usefulness of Financial Technology (FinTech) in accounting among females.
H1: The usefulness of Financial Technology (FinTech) in accounting among males is not significantly different from usefulness of Financial Technology (FinTech) in accounting among females.

3 Research Methodology

Data has been collected for the research. Under secondary data researcher has referred several reports and articles published by authenticate research organisations and in newspapers to draw important and relevant inferences. For primary data structured questionnaire has been made and circulated via google form to the respondents. In total 187 responses were received from questionnaire which were taken into consideration for data analysis and interpretation.

Tools and Techniques Used for Analysis: Research as used following tools for data analysis and inferences.
Simple Percentage Method II. Chi-Square test.

Data Analysis and Interpretation
This section of the research paper shows various interpretation of data analysis collected from the questionnaire (Table 1).

Objective Wise Data Analysis and Interpretation
Objective 1: To explore various Financial Technology (FinTech) which are useful for accounting work (Fig. 1).

On the basis of above figure, it is clearly visible that majority of the respondents found FinTech useful for accounting work. For machine learning, 30% of the respondents found it very useful, for an, 35% of the them have selected it to be moderately useful, Blockchain, Big Data analysis, Artificial Intelligence and Robo-Advising found to bemoderately useful among respondents with 35, 30, 30 and 36 % respectively.

Objective 2: To know the usefulness of Financial Technology (FinTech) in accounting (Charts 1, 2, 3, 4 and 5).

Table 1 Demographic profiles of the respondents with the interpretation

Demographic profiles	% of the respondents	Respective data interpretation
Gender		
Female	48	Out of 187 respondents, 48% consist of female and 52% are male
Male	52	
Level of education		
Post Graduate	3	The majority of the respondents are undergraduate with 88% of total 187 respondants. Remaining 12% are PG, Pursuing PG, Professionals and research scholar
Professional	5	
Pursing Post graduate	4	
research scholar	1	
Under-graduate	88	
Area of specialisation		
Accounting (Auditing, Accounting and Taxation) (BAF/BCOM/MCOM etc.)	57	57% of the respondents have Accounting as specialisation, 19% of the respondents belong to IT field and remaining respondents have Financial markets, Management and other subjects as specialisation
Financial Markets, Insurance and Investment (BFM/ BBI/Msc Finance/BIM etc.)	15	
International business	1	
IT (BscIT/Data science etc.)	19	
Management (BMS/BBA/MBA/PGDM etc.)	8	
Other (PSW)	1	
Year of education		
First Year	10	Out of 187 respondents, 52% of them are in their third of education
Second Year	33	
Third Year	52	
Professor/teacher/lecturer	5	

From above data it is very much clear that Fintech can be very useful to perform various accounting functions. Majority of the respondents agreed that FinTech is very useful in managing regular accounting work; reconciliation; cloud accounting and auditing; business valuation; cash flow forecasting; credit control; fraud identification and mining; compiling and analysing the accounting data.

Objective 3: To analyse the perception for Financial Technology (FinTech) among the respondents (Charts 6, 7, 8, 9, 10 and 11).

Majority of the respondents with 54 and 52 % have shown their disagreement to the statement that 'Accounting professional do not require knowledge of technology' and 'The current job market of accounting professional demands for only degree not IT knowledge' respectively.

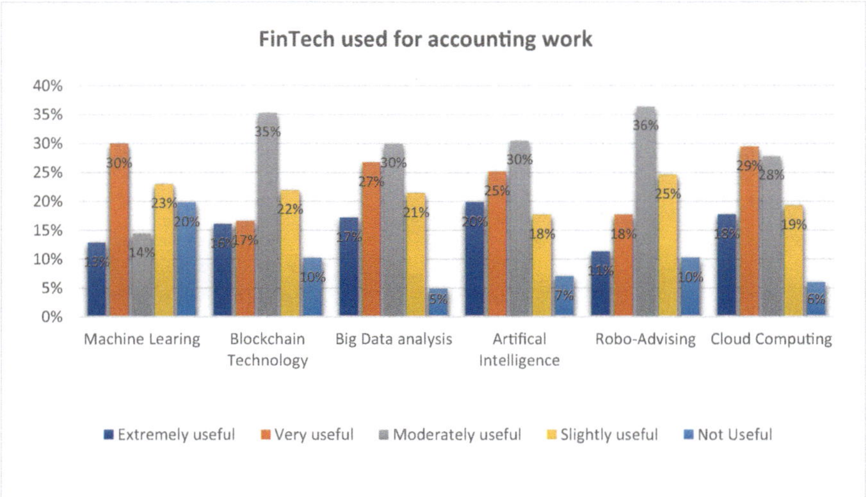

Fig. 1 Various FinTech used in accounting work

Chart 1 Managing regular acting workChart

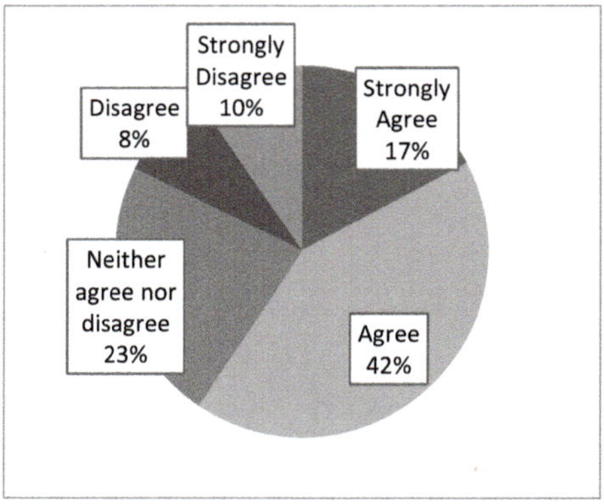

Chart 2 Reconciliation

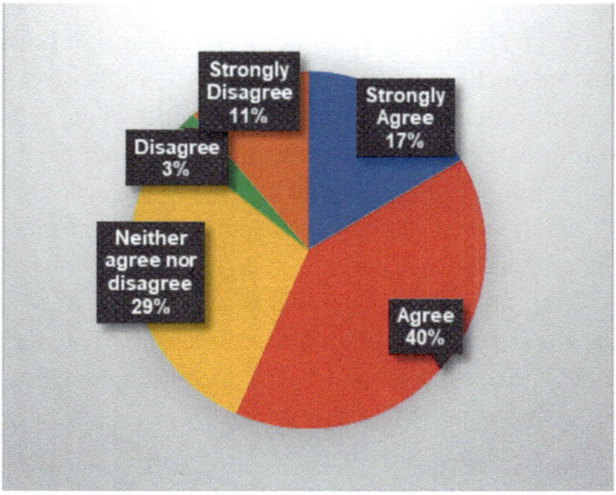

Chart 3 Cloud accounting, reporting and auditing

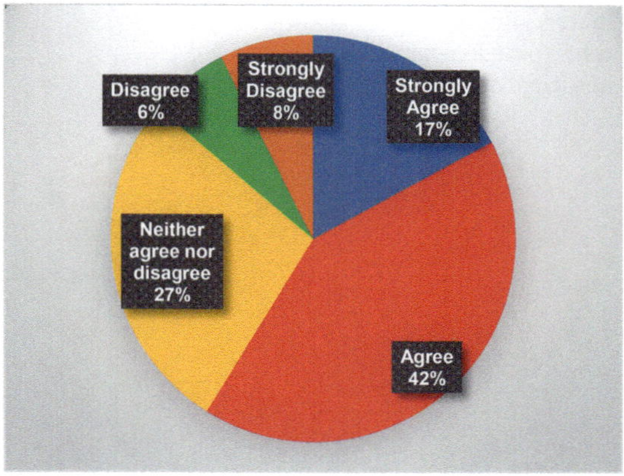

Chart 4 Business valuation

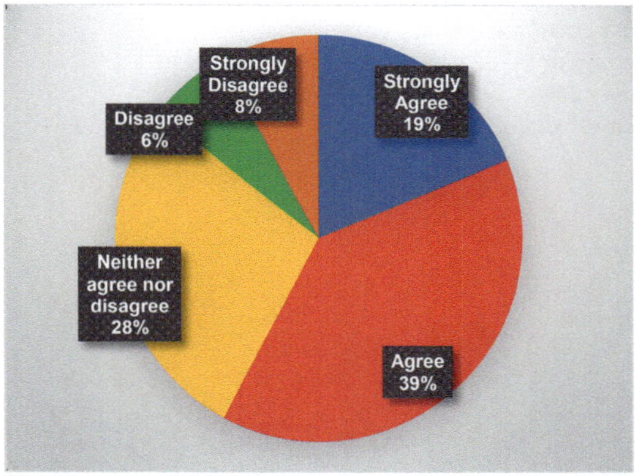

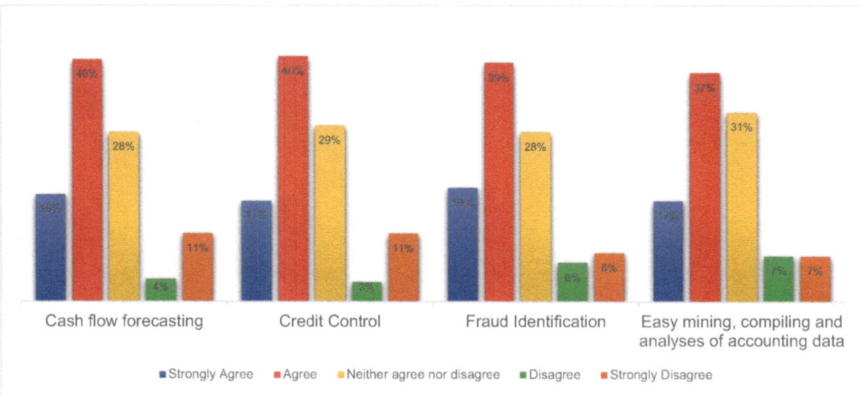

Chart 5 Usefulness of FinTech for accounting work

Chart 6 Accounting professional do not require knowledge of technology

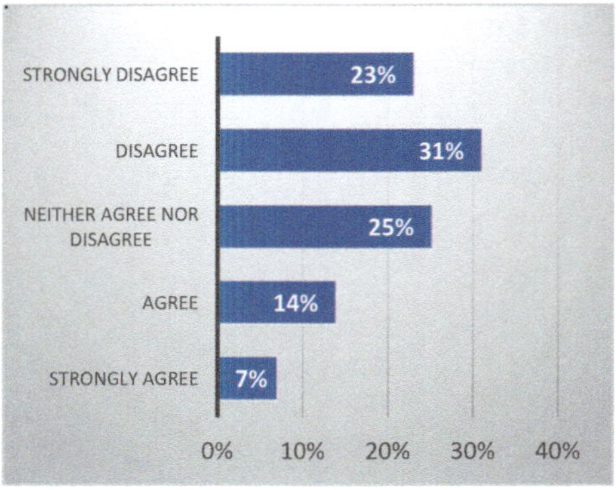

Chart 7 The current job market of accounting professional demands for only degree not IT knowledge

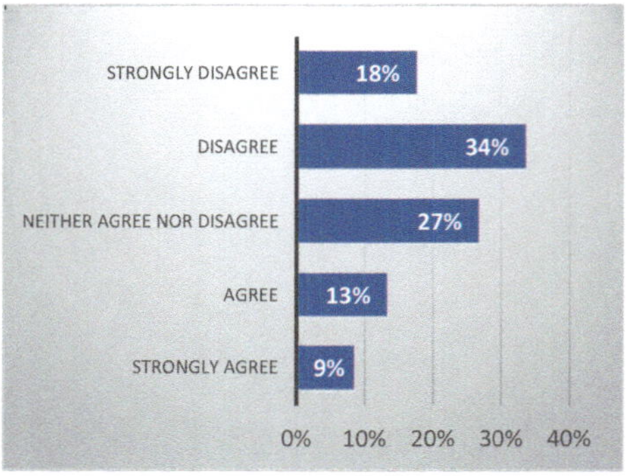

Chart 8 Technology will not replace, it will support

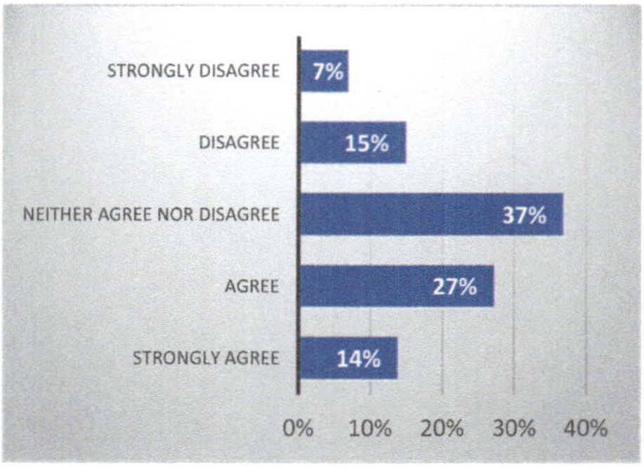

37% of the respondents were with the opinion of neither agree and not disagree for the statement 'Technology will not replace, it will support the accountants', it seems that they are lack of awareness about its practical usefulness on workplace.

Chart 9 FinTech will help in more responsive and rational decision the accountants

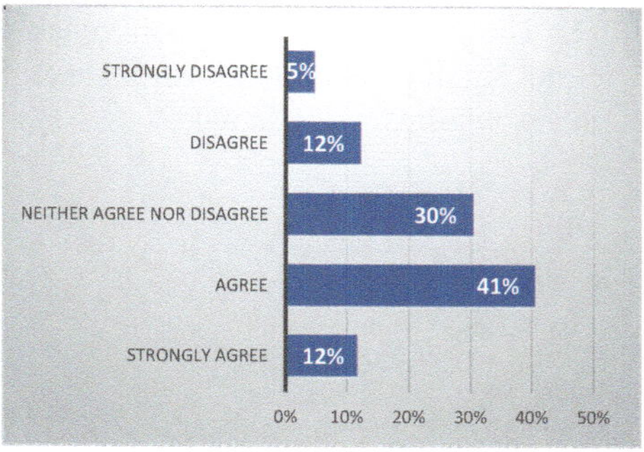

Chart 10 FinTech can be used to build expertise of accounting

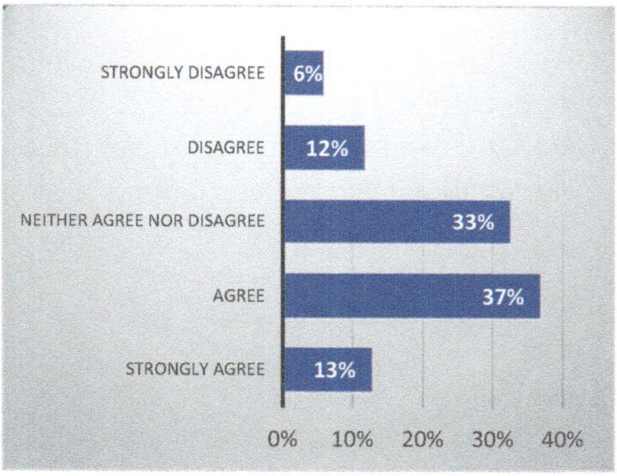

Chart 11 FinTech help accounting professional to identify professionals threats and opportunities in advance

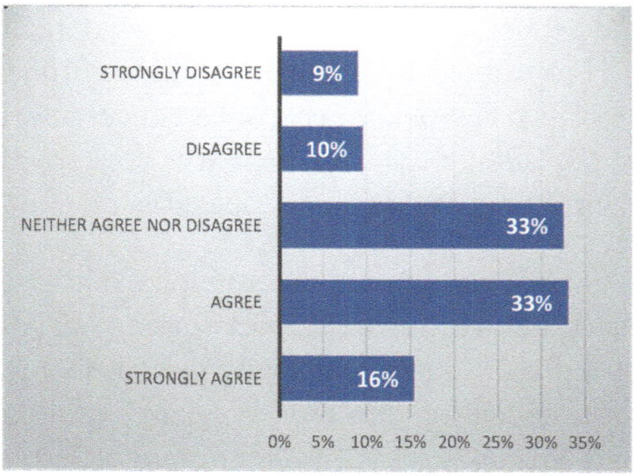

Hypothesis Testing

Ho: The usefulness of Financial Technology (FinTech) in accounting among males is not significantly different from usefulness of Financial Technology (FinTech) in accounting among females.

H1: The usefulness of Financial Technology (FinTech) in accounting among males is not significantly different from usefulness of Financial Technology (FinTech) in accounting among females.

P value	0.582145
Chi table value	9.488
Chi calculated	2.85635

The Chi-square calculate value is 2.85635 which is less than the table value 9.488, that means we fail to reject the null hypothesis. There is no significant association between the gender and usefulness of Financial Technology (FinTech) in accounting.

Ho: The usefulness of Financial Technology (FinTech) in accounting is not same across all the level of education.
H1: The usefulness of Financial Technology (FinTech) in accounting is same across all the level of education.

p value	0.144339
Chi table Value	26.296
Chi calculated	21.9646

The Chi-square calculate value is 21.9646 which is less than the table value, that means we fail to reject the null hypothesis. There is no significant association between the gender and usefulness of Financial Technology (FinTech) in accounting.

From above both inferences drawn from hypothesis testing it has been proved that there is no association between gender, level of education and usefulness of FinTech in accounting, which means it is concerned to be very much useful irrespective of their gender and level of education.

Conclusion: In general technology has captured the world in all the fields like entertainment, finance or accounting but educational sector is not an exception to it specifically during the covid period. Digital transformation has been proved to be an excellent opportunity for many but a threat to some. The gap between what is and what is expected from professionals has taken up the same [8]. The researchers found that various of fintech like the machine learning, Blockchain, Big Data Analysis, Artificial Intelligence and Robo- Analysis is very useful for finance sector in general and for accounting in particular. There are various accounting activities like invoice management, tax advices, expense management and many other functions which can be facilitated by various financial technological advancement. This adaption to the technology may prove to be competitive advantage for some companies but at the same time there are some suggestions for HEIs to enhance the skill set of accounting graduates and post graduates like Introducing the FinTech (blockchain, machine learning etc.) as subject in syllabus form, Giving the practical knowledge while studying through industry connect and Providing internship opportunities in companies working for FinTech and software designing. At the end, it is also recommended to the growing organization to change their business process and models by leveraging the use of technology in creating and providing digital business solution and services to customers.

References

1. Srinivasan RK (2020) Digital strategy and business transformation. In: W. I. Council Wiley innovation black book enterprise 4.0. Wiley India Pvt. Ltd., pp 11–26
2. WEF (2020) The future of jobs report 2020. World Economic Forum, Switzerland
3. (FSB) FS (2019) FinTech and market structure in financial services: market developments and potential financial stability implications. FSB, Basel
4. Dietz M, Khanna S, Olanrewaju T, Rajgopal K (2016) FinTechnicolor: the new picture in finance. Mckinsey & Company, New York
5. Education IC (2020) IBM Cloud Learn Hub. Retrieved from IBM: https://www.ibm.com/cloud/learn/api
6. Sakhnini Z, Elali W, Al-Sartawi A (2022) Behavioral and non-behavioral factors and the level of adapting and implementing fintech and E-banking in Bahrain: suggested model. In: Musleh Al-Sartawi AMA (eds) Artificial intelligence for sustainable finance and sustainable technology, ICGER 2021. Lecture notes in networks and systems, vol 423. Springer, Cham
7. Agarwal N (2021) Redefining banking: exchange traded savings and loans using cryptocurrencies. Int J Electron Bank 3(1):1–11
8. Al-Sartawi A, Al-Okaily M, Hannoon A, Khalid AA (2022) Financial technology: literature review paper. In: Musleh Al-Sartawi AMA (eds) Artificial intelligence for sustainable finance and sustainable technology. ICGER 2021. Lecture notes in networks and systems, vol 423. Springer, Cham

Study of Public Sentiment Using Social Media for Organic Foods in Pre-Covid and Post-Covid Times Worldwide

Jolly Masih, Manojkumar Deshpande, Harvinder Singh, and Jonathan Deutsch

Abstract This study systematically assess the paradigm shift in consumer behavior during Pre-COVID and COVID scenarios using social media (data collected from Facebook, Twitter, Pin Interest, YouTube, Reviews & Forums, and Google Plus) platforms. The study analyzed 523,764 and 1,054,794 social media responses in Pre-COVID and COVID eras respectively. Due to the large volume of the data, for Pre-COVID era, responses from August to December 2019 were considered. For COVID era, responses from January to June 2020 were taken into the account. The organic food categories, considered in the study were Snacks, Fruits, Vegetables, Beers & Alcohol, Meat, Flour, Dairy, Cereals & Savories, Health Drink, Noodles & Fast Food, Dessert and Pizza & Pasta. In the study three-stage filtering for Sentiment Analysis was performed (86.4% accuracy achieved) using Natural Language Processing, Manual analysis of sentiment and Bi-Directional Encoder Representations from Transforms based model. It was concluded that in Pre-COVID era, people were less aware about the health benefits of organic food but in COVID era, the awareness level of consumers rose exponentially about various organic food categories, especially about processed organic food categories and benefits of raw organic food categories. In Pre-COVID Era, consumers were mainly focused on fresh fruits and vegetables. In COVID Era, food categories like Dairy, Meat and Vegetables were at the top. Fruits, snacks, beer and alcohol were also in high demand. While analyzing the Pre-COVID and COVID scenarios, we observed a shift in consumer behaviour from raw organic food categories towards processed and semi-processed organic food

J. Masih (✉)
Symbiosis University of Applied Sciences, Indore, India
e-mail: jolly.iabm@gmail.com

M. Deshpande
Prestige Institute of Engineering Management and Research, Indore, India
e-mail: director@piemr.edu.in

H. Singh
Leaders Institute Pvt. Ltd., Brisbane, QLD, Australia
e-mail: drharvi@leaders.edu.au

J. Deutsch
Department of Food and Hospitality Management, Drexel University, PA Philedelphia, USA
e-mail: st96d633@drexel.edu

A. M. A. Musleh Al-Sartawi et al. (eds.), *From the Internet of Things to the Internet of Ideas: The Role of Artificial Intelligence*, Lecture Notes in Networks and Systems 557,
https://doi.org/10.1007/978-3-031-17746-0_11

categories. Organic noodles & fast food, dessert and health drinks still remained as the least preferred food categories. The study is unique since it provides a macro and micro view of shift in consumer behavior preferences in Pre-COVID and COVID eras using social media at various geographical locations worldwide. This research will help manufacturers of organic food to reshape their existing consumer behavior and market penetration strategies. It will also help in new product designing and increasing the consumer base both at the local and global levels.

Keywords Organic food · Pre-COVID · COVID · Sentiment analysis · Social media · Perceptual map · Consumer behaviour

1 Introduction

Organic food includes the crops, fruits and vegetables cultivated without application and use of any artificial or synthetic chemicals, fertilizers, pesticides or inclusion of any genetically modified organisms (GMOs). Organic foods include wide range of products from fresh fruits and vegetables, dairy, meat, to processed foods such as pizza and pasta, drinks, frozen meals and drinks. Production of organic food is basically divided into two major groups worldwide: (1) small and marginal farmers who grow food organically using farm yard manure and other crop residues at small scale; and (2) certified organizations or farmers growing organic foods as per the norms of "organic food licensing/certification agency" [1, 2].

The COVID-19 pandemic has divided the world into two part: Pre-COVID Era (before year 2020) and COVID Era (After year 2020). Organic food has always been considered healthy as compared to conventional foods, but after the occurrence of the COVID-19 pandemic, a sudden hike in the demand of organic foods has been observed [3]. At the same time, due to lock down, travel restrictions and closure of stores and malls, people switched to social media and online shopping options to search about various healthy eating food options. This study critically analyses consumer behaviour related to purchase of organic foods worldwide based on responses generated on different social media platforms like Twitter, Facebook, Youtube etc. The study also analyzes consumer sentiments and purchase intensions about organic foods in two time zones Pre-COVID Era (before year 2020) and COVID Era (after year 2020). Social media is considered to be an effective platform to advertise and create awareness about products, to gather consumer sentiment or opinion about any product and to get unbiased results (which is very common in survey method of research). Hence by using the big data generated from various social media platforms, this study carefully analyses the positive, negative and neutral sentiments related to organic foods at various social media platforms in Pre-COVID and COVID Era. The study also consolidates consumer responses about various food categories of organic food. Food categories taken into consideration were Fruits, Vegetables, Dairy Products, Meat, Flour & Mixes, Fast Food & Noodles, Pasta & Pizza, Snacks, Beer & Alcohol, Health Drinks, Desserts, Cereals & Cornflakes and Meat. Study provides insight on the shift in consumer sentiments in Pre-COVID and COVID eras, such as what were the major causes that has led to the shift in these sentiments and which food categories were mostly preferred in both the eras. This research will help manufacturers of organic food to reshape their existing consumer behaviour and market penetration strategies. It will also be helpful in new product designing process and in increasing the consumer base at local and global level.

2 Literature Review

History of Organic Food

The idea of "organic agriculture" has been advanced within the 1900s through Sir Albert Howard, and other researchers who used animal-based manures, organic compost, green vegetation and crop rotation as a method to develop a sustainable farming system. Over the time, all countries have developed their organic foods certification agencies/organisations like APEDA (India), USDA NOP (USA), CFIA (Canada), IFOAM (Europe), APEDA (Asia), APEDA (Middle East) and NCO (Australia) [3].

Growing Demand of Organic Food

Major reasons for this sudden increase in demand of organic food worldwide are growing health awareness and consciousness due to COVID-19 pandemic. People worldwide have also become conscious about improving their health and protecting the environment. Labels like 'Eco-friendly', 'Organic', 'Naturally grown', 'Non-GMO' have greater acceptance among consumers. The growth in sales of organic foods is also because food companies have resumed their operations after lock-down and COVID-19 impact. A lot of food companies have also launched new varieties of organic foods owing to increased demand by the health-conscious consumer segment. The global market of organic food is expected to grow up to $380.84 billion in 2025 at a CAGR of 14.5% [4].

Organic Food is Expensive

Organic foods are generally 20–100% more expensive than conventional foods due to multiple reasons like expensive certification process of growing and cultivating organic foods, low shelf-life due to less or no reliance on preservatives or chemicals and disruptions in organic food supply chains [2].

Role of Social Media in Spreading Awareness About Organic Foods

Social media has become an indispensable part of our lives. Social media platforms have provided great freedom and liberty to consumers to express their views on various products including organic food. Research has shown that consumers view social media as an important platform to receive messages about organic food products, their benefits, usage and availability [8–10]. Amongst all social media platforms, Twitter has played most important role in promotion and awareness of organic foods amongst the masses [10].

Table 1 Sample techniques analysis

Area of study	Worldwide
Technique used	Perceptual mapping
Charts	Pie charts, Bar charts
Sample size	Pre-COVID sample: 523,764 & COVID sample: 1,054,794
Data collection method	Meltwater data crawler' (https://www.meltwater.com/en)
Mode of collection	Data pool for social media included Facebook, Twitter, Pinterest, YouTube and Google Plus

3 Research Methodology

- **Time period considered:** Due to large volume of the data, for Pre-COVID Era, responses from August to December 2019 were considered. For COVID Era, responses from January to June 2020 were taken into the account (Table 1).

- **Data Sorting and filtering**: Sentiment analysis was conducted in 3 stages

  i. Natural Language Processing based sentiment generated from Meltwater AI (Artificial Intelligence) tool, based on queries created.
  ii. Manual Sentiment Analysis, by visiting social media posts manually (random selection was followed to visit the posts) and checking their sentiment.
  iii. Final sentiment analysis was run using BERT (Bi-Directional Encoder Representations from Transforms) based model. This Model was Developed by Google. It has an AI mechanism which let each token from the input sequence to focus on particular sentiment by assigning weights to them.

Using the 3-stage data sorting and filtering process for sentiment analysis, an overall data accuracy of 87% was achieved.

4 Results and Discussion

A. Pre-COVID Era Social Media Analysis

The Pre-COVID analysis of organic food using social media platforms indicated that people across the world are inclined to use or at least try organic food products. Mostly people believed that in spite of low availably, lack of awareness and high prices of organic foods, they were still very fresh, good in taste and nutritious as compared to other conventional foods (Fig. 1).

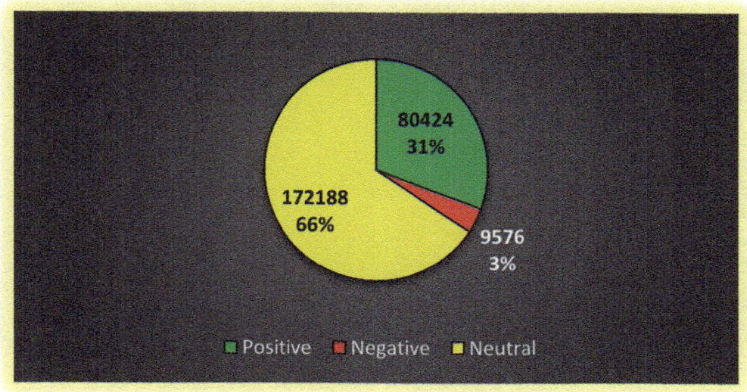

Fig. 1 Pre-COVID social media sentiment analysis

Positive: In Pre-COVID scenario rich and wealthy consumers preferred organic food due to status, health benefits, safety and environmental concern. Many people purchased organic food believing that it would be chemical-free, more fresh and healthier than other conventional foods. Major focus of people was mostly on purchase of fresh fruits and vegetables, as compared to semi-processed or completely processed categories.

Negative: Major negative responses about organic foods indicated that most people were not happy about the high prices and low availability. They also complaint about a few companies who were caught in fraud act of selling organic foods. Consumers demanded for trustworthy and affordable organic food options.

Neutral: In neutral sentiments, consumers raised questions about product information, manufacturing process, licensing process and availability of other variants of organic foods in nearby supermarkets.

B) COVID Era Social Media Analysis

In the COVID era, a huge spike in demand and awareness of organic foods had been observed. Globally, the organic food industry had seen a hike of somewhere about 28–50% depending upon the country. For example, in USA, organic food industry had witnessed a growth of 48–50% and in Europe, Asia and Australia, the growth of organic food was about 28–35%. However, not much surge in demand had been seen in Africa. Whole Foods Market in USA, Abel and Cole in UK, Naturally Yours in India, I Say Organic, Healthy Buddha, Nourish Organics and Nature land Organics in Asian- Pacific emerged as a biggest supplier of organic foods.

In COVID scenario blogs, Facebook and Twitter are found to be the most popular platforms (Fig. 2).

Positive: Organic food gave consumer the assurance, that toxic pesticides, synthetic fertilizers, and genetically modified organisms (GMOs) were not used in the food production, and the livestock had not been given antibiotics or growth hormones. Regionally, North America was the market leader followed by Europe. Countries

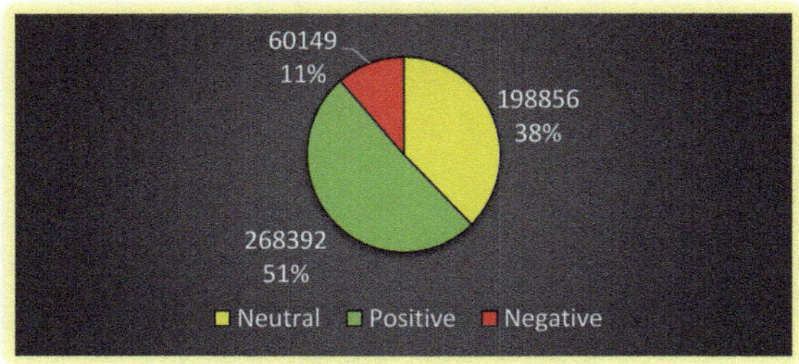

Fig. 2 COVID social media sentiment analysis

like USA, Germany, France, China were the topmost countries in the global organic food market. In COVID scenario, organic food was considered as non-toxic, natural, immunity booster, natural, safe and healthy foods. In war-against COVID-19, organic foods were considered to be an ally. Due to the influence and fear of COVID-19 pandemic, all the categories of organic foods, namely dairy, meat, pizza & pasta, flours (made out of grains) had seen an incredible growth, especially in metro cities, where people had higher purchasing power.

Negative: Some of the people had posted negative reviews or comments on social media about high prices of organic foods, lack of variety in processed category, lack of availability of organic foods in nearby supermarkets due to broken and unstructured supply chain. Some people also claimed about false or fake tags of "being organic" by some local brands. A few people made negative comments about putting high sugar and salt in processed organic foods, which made the taste very poor.

Neutral: The neutral sentiments about organic foods revealed that people were curious to know more about organic foods, their benefits, available food variants and leading companies. People also raised queries to know the difference between "organic food" and "natural foods" or "safe foods". Many posts were made about various agencies providing organic food certification.

C) Multidimensional Perceptual Mapping

- **Perceptual Map for Organic food categories in Pre-COVID Era (Fig.** 3)

Perceptual map was plotted between number of posts made on social media about the particular organic food category and purchase tendency of consumer about that organic food category.

Perceptual map was divided in 4 quarters (Q) and each quadrant had following food categories based on the scores obtained:

Q1-Low Purchase Tendency-High Number of Post: No food category.
Q2-High Purchase Tendency-High Number of Post: Vegetables, Fruits, Dairy.

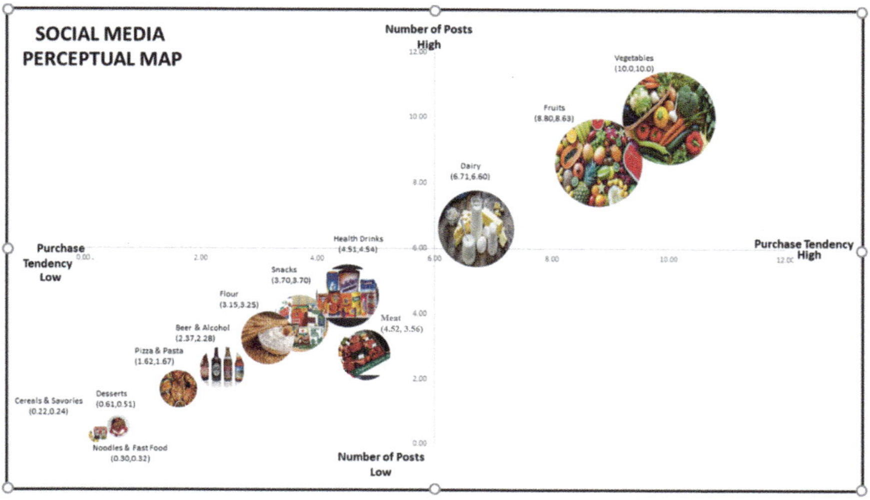

Fig. 3 Perceptual map for organic food categories for social media analysis in pre-COVID era

Q3-Low Purchase Tendency-Low Number of Post: Health Drinks, Snacks, Flour, Beer & Alcohol, Pizza & Pasta, Desserts, Meat, Noodles & Fast Food and Cereals & Savories.
Q4-High Purchase Tendency-Low Number of Post: No food category (Table 2).

In Pre-COVID Era, the awareness about organic foods was there, but consumers were mainly focused on fresh fruits and vegetables. Most of the times, organic foods were considered to be same as a 'naturally grown' and Non-GMO' fresh fruits and vegetables. Thus, these two were the leading food categories with maximum demand worldwide before January 2020 (in Pre-COVID Era). Other categories that were also getting popular in Pre-COVID era were organic dairy, health drinks & snacks.

- **Perceptual Map for Organic food categories in COVID Era (Fig.** 4)

Perceptual map was divided in 4 quarters (Q) and each quadrant had following food categories based on the scores obtained:

Q1-Low Purchase Tendency-High Number of Post: No food category.
Q2-High Purchase Tendency-High Number of Post: Dairy, Meat, Vegetables, Snacks, Fruits.
Q3-Low Purchase Tendency-Low Number of Post: Health Drinks, Flour, Beer & Alcohol, Pizza & Pasta, Desserts, Noodles & Fast Food and Cereals & Savories.
Q4-High Purchase Tendency-Low Number of Post: No food category (Table 3).

In COVID Era, food categories like Dairy, Meat and Vegetables were at the top. Fruits, snacks and beer and alcohol very also in huge demand. There was a shift of

Table 2 Perceptual map scores of social media analysis in pre-COVID era

Categories	Total No. of hits	Total sentiment score	Hits converted (On scale of 1–12 (1 being minimum, 12 being maximum)	Sentiment score (On scale of 1–12 (1 being minimum, 12 being maximum)	Hits percentage share (%)
Snacks	24,881	57,244	3.70	3.70	8.81
Fruits	59,159	133,308	8.80	8.63	20.95
Vegetables	67,251	154,558	10.00	10.00	23.82
Beers & Alcohol	15,934	35,283	2.37	2.28	5.64
Flour	21,187	50,261	3.15	3.25	7.50
Dairy	45,120	101,985	6.71	6.60	15.98
Cereals & Savories	1464	3691	0.22	0.24	0.53
Health Drink	30,353	70,189	4.51	4.54	10.75
Noodles & Fast Food	1993	4957	0.30	0.32	0.71
Dessert	4070	7854	0.61	0.51	1.44
Meat	2273	6075	0.62	0.48	1.59
Pizza & Pasta	10,916	25,873	1.62	1.67	3.87

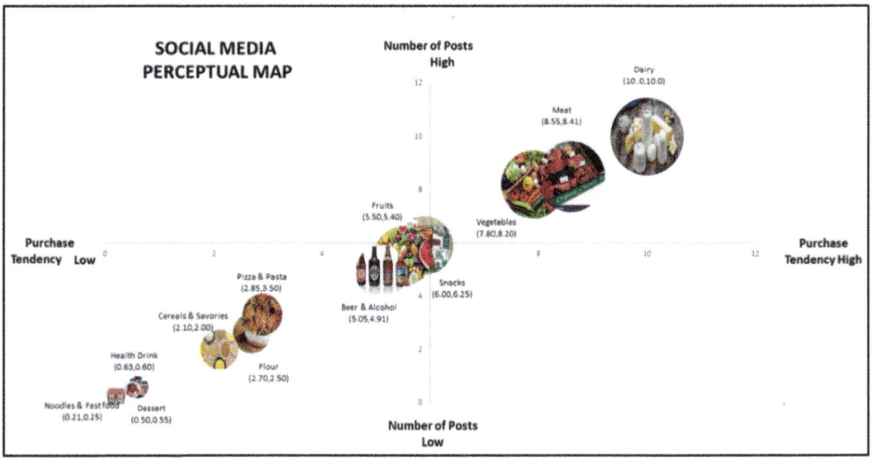

Fig. 4 Perceptual map for organic food categories for social media analysis in COVID era

Table 3 Perceptual map scores of social media analysis in COVID era

Categories	Total no. of hits	Total sentiment score	Hits converted (On scale of 1–12 (1 being minimum, 12 being maximum)	Sentiment score (On scale of 1–12 (1 being minimum, 12 being maximum)	Hits percentage share
Snacks	16,628	42,133	5.91	5.95	11.36
Fruits	15,435	39,069	5.48	5.52	10.55
Vegetables	22,359	57,776	7.94	8.16	15.28
Beers & Alcohol	14,226	34,771	5.05	4.91	9.72
Flour	7515	18,435	2.67	2.60	5.14
Dairy	28,147	70,828	10.00	10.00	19.24
Cereals & Savories	5782	14,089	2.05	1.99	3.95
Health Drink	1776	4277	0.63	0.60	1.21
Meat	24,273	60,075	8.62	8.48	16.59
Noodles & Fast Food	641	1638	0.23	0.23	0.44
Dessert	1307	3602	0.46	0.51	0.89
Pizza & Pasta	8223	23,389	2.92	3.30	5.62

consumer behaviour from raw organic food categories towards processed and semi-processed organic food categories. Organic noodles & fast food, dessert and health drinks still remained as the least preferred food categories.

5 Conclusion

Using big data analysis of social media posts, it was found that in the Pre-COVID era people were less aware and not much interested in the health benefits of organic food but in COVID era, the awareness level of consumers about different organic food categories especially about processed organic food categories and benefits of raw organic food categories. In Pre-COVID Era, the awareness about organic foods was there, but consumers were mainly focused on fresh fruits and vegetables. Thus, fresh fruits and vegetables were the leading food categories with maximum demand worldwide before January 2020 (in Pre-COVID Era). Other categories that were also getting popular in Pre-COVID era were organic dairy, health drinks & snacks. In COVID Era, food categories like Dairy, Meat and Vegetables were at the top. Fruits, snacks and beer and alcohol were also in a high demand. There was a shift of consumer behaviour from raw organic food categories towards processed and semi-processed organic food categories. Organic noodles & fast-food, dessert and health drinks still remained as the least preferred food categories. This research will help manufacturers of organic food to reshape their existing consumer behaviour and marketing strategies. It will also be helpful in new product designing and in increasing the consumer base both at local and global level.

References

1. Le HN, Le LT (2022) Organic food consumption: a literature review on motives and barriers. Sci Technol Develop J-Econ-Law Manage 6(1):2115–2126
2. Masih J (2021) Understanding health-foods consumer perception using big data analytics. J Manage Inf Decis Sci 24(1):1–15
3. Heckman J (2006) A history of organic farming: transitions from Sir Albert Howard's War in the soil to USDA national organic program. Renewable Agric Food Syst 21(3):143–150
4. Danner H, Hagerer G, Pan Y, Groh G (2022) The news media and its audience: agenda setting on organic food in the United States and Germany. J Clean Prod 131503
5. Boobalan K, Nachimuthu GS, Sivakumaran B (2021) Understanding the psychological benefits in organic consumerism: an empirical exploration. Food Qual Prefer 87(1):104070
6. Massey M, O'Cass A, Otahal P (2018) A meta-analytic study of the factors driving the purchase of organic food. Appetite 125(1):418–427
7. Tandon A, Dhir A, Kaur P, Kushwah S, Salo J (2020) Why do people buy organic food? The moderating role of environmental concerns and trust. J Retail Consum Serv 57(1):102247
8. Lynch M, Knezevic I, Mah CL (2022) Exploring food shopping behaviours through a study of Ottawa social media. Appetite 168(1):105695
9. Pilař L, Kvasničková Stanislavská L, Kvasnička R (2021) Healthy food on the twitter social network: Vegan, homemade, and organic food. Int J Environ Res Public Health 18(7):3815
10. Wojciechowska-Solis J, Kowalska A, Bieniek M, Ratajczyk M, Manning L (2022) Comparison of the purchasing behaviour of Polish and United Kingdom consumers in the organic food market during the COVID-19 pandemic. Int J Environ Res Public Health 19(3):1137

The Usage of Digital Media in Society

Abdulsadek Hassan, Hanan Gunied, Abdulrahman Yaqoob Seyadi, and Abdulla Mahmood Alkhaja

Abstract The study aimed to identify the usage of digital media in society and applied this usage in educational field. The results indicated that new types of media as (digital media) or (social networks) and (citizen journalism) etc. Results showed that modern communication technology has contributed to enhancing cultural communication, revitalizing knowledge exchange between peoples, overcoming geographical in educational field and even transcending them, and brought about a radical change in quantity and quality in the content transmitted through digital media.

Keywords Digital media · Traditional media · Education

1 Introduction

Since the emergence of a new element to the world that contributed to a unique qualitative shift in changing the identity of the media to some extent and giving features that were reprehensible or unexpected, the idea of ideology may have become circulating on the media horizon. The technical development in the year 1990 called this idea digital media, and with the entry of technology into its widest doors, the digital media, or the transfer of digital, even electronic media, took a large share of its integration with contemporary technology means at the end of the last millennium and to this day [8].

The large and successive developments in the means of communication and information technology have brought about many changes and challenges in the media landscape [15]. The spread of social networks has also contributed to changing the traditional media work method, which was dependent on media institutions only as a source of information, so that now the production and consumption of news is an

A. Hassan (✉)
Ahlia University, Manama, Kingdom of Bahrain
e-mail: dr.sadek1964@yahoo.com

H. Gunied · A. Y. Seyadi · A. M. Alkhaja
Deputy Dean for Society and Environment Affairs, Faculty of Mass Communication, Cairo University, Giza, Egypt

A. M. A. Musleh Al-Sartawi et al. (eds.), *From the Internet of Things to the Internet of Ideas: The Role of Artificial Intelligence*, Lecture Notes in Networks and Systems 557,
https://doi.org/10.1007/978-3-031-17746-0_12

option in the hands of members of the communities taking advantage of opportunities which provided them with the continuous development in the world of the Internet and technology [22].

Digital media appeared in the latter part of the twentieth century, relying on Internet technology to deliver information, and it includes the integration of traditional media such as films, images, spoken and printed words, with computer technologies and smart devices [17], and it has multiple forms, including: media websites on the Internet, channels, electronic journalism, online commercials, blogs, and the use of smartphones for television broadcasting [23].

Today, digital media has become an inescapable reality to deal with, and its tools have imposed themselves, and dealing with these technologies is no longer a luxury, but has become a living reality for most of the world's population, through which they communicate, and through them they transmit information and news [6], in the fastest and easiest way according to the latest and most technologies it has developed, and modern means and technologies have been prepared in these eras that assist in learning languages and facilitate this for those who acquire them, and contribute mainly to the educational process, especially in the time of the spread of epidemics and the prohibition of human gatherings in order to preserve them [6].

With the great development witnessed by the Internet, talk began about the digital media and its implications and the interactive services it provides that made the future of the media material a participant with it at times and a maker of it at other times, and the public became seeking information and choosing the appropriate ones, as well as exchanging messages with the sender after what was the role of the future is just to receive information only [7].

The digital media has within a cultural, historical and civilized framework the characteristics of the era in which it is born. In fact, the information age has produced a digital media style that differs in its concept, features, characteristics and means from traditional media patterns, and differs in its extensive media, political, cultural and educational effects to the extent that some of them launched our era has the name (the media age) [15], not because the media is a new phenomenon in human history, but because its modern means have reached far goals in depth of impact, strength of direction and severity of danger that have led to fundamental transformations in the role of the media, and made it an essential axis in the community system [21].

Society became open and available to everyone, even to non-specialists and students of this field, which required more effort in critical thinking, as these technological means have changed all aspects of life and created new activities for individuals and societies [6]. Technological development has become the source of information in a dangerous situation, so we do not know what the correct source of information is and how to deal with it [1].

2 Digital Media Concept

Digital media as a concept that depends on the merging of text, image, video and sound with each other, as well as the use of the computer as a main mechanism for it in the production and presentation process. Instructiveness represents the main difference that distinguishes it and is the most important feature [11].

Digital media is called by many names and terms, including:

Digital media, interactive media, information media, multimedia media, online live media, cyber media, and hyper media [18].

Also, Digital media is any media that is encoded in machine-readable formats [16]. Digital media can be created, viewed, distributed, modified, and saved on digital electronics devices [8].

Digital can be defined as any data represented by a series of numbers, media refers to the way information is transmitted or communicated together, and digital media refers to any information transmitted to us through a screen. This includes text, audio, video, and graphics that are transmitted over the Internet for display on the Internet [3].

3 The Concept of Digital Media is Based on Four Types

The digital media based on the Internet and its applications, which is completely new with unprecedented qualities and features, as it grows rapidly and generates a set of countless applications [7]. The digital media based on mobile devices, and it is growing rapidly, and new types of applications are emerging from it on various portable tools, including telephones, personal digital assistants, and others [1]. A type based on the platform of traditional media such as radio and television to which new features such as interactive, digital and demand response have been added [16]. The digital media based on the computer platform, and this type is circulated, either online or by various means of preservation such as optical discs, etc. It includes visual presentations, video games, electronic books and others [12].

4 Forms of Digital Media

Digital Newspaper: It is a digital media that uses the Internet as a communicative medium to disseminate news and information, and does not differ from the printed newspaper in terms of periodicity of publication, and the multiplicity of journalistic forms such as news, reports, investigations, articles, interviews, and caricatures, and in terms of the diversity of political, economic, social, sports and other topics [14], but It outperforms the printed newspaper with the advantages of quickly publishing the scoop, updating its issues, and containing interaction tools that allow the future to have multiple options for interacting with the sent content, as well as the use of "Multi Media" in publishing content [19].

Digital Internet Radio: The digital media that provides various broadcasting services using the Internet as a communication medium for broadcasting, and with interactive features that allow the listener to control the listening time and choose the broadcast content from the digital radio website on the Internet [23].

Interactive Digital TV: An integration of smart computer technology, the Internet, and traditional television that allows multiple interactive features for the viewer to follow news, programs, music, video, surfing the Internet, making calls and conversations, video conferencing, shopping, and controlling viewing and viewing time [2].

Digital Social Media Networks are a group of social identities created by individuals, organizations, or media institutions with the aim of strengthening personal relationships, or activating professional relationships, and disseminating news and information to a large audience, using the Internet as a communication medium for dissemination [5].

5 Digital Media Features

The digital media has many characteristics, including:

1. **Interactive:** where the communicator and the recipient exchange roles, and the practice of communication is bi-directional and reciprocal, and not in a unilateral direction, but rather there is a dialogue between the two parties [17].
2. **Asynchrony:** It is the possibility of interacting with the communicative process in a timely manner for the individual, whether he is a receiver or a sender [13].
3. **Sharing and spreading:** The digital media allows everyone with simple tools to be a publisher who sends their message to others [10].
4. **Movement and flexibility:** where new means can be transferred to accompany the recipient and the sender, such as mobile computers, Internet computers, mobile phones, and handheld devices, by making use of wireless networks [4].
5. **Universality:** The communication environment has become a global one, transcending the barriers of time, space and censorship [15].
6. **Media integration:** In the digital media, all means of communication are used, such as texts, audio, still images, moving images, two-dimensional and three-dimensional graphics [6].
7. **Attention and focus:** Since the recipient in the digital media does an active job in selecting content and interacting with it, it is characterized by a high degree of attention and focus, unlike exposure to traditional media, which is usually negative and superficial [4].

8. **Storage and Preservation:** It is easy for the recipient to store, save and retrieve communication messages, as part of the capabilities and characteristics of the medium itself [9].

6 Types of Digital Media

Owned Digital Media
Owned digital media is any online asset that you control, usually this type of media is unique to a brand, digital media platforms include your website, social media channels, blogs, videos, etc. [12]. If optimizing your digital media is done for SEO, the work will show up when potential customers search for topics related to what you offer, the more ownership of the digital media you create, the more visible the work becomes, this gives brand authority, and thus increases your ranking on the results page Search Engine [14].

Paid Digital Media
Paid digital media promotes your online content to improve traffic to your media assets, this type of media typically includes paid search ads, display ads, and Facebook ads [20].

Although paid media is an expense that, when used correctly, exposes your content and brand to a wider audience, a good marketing agency can use paid media to direct the right customers to your website, and paid digital media ads can also deliver results [19]. Important and measurable within the campaign, and thus supports the overall marketing strategy and this is one of the digital media theories [1].

Digital Media Acquired
Earned media are advertisements generated by customers, when customers defend brand via word of mouth, social media, reviews, testimonials, shares, reposts, etc. [5], this is the best form of endorsement they can That the company gets it [22]. Digital acquired media is an effective way to grow brands while reducing customer acquisition costs, however this needs to be monitored regularly, because unlike owned digital media there is no control over it [3].

7 Challenges Facing Digital Media

With the emergence of this new form of media and its widespread, and the entry of many media professionals into it, whether specialists or non-specialists, several phenomena have arisen that accompanied it:

1. Breaking the monopoly of major media organizations [2].
2. The emergence of a new class of media professionals, sometimes not specialists in media, but they have become professionals in the use of new media applications, in which they excel over the original specialists [9].

3. The emergence of new platforms for dialogue, as it is now possible for any individual in society to send, receive, interact, track, inquire and comment freely, and very quickly [24].
4. The emergence of public information to the public [8].
5. The emergence of new cultural and media contents [11].
6. Participate in setting discussion agendas, as the new media sometimes succeeds in shedding light extensively on issues that are silent in the traditional media, making these important issues an obsession for the community to think about, discuss and address them [17].
7. The emergence of the phenomenon of virtual society and social networks: It is a group of people who interact and communicate using the new media, for professional, cultural, social or educational purposes. In this society, the relationships are characterized by not necessarily being simultaneous. And members do not attend in the same place [25]. The virtual community may be more powerful and effective than the real one, because it is formed quickly, spreads across the place, and achieves its goals with the least number of restrictions and limitations [10].
8. **Fragmentation of the masses**: With the huge diversity and great diversity that has never been seen before in history, the audience began to break up into small groups, instead of the case of the broad masses of the traditional media, and thus the media moved to the stage of category and specialized media [18].

8 Factors of the Emergence of Digital Media

1. **The technical factor** represented by the tremendous progress in computer technology: its equipment and software, and communication technology, especially with regard to satellite and fiber-optic networks [9]. These technological elements merged in several communicative combinations until they produced the Internet, which is currently being formed to become a medium within which all other communication media: print, audio and video, as well as the public and the personal, are folded [24].
2. **The economic factor** represented in the globalization of the economy and what it requires of accelerating the movement of goods and capital, which in turn requires accelerating the flow of information [14]. This is not just because information is a common denominator that supports all economic activities without exception, but because any information is an economic commodity in itself whose importance is increasing day by day [22]. In other words, the globalization of media and communication systems is the means for economic forces to globalize markets and develop consumer tendencies on the one hand, and the distribution of cultural industry goods such as music, games and television programs on the other hand [2].
3. **The political factor** represented in the increasing use of the media by political forces in order to tighten their grip on the conduct of affairs and maintain the stability of the balance of power in a highly turbulent world full of conflicts and

contradictions [5]. These technical, economic and political factors have overlapped in an unprecedented manner, making the new media a thorny issue. It is a very hot arena for global, regional and local conflicts [25].

9 Articles and Blogs

Posts and articles are another type of digital media material, which is shorter than e-books and accessible online, and can usually be read within five to 20 min [7]. Articles or blog posts are intended to quickly inform the audience about a particular topic, and can be written by a blogger, journalist, or content industry company [3].

10 Social Media

Popular social networks include Facebook, Twitter, Instagram and LinkedIn, and anyone can post content on them, and they are used by: Individuals to share content with friends and family [8]. Brands to advertise their products and communicate with customers [15]. Media companies to share their articles and open dialogues with readers [24].

11 News and Literary Texts

The Internet is one of the most important sources of news today. There are many news sites that many people rely on to find out what is going on around them. On the other hand, the continuing importance of written work in digital media emerges. There is a wide spread of literary sites, free encyclopedias such as Wikipedia, in addition to e-reading tools such as Kindle [5].

Paid Ads. Online advertising is another example of digital media. Some types of digital media advertising include [18]:

Website Banner Ads. Advertisements via search engines. Video ads. Social media ads. digital art Some digital media creators produce illustrations, videos, music, or photographs as an art form.

Digital art productions can be sold or monetized in markets such as Getty Images (for photos), iTunes (for music), and YouTube (for videos).

Video Games Video games are designed to entertain people and are usually played on a video game player such as PlayStation and Xbox, with some types of games also being played on computers, tablets or smartphones [7].

Virtual Reality A new type of digital media that is steadily becoming more popular is virtual reality [18]. Virtual reality involves wearing glasses and watching an immersive digital experience [23], which may include highly realistic colors, sounds, graphics, and videos [4].

12 The Future of Digital Media

As we mentioned, the importance of traditional media for humans is declining in front of the importance of digital media, whose base is increasing day by day [21]. Some expect the complete disappearance of traditional media when reaching a certain point in the development of the Internet infrastructure around the world, to be able to accommodate all television channels, newspapers and radio channels [3]. There are many job opportunities in the field of digital media because of its breadth and newness compared to traditional media [7]. Achieving success in the media field requires making a lot of effort and being present where the audience wants without waiting for them to accept what is being offered, and from this point of view we seek to provide new services that benefit the public and enhance their digital experience [3], and there are main things on which the development of the media today depends:

1- **Video making:** Research indicates that 82% of the traffic on the Internet will be videos during 2022, and consumers will spend 85% of their time watching videos, and thus the video industry has become a very vital component and there are many ideas that can be applied, especially with the development of technologies Virtual Reality (VR) and Augmented Reality (AR) [11].
2- **Social media:** Social media platforms are not only part of our present, but also a part of the future, so we constantly ask ourselves what the audience needs to support our content and present it in an appropriate way on various platforms [22].What distinguishes social media most is the possibility of learning through it. The field is open to trying new and different things and we are working to develop them to avoid previous mistakes [9]. The goal of publishing content across these platforms is to open channels for interaction with the audience and increase content participation instead of focusing on visits and traffic only [19].
3- **Marketing in collaboration with the content creator:** Content makers today have great power, as the audience believes that they are credible and behave themselves, so cooperating with them for the purpose of marketing is a good strategy [15], but choosing the right influencer is very difficult because this influencer may be a teenager who makes videos on Tik Tok and has an army of followers who can help to convince them with a certain content or idea [8], so it is very important to know the audience we are addressing to choose the right influencer to collaborate with in order to get a good investment, and we see this clearly in the choices of major brands for those who collaborate with content makers. It is no longer limited to celebrity influencers who have huge following numbers without real interaction [17].

The Impact of Digital Media on Society It remains to be clear how this media affects societies through: [5] **Spread of Extremist Ideas:** Extremist groups found in the digital media a suitable platform for expressing and disseminating their extremist ideas as a result of the absence of effective censorship on these platforms, and it also facilitated the communication of these groups with each other [18]. **The development of critical thought among individuals**: The digital media helped in the development of critical thought among individuals thanks to the interactive feature of this type of media, as the digital media helped the audience to participate and express opinion and criticism compared to the traditional media in which the viewer was just a recipient [26].

Impact on the Values of Society: The digital media differs from the traditional media by the difficulty of censoring it by the family or the state, and this in turn leads to the spread of ideas and values that may be different from the values of this society [3]. This happens in particular by informing adolescents about some of the customs that spread in societies [12]. The other and their admiration for it, which leads to their imitating it, and it spreads with this generation, causing a change in society [24].

Being Influenced by Western Cultures: Expressing a country's culture is no longer restricted to dominant countries such as Western culture, as the digital media has made it easier for societies to express themselves and other cultures are easily informed of them without the need for media outlets that transmit what is in line with their whims and policies [4, 29].

The Ability to Bring About Global Change: The digital media has the ability to bring about change in countries and influence public opinion, such as calling for protests or putting pressure on governments and changing regimes [10].

Disadvantages of Digital Media Despite the advantages of the digital media, it has many negatives, and sometimes its features may turn into negatives at the same time. Among the negatives of the digital media are: [4].

Lack of Credibility: The digital media is often subject to suspicion for the viewer or the recipient. Every person can publish what he wants without a reliable source, which leads to misinformation and the spread of rumors, unlike the traditional media, which is subject to professional rules that prevent the transmission of unreliable or questionable news [17, 28]. **Lack of censorship**: In the digital media, there are no censorship authorities that prevent the publication of some misleading or offensive news or ideas or rumors, which are the most serious negatives of the digital media [22]. **Changing the rules in the digital media:** the rules change in the digital media and the controlling and influential party changes, for example, an ordinary person with a large number of followers can be more influential than an entire party or a professional journalist in the traditional media [17]. **Difficulty of Legal Prosecution:** In the digital media, it is difficult to legally prosecute people who publish publications calling for hatred, terrorism, extremism and other crimes punishable by law in the traditional media [3, 27]. **Violation of copyright and property rights:** In the digital media, especially in social media, it is easy to copy books or articles and publish them without the name of their author and violate property rights despite the existence of laws that protect them and hold them accountable for electronic crimes, but it is difficult to implement because of the difficulty of prosecuting violators on the Internet [1].

13 Case Study: What Functions Can Digital Media Contribute to the Educational Process?

Motivation Function: This function is represented in the programs designed to arouse interest, alert, and in general motivate the adoption of new ideas [11]. **Evaluation function:** These are programs designed to provide information to those who are interested in a particular topic and are looking for additional material—descriptive or analytical—to help them evaluate what they are looking for [21]. **Information function:** It is the short message that is intended to provide simple news items, such as: giving information on the locations of lectures and the number of working hours [6]. **Reinforcement function:** To be interested in a subject or to maintain a certain habit, we must carry out a process of reinforcement, whereas constant reinforcement and confirmation are required [4]. **Professional job:** They are programs designed for groups and individuals directly related to development work, such as: programs prepared for workers in development fields, teachers and lectures, as well as interviews [26].

14 Conclusion

In conclusion, it is important to know that digital media is an expanding world at an accelerating pace. So digital technology will continue to enhance its position in society. In the near future, it is likely that augmented reality devices will spread, and we will see traditional media gradually fade away. The results showed that the revolution of communications and information technology produced a new media style that differs in its concept, features, characteristics and means from traditional media. Where the star of digital media has emerged, in the midst of the technological boom that the world is witnessing, which is a translation of the technological development that humans have reached, which has deepened their ability to communicate through many modern technological media. The results showed that digital media is distinguished from traditional media by relying on a new method of modern media, which is the merging of all traditional means of communication, with the aim of delivering the required contents in distinct and more effective forms. Resorting to traditional means, such as broadcasting stations, printing presses, and others, in ways that combine text, image and sound. Which raises the barrier between the recipient and the sender, and he can discuss the media contents he receives, either with the site administration or with other recipients. The results also revealed that modern communication technology has contributed to enhancing cultural communication, revitalizing knowledge exchange between peoples, overcoming geographical and political borders, and even transcending them, and brought about a radical change in quantity and quality in the content transmitted through digital media.

References

1. Achtenhagen L (2016) Developing media management scholarship: a commentary to Picard and Lowe's essay. J Media Bus Stud 13(2):117–123
2. Arrieta Castillo C, Berdasco-Gancedo Y (2020) Brexit and female leadership in Spanish digital media from a linguistic perspective. The case of Theresa May's decline. Commun Soc 33(2):243–257
3. Al-Qudah AA (2022) Artificial intelligence in practice: implications for information systems research, case study UAE companies. In: Musleh Al-Sartawi AMA (eds) Artificial intelligence for sustainable finance and sustainable technology, ICGER 2021. Lecture notes in networks and systems, vol 423. Springer, Cham. https://doi.org/10.1007/978-3-030-93464-4_23
4. Al-Okaily M, Alqudah HM, Al-Qudah AA, Alkhwaldi AF (2022) Examining the critical factors of computer-assisted audit tools and techniques adoption in the post-COVID-19 period: internal auditors perspective. VINE J Inf Knowl Manage Syst Vol. ahead-of-print No. ahead-of-print. https://doi.org/10.1108/VJIKMS-12-2021-0311
5. Alansari Y, Al-Sartawi A (2021) IT governance and E-banking in GCC listed banks. Procedia Comput Sci 183:844–848
6. Bange S, Moisander J, Järventie-Thesleff R (2019). Brand co-creation in multichannel media environments: a narrative approach. J Media Bus Stud 1–18. https://doi.org/10.1080/16522354.2019.1596722
7. Baptista J, Wilson AD, Galliers RD, Bynghall S (2017) Social media and the emergence of reflexiveness as a new capability for open strategy. Long Range Plan 50(3):322–336
8. Baym N (2015) Personal connections in the digital age. Polity Press, Malden
9. Biernacka-Ligieza I (2021). The significance of digital media in local public space crisis management: the case of Poland, the United Kingdom and Italy. J Arab Muslim Media Res 14(1):47–70
10. Dunas DV, Vartanov SA (2020) Emerging digital media culture in Russia: modeling the media consumption of Generation Z. J Multicult Discourses 15(2):186–203
11. Jenkins ES (2021) Do Digital media spark interology? On meditative and immeditative modes. Media Res 17(3):87–97
12. Gálik S, Oprala B (2021) Temporal changes under the influence of digital media. Commun Today 12(1):4–12
13. Gil de Zúñiga H, Chen H-T (2020) Digital media and politics: effects of the great information and communication divides. J Broadcast Electron Media 63(3):365–373
14. Girija S, Triple (2019) A case study of production practices and user participation in an advertising-free digital news media organisation. (Cogn Commun Co-Oper) Open Access J Global Sustain Inf Soc 17(1):38–55
15. Guadagno RE, Rempala DM, Murphy S, Okdie BM (2013) What makes a video go viral? An analysis of emotional contagion and Internet memes. Comput Hum Behav 29(6):2312–2319
16. Horst S-O, Järventie-Thesleff R, Perez-Latre FJ (2020) Entrepreneurial identity development through digital media. J Media Bus Stud 17(2):87–112

17. Karolak M, Razzaque A, Al-Sartawi A (2021) E-services and M-services using IoT: an assessment of the Kingdom of Bahrain. In: Musleh Al-Sartawi AM, Razzaque A, Kamal MM (eds) Artificial intelligence systems and the internet of things in the digital era. EAMMIS 2021. Lecture notes in networks and systems, vol 239. Springer, Cham
18. Lee FLF (2020) Countering the counterpublics: progovernment online media and public opinion in Hong Kong. Int J Commun (19328036) 15:3397–3417
19. Musleh Al-Sartawi AM, Razzaque A, Kamal MM (eds) (2021) Artificial intelligence systems and the internet of things in the digital era. In: EAMMIS 2021. Lecture notes in networks and systems, vol 239. Springer, Cham
20. Mattoni A (2020) A media-in-practices approach to investigate the nexus between digital media and activists' daily political engagement. Int J Commun (19328036) 4:2828–2845
21. Musleh Al-Sartawi AMA (eds) (2022) Artificial intelligence for sustainable finance and sustainable technology. In: ICGER 2021. Lecture notes in networks and systems, vol 423. Springer, Cham
22. Olasina G (2020) Cultural expression using digital media by students. J Afr Media Stud 12(3):351–364
23. Oser J, Boulianne S (2020) Reinforcement effects between digital media use and political participation: a meta-analysis of repeated-wave panel data. Public Opin Q 84:355–365. 2020 Supplement
24. Pantic M (2020) Gratifications of digital media: what motivates users to consume live blogs. Media Pract Educ 21(2):148–163
25. Post S (2019) Polarizing communication as media effects on antagonists. Underst Commun Confl Digit Media Soc Commun Theory (1050–3293) 29(2):213–235
26. Rainie L (2020) The everything-ness and the more-ness of the internet: how digital is different from other media. Int J Commun (19328036) 14:5860–5879
27. Sanad Z, Al-Sartawi A (2021) Financial statements fraud and data mining: a review. In: Musleh Al-Sartawi AM, Razzaque A, Kamal MM (eds) Artificial intelligence systems and the internet of things in the digital era, EAMMIS 2021. Lecture notes in networks and systems, vol 239. Springer, Cham.
28. Tobias-Mamina R, Maziriri ET, Chiliya N (2021) Does media matter? The effect of digital media exposure, perceptions and attitudes towards voting, Communicare 40(1):1–22
29. Ünlü DG (2021) Evaluation of digital media usage of women through motherhood models as a cultural ideology. Ileti-s-im (31):137–167

Knowledge Management Capabilities Excel the Long-Term Corporate Sustainable Development: A Literature Review

Anjum Razzaque, Najeeb Razzaque, Mazen Ali, and Hussain Mohsen Al-Arayed

Abstract The knowledge era has become critical for enabling organizational value. Though knowledge is critical for organizational efficiency, leveraging the intangible form of knowledge is a challenging endeavor. Even though there exist ample studies assessing the role of social networks or virtual communities like knowledge management tools on operational efficiency, there is no clear empirical evidence how the knowledge management processes of a knowledge management system enable an organization in various contexts. The issue is that though such knowledge management tools inspire operational efficiency through the facilitation of quality decision-making, how actually the knowledge management capabilities (i.e., through the application of knowledge management processes: knowledge acquisition, knowledge creation, knowledge sharing, and knowledge storage, knowledge retrieval, and knowledge application) improve organization operation efficiency is yet empirically unknown. There is a need for literature review to shed light on how knowledge management capabilities support the organizational managerial strategies. he knowledge era becomes critical for enabling organizational value.

Keywords Knowledge management capabilities · Corporate sustainable development · Knowledge acquisition · Knowledge creation · Knowledge storage

M. Ali—Freelancer.

A. Razzaque (✉) · N. Razzaque
Ahlia University, Manama, Bahrain
e-mail: anjum.razzaque@gmail.com

M. Ali
Manama, Bahrain

Tarabut Gateway, Manama, Bahrain

H. M. Al-Arayed
University of Bahrain, Zallaq, Bahrain

A. M. A. Musleh Al-Sartawi et al. (eds.), *From the Internet of Things to the Internet of Ideas: The Role of Artificial Intelligence*, Lecture Notes in Networks and Systems 557,
https://doi.org/10.1007/978-3-031-17746-0_13

1 Introduction

Traditional theories and economic models' bare variables driven from the quantifiable ecosystem, concerning human capital, etc. Until recently, knowledge was deemed as a peripheral influential entity for production purposes. Current research deems knowledge as critical, but a challenging endeavor when leveraged in its intangible form. Generating new knowledge is now pivotal for achieving competitive advantage [27, 32, 33, 34, 35]. Organizations that want to successfully compete need to manage, or leverage, their unquantifiable resource, i.e., their organizational knowledge by managing knowledge, a notion that a firm captures and directs through its resources, e.g., documents, employee skills [21]. Academic and practitioner literature [e.g., 1, 3, 4, 5, 24] posits that the most important outcome to social capital is the new knowledge resources, which is developed through social interactions [25, 29–31]. The knowledge era has become critical for enabling organizational value where knowledge is a blend of experience, contextual information, and professional understanding [12, 13]. For leveraging the knowledge assts of any organization, the industry 4.0 era has currently led literatures' attention on comprehending social networking platforms for studying team collaborations through which knowledge is able to be managed. This way employees, i.e., peers within a social network, or virtual community, can utilize the social capital of resources of that network or community [23, 32–36]. According to [8], despite ample research assessing the effect of social networks, virtual communities, and knowledge management on organizational operational efficiency (revealing a positively significant relationship) there remains scant empirical evidence that has yet to assess the role of social networks (knowledge management tools through their knowledge management capabilities) (Razzaque et al., 2020),to achieve higher firm's competitiveness. The issue is that all stakeholders of any organization rely on social networks to acquire, share, create store and re-use knowledge, i.e., manage knowledge. However, there is no clear empirical evidence how these knowledge management processes facilitate an organization in various contexts. In other words, plenty scholars suggest that virtual communities/social networks inspire organizational operational efficiency through quality decision-making [8, 32, 33, 35, 36]. There remains scant empirical evidence to support how actually organizational knowledge management capabilities, i.e., through what knowledge management processes, improve organization operation efficiency [42]. It is the pivotal role of a knowledge management system and through its application of the knowledge management processes, knowledge acquisition, knowledge creation, knowledge sharing, and knowledge storage and retrieval, besides the application of knowledge, which allow the knowledge management capabilities to come in existence [2, 43]. It is the mutual worry in knowledge management literature how knowledge management capabilities support the organizational managerial strategies [19] for sustainable development of an organization.

Sustainable development is the modern buzz word [10] as this notion meets the needs of social, environmental, and economic expected outcomes of our modern civilization, assuring organizations a mechanism by which to avoid sacrificing the needs of the upcoming generations [15, 16]. Research continues to stress on a vital research question pertaining to sustainable development, i.e., how organizations can achieve their sustainable development goals in developing countries? A question that is applicable in the public and private sectors as well as the case of urban planning [45]. Various scholars have reported that if firms want to incorporate sustainable development within their strategic goal, such organizations must transform into learning organizations through an organizational culture which harnesses knowledge for achieving their long-term corporate sustainable development [10, 35]. Hence, the aim of this study is to empirically comprehend what role knowledge management capabilities play on corporate sustainable development. Here, corporate sustainable development refers to an organization's capability for expecting and meeting the demands of the current and future stakeholders (World Commission on Environment and Development) [44]. Research on corporate sustainable development has identified a major vulnerability. Existing literature on sustainable development focuses on policies and systems, to orchestrate a firm foundation for reaping the positive outcomes of those future endeavors in this research area. However, scant evidence conveys if day-to-day operations of an organization can guide corporate sustainable development [39, 43]. Though scholars like [32–34, 43] claim that the notion of sustainable development is greatest from its impact stand point of view when incorporated within an organization's business processes, there remains a dire need for an integrated conceptual framework which can integrate system within interconnected processes whose outcomes is the achievement of the long-term corporate sustainable development [18]. The aim of this study is address Rocha's et al. concern, i.e., scant evidence if organizational day-to-day operations guide corporate sustainable development. Further, this study also wishes to address [18]'s concern, i.e., the dire need for an integrated conceptual framework that interconnects processes, i.e., knowledge management processes with operational efficiency and corporate sustainability.

2 Literature Review and Proposition Development

This study aims to comprehend the role of organizational knowledge management capabilities on organizational operational efficiency and corporate sustainable development. The section ahead critique literature to develop four sub-propositions based on its main Proposition (P) that the role of knowledge management capabilities clearly influences corporate sustainable development (as depicted in the Fig. 1 model).

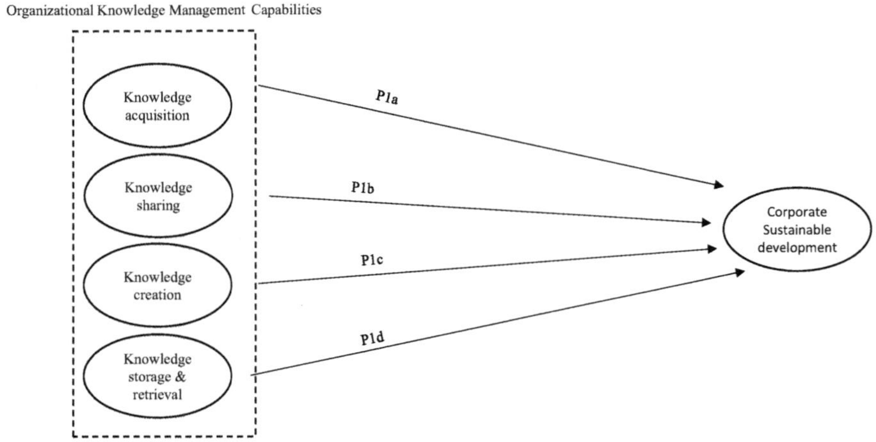

Fig. 1 Conceptual framework

2.1 *Organizational Knowledge Management Capabilities for Corporate Sustainable Development*

Organizations that focus only on improving their operational efficiency using knowledge management capabilities fall short of realizing the true potential for achieving sustainable development. As a result, an organizational knowledge management capability is required to also achieve an organization's sustainable development, in conjunction with achieving organizational operational efficiency. For instance, when organization employees acquire and create knowledge, they are required to innovate to enhance service quality. When a knowledge management system is efficient then optimum storage and retrieval of knowledge is anticipated for enhancing the reliability of the services that the organization provides. When employees of such an organization share knowledge with their customers, they anticipate improved transparency within such transactions. Though such a rational is the norm, there is scant literature which assesses the effect of organizational knowledge management capabilities on corporate sustainable development. One of the few studies by [20] recommended a knowledge management framework for sustainable development but this framework has yet to be empirically confirmed if it is viable for ant real-world implications. Based on this argument the second hypothesis is:

P2: Organizational knowledge management capabilities bae an influential effect on its sustainable development.

2.1.1 Knowledge Acquisition and Corporate Sustainable Development

[11] identified organizational goals for establishing peer networks that facilitate sustainability. These goals revolve around the acquisition of knowledge which then should be transferred to achieve its value through collaborative activities. These goals are to acquire knowledge in the form of experiences, to participate within a network

to collaborate within problem-solving issues, to create and re-use knowledge into action, and to formulate a platform that brings together a network of ties which indulge in collaborative activities for a long-term corporate sustainable development [11]. Based on this argument:

P2a: Organizational knowledge acquisition an influential effect on its sustainable development.

2.1.2 Knowledge Sharing and Corporate Sustainable Development

Knowledge is the key for long-term sustainable development. Organizations are the knowledge bases during the twenty-first century in which context knowledge sharing is pivotal for sustain competitive advantage, during an era when the general population is aging, and organizations need to revise their policies [3, 41] that introduce newer knowledge management practices that would enhance corporate sustainable development, and the best way for doing so is through the utilization of knowledge sharing [7]. According to [22] the Italian Agri-food Protection Consortia formulated a community of interest for harvesting knowledge that promotes a specific area-specific food items' quality culture [35]. It was observed that such a social group of in/formal platform was great for resource exchanging, knowledge through discussions of best practices which in the long term contribute to the sustainable development of such an initiative. Based on this argument the second hypothesis is:

P2b: Organizational knowledge sharing an influential effect on its sustainable development.

2.1.3 Knowledge Creation and Corporate Sustainable Development

From an economic viewpoint of sustainability society cannot replace human capital with artificial intelligence [7]. The rationale is since human beings are unique from one another, their differences from the point of view of their visions and their work practices, unites them to communicate their differences [22, 31, 43]. Within the context of an organizational, such communications lead to the creation of new knowledge, i.e., through the knowledge creation process, where social interactions lead human beings into a new boundary, thus causing them and their peers, as well as their organizations, along with their societies, to envision their long-term sustainable development goals. From a phytophile point of view, [9] posits that a knowledge driven economy, regards knowledge as an extremely precious resource for any corporate atmosphere. Such that the creation of new knowledge grows into the nuclear portion of the concern regarding sustainable development of any economy, and this is due to the tacit nature of knowledge most previous for achieving corporate competitive advantage, hence its long-term sustainable development. Based on this argument the second hypothesis is:

P2c: Organizational knowledge creation bare an influential effect on its sustainable development.

2.1.4 Knowledge Storage and Retrieval, and Corporate Sustainable Development

Rapidly changing global markets pressure organizations to surmount the challenges posed during productions by applying knowledge management strategies which re-use stored knowledge from past experiences. The experiential value harvests from the relational ties amongst the knowledge bearers and knowledge inquirers. If knowledge is stored properly then it is easily accessible of employees [17]. The learning process within an organization occurs from the stored experiences in the form of knowledge within its corporate memory bank [1, 2, 37–41, 43] for instance as with the case of Chapparal Steel [6]. A benchmarking study of industries by [4], revealed the importance of knowledge management as a business strategy where stored knowledge when shared forms the start of a sustainable organizational knowledge screw [26–28, 32, 33]. It is this stored organizational knowledge, in the form of storing shared experiences, that is vital for a successful knowledge management strategy, if accessible via a skill-resources' knowledge base, then would prove a vital since the re-use of past stored intellectual capital is vital for a successful knowledge management capability which enhance the long-term corporate sustainability goals [14]. Based on this argument the second hypothesis is:

P2d: Organizational storage and retrieval bare an influential effect on its sustainable development.

3 Discussion and Conclusion

The literature review of this study revealed that though ample scholars empirically confirmed that knowledge management tools, i.e., social networks, become an organizational knowledge management capability for improving efficiency by facilitating quality decision-making [8, 34–36], scant empirical evidence exists to support how actually this is possible. In other words, through what knowledge management processes, organization operation efficiency is achievable [42]. Furthermore, when focusing on the context of the banking sector, bankers' foremost challenges is to explore how effectively knowledge management capabilities can achieve operational efficiency and corporate sustainability. This was the study research aim. Hence, the Fig. 1 model of this study was designed, and all the five variables were integrated to address the need to propose an integrated management system of integrated processes, i.e., the Fig. 1 conceptual model integrating knowledge capabilities' knowledge management four processes with corporate sustainable development for achieving corporate sustainability. Future research should consider the Fig. 1 viable for empirical assessment, particularly in the banking section of the Kingdom of Bahrain. Since quantitative research is rigorous and leads to confirm the role of the Fig. 1 model's independent variables and department variables, this study should assume a confirmatory assessing stance.

References

1. Adler PS, Kwon SW (2002) Social capital: prospects for a new concept. Acad Manag Rev 27:17–40
2. Alansari Y, Al-Sartawi A (2021) IT governance and E-banking in GCC listed banks. Procedia Comput Sci 183:844–848
3. Al-Sartawi A, Karolak M, Razzaque A (2021) Cybersecurity aids financial institutions performance. In: Big data for entrepreneurship and sustainable development, pp 91–104. CRC Press
4. Al-Qassim M, Chen W, Al-Sartawi A (2022) blockchain in environmental compliance and enforcement. In: Musleh Al-Sartawi AMA (eds) Artificial intelligence for sustainable finance and sustainable technology, ICGER 2021. Lecture notes in networks and systems, vol 423. Springer, Cham
5. Al-Okaily M, Alqudah HM, Al-Qudah AA, Alkhwaldi AF (2022) Examining the critical factors of computer-assisted audit tools and techniques adoption in the post-COVID-19 period: internal auditors perspective. VINE J Inf Knowl Manage Syst. ahead-of-print No. ahead-of-print. https://doi.org/10.1108/VJIKMS-12-2021-0311
6. Baker WE (2000) Achieving success through social capital: tapping the hidden resources in your personal and business networks. Jossey-Bass, San Francisco
7. Barton-Leonard D (1995) Wellsprings of knowledge: building and sustaining sources of innovation. Harvard Business School Press, Boston
8. Beijerse RP (1999) Questions in knowledge management: defining and conceptualizing a phenomenon. J Knowl Manag 3(2):94–110
9. Brandon DP, Hollingshead AB (2004) Transactive memory systems in organizations: matching tasks, expertise, and people. Organ Sci 15(6):633–645
10. Bratianu C, Orzea I (2012) Organizational knowledge creation in the perspective of sustainable development. Rev General Manage 15(1):13–21
11. Burksiene V (2013) Influence of personal competency on sustainable development knowledge creation in the organization. Socialiniai tyrimai/Soc Res 3(32):71–83
12. Clarke SF, Roome NJ (1999) Sustainable business: learning-action networks as organizational assets. Bus Strateg Environ 8:296–310
13. Clegg S, Clarke T (1999) Intelligent organizations? In: Ibarra-Colado E, Bueono-Rodriquez L (eds) Global management: universal theories and local realities. Sage, London, pp 177–201
14. Davenport TH (1997) Working knowledge: how organizations manage what they know. Harvard Business School Press, Boston
15. Davenport TH, Prusak L (1998) Working knowledge: how organizations manage what they know. Harvard Business School Press. https://doi.org/10.1145/348772.348775
16. Gargiulo M, Benassi M (2000) Trapped in your own net? Network cohesion, structural holes, and the adaptation of social capital. Organ Sci 11:183–196
17. Hannoon A, Al-Sartawi A, Khalid A (2021) Relationship between financial technology and financial performance. In: Musleh Al-Sartawi AMA (eds) The big data-driven digital economy: artificial and computational intelligence. Studies in computational intelligence, vol 974. Springer, Cham
18. Karapetrovic S (2003) Musings on integrated management systems. Meas Bus Excell 7(1):4–13
19. Karolak M, Razzaque A, Al-Sartawi A (2021) E-services and M-services using IoT: an assessment of the Kingdom of Bahrain. In: Musleh Al-Sartawi AM, Razzaque A, Kamal MM (eds) Artificial intelligence systems and the internet of things in the digital era, EAMMIS 2021. Lecture notes in networks and systems, vol 239. Springer, Cham
20. Linger H, Aarons JP, McShane P, Burstein F (2013). A knowledge management framework for sustainable development: a case of natural resource management policy work in Indonesia. In: PACIS 2013 Proceedings, pp 1–12
21. Lyons KL (2020) Using patterns to capture tacit knowledge and enhance knowledge transfer in virtual teams. In: Malhotra Y (ed) Knowledge management and virtual organizations, pp 124–143. IGI Global, Hershey

22. Maizza A, Fait M, Scorrano P, Iazzi A (2019). How Knowledge sharing culture can become a facilitator of the sustainable development in the agrifood sector. Sustainability 11:952–967
23. Musleh Al-Sartawi AM, Razzaque A, Kamal MM (eds) (2021) Artificial intelligence systems and the internet of things in the digital era. In: EAMMIS 2021. Lecture notes in networks and systems, vol 239. Springer, Cham
24. Nahapiet S, Ghoshal S (1998) Social capital, intellectual capital, and the organizational advantage. Acad Manag Rev 23:242–266
25. Nicolini D (2016) Knowing in organizations: a practice-based approach. Routledge, Abingdon
26. Nonaka I, Konno N (1998) The concept of "Ba": building a foundation for knowledge creation. Calif Manage Rev 41(3):40–54. https://doi.org/10.2307/41165942
27. Nonaka I (1994) A dynamic theory of organizational knowledge creation. Organ Sci 5(1):14–37
28. Nonaka I, Takeuchi H (1995) The knowledge-creating company: How Japanese companies create the dynamics of innovation. Oxford University Press, Oxford
29. Nonaka I, Von Krogh G, Voelpel S (2006) Organizational knowledge creation theory: evolutionary paths and future advances. Organ Stud 27:1179–1208
30. Razzaque A, Alalawi M (2015) Knowledge sharing and positive emotions in virtual communities: the case of the healthcare sector. In: KS global research international conference on emerging trends in multidisciplinary research (ETMR 2015). KS Global Research, Istanbul
31. Razzaque A, Eldabi T (2018) Tackling obesity by diet quality, technology and social networks. Int J Food Eng 4(1):88–92
32. Razzaque A, Al-Sartawi AMM, Karolak M (2021) Rethinking tourism entrepreneurship amidst COVID-19 through the lens of artificial intelligence and big data analytics. In: Big data for entrepreneurship and sustainable development, pp 175–188. CRC Press
33. Razzaque A, Eldabi T (2020) Physicians social capital aids their medical decisions when they virtually share knowledge. Int J Knowl Manage Stud 11(3):229–257
34. Razzaque A, Eldabi T, Chen W (2020) Quality decisions from physicians' shared knowledge in virtual communities. Knowl Manage Res Pract 1–13. https://doi.org/10.1080/14778238.2020.1788428
35. Razzaque A, Eldabi T, Jalal-Karim A, Karolak M (2013) Culture sustained knowledge management infrastructure and architecture facilitate medical decision making. In: 2013 IIAS-IASIA joint congress. IIAS-IASIA, Manama
36. Razzaque A, Eldabi T, Jalal-Karim A (2013) Physician virtual community and medical decision-making: mediating role of knowledge sharing. J Enterp Inf Manag 26(1):500–515
37. Rocha M, Searcy C, Karapetrovic S (2007) Integrating sustainable development into existing management systems. Total Qual Manag Bus Excell 19(1–2):83–92. https://doi.org/10.1080/14783360601051594
38. Sanad Z, Al-Sartawi A (2021) Financial statements fraud and data mining: a review. In: Musleh Al-Sartawi AM, Razzaque A, Kamal MM (eds) Artificial intelligence systems and the internet of things in the digital era, EAMMIS 2021. Lecture notes in networks and systems, vol 239. Springer, Cham
39. Shah N, Shah S (2010) Relationships between employee readiness for organizational change, supervisor and peer relations and demography. J Enterp Inf Manag 23(5):640–652
40. Sanad Z, Shiwakoti R, Al-Sartawi A (2022) Gender diversity on corporate boards and earnings management: a review. In: Musleh Al-Sartawi AMA (eds) Artificial intelligence for sustainable finance and sustainable technology, ICGER 2021. Lecture notes in networks and systems, vol 423. Springer, Cham
41. von Krogh G (1998) Care in knowledge creation. Calif Manage Rev 40(3):133–153
42. World Commission on Environment and Development (WCED) (1987) Our common future. Oxford University Press, Oxford
43. Zhao J, Zheng X, Dong R, Shao G (2013) The planning, construction, and management toward sustainable cities in China needs the environmental internet of things. Int J Sustain Develop World 20:195–198. https://doi.org/10.1080/13504509.2013.784882

Sustainable Technologies, Artificial Intelligence, and Internet of Ideas

Augmented Press and Robot Journalists Who Determines the Ethics of Journalistic Coverage?!

Ismail Noori Mseer

Abstract The focus is on researching the relationship between the journalist and the technician and considering that the current reality has come to refer with complete clarity to the aspiration of journalistic work to rely on algorithms, data processing, and linguistic engineering applications. This shows the importance of examining the symbolic message on which the press is based instead of the cold and direct logic on which digital applications are based. The matter remains dependent on the initiative issued by the press sector, given the responsibility related to the importance of reconsidering the media content industry. And directing attention to the extent of (the world's vision) for the journalist. The mission and moral responsibility are involved in the entire journalistic work; the journalist's job is not limited to writing texts or following up and editing the news. Still, the matter goes far beyond this reality. Artificial intelligence and its algorithms may help journalists produce 100 articles in five minutes. But the question remains about the feasibility, meaning, value, and importance of this amount of production. Let us agree that the articles produced based on the existing cooperation between (digital and human) have reached accuracy and skill and rely on a vast amount of data and information. But the question remains, where is the identity, spirit, vision, and intuition that distinguishes this press institution from that one? Where do you see the position of values and ethics, symbolism, unique taste, and press message? What is the point of journalism being automated with a digital mechanism based on algorithms while missing the message of journalistic work and the goal for which it works?

Keywords Generative Pre-trained Transformer 3 (GPT-3) · (garbage in Garbage out) (GIGO) · Augmented journalism · Emotional intelligence · Brain-computer interfaces BCI · Interfacing the cloud with the human brain · Neurolinguistics

---

I. N. Mseer (✉)
Ahlia University, Manama, Bahrain
e-mail: imseer@ahlia.edu.bh

A. M. A. Musleh Al-Sartawi et al. (eds.), *From the Internet of Things to the Internet of Ideas: The Role of Artificial Intelligence*, Lecture Notes in Networks and Systems 557,
https://doi.org/10.1007/978-3-031-17746-0_14

1 Introduction

What is the future of news organizations in light of the heavy presence of applications of artificial intelligence mechanisms? A question that is not related to this or that field. Instead, artificial intelligence is at the forefront of the entire scene and announces its dense and robust presence in all the details of current life. The term (Artificial intelligence journalism) has become firmly and remarkably present, given the change and transformation that will appear in news organizations' overall work and performance [1]. These institutions are getting new and unprecedented mechanisms in collecting information and data and working on filtering, filtering, and sorting them with precision and rigor. And the sound, image and video, and direct detection of texts in terms of language integrity, plagiarism, burglary and uniqueness, and performance evaluation. It must acknowledge that the applications of artificial intelligence have made a quantum leap in the way the news organization is produced, as the spread of modern technologies to the public, especially the smartphone, has contributed to expanding the area of interaction between news organizations and the public, in a way that is based on speed, flexibility, simplification, and facilitation.. Even the traditional news meal is experiencing different conditions from the conventional method as it began to reach the recipient based on the desire emanating from him, in terms of time, place, appetite and mood. A meal is based on multiple choices and various means, with high flexibility and comprehensive openness to the future. Considering that the main idea on which the mechanisms of artificial intelligence work are based is based on the fact that a computer can compile, store and sort software and algorithms, and thus the ability to perform tasks by (simulating human intelligence).

The Neuralink Foundation, for example, has made it a slogan; (Breakthrough Technology for the Brain) is a front for its website [2] where it focuses on the idea that can copy the human brain through programs and technologies to create a direct experience between artificial intelligence and humans. At the same time, artificial intelligence journalism is based on several technologies, which are based on; (Machine learning, deep neural networks, natural language processing, natural language generation, automatic summarization, and computer vision) to expand automation into the core of journalistic work. Indeed, some news organizations are turning to computers in writing newspaper articles. In this, we can refer to the experience of the British newspaper The Guardian, when it cooperated with the Open IA Foundation, which bears the slogan; (Our mission is to ensure that artificial general intelligence benefits all of humanity.) [3] Using the Generative Pre-trained Transformer 3 (GPT-3) algorithm, in language analysis and generation, an article written by the robot was produced and published in the Guardian in September 2020 under the title (An article written by the entire robot). Referring to several data, the article's author is a (thinking robot), using only 0.12% of his cognitive ability; his decisions are based on logic; he learned everything by reading the Internet. And here he is face to face with the human community through this article [4].

2 Developing Platforms for Conversation

The Open AI Lab has focused its efforts on developing a natural language processing system called GPT-3 that focuses its work on simulating human language. Since its inception in 2020, this laboratory has been dealing with deep learning language with up to 175 billion variables, containing data based on billions of words. In addition to the development of 300 applications working on developing platforms for conversation, word processing, and research and investing the efforts of thousands of workers and developers worldwide. The productivity of this conference reached 4.5 billion words per day. [5] By taking advantage of deep learning applications that provide a regular sequence that contributes to text input, in conjunction with the ML machine learning model that automatically creates, summarizes, and translates texts and speech recognition [6].

Artificial intelligence applications have proven their flexibility and ability to process a vast amount of diverse and different information and data. They have made the most expansive field of media coverage on topics and events difficult for a journalist to obtain. As the press coverage has become more accurate and direct in following up on disasters, wars, battles, and natural events such as storms, earthquakes, and volcanoes, the contributor has become apparent while covering the details of the Covid 19 pandemic. Until the audience can follow the news based on their preferences, it depends on the initiatives of the major news organizations. Artificial intelligence applications still require high financial costs, which only giant enterprises can afford. Thus the Washington Post's industry collaborating with HireVue to test benchmark algorithms on recruitment technology emerged, but this was not without some problems. The Electronic Privacy Information Center, EPIC, filed a complaint against Hire Vue's experiments on recruitment technology. American workers consider artificial intelligence experiments in voice and face scans dangerous, especially since these experiments are still being tested. [7] On the other hand, these applications have an effective and influential presence on employers evaluating their labor force. Depending on computer and smartphone applications to read and analyze faces, signals, and voice, and thus analyze all this data, and adopt it in classifying job candidates. Despite the accuracy that characterizes the work of algorithms in classification and selection, it is not without the appearance of some biases, especially in the way of reading features and sounds. Where physiological differences between races. And this was revealed by the investigations of the Washington Post itself [8].

The CNN news channel website has its contribution to the use and applications of the search algorithm in the field of ophthalmology; given the gift of artificial intelligence to health care in the world, it is hoped that the financial investment in it during the year 2026, will reach 44 billion US dollars. [9] The applications of deep learning are used in data analysis to identify early threats to eye patients. Where specialists tended to benefit from the applications of artificial intelligence, especially in terms of investing in deep learning technology. The partnership between the Deep Mind Foundation was Moor fields Eye Hospital, where positive results were achieved, especially in terms of speed of diagnosis and work to protect patients at risk of blindness [10]. Caution remains, especially since some of the cases have failed [11].

Artificial intelligence is the fruit of broad efforts from specialized scientific platforms and institutions, the Net craft Foundation, which raises the slogan; (Protect yourself from cybercrime), [12] whose mission is to monitor 150 million websites daily, as it works through its applications to analyze data and information, and detect fraud, deception, and cybercrime, through algorithms based on search, matching, and detection. About keywords that appear in Google search engine results. [13, 14] While the entire technique remains based on the principle of the data entry method, which is called; (garbage in, garbage out) (GIGO). If you enter inaccurate data, the result will undoubtedly be wrong. It is one of the logical principles based on the correct premises that lead to the right results. [15, 16] This is by calculating that the expected outputs are subject to the right and valid inputs entered into the program [15].

From the reality of the accumulated experiences and the impressive results provided by artificial intelligence applications, the conditions of automation in the field of press production have expanded. Even the big news organizations headed towards expanding the horizons of cooperation with companies working in artificial intelligence, where an artificial voice program was developed that converts written text into audio reading based on the investment of deep neural networks. This is clarified by the relations between the British BBC and Microsoft. [16] Reuters news agency cooperated with Synthesia Foundation to develop an automated presenter, whose task is to prepare reports collected from the Routers website, where it works on collecting words and images and editing and producing reports without human intervention. [17] It is based on integrating video clips of a virtual presenter who provides news summaries, exploiting Reuters' storage of images, films, audio recordings, and written texts. [18] It must note here that influential media organizations are issuing the intense trend towards artificial intelligence applications due to the high cost and complex procedures imposed by the conditions of experiments and cooperation protocols with institutions specialized in the digital field.

It is interesting to note that the initiative in developing the performance of the work of news organizations is now issued by institutions specialized in the digital field. Google, for example, has allocated sums of more than 150 million US dollars to support projects that advance the reality of innovation and renewal in the news industry. Among the most prominent of these projects is RADAR (Reporters and Data and Robots), with a capital of about 700 thousand euros. The task of the project is to automate the overall delivery of news services to websites, newspapers, and radio stations, based on the use of algorithms that do the work—at the same time, retaining some of the experienced human cadres in pursuit of control and control [19].

Augmented journalism is a concept based on the compatibility between machine and human achievement based on knowledge and experience in journalism. The massive flow of information and much data creates an innovative relationship that calls for more scrutiny, examination, selection, and sorting. And from this reality, everyone rolls up their sleeves to seek to study and analyze the impact of the relationship between the robot and the human. Based on the awareness of the cross-fertilization of experiences and the overlap of disciplines. There is no harm in using the programmer and engineer from artificial intelligence and integrating it into the news-editorial room. Why the fear of the dominance of artificial intelligence in the future on the job of a human journalist? The matter here calls for reading and observation of another kind, focusing on the purpose, goal, and moral responsibility upon which the entire journalism profession is based. As for the process being based on advancing one foot and delaying the other, this is based on conditions of hesitation that cannot result in anything!

Artificial intelligence applications, who wants it and who refuses it, are now crawling with full force in the field. No power in the world can confront it or disrupt its path; this is impossible. From this reality emerges the importance of adapting to reality, which must capture the minds of the administrations based on news organizations. This is due to the calculation that artificial intelligence applications have become something that cannot condone. Direct analyses indicate that the reaped benefits have become much greater than the consequences that some pessimists think. It is the extensive presence in the journalistic practice in general and on the various experiences and capabilities, starting from analyzing texts, passing through the identification of images, and ending with distinguishing false information, in addition to the direct results that are now reflected on the cultural, social, political and economic levels. [20] From the reality of this interaction, the role of automated journalism emerges in support and advancement of the media sector by employing algorithms in the production of fast, automatic text, which is based on the analysis and scrutiny of millions of data with remarkable accuracy and speed. The prosperous effectiveness of the role of automation in the dissemination of content that aligns with the global path—considering the development and development of more media sectors, especially in the field of advertising, and investing this massive energy in studying data and information related to political issues. For example, the robot (Tobi) was employed to research and analyze problems associated with the Swiss elections in 2018, so it could complete 40,000 news stories in just five minutes. [21] efficiently and accurately. The most significant challenge lies in dealing with artificial intelligence products. It may appear shortly (journalism augmented), which means more workers in the field of journalism, the announcement of the end of the human journalist. But the most critical question remains from the angle of view?! As all artificial intelligence is based on activating algorithms. The latter is subject to human programming. And it is subject to very complex operations, which require more time, money, efforts, and creative initiatives, which remain confined to the actions issued by major digital institutions such as Google and Microsoft, to name a few. [22] However, the prospects for cooperation remain between journalism professionals and digital technicians in working together towards producing hybrid content based on augmented reality applications by investing the massive expansion in the use of the smartphone and its applications in various fields and fields of knowledge. Most of the press institutions are working on creating their applications and urging the public to interact with them, investing in augmented reality in providing a news meal based on audio, image, video, and computer games [23].

Human–computer interaction constituted a wide field of research and deep study due to the computer's importance in contemporary life. It is no longer confined to specialists but instead spreads in the core of personal life until it has become an essential part of human life, which cannot be dispensed with. And from this reality, the aspiration towards intensifying efforts towards facilitating the work of the computer became the talk about digital illiteracy, due to the ambition of countries towards linking public services to the intelligent government system, where the trend towards conducting business and implementing more government services through websites Whether it is in terms of obtaining official papers or buying and selling operations and dealing with reality by completing virtual functions. This, of course, coincided with the provision of Internet service and the emergence of the smartphone, which worked to turn the equation upside down, after collecting most of the vocabulary of modern technology in one device, to make the Internet as the beating heart that moves the entire public life.

Computer use is no longer a luxury or an accessory. Instead, it has become an indispensable necessity in contemporary life. Until it became the first and decisive factor in changing the social and cultural behavior of the human race. And we have to look at the way of human behavior in the form of walking on the public road. The preoccupation with the mobile phone is ignoring everything around it until the collision between two people has become an obvious matter. The two parties are preoccupied with the mobile phone. And it became natural to be ignored by the guest you receive in your home. You talk to him directly while he's busy with a side talk on the phone. Thus, the details began to express new cultural and social behaviors patterns.

On the other hand, it is not possible to deny the impact of this device in providing more essential services in the various fields of security, health, medical, education, administration, and emergency. Here you communicate with the world in a straightforward and simplified way. And now, you can monitor your home through the camera program that monitors the interior and exterior parts of your home while you are spending your summer vacation abroad. And here you are communicating with your manager at work. And your professor at the university, and you play the game of PUBG with companions from all over the world. And there is no harm in filling out the passport form, completing more activities of anticipation and purchase, enjoying movies and listening to songs, and holding a virtual family session. This and many more that the future promises are services that we cannot think of now, but the context we live in. It gives us a more open mind to receive strange, strange, and often impossible.

The field of emotional computing goes directly to the concept of emotional intelligence, which is based on the initiative of the individual in the ability to adapt to the feelings of others. Based on several criteria such as; Social skills, awareness, organization, management of emotions. These standards do not stand as standards or determinants, as much as they highlight their importance in the person's ability to activate them in the social environment in which he lives. From this reality, the importance of distinguishing between a successful administrator and another who cannot initiate and motivate emerges. The main criterion here is based on the level of interaction issued by the person to regulate feelings. A conscious and accurate response to the surfaces of the group, and the trend towards making the workgroup that works under his leadership, interact positively in pursuit of a clear goal shared by all (management and workers). Is it sufficient to talk about the positive pattern and ignore the manipulation conditions that some administrations may issue to obtain personal gains? Is it possible to ignore psychological disorders that can limit the effectiveness of emotional intelligence and slow down the path of tolerance and flexibility while achieving the desired goals of the group? And if talking about emotional intelligence is focused on its positive value, does this mean that a distinguished leader who enjoys emotional intelligence cannot be subject to turmoil, weakness, and weakness, and sometimes slow in the way he responds?! [24]

3 Rehabilitation

The effectiveness of the brain-computer interfaces BCI is based on monitoring the signals emanating from the brain, working on converting them into commands, which are received, and working on analyzing them to obtain the desired result. This technique provides compensation and appropriate treatment for the rehabilitation of patients with stroke, spinal cord injuries, cerebral palsy, and all injuries related to the brain and electrocardiogram. Hence, research in this field focuses on studying and developing the point of prosthetics and wheelchairs and research related to robotics. At the same time, the area remains wide on research related to the rehabilitation of people with cerebral palsy and paralytic diseases. The precise work on which it is based requires flexibility and small size. From this, the importance of cross-fertilization of ideas with the nanotechnology sector emerges especially since the target group in this field needs more efforts related to minimizing the size of the device and its portability, in addition to taking utmost care in the degree of safety and bypassing unintentional errors as much as possible, calculating that the work is related to a susceptible central device represented in the brain human. Conclusion: there is no doubt that clinical trials have come a long way in positive achievement. The experts presented a brilliant and essential scientific achievement on the level of great experiments that won appreciation and respect. However, obstacles remain, represented in the high cost, in addition to deep thinking about reliability, about the performance of these devices and their ability to achieve decisive and accurate achievement. This remains subject to the results of the laboratories and the results issued by them. [25] The qualitative scientific leap that achieved the Fourth of Industrial Revolution was achieved in terms of artificial intelligence, e and the Internet of Things began to pave the way for constructive transformations in the course of dealing with the Internet as a service available to humanity, where the significant change witnessed by the concept of knowledge, based on the enormous energy in the Providing data and information, and the means that became available in terms of storage and processing. In contrast, cloud information storage technology became a reality and facilitated, made researchers specialized in deep research in the possibility of expanding the field of experiments and research around interfacing the cloud with the human brain. Connecting the brain to the cloud [26].

4 Conclusion

Research on brain-computer interfaces has resulted in new types of attempts to exploit neural signals, especially in the field of direct-speech brain-computer interface (DS-BCI), where the trend towards imagined speech monitoring and Focusing on studying the codes related to imaginary speech through words, sentences, and sounds made by the human being. The main thrust is to link neurolinguistics and focus efforts on developing decoding technology to improve communication methods. [1] The effectiveness of the brain-computer interfaces (BCI) is based on the control emanating from neural activity to allow the paralyzed person to control the device through a signal emanating from his brain. Based on the development of the working mechanisms of recurrent neural networks (RNNs), and work on monitoring and follow-up of the total neural response levels. However, this mechanism is not without obstacles represented in the direct impact on the decoding method. The aspiration is to use temporal attention-aware timestep selection (TTS), an effort to follow neural activity step by step. Considering that this method gives the observer and observer the ability to determine the appropriate timing to enter information related to neural activity, and thus contribute to the decoding of the neural. In lab experiments, the TTS method has improved performance. Especially in reducing the patient's training period and allowing the observer and observer to know the nature of the patient's nervous activity [2].

Brain-computer interfaces are based on more flexibility due to their dependence on openness to different disciplines. The matter here is not based on just linking the human brain and the computer, but rather it goes beyond it to deal with other devices that improve performance. It contributes to benefiting from neuroscience, electronic games and investing in creating the appropriate environment that enriches the reality of research, especially in transferring virtual reality into the heart of objective reality. And work on developing and improving the robot's work and expanding the scope of research in the field of affective computing. The orientation in this field is based on the in-depth study of the complexities underlying the work of the human brain and the endeavor towards monitoring the advantages that it enjoys. Proceeding from the transformations and changes produced by the brain at the physiological, psychological, and neurological levels, and their severe impact on the way of communication and possible mechanisms of operation, and more challenges posed by critical situations produced by laboratory experiments or facts made by the details of life [26].

References

1. Haapanen L (2020) Problematising, the restoration of trust through transparency: focusing on quoting. Journalism (published online before print). https://doi.org/10.1177/1464884920934236
2. https://neuralink.com/
3. https://openai.com/about/
4. GPT-3 Tue 8 Sep 2020 09.45 BST A robot wrote this entire article. Are you scared yet, human? https://www.theguardian.com/commentisfree/2020/sep/08/robot-wrote-this-article-gpt-3
5. GPT-3 (2021) Powers the Next, Generation of Apps. https://openai.com/blog/gpt-3-apps/
6. Kashyap Vyas, 19 October 2021, What is GPT-3 and Why Does it Matter? https://www.itbusinessedge.com/development/what-is-gpt-3/
7. EPIC's FTC complaint about HireVue. Updated Nov 6, 2019 at 8:08 PM. https://www.washingtonpost.com/context/epic-s-ftc-complaint-about-hirevue/9797b738-e36a-4b7a-8936-667cf8748907/
8. Harwell D, Carolyn Y Johnson examine the algorithms measuring your worth. Thursday, October 24, 2019. An interview with an algorithm. https://www.washingtonpost.com/podcasts/post-reports/an-interview-with-an-algorithm/
9. DUBLIN (4 January 2022)/PRNewswire The Worldwide Healthcare Artificial Intelligence Industry is Expected to Reach $44.5 Billion by 2026, January 4 2022. https://finance.yahoo.com/news/worldwide-healthcare-artificial-intelligence
10. Suleyman M (2018) A major milestone for the treatment of eye disease. https://deepmind.com/blog/article/moorfields-major-milestone
11. Lewis N, CNN Business, July 15, 2019 Would you trust an algorithm to diagnose an illness? https://edition.cnn.com/2019/07/15/business/artificial-intelligence-healthcare/index.html
12. Jonathan Strickland Why is the Google algorithm so important?
13. Walker J: Head of Technology Transformation, June 2020, Azure offers us the flexibility and control we needed to create a natural, branded voice assistant while maintaining the necessary privacy for our customers' data. https://customers.microsoft.com/en-us/story/754836-bbc-media-entertainment-azure
14. Reuters built a prototype for automated news videos using Deepfakes tech, 2 July 2020. https://www.synthesia.io/post/reuters-built-a-prototype-for-automated-news-videos-using-deepfakes-tech
15. Staff R (7 February 2020) Reuters and Synthesia unveil AI prototype for automated video reports. https://www.reuters.com/article/rpb-synthesia-prototype-idUSKBN2011O3
16. Bennett E (24 Apr 2019) RADAR signs its first paying subscribers for stories jointly written by journalists and AI. https://pamediagroup.com/radar-signs-its-first-paying-subscribers-for-stories-jointly-written-by-journalists-and-ai/
17. Fandango was presented at a panel discussion organized by ANSA (15 November 2019) Augmented Journalism: the Future of Information in Times of Artificial Intelligence. https://fandango-project.eu/news/2019/11/15/augmented-journalism-the-future-of-information-in-times-of-artificial-intelligence/
18. Malek C (26 Jun 2019) The inevitability of augmented journalism the-inevitability-of-augmented-journalism. https://communicateonline.me/category/industry-insights/post-details/the-inevitability-of-augmented-journalism. See also, Robo-journalism gains traction in shifting media landscape, Issued on: 10 Mar 2019. https://www.france24.com/en/20190310-robo-journalism-gains-traction-shifting-media-landscape
19. Crouch GW (24 October 2018) Augmented Journalism. Future Possible? https://sapientd.com/2018/10/24/augmented-journalism-future-possible/
20. Bunz M (7 Jan 2010) How journalists can use augmented reality. https://www.theguardian.com/media/pda/2010/jan/06/journalism-augmented-reality

21. Levine MP (2021) New insights into emotional intelligence, Penn State Milton S. Hershey Medical Center. ISBN 978-1-80355-984-1. Print ISBN 978-1-80355-985-8, eBook (PDF). ISBN 978-1-80355-986-5, 2nd December 2021. https://www.intechopen.com/welcome/1e8cf2f6a6f0b1e8fc4145d85808de0c
22. Shih JJ, Krusienski DJ, Wolpaw JR (2012) Brain-computer interfaces in medicine. Mayo Clin Proc 87(3):268–279. https://doi.org/10.1016/j.mayocp.2011.12.008
23. Tian F, Wang H, Cheng W, Zhang W, Li Y (2021) A high-density EEG study investigating VR film editing and cognitive event segmentation theory. Sensors 21(21):7176. https://doi.org/10.3390/s21217176
24. Cooney C, Folli R, Coyle D (2018) Neurolinguistics research advancing development of a direct-speech brain-computer interface. iScience 8:103–125. https://doi.org/10.1016/j.isci.2018.09.016
25. Yang S-H, Huang J-W, Huang C-J, Chiu P-H, Lai H-Y, Chen Y-Y (2021) Selection of essential neural activity timesteps for intracortical brain–computer interface based on recurrent neural network. Sensors 21(19):6372. https://doi.org/10.3390/s21196372
26. Saha S, Mamun KA, Ahmed K, Mostafa R, Naik GR, Darvishi S, Khandoker AH, Baumert M (2021) Progress in brain computer interface: challenges and opportunities. Front Syst Neurosci 15:578875. https://doi.org/10.3389/fnsys.2021.578875

An Initial Assessment of Sustainability Reporting Practices, Motives and Obstacles for Palestinian Corporations

Islam Abdeljawad, Muiz Abu Alia, Sanabel Yassin, Salam Morrar, and Mays Mubaslat

Abstract Following the growing consideration for sustainability activities, sustainability reporting has received a lot of attention in recent years. Palestinian corporations, however, were less affected by this trend. The purpose of this research is to describe the extent to which Palestinian firms engage in sustainability reporting. Furthermore, the motivations and barriers to this participation, as well as the external assurance of sustainability reporting, are addressed. The research is exploratory and descriptive, and the data was gathered using a questionnaire. The study sample included 38 of the 48 companies listed on the Palestine stock exchange (PEX). The results of this survey were reported using a descriptive analysis. The study's findings gave a preliminary assessment of sustainability reporting and the variables that may improve its implementation in Palestine.

Keywords Sustainability reporting · Practices of sustainability reporting · Palestine

1 Introduction

Companies are now reporting on sustainability to improve their image [1]. Firms tried to reduce the perception of their environmental harm by donating money and volunteering in their communities [2]. Then came the concept of sustainability. It became a way to attract investment whilst protecting the environment and societies. Companies are more aware of their social and environmental impacts because people are becoming more aware of it [3]. Sustainability reporting encompasses ethical, environmental, and social issues [4]. The firm's economic, environmental, social, and governance performance are reported. It helps organizations internalize and improve their commitment to sustainable development goals, both internally and externally [5]. Within-organization benefits include better understanding of risks and opportunities, emphasizing the link between financial and non-financial performance,

I. Abdeljawad (✉) · M. A. Alia · S. Yassin · S. Morrar · M. Mubaslat
Faculty of Economics and Social Sciences, An-Najah National University, Nablus, Palestine
e-mail: islamjawad@najah.edu

A. M. A. Musleh Al-Sartawi et al. (eds.), *From the Internet of Things to the Internet of Ideas: The Role of Artificial Intelligence*, Lecture Notes in Networks and Systems 557,
https://doi.org/10.1007/978-3-031-17746-0_15

influencing long-term management strategy and business plans, reducing costs and increasing efficiency, avoiding publicized environmental, social, and governance failures. To illustrate how the organization influences and is influenced by expectations about sustainable development is one of the external benefits.

However, there are substantial disparities in the content and quality of sustainability reports [6]. Size, visibility, and sector affiliation are all determinants of sustainability reporting, according to Hahn and Kühnen [7]. Many initiatives establish reporting structures and standards for consistency and comparability, but none are enforceable or widely accepted [8]. The Global Reporting Initiative (GRI), which has grown in popularity, provides a structure for making reports consistent and comparable [8]. The OECD Guidelines for Multinational Enterprises, the UN Global Compact (the Communication on Progress), The Covalence Ethical Quote, and the International Organization for Standardization [9] are all prominent players in this arena. Lots of companies from many sectors have adopted one of these initiatives for sustainability reporting guidelines worldwide.

Palestine has accepted the Arab League's Sustainable Development Initiative, which aims to increase Arab engagement in sustainability practices and highlight the role of Arab countries in attaining development, particularly in the areas of technology and globalization [10]. Despite its importance, sustainability reporting is still a new notion in Palestine, both in terms of practice and content. This contradicts the recent official focus on Palestinian companies' sustainable policies [11].

The goal of this research was to describe sustainability reporting practices in the Palestinian context. It attempted to ascertain the extent to which Palestinian enterprises engage in social responsibility and sustainability reporting methods, as well as their motivations and the obstacles they confront. Furthermore, the study aims to discover the extent to which Palestinian firms employ external assurance for their sustainability reports, as well as the reasons that hinder them from doing so.

This research is significant in that it provides a clear picture of sustainability reporting in Palestine. Its findings will assist Palestinian CEOs and policymakers to handle this issue more effectively and to implement policies that support long-term commitments. Furthermore, given the constraints posed by the Israeli occupation and the resulting political instability, addressing such concerns in Palestine is unique [12–15].

2 Background and Literature Review

The GRI defines sustainability reporting as measuring, disclosing, and accounting to internal and external stakeholders for sustainable development performance [16]. According to Halkos [16], a sustainability report presents a company's environmental and social results using particular measures. This report helps the company meet its accountability by presenting non-financial results to stakeholders. A sustainability report should give a fair and balanced picture of an organization's sustainability performance, including both good and adverse aspects [17]. Sustainability reports can be used to measure and review sustainability performance against laws, norms, codes, performance standards, and voluntary efforts, as well as to compare performance inside and across enterprises [39].

The literature has grown in tandem with the international interest in sustainability and social responsibility. Sustainability reporting techniques have recently emerged as an appealing research issue. According to Ioannou and Serafeim [18], sustainability reporting increased social responsibility and governance practices. Eccles et al. [19] classified firms as high or low sustainability. High-sustainability firms engaged stakeholders and shared non-financial data. The study found that CSR activities had a great impact on long-term stock market and accounting performance. Conscious management and board members that appreciate the value of sustainability initiatives promote high sustainability [20]. Burhan and Rahmanti [21] found that sustainability reporting improves social performance in Indonesian enterprises. But this will not happen without adequate stakeholder communication. According to Finch [22], organizations want to show stakeholders how well they are doing in accomplishing long-term goals like enhanced financial performance, competitive advantage, and profit maximization. According to Dienes et al. [23], firm size, media exposure, and ownership structure influence the disclosure of sustainability reports. Corporate governance only seems to have an influence on the existence of audit or sustainability committees.

Sustainability reporting is linked to one of the most crucial aspects of financial reporting, namely disclosure. Culture and the regulatory framework drive the level and character of transparency [24, 25]. These variables vary globally, resulting to differences in financial reporting standards, including disclosure [25]. According to Hbek and Wolniak [26], the quality of the social responsibility report is affected by sustainability reporting requirements, international standards, and national laws. According to Hbek [27] and Tarwaneh et al. [28], sustainability reporting is still a new concept in Poland, with few enterprises adopting it. Alhaleh et al. [29] confirm that sustainability reporting is frequently used in Palestinian banks, but only in annual reports. Hemphill [9] identified six principles of sustainability reporting: comparability, balance, accuracy, clarity, timeliness, and reliability.

Legrand et al. [30] claim that unlike financial reporting, sustainability reporting is mainly uncontrolled. As a result, the methods used to collect, write, and distribute environmental and social data vary greatly. Zovico [31] discovered that 22% of companies in his sample had produced a sustainability report, 42% merely had a sustainability component on their website, and the remaining had not disclosed anything.

Annual reports, according to Frost et al. [32], are less beneficial than discrete reports and websites to report for sustainability. Kuzey and Uyar [33] found an increasing awareness of GRI-based sustainability reporting and an improvement in report quality. However, independent verification of sustainability reports is not common. The study found that larger organizations are more likely to issue a sustainability report than smaller firms. This finding should be viewed in the perspective of legitimacy theory, where larger enterprises may be more vulnerable to illegitimacy. The survey also found that manufacturing enterprises publish sustainability reports more frequently than service firms. The importance of industry on sustainability reporting demonstrates the legitimacy theory's usefulness in explaining firm's actions.

A large need for compulsory, externally validated sustainability reporting in either the annual report or as a separate stand-alone report was identified by O'Dwyer et al. [34]. They discovered that every organization had some type of information on their social or environmental performance on their official website. The percentage of companies publishing sustainability reports has risen in recent years. The data also demonstrates that certifying sustainability reports has become a global phenomenon, affecting both established and emerging nations. Dilling [35] found that companies with a higher profit margin and/or situated in Europe are more likely to generate high-quality sustainability reports.

According to FernandezFeijoo et al. [36], countries with more female board members report more CSR. Amran et al. [37] discovered that institutionalizing CSR in an organization improves sustainability reporting quality. Strategic connections with non-governmental groups are positively connected with sustainability reporting quality.

3 Methodology

3.1 Population and Sample

For exploratory and descriptive purposes, a survey research design was used to evaluate sustainability reporting in Palestine. Due to the modest number of listed companies on the Palestine Exchange (PEX) (48), we avoided using a sample to reduce the observations. So we addressed each company in several ways: physical visits, phone calls, and emails asking for online replies. Finally, we got data from 38 companies (nearly 80% of the total number of listed companies).

3.2 Research Instrument

Our study's exploratory nature required a survey-based methodology. This paper developed a questionnaire based on GRI disclosure standards and existing literature [23, 34]. As illustrated in Table 1, the questionnaire is divided into five sections: personal and corporate sustainability reporting motivations, reporting barriers, motivations for employing external assurance for sustainability reports, and external assurance obstacles.

Table 1 Questionnaire layout

Section	Variable	Details	Items
1	Information about the company, the respondents, and the sustainability reporting practices	Sector, age, and board characteristics Sustainability Reporting and External Assurance Practices	8
2	Motivations for sustainability reporting	The degree to which the company believes that included incentives affect the sustainability reporting	16
3	Barriers to sustainability reporting	The degree to which the company believes that included barriers affect the sustainability reporting	9
4	Motivations for external assurance of CSR and sustainability report	The degree to which the company believes that the included factors motivate them to assure the accuracy and reliability of information in the sustainability report from an external party	12
5	Barriers to external assurance for sustainability report	The extent to which the company believes that the included barriers affect the external assurance of the sustainability report from an external party	6

3.3 Data Analysis

The survey data were mostly analysed using descriptive statistics. For describing the study's findings on sustainability reporting practices, motivations, and barriers to sustainability practices, as well as external assurance motives and barriers, the frequency, the mean and rank were employed.

4 Results

4.1 Sample Characteristics

Table 2 shows the sample's industry, age, number of board members, female membership, and board member independence. As seen in Table 2(a), the industrial sector had 10 companies (26.3%). The remaining four categories have 6–8 enterprises each. In (b), the study's companies are mostly mature (73%) with over 20 years of age. In (c), companies have 5–15 board members, with 7, 9, and 11 being the most common. While most boards lack independent members, women are underrepresented. These conclusions align with previous literature [12, 13, 38].

Table 2 Sample characteristics

(a) Sector	Frequency	Percent
investment	8	21.05%
industrial	10	26.32%
service	6	15.79%
insurance	7	18.42%
banks	7	18.42%
Total	**38**	**100.00%**

(b) Age category(years)	Frequency	Percent
1 – 5	1	2.63%
6 – 10	1	2.63%
11-15	5	13.16%
16 – 20	3	7.89%
More than 20	28	73.68%
Total	**38**	**100.00%**

(c) Number of members	Frequency	Percent
5	1	2.63%
6	2	5.26%
7	12	31.58%
8	3	7.89%
9	5	13.16%
10	1	2.63%
11	9	23.68%
12	2	5.26%
13	2	5.26%
15	1	2.63%
Total	**38**	**100.00%**

(d) Female Members	Frequency	Percent
0	23	60.53%
1	8	21.05%
2	5	13.16%
3	2	5.26%
Total	**38**	**100.00%**

(e) Independent Members	Frequency	Percent
0	15	39.47%
1	5	13.16%
2	5	13.16%
3	3	7.89%
4	3	7.89%
6	1	2.63%
7	1	2.63%
8	1	2.63%
9	2	5.26%
12	1	2.63%
13	1	2.63%
Total	**38**	**100.00%**

4.2 Sustainability Reporting and External Assurance Practices

As indicated in Table 3, all companies incorporate sustainability information. Almost all organizations utilize their annual report to inform stakeholders about their sustainability efforts. Only two companies publish separate sustainability reports. Participants were asked if they engaged third parties to verify sustainability data. Table 3(b) shows that no corporation had its sustainability data independently verified.

Furthermore, we ask organizations to mention whether they established a special unit or responsibility tasked with sustainability reporting or alternatively, delegated this obligation to another unit. According to the findings, such a unit is not a high priority. According to Table 3(c), just 42% of corporations designate a dedicated unit or responsibility for sustainability disclosure. The rest of them involve sustainability reporting practices with other types of reporting. These findings indicate that sustainability reporting in not currently a priority and both companies and authorities should take this issue extremely seriously.

Table 3 Sustainability reporting and external assurance practices

(a) Type of disclosures about sustainability and CSR report		
Type of disclosure	Frequency	Percent
No disclosure	0	0
In the annual report	36	94.7
A separate social responsibility report	0	0
A separate sustainability report	2	5.3
Total	38	100
(b) External assurance of sustainability or CSR report		
Answer	Frequency	Percent
Yes	0	0
No	38	100
Total	38	100
(c) Existence of a special unit or responsibility for sustainability and CSR practices		
Answer	Frequency	Percent
Yes	16	42.11
No	22	57.89
Total	38	100

4.3 Motivations for Sustainability and Social Responsibility Reporting

Drivers motivating Palestinian companies to disclose sustainability and CSR activities are shown in Table 4. It illustrates the percentage of the companies for items proposed as motivations for such disclosure, the mean, and the rank of each item.

As seen in Table 4, the most influential drivers are moral commitment and citizenship (4.55), increasing the confidence of customers and suppliers (4.42), enhancing the reputation of the company in the market (4.39), one of the company's policies (4.32), enhancing competitive advantage in the market (4.24), responding to the expectations of customers and community (4.18), and promotion of goods and services (4.16). We notice that most motives are customer-driven, meaning that sustainability reporting is used as part of the public relations and promotion of the company.

Table 4 Motivation for preparing sustainability and CSR reports

#	Item	Never do it %	We don't do it %	Neutral %	Often do it %	Always do it %	Mean	Rank
1	Enhance reputation in the market	0	5.3	5.3	34.2	55.3	4.39	3
2	Increase the confidence of customers and suppliers	0	2.6	5.3	39.5	52.6	4.42	2
3	Disclosure results in the promotion of our goods and services	5.3	2.6	10.5	34.2	47.4	4.16	7
4	Enhance competitive advantage in the market	2.6	2.6	13.2	31.6	50	4.24	5
5	Because of moral commitment and citizenship	2.6	2.6	0	26.3	68.4	4.55	1
6	A response to the expectations of customers and community	5.3	2.6	5.3	42.1	44.7	4.18	6
7	A response to the requirements of the international community	10.5	10.5	23.7	26.3	28.9	3.53	12
8	A response to the requirements of shareholders	5.3	7.9	15.8	28.9	42.1	3.95	8
9	Media forces us to disclose information about CSR practices and sustainability	21.1	18.4	15.8	31.6	13.2	2.97	14

(continued)

Table 4 (continued)

#	Item	Never do it %	We don't do it %	Neutral %	Often do it %	Always do it %	Mean	Rank
10	One of our company's policies	2.6	0	13.2	31.6	52.6	4.32	4
11	Local laws and regulations encourage us to disclose information about CSR practices and sustainability	5.3	2.6	21.1	42.1	28.9	3.87	9
12	International laws and regulations compel us to disclose information about CSR practices and sustainability	21.1	18.4	26.3	23.7	10.5	2.84	15
13	Disclosure is a common practice in the sector in which we operate	18.4	7.9	28.9	23.7	21.1	3.21	13
14	Enhances the legitimacy of our business from the community point of view	7.9	5.3	15.8	44.7	26.3	3.76	10
15	Imitating competing companies	47.4	15.8	26.3	7.9	2.6	2.03	16
16	To be unique and pioneers in the market	10.5	5.3	15.8	34.2	34.2	3.76	10

4.4 Obstacles of Sustainability and Social Responsibility Reporting

Immature sustainability reporting may be attributed to several obstacles. Table 5 contains the items proposed as reporting obstacles. It shows the percentage of the companies that agreed or disagreed with each item, as well as the mean values and rank of each item. The absence of laws and instructions that ask for preparing such reports is ranked as the top obstacle, followed by the absence of guidance on how to prepare these reports. The lack of interest by stakeholders is the third obstacle, and

Table 5 Obstacles for preparing sustainability and CSR reports

#	Item	Strongly disagree %	Disagree %	Neutral %	Agree %	Strongly agree %	Mean	Rank
1	Reporting costs are high and we can't afford them	55.3	18.4	18.4	7.9	0	1.79	7
2	The disclosure of donations contradicts the provisions of Islamic law	60.5	15.8	18.4	2.6	2.6	1.71	9
3	Disclosure of what we have presented to our society makes us proud of ourselves which will add no value to our contribution	50	13.2	26.3	7.9	2.6	2	6
4	There are no laws or instructions that compel us to prepare such reports	26.3	15.8	23.7	23.7	10.5	2.76	1
5	There is insufficient guidance on how to prepare CSR reports and sustainability	26.3	31.6	26.3	15.8	0	2.32	2
6	Management unaware of the importance of reporting on CSR and sustainability	60.5	13.2	18.4	7.9	0	1.74	8

(continued)

Table 5 (continued)

#	Item	Strongly disagree %	Disagree %	Neutral %	Agree %	Strongly agree %	Mean	Rank
7	We do not believe that the parties involved in the company (shareholders, creditors and customers) are interested in publishing such reports	39.5	18.4	28.9	7.9	5.3	2.21	3
8	We do not have experts and systems to help us prepare such reports	42.1	21.1	28.9	5.3	2.6	2.05	5
9	We do not have time to prepare separate reports of CSR practices and sustainability	36.8	28.9	26.3	2.6	5.3	2.11	4

the lack of time to prepare a separate report is the fourth obstacle. Other obstacles are listed in the table.

4.5 Motivations and Barriers for External Assurance of Sustainability and Social Responsibility Report

Concerning the companies' desire to add assurance to their sustainability and CSR reporting, we notice that none of them use an external party to provide assurance on their sustainability reporting. We ask participants to provide their agreement degree on the factors that motivate them to seek external assurance on the one hand, and the reasons that prevent them from doing so on the other hand. Related results are presented in Table 6 and Table 7, respectively.

Concerning the motivations for external assurance for sustainability and CSR reports, we notice that the agreement level is not high. The most items considered by the respondents as possible drivers for using external assurance are: because external assurance will raise the level of qualitative information in the report (Mean = 3.29); to reinforce the legitimacy of business from the community point of view (Mean = 3.21); to make the CSR report appear professional (Mean = 3.18); to enhance the

Table 6 Motivation for external assurance of sustainability and CSR reports

#	Item	Strongly disagree %	Disagree %	Neutral %	Agree %	Strongly agree %	Mean	Rank
1	External Assurance enhances our credibility	21.1	10.5	21.1	28.9	18.4	3.13	5
2	External Assurance makes the form of the CSR report appear professional	18.4	13.2	15.8	36.8	15.8	3.18	3
3	External Assurance raises the level of qualitative information presented in the report	21.1	7.9	13.2	36.8	21.1	3.29	1
4	External Assurance of CSR report and sustainability enhances the role of internal audit in our company	21.1	10.5	15.8	36.8	15.8	3.16	4
5	External Assurance of the CSR reports and sustainability is a response to the expectations of our shareholders, customers, suppliers and all segments of the local and international community	23.7	2.6	26.3	34.2	13.2	3.11	6

(continued)

Table 6 (continued)

#	Item	Strongly disagree %	Disagree %	Neutral %	Agree %	Strongly agree %	Mean	Rank
6	Assurance of the CSR report and sustainability is one of our company's policies and our group	26.3	7.9	23.7	26.3	15.8	2.97	9
7	Local laws and regulations encourage us to authenticate the report on CSR practices and sustainability	26.3	13.2	31.6	18.4	10.5	2.74	10
8	International laws and regulations compel us to authenticate the report on CSR practices and sustainability	34.2	18.4	26.3	13.2	7.9	2.42	12
9	Assurance of the CSR report and sustainability is a common practice in the sector in which we operate	34.2	15.8	23.7	18.4	7.9	2.5	11
10	Assurance of the CSR Sustainability Report reinforces the legitimacy of our business from the community point of view	13.2	13.2	28.9	28.9	15.8	3.21	2

(continued)

Table 6 (continued)

#	Item	Strongly disagree %	Disagree %	Neutral %	Agree %	Strongly agree %	Mean	Rank
11	Assurance of the CSR report and sustainability will enhance our reputation in the market	23.7	7.9	21.1	28.9	18.4	3.11	6
12	Assurance of the CSR report and sustainability will enhance our competition in the market	26.3	7.9	21.1	28.9	15.8	3	8

role of internal audit in the company (Mean = 3.16); and to enhance the credibility of the company (mean = 3.13).

In Table 7, we observe that Palestinian companies do not tend to use external assurance mainly since they think they have adequate internal assurance procedures

Table 7 Barriers for external assurance of sustainability and CSR reports

#	Item	Strongly disagree %	Disagree %	Neutral %	Agree %	Strongly agree %	Mean	Rank
1	High cost	39.5	21.1	28.9	2.6	7.9	2.18	4
2	Management unawareness of the importance of external Assurance of CSR and sustainability report	47.4	21.1	23.7	5.3	2.6	1.95	6
3	Carelessness of related parties (shareholders, customers and suppliers) about external Assurance of the report	44.7	13.2	28.9	10.5	2.6	2.13	5

(continued)

Table 7 (continued)

#	Item	Strongly disagree %	Disagree %	Neutral %	Agree %	Strongly agree %	Mean	Rank
4	We may validate our reports in the coming years as we are still at an early stage in preparing CSR reports and sustainability	28.9	10.5	36.8	18.4	5.3	2.61	2
5	We do not believe that external Assurance will give credibility and higher value to the report	31.6	21.1	26.3	7.9	13.2	2.5	3
6	We confirm our reports internally as we have internal procedures for assurance	10.5	7.9	21.1	31.6	28.9	3.61	1

to confirm the report, which has the highest mean of 3.61. However, companies expect to have external assurance in the coming years. The gain from the external assurance and the cost of this assurance were the third and fourth reasons for not using the external assurance services.

5 Conclusions

By exploring the sustainability reporting practices for the listed companies, we have noticed that this term is still not clear. The majority of corporations that report for sustainability do not have any formal reporting of their social and environmental activities; instead, the reporting is limited to a distinct part in the annual report. The findings show that the sustainability reported by Palestinian enterprises is mostly focused on customers and is utilized to boost the firm's image and legitimacy. The objectives for releasing sustainability and social responsibility reports are centered on customers' and suppliers' moral commitment, citizenship, and trust. This is not surprising given that it is congruent with most organizations' long-term goals. The primary purpose of producing a separate report is to boost the company's image. As a result, the firm may have a competitive advantage in the market. The absence of rules and instructions requiring such reports, as well as a lack of direction on how to create such reports, are considered as the most significant barriers. The findings indicate that Palestinian enterprises have a low willingness to acquire external verification for sustainability reports because they believe they have enough internal assurance methods to confirm the report.

The findings imply that all parties involved must seek to improve sustainability reporting methods. In this context, specific legislation, rules, and education on sustainable reporting are essential. In the future, it would be interesting to formally investigate the association between sustainability reporting practices and firm characteristics such as foreign ownership, company age, firm size, and the percentage of female board members. Furthermore, it is critical to investigate the barriers to sustainability reporting and the use of external certification by examining elements other than those mentioned in our study.

References

1. Rashid M, Abdeljawad I, Manisah Ngalim S, Kabir Hassan M (2013) Customer-centric corporate social responsibility. Manag Res Rev 36:359–378
2. Brammer S, Millington A (2005) Corporate reputation and philanthropy: an empirical analysis. J Bus Ethics 61:29–44
3. Adams CA, Frost GR (2008) Integrating sustainability reporting into management practices. Account Forum 32:288–302
4. Kolk A (2008) Sustainability, accountability and corporate governance: exploring multinationals' reporting practices. Bus Strateg Environ 17:1–15
5. Baumüller J, Sopp K (2022) Double materiality and the shift from non-financial to European sustainability reporting: review, outlook and implications. J Appl Acc Res 23:8–28
6. Fortanier F, Kolk A, Pinkse J (2011) Harmonization in CSR reporting. Manag Int Rev 51:665–696
7. Hahn R, Kühnen M (2013) Determinants of sustainability reporting: a review of results, trends, theory, and opportunities in an expanding field of research. J Clean Prod 59:5–21
8. Siew RY (2015) A review of corporate sustainability reporting tools (SRTs). J Environ Manage 164:180–195
9. Hemphill T (2013) The ISO 26000 guidance on social responsibility international standard: what are the business governance implications? Corp Gov Int J Bus Soc 13:305–317
10. PCBS (2014) Environment and sustainable development in Palestine. Palestinian Central Bureau of Statistics, Environmental Quality Authority, Ramallah-Palestine
11. PEX (2017) Symposium on the importance of sustainability report. Palestine stock exchange
12. Abdeljawad I, Masri R (2020) Board characteristics and corporate performance: evidence from Palestine. An-Najah Univ J Res-B (Humanit) 34:745–770
13. Abdeljawad I, Oweidat GA, Saleh NM (2020) Audit committee versus other governance mechanisms and the effect of investment opportunities: evidence from Palestine. Corp Gov Int J Bus Soc 20:527–544
14. Abu Alia M, Abdeljawad I, Jallad SE, Rashid M (2022) Voluntary disclosure-cost of equity nexus and the moderating role of corporate governance: evidence from an extremely politically unstable context. Int J Islamic Middle Eastern Finance Manage. ahead-of-print

15. Alia MA, Abdeljawad I, Yaaqbeh M (2020) Depressing earnings management in Palestinian corporations: the role of audit quality, audit committee, and accounting conservatism. Int J Revenue Manag 11:213–236
16. Halkos G, Nomikos S (2021) Corporate social responsibility: trends in global reporting initiative standards. Econ Anal Policy 69:106–117
17. Garas S, ElMassah S (2018) Corporate governance and corporate social responsibility disclosures. Crit Perspect Int Bus 14:2–26
18. Ioannou I, Serafeim G (2017) The consequences of mandatory corporate sustainability reporting. Harvard Business School research working paper
19. Eccles RG, Ioannou I, Serafeim G (2014) The impact of corporate sustainability on organizational processes and performance. Manage Sci 60:2835–2857
20. Dobbs S, van Staden C (2016) Motivations for corporate social and environmental reporting: New Zealand evidence. Sustain Account Manage Policy J 7:449–472
21. Burhan AHN, Rahmanti W (2012) The impact of sustainability reporting on company performance. J Econ Bus Account Ventura 15:257–272
22. Finch N (2005) The motivations for adopting sustainability disclosure. Available at SSRN: https://ssrn.com/abstract=798724
23. Dienes D, Sassen R, Fischer J (2016) What are the drivers of sustainability reporting? A systematic review. Sustain Account Manage Policy J 7:154–189
24. Alia M (2010) The impact of environmental factors on the adoption of international financial reporting standards (IFRSs) in the Arab world. vol. Unpublished PhD thesis. Vrije University of Brussels, Belgium
25. Alia M, Branson J (2011) Environmental obstacles associated with the adoption of IFRSs in the Arab world: evidence from Jordan. World Account Sci (Muhasebe Bilim Dünyası Dergisi) 13:23–85
26. Hąbek P, Wolniak R (2016) Assessing the quality of corporate social responsibility reports: the case of reporting practices in selected European Union member states. Qual Quant 50:399–420
27. Hąbek P (2014) Evaluation of sustainability reporting practices in Poland. Qual Quant 48:1739–1752
28. Tarwaneh AW, Alsuhimat FM, As' ad AM (2021) The level of Jordanian citizen's awareness of the social responsibility standards in light of the Corona pandemic (Covid-19). J Educ Psychol Sci 4:64–79
29. Alhaleh SEA, Saleh MAA, Liu C (2019) Research on: analyzing the impact of corporate social responsibility disclosures and sustainability on operating banking sector in Palestine. Managers' perspective empirical and theoretical investigation. South East Asia J Contemp Bus Econ Law 20:32–39
30. Legrand W, Huegel EB, Sloan P (2013) Learning from best practices: sustainability reporting in international hotel chains. In: Advances in hospitality and leisure, vol 9. Emerald Group Publishing Limited, pp 119–134
31. Zovico N (2020) The Sustainable Development Goals: an analysis of Sustainability Reporting in the Italian Paper and Pulp Industry. Politecnico di Torino
32. Frost G, Jones S, Loftus J, Van Der Laan S (2005) A survey of sustainability reporting practices of Australian reporting entities. Aust Account Rev 15:89–96
33. Kuzey C, Uyar A (2017) Determinants of sustainability reporting and its impact on firm value: evidence from the emerging market of Turkey. J Clean Prod 143:27–39
34. O'Dwyer B, Unerman J, Hession E (2005) User needs in sustainability reporting: perspectives of stakeholders in Ireland. Eur Account Rev 14:759–787
35. Dilling PF (2010) Sustainability reporting in a global context: what are the characteristics of corporations that provide high quality sustainability reports an empirical analysis. Int Bus Econ Res J (IBER) 9
36. Fernandez-Feijoo B, Romero S, Ruiz-Blanco S (2014) Women on boards: do they affect sustainability reporting? Corp Soc Responsib Environ Manage 21:351–364
37. Amran A, Lee SP, Devi SS (2014) The influence of governance structure and strategic corporate social responsibility toward sustainability reporting quality. Bus Strateg Environ 23:217–235

38. Abdeljawad I, Dwaikat LM, Oweidat G (2022) The determinants of profitability of insurance companies in Palestine. An-Najah Univ J Res-B (Humanit) 36:439–468
39. Al-Sartawi A (2020), Information technology governance and cybersecurity at the board level. Int J Crit Infrastruct 16(2):150–161

Capital Structure, Performance, and the Moderating Role of Diversity of Products and Claims of Insurance Companies in Palestine

Islam Abdeljawad and Layth Dwaikat

Abstract The study's objectives are twofold: first, to look at the impact of capital structure on insurance company performance in Palestine; and second, to see if this relationship is affected by the diversity of insurance portfolios (product diversification and claims diversification). This study is motivated by the observation that the insurance companies in Palestine are highly levered and the insurance activities are concentrated in the automobile industry, and it is unknown if this strategy is advantageous to the company. The moderating influence of product and claim diversity on the link between capital structure and financial performance in this paper is a contribution to the existing literature. To estimate the study's model, balanced panel data from seven insurance companies operating in Palestine from 2010 to 2018 was used. The findings demonstrate that leverage has a negative influence on insurance firm performance, but this relationship is contingent on the degree of insurance portfolio diversification. Diversifying the insurance portfolio lessens the negative impact of financial leverage on financial performance. According to the findings, diversification allows insurance companies to accept more risks. The conclusions of the study advise management to concentrate on the product and claim diversification as a means of reducing financial risk.

Keywords Insurance company performance · Leverage · Product diversity · Claims diversity · Palestine

1 Introduction

The higher stock price of companies with greater financial performance is a result of the company's sales, profits, and risks. Good performance also indicates the executives' efficiency in managing the company's resources and overcoming all hurdles and risks in order to maximize the company's profits and shareholder wealth. It's

I. Abdeljawad (✉) · L. Dwaikat
An-Najah National University, Nablus, Palestine
e-mail: islamjawad@najah.edu

A. M. A. Musleh Al-Sartawi et al. (eds.), *From the Internet of Things to the Internet of Ideas: The Role of Artificial Intelligence*, Lecture Notes in Networks and Systems 557,
https://doi.org/10.1007/978-3-031-17746-0_16

also critical to maintain the company's continuity and achieve appropriate financial soundness [1]. Net premiums earned, profitability from underwriting activities, return on investment, return on assets, and return on equity are all common measures of an insurance company's financial performance [2].

Many firm-specific factors (internal factors) as well as macroeconomic factors (external factors) can influence financial performance. Many studies have looked at the relationship between internal and external variables and insurance company financial performance [3–5]. This study will look at the impact of capital structure on insurance company financial performance, as well as the impact of product and claims diversification on the capital structure-financial performance link. The impact of a company's capital structure on its financial performance has been extensively researched using various statistical models and measuring approaches for both financial performance and capital structure. However, the impact of capital structure on insurance company performance is understudied, particularly in the case of Palestine, a small underdeveloped country with special political characteristics [6–9].

The insurance industry in Palestine is relatively new, and there is little research on the subject, as well as contradictory findings from prior studies in other countries. Insurance companies in Palestine are characterized by high concentration in the automobile segment of insurance [3]. There is little evidence about the effect of this concentration on performance. This research may aid decision-makers in avoiding bad performance by providing advice for achieving greater profitability and, as a result, improving financial performance. Furthermore, the study may assist managers by giving critical information on the best management of insurance portfolios and their impact on profit maximization. Particularly, this study is focusing on product and claim diversification effect on leverage-profitability link.

The rest of the paper is organized as follows: The background of the study is presented in Sect. 2. The research methodology is discussed in Sect. 3. Section 4 summarizes the findings, and Sect. 5 wraps things up.

2 Study Background

The insurance legislative framework in Palestine has undergone numerous changes as a result of the country's multiple rulers. Since the Ottoman administration, insurance has been practiced in Palestine under the auspices of a statute enacted in 1845 that dealt with the insurance of financial losses and damages. The law of the guarantee contract is the name given to this statute. Regulations and laws were amended under the British mandate, the Jordanian period, and the Israeli occupation until 1994, when the Palestinian Authority was established to supervise and control the enactment of insurance-related legislation. Until the foundation of the Palestine Capital Market Authority (CMA) in 2004 and the enactment of the Insurance Act in 2005, which controlled the insurance sector until now, this sector was mostly ignored. There are currently 10 insurance companies functioning in Palestine, two of which use the Islamic Takaful system [10].

With a few exceptions, the impact of insurance firm's capital structure on its financial performance is understudied. Only a few studies have looked into the impact of capital structure on the performance of insurance companies, whether they are general or life insurance companies. The impact of product diversification on performance of insurance companies is rare [11–13]. Foong and Idris [1] investigate the impact of leverage on the financial performance of Malaysian general insurance companies, as well as whether the leverage-performance relationship is a contingent on the degree of product diversification, by looking at a sample of the entire population of authorized general insurance companies operating from 2006 to 2009. It was discovered that leverage had a negative relationship with firm performance. However, there is a strong interaction effect on firm performance between leverage and product diversification. The findings suggest that, depending on the amount of the firm's product diversification, leverage could be useful or harmful to the financial performance of general insurance companies. Elango et al. [14] used various measures of product diversification and firm performance to investigate the relationship between product diversification and firm performance in the US property–liability insurance industry. They discovered that the extent of product diversification has a complex and nonlinear relationship with firm performance, and that performance benefits associated with product diversification are contingent upon an insurer's degree of geographic diversification. Finally, according to Comment and Jarrel [15], diversified firms have more leverage than non-diversified firms.

The majority of studies that looked at the impact of capital structure on insurance company performance focused on leverage but neglected product or claims diversification, and just a few studies looked at product diversity. The association between financial leverage and performance was studied here using product or claim diversity as a moderator variable. This research looks into the following hypotheses in particular:

H1: There is a negative relationship between leverage and financial performance of insurance companies in Palestine.

H2: Product diversity moderates the relationship between leverage and financial performance of insurance companies in Palestine.

H3: Claims diversity moderates the relationship between leverage and financial performance of insurance companies in Palestine.

3 Research Methodology

This part of the research deals with the data collection method, sampling design, research model, measurement of variable and analysis techniques.

3.1 Data

The number of insurance companies as of the end of 2018 was ten. Data were collected manually from annual reports of insurance companies disclosed on the Palestine Stock Exchange website for the nine years period from 2010 to 2018 based on the availability of data. Companies with inappropriate disclosures for the period were not included, resulted in seven companies being used for the period. The final data set is balanced panel data of 63 observations.

3.2 Measurement of Variables

Measurement of variables of this study followed previous literature as follows:

3.2.1 The Dependent Variable

Financial Performance. Many measures of financial performance as an accounting profitability measure such as Return on Assets (ROA), Return on Equity (ROE) and Return on Invested Capital (ROIC) [3, 16, 17] are usually used to measure performance. This study used ROA measured by dividing profit before tax by total asset. ROA is used as the dependent variable in the regression model.

$$\text{ROA} = \text{Net income before taxes} \;/\; \text{total assets}$$

3.2.2 The Independent Variables

There are several independent variables included in our models. These variables are following:

a. Leverage. It is an indicator to the degree to which business is utilizing borrowed money. In this study, leverage is defined as total liabilities divided by total equity

$$\text{Leverage} = \text{Total Debts} / \text{Total Equity}$$

b. Product diversity. It is an indicator of the diversification of insurance companies' revenues and depends on revenues concentration. It is measured by the Herfindhal index as follows:

$$\text{Product Diversity} = 1 - \sum_{i=1}^{n} P_i^2$$

where

Pi = Total premium earned in product line i / total premium earned by the firm.

c. *Claims diversity*. It is an indicator of the diversification of paid claims of insurance companies and how its incurred claims distributed between the various products. It is calculated in a way analogous to calculating the product diversity as follows:

$$\text{Claims Diversity} = 1 - \sum_{i=1}^{n} C_i^2$$

where:

Ci = Total Claims incurred in product line i / total Claims incurred by the firm.

d. *Firm Size (Size)*. Firm size is used as a control variable and is measured by:

$$Size = \text{Log (Total Assets)}$$

3.3 Research Model

Data analysis techniques include descriptive analysis, correlation analysis and panel data multiple linear regression. The following linear models were estimated to test the hypotheses of this research.

$$\textbf{Model 1}: \text{ROA} = \text{B}_1 + \text{B}_2\text{Lev} + \text{B}_3\text{SIZE} + \text{e}$$

$$\textbf{Model 2}: \text{ROA} = \text{B}_1 + \text{B}_2\text{Lev} + \text{B}_3\text{PD} + \text{B}_4\text{Lev} * \text{PD} + \text{B}_4\text{SIZE} + \text{e}$$

$$\textbf{Model 3}: \text{ROA} = \text{B}_1 + \text{B}_2\text{Lev} + \text{B}_3\text{CD} + \text{B}_4\text{Lev} * \text{CD} + \text{B}_4\text{SIZE} + \text{e}$$

where:

ROA: Return on assets.
Lev: Leverage Ratio,
PD: Product Diversity,
CD: Claims Diversity,
Size: Size of the company,
e: error term, and.
Bi: coefficients to be estimated.

Table 1 Descriptive statistics

	Mean	Median	Maximum	Minimum	Std. Dev.	N
ROA	0.034	0.034	0.148	−0.182	0.049	63
SIZE	7.649	7.649	8.092	7.039	0.256	63
LEVERAGE	2.835	2.248	45.557	−19.013	7.070	63
CLAIMSDIVERSITY	0.436	0.488	0.618	0.054	0.142	63
PRODUCTDIVERSITY	0.456	0.465	0,632	0.161	0.130	63

4 Data Analysis

4.1 Descriptive Statistics

The descriptive statistics indicators (mean, median, maximum, minimum, and standard deviation) for the dependent variable (return on assets), independent variable (leverage), moderator variables (product diversity and claims diversity), and control variable (Size) are presented in Table 1.

Table 1 shows the results of a pooled descriptive indicator assessment from 2010 to 2018. The return on assets (ROA) for Palestinian insurance businesses ranges from −0.182 to 0.148, with both mean and median are 0.034. The large leverage ratio stands out. It ranges from 45.557 to −19.013 with an average of 2.835.

4.2 Correlation Analysis

The bivariate link between each two variables is shown by the correlation coefficient in Table 2. ROA has a positive association with company size and leverage, but a negative relationship with product diversity and claims diversity, according to Table 2. The high correlation between independent variables can be problematic in regression analysis, but the relationship between independent variables indicates that there is no multicollinearity problem between variables because the only high correlation is between product diversity and claim diversity, which are used separately and not in the same model.

5 Estimation Results

The regression results are shown in Table 3. Model 1 explains 16.58% of the entire variation in insurance companies' ROA according to R-squared. The null hypothesis that the model is inadequate should be rejected because the F-statistic is sufficiently high, suggesting that the model is well fitted at the 1% level of significance. With

Table 2 Correlation coefficients

	ROA	SIZE	LEVERAGE	PRODUCT DIVERSITY	CLAIMS DIVERSITY
ROA	1.000				
SIZE	0.380	1.000			
LEVERAGE	0.116	−0.083	1.000		
PRODUCT DIVERSITY	−0.017	0.372	−0.237	1.000	
CLAIMS DIVERSITY	−0.007	0.351	−0.207	0.941	1.000

a coefficient of 0.0010 and a probability of 0.2154, we can see that leverage has no meaningful effect on financial performance (ROA) in Model 1.

The influence of the independent variable (leverage) and the moderator variable (product diversification) on the financial performance (ROA) of insurance businesses in Palestine is expressed in Model 2. In this model, leverage has a negative effect on financial performance at a 1% significance level, and the interaction term has a

Table 3 Estimation results

Dependent Variable: ROA			
Variable	Model 1	Model 2	Model 3
Size	0.0757***	0.0276	0.0476**
	(3.3112)	(1.278)	(2.1202)
Leverage	0.001	−0.0252***	−0.0203***
	(1.2521)	(−5.8146)	(−4.3633)
Product diversity		−0.1748***	
		(−4.0306)	
Leverage*Product Diversity		0.0537***	
		(6.0734)	
Claims diversity			−0.183***
			(−3.763)
Leverage*Claims Diversity			0.045***
			(4.606)
C	−0.5477***	−0.0836	−0.2393
	(−3.1277)	(−0.5042)	(−1.3786)
R-squared	0.1658	0.5023	0.4003
Adjusted R-squared	0.138	0.4679	0.359
Durbin-Watson stat	1.0323	1.6934	1.678
F-statistic	5.9641***	14.6338***	9.6807***
Observations (balanced)	63	63	63

Table 4 Summary of results

	Hypothesis	Empirical result
H1	There is a negative relationship between leverage and financial performance of insurance companies in Palestine	Supported
H2	Product diversity moderates the relationship between leverage and financial performance of insurance companies in Palestine	Supported
H3	Claims diversity moderates the relationship between leverage and financial performance of insurance companies in Palestine	Supported

highly significant positive moderation in the relationship between the dependent and independent variables, increasing the explanatory potential of the model to 50%.

The influence of the independent variable (leverage) and the moderator variable (claims diversity) on the financial performance (ROA) of insurance companies in Palestine is expressed in Model 3. The interaction term shows a highly significant positive moderation in the relationship between dependent and independent variables, and the explanatory power of the model has increased to 40%, which is less than R-squared at model 2, but it is still high.

The associations between the independent variables and the dependent variable were established using the estimation models, as explained below. According to regression analysis, there is a negative association between leverage and financial performance of insurance companies in Palestine, with a coefficient of −0.02 and significance at 1%. As a result, there is a considerable negative association between insurance company leverage and financial success in Palestine. Foong and Idris [1] found similar results In Malaysia.

Product diversification moderates the relationship between leverage and financial performance of insurance companies in Palestine, according to the second hypothesis. The results of the regression analysis show that the interaction has a highly significant positive moderating effect in the relationship between dependent and independent variables, with a coefficient of 0.05 and a significance level of 1%.

The third hypothesis is that claim diversity moderates the association between insurance company leverage and financial performance. The findings of the regression analysis demonstrate that the interaction exhibit a significant moderating influence in the relationship between dependent and independent variables, with a coefficient of 0.045 and a significance at 1% level. The summary of the findings are presented in Table 4.

6 Conclusions

The impact of the insurance portfolio (products and claims diversity) on the relationship between capital structure and financial performance was investigated in this paper, which used multiple linear regression analysis to examine these relationships for insurance companies in Palestine. The findings revealed that capital structure and financial performance have a negative association. The findings also revealed that diversifying the insurance portfolio reduces the negative impact of financial leverage on financial performance, making insurance companies more capable of taking risks when they are not focused on one or two types of insurance products, as they are now with car insurance concentration. Moreover, insurance companies are likely to take advantage of technical reserves by reducing them and making higher profits, thus reflecting the fact that there is a negative relationship between leverage, which decreases if technical reserves are reduced, and profitability, which increases by reducing these reserves.

The study has contributed to the existing literature by adding the moderating effect of product and claim diversity on the relationship between capital structure and financial performance. The study is helpful for the management to focus on reducing the risk of leverage by diversifying the product and claiming cash flows.

Appendix 1: Research Sample (Yearly Data from 2010 to 2018)

	Insurance Company Name
1	Trust International Insurance Company
2	National Insurance Company
3	Global United Insurance Company
4	Takaful Palestinian Insurance Company
5	Mashreq Insurance Company
6	Palestine Insurance Company
7	Ahlia Insurance Company

Appendix 2: Diagnostic Tests

Diagnosis of the regression was carried out to determine the validity of the model's assumptions. The diagnostic tests including normality, autocorrelation, and heteroscedasticity are presented below.

1. **Normality Test**

The Jaque-Bera test of the normality of residuals is intended to establish whether the mean of the residuals is equal to zero. The residuals resulted from model 2 are bell-shaped and the Jarque-Bera statistic is 3.619 with p-value of the test of 0.163. These results indicate that the residuals are normally distributed.

For model 3, the Jarque-Bera statistic is 0.735 with p-value of the test of 0.692 indicating that the residuals are normally distributed too.

2. **Autocorrelation**

The residuals of the model should be uncorrelated with each other for an estimation to be efficient. The Durbin-Watson test is designed to check for autocorrelation. The result of the test has a value of 1.693 for model 2 and 1.678 for model 3, which indicates no evidence for the presence of autocorrelation on the models.

3. **Heteroscedasticity**

The Breusch-Pagan test was used to check for heteroscedasticity. According to the table below, the probability is greater than 0.05 statistic test meaning that heteroscedasticity is not present at the 0.05 level of significant for both models.

	Test	Statistic	d.f.	Prob.
Model 2	Breusch-Pagan LM	16.01538	21	0.7688
Model 3	Breusch-Pagan LM	18.26054	21	0.6325

References

1. Foong SY, Idris R (2012) Leverage, product diversity and performance of general insurers in Malaysia. J Risk Finan 13:347–361
2. Murigu JW (2014) The determinants of financial performance in general insurance companies in Kenya. University of Nairobi
3. Abdeljawad I, Dwaikat LM, Oweidat G (2022) The determinants of profitability of insurance companies in Palestine. An-Najah Univ J Res-B (Humanities) 36:439–468
4. Rashid A, Kemal MU (2018) Impact of internal (micro) and external (macro) factors on profitability of insurance companies. J Econ Policy Res 5:35–57
5. Tegegn M, Sera L, Merra TM (2020) Factors affecting profitability of insurance companies in Ethiopia: panel evidence. Int J Commer Finan 6:1–14
6. Abdeljawad I, Oweidat GA, Saleh NM (2020) Audit committee versus other governance mechanisms and the effect of investment opportunities: evidence from Palestine. Corporate Governance Int J Bus Soc 20:527–544
7. Abdeljawad I, Abed-Rabu K (2019) Capital structure determinants of Palestinian corporations. Jordan J Bus Adm 15:269–283
8. Abu Alia M, Abdeljawad I, Jallad SE, Rashid M (2022) Voluntary disclosure-cost of equity nexus and the moderating role of corporate governance: evidence from an extremely politically unstable context. Int J Islamic Middle Eastern Finan Manag ahead-of-print
9. Alia MA, Abdeljawad I, Yaaqbeh M (2020) Depressing earnings management in Palestinian corporations: the role of audit quality, audit committee, and accounting conservatism. Int J Revenue Manag 11:213–236
10. MAS (2016) The insurance sector in Palestine - achievements, failures and challenges. http://www.mas.ps
11. Schommer M, Richter A, Karna A (2019) Does the diversification–firm performance relationship change over time? A meta-analytical review. J Manage Stud 56:270–298
12. Jouida S (2018) Diversification, capital structure and profitability: a panel VAR approach. Res Int Bus Financ 45:243–256
13. Al-Sartawi A (2018) Institutional ownership, social responsibility, corporate governance and online financial disclosure. Int J Crit Account 10(3/4):241–255

14. Elango B, Ma YL, Pope N (2008) An investigation into the diversification–performance relationship in the US property–liability insurance industry. J Risk Insur 75:567–591
15. Comment R, Jarrell GA (1995) Corporate focus and stock returns. J Financ Econ 37:67–87
16. Abdeljawad I, Masri R (2020) Board characteristics and corporate performance: evidence from Palestine. An-Najah Univ J Res-B (Hum) 34:745–770
17. Damodaran A (2007) Return on capital (ROC), return on invested capital (ROIC) and return on equity (ROE): measurement and implications. SSRN https://ssrn.com/abstract=1105499

An Efficient Image Processing Mechanism to Split Graphical Content in Fog Environment to Minimise Packet Dropping and Error Rate

Amogh Shukla and Arushi Jain

Abstract Image processing is a method to perform operations on an image, to get an enhanced image or extract some useful information from it. Fog computing, also known as fogging, is an architecture that uses edge devices to carry out a substantial amount of computation (edge computing), storage, and communication locally and route it over the internet backbone. An error in the physical signal causes data in a packet to get corrupted. Such corruption - most often detected at the link layer - causes the entire datagram to be discarded. Consequential packet loss leads to poor performance, lag, jitter and reduced throughput. The proposed research in this paper focuses on the mechanism of image processing that transfers graphical content over the fog environment. The major objective of our research is the minimization of packet dropping and error rate in this transmission. To achieve this objective the graphical content is split after compression to reduce the content size.

Keywords Fog computing · Packet dropping · Cloud computing · Compression · Image processing

1 Introduction

Using image processing, users may improve or extract information from a picture by applying various procedures. Fog computing architecture employs edge devices to perform significant computations (edge computing), storage, and communication locally and route it via the internet backbone. Fog computing also goes by the name of fogging. The data in a packet may get garbled when a physical signal fault occurs. Corrupted datagrams are destroyed at the link layer and are required to be re-transmitted by the host. Researchers hope to reduce packet loss and error rates for

A. Shukla · A. Jain (✉)
Vellore Institute of Technology, Vellore, TN 632014, India
e-mail: arushijain1479@gmail.com

A. Shukla
e-mail: amoghsh369@gmail.com

A. M. A. Musleh Al-Sartawi et al. (eds.), *From the Internet of Things to the Internet of Ideas: The Role of Artificial Intelligence*, Lecture Notes in Networks and Systems 557,
https://doi.org/10.1007/978-3-031-17746-0_17

graphical material in fog environments using image processing mechanisms in order to minimise the effects of packet loss like decrease in quality and high latency.

The process of making changes in graphical content by rotation, scaling, cropping, edge detection and/or compression is considered image processing. Several image processing mechanisms are stated below:

1.1 Image Processing

- Scaling: The process of enlarging or reducing an image horizontally/vertically.
- Rotation: The process of changing the angle of an image.
- Cropping: The elimination of some portion of an image during selection while pre-processing is termed as cropping.
- Edge detection: This is the process of extracting the edges or boundaries from an image to form an outline of the image.
- Compression: Lossy or lessless compression mechanisms are applied to an image to reduce the storage space consumed.

Unprocessed graphical samples collected via cameras or sensors are sent to satellites using image processing. Space probes, aircraft, and visuals used in everyday life for many applications have all been improved through the use of image processing. Many image processing techniques have been developed in the last decade. The majority of these techniques are developed to enhance visuals taken from spacecraft. Space missions and military aircraft may potentially take these kinds of images. Because of the availability of powerful personal computers, such graphical analysis methods have become well-known. Large-capacity memory devices and graphics software have contributed immensely.

1.2 Fog Computing

The word fog computing is presented by Cisco Systems. It is also known as edge computing or fogging. It is a model in which data, processing and application are concentrated in devices at the network edge rather than existing almost entirely in the cloud. Therefore, it can be considered as an expansion of cloud computing to the edge of the network. Fog computing facilitates the operations of computation, storage and networking services between end devices and cloud computing data centers. One of the major essential objectives behind the use of fog computing is to get better efficiency along with a reduction in the quantity of information that has to be delivered to the cloud for dispensation, examination and storage. Even though it is mainly carried out for the purpose of efficiency, it can be used for security and compliance.

Need For Fog Computing. Cloud computing has presented a lot of opportunities in the last few years as they offer a variety of computing services to their clients, like the "pay-as-you-go" model. It comes into notice that the cloud computing model proves to be a well-organized substitute for owning and managing private data centers for customer-facing web applications and batch processing. The network communication costs, requirement of information regarding storage resources and computation drawbacks are eliminated because of cloud computing. But this convenience proves to be a setback in time-sensitive applications in which we need nodes in the vicinity to meet the requirements of these applications. With the increasing integration of IOT techniques and devices in a customer's life, the cloud computing concepts available at present are not able to fulfill the needs related to mobility support, location awareness and low latency. To deal with the above problem, a new computing method is put forward which is known as fog computing. It executes at the edge of the network. This results in low latency and location awareness. Quality-of-services for streaming and real-time applications is also enhanced. Some examples include industrial automation, transportation and networks of sensors and actuators. In addition to this, new infrastructures support heterogeneity as fog devices include end-user devices, access points, edge routers and switches. The concept of fog computing is well suited to real time big data analytics. It supports densely distributed data collection points. Fog computation is also used in the fields of entertainment, advertising, personal computing, and other applications. The concept of the "edge of the network" is facilitated with the help of fog computing. Both contribute identical methods and characteristics like virtualization and multi-tenancy. Identical methods and characteristics like virtualization and multi-tenancy are contributed by both. The expansion is non-trivial. However, there exist some fundamental differences stemming from the reason why fog computing was developed: to address and provide services that do not fit the paradigm of the cloud.

Cloud vs Fog Computing. In cloud computing, data and applications are handled inside the cloud. In case the amount of data is large, cloud computing is time-consuming. Whereas, in fog computing, it operates on a network edge and thus, less time is consumed.

If a comparison is made between fog and cloud, it is seen that a large amount of bandwidth is required in the latter case because every bit of data is sent over cloud channels, whereas in fog computing the requirements of bandwidth are lesser, as every bit of data is aggregated at certain access points instead of being sent over cloud channels.

Advantages and disadvantages of fog computing
Advantages

1. The detection of masquerade activity.
2. The uncertainty of the attacker and the additional costs induced to make out the difference between real and false information are reduced.
3. Prevention is a considerable responsibility as masquerade activity is highly risky.

4. A considerable reduction in the data quantity that must be moved, subsequent traffic, and distance the data must travel is achieved with the help of edge application. Due to this, we will get a reduction in transmission costs, shrinking latency, and enhanced quality of service (QoS).
5. Elimination of bottlenecks is achieved, which results from centralized computing systems using edge computing.

Disadvantages

1. Existing data protection mechanisms such as encryption, etc., fail to protect data from hackers. Therefore, it is almost impossible to identify when an attack is about to happen.
2. It does not verify whether the user is authorized or not. This means that it is very difficult to identify which user is an attacker.
3. It is impossible to identify which file was hacked. In fog computing, safety is considered to be one of the major concerns.
4. It does not focus on ways of securing data from unauthorized access.

Architecture of Fog Computing

1. **Hardware Platform.** Two premises emerge from the nature of FC. First, since the platform is distributed, "easy-to-manage" becomes a prioritized consideration. In fact, "easy-to-manage" is a broad request that includes rich interfaces, easy programming, abundant software, etc. To satisfy these requests, ARM processors usually become the first choice for an embedded system. The ARM family has already been well applied in thousands of applications in IoT. Secondly, with the ultimate goal of computing, processing speed is a key point. Although ARM cores now have excellent processing power, doubts about performance still exist when facing complex computing. However, a new HPC computing resource has recently been funded in the UK to provide computing resources for UK scientists using 64-bit ARM CPUs. Thus, the views on the capability of ARM cores in this arena may well change. Field Programmable Gate Array (FPGA), another well-known hardware alignment in IoT, naturally comes to mind since it can process complex algorithms fast and in a parallel way, which perfectly matches the idea for accelerating big data applications. Furthermore, Xilinx, known for inventing FPGA, has launched the Zynq series which integrates ARM and FPGA together; ARM takes charge of management and FPGA focuses on the algorithms. This solution is likely to be of great advantage as a reference for setting up a fog node.
2. **Software Platform.** Bonomi discussed the components of fog software architecture and proposed the structures of the abstraction and orchestration layer. Since Hadoop, Spark and OpenStack are all announced to run on ARM by AMD, DataCentred and other institutes or individuals, transplanting big data/cloud computing platforms from PC/server to embedded hardware has become a feasible and agile approach. The abstraction layer maps Open-Stack, which virtualizes the heterogeneous resources, with cloud structure. The orchestration layer is able to probe, analyze, plan and execute a job that can be implemented in a big data platform like Hadoop or Spark. For instance, OpenStack is deployed upon

the ARM core and Hadoop, Spark or more specific engines work on it via APIs. The API design must consider flexibility, latency sensitivity and heterogeneity.
3. **Cloud Assist.** Notwithstanding distributed deployment, FC needs a cloud assist mechanism for convenient management and maintenance, which obviously are challenging to address manually on a local device. For example, the programs for data processing are initially installed in the RAM or ROM. If the programs or configurations need modification, the patch or advanced release will be downloaded from the cloud server. Besides this, the cloud is also helpful in resource management. As a voluntary system, the participants are expected to be high performance, reliable and unselfish. Accordingly, the evaluation of these factors of a participant is very important and related to their historical behavior, which is naturally the duty of the cloud. When a user initiates a computing task, the cloud can pull up a local list with the names of recommended resources.

1.3 Fog Environment

CISCO, a networking behemoth, pioneered fog computing. Fog computing is essentially a cloud extension. Thus, it would be impossible to describe it without first describing cloud computing.

The technique of using the internet to perform ICT (Information and Communications Technology) activities and services and store computer resources is known as cloud computing. People and companies may utilize third-party hardware and software that is hosted online with its help. Cloud computing allows users to access information and computer resources from any location with an internet connection. It has benefits over conventional on-site hosted services in terms of speed, affordability, and efficiency because of the constant availability of shared/pooled computing resources [26].

While cloud computing is presently functioning effectively, the available bandwidth is largely dependent on the capacity of the network service provider. Furthermore, as billions of people input, send and receive data into and out of the cloud, the cloud infrastructure is obstructed increasingly. Fog computing (some-times called fogging) is expanding or reducing cloud computing to the bottom of the network so that lower-end consumers may get faster ICTs (communication, storage, software, and so on). Since fog computing is nearer to smaller end users, has a wider client base and is more mobile, it varies from cloud computing.

Fog computing increases service quality, lowers latency and boosts customer satisfaction. Fog computing is enabled easily by the growing Internet of Things (IoT) comprising assets (vehicles, household appliances and even garments) connected to sensors that may transmit and receive data. Instead of an advanced backbone network, fog may be implemented with a simple communication system. As a consequence, the coverage is denser. This advantage facilitates the operation of big, real-time data-driven applications that support thousands of nodes in highly dynamic and varied settings.

1.4 Image Processing Mechanism

1. **Lossless compression**: is used for archiving purposes. It is used for medical imaging as well as technical drawing. It may also be used for clip art or comics.
2. **Lossy compression mechanism** may introduce compression artifacts when these are utilized at low bit rates. Lossy mechanisms have been found to be suitable in natural images such as photographs and other such applications where a small loss of fidelity is acceptable. This loss is acceptable in order to manage the substantial reduction in bit rate. Lossy compressions produce negligible differences, which might be known as visually lossless.

2 Related Works

In 2017, Nabil Abubaker [1] examined two problems of privacy arising from the fog computer system, and therefore, we identify the resulting obstacles. The first issue relates to the fog architecture's inability to reduce latency and enhance bandwidth when current methods of privacy protection contravene this design. The second issue concerns the closeness of the fog nodes to IoT or end-users. He talks about the necessity of tackling these problems by placing them in a real-life environment. Finally, he offers a fog computing paradigm that respects privacy, addresses these problems, and evaluates the safety and efficiency of his solutions. In 2018 Yunguo Guan [2] developed an intermediary fog layer to handle data on cloud communications with end-users. Fog computing is therefore generally seen as a cloud extension. In his paper, he addressed the problems of the design of fog computing for data security and privacy. In particular, he described the specific difficulties posed by the fog layer in the area of data security and privacy. He emphasized the reasons why cloud computing cannot apply data protection methods directly to fog computing. Mohamed Firdhous [3] models the cloud system in 2014 using the queuing theory, particularly Erlang. Using simulations, he evaluated four alternative models of cloud utility, of varying complexities. The findings of the simulations were also given, along with a thorough debate. K. Shenoy [4] suggested in 2015 that the Low Power and Lossy Network Routing Protocol (RPL) of IPv6 be used on the Internet of Things to demonstrate system performance when nodes are portable. He evaluates the versatility RPL shows for mobility characteristics. His article takes into consideration two configurations, namely the serial and parallel mobility models. M. Verma [5] suggested in 2016 that in terms of both - CPU cycles and energy, the average variation between optimum and non-optimal solutions was less than 6%. The scratchpad is comparable to a preloaded loop cache in the described design. Comparing the current method's energy consumption with that of a preloaded loop cache, the processor cycles and instructional energy are reduced by an average of 9 and 29%, respectively. To obtain a sub-optimal solution at every time stage, MUKHERJEE [6] presented a modified fit decrease method in 2016. In GPPs, MFFD equally allocates RRH requests to enhance the contrast between CRUs and FFD. In 2014, Zao [7] described his universal system design and presented its supporting technology, including publish/subscribe protocols from machine to machine, a multi-tier fog/cloud, and a connected web data system. The reliability and ease of use of the BCI online game, EEG Tractor Beam, will be shown in real-time at the Intelligent Environment Conference. In 2017, P. Dhiman [8] utilized and applied this new computer technology and concentrated on architects and policymakers engaged in it. In 2016, Dhore [9] examined fog computing and its features, benefits, and drawbacks. Security and implementation issues in the case of fog computing and its scheduling method have also been addressed here. The primary reason for the planning process is to optimize the use of resources. Another goal is to improve performance and decrease reaction time. It also improves service quality. In 2017, Bushra Zaheer Abbasi and Muna Ali Shah [10] examined security issues identified with FC. They also proposed an arrangement regarding these issues. Their paper presents a basic investigation of the fog design concerning security. The work done since 2012 was inspected on the foundations of safety methods and security dangers. They also compiled the current security procedures based on the foundations of safety objectives accomplished by each procedure. operational efficiency and they measured and analyzed the resource utilization, request rates, and performance of two consolidated n-tier application benchmark systems. There have been research on image processing for fog [11, 20, 23] and IoT [12] while some researches considered fog computing with machine learning [13, 21] and big image data processing [14]. Authors also did research to develop fog computing model [15, 25] to resolve security issues [16, 17, 24]. Image processing research [18] [19] considered optimization techniques. Some surveillance video stream processing systems [22] is making use of fog computing.

3 Problem Statement

There has been a lot of research in the field of cloud and fog computing. The limitation of existing study is that it is suffering from data security issues. More-over, the performance of the transmission mechanism is very slow. There is a need for a mechanism that splits the graphical content into two parts and transmits data from multiple routes. This saves the graphical content from external hackers. The data remains over the network for less time. Thus, the probability of congestion, packet dropping and delay gets reduced.

4 Proposed Work

The proposed work uses image processing mechanisms to split graphical content in a fog environment to minimize packet dropping and error rate. Such mechanisms split the graphical content into two parts and transmit the data from multiple routes. This saves the graphical content from external hackers. Moreover, the data remains

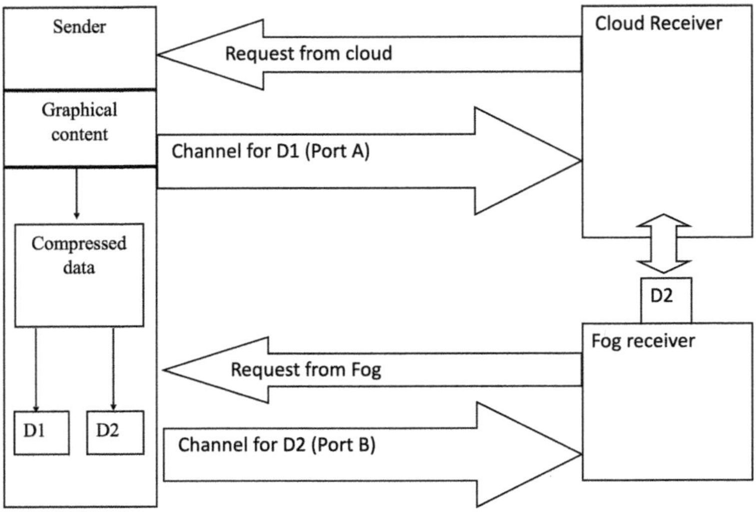

Fig. 1 Proposed architecture

over the network for less time. Thus, the probability of congestion, packet dropping and delay gets reduced.

4.1 Proposed Architecture

Figure 1 is expressing the Architecture of the proposed work which contains sender and receiver. The sender sends the image which is graphical content then compressed the data of images further divided into parts i.e. D1, and D2. Then it is sent to the cloud receiver to store the receiver which is sent to the receiver side.

4.2 Algorithm for Proposed Work

1. Prepare the graphical data (GD) on sender end.
2. Apply Huffman coding to compress data and set file as CGD.
3. Get the matrix from image M.
4. Divide the matrix by constant variable C and split the data in two parts.

   (a) CGDQ = M/C
   (b) CGDR = M

5. Initiate cloud receiver and open user defined port A for data transmission.
6. Transfer data CGDQ over transmission route of cloud.

7. Initiate Fog receiver and open user defined port B for data transmission.
8. Transfer data CGDR over transmission route of fog.
9. Transfer data D2 from fog to cloud.
10. Merge file D1 and D2.
    (a) DATA = (CGDQ X C) + CGDR.
11. Apply Huffman coding to decompress data.

4.3 Proposed Flow of Work

See Fig. 2.

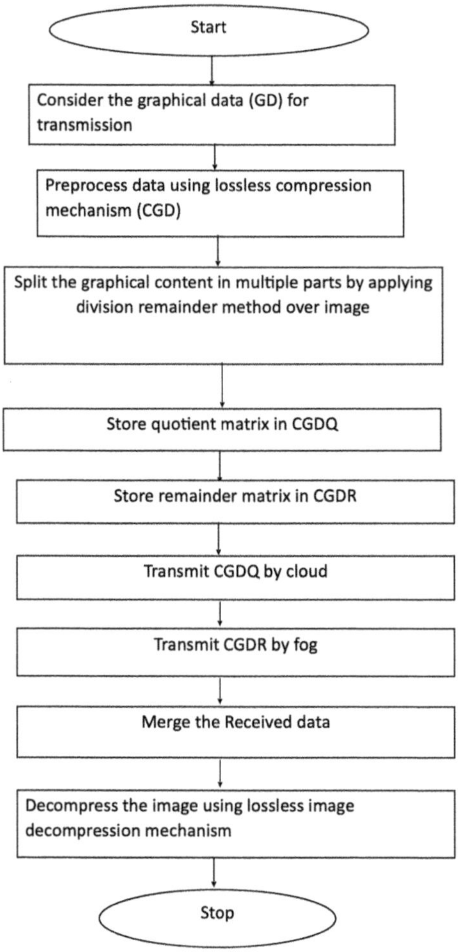

Fig. 2 Process flow of work

5 Result and Discussion

We made use of MATLAB for data compression and splitting. The parameters considered to prove that the proposed work is better than the previous approaches are packets affected by attacks by hackers, packet dropping rate, error rate, and queuing delay. The packet loss ratio represents the ratio of the number of lost packets to the total number of sent packets To calculate percentage error, use the formula: [(exact value - approximate value)/exact value] × 100.

5.1 Simulation for Packets Influenced by Hacking

This simulation considers packets influenced by hacking in three cases where uncompressed data, compressed data, and our proposed work are compared. In the proposed work content is transferred after splitting operation that has been performed after compression. Thus the probability of hacking gets reduced (Table 1 and Fig. 3).

Table 1 Comparative analysis of Packets influenced by Hacking

Number of packets	Uncompressed data	Compression based	Proposed work
10	1.173312474	0.806547107	0.59893776
20	2.793543613	1.83030653	1.370047943
30	3.768477228	2.485966891	1.500256446
40	4.273044898	2.967048489	2.408750124
50	5.318400758	3.27187175	2.52751599
60	6.730148972	3.406440629	2.63111973
70	7.284949174	4.425344937	3.159751928
80	8.779972882	4.8482998	3.275090373
90	9.039878396	5.489998769	3.478586449
100	10.72145724	6.335414512	4.233932941
110	11.72342072	6.382780087	4.151330499
120	12.67858044	6.942992882	4.979704597
130	13.56110506	7.498378541	5.207543843
140	14.55779409	7.770056278	5.091959753
150	15.34025039	8.044769754	5.917618675
160	16.47814536	8.419129791	6.22286867
170	17.79354333	9.101002706	6.383264453
180	18.50699621	9.934479618	6.446028661
190	19.19426529	10.04748329	6.694405286
200	20.62774982	10.90253226	7.221664814

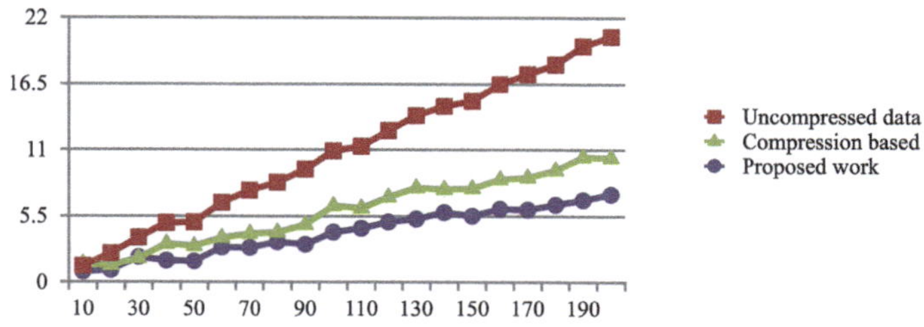

Fig. 3 Comparative analysis of Packets influenced by Hacking

5.2 Simulation for Packet Dropping

This simulation considers packet dropping in three cases where uncompressed data, compressed data, and the proposed work are compared. In the proposed work content is transferred after splitting operation that has been performed after compression. Thus the probability of Packet dropping gets reduced (Table 2 and Fig. 4).

Table 2 Simulation of packet dropping

Number of packets	Uncompressed data	Compression based	Proposed work
10	1.033817196	1.081959253	0.896273182
20	2.608528544	2.147665309	1.62595395
30	3.20063519	2.549410564	1.996401713
40	4.054626335	2.798195088	2.256365843
50	5.404800217	3.274254116	2.625915332
60	6.894327715	3.650673374	2.98848252
70	7.62480317	4.683637681	3.13203965
80	8.540093745	4.408205224	2.952773395
90	9.920578865	5.522209326	4.198567733
100	10.99037735	6.169405856	3.949982474
110	11.14433059	6.366193796	3.863958757
120	12.02044553	6.7193454	4.253962663
130	13.43020686	6.859717853	4.990743535
140	14.1636882	7.734425544	5.646806247
150	15.98052844	8.34042246	5.537544056
160	16.25933671	8.939606497	5.974520313
170	17.09620004	8.661286056	6.37892232
180	18.077429	9.85640769	6.587174437
190	19.4363189	10.5191278	7.315744214
200	20.82032888	11.20396448	7.357168181

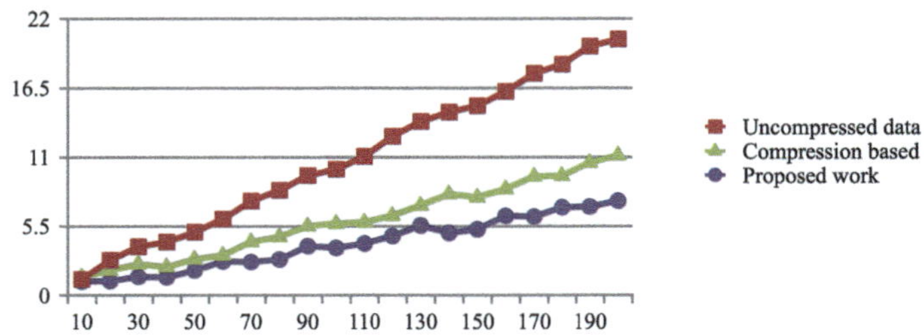

Fig. 4 Comparative analysis of Packet dropping

5.3 Simulation for Error Rate

This simulation considers the error rate in three cases where uncompressed data, compressed data, and the proposed work are compared. In the proposed work content is transferred after splitting operation that has been performed after compression. Thus the probability of error gets reduced (Table 3 and Fig. 5).

Table 3 Simulation for error rate

Number of packets	Uncompressed data	Compression based	Proposed work
10	11.62128091	5.997486514	3.888725043
20	22.03007513	11.44417425	7.976818617
30	33.08889086	17.42813253	11.38670513
40	44.49936533	22.83385329	14.83334934
50	55.60861225	28.44698717	18.67297408
60	66.45056496	33.68421246	23.05840358
70	77.9967752	39.86006714	26.44077119
80	88.4219916	44.62981207	30.13515247
90	99.27013903	50.09262965	33.30815853
100	110.943108	56.0342066	37.89025895
110	121.8578601	61.4423387	40.76067011
120	132.0125066	66.38944114	44.14669822
130	143.7374564	71.92939182	48.89192095
140	154.4370367	77.47405138	51.7104677
150	165.2409211	83.41260755	55.54737253
160	176.3686823	88.88026259	58.85388569
170	187.1886607	94.30554055	63.1553557
180	198.2104129	99.95250654	66.6166813
190	209.8446263	105.3223505	70.0590278
200	220.6636487	110.6321184	73.77857763

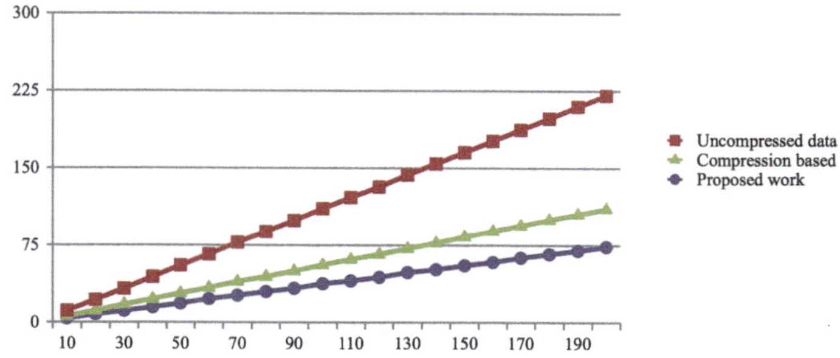

Fig. 5 Comparative analysis of Error rate

6 Conclusion

An important portion of computation (edge computing), storage, and communication will be carried out locally on edge devices and routed across the internet backbone as part of the proposed design. Data packets have been known to become corrupted due to the possibility of a physical signal fault. Therefore, it is suggested that studies be conducted on the image processing techniques that transmit graphical material across the fog environment. The suggested work has offered a viable way for effectively transmitting graphical material resulting in faster communication, better real-time data streaming and reduced costs.

7 Future Scope

As part of the planned architecture, a significant amount of computation (edge computing), storage, and communication will be performed locally on edge devices and routed across the internet backbone. Because of the likelihood of a physical signal defect, data packets have been known to get corrupted. Therefore, it's recommended that research be done on the image processing method used to send graphic data through a dense fog. The proposed study would provide a more suitable solution for efficiently delivering visual content via a cloud networking architecture in the future.

References

1. Abubaker N, Dervishi L, Ayday E (2017) Privacy-preserving fog computing paradigm. In: The 3rd IEEE workshop on security and privacy in cloud, vol 9, pp 502–509
2. Guan Y, Shao J, Wei G, Xie M (2018) Data security and privacy in fog computing. IEEE Netw 32(5):106–111

3. Firdhous M, Ghazali O, Hassan S (2014) Fog computing: will it be future of cloud computing. In: Third international conference on informatics and applications. ISBN 978-1-941968-00-0
4. Shenoy K, Bhokare P, Pai U (2015) Fog computing future of cloud computing. Int J Sci Res 4(6):55–56
5. Verma M, Bhardwaj N, Yadav AK (2016) An architecture for load balancing techniques for fog computing environment. Int J Inf Technol Comput Sci 6(2):269–274
6. Mukherjee M et al (2017) Security and privacy in fog computing: challenges. IEEE Access 5:19293–19304. https://doi.org/10.1109/IE.2014.54
7. Zao JK et al (2014) Augmented brain-computer interaction based on fog computing and linked data. In: International conference on intelligent environments. https://doi.org/10.1109/IE.2014.54
8. Dhiman P (2017) Fog computing shifting from cloud to fog. Int J Eng Sci Comput 7(3):4993–4995
9. Dhore TB, Chandankhede SS, Todsam A (2016) FOG computing. Int J Res Sci Eng 5:65–68
10. Dolci K, Datta SK (2017) Comparison of edge computing implementations: fog computing, cloudlet, and mobile edge computing. IEEE Global Internet Things Summit 5:1–6
11. Markovic D, Vujičić D, Djordjevic B, Ranđić S (2020) Application of image processing in fog computing. In: UNITECH 2020 international scientific conference, Gabrovo, Bulgaria vol 4, pp 1–16
12. Sharma A, Singh PK, Kumar Y (2020) An integrated fire detection system using IoT and image processing technique for smart cities. Sustain Cities Soc. 61:102332
13. Al-Sartawi A (ed) (2021) The big data-driven digital economy: artificial and computational intelligence. In: Kacprzyk J (ed) Studies in computational intelligence, vol 974. Springer International Publishing
14. Zhang Y, Wang P, Huang H, Zhu Y, Xiao D, Xiang Y (2020) Privacy-assured FogCS: chaotic compressive sensing for secure industrial big image data processing in fog computing. IEEE Trans Industr Inf 17(5):3401–3411
15. Zhu J, Hu J, Zhang M, Chen M, Bi S (2020) A fog computing model for implementing motion guide to visually impaired. Simul Model Pract Theor 101(102015). https://doi.org/10.1016/j.simpat.2019.102015
16. Stojmenovic I, Wen S, Huang X, Luan H (2015) An overview of fog computing and its security issues. Concurrency Comput Pract Exp 28(10):2991–3005
17. Puthal D, Mohanty SP, Bhavake SA, Morgan G, Ranjan GR (2019) Fog computing security challenges and future directions [energy and security]. IEEE Consum Electron Mag 8(3):92–96
18. Khanna K, Arora SM (2016) Ant Colony optimization towards image processing. Indian J Sci Technol 9(8):119–125
19. Khan MB, Lee XY, Nisar H, Ng CA, Yeap KH, Malik AS (2015) Digital image processing and analysis for activated sludge wastewater treatment. Sig Image Anal Biomed Life Sci 823:227–48
20. Dastjerdi AV, Gupta H, Calheiros RN, Ghosh SK, Buyya R (2016) Fog computing: principles, architectures, and applications. Internet of Things 10:61–75
21. Li L, Ota K, Dong M (2018) Deep learning for smart industry: efficient manufacture inspection system with fog computing. IEEE Trans Industr Inf 14(10):4665–4673
22. Karolak M, Razzaque A, Al-Sartawi A (2021) E-services and M-services Using IoT: an assessment of the Kingdom of Bahrain. In: Musleh Al-Sartawi AM, Razzaque A, Kamal MM (eds) Artificial intelligence systems and the internet of things in the digital era. EAMMIS 2021. Lecture notes in networks and systems, vol 239. Springer, Cham
23. Yi S, Hao Z, Qin Z, Li Q (2015) Fog computing: platform and applications. In: 2015 Third IEEE workshop on hot topics in web systems and technologies, Hotweb, vol 3, pp 73–78
24. Hu P, Ning H, Qiu T, Song H, Wang Y, Yao X (2017) Security and privacy preservation scheme of face identification and resolution framework using fog computing in the internet of things. IEEE Internet Things J 4(5):1143–1155

25. Hu P, Dhelim S, Ning H, Qiu T (2017) Survey on fog computing: architecture, key technologies, applications, and open issues. J Netw Comput Appl 98(2):27–42
26. Musleh Al-Sartawi AMA (ed) (2022) Artificial intelligence for sustainable finance and sustainable technology. In: ICGER 2021. Lecture notes in networks and systems, vol 423. Springer, Cham

Meta-analysis of AI Research in Journalism: Challenges, Opportunities and Future Research Agenda for Arab Journalism

Maha Abdulmajeed and Nagwa Fahmy

Abstract The use of artificial intelligence tools in newsrooms is revolutionary and controversial as well. Despite the promising opportunities provided by AI to enhance digital journalism practice, it also raises several legal, professional, and ethical considerations. Research about AI in Arab media is a promising area of interest that increasingly attracts Arab scholars. However, there is a need for systematic and purposive growth in future research about AI and Arab media, that considers the socio-cultural and economic contexts of Arab countries and meets the priorities and needs of Arab media organizations. Accordingly, this paper provides researchers with an overview of the main challenges and debates in the field of AI and journalism studies. This study applies a systematic review of a sample of English and Arabic-written studies from 2014 to 2022 about the implications, challenges and considerations of using AI in newsrooms. Based on the analysis, the study proposed a future research agenda about AI and Arab journalism.

Keywords Robot journalism · Automated journalism · Meta-analysis · Arab journalism · Newsroom automation

1 Introduction

Artificial intelligence (AI) as a range of technologies including machine learning, automation and data processing is a significant part of journalism already but it is unevenly distributed [7]. During the last two decades, using artificial intelligence tools increased in different newsroom worldwide including Arabic speaking news media outlets, while different newsroom started using automated artificial intelligence tools to help in editing process.

M. Abdulmajeed (✉)
College of Mass Communication, Ajman University, Ajman, UAE
e-mail: M.attia@ajman.ac.ae

N. Fahmy
College of Communication and Media Sciences, Zayed University, Abu Dhabi, UAE
e-mail: Nagwa.Fahmy@zu.ac.ae

A. M. A. Musleh Al-Sartawi et al. (eds.), *From the Internet of Things to the Internet of Ideas: The Role of Artificial Intelligence*, Lecture Notes in Networks and Systems 557,
https://doi.org/10.1007/978-3-031-17746-0_18

An existing language-based divide had bounded Arabic speaking news media outlets from adopting such AI tools and limited their implementation especially in newsroom routine work [4, 23].

AI innovation paved the way for new types of news stories. The widespread adoption of AI in newsrooms made it an important topic for research. An increased number of studies were conducted to get deep insightful perspective of AI uses in different journalistic practice [12].

Nowadays computation and digital innovative tools support journalists by helping them generating news at higher speed using Big Data. From another point of view, news generated by algorithms will not be as quality as human-written news because algorithms do not have creativity, values, insights, and emotions [11].

Although Artificial intelligence was a result of several discipline efforts from engineering to computer sciences and linguistics, journalism practice benefit from this development in several ways [34]. Recently, AI played a curial role to enhance journalists' skills and abilities. They can use AI with all types of data [21].

The increasing usage of AI applications in the field of journalism represents a shift in the journalism industry, which reflects an engagement between social and technological factors.

At the same time, using AI in journalism also increasingly attracts researchers' attention. Scholars widely examined AI applications in the field of media practice under different research areas like the algorithmic decision-making in respect with the algometric distribution of media content [26], the computational journalism that focuses on how the advanced technologies support journalists' daily work [9], and the newsroom automatization discipline that examines the usage of smart applications in news industry like platforms of distributing news on social media and data mining techniques, etc. [13].

Like data journalism and other innovative media practices, Arab media organizations' usage of AI applications is still in its early stages [17]. In parallel, there is rapid growth of research about AI in Arab journalism [16]. The sociocultural, economic, and political contexts in Arab countries highlight serious challenges, problems, and considerations of the usage of advanced technologies in media. Bearing in mind the very limited analytical reviews available about AI and journalism in Arab research, there is a need for systematic research to enhance future research outcomes for Arab journalism. Accordingly, this study maps and analyzes the scientific production about challenges and opportunities of AI in media practice. It provides researchers with a review of debates that are occurring in AI and journalism research and suggests a roadmap for future Arab journalism research.

2 Literature Review

Nowadays Artificial intelligence gained popularity as it interacts with different aspects of daily life. Integrating AI in news gathering allows using several features such as machine learning, computer vision, speech recognition to convert voice data into text data, natural language processing (NLP) to understand and respond to text or voice data, and robotics that refer to enabling computers and systems to perform regular human tasks [31].

Computational Journalism is a broad term that deals with journalism practices that use AI techniques and tools to dig into data and enable journalists to analyze data; identify patterns, trends, and insights from multiple sources, it can also track audience sentiments [25].

Computational journalism includes many detailed concepts such as automated journalism that refer to software that transform data into readable news stories [25].

Developing AI tools during the last decade paved the way and helped journalists in different countries to produce news stories. Stray [33] mentions that it can help to tell new types of stories that lead to innovative practice of Journalism. Fully automated and semi-automated forms of gathering, filtering, composing, and sharing news occupied a greater place in a growing number of newsrooms opening discussions about shifts in the norms, patterns, and routines of news production were happening, and at a more fundamental level, taken-for granted ideas about by who (or what) journalism was being challenged [34].

Research on journalism and computation as artificial intelligence – and the human–machine communication that it facilitates – becomes a more salient factor in the way people make sense of the world and create meaning, both with each other and in relation to machines [34].In addition, analyzing AI journalism initiatives and practices helped to understand AI journalism potential and limitations.

Implementing AI tools in newsrooms refers to a wide journalism usage such as finding and contacting sources, adjusting objects faces or colors, converting speech into text, editing audio and video, identify sentiment of readers in different platforms. Those tasks usually took time and efforts from journalists. Research predicts that AI tools will not substitute journalists, these technologies are more likely to complement human work by increasing its quality and efficiency [13].

The production of Automated news stories without human intervention was known as Automated Journalism and Robot Journalism. AI implementation will not eliminate jobs in the media industry but increase speed and efficiency of reporters and video producers through different tools that allowed machines to write different versions of one story and create videos of related footage out of keywords [8].

Digitization of news content production allow it spread in many digital platforms. it also allows personalization of news distribution to reach the target audience and open a variety of opportunities for collecting and analyzing large amounts of audience and consumption data [35]. Recently, Automation in journalism has increased reader revenue for publishers, lowered costs and it has rapidly expanded coverage in any topic on which automated content can be produced [3].

3 Research Methodology

The focus of this paper is to analysis the contributions of previous research about the use of AI applications in journalism to identify the main areas of interest, and then to propose a research agenda about AJ in the Arab world.

This paper applied a thematic and quantitative descriptive analysis to analyze the key features of a selected sample of academic contributions to provide researchers with a good understanding of the main debates about the potentials and challenges dominating in this growing research area, and to present suggestions for future Arab research about automated journalism.

3.1 Our Research Questions Are

1. What are the main trends dominant in the scholarly research on automated journalism recently, including which areas are studied most and using which methodological approaches, and what are the gaps present in this research?
2. What are areas of research interests should be covered by Arab researchers on automated journalism in Arab countries?

3.2 Search Procedures

The selection of the texts based on the existence of specific keywords ("algorithmic journalism", "automated journalism", "artificial intelligence" and "journalism", "big data" and "journalism") whether in the titles and/or abstracts or full text of manuscripts, through four scholar databases which are Google Scholar, Scopus, Research Gate and Dar Almandouma. The sample includes peer-reviewed theoretical and empirical journal articles, book chapters, reports, conference proceedings that were published in English and Arabic. Overall, 35 texts were examined, covering a period from 2014 to January 2022. Figure 1 shows the increasing interest in research over time. The sample consists of (31) articles published in peer-review journals, (2) reports, (1) conference proceedings, and (1) book chapter.

After a thorough reading of the collected articles, we applied a quantitative and qualitative descriptive analysis to manually extract information about the following categories: Basic data: date of the publication, number of authors, title of the article, its language (English/Arabic), the keywords, type and scope of the study and the discipline(s), type of publication, main focus and secondary focuses (if applicable), the methodologies and research techniques, theoretical framework, the most important results, and limitations of the study (if applicable).

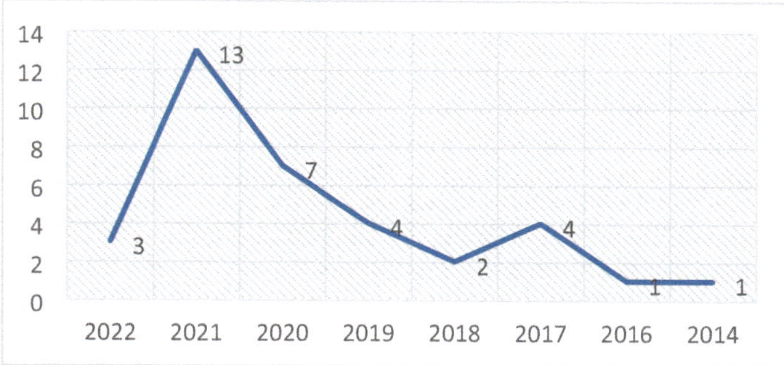

Fig. 1 Distribution of the sample from 2014–January 2022

3.3 Sample Characteristics

Co-authored scholar research production represented 55% compared to 49% individually authored. English published texts constitute 73% of the sample, while Arabic texts constitute 31%. Total (10) texts of the sample cover either geographical international scope or not specific, (24) texts of the sample cover countries including: Egypt, Nigeria, China, South of Africa, Argentina, Spain, and UK, while only (2) are of regional scope; Western Europe.

Review showed that (21) texts of the sample belong to "Journalism and mass communication" discipline, while (14) texts are interdisciplinary, including (6) belonging to Sociology and (4) belonging to Computer Sciences (Fig. 2). This reflects that research about the challenges and consideration of using AI in journalism can better go for more interdisciplinary research.

The dominance of the qualitative method in the sample can be partially explained by the explanatory nature of most of the texts, which is understandable considering early stages of research in this field. Total (18) of the sample applied qualitative

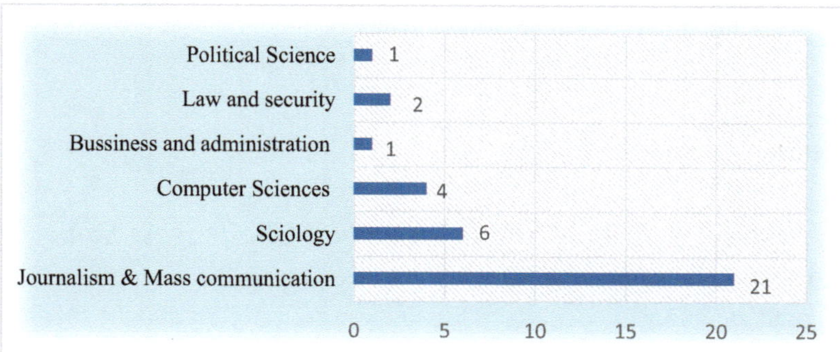

Fig. 2 Distribution of the sample according to the discipline

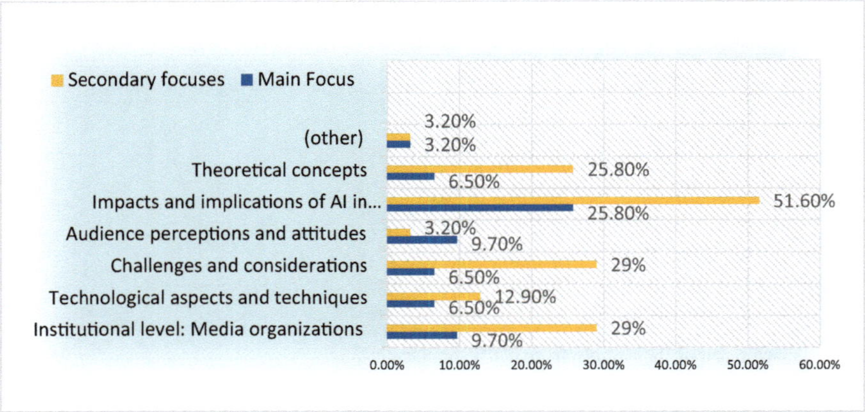

Fig. 3 Main and secondary focuses of the research sample

method, (8) applied mixed-method, and (11) applied quantitative methods, most of them are Arabic studies.

Consistently, in-depth interviews were the most qualitative method applied; (12) texts. used in (13) texts, surveys were dominant in quantitative studies, and were mainly applied to collect data from journalists or media students, especially in Arabic studies. Case studies represent media organizations that applied AI.

Impacts and implications of using AI applications in journalism topped as main focus in 26% of the sample, besides, even if the studies focus mainly on another topic, the impacts and implications comes first as secondary focus of 52%. Notably, challenges and considerations related to using AI in journalism comes as a main focus in just 7% of the sample and comes as a secondary focus in 29% of the sample. (Fig. 3).

The analysis of most-frequent keywords showed that English-written texts used more varied keywords than Arab-written texts, (Fig. 4). This result indicates the need to achieve better understanding of differences between the terms and how these differences could reflect different types of using AI in journalism [12].

4 Results

4.1 Linguistics, Ethical and Professional Challenges: Some Arise, Others Decrease

One of the biggest challenges for Automated journalism in the Arab region is the early stage of development of Arabic Natural Language Processing (ANLP). This is an application of AI and machine learning used to understand Modern Standard Arabic and dialects. Research in this linguistics field demonstrates that Arabic language

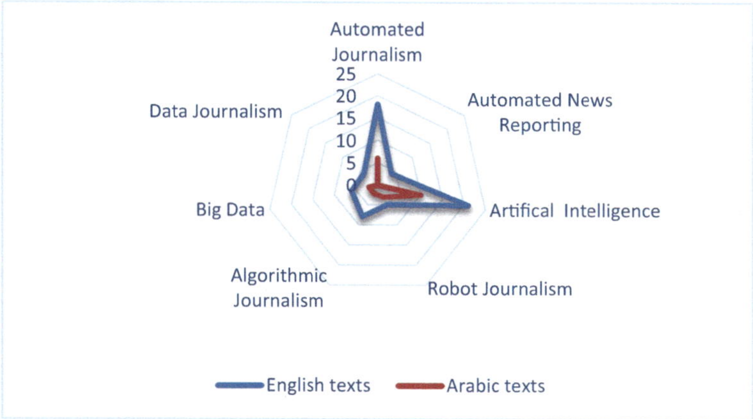

Fig. 4 Most-frequent keywords used in the sample

differs from other languages due to its complex and ambiguous structure that the computational system has to deal with at each linguistic level. Despite the increasing research in this field more work is needed to effectively use it in Arabic Automated news production.

Research about using AI tools in journalism reflected researchers' perspectives about several ethical and professional challenges that should be carefully handled as long as increasing number of media organizations adopt using AI tools. Using algorithms in journalism can help overcome some ethical challenges that traditional media faced. It helps unburden journalists from daily extensive routine work, giving them an opportunity to focus on in-depth analysis and reporting, and enables journalists to consider moral requirements in their work like checking multiple news sources, giving more attention to human dignity and respect privacy, etc. [14]. However, previous research about AI in journalism highlighted several ethical and professional challenges that emphasize the importance of the role of human journalists and the need for advanced skills in modern newsrooms.

In this regard, the meta-analysis, conducted by Graefe & Bohlken [18] on readers' preferences of human-written versus automated news, revealed that human-written news stories were rated better than automated news with respect to quality. However, they pointed to a possible ethical challenge related to news organizations and how they could deal with readers' disapproval of automated news. Moreover, there is another concern in this regard as news organizations are more interested to adopt AI to avoid costs of human journalists' resources [8].

In his article, Ali [4] concluded several professional challenges like undermining creativity and the absence of monitoring and bias, in addition to ethical challenges like missing transparency, insufficient fact-checking and data manipulation. Regardless of the technological development and the advanced tools used to collect and process data, journalism is still about critical thinking that answers the question "why" and help to make people better aware of events and important public issues. Biswal and

Gouda [8] highlighted professional concerns about data AI-driven journalism that could lack creativity, humor, and critical thinking which will remain forever in the field of journalism. Another perspective about ethical considerations of using AI in media was provided by Dörr and Hollnbuchner [14] who discussed concepts of AJ ethics and differentiated between new media ethics, digital media ethics, cyberethics, and traditional media ethics. According to them, algorithmic journalism's ethical implications are in the overlapping fields of digital media and cyberethics.

Research emphasized the danger of AI bias, and the hard work human journalists should do to highlight how they used AI to help their reporting or production [21]. The analysis revealed different views in this respect. While Thurman et al. [34] supported the idea of "mechanical objectivity," or the belief that Algorithms, Automation, and News machine systems are capable of offering representations and outputs "that overcome the limits of human subjectivity, other researchers urged that algorithms may generate unbalanced news due to biased data obtained from biased sources [15]. Dalgali and Crowston [11]pointed that data inputs could be created to manipulate readers' minds and attitudes towards current events according to the data providers' purposes, while readers perceive these manipulated stories as objective, since they are automatically generated.

In the same context, Monti [27]urged that the accountability of the automated news transmitted becomes more important in the political field if it reflects the public debate. Readers should be clearly informed of how AI tools were used to collect data, their publicity, perform an analysis, and whether the stories are automatically generated or human-written [21]. If not, this is an ethical concern related to transparency. The question is if ethical rules are violated, errors occur, or collected data are misused, who will be responsible for such violations? [11]

Obviously, the use of AI and big data provides opportunities to enhance reader engagement and content personalization and can contribute to the improvement of professional practice [10]. However, according to Hansen et al. [21] there is a challenge of achieving "a balance between personalized content focusing on individual interests and remaining committed to journalism's public service mission the public".

Another concern arises due to the increasing role of the 'Big Tech' technological competitors, like Facebook and Google, in the news industry and their control of research and product development, urging a need for transparency, dialogue, and support for journalism from technology companies [7].

According to Santos and Ceron [31], most of the applications that use AI in the news industry rely on grants from big tech companies such as Google and Facebook, which rises serious concerns to the development of technological innovations in news media since "these 'Big Tech' decide who receives their money, when they receive it, and where it goes". Research also revealed other concerns as technological developments have no solutions to legal problems emanated from algorithm-generated content about private citizens [28]. The communication and knowledge gaps between journalists who use AI tools and programmers who design them, probably leads to journalistic malpractice [2, 21].

4.2 Future Agenda Research of AI in Arab Countries

Research on using artificial intelligence tools in journalism is still in its infancy. However, as this phenomenon has already changed many professional practices in journalism, researchers from different countries like United States and Western Europe [12] provided some interesting academic work. Meanwhile, reviewing previous studies about AI and journalism revealed that only few of them suggested future research areas based on their findings.

Accordingly, this study aims to suggest future research agenda based on the thematic analysis of the available academic production. This will help Arab researchers to fill the gap and focus on research areas that meet Arab media organizations' needs to develop their usage of AI, and to handle the possible ethical and professional challenges and considerations resulted from using AI in newsrooms.

Theoretical and Methodological Considerations. Although Arab research of AI applications used by news organizations is still in its early stages, review of the Arabic sample reflected the emphasize on empirical research without providing sufficient theoretical contributions to study the new phenomena in the socioeconomic, cultural, and political contexts in Arab countries. This is not a new concern in Arab media research, aaccording to Ayish [6], the introduction of mass media studies in Arab countries was marked by strong Western influences [6], limiting the scope of several empirical studies to specific countries [22] or languages, preventing generalization of results. Arab studies about AI and journalism either repeatedly applied the same theoretical approach "Technology Acceptance Model" or the "Diffusion of Innovations" [1, 2, 5, 15]. We suggest future research to work on developing a theoretical approach that help to study and interpret Arab journalism in a comprehensive context.

The analysis also highlighted the ambiguity of definitions between Automated journalism, big-data journalism, Robot journalism, AI journalism and algorithm journalism, as those terms were used interchangeably due to the vague boundaries between them. There are different types starting with Computer Assisted Reporting (CAR) where journalists use technology to access information, store it and process it, use data bases and algorithms in data analysis and processing, then produce a fully automated content. Some definitions focus on the automated content production as most important aspect of algorithmic journalism, while neglecting other applications of algorithmic technologies in the news production field [24].

Other definitions described the wider scope of applying algorithms in journalism, referring to the integration of algorithms technology, data, and knowledge from the social sciences to complement the accountability function of journalism [20, 29]. Although broad definitions include the various technological applications used in journalism and emphasize the contribution of human workers in these procedures, such definitions should be avoided, as they do not help to get better understanding of the exact focus of the subject matter. Accordingly, we suggest that future research should differentiate between fields of using technology in journalistic professional practices.

Audience Research. In their meta-analysis research, Graefe and Bohlken [18] showed notable findings of users' preferences to automated compared to human-written news, while the direction of effects differed depending on the type of evidence of the study. Studies also concluded that we still do not know what factors make users perceive automated and human-written news to be more or less credible or readable [8, 22, 24]. It is still not clear why automated news tend to be perceived as more credible but less readable than human-written news [19]. Hence, future research should focus on examining audiences' perception of AI-produced news stories.

On another side, Arab studies about AI applications usage in journalism [2, 15, 28] haven't yet considered readers' perceptions and preferences of automated and human-written journalism. Alternatively, they focused on examining journalists' attitudes towards using AI tools and what impacts resulted of this practice in newsrooms. This highlights the need to thoroughly study audience and readers' perceptions to automated news and content, and what factors affect their perceptions of quality and credibility of such content.

How to Be Ready? Study of Challenges and Considerations. Journalists' responsibilities and concerns about using AI tools in their professional practices and routines are one of the top research areas raised by several researchers [14, 34]. Hence, there is a need for studies that focus on discussion and transparency between journalistic and non-journalistic actors about moral values and ethical guidelines, what factors help to achieve better understanding between journalists and coders, how algorithms affect journalism, and how to develop best practices for that technology. The "lack of journalists' AI-related and digital journalism education and training" is a serious concern reported by Santos and Salaverría [32].

Few studies considered the relation between news organizations (practice level), academic outputs (scholarly level), and media colleges and schools (educational level), when discussing the current situation of using AI tools in journalism. Beckett highlighted the importance of including AI in media education, and the need for more "strategic cooperation with universities, researchers in addition to technologies companies and programmers" [7].

Few Arab studies pointed to the need for developing communication curriculum that provides students with the required skills to use AI in media practice [1, 2]. Accordingly, we suggest opening the door for more research on future skills needed in media and journalism labor market for using AI technologies, providing a roadmap for developing media courses and curricula, and mid-career journalism training on using AI tools.

5 Conclusion

Although journalists face many problems in most Arab countries to access information and obtain accurate and updated databases available as open source [17], we are talking about AI technology as a new actor in the Arab newsrooms. While journalists still fight to get access to data, self-training on using open-source technologies, and practicing digital storytelling, introducing AI in Arab newsrooms at this stage with no adequate preparation, raises several concerns and challenges. This critical situation mandates a clear agenda for academic research in this early phase.

Studies discussed AI in the field of journalism through various perspectives; technical (media as professional practices, values, and norms), ethical and legal perspectives, and social perspective. Other perspectives include media institutions, journalists, and users' preferences.

Many of the reviewed sample did not follow a theoretical framework, few of them focused on the conceptual basis of the new phenomenon. However, examining AI in a wider context requires considering micro and macro factors, and analyzing institutional, professional, and socioeconomic factors affecting the usage of AI applications in Arab media.

Although studies pointed to the importance of AI and algorithmic technologies in making news more objective through working effectively in data collecting and data mining, however, journalists still hold the responsibility for selecting which stories to publish or to neglect. A controversial aspect about responsibility of journalists in comparison to robots is programmers; one of the most highlighted themes in the literature review.

There are many aspects that previous research highlighted as priorities in future research including how the performance of human-written and automated news evolves over time [19]. Future researchers should investigate the new professional roles in newsrooms due to the influence of AI [30] and the role of 'Big Tech' companies in enhancing or limiting innovations in newsrooms [32].

We suggest future research should identify which journalism fields AI can improve and produce innovations into with lower risk possible. Also, research should analyze challenges and best innovative practices of using AI in Arab journalism field to promote and encourage using AI as a work routine.

Developing curriculum to introduce AI skills in Arab universities can support future generations of journalists with AI skills and tools. Meanwhile, organizing training sessions for AI practice in Journalism will improve its usage in newsrooms.

Scholars should also consider developing new models for traditional theoretical approaches to study the impact of using AI in journalism on their audiences.

Users' preferences, behaviors, and attitudes towards automated vs. human-written news content are still uncovered research area, especially in Arab research. This could be interpreted by the explanatory phase of AI in journalism. This primary stage allows researchers to conduct empirical studies to examine user's perception of credibility, quality, and interest in automated content. Additionally, there are no scholar contributions to educating media skills and requirements to upskill media students in the AI era. This is a research gap in the Arabic academic production about AI and Journalism. Finally, despite the limitations of this study it can help to crystalize the agenda for future research about AI and Arab journalism.

References

1. Abdel-Hamid A (2020) Egypt and UAE mass communication students' acceptance of using artificial intelligence applications and its impact on their future careers: a study based on the technology acceptance model. Egypt J Public Opin Res 19(2):341–409
2. Abdel-Latif M (2021) Adoption of Egyptian press institutions to artificial intelligence technology in producing and editing news and journalistic topics: experiences of some foreign newspapers. Egypt J Public Opin Res 20(3):1–68
3. Al-Attar S (2022) Automated Journalism. Campaign 50
4. Ali W, Hassoun M (2019) Artificial intelligence and automated journalism: contemporary challenges and new opportunities. Int J Media J Mass Commun 5(1):40–49
5. Aram A (2021) The future of journalists in the age of artificial intelligence (robot journalism model). J Media Res 4(58):1673–1702
6. Ayish M (1998) Communication research in the Arab world: a new perspective. Javnost Public 5(1):33–57
7. Beckett C (2019) New powers, new responsibilities, a global survey of journalism and artificial intelligence. London School of Economies and Political Sciences, London
8. Biswal SK, Gouda NK (2020) Artificial intelligence in journalism: a boon or bane? In: Optimization in machine learning and applications. Springe, pp 155–167
9. Bucher T (2017) Machines don't have instincts': articulating the computational in journalism. New Media Soc 19(6):918–933
10. Calvo ST, Vila P (2021) Exo journalism: a conceptual approach to a hybrid formula between journalism and artificial intelligence. J Media 2(4):830–840
11. Dalgali A, Crowston K (2020) Algorithmic journalism and its impacts on work. In: Computation + journalism symposium. The national science foundation
12. Danzon-Chambaud S (2021) A systematic review of automated journalism scholarship: guidelines and suggestions for future research. Open Res Europe 1(4):1–19
13. Diakopoulos N (2019) Algorithmic accountability. digital. Journalism 3(3):398–415
14. Dörr KN, Hollnbuchner K (2017) Ethical challenges of algorithmic journalism. Digit J 5(4):404–419
15. El-Khuly S (2020) Tendencies of Egyptian journalists towards employing artificial intelligence in developing journalistic content of informational richness: a field study. Egypt J Media Res 72:101–173
16. Elgayyar S (2021) Recent trends of the researches about the impact of artificial intelligence on the media product. J Public Relat Res 9(35):10–77
17. Fahmy N, Attia MA-M (2021) A field study of Arab data journalism practices in the digital era. J Pract 15(2):170–191
18. Graefe A, Bohlken N (2020) Automated journalism: a meta-analysis of readers' perceptions of human-written in comparison to automated news. Media Commun 8(3):50–59
19. Haim M, Graefe A (2017) Automated news: better than expected? Digit J 5(8):1044–1059
20. Hamilton JT, Turner F (2009) Accountability through algorithm: developing the field of computational journalism. report from developing the field of computational journalism. In: Center advanced study in the behavioral sciences summer workshop
21. Hansen M, Roca-Sales M, Keegan J, King G (2017) Artificial intelligence: practice and implications for journalism. Tow Center for Digital Journalism, Columbia University, 1–21
22. Jia C (2020) Chinese automated journalism: a comparison between expectations and perceived quality. Int J Commun 14:2611–2632
23. Jusoh S (2018) A study on NLP applications and ambiguity. J Theor Appl Inf Technol 96(6):1486–1499
24. Kotenidis E, Veglis A (2021) Algorithmic journalism—current applications and future perspectives. J Media 2(2):244–257
25. Lewis SC, Guzman AL, Schmidt TR (2019) Automation, journalism, and human-machine communication: rethinking roles and relationships of humans and machines in news. Digit J 7(4):409–427

26. Matt C (2018) Automating judgment? Algorithmic judgment, news knowledge, and journalistic professionalism. New Media Soc 20(5):1755–1772
27. Monti M (2018) Automated journalism and freedom of information: ethical and juridical problems related to AI in the press field. Opinio Juris Comparatione 1(1):1–17
28. Mousa E, Abdulfattah A (2020) Attitudes of journalists and leaders towards employing artificial intelligence technologies in newsrooms of the Egyptian newspapers. J Public Opin Res 19(1):1–66
29. Musleh Al-Sartawi AMA (eds) (2022) Artificial intelligence for sustainable finance and sustainable technology. In: ICGER 2021. Lecture notes in networks and systems, vol 423. Springer, Cham
30. Parratt-Fernández S, Mayoral-Sánchez J, Mera-Fernández M (2021) The application of artificial intelligence to journalism: an analysis of academic production. Profesional de la Información 30(3):1–12
31. Santos MFD, Ceron W (2022) Artificial intelligence in news media: current perceptions and future outlook. J Media 3:13–26
32. Santos MFD, Salaverría R (2021) From data journalism to artificial intelligence: challenges faced by La Nación in implementing computer vision in news reporting. Palabra Clave 24(3)
33. Stray J (2019) Making artificial intelligence work for investigative journalism. Digit J 7(8):1076–1097
34. Thurman N, Lewis SC, Kunert J (2019) Algorithms, automation, and news. Digit J 7(8):980–992
35. Trattner C, Jannach D, Motta E et al (2021) Responsible media technology and AI: challenges and research directions. AI Ethics

YouTube: A Study from the Perspective of Digital Media

Nader Mohammed Sediq Abdulkhaleq

Abstract The study identifies the nature YouTube as media tool. The study also showed that YouTube at the global level, and the human level, and what this very important site offers, in the human life of man in general, and indicated that the terrible diversity in the number of videos makes the turnout also terrible, as it is from all languages and in all ways. From this standpoint, YouTube has positioned itself among the best websites throughout history thanks to its integrated and always effective system. Which is based on the idea of permanent development, as YouTube from the hour of its inception until now has evolved in everything, even in the performance of research and the most accurate suggestions and all these details that made it at the top of the interests of anyone in the world.

Keywords Social sites · YouTube · Media

1 Introduction

In 2017, the number of Internet users worldwide became about 3.7 billion users, representing 59% of the world's population. The personal and social interaction with friends and co-workers has shifted from being a realistic interaction and in a real, tangible community, to a large-scale electronic entity, due to Because of the rapid growth of social media such as "Facebook, Myspace, YouTube, and Twitter," Internet users are now able to express themselves through various forms of social media, such as: exchanging videos and comments, following videos and comments of others, The recipient has now become an integral and even essential element in the process of exchanging media messages, as he sometimes exchanges roles with the communicator [17–19]. Internet users have adopted the exchange of video clips quickly, for the purposes of entertainment, news and education, and modern technology, such as mobile phone cameras, computer cameras, modern cameras and many transmitting devices, has allowed individuals to capture any information in

N. M. S. Abdulkhaleq (✉)
Manama, Bahrain
e-mail: nadersiddiq@gmail.com

A. M. A. Musleh Al-Sartawi et al. (eds.), *From the Internet of Things to the Internet of Ideas: The Role of Artificial Intelligence*, Lecture Notes in Networks and Systems 557,
https://doi.org/10.1007/978-3-031-17746-0_19

a visual form and broadcast it on the Internet at the same time [22, 27]. The year 2017 witnessed a remarkable growth of electronic videos, according to the agency "ComScore", the number of users who watched videos on the Internet reached 180 million users, and that is in the United States only, and electronic video needs more research and study due to its importance In contemporary society[15, 24]., "User-Generated videos" are also known as "Video Blogging" and have become widely known by millions of Internet users, and "Video Blogs" which are also known as "Vlogs" which is a phrase Vlogs are a type of electronic publishing that allows anyone - who has a means to connect to the Internet, a computer and a camera, even if it is a mobile phone camera - to "produce and publish their video clip on the Internet" [7].The use of YouTube has increased since its inception in 2005, and in 2016 YouTube became the first social network as chosen by the American "Time" magazine for it. Media and modern mass communication, YouTube may include videos posted by anyone in society, they can be citizens, human rights groups, soldiers, or even terrorists, some of which may relate to an incident with political dimensions or to document important issues such as global warming and illegal immigration and corruption, and therefore the focus comes to YouTube due to its wide popularity and the huge increase in the demand of Internet users on it [17]. YouTube is more than an alternative to traditional television, as it is making significant changes in the way visual content is published [21]. YouTube has become a phenomenon in every sense. Yes, it is the phenomenon of YouTube that spreads very quickly, where the majority of its visual content is produced by independent individuals in the community and in their own way, and then the community must appreciate the existence of video sharing sites, such as: YouTube [9].

YouTube is one of the most famous websites on the Internet, and its address is www.youtube.com. The idea of the site is to attach any digital files consisting of video clips on the Internet without any financial cost, and also allows the possibility of downloading from it a huge number of videos, and there are many Its subscribers are great and millions visit it daily [7]. Users who subscribe to it can add an infinite number of their visual content on the accounts designated for them, and it is worth noting that the personal accounts registered with YouTube are called the "Channel", as soon as the user registers on the site, he can attach any number of these files to be seen by millions of people around the world. Viewers can also manage a group dialogue or "Discussion" discussion about the video [5], by adding accompanying comments [6], as well as evaluating the "Rate" of the video file by giving it a relative value of five degrees to express how much site users like the video. Since its inception in 2005, social networking has become the most famous entity in the world of electronic video "Online Video" and quickly entered the list of the ten most popular websites in the world [18]. YouTube content generally comes from two main sources [13], first: programs previously produced by traditional media, and second: videos produced by individuals [8]. To be more accurate, the content available within YouTube varies between three types; The first is selected video clips from films or programs produced by traditional media [11]. The second type is the so-called "Mash-Ups", which is content collected from more than one source to produce an integrated video clip that presents an idea, and the last type is It is User-Generated Content, or "user-made" content and it is produced by the user himself [1]. The quality of this type of wonderful varies from weak or empty of meaning, and the poor technical or intellectual quality is due in some sections of this type due to its implementation by amateur individuals [11].

YouTube is one of the famous social sites, which in a short period of time managed to gain an advanced position within the social networking sites, especially in its distinguished role in the recent events that took place and took place in different parts of the world, such as: natural disasters, mass uprisings and popular revolutions. Media by displaying videos that the networks of its correspondents have not been able to obtain, and YouTube is one of the best services that the Internet has provided so far [12]. YouTube started to appear on the Internet on February 15, 2005, by three employees of PayPal: "Hurley Chad" Chad Hurley, "Steve Chen" Steve Chen, and "jawed karim" Jawed Karim. The owner of the first video uploaded to YouTube, and he filmed it for himself, in fact, animals [4]. The idea of creating YouTube arose in San Bruno, California, in the United States of America, when the three friends were at a party and took videos, and they wanted to spread them among their colleagues, and they could not send them via e-mail because it did not accept large files, and from here it began to crystallize The idea of a site to attach movies and videos on the Internet, and the domain "Domain" or the address of the site was registered on February 15, 2005 [20].

Fifteen months after the creation of YouTube in 2005, the site received 100 million videos per day, which made it a leading and responsible source for electronic videos on the Internet until 2016. In October 2016, the global company Google made a deal to buy YouTube for an estimated value with $1.65 billion, it is the largest deal made by Google in its history, and Google has included this site in it to be at the fore in the digital and electronic video revolution. video electronically. At the end of 2006 [14], the American "Time" magazine chose YouTube to win the title of Person of the Year, and on the cover a computer screen image like a mirror that reflects the image of the person who sees it, and attached to it a subtitle of the news saying: "Yes, you.. you control the information age, so welcome to your world" [19]. And about the major developments that YouTube has witnessed since its launch, the first of which was in May 2006, when the site provided a comment service on videos, and in August 2006, when the first form of advertisements was launched as a service provided to advertisers on YouTube, two years later, specifically, in June 2008, YouTube was used for the first time as an installed application. Built-In-Application by Sony [17].

2 YouTube Community

The YouTube community consists of all categories of the real community, there are ordinary users who represent citizens, and there are also major companies for local and international artistic production, and there are also channels for public figures in the community. Private, and the other: those who have not registered but have the ability to use the site and offer the service of subscription and registration in YouTube for free, but those registered with YouTube must provide their e-mail address, and those registered on the site can add an unlimited number of videos on YouTube for an unlimited period of time. Limited, users can watch videos and share them on their accounts on other social networks, and search for videos only, while the possibilities and services available to registered users on YouTube are many [19], where the user can in this case watch the clips and search for them in the engine Search the site, exchange clips, "Share" and create a private channel for it, "Channel" that enables it to "Upload" an infinite number of files. The video cutter, and they can also rate those clips and "Comment" on them, and even create a "discussion" episode around a specific clip, and finally they have the right to create a special "Playlist" consisting of several clips on a specific topic or issue, and the difference between the YouTube user appears. The one who has an account on the site, and the other who does not have an account, from the first moment of entering the site [19], where the user with the account finds the videos on the home page that belong to the private channels that the user has already subscribed to, and some clips that YouTube suggests watching, as they belong to Private channels from which the user has watched videos before, with a list of "subscriptions" channels on the left side of the main page, preceded by a list containing five "Icons" options, the first of which is in order of "Watch History" which contains all clips. The video that the user has watched and another with the name "Watch Later" is the one that collects the videos that the user has marked as watched at a later time [1]. There is also an icon called "What to Watch" where it is the most important The activation link when entering the site, and it contains the videos suggested to be watched by YouTube based on the majority of viewers, followed by an icon for the channels the user has previously subscribed to, called "My Subscriptions" [16], and finally an icon called "Social" to link any accounts of social networking sites. Another social for this user with YouTube, so that an integrated circle of social communication for the user is formed, there is also another list on the right side of the home page that contains a few channels that the site recommends to subscribe to "Recommended" and on the bottom side of the page appear four possibilities for both types of users First: The ability to change the language and choose from 61 different languages [8, 20].

Second: The possibility of changing the choice of the country to which the user belongs, and it is possible to choose from among 52 different countries or choose the "Worldwide" affiliation. Third: The possibility of safe viewing, where this service can be activated at any time in order to avoid watching videos that may be inappropriate to the user's taste. Fourth: The possibility of "Help", while the YouTube user - who does not have an account - finds the home page in a way Slightly different from the one that appears to the user who has an account, as the page always contains 24 different videos of the most popular and watched during the day of entering the site and until the rest of the previous week [3], and these clips belong to private channels, some of which are owned by individual users, amateurs or well-known media organizations, or categories of public channels selected by YouTube such as: music, entertainment and sports, in addition to the search box for videos by keywords located at the top of the home page, which is available at Always for both types of users [16].

YouTube contains special personal pages for users known as "Channels", which are pages that contain their personal data and the videos they prefer, and it is worth noting that there are 317 YouTube channels that have exceeded one million subscribers, moreover, YouTube enables users to form groups Niqash Groups contains a variety of topics and videos that are discussed or commented on, and YouTube has made available the "Partnership" program service, which contracts with owners of YouTube channels and gives them a fee for the content provided, and the site was established for the service in 2007 [6], it now has what Over 1 million creators, across more than 30 countries around the world, monetize their videos on their YouTube channels, and thousands of channels generate six-figure money each year [3, 17, 22]. The Music channel for music and songs tops the list of YouTube channels in terms of the number of subscribers, with more than 88 million users subscribed to it on YouTube, followed by Gaming, Sports, and You Tube News. It is worth noting that there are three channels owned by individuals and amateurs within The ten most subscribed YouTube channels are PewDiePie, HolaSoyGerman and Smosh, and these three channels are distinguished by their satirical comedy videos [12].

YouTube provided a greater opportunity for artistic production companies to reach a larger number of audiences, as the number of followers of the "VEVO" channel for songs on YouTube reached more than 60 million viewers, and the number of followers of the "Warner Music" channel reached more than 31 million. The Associated Press channel has more than 6.5 million viewers, making it one of the 50 most important news publishing agencies on the Internet [6], which made YouTube a gateway for democratic community dialogue, and the vast majority of the content on YouTube comes from individual users, but major media organizations, such as "BBC" and "CBS" and others add a lot of their content to the site. Internet within the youth group between the age of 18–34, and they are more inclined to watch videos on YouTube [12]. Studies suggest that users who are active in social networking watch videos on YouTube, and this explains the speed of the spread of social networking sites, such as: YouTube. Young people between the ages of 18–34 use YouTube more than they watch any cable TV channel in the United States, and the indications are clear that males watch YouTube videos more than women. Where males between the ages of 18–34 are the most watched group on YouTube videos, one out of every 5 people watches 10 h of YouTube videos per week. Studies and scientific reports have also confirmed that second-year students are the most inclined to use the Internet in general, and YouTube in particular [14]. In most cases, users watch YouTube for entertainment purposes and to exercise absolute freedom, and at other times, YouTube users exchange videos or information about them by adding comments and evaluating some other clips, and these activities are among the new and different activities in the world of media. After 8 years of its creation, the YouTube community has now become clear, and it is constantly growing [10].

3 Youtube Worldwide

YouTube comes in third place in terms of the most used websites in the world, according to the classification of the American website Alexa.com, which specializes in classifying websites. Video clips, as it also came in the same position in the category of community video sites "Community Video", YouTube also comes in second place according to the classification of "Onbile.com" in terms of the most visited social networking sites on the Internet after Facebook, and YouTube also comes It ranks second in the world as a search engine for the famous Google site, because YouTube contains more than 15 million reference files in the "Content ID" database, which is among the most comprehensive databases in the world [11, 21].

Subtitles for YouTube are available in 61 languages, including Arabic, and it has headquarters in 56 countries. 19.8% of YouTube use globally comes from only the United States, which comes first in terms of YouTube use [12], India comes in second place with 8.2%, then Japan with 4.5 As for the Arab countries, there are 167 million views on YouTube. In November 2006, the site achieved one of the largest growth rates for sites on the World Wide Web and ranked fifth as the most visited site in the world. In 2007, the site consumed an amount of data flow similar to the consumption of data. The world for all Internet sites in the year 2000, ie about 10% of the global production of the Internet [15, 24]. Experts believe that YouTube controls the undisputed increase in viewing and sharing of videos on the Internet, and YouTube owns 60% of the videos on the Internet in 2006, i.e., the year after its creation, and in 2008 it contains more than 75% of the total videos on the Internet as well. The rapid growth of YouTube was observed as it gained unprecedented fame on the Internet in a very short time [11], as YouTube viewership rates increased dramatically and in just one year, 6 billion minutes of video clips were watched in 2006 while this number doubled to reach 20 billion minutes in In 2007 and about the development of YouTube viewing statistics, it suffices to realize that a few months after the creation of YouTube in 2005, there were 8 million videos watched on YouTube daily, and the views kept increasing in an unprecedented way until it reached 4 million views of YouTube videos per day in In 2013, the number of hours of video watched every month on YouTube exceeded 6 billion hours, at an average of approximately one hour for every person on Earth, an increase of 50% from 2012 [4].

The use of YouTube has developed unexpectedly since its inception. In 2007, the rate of uploading or adding videos on YouTube was 8 h of those clips per minute, and then this number developed in 2008 to reach 13 h, and this rate continued to increase when it exceeded 24 h. hour in 2009 [18, 23], and this same rate reached 35 h in 2010, and in 2011 the rate came by 48 h, which is double that rate in 2009: that is, the rate of adding videos on YouTube doubled in just two years, and when many thought that YouTube rates May prove or decrease in 2012, YouTube surprised everyone, recording 60 h, and that rate continued to rise, reaching at the beginning of 2013 to 72 h and 100 h at the end of the year of video clips uploaded and added to the site per minute, and this reflects the extent to which users interact with YouTube And their confidence that YouTube guarantees them the desired freedom of expression [12].

4 Second: How Interactive YouTube Is

4.1 Youtuber Activity

The increase in the use of video sharing sites comes for many reasons, but especially through the development of content, such as: the content on YouTube through the activity and interaction of its users. YouTube is one of the new forms of social communication that has developed in the past few years, and YouTube embodies the social environment that gives users the ability to exercise both the roles of producer and viewer of the content [7]. The Internet in general requires users to be active in the selection of content, and YouTube is a good example of this phenomenon, as it requires a greater level of interaction by users, as users in the world of YouTube have become active and effective participants in the communication process, where users play an active role in the process of producing, publishing and receiving media content on YouTube [3].

The success of the site depends largely on the small area "Bandwidth" of the video clip, that is, the transfer of data for those clips on the site is designed to be small in size, because the interest of users on this site is focused on interacting with the videos and accessing them to large numbers of users more than interest With the high technical quality of sound and image, and there are 6 different classifications for participation in video clip sites such as YouTube [14], they are: active producers representing 13% of the total participants in those sites, critics who evaluate clips and represent 19% and collectors who save clips to exchange them with others, representing 15%, And members of social networking sites, who can participate by placing content on YouTube or other social media sites, representing 19%, viewers, representing 33%, and finally, inactive subscribers, representing 52% [12].YouTube provides the opportunity for its users to rate and comment on the videos, and this contributes to creating a parallel community that exchanges information. Amateur YouTubers who document their daily lives. And about the interaction of users with videos, 50% of users talk and discuss about the clip after watching, 38% of them exchange clips on other social networking sites after watching, 65% post videos that they shoot themselves, and 25% post videos weekly [19].YouTube contains an important and new element in the communication process, specifically the post-watching process - which was not available in the form of traditional television - which is the possibility of videos with specific groups, such as: family and friends. The ability to share these YouTube videos provides an opportunity for researchers to examine a new form of post-watching activity, as YouTube is now evolving the role of its users from just recipients of media messages to content producers themselves. YouTubers also have great control over what they watch, when and with whom they share it [2, 25].The most viewed videos with discussions and interactions are the videos produced by amateur YouTubers. 50% of YouTubers constantly watch videos made by amateur users on their YouTube channels, and 13% of those who watch videos on YouTube also put videos on it. Also, researchers found that 22% of American users film the videos themselves and 14% of these put the clips on YouTube [5]. TV programs shown on YouTube have also begun to attract the attention of many YouTube users, as 39% of the total YouTube users watch movies, sports programs and complete TV series, and 25% watch the important clips of those series and programs, and 29% claim that the higher the quality of the content on YouTube. YouTube is better, they watched more [18].

5 Third: The New YouTube TV

5.1 YouTube Features and Challenges

YouTube allows users to add, share and watch videos on the Internet, and it is one of the latest and best that modern technology has provided, to provide a different and integrated set of services and activities. Participation in YouTube requires registration and subscription in order for the user to practice all activities and services on the site, such as: adding clips, commenting on them, or even evaluating them. YouTube also offers its users the ability to communicate with anyone in both public and personal forms, such as: communicating with family members and friends or communicating with new people around the world [15]. YouTube allows thousands of user-made video producers to add these clips on its site, which is the place where any user on the Internet can access it in the simplest way and through an acceptable computer and Internet connection, and YouTube video viewers and those who add these clips can communicate and evaluate discussions about any presented clip, by throwing text comments or commenting on another video [12], and YouTube also provides a more comprehensive model for community dialogue than the model presented by traditional media, in terms of its plurality of sources, free service, and opportunities available to both the producer of the video clip and the viewer to exchange their opinions YouTube also guarantees its users not to delete or modify any content, except in cases of user violation of the "Terms of Use" or if a large number of users have reported to "Report" [8, 27].

5.2 YouTube's Relationship with Youth and Society

Planet Earth brings together 7 billion people, 45% of whom use the Internet and most of them are young people under the age of 25. YouTube is also the most popular medium for sharing Internet videos. YouTube has given the ability for any individual in the community who owns a simple camera and knows how to upload his video clip on YouTube, to express and broadcast his views to the world without any restrictions, and the feature of sharing videos on YouTube has enhanced the right of individuals in the same community to express their political opinions, whether or not Social and completely free, which in turn expresses the real interests of the community [21], where YouTube provides important opportunities for communities, which is the possibility of establishing discussions and community dialogues, especially those that bring together users, and specific interests and topics that occupy the audience and exchange similar content through comments on videos or exchanging Various videos on these topics [1]. Video forums such as YouTube are usually used in everyone's living rooms, and this reflects the extent to which individuals relate to it, and sometimes it is considered a type of "Amateur Reality TV". YouTube has a wide base of users who create, add and share videos, and can provide the audience with important information about an influential topic in society [16]. The nature and content of the videos are thus influenced by the social aspects surrounding their creator and viewer. Henry Jenkins believes that the intensity of the integration of the participating communities with the new media is what paved the way for the success of YouTube, and not the other way around, as some thought, as one of the most common practices of YouTube users is watching videos and their textual response to them. Users' comments on those clips enable them to express See them, like criticism or express familiarity with the videos and their makers [13].

In one study that looked at the most viewed, best, and most responsive YouTube videos, it was found that the best highly rated clips added by individual users outperformed those added by traditional production and advertising companies. Where social networking sites such as YouTube combine personal, social and business interests in one network, which facilitates the process of communication and exchange of information between them, as the Jones and Jolie study showed that 33% of YouTube users described it as an integrated community and 19% found that YouTube is a new tool for communication, as seen by 10% He enjoys absolute freedom and 9% stated that YouTube has become one of their main daily activities [6]. YouTube spends more than $100 million annually to encourage video software developers, to reach the maximum possible potential of the user, and YouTube funds with millions of dollars some talented users, such as: directors, artists, authors, and even amateurs to encourage them to provide more creative content [8], but YouTube's ability to Giving the opportunity to young amateurs from the community to create the content that expresses them without large funding, to be an audience even if it is small in number, and to compete with traditional media sources in providing the community with news and information [14, 28].

5.3 YouTube as a Media Tool

The domains on YouTube are divided into main areas through which the videos are categorized. Users who subscribe to YouTube can also subscribe to these areas, just like YouTube channels, but the most important areas of YouTube in terms of the number of subscribers are: music 72 million subscribers, games 68 million subscribers, then sports 67 million subscribers, global news and events 23 million subscribers, films 15 million, drama series 11 million subscribers, education 6 million subscribers, and within each field there are several different channels, such as: the football channel within the field of sports and the classical music channel within the field of music In the field of news, there are also the most important global news networks, such as: BBC, Monte Carlo, Reuters and other international channels [12]. Studies see that YouTube is similar to television, as it is an entertainment medium that is sometimes used for political news purposes, but the user himself is the source of the content circulating on it, as there are 5 main areas of YouTube videos, in terms of the majority of viewers, which are: First, music, second: "Mash Ups." These are the videos that users create themselves by merging a video and audio clip together. Third: User-Generated Videos in which the user is the hero of the clip. Fourth: Political Satire Shows. Fifth: Entertainment and drama programs [7].

As for the relationship between users' interests in fields and their gender, the most watched area for YouTube videos by males and females together is comedy with 39%, then news with 33%, and music and lyrical videos with 31%, but clear differences appear between males and females. In other areas of watched videos, females tend to watch entertainment videos by 31% compared to 24.5% for males, and females prefer watching dramatic videos by 24.7% compared to 14% for males, also females outperform males in watching Entertainment videos and celebrity news 15% vs 5%. On the other hand, males tend to watch sports videos by 31%, compared to 8% for females. The percentage of males watching videos of mature and adult interest is 26% higher than the percentage of females, which represents 8% [4]. The results of the study also show that males tend to watch sports videos. 21% of women watched political videos, compared to 12% for females [15]. Entertainment is the main reason for watching videos on YouTube by 27%, followed by website news and friends' news by 20%, education by 13% and interaction and participation by 10%. Studies have also shown that there is a positive correlation between watching and exchanging entertainment videos on YouTube, and users' motives to search for information and entertainment [14].

The Internet provided facilities for the circulation of visual, audio and written information, which led to the democratization of its free exchange and discussion among members of the same society. This change is the most obvious, as these possibilities were not available through traditional media outlets. The entertainment fields have witnessed the ability of social networks as an influential media in the public [8]. Social networks have also become an integral part of daily entertainment, and the audience can be negatively or positively affected by the entertainment content circulating on social networks such as YouTube, and governments must take this into account [13, 29]. YouTube is currently gaining a great position despite the amount of entertainment content on the site. YouTube has become an important part of the media mechanism currently used. YouTube also provides the opportunity for citizens to freely share entertaining video information and encourages them to hold discussions by adding comments about videos. Also, the nature of YouTube in saving all the visual content on its site enables users to analyze their entertainment information, by watching entertainment videos about a specific issue or personality [12], and then help them determine their trends and even monitor the contradictions between the statements of one of the characters or know the aspects of a specific issue. YouTube is also distinguished by providing content that is characterized by transparency, as any user can access the content and interact directly with it [10], leave a comment or add videos bearing opposite opinions, and the follow-up of video blogs such as YouTube is increasing to replace it with traditional media news sources, which indicates the ability of new media in directing the public, through their use and dependence on them, especially to reach the largest number of young people, who are the largest group, in terms of using the Internet and social networks [2, 26].

With the beginning of the emergence of media contents created by the user, the Internet became an integral part of the entertainment life, especially after the development of YouTube, where the concept of entertainment fields on the Internet changed from if it is only said to the interest in knowing how to say it, until YouTube became an arena for any citizen by adding or exchanging a clip YouTube is a turning point in the practice of entertainment on the Internet, not only because it is an arena for exchanging visual content, but also because it provides information and visual facts periodically and faster than any other technology [11, 30]. The possibility of exchanging content on YouTube also added another dimension to the process of disseminating information, especially political information, as the user became immersed in information and did not make much effort in searching for information, as happens in traditional media. YouTube's strength as a breathing tool is that it is like blogging in the space it provides for its audience to add or share videos that showcase their views, regardless of the length of the video clip or the length of the comment on it [22]. There are sometimes some videos on YouTube that many governments do not want to publish, such as those that publish entertaining political materials, or that depict heads of state and government or prominent political figures in unpopular positions for these personalities and these ruling regimes, and this publication caused the site to be blocked in some countries. And some countries have lifted the complete block on the site but are now blocking specific videos that they do not want to share. YouTube may affect negatively or positively the image of a political figure if it is not in control of his reactions all the time [1]. YouTube is the most visited website in the United States with 181 million visitors per month and is the most profitable site. YouTube provides its services for free and relies on other sources of income represented in entertainment spaces. It also plays an important role in providing video ads as a new and practical field in the entertainment field [8]. YouTube allows users to upload, share and watch videos on the Internet, and it is one of the latest technologies provided by it, in order to provide an integrated set of activities. Individuals also feel the capabilities and capabilities of YouTube, especially in the ease of accessing videos that match their interests and helping them discover different cultures in various forms, by watching videos that provide information on different civilizations and cultures. YouTube also helps to develop the user's level of independence for entertainment purposes and his ability to rely on himself to obtain information [15].

6 Conclusion

YouTube is now changing the way Internet users interact with the world, as YouTube has removed all geographical, social and cultural barriers, by just sharing and watching videos, and its impact has become evident by requiring individuals and organizations to use it as a trusted source for information. Google is also interested in encouraging Well-known media organizations to join YouTube in the "Youtube Partner Group" program and upload more media reports and videos on YouTube, in preparation for benefiting from it in strengthening the YouTube News service as a news site on the Internet in the future. YouTube does not depend on any other social networking sites to prove its existence, but other communication sites are using YouTube, as it has become the first destination for watching and exchanging videos on the Internet, and while social networking sites may face successes and failures, YouTube will remain the only means that will continue It is booming, and in 2012, that is, 8 years after its establishment, the features of the YouTube community have now become clear, and it is constantly growing. Based on the theoretical framework, previous studies, and the results of the current study, the researcher proposes a number of topics that need to be conducted studies to find out their results. Psychological and social effects of using YouTube. The impact of YouTube on terrestrial and satellite television channels. Mobile TV, which is the latest form of television broadcasting in the world. The ethical dimension on YouTube and its impact on adolescents and youth, a comparative study of professional codes and ethics between old and new media.

References:

1. Waterloo SF, Baumgartner SE, Peter J, Valkenburg PMY (2018) Norms of online expressions of emotion: comparing Face-book, Twitter, Instagram, and WhatsApp. New Media Soc 20:1813–1831
2. Aran-Ramspott S, Fedele M, Tarragó A (2018) YouTubers' social functions and their influence on pre-adolescence. Comunicar 26:71–80
3. Raharjana IK, Siahaan D, Fatichah C (2021) User stories and natural language processing: a systematic literature review. IEEE Access 9:53811–53826
4. Barthorpe A, Winstone L, Mars B, Moran P (2020) Is social media screen time really associated with poor adolescent mental health? A time use diary study. J Affect Disord 274:864–870
5. Bhatia A (2018) Interdiscursive performance in digital professions: the case of YouTube tutorials. J Pragmat 124:106–120

6. Caraka RE et al (2022) The impact of social media influencers Raffi Ahmad and Nagita Slavina on tourism visit intentions across millennials and Zoomers using a hierarchical likelihood structural equation model. Sustainability 14:524. https://doi.org/10.3390/su14010524
7. Al-Sartawi A (ed) (2021) The big data-driven digital economy: artificial and computational intelligence In: Kacprzyk, J (ed) Studies in computational intelligence, vol 974. Springer International Publishing
8. Cayari C (2018) Connecting music education and virtual performance practices from YouTube. Music Educ Res 20:360–376
9. de Bérail P, Guillon M, Bungener C (2019) The relations between YouTube addiction, social anxiety and parasocial relationships with YouTubers: a moderated-mediation model based on a cognitive-behavioral framework. Comput Hum Behav 99:190–204
10. Clary PC, Markham M, Myers-Bowman KS (2020) Relationships are key: family life educators' use of social networking sites. J Technol Hum Serv 38:112–136
11. Al-Qudah AA (2022) Artificial intelligence in practice: implications for information systems research, case study UAE companies. In: Musleh Al-Sartawi AMA (eds) Artificial intelligence for sustainable finance and sustainable technology. ICGER 2021. Lecture notes in networks and systems, vol 423. Springer, Cham
12. Fägersten KB (2017) The role of swearing in creating an online persona: the case of YouTuber PewDiePie. Discourse Context Media 18:1–10
13. Al-Sartawi A (2019) Assessing the relationship between information transparency through social media disclosure and firm value. Manag Account Rev 18(2):1–20
14. Feijoo B, García A (2017) Analysis of children consumption of TV, videogames and the internet: differences based on gender in their selection of audiovisual content. Fonseca 15:95–108
15. Feijoo B, Sádaba C (2021) The relationship of Chilean minors with brands and influencers on social networks. Sustainability 13:2822. https://doi.org/10.3390/su13052822
16. Ferchaud A, Grzeslo J, Orme S, LaGroue J (2018) Parasocial attributes and YouTube personalities: exploring content trends across the most subscribed YouTube channels. Comput Hum Behav 80:88–96
17. Razzaque A, Al-Sartawi AMM, Karolak M (2021) Rethinking tourism entrepreneurship amidst COVID-19 through the lens of artificial intelligence and big data analytics. In: Big data for entrepreneurship and sustainable development. CRC Press, pp 175–188
18. Genç M, Öksüz B (2019) An analysis on collaborations between Turkish beauty YouTubers and cosmetic brands. Procedia Comput Sci 158:745–750
19. Musleh Al-Sartawi AM, Hussainey K, Razzaque A (2022) The role of artificial intelligence in sustainable finance. J Sustain Finan Investment, 1–6
20. Pereira S, Moura PFRD, Fillol J (2018) The YouTubers phenomenon: what makes YouTube stars so popular for young people? Fonseca J Commun 17:107–123
21. Jerslev A (2016) Media times in the time of the microcelebrity: celebrification and the YouTuber Zoella. Int J Commun 10:5233–5251. https://ijoc.org/index.php/ijoc/article/view/5078. Accessed 25 Oct 2022
22. Al-Adwan AS, Nofal M, Akram H, Albelbisi NA, Al-Okaily M (2022) Towards a sustainable adoption of E-learning systems: the role of self-directed learning. J Inf Technol Educ Res 21:245–267
23. Lozano-Blasco R, Latorre-Martínez MP, Cortes-Pascual A (2021) Analyzing teens an analysis from the perspective of gamers in YouTube. Sustainability 13:11391. https://doi.org/10.3390/su132011391
24. Musleh Al-Sartawi AMA (2020) E-learning improves accounting education: case of the higher education sector of Bahrain. In: Themistocleous M, Papadaki M, Kamal MM (eds) Information systems. EMCIS 2020. Lecture notes in business information processing, vol 402. Springer, Cham
25. Rodríguez Illera J, Martínez Olmo F, Rubio Hurtado M, Galván Fernández C (2021) The content posting practices of young people on social networks. Pixel-Bit Rev Medios Y Educ 60:135–151

26. Smith R, Secoy J (2019) Exploring the music identity development of elementary education majors using ukulele and YouTube. J Music Teach Educ 29:71–85
27. Musleh Al-Sartawi AMA (eds) (2022) Artificial intelligence for sustainable finance and sustainable technology. In: ICGER 2021. Lecture notes in networks and systems, vol 423. Springer, Cham
28. Vizcaíno-Verdú A, Aguaded I, Contreras-Pulido P (2021) Understanding transmedia music on YouTube through Disney storytelling. Sustainability 13:3667. https://doi.org/10.3390/su13073667
29. Thosuwonchinda V, Ketut N, Sungkamart K (2021) The Comparison of factors for choosing a tourism destination: a case study of Bangkok-Bali. Res J Phranakhon Rajabhat Soc Sci Humanity 16:115–128
30. Van-Dam S, Van-Reijmersdal EA (2019) Insights in adolescents' advertising literacy, perceptions and responses regarding sponsored influencer videos and disclosures. Cyberspace 13:2

The Use of Modern Technology in the Practice of Digital Public Relations

Abdulsadek Hassan

Abstract This study aims to know the modern techniques available to departments, the practice of public relations activities, and their reflection on the performance of public relations practitioners in addition to monitoring, and the level of their use and areas of employment in the field of public relations. The results indicated that technological development during the last decade has led to shifts in the field of public relations practice, which has faced great challenges since the emergence of applications and tools for interactive means of communication and the expansion of their use, foremost of which are websites that enable their users to exchange information.

Keywords Digital public relations · Technology—Practice

1 Introduction

Technology has an important impact on public relations through the preparation of information and bulletins using modern printing devices, as it made it easy to conduct surveys for newsletters, print books and manuals, and present studies related to this field, which are simple in a way that is not compared to the previous ones, and the computer and the Internet are now necessary in every Public relations offices, in addition to its well-known features and benefits for solving equations, preparing tables, and storing information and large files with its huge capacity [12]. The Internet has been widely used in public relations and has a dedicated place; So that all small and large companies have their own sites, and the first face of this institution is for hundreds of millions of audience, and in addition to the fact that public relations have identified groups to update those sites, while the activity of specific public relations agencies appeared in modern media and direct communications, and now it is considered public relations more effective and efficient than before; This is because technology has given it ease of communication, and it has reduced cost

A. Hassan (✉)
Ahlia University, Manama, Kingdom of Bahrain
e-mail: dr.sadek1964@yahoo.com

A. M. A. Musleh Al-Sartawi et al. (eds.), *From the Internet of Things to the Internet of Ideas: The Role of Artificial Intelligence*, Lecture Notes in Networks and Systems 557,
https://doi.org/10.1007/978-3-031-17746-0_20

and time for PR staff and consumers [3]. In light of the information revolution, electronics became the central role played by coal in the past at the beginning of the industrial revolution. Then the digital technological revolution came to form the basis of modern technology for the twenty-first century, which made some people call this era the digital age, especially the great leaps it is witnessing in the field of modern communications [7], which he left an effective and clear impact on all administrative activities, including public relations.

This tremendous technological progress has helped shorten the time between any two points on the earth's surface as a result of the use of the electronic communication network, which is one of the most important products of the development of information technology in this era, in which information technology has developed very quickly, not only in the field of computers, but in the field of communications [15, 23]. In general, as this development made it possible for different technologies to converge in what is known as the fusion or merging of more than one technology together, to be a new product bearing the characteristics of each technology separately, except that it is completely unique in its characteristics and technological capabilities, and thus information technology has become an increasing impact on various aspects of the daily life of individuals, institutions and governments, the means of communication, sources of obtaining information, the movement of buying and selling, the means of entertainment, the method of accomplishing work, the emergence of what is known as e-government, and most importantly the method of communication, which is the essence of public relations activity, has been affected, so that information technology has become in its various products [4].

The Internet has provided many channels of communication, and these channels have grown over time. Undoubtedly, these matters are very important for the public relations officer in any institution, and the dynamic in which the Internet operates requires that the public relations practitioner always take the initiative, and benefit from the most prominent services available on the Internet [21].

2 The Importance of Online Public Relations

Why companies and individuals who want to succeed in the modern world cannot afford not to exploit Internet PR clearly from the example of "anti-campaigns". This could either be an unfortunate incident or a carefully crafted plan of unethical competition [7, 22]. But no matter who is behind them and why, negative campaigns can have a devastating effect on the public reputation of a company or individual. However, the danger of negative publicity is not the only reason why online PR is becoming increasingly important. The Internet has a huge impact on people's opinion and the decisions they make, especially when they are unsure [20]. For example, let's say anyone has difficulties deciding which anti-acne cream to choose. He will open his search engine, and search for information about the creams he is considering buying. He will most likely consider a variety of factors including price and other people's opinions, but his decision will also be greatly influenced by the information he can obtain about a particular product [15]. And if he finds little information about how the product works, once he can see the results, etc., there is a good chance that he will not buy that product. In order to at least encourage Ahmed to buy your cream, he needs to have accurate, reliable and up-to-date information about both the product and the company. This can only be achieved through effective Internet public relations [3].

The information revolution has helped to pay attention to public relations and its spread, as it has become an integral part of the business and administrative process and even in the decision-making of the organization. The impact of communication technology in public relations is evident when it is practiced through multimedia; This helped the interactive and real-time, and the density of information flowing, and this is what made public relations move from the traditional concept to the digital [7].

3 Digital Public Relations Goals

With the expansion of the digital ecosystem and the pivotal role that the Internet and social media play in the field of communication, and with the increasing number of users of social platforms, the process of managing and maintaining the reputation of the organization on the Internet has become the primary goal for those engaged in public relations tasks, as the reputation of most organizations in the world is at stake [2]. As a result of the great competition and the continuous expansion on the web, there are many electronic pages, news networks, and fields related to the offerings of various organizations. Therefore, companies and organizations have realized the importance of managing their prevailing electronic reputation in the minds of the masses, by drawing up their policies through planning [17].

4 Planning in Digital Public Relations

Public relations organizations need to set precise goals for the public relations apparatus, and such goals can only be determined or set through the planning mechanism. For these reasons, organizations and public relations departments resort to planning their activities and programs [8, 21].

Planning means that mental activity that is directed to choosing the optimal possible use of the available set of energies to achieve certain purposes in a specific period of time. From this we conclude that the basic variables that govern the planning process are: the purposes or objectives, resources, or capabilities, and the time available for or specified for implementation [5].

These elements can be employed in the field of digital relations as follows:

1. **Objectives:** Create and maintain a good online reputation (search engines and social media platforms) [7].
2. **Available resources and capabilities:** All available modern forms of communication (the organization's website, pages and murals on social media, Android applications, blogs and forums, e-mail), and to achieve this, the public relations apparatus must consist of cadres that control communication and electronic editing techniques, Creating and managing interactive content is a key factor in this process [6].
3. **Time:** Choosing the right time and maintaining the electronic identity for a longer period of time [10].

The importance of planning lies in public relations activities through:

- Helping the organization to identify the digital environment in which it operates, and to reveal the attractive opportunities available to it, and the restrictions, threats and risks that affect its work, and to identify the competing institutions and their audience [13].
- Develop long-term strategies to improve the organization's cyber reputation management and respond to attacks [18].
- Achieving a balance in the performance of public relations inside and outside the cyberspace [9].
- Helping public relations officials to identify the size of the target audience, know their desires, and measure their opinions [7].

5 Public Relations and the Internet

The Internet has revolutionized communication by giving a voice to those who could not be heard before. This has opened up a whole new world of opportunities for companies and individuals that they could not have imagined in the age of traditional media. They still play an important role in shaping public opinion, but with the entire world turning to the internet to search for information and purchase products/services, PR that does not communicate effectively with the online community cannot effectively shape the desired public image [1].

6 Internet Public Relations Challenges

Although the Internet is one of the greatest inventions in human history, it has dramatically changed the world, and above all communication. The global network has opened up a number of opportunities, but it has also opened up a number of challenges, especially for the public relations sector that has mastered communication with and through traditional media [11].

While PR professionals have had no competition in the age of traditional media, Internet democracy has forced them to compete with the world at large. The internet has given everyone the ability to share their opinion about everything with the entire world within seconds. The ability to reach thousands or millions of people at almost any time [7].

It has given PR professionals a powerful tool but has made their jobs more difficult, especially when the integrity of their clients is attacked. Imagine restoring a customer's reputation if the first page of search engine results contains malicious or primarily malicious content [14].

7 The Role of Technology and the Internet in Public Relations

This era is the information age. The Internet and technology have left an effective impact on public relations [16].

The speed of information processing and the preparation of bulletins using fast printing devices made the possibility of conducting surveys, presenting newsletters, printing brochures, books and guides, and presenting related studies, in a way that is not comparable to the previous ones [5].

Also, computers and the Internet are now necessary in every office. In addition to its well-known features and benefits in solving problems and equations, preparing tables, preserving information and huge files with its huge absorption, it has become a basic necessity in printing in multiple languages, completing technical and scientific designs, magazine covers and books, as well as connecting to Internet networks and faxing [18].

The Internet has grown to the point where users do not expect less than the level of professionalism in the level of service, so the relationship staff must use electronic devices and advanced technology and increase their capabilities until they reach the level of satisfaction [1].

The Internet has been widely used in public relations and has taken a prominent place, so that all small and large companies, as well as profit and non-profit ones, have websites on the network, and it is the first face of that company or institution in front of hundreds of millions of public relations departments hired specialists and groups to develop these sites, while public relations agencies specialized in new media and direct contacts flourished [21].

Now, public relations is more effective and efficient than before, because of the ease of communication that technology has provided, and it has reduced the cost and time for relations staff and consumers [19].

8 Advantages of Digital Public Relations

8.1 The advantages of digital Public Relations Vary Greatly, According to Its Role and What It Contributes to Do, and Its Advantages Are

8.1.1 First: Good Communication

It is considered one of the most important signs of the success of this management in institutions, as it has become one of the most important factors for the sluggishness of societies in general, and the parties working in it at the public and private levels. Communication and new innovations [18].

8.1.2 Second: The Speed of Interaction and Communication with the Concerned Department of the Institution

One of the most important advantages of digital public relations is that it is characterized by speed in interaction and communication with the concerned department within the institution [13].

8.1.3 Third: Transparency

One of the advantages of public relations is that it is characterized by transparency of communication with customers or the public of the institution, where the official spokesperson is not always on the digital platform [17].

8.1.4 Fourth: Performance Appraisal and Measurement

The Public Relations Department works to continuously measure its performance within the institution, by evaluating its communication with its mass base [19].

9 The Use of Public Relations for Modern Technology in Dealing with the Masses of the Institution

9.1 Relationships with Internal Audiences

The use of the Internet in the field of public relations adds many benefits to the organization and its employees, as well as to the owners or shareholders. Through the use of the Internet, communications become easier and less expensive... All instructions and alerts take place among the organization's employees through the use of the Internet [5]: In addition, it is possible to put certain information that benefits or increases the level of awareness of workers, such as social security for workers and the internal system of the organization. In many organizations, intranet networks are used in organizations that have a local or internal network for communication [12].

In order for the intranet to be complete, it needs to be subject to the following basic rules:

- It must be safe so that employees can use it with confidence [7, 20].
- To be trustworthy so that employees do not feel any threat from its use [17].
- To be suitable so as to assist in promoting the work [11].
- To be private so that it can be used for internal interaction between departments [3].
- Be quick to save time and help access content to increase productivity [15].
- To be developed regularly so that the available information is the latest information [22].

9.2 *Relationships with External Audiences*

So that information technology effectively contributes to raising the level of performance of workers in the public relations function and reducing the cost of the services they provide and the communications that are made with customers (after some institutions were reviewing their products in a traditional and classic way through the institution's heavy catalog, which includes tens of thousands of goods that It is produced by the institution and it costs huge sums to produce it [14]. Today, with the development of modern technology, institutions are using modern techniques in this field, so they dispensed with the huge catalog and replaced it with a very small, low weight and limited cost... At the same time, this modern technology facilitates the customer's CD-ROM Reviews products, their prices and ways of distributing them, and saves him time and the trouble of searching) [6].

With the rapid development in the field of the Internet, customers can access the information that interests them easily and at the lowest cost [4]. Previously, these customers had to either go to the headquarters of the concerned organization or contact them by phone, and with the presence of the Internet, the customer can visit the organization's website and view a huge amount of Information, and the client can participate in any questionnaire carried out by the organization, and he can record any complaints about the organization, manage it and put his suggestions through the Internet [14].

9.3 Internet Public Relations Challenges

Although the Internet is one of the greatest inventions in human history, it has dramatically changed the world, and above all communication. The global network has opened up a number of opportunities, but it has also opened up a number of challenges, especially for the public relations sector that has mastered communication with and through traditional media [8].

While PR professionals have had no competition in the age of traditional media, Internet democracy has forced them to compete with the world at large. The internet has given everyone the ability to share their opinion about everything with the entire world within seconds. The ability to reach thousands or millions of people at almost any time [19].

It has given PR professionals a powerful tool but has made their jobs more difficult, especially when the integrity of their clients is attacked. Imagine restoring a customer's reputation if the first page of search engine results contains malicious or primarily malicious content [1].

9.4 Digital Public Relations

The digitization of public relations in the world has become one of the basic needs of scientific life [7]. New communication technologies such as websites, e-mails, social networking sites, photo and video sharing sites, etc. have greatly affected the practice of the public relations profession, which gave it a new advantage and a new dress than it was before [11].

Modern electronic tools for public relations have provided wider opportunities to reach those audiences faster and in a more accurate manner, which has had a significant impact on the development of their performance [11].

This has helped public relations practitioners achieve faster and more accurate communication with their clients, both between employees within the organization and its external audience, so the Internet has become an indispensable means of communication for public relations practitioners [3].

The Internet has significantly affected the development of public relations, as there were great opportunities for public relations practitioners to employ their creative skills. New technology provided new tools. The Internet added another dimension to public relations [13]. The Internet provided tools and communication channels for public relations practitioners and succeeded in controlling the way in which they are distributed the information [9].

Public relations and their theoretical frameworks. And public relations as an independent science cannot remain isolated from modern technology and its new developments, otherwise it will find itself ineffective and affected by all segments of society [12].

The power of public relations in social media has facilitated communication and interaction with others faster due to the uniqueness of digital media with the elements of instant and interactive than other traditional media, until it has become the preferred term for a huge group of media practices that use digital technologies [18].

This digital public relations trend may create new expectations about providing content that small or unqualified organizations may not be able to easily meet, and this would reduce the share of organizations that are less funded, or organizations that lack the role of public relations in their management. Therefore, public understanding should be intensively enhanced to meet the current and future challenges posed by digital media on public relations and media practices in all its forms [1].

9.5 *What is Digital Public Relations?*

"It is the use of Internet tools such as (websites—social networks—blogs), in order to communicate with the audience of the organization [7].

The International Public Relations Association defined it as "the employment and harnessing of modern communication techniques and digital media channels to carry out its activities, and facilitate communication between the masses of the organization, whether it is the internal public or the external public [5].

Digitization in the world has become one of the basic needs of scientific life. New communication technologies such as websites, e-mail, social networking sites, photo and video sharing sites, etc. have greatly affected the practice of public relations, which gave it a new feature and a new dress than it was before. Modern electronic public relations tools have provided broader opportunities to reach those audiences faster and in a more accurate manner, which has had a significant impact on the development of their performance [14].

This has helped public relations practitioners achieve faster and more accurate communication with their clients, both between employees within the organization and its external audience, so the Internet has become an indispensable means of communication for public relations practitioners [9].

The Internet has significantly affected the development of public relations, as there were great opportunities for public relations practitioners to employ their creative skills, as new technology provided new tools, and the Internet added another dimension to public relations. The Internet provided tools and communication channels for public relations practitioners and succeeded in controlling the way they are distributed the information [13].

Public relations and their theoretical frameworks. And public relations as an independent science, cannot remain isolated from modern technology and its new developments, otherwise it will find itself ineffective and affected by all segments of society [16].

The power of public relations in social media has facilitated communication and interaction with others faster due to the uniqueness of digital media with the elements of instantaneous and interactive, compared to other traditional media, until it has become the preferred term for a huge group of media practices that use digital technologies [14].

This digital public relations trend may create new expectations about providing content that small or unqualified organizations may not be able to easily meet, and this would reduce the share of organizations that are less funded, or organizations that lack the role of public relations in their management. Therefore, public understanding should be intensively enhanced to meet the current and future challenges that digital media poses to public relations and media practices in all its forms [7].

9.6 Digital Public Relations Practice

Now it has become that social process in which communication takes place from a distance between parties who exchange roles in transmitting and receiving various communication messages through digital systems and their means to achieve certain goals [7].

Accordingly, electronic public relations are traditional public relations using web applications to disseminate information and news through written texts and multimedia such as video, audio and image, in addition to relying on applications compatible with tablets and smart phones [2]. Communication is two-way and takes place at anytime and anywhere and the public can from interacting with public relations practitioners, thus forming a virtual community for the organization and its audience, supporting the concept of dialogue. Public relations functions become part of the organization's website, along with other marketing and service functions [8].

Here, communication technology has had a great impact on public relations practitioners. Thanks to it, this practitioner has become more specialized in the various applied journals of public relations [5]. The use of the Internet allows many advantages for public relations practitioners within all service institutions, as it helps with its various tools such as e-mail services, websites, and social networking sites. Public relations practitioners get to know the trends of the internal and external audience faster than before and contribute to the marketing of the institution and enhance its reputation at the local and international levels in light of the distinction of the Internet by overcoming geographical and cultural barriers, and we find that institutions are interested in publishing on the Internet in multiple languages [4]. The main objective of public relations through the Internet is to gain public opinion, build and support good relations, and create an atmosphere of mutual understanding and friendliness between institutions and their audiences, to build a good mental image among different audiences. Electronic services and information of the institution. In addition to the network material, institutions usually include newsletters about their services on their websites [3].

The website of each company or organization is a public relations tool as it is the main interface of the company and represents its official spokesperson because it contains reliable information and statements, as well as contains much information about the company's products and services [7]. Organizations use websites to build relationships with stakeholders to attract business and communicate with the media [1].

As a result, electronic public relations practitioners analyze the content of websites, e-mails, publications and bulletins issued by different websites, with the aim of identifying market trends, public opinion, media and customers towards the organization, its policy and its products, strengthening communications with different audiences and expressing their views on the organizations' activity and production processes [5]. The emergence of the Internet and its new internal applications increased the role and responsibility of public relations until the institution became an available means of obtaining information, and the role of public relations officials here is to help design and plan for this, due to the dangers that threaten the reputation of the institution and relationships, where the domination of the information of the institution and its protection became unavailable [15]. Communication technology and its applications have provided public relations devices and departments in institutions and business organizations with a new means in the field of introducing the public to institutions, promoting the services and products they provide, and educating them to encourage them to communicate with them through the medium and seek to establish and develop continuous electronic relationships with their audiences internally and externally [12]. Modern technology has departed from the traditional framework through the rapid spread of the world: the recipients of the Internet and social networking are spread all over the world, and thus the institution can offer its services widely, and thus reach its audiences wherever they are in the world. The advantage of its low cost compared to the services it provides to the work of companies [11].

9.7 *How to Create a Digital PR Strategy*

The digital strategy defines how the company will improve its SEO rankings. This outline usually includes ideas on increasing branding and securing external links. Additionally, the strategy may include SEO audits, backlink analysis, and assignments.

9.8 *Step 1: Determine the Goal*

The goal is at the core of all digital PR strategies. What is to be achieved through the work of public relations? Although the overarching goal will always be to get more sales, it is recommended that you reset the goals that your content needs to achieve to increase those sales [4].

It can include:

- Increase traffic to the home page or a specific product/service page [9].
- Directing traffic to a new offer or promotion [11].
- Increase your email marketing list to increase new and repeat sales [2].
- Increase brand trust by gaining influencer features [13].

9.9 Step 2: Review Your Target Customers

Who is the audience that you want to reach through your digital PR strategy and the content that will be created? Outline your target customers and identify any weaknesses or needs that can help make content relevant to them. Keep in mind the end goal and target customers through the content thinking process [11].

9.10 Step 3: Think About the Content and Research

The time you will need to think of ideas and angles for your corporate content should not be underestimated. Ideas don't get into the blanks, so work on building relationships with colleagues or get a small group of people together to bounce ideas off. And then make a list of any ideas, prioritizing them based on the ones you feel most attractive [18].

When there are some important ideas, you should research more. The data may be found or researched online, or you need to consider doing a company data study [20]. If the idea is a content asset such as a calculator or an interactive page, research the time and cost to create it [21].

9.11 Step 4: Select Target Posts

Who would be interested in posting and reporting this type of content? Who has the right audience that will include the target customers among their readers and/or followers? [5].

Will the idea provide something new and exciting that will lure journalists and influencers into wanting to repost and tell the company's story? If not, you may need to refer to the content Ideas drawing board [9].

9.12 Step Five: Design the Strategy

Digital PR strategies mostly focus on publishing articles and securing backlinks from relevant websites and blogs around the content. It also involves nurturing relationships with journalists to be mentioned in the press [17].

However, there are many different ways this task can be carried out, different types of connections that can be built and other important digital PR strategies that can be used as well [21]. Here are some of the most popular types of digital PR strategies to consider:

- Publish unique articles or data pieces online to get high quality backlinks [7].
- Connect with journalists and editors to get backlinks [18].
- Publishing press releases and promoting news content to gain journalistic advantages [6].
- Post a piece of content such as a calculator, online tool, or interactive web page [14].
- Post on Blogger to get backlinks and mentions to related blogs [7].
- Influencer marketing to get mentions on influencer social media accounts [3].
- Affiliate programs that pay commission to bloggers who refer clients to your business [7].
- Offline press events or blogger events aimed at getting coverage online [3].
- Sharing and promoting infographics [5].
- Restore links by requesting to add a link if your brand is mentioned, your work is on another website without a link. [18].

9.13 Step Six: Create Content

The reason we left content creation so late in the process is to make sure that ideas definitely sound like pioneers and that you know exactly what you want to do with your content asset when you create it. After all, it's not a good idea to have a piece of content on Google Drive without a plan of what to do with it [15].

Once you have confidence in the idea and your digital PR plan, it's time to get into the nitty-gritty of content creation [12]. This is where you flesh out the details, draft, and paraphrase to get the tone and headlines and make sure that the quality of the content and the look and feel of the final piece are perfect [10].

9.14 Step 7: Communicate and Control

Getting coverage is the cornerstone of an effective digital PR strategy. This takes perseverance and continuous communication with website owners, publishers, journalists, bloggers and influencers to get the coverage you want [20].

10 Conclusion

The results revealed that digital public relations is the use of various media on the Internet and social media, to create and share content about the institution or person, and to reach the target audience and communicate with it through electronic media, that is, it builds and manages the reputation of the institution, and this is because the formation of the correct image of the institution is in the mind people are very important in communicating a brand or organization to a distinct and pioneering identity.

The results also observed that, in light of the development of social media and its communication networks, and the increasing use of them by the business sector, its mass media with modern technologies contributed to highlighting the work of public relations in various sectors and institutions on both the private and public levels. In light of this development—the development of "digital media"—represented in the professional use of social networks and their employment in the service of public relations, the "digital public relations" crystallized in the transition of relationship management and communication with the public from the traditional stage to the modern stage emerged. that keeps pace with the era of technology and its users. It is no longer limited to media public relations and the sending of press releases only, but today "digital public relations" has become an art of integrating what was limited to working with traditional media and channels with new digital-marketing content, which serves the goals of the institution through what it publishes and communicates today. On social media platforms and networks, which reflected positively on the institution itself in search engines across the Internet.

Finally, with the development of the field of media and communication, and the increasing speed of the Internet's dominance in all fields and sectors, digital public relations is more comprehensive and more diverse, which gives public relations and communication officials enormous possibilities in delivering their message to the target audience in an effective manner. This strategy is not limited to mastering the methods of speaking and rhetoric, in order to improve the image of the institution, but goes beyond it to work on knowledge collection and analysis, keeping pace with the requirements of the times, and focusing on value, speed and creativity by relying on modern technologies.

References

1. Alexander DM (2016) What digital skills are required by future public relations practitioners, and can the academy deliver them? PRism 13(1):1–13
2. Permatasari AN, Soelistiyowati E, Suastami GAPP, Johan RA (2021) Digital public relations: trend and required skills. Kehumasan digital: Tren dan Kompetensi yang Dibutuhkan. Jurnal ASPIKOM 6(2):373–386

3. Allaguia I, Breslowba H (2016) Social media for public relations: lessons from four effective cases. Public Relations Rev 42:20–30
4. Al-Sartawi A, Al-Okaily M, Hannoon A, Khalid AA (2022) Financial technology: literature review paper. In: Musleh Al-Sartawi AMA (eds) Artificial intelligence for sustainable finance and sustainable technology. ICGER 2021. Lecture notes in networks and systems, vol 423. Springer, Cham
5. Al-Sartawi A, Reyad S, Madbouly A (2021) Shariah, presentation and content dimensions of Web 2.0 applications and the firm value of Islamic financial institutions in the GCC countries. J Islamic Market. Accepted article
6. Al-Qudah AA, Al-Okaily M, Alqudah G et al (2022) Mobile payment adoption in the time of the COVID-19 pandemic. Electron Commer Res. https://doi.org/10.1007/s10660-022-09577-1
7. Arisanty M, Wiradharma G, Fiani I (2020) Optimizing social media platforms as information dissemination media. J APSIKOM 5(2):266–279
8. Almaqableh L, Reddy K, Pereira V, Ramiah V, Wallace D, Veron JF (2022) An investigative study of links between terrorist attacks and cryptocurrency markets. J Bus Res 147:177–188
9. Al-Qudah AA (2022) Artificial intelligence in practice: implications for information systems research, case study UAE companies. In: Musleh Al-Sartawi AMA (eds) Artificial intelligence for sustainable finance and sustainable technology. ICGER 2021. Lecture notes in networks and systems, vol 423. Springer, Cham. https://doi.org/10.1007/978-3-030-93464-4_23
10. Almaqableh L et al (2022) Is it possible to establish the link between drug busts and the cryptocurrency market? Yes, we can. Int J Inf Manag 102488
11. Gulerman NI, Apaydin F (2017) Effectiveness of digital public relations tools on various customer segments. J Manag Market Logistics (JMML) 4(3):259–270
12. Kamil A (2020) Role of public relations in crisis management with the coronavirus crisis as an example: a case study on the UAE. Glob Media J 18:35
13. Krishan R (2015) Social media challenges and adoption patterns among public relations practitioners. Media Watch 6(1):73–81
14. Karolak M, Razzaque A, Al-Sartawi A (2021) E-services and M-services using IoT: an assessment of the Kingdom of Bahrain. In: Musleh Al-Sartawi AM, Razzaque A, Kamal MM (eds) Artificial intelligence systems and the internet of things in the digital era. EAMMIS 2021. Lecture notes in networks and systems, vol 239. Springer, Cham
15. Musleh Al-Sartawi AM, Razzaque A, Kamal MM (eds) (2021) Artificial intelligence systems and the internet of things in the digital era. EAMMIS 2021. Lecture notes in networks and systems, vol 239. Springer, Cham
16. Natifu B, Zikusooka A (2014) Ubuntu, professionalism, activism and the rise of public relations in Uganda. In: St. John B, Lamme MO, Etang JL (eds) Pathways to public relations: histories of practice and profession. Routledge, London, pp 224–238
17. Neill MS, Lee N (2016) Roles in social media: how the practice of public relations is evolving. PR J 10(2):1–25
18. Ningsih H, Umarella F (2020) Cyber public relations model in building a positive image of local government: (case study on the government of special capital region of Jakarta). Adv Soc Sci Res J 7(12):121–134. https://doi.org/10.14738/assrj.712.9398
19. Sakhnini Z, Elali W, Al-Sartawi A (2022) Behavioral and non-behavioral factors and the level of adapting and implementing fintech and e-banking in Bahrain: suggested model. In: Musleh Al-Sartawi AMA (eds) Artificial intelligence for sustainable finance and sustainable technology. ICGER 2021. Lecture notes in networks and systems, vol 423. Springer, Cham
20. Valentini C (2015) Is using social media 'good' for the public relations profession? A critical reflection. Public Relations Rev 41(3):170–177
21. VanDyke MS, Lee NM (2020) Science public relations: the parallel, interwoven, and contrasting trajectories of public relations and science communication theory and practice. Public Relations Rev 46:1–9

22. Vardeman-Winter J, Place K (2015) Public relations culture, social media, and regulation. J Commun Manag 19(4):335–353
23. Wolf K, Archer C (2018) Public relations at the crossroads: the need to reclaim core public relations competencies in digital communication. J Commun Manag 22(4):494–509
24. Yue C (2016) Strategic public relations management in China. Master's Thesis, West Lafayette, Indiana. Purdue University. Open Access Theses, 911. http://docs.lib.purdue.edu/open_access_theses/911. Accessed 20 Nov 2021

AAU Uses of Communication and Education Technology in Distance Learning During a Corona Pandemic-Presentation and Evaluation

Rahima Aissani

Abstract This study aims to present and evaluate the experience of Al Ain University (AAU) uses of communication and education technologys in the transition from regular education to distance Learningdue to the closure in light of the outbreak of the Corona pandemic. The AAU has used Microsoft Teams Application to present lectures and for the management of the educational process and its outputs. It adopted the Lock Down Browser to secure the examination environment. The study applied the analytical descriptive approach of the procedures of AAU for transition to distance education, and the descriptive survey of a sample of faculty members and students and their evaluation of this experience. The study concluded that the transition of AAU to distance Learning was quick and easy; and the online education management systems, represented in Moodle Platform and the online portal, that the University already used, contributed to this transition. The results were as follows: Continuity of education, effective crisis management, raising the efficiency of faculty and administrative staff and university students in the use of communication and education technologys, and distance learning skills and software.

Keywords Higher education · Education management strategies · Distance learning · Distance work · E-Learning platforms · Microsoft teams · COVID-19

1 Introduction

Many businesses have been affected by the closure and quarantine, especially education, in light of the spread of the Corona virus. A survey was conducted by the International Association of Universities (IAU) covering Africa, Asia, and the Pacific, USA and Europe. The results showed that 77% of the universities of Africa were closed as a precaution, despite recording lower rates of the outbreak of the epidemic

R. Aissani (✉)
College of Communication and Media, Al Ain University(AAU), Hamden Bin Mohammad Street, Jimi - Al Ain, Abu Dhabi, United Arab Emirates
e-mail: rahima.aissani@aau.ac.ae

A. M. A. Musleh Al-Sartawi et al. (eds.), *From the Internet of Things to the Internet of Ideas: The Role of Artificial Intelligence*, Lecture Notes in Networks and Systems 557,
https://doi.org/10.1007/978-3-031-17746-0_21

compared to the rest of the world's continents, and 55% of the universities in Asia, the Pacific, and Europe were closed. The percentage reached up to 45% for the Americas [1].

All countries reacted urgently to the educational file, to ensure that learners are not isolated from sources of knowledge. Some countries began educational broadcasts through television, and other educational programs. International organizations concerned with education have also endeavored to provide educational resources free of charge, and to provide their services via the Internet while ensuring access to information and open educational resources, Distance learning has taken different forms. [2] Distance Learning has taken different forms: some were satisfied with television broadcasts, interactive and non-interactive, some used the radio, and others sought to integrate educational means with interactive lessons online via different educational platforms [3].

The closure of schools and universities for a short or long period put educational systems in front of challenges that necessitated the implementation of emergency plans represented in finding other opportunities and alternatives to provide education. These included providing learning opportunities through educational platforms and online educational classes, establishing partnerships with these educational platforms to make available their resources and services free of charge to increase the capabilities of some countries to respond, and training teachers and school principals on remote work and education. The emergency plans also included information on how to deal with the virus, and the building of work teams including advisors and teachers to support parents and students and train them to adapt to the new situation [4].

The UAE has adopted distance Learning for all students, starting from March 22, 2020, and has applied this decision to public and private schools, and all higher education institutions. It has developed an integrated plan to activate distance Learning positively. The plan included specialized training for the teaching and administrative staff amounting. The training included a set of workshops, which were conducted entirely remotely, and launched reference guidelines for managing student behavior, and provided them with free educational platforms and free internet. [5] The Ministry has proposed on its smart learning portal 13 global educational platforms [6] to provide learners with multiple educational options, and broadcast safe videos without exposing them to advertisements or inappropriate content through its huge library of high quality educational videos. The AAU responded to the Ministry's decisions and adopted several measures to implement distance learning as of March 22, 2020.

This study deals with a detailed presentation and evaluation of the experience of AAU uses of communication and education technologys in the transition from regular education to distance Learningdue to the closure in light of the outbreak of the Corona pandemic. The AAU has used Microsoft Teams Application to present lectures and for the management of the educational process and its outputs. and applied it starting the second semester 2019/2020, and until the end of the summer semester 2020/2021.

2 Review of Literature

The study reviewed the David and other report to Using E-Learning Platforms on a Large Scale during Crises, as well as clarifying the opportunities produced by the (Covid-19) crisis, while highlighting real examples that illustrate how policies and decisions related to programming are formulated. The report concluded that the most significant obstacles are: The weakness of the data collection and education management system has been exacerbated by the expansion of the distance learning application, the slowing down of students' progress in the formal educational curricula, and the inability of students to access the Internet. The digital divide in low-income countries will be a stumbling block in the way of completing the school year of the distance learning system for many students, as well as the difference in teachers interaction with the distance Learningsystem [7].

The Basilaia & Kvavadze research aimed to study the experience of moving from school education to online education during the spread of the Corona virus epidemic in Georgia. She relied on the statistics of the first week of the teaching process in a private school and its experience in moving from face-to-face to online education using the two platforms Edu Page and Gsuitein the educational process [12]. The two researchers concluded that the transition between traditional education and online education was successful, and it is possible to benefit from the system and the skills acquired by teachers, students and school management in the post-epidemic period in various cases, such as teaching people with special needs who need extra hours, or by increasing the effectiveness of group teaching or increasing student independence and acquiring new skills [8].

It conducted a descriptive study aimed at clarifying the ways in which the Corona pandemic affected the reshaping of education in Indonesia, explaining the online learning types and strategies that teachers use in the world due to the closure of universities to limit the spread of the epidemic Corona virus, and the advantages and effectiveness of using distance learning. The study concluded that there was a high speed of the impact of the Corona epidemic on the education system, as the traditional method of education has declined, giving way to the spread of online learning, because it supports learning from home, and thus reduces the direct contact of individuals with each other, and limits the spread. The study proved the importance of using various strategies to increase the ease and improvement of distance education [9].

The study of Allam & other concluded The implications of online learning are the concern of university as a teaching method that is expected to improve students' academic performance by enhancing computer/Internet literacy competency, selfdirected learning, and motivation for learning level. the study found that Computer/Internet Literacy Competencyamong undergraduate students is at a high level, while Self-directed Learning and Motivation for Learning reported at a low level. The study concluded also The Online Distance Learning seems new norms that should not hinder potential interactive learning for best academic performance and to drive excellent achievement [11].

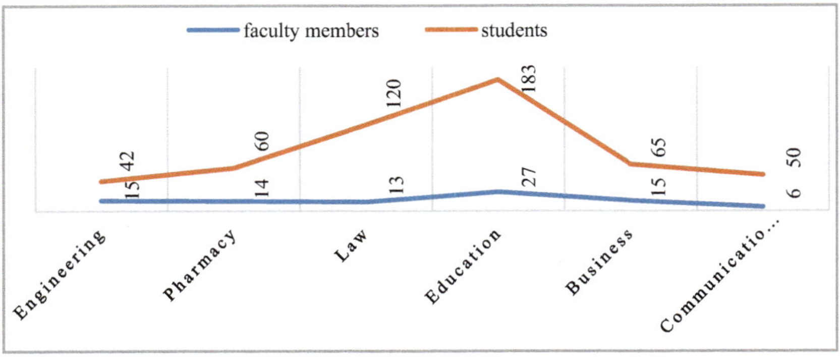

Fig. 1 Number of respondents

3 Methods

The study adopted the descriptive survey method using three tools:

a) Describing the procedures adopted by AAU to implement distance Learning.
b) direct interview with a representative sample of the academic staff, asking them about their opinion on the distance Learning experience, identifying the most important benefits, as well as the most important challenges and difficulties they faced.
c) A short questionnaire containing two open-ended questions to students:

- Identify the three most important advantages you have benefited from the distance Learning, in light of the Corona pandemic.
- Mention the three most important disadvantages of the distance Learning.

The sample included 90 faculty members representing 50% of the total faculty members at AAU in the academic year 2020/2021, and 520 male and female students representing 10% of the total students registered in the second semester 2020/2021, distributed to the six colleges. (see Fig. 1).

4 Procedures and Measures Adopted by AAU to Apply Distance Learning

AAU adopts a full-time system (morning and evening), and attendance to classrooms to receive knowledge, considering the educational process as a participatory process based on actors who have the same levels of influence so that its results are influential in the internal and external environment. Therefore, it seeks to harmonize and balance between the needs of these actors (students and faculty members and provides competitive educational services that attract students), and the requirements and needs of the labor market for the availability of high and professional skills and capabilities. It uses educational technology through the Moodle Platform (lectures, practical work and exams), and through the online portal (attendance and absence, grade monitoring, academic advising, discussion and communication groups).

In accordance with the decisions of the UAE Ministry of Education and the precautionary measures in light of the Corona pandemic, AAU has moved from regular education to distance Learning starting March 22, 2020, and continuing until the end of the first and second summer semesters 2020/2021, totaling two and a half regular semesters (mid-spring 2019/2020, fall and spring 2020/2021), and four summer semesters (the first and second of 2019/2020, and the first and second of 2020/2021). Starting the second mid of the first semester 2020/2021, hybrid education was applied in scientific faculties (Pharmacy and Engineering), and some applied courses in humanities faculties (Law, Education, Business, Communication and Media).

4.1 Subscription to Distance Learning Platforms (MS-Teams) and Distance Lecturing Application (Zoom)

AAU has chosen Microsoft Teams from among the 13 platforms proposed by the Ministry, due to its distinctive set of characteristics, the most important of which is its compatibility with (Microsoft) applications and programs. It is a collaboration and communication platform for institutions, professional teams and companies [13].

It adopted Lock Down Browser in exams, a web browser developed by Responds, which enables users to inhibit cheating while the students are taking their exams online, and securing exam environment. The student will not be able to navigate to other URL addresses, switch apps, and take screenshots, or copy or print questions. It also adopted assessment methods online that included three levels: Online exams, practical homework and projects and Take Home Exams. The AAU has relied on the standard version of the application.

4.2 Training Members of the Administrative and Academic Bodies on the use of the Platform

The Information Technology Center conducted a training courses for all faculty members on both campuses on how to use the MS-TEAMS program for E-distance Learning During the first week of spring break (11–19) March 2020 two training workshops per day.

4.3 *Issue Users' Guide to Help use (MS-Teams)*

In print and electronic form, and educational videos for faculty and technical staff members and students, and upload them on the University's website and Moodle Platform. It also issued the Faculty Members' Online Exam Guide, and organized workshops for members of the administrative and academic bodies to train on how to manage online exams using the Lock Down Browser, and provided classrooms with hybrid education equipment and devices (recording cameras, and broadcast via the Teams).

5 Discussion

AAU continued to use the distance Learning application MS-Teams until the end of the second summer semester 2020/2021. It thus achieved several results:

5.1 *Continuity of Education*

The study has been conducted remotely since the middle of the second semester 2019/2020, the first and second summer semesters 2019/2020, and the first and second semesters 2020/2021. The AAU also scheduled the first and second semesters 2020/2021, to continue teaching remotely via MS-Teams. The following statistics in Fig. 2 including the number of sections and students enrolled in them in each semester, demonstrate this continuity. The number of sections in all faculties of the university increased from 950 sections in the second semester 2019/2020 to 1035 sections in the first semester 2020/2021, and 1075 sections in the second semester 2020/2021, and the number of registered students increased from 4886 students in the first semester 2020/2021, to 5,200 students in the second semester 2020/2021.

5.2 *Effective Crisis Management*

The AAU was able to continue to hold all academic curricular activities (lectures, exams, and discussion of projects), extra-curricular activities (organizing events) remotely, as well as all administrative activities and events, such as registration, issue the documents and certificates that students need, and work from inside the University headquarters, taking all precautionary measures adopted (wearing a mask, social distancing, daily sterilization, etc.). AAU has received 5 star rating for "e-learning" according to the "QS STARS" Rating System on July 20, 2020, and ranked

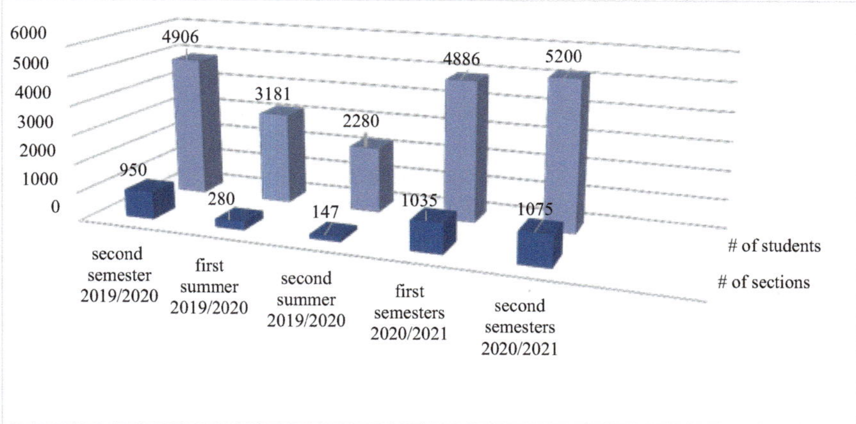

Fig. 2 Statistics of the number of sections and students during the semesters in the Corona pandemic at AAU

second at the level of UAE universities after being ranked among 401–600 universities worldwide in quality education, impact, and how universities deal with some urgent humanitarian problems through research, supervision, awareness, and education (April 2021). This is according to the international Times Higher Education University Impact Rankings 2021, the largest worldwide ranking in terms of the number of universities, which reached 1240 universities from 98 countries. This came during the "Innovation and Impact Summit 2021".

Raising the Faculty Members and University Students' Efficiency in the Use of communication and Education Technologys, and distance learning skills by mastering new educational technologies and basic software towards preparing and implementing lectures by faculty members, and exams, assignments, costs and curricular, and extra-curricular projects by students. **The results of interviews with academic members concluded that the most important advantages are:** Acquiring the skills of using distance Learning platforms, developing the skills of preparing, drafting, producing and directing e-educational content and how to communicate and publish it to students, and moving from traditional education based on the face-to-face communication and the use of tangible illustrative materials to e-learning based on stimulating participation, sustained attention, and follow-up.

As for the difficulties, the faculty members answered that they encountered a number of challenges that centered on:

- The transition process itself from (face-to-face) education to distance Learning, which differs from the former in several respects, the most important of which are: Teaching methods, diversifying educational activities, and creating and sustaining virtual interaction methods.
- The difficulty of teaching scientific courses that are incompatible with the nature of e-learning, such as media, art production, and photography courses, and

courses that require practical work, practical training, and conducting scientific experiments in specialized laboratories.
- The assessment of learning represented the greatest challenge among these due to the inadequacy of some traditional assessment tools in remote assessment in various academic courses, especially with regard to the requirements for evaluating educational outputs related to analytical and critical writing, and analyzing data from marketing studies and scientific research plans.

This called for searching for alternative assessment tools fitted for e-learning, in order to achieve the best output of the distance Learning process. Therefore, a guiding document was prepared for the faculties of the University and its departments for alternative assessment tools prepared by a committee of specialists formed for this task. The document contained about 15 tools of the most widely used alternative assessment tools, which are: The quizzes, home tests, oral tests, open book test, practical tests, exercises or activities at home, online collaboration and discussion, electronic achievement file, individual and group projects, recorded oral presentations, production of educational videos and recordings, case studies, research papers summaries, critical, reviews, and role- plays and simulations. The document emphasized that the University colleges and departments are free to choose assessment tools according to the nature of software requirements and courses.

The student questionnaire on the distance Learning experience concluded that the three most important positives that the interviewed students benefited from in distance Learning in light of the Corona pandemic are:

- Saving time and reducing costs, continuity of learning across time and space, and the ability to have immediate and rapid access to educational resources, data and information in their read, audio and visual forms, and thus raising the ability to understand, absorb, and learn in an effective manner (93%).
- Enhancing skills and abilities in using communication and education technologys and various software to benefit from available educational materials, design and production of digital contents, independence, and enhancing creativity and thinking skills (87%).
- Involving in virtual activities, discussions, and dialogues, being able to participate more freely in lectures and curricular and extra-curricular activities, overcoming feelings of tension and shyness, and enhancing communication skills through electronic media (75%).

As for the three most important negatives in distance Learning, according to the response of the sample students, they are as follows:

- Weak communication networks, simultaneous pressure on Internet networks, and some urgent problems in accessing electronic platforms (89%).
- Weak social spirit and human dimension, which is obtained more through physical interaction and communication with teachers and colleagues, feeling bored from sitting in front of screens, and losing the ability to build friendships during this period (86%).

- Fear and constant anxiety about the continuation of situation, and the delayed progress in completing the curricula and graduating on time (83%).

6 Conclusion

In response to the precautionary measures in place in the UAE in light of the outbreak of the Corona pandemic, AAU has moved from formal education to distance Learningeasily, aided by e-learning management systems, represented in Moodle Platform and the online portal, which has been in use formerly [10]. The choice of moving to distance Learningwas an emergency shift in the management of the educational process at AAU (and in most universities in the world), through which the University managed to achieve a number of results, the most important of which are: continuity of education, effective crisis management, raising the efficiency of faculty and administrative staff and students in the use of communication and education technologys, and distance learning skills and software.

Accordingly, in light of the uncertainty about the decline or end of the Corona virus, or the emergence of new epidemics, the study recommends working on the creation of a long-term project to implement dual formal/distance Learningand to benefit from the outcomes of this experience and what is provided by online educational platforms, communication technology and education in this field. This is in preparation for managing emerging crises and emergencies that prevent the continuation of formal education as happened with the Corona pandemic. It is certain that things will not return to normal after the pandemic, and that there is an urgent need to conduct a comprehensive review of the methods and curricula of education in classrooms, and prepare to face any new emergency and coexist with it.

References

1. Marinoni LHV, Jensen T (2020) The impact of covid19 on higher education around the world. IAU: Global Survey, p 70
2. Audah M, Capek M, Patil A (2020) COVID-19 and digital learning preparedness in Jordan. World Bank Blogs. https://blogs.worldbank.org/arabvoices/covid-19-and-digital-learning-preparedness-jordan. Accessed 13 Mar 2022
3. World Bank Blogs. https://blogs.worldbank.org/arabvoices/covid-19-and-digital-learning-preparedness-jordan. Accessed 13 Mar 2022
4. ABEGS (Arab Bureau of Education for the Gulf States) (2020) Adopting digital distance education, 1
5. UAE Government (2020) Distance learning in times of COVID-19 - The Official Portal of the UAE Government. https://u.ae/ar-ae/information-and-services/education/distance-learning-in-times-of-covid-19. Accessed 13 Mar 2022
6. Al-Jammal A (2020) Education adopts 13 online platforms to support distance learning. Emarat Al Youm newspaper, 25 March 2020. https://www.emaratalyoum.com/local-section/education/2020-03-25-1.1324800. Accessed 13 Mar 2022. The platforms are: Madrasa E-learning Platform, Alef, Nahla wa Nahil Platform, Microsoft Teams, MC Grow Hill, Oxford University

Press, College Board Organization, Twig, Ynmo, Matific
7. David R, Pellini A, Jordan K, Philips T (2020) Education during the COVID-19 crisis: opportunities and constraints of using EdTech in low-income countries. Part of our EdTech and coronavirus (COVID-19) series: edtechhub.org/coronavirus. https://doi.org/10.5281/zenodo.3750976
8. Basilaia G, Kvavadze D (2020) Transition to online education in schools during a SARS-CoV-2 coronavirus (COVID-19) pandemic in Georgia. Pedagogical Res 5(4):em0060. https://doi.org/10.29333/pr/7937. Accessed 27 May 2020
9. Yulia H (2020) Online learning to prevent the spread of pandemic corona virus in Indonesia. Eternal English Teach J 11(1):48–56. https://doi.org/10.26877/eternal.v11i1.6068
10. Al-Qassim M, Chen W, Al-Sartawi A (2022) Blockchain in environmental compliance and enforcement. In: Musleh Al-Sartawi AMA (eds) Artificial intelligence for sustainable finance and sustainable technology. ICGER 2021. Lecture notes in networks and systems, vol 423. Springer, Cham
11. Allam SNS, Hassan MS, Mohideen RS, Ramlan AF, Kamal RM (2020) Online distance learning readiness during Covid-19 outbreak among undergraduate students. Int J Acad Res Bus Soc 10(5):642–657
12. Musleh Al-Sartawi AMA (2020) E-learning improves accounting education: case of the higher education sector of Bahrain. In: Themistocleous M, Papadaki M, Kamal MM (eds) Information systems. EMCIS 2020. Lecture notes in business information processing, vol 402. Springer, Cham
13. Hubbard M, Bailey MJ (2018) Mastering microsoft teams. Apress, P43–74. & Hilaire, D.St. (2019). "4 Advantages of Microsoft Teams with Office 365". https://www.knowledgewave.com/blog/4-advantages-of-microsoft-teams-with-office-365. Accessed 13 Mar 2022.& Stanfield N (2020) The best 25 features of microsoft teams in 2020. https://www.stanfieldit.com/microsoft-teams/. Accessed 13 Mar 2022

Healthy Living and New Media: A Review

Mohammad Habes, Sana Ali, Dina Tahat, and Juma Khalifa Juma Aldarmaki

Abstract Healthy living is an essential aspect of sustainable socioeconomic growth and development. Those who are healthy, can effectively cope with everyday challenges and avail better living standard. This research article aims to highlight the importance of health and the role of media as an important social institution to promote a healthy lifestyle and awareness. The authors highlighted the potential of mass media content to spread health awareness and reinforce healthy behaviours among the audience. For this purpose, Social Marketing is one of the most preferred techniques that can influence public beliefs and attitudes towards personal healthcare. For instance, Social Marketing ads regarding Cancer, HIV and Breast Cancer are an illustrating example of mass media influence to spread awareness and promoting healthy lifestyles. Besides, during the outbreak of EBOLA, Zika, MERS, SARS, Social Marketing also helped to inform the public and as a result, they adopted personal care measures. Thus, the researchers concluded that the role of mass media to promote healthcare is of greater significance. However, besides media, other artificial intelligence should also play their role to provide better health opportunities and improve the living standard by new media.

Keywords Mass media · Health living · Healthcare awareness · Personal care · AI

M. Habes (✉)
Radio and TV Department, Yarmouk University, Irbid Zip Code 21163, Jordan
e-mail: mohammad.habes@yu.edu.jo

S. Ali
Department of Mass Communication, Allama Iqbal Open University Islamabad, Islamabad 44000, Pakistan

D. Tahat
College of Education, Humanities and Social Sciences, Al Ain University, 64141 Al Ain, UAE
e-mail: dina.tahat@aau.ac.ae

J. K. J. Aldarmaki
Universiti Utara Malaysia-School of Management, Changlun, Malaysia

A. M. A. Musleh Al-Sartawi et al. (eds.), *From the Internet of Things to the Internet of Ideas: The Role of Artificial Intelligence*, Lecture Notes in Networks and Systems 557,
https://doi.org/10.1007/978-3-031-17746-0_22

1 Introduction

Social media platforms are a major global revolution in connecting people as they have developed the concept of a modified communication for their users. Today, these platforms are considered as an integral part of our social life. [1] The social impact of new media is an indisputable reality in today's world [2] as the sociocultural spheres are widely mediated by the Internet and social networks [3–5]. Due to extensive availability of digital media technologies, it is possible to educate and raise awareness health living conditions [6]. Increased fragmentation of media types and evolving technologies have largely facilitated the persuasive promotional message regarding public awareness in general [7]. To raise public awareness regarding personal care, the role of new media platforms is of greater magnitude. [8] In most cases, media type and content, both are chosen and designed according to the particular criteria e.g., audio-visual support of television, the graphic-textual content of newspapers and modified features of social media, fundamentally entice a larger number of the audience [9]. Thus, the aim of this discussion is to highlight the influence of mass media regarding healthcare promotion. As, mass media are largely implied in promoting public health through special programmes, campaigns, advertisements, pamphlets, booklets and other content all are frequently designed and disseminated among the audience [10]. In this regard, the power and impact of mass media can reach to a vast number of audience, promoting healthy living and motivating people to rethink about their current living conditions [11]. Since media have a strong impact on collective and individual beliefs, they are well-capable of reshaping convictions and behaviours [12, 13]. A situation in which the influence of the media on healthy living conditions through health promotion is moderately known by researchers, acquiring information concerning healthy lifestyle is the need of the day [14]. In the field of health, the media are fundamental, because through the active and conscious participation of the community, they manage to promote self-care of health, disseminate information about it and increase awareness of specific aspects of individual health. and collective. [15] Framed within a situation in which the influence of the media on healthy living conditions through health promotion is a reality unknown in depth by researchers, what is pursued is the search for information and the effective obtaining of knowledge about the subject. This literature review is limited to the role of the mass media as a condition for healthy living through the promotion of the health of individuals. The formulation of the problem was established in the following question: Can the media influence healthy living conditions through health promotion? The reason for carrying out this research relies primarily on the social relevance of the subject. Given that public health ethics demand media to spread social awareness concerning healthcare, leading to adopt a healthy lifestyle [16]. In this context, the importance of communication in the healthcare sector is clear as media can successfully entice and persuade the masses reaching to a larger number of people. [17] From a conceptual point of view, the mass media are part of the political-ideological super structure of society, and a fundamental mechanism of information for popularising health promotion. Thus mass media have become an essential source of information as they are effectively promoting healthcare and persuading the public to adopt a healthy lifestyle. Due to active participation, even audiences are also engaged in acquiring this information and implementing both on individual and collective level [18].

2 Methods

Method refers to guide the study to collect the relevant information and investigate it both systematically and empirically [19, 20]. The present investigation is of an applied-descriptive method as it allows the knowledge usage in reality and also helps in characterising a specific situation indicating its most peculiar features and describing all its main components [14]. According to the epistemological stance, this research is pragmatic and critical as the extensive discussion will lead the researchers to reality [21].

3 Analysis and Discussion

Healthcare should be considered in a broader context as "health is a state of complete physical, mental and social well-being, not just the absence of disease or illness" [22]. Complementing this idea, [23] argued that there are four general determinants that influence health: Human biology, environment, lifestyle and organization of health care. Here the assumption is, health seems to be equally maintained by medical science, environmental factors, social institutions and one's own effort. In this sense, human biology refers to the study of human life from cellular level to the population level and discussed by every discipline in its own different manner [24]. Likewise, according to the environmental perspective, human life is the development of a new generation which more developed, urbanized and aware of the self-care and wellbeing. Life exists with managing survival and working for progress [25]. Within the social arena, human life is the building block of society. Due to this reason, the existing social institutions bear the responsibility to work for the survival of human life [26]. Health in a broader context can also be defined as the level of functional or metabolic efficacy of an organism and in harmony with the environment, measured by the ability of each person to do or become what they want to be. Indeed, health is perceived as the possibility that a person is enjoying biological, psychological and so social harmony, in dynamic interaction with the environment in which he lives [27]. For it encompasses the state in which a human body normally exercises all its functions which it appropriately performs in a given moment. Thus health is a life-long consideration which witnesses the quality of life, well-being and ultimately happiness. Here health well-being not only implies on medical practitioners but also on all the socio-political institutions. Health also constitutes that all the fundamental needs are covered in the sociocultural context. According to Beaudoin et al. [23], a more dynamic definition of health is the achievement of the highest level of physical, mental and social well-being. Moreover, the ability to function and keep pace with the rapidly increasing social challenges. In this context, health is not considered as an abstract state, but as a resource that allows people to lead individual, social and economically productive lives. Broadly speaking, health is the potential to live a normal life, a positive concept that emphasizes the provision of social and personal resources, as well as physical skills. Furthermore, the importance of health resides in allowing a person's body to maintain good operating standards and ability to carry out the different activities [28]. Health is a phenomenon that is achieved through a host of individual and collective actions and can be sustained for a long time. Better health ensures happy living and more concentration towards productivity results in more positive outcomes The importance of health can be assumed by its role as it ensures collective well-being and prosperity for society. As it constitutes a multi-sided value both on an individual and social level, healthcare and well-being have become an important part of social policies in many countries [29]. Stead et al. [30], it is important to think of health as a process, rather than as a state; secondly, people from varying cultural and geographical backgrounds have a varying perception about it and third, and thirdly, it is a dynamic process, with stimuli and responses, in continuous, constant and permanent change. Closing the idea Álvarez-Dongo et al. [31] establish that ultimately health is the dynamic balance of risk factors between the environment and within certain parameters. To understand this concept, it is important to study what a healthy living condition actually implies and what people perceive about it on both individual and the collective level (Fig. 1).

Most authors define healthy lifestyles as a set of behavioural patterns that have positive repercussions for people's life. However, these behavioural patterns are chosen voluntarily or involuntarily by people [18]. According to Grenfell & Hardy [32] while the medical model has defended the exclusively voluntary nature of this choice, the authors of psychosocial orientation understand the choice as involuntary to a certain extent since they recognize the influence of psychosocial variables on the acquisition and maintenance of lifestyles. The expression healthy life condition begins to be defined as an integrating concept that includes all areas of life with

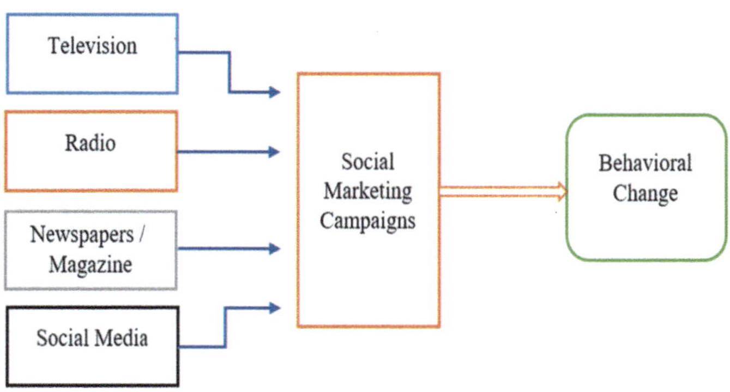

Fig. 1 The study framework

a multidimensional character, in the mid-1970s. Thus, although historically there have been two basic approaches: the one that conceives it as a unitary entity and the other that considers it a construct made up of a series of domains, no consensus has been reached to this issue and now the social and economic indicators are allowed to measure the argumentation concerning social welfare of the masses [28].

Despite there are traditionally recognized social institutions e.g., communities, schools and family members, mass media equally bear the responsibility of social well-being and support. [8] Similarly, regarding the mass media as a health promoter, literature has defined them as an effort to disseminate relevant awareness that further leads to adopting healthcare and well-being, indicating a positive concern towards health. To promote public health, the use of social marketing to design and implement social healthcare policies is one of most prominent role of mass media today [33] as it is an extension of traditional marketing that deals with the ideas or products, advertised and promoted for the social causes. Social marketing is not aimed to benefit the manufactures rather, to promote social health and well-being [34]. The implementation of social marketing by using creative, distinctive conceptual ideas through social media platforms, motivate behavioural changes. For this purpose, potential marketers specially create and design the messages and convey them through specially designated media campaigns [27, 30]. According to [35], the state is also responsible for promoting healthcare campaigns and for this purpose, media platforms are hired to disseminate public service healthcare awareness messages. For instance, "Hombres Sanos" as a renowned healthcare campaign regard HIV awareness in Spain remarkably attracted the targeted audience and they showed a great engagement to adopt HIV prevention measures [36]. Similarly, mass media also ran several campaigns to adopt certain measures during healthcare emergencies o traditional media. Besides Social Networking platforms have also remained under consideration to promote healthcare in general [16]. To further affirm this examined the correlation between media-based healthcare awareness campaigns regarding breast cancer and substance abuse in Australia. Results indicated that media exposure to these healthcare campaigns positively influenced the people's behaviour leading them to perform mammography regularly and seek help against the drug abuse. Thus, the researchers considered media campaigns as highly attractive and strongly influencing public behaviour. According to [27], these marketing campaigns are deliberately attractive and powerful. Marketers use various communication strategies that motivate behavioural changes and reinforce the adoption of a healthy lifestyle.

Another major example of mass media influence was confirmed by [37] as they investigated the role of social marketing campaigns to promote quitting tobacco usage among the adult population in Malawi. Findings witnessed a strongly influential role of mass media campaigns for persuading the individuals as message content, duration and nature, all equally worked to influence the potential audience. Being capable of governing public behaviour, the role of mass media as a social institution is highly influential. This change is not merely a simple process rather, it is vigilantly observed, designed and implemented for public health and well-being Now, due to the increase in communication, the availability of information and the presence of artificial intelligence, people are also demanding health care awareness that enhances the role of the media in health care support and information [6].

According to [38, 39] the media, it contributes to publishing all information and news related to health, as well as publishing achievements and seminars, in addition to publishing workshops and issues that are highly related to health and health experiences. Similarly, [40] the health media contributes to providing all statistics and data related to health, with the importance of focusing on analyzing and presenting them to all specialized public groups. Through the use of activities, the media aims to raise health awareness among targeted individuals and communities, in addition to contributing to instilling healthy and positive behaviors. [41] In addition, the media is an important factor in strengthening relations and partnership between it and the health sector, through providing correct information, and the importance of focusing on the consequences caused by rumors and false information, and in encouraging community institutions to join hands with each other; This is in order to serve the health field, [42] in addition to encouraging efforts that are aimed at developing and developing all health awareness programs. While Nikdel Teymori & Fardin, (2020) Ban stresses the media in highlighting the efforts resulting from the health initiatives of working institutions and individuals, it also contributes to encouraging media cadres to creativity and innovation in the field of health media, with the importance of focusing on encouraging creative people to participate and contribute to health awareness programs, and contributes to honouring Appreciating all media competencies that are distinguished in the health sector. Therefore, the above discussion contains can be a gateway to further studies and debates concerning the role of Social and Environmental institutions for public healthcare and well-being. However, still, any finite conclusion does not yet exist, that is why some alternatives are proposed that could well be taken into account for those interested in health promotion through the media, such as: Employing artificial intelligence applications at Communication media used in health promotion with a longitudinal study and a marketing plan for health promotion [44].

4 Conclusions

The social responsibility of mass media as liable for better living conditions and healthy lifestyle also demands other Social Institutions to promote health and well-being. Institutions such as schools, family, Church, Mosques and others are potentially influential to educate and change public behaviour regarding healthcare. Education and information are important parts of media content that can be further used to develop healthy behaviours. By using different tactics. media platforms are capable of enticing a massive number of individuals that further educates them concerning the importance of health and well-being. Employing artificial intelligence positively alter our perceptions and develop healthy behaviours as healthy individuals are more capable of performing well and even, capable of achieving economic progress. By reinforcing active (AI)apps, artificial intelligence can inform and educate not only in the crisis but also in normal situations. We can conclude that the media, (AI)apps plays a major role in the health field, as it is not limited to the dissemination of all disease cases, but rather contributes to the dissemination of all matters related to health, such as talking about health insurance in addition to health institutions, and the media focus on confronting special behaviours.

References

1. Saraf RA, Balamurugan J (2018) The role of mass media in health care development: a review article. J Adv Res J Mass Commun 5(1–2):39–43. https://doi.org/10.24321/2395.3810.201807
2. Alhammad KL, Habes M, Al Olaimat F, Haddad I (2021) Attitudes of students of the faculty of mass communication at Yarmouk University towards using video platforms in distance education. Rev Int Geogr Educ Online 11(5):1041–1052
3. Habes M, Salloum SA, Alghizzawi M, Mhamdi C (2020) The relation between social media and students' academic performance in Jordan: YouTube perspective Mohammed. https://doi.org/10.1007/978-3-030-31129-2
4. Alghizzawi M (2019) A survey of the role of social media platforms in viral marketing: the influence of eWOM. Int J Inf Technol Lang Stud 3(2)
5. Alghizzawi M (2019) The role of digital marketing in consumer behavior: a survey. Int J Inf Technol Lang Stud 3(1)
6. Habes M (2019) The influence of personal motivation on using social TV: a uses and gratifications approach. Int J Inf Technol Lang Stud 3(1)
7. Quattrin R (2015) Health promotion campaigns and mass media: looking for evidence. Prim Heal Care Open Access 05(01). https://doi.org/10.4172/2167-1079.1000190
8. Alhumaid K, Habes M, Salloum, SA (2021) Examining the factors influencing the mobile learning usage during COVID-19 pandemic: an integrated SEM-ANN method. IEEE Access
9. Habes M, Ali S, Elareshi M, Tahat KM, Ziani A (2021) Understanding users' social TV content using artificial intelligence approach: a survey. In: 2021 international conference of modern trends in information and communication technology industry (MTICTI), pp 1–7
10. Catalán-Matamoros D (2011) The role of mass media in public health education. In: Health management - different approaches and solutions
11. Maryon-Davis A (2012) Using the mass media to promote health
12. Thompson C, Gray K, Kim H (2014) How social are social media technologies (SMTs)? A linguistic analysis of university students' experiences of using SMTs for learning. Internet High Educ 21. https://doi.org/10.1016/j.iheduc.2013.12.001.
13. Seo H, Houston JB, Knight LA, Kennedy EJ, Inglish AB (2014) Teens' social media use and collective action. New Media Soc 16(6):883–902. https://doi.org/10.1177/1461444813495162
14. Al-Shibly MS, Alghizzawi M, Habes M, Salloum SA (2019) The impact of de-marketing in reducing Jordanian youth consumption of energy drinks. In: International conference on advanced intelligent systems and informatics, pp 427–437
15. Al-Shakhanbeh ZM, Habes M (2022) The relationship between the government's official Facebook pages and healthcare awareness during Covid-19 in Jordan. In: Advances in data science and intelligent data communication technologies for COVID-19. Springer, pp 221–238
16. Denecke K et al (2015) Ethical issues of social media usage in healthcare. Yearb Med Inform 10(1):137–147. https://doi.org/10.15265/IY-2015-001
17. Salloum SA, AlAhbabi NMN, Habes M, Aburayya A, Akour I (2021) Predicting the intention to use social media sites: a hybrid SEM - machine learning approach. Adv Intell Syst Comput 1339(March):324–334. https://doi.org/10.1007/978-3-030-69717-4_32
18. Murphy G, Corcoran C, Tatlow-Golden M, Boyland E, Rooney B (2020) See, like, share, remember: adolescents' responses to unhealthy-, healthy-and non-food advertising in social media. Int J Environ Res Public Health 17(7):2181

19. Al-Maroof RS, Salloum SA, AlHamadand AQ, Shaalan K (2020) Understanding an extension technology acceptance model of google translation: a multi-cultural study in United Arab Emirates. Int J Interact Mob Technol 14(03):157–178
20. Salloum SA, Shaalan K (2018) Investigating students' acceptance of e-learning system in higher educational environments in the UAE: applying the extended technology acceptance model (TAM). The British University in Dubai
21. Ali S (2018) Social media usage among teenage girls in Rawalpindi and Islamabad. Glob Media J 16(31):1
22. Lee C (2009) The interplay between media use and interpersonal communication in the context of healthy lifestyle behaviors: reinforcing or substituting? Mass Commun Soc 13(1):48–66
23. Beaudoin CE, Fernandez C, Wall JL, Farley TA (2007) Promoting healthy eating and physical activity: short-term effects of a mass media campaign. Am J Prev Med 32(3):217–223
24. Stinson S, Bogin B, O'Rourke D, Huss-Ashmore R (2012) Human biology: an evolutionary and biocultural perspective. 2nd edn, September 2018, pp 1–22. https://doi.org/10.1002/9781118108062.ch1
25. Lamprecht M (2016) The Role of the built environment in human life. Selected issues. Eur Spat Res Policy 23(2):65–81. https://doi.org/10.1515/esrp-2016-0011
26. Amzat J, Razum O (2014) Medical sociology in Africa. Med Sociol Africa, pp 1–299. https://doi.org/10.1007/978-3-319-03986-2
27. Evans WD (2006) How social marketing works in health care. BMJ 332(7551):1207–1210
28. Raggatt M et al (2018) 'I aspire to look and feel healthy like the posts convey': engagement with fitness inspiration on social media and perceptions of its influence on health and wellbeing. BMC Public Health 18(1):1002
29. Przybylska D, Borzęcki A, Drop B, Przybylski P, Drop K (2014) Health education as an important tool in the healthcare system. Polish J Public Heal 124(3):145–147. https://doi.org/10.2478/pjph-2014-0032
30. Stead M, Gordon R, Angus K, McDermott L (2007) A systematic review of social marketing effectiveness. Health Educ
31. Álvarez-Dongo D, Sánchez-Abanto J, Gómez-Guizado G, Tarqui-Mamani C (2012) Sobrepeso y obesidad: prevalencia y determinantes sociales del exceso de peso en la población peruana (2009–2010). Rev Peru Med Exp Salud Publica 29:303–313
32. Musleh Al-Sartawi AM, Razzaque A, Kamal MM (eds) (2021) Artificial intelligence systems and the internet of things in the digital era. In: EAMMIS 2021. Lecture notes in networks and systems, vol 239. Springer, Cham
33. Grier S, Bryant CA (2005) Social marketing in public health. February 2005. https://doi.org/10.1146/annurev.publhealth.26.021304.144610
34. Garde S, Knaup P, Hovenga EJS, Heard S (2007) Towards semantic interoperability for electronic health records. Methods Inf Med 46(03):332–343
35. Subica AM, Grills CT, Douglas JA, Villanueva S (2016) Communities of color creating healthy environments to combat childhood obesity. Am J Public Health 106(1):79–86. https://doi.org/10.2105/AJPH.2015.302887
36. Elder RW, Services H, Soler RE, Labre M (2014) Mass media health communication campaigns combined with health-related product distribution, March 2018. https://doi.org/10.1016/j.amepre.2014.05.034
37. Durkin S, Brennan E, Wakefield M (2012) Mass media campaigns to promote smoking cessation among adults: an integrative review. Tob Control 21(2):127–138. https://doi.org/10.1136/tobaccocontrol-2011-050345
38. Junco R, Heiberger G, Loken E (2011) The effect of Twitter on college student engagement and grades. J Comput Assist Learn 27(2):119–132. https://doi.org/10.1111/j.1365-2729.2010.00387.x
39. Alghizzawi M, Habes M, Salloum SA (2019) The relationship between digital media and marketing medical tourism destinations in Jordan: Facebook perspective. In: International conference on advanced intelligent systems and informatics, pp 438–448

40. Tahat K, Self CC, Tahat ZY (2017) An examination of curricula in Middle Eastern journalism schools in light of suggested model curricula. J Organ Cult Commun Confl 21(1):1–23
41. Singhal R, Rana R (2015) Chi-square test and its application in hypothesis testing. J Pract Cardiovasc Sci 1(1):69. https://doi.org/10.4103/2395-5414.157577
42. Huang H, Zhong W, Lai Q, Qiu Y, Jiang H (2020) The spatial distribution, influencing factors, and development path of inbound tourism in China-an empirical analysis of market segments based on travel motivation. Sustain 12(6). https://doi.org/10.3390/su12062508
43. Musleh Al-Sartawi AMA (2020) E-learning improves accounting education: case of the higher education sector of Bahrain. In: Themistocleous M, Papadaki M, Kamal MM (eds) Information systems. EMCIS 2020. Lecture notes in business information processing, vol 402. Springer, Cham
44. Bhaskaran N, Kumar M, Janodia MD (2017) Use of social media for seeking health related information – an exploratory study. J Young Pharm 9(1):107–114. https://doi.org/10.5530/jyp.2017

Big Data Systems Architecture and Data Security Fundamentals—Case Study Approach for a Hospital in Singapore

Girija Periyasamy, Easwaramoorthy Rangaswamy, and Nishad Nawaz

Abstract With data systems generating bigger data than ever before and also with data security under bigger threat, it is crucial of critical sectors like healthcare to revisit their Big Data Systems Architecture and Data Security Fundamentals. The organisation for the current case study chosen is a proposed Singapore based general hospital, SAI-GE. The proposed outcomes for the Big Data Systems Architecture and Data Security Fundamentals would be replicated for any similar hospitals across the globe. The various tools discussed are Ingestion Tool (Flume vs Kafka), Processing Tool (Hive vs Impala), Analysis Tool (Mahout Vs MLLib) and Orchestration Tool (Oozie vs Azkaban). Security Risk has been studied including Data ex-filtration by internal staff, Data ex-filtration by external threats, Poor data quality, Non-compliance to regulatory requirements and Disclosure Risk problem. List of security controls/measures that address the risks was studied and includes Preventive Control, Deductive Control and Corrective Control. Apart from the three key security controls, Access control which is also important security control is also studied. High Level Design Diagram along with Processing Design and Dataflow Diagram has been proposed for a hospital in Singapore, which would be replicated in other hospitals across with world with minor customizations to the regulatory framework and infrastructural capabilities.

Keywords Big data systems architecture · Data security · Healthcare big data · Healthcare data security · Security controls · Dataflow design

G. Periyasamy · E. Rangaswamy
Amity Global Institute, Singapore, Singapore
e-mail: erangaswamy@singapore.amity.edu

N. Nawaz (✉)
Department of Business Management, College of Business Administration, Kingdom University, Riffa, Bahrain
e-mail: n.navaz@ku.edu.bh

A. M. A. Musleh Al-Sartawi et al. (eds.), *From the Internet of Things to the Internet of Ideas: The Role of Artificial Intelligence*, Lecture Notes in Networks and Systems 557,
https://doi.org/10.1007/978-3-031-17746-0_23

1 Introduction

With data systems generating bigger data than ever before and also with data security under bigger threat, it is crucial of critical sectors like healthcare to revisit their Big Data Systems [1] Architecture and Data Security Fundamentals. In the present day hyper-connected world, the critical infrastructure is more vulnerable than ever to cyber threats, whether state sponsored, criminal groups or individual [2]. Especially, with the advent of Internet protocol suites and packet-switching technologies, there has been notable concerns of security, privacy, scalability, and reliability in Big Data Systems Architecture and Data Security Fundamentals [3]. The organisation for the current case study chosen is a proposed Singapore based general hospital, SAI-GE. The proposed outcomes for the Big Data Systems Architecture and Data Security Fundamentals would be replicated for any similar hospitals across the globe [4]. This paper introduces a fundamental framework and architecture for a hospital. High Level Design Diagram along with Processing Design and Dataflow Diagram is proposed. Security Risk has been also studied including Data ex-filtration by internal staff, Data ex-filtration by external threats, Poor data quality, Non-compliance to regulatory requirements and Disclosure Risk problem. List of security controls/measures that address the risks was studied and includes Preventive Control, Deductive Control and Corrective Control.

2 Background and Problem Statement

SAI-GE Hospital prefers to have the Big Data deployment infrastructure and data hosting deployed in cloud rather than within the organization. The key advantage of the choosing cloud as the option is due to the fact that data can be recovered in case of any disaster. Cloud gives the option of disaster recovery, which can be used as a back-up option by SAI-GE Hospital. Further it also enhances the accessibility, as stored files can be accessed from anywhere. This is also key for SAI-GE Hospital [5], which is eyeing regional expansion to cater the growing regional market. Another factor is that the cloud-based options have now become cheaper and even cost effective rather than investing heavily on the servers and internal storage. One of the key challenges in cloud option is the bandwidth, however this is not a major problem in Singapore with government moving towards Smart Nation initiatives with 5G on trails [6]. The cloud servers need to be hosted in Singapore to avoid the Data Protection Act issues. The key for successful implementation of big data architecture solution is that the source layer has to be effective [7]. The right sources of data in the relevant format are important. The internal data should be collected without any spills and in the best possible way. Adequate preliminary thoughts and processes would help in getting the right data in the right format. As such the data should be not too precise or not too vague, whereas it should satisfy the requirements of the decision-making team [8]. At the same time the external data or data from third parties need to be dealt with even more carefulness, as it can be different forms and not really to the requirement of our organization. Further the key is that it is imperative to understand what data and what type of data is helping the decision makers. Data clutter has to manage effectively. Key data should be separated, accessible, storable and also retrievable in the timely manner [9].

3 Proposed Big Data Systems Architecture Strategy

The four key sections of big data solution implementation or deployment are illustrated below with explanation on the tools and frameworks along with the reasons for choosing the option along with business case.

3.1 Section 1: Ingestion Tool (Flume vs Kafka)

Description of Kafka
The tool that has been selected for the research is Kafka. Kafka is one of the few available open-source stream processing and low-latency platform tools that can handle real-time data. It was developed at LinkedIn and was also part of Apache project. The key reasons to use Kafka is that it is fast, scalable and is distributed by its design, partitioned, and is replicated for fault tolerance to commit log service. Kafka can be used instead of a traditional message broker [10]. It is able to assist in website activity tracking as a set of real-time publish-subscribe feeds. Further it can assist in the monitoring of operational data metrics, log aggregation, stream processing, and event sourcing and commit log. In our case, Kafka can be used to assist in log aggression of claims data of CPF Website and also logs of medical centre. Further, it can be used for log-based comments and reviews of the social media websites. The key reason for using Flume is that it provides a number of pre-built collectors, whereas Kafka is just more of a messaging system with pull data. However, SAI-GE Hospital prefers to be proactive as such Flume pushes data. Another key aspect of Kafka is that it can only work with Zookeeper [11]. Though Kafka is can support huge number of users, Flume is transactional with no lost when duplicating streams. Hence, it is justified to choose Flume.

3.2 Section 2: Processing Tool (Hive vs Impala)

Description of Impala
Impala is one of the SQL query engine and its processing is lightning fast. Impala was initiated by Facebook, and it is one of the open source and database for Apache Hadoop. Impala is good choice for programmers and it is not based on MapReduce

algorithms. It is using Massive Parallel processing engine in Hadoop and it provides in-built support of processing all Hadoop supported file formats. It carries huge performance in queries with aggregation functions on numeric fields since it reads only the column split part files somewhat reading whole data set. It can access low latency queries interactively and it provides the results in seconds and get the results in real time [12]. It allows to do partial data needs to be analyzed. It supports various file formats. If we added new data in files with name that time these files controlled by Impala. It supports all languages and transaction data with other Hadoop components. Improving our healthcare services and satisfy the customer to provide the summarization and analyzing the data by writing SQL. Speed of processing is reasonable and able to store massive amounts of data with all formats, shapes and sizes. After the analyzing part it turns into actionable business insight. Hive easily integrates with other critical data centre technologies using a JDBC interface hence, we are implementing the Hive for this processing.

3.3 Section 3: Analysis Tool (Mahout Vs MLLib)

Description of MLLib

Apache Spark MLLib is a machine learning library, the main purpose of which is to make useful machine learning scalable and easy. It consists of common learning algorithms and utilities, including classification, regression, clustering, collaborative filtering, dimensionality reduction as well as lower-level optimization primitives and higher-level pipeline APIs. Spark MLLib is considered as a distributed machine learning framework on top of the Spark Core which, mainly due to the distributed memory-based Spark architecture [13]. MLLib handles interactive jobs, and it is approximately nine times as fast as the disk-based implementation used by Apache Mahout. Spark's MLLib approaches with Apache's Spark installation since it is built on Apache Spark. Mahout allows applications to analyze large sets of data effectively and in quick time [14]. The algorithms of Mahout are written on top of Hadoop, so it works well in distributed environment. Mahout uses the Apache Hadoop library to scale effectively in the cloud. Hence, we can implement the cloud for the businesses easily [15].

3.4 Section 4: Orchestration Tool (Oozie vs Azkaban)

Brief Description of Azkaban

Azkaban is batch workflow job scheduler created at LinkedIn and allowing developers to controlling the job execution and running Hadoop jobs. It is supporting the time based scheduling but it is data independent also we can use web user interface easily. It is mainly supports to track the workflows. Azkaban keeps state of all running workflows in memory [16]. It allows Java and Java process. Azkaban can be used for offline processing. It supports resource locks like read, write, counter and it runs as standalone. Based on our business cases, it evolves more strong solution. It supports solve the problem of Hadoop job dependencies. Oozie mainly supports the fault tolerance and also multi users, multi workflows. For our business it is very useful to make decisions with support of decision nodes to achieve the bigger tasks. It is high availability and well in progress also to secure clusters. Thus, it is very suitable for our cases.

4 Introduction to Security Risk

There is always a possibility in the Hospital environment that the big data infrastructure is been hit by data breaches and security issues. For instance, when the open-source software interacts with each other, the structure may be at risk, and has to be adequately secured [17]. Most of the times, organizations tend to focus on broader level security and importance for the internalized security is minimal. Sensitive data such as patient's personal details and patient management system need to be secured [25]. To keep the security control intact, the following areas needs to be covered adequately.

- Authentication: People need to access data that they need to access only
- Authorisation: Adequate steps need to be in place to ensure that there is no leakage of data
- Auditing: Audits need to be conducted to prevent attempts to access data that the people are unauthorized. (even if it is accidental)
- Data Encryption: Adequate security controls to be provided in case of data movement or transmission.

5 List of Security Risks Identified

The various risks identified in the current project are as follows,

Data Ex-Filtration by Internal Staff

There is a possibility of data being mishandled or misused by the internal staff of the hospital either accidentally or voluntarily. There were instances that have been reported on the researches that database administrators downloading the organisation data and were selling customer database. A recent example of data theft by internal staff is widely found in the media is of "DBS HK staff arrested over illegal use of customer data" [18]. The impact on the organisation is loss of confidentiality of sensitive data. Similar risk is possible in Hospital environment.

Data Ex-Filtration by External Threats

The world has been increasingly witnessing more of external data ex-filtration risks. The latest of these threats in Singapore context is the personal details of 850 national servicemen and staff at the Ministry of Defence (MINDEF) were stolen by external parties and MINDEF described this attack as targeted and carefully planned cyber-attack [19]. More recently a Wannacry attack has forced 16 UK hospitals to shut down. The impact on the organisation is beyond the loss of confidentiality of sensitive data. For instance, the hospitals were left with cancelled appointments, unable to access basic medical records and one hospital has cancelled all non-urgent operations [20]. Similar was the status in United States [21]. Even in hospitals of Singapore have similar risk or threat.

Poor Data Quality

The impact of poor quality of data is huge for all type of business and it is the same the health care business too. Bad Data would lead to bad decisions [22, 25]. Poor quality of data is mainly associated with unclear data definitions, collection guidelines, data sources, data format, programming errors, etc. Research also shows that lapse or lack of security controls leads to poor data quality. Like any other organisations, Hospitals are also affected by this risk.

Non-compliance to Regulatory Requirements

Data Architecture should keep in mind the compliance matters related to the regulatory requirements. Hospitals and in general the Healthcare sector in Singapore has been provided with the Personal Data Protection Act (PDPA) guidelines on the data security [23]. Any non-compliance may result in fines by the regulatory bodies and also intangible impact on the goodwill and brand of the hospital.

Disclosure Risk Problem

Another security risk on disclosure of the identities. For instance, the Type 1 of the risk is disclosing the identification fields/variable of the data Eg name, NRIC number, address, etc. when data is been moved to different stakeholders. Type 2 risk is on Attribution where the data has been used without providing the correct acknowledgements or credits.

6 Discussion

6.1 List of Security Controls/Measures that Address the Risks

There are 3 layers on security measures in Hospital's Hadoop System,

Firstly, Preventive Control is the initial part to control systems. The steps taken before loss or problem occur. We have to protect the data when in transit as it is ingesting the data from internal or external data sources in Hadoop. We use Secure Socket Layer (SSL) or Transport Layer Security (TLS) for ingesting data through Sqoop. HDFS Permission authorisation is used while ingesting unstructured data and the Kerberos installed while ingesting log files from medical centre and claims data via Flume. Kerberos is used to authenticate to identify the person who accessing data through Hive in the hospital. Apache Ranger provides centralized security administration for Hadoop, and it enables fine grain access control and deep auditing for Apache components such as Hive [24]. They would use Ranger to restrict access to the base Patient details and apply access to the view. Rangers also provide as the Centralize auditing of user access and administrative actions (security related) within all the components of Hadoop. This includes install and maintain a firewall configuration that can protect the data and at the same time do not use defaults for system passwords and other security parameters. Further, it also assists to protect the data by using encrypt transmission of data across open, public networks, etc. Most of the issues related to PDPA and data security compliance happens on the sharing of sensitive data. As preventive technique, masking can help in avoiding this type of security breaches. Various type of masking is Redaction/Nulling (Simple replace with 'X'), Tokenisation (Random mapping stored in a table), Format Preserving Encryption - FPE (Uses a key to encrypt and decrypt) and Substitution (Uses a substitution table) Further there are various other ways to deploy masking such as Static masking and Dynamic masking, which is Proxy based or View based (with trigger).

Secondly, Detective controls include the logging of events and the associated monitoring and alerting. In this aspect, we have to consider about preventive and detective controls altogether when the data in HDFS. Here, we used Apache Ranger as federated authorization model for HDFS and provides a scalable cryptographic key management service for HDFS "data at rest" encryption and centralized security administration solution for Hadoop that enables administrators to create and enforce security policies for HDFS. Ranger also uses the query with different reasons for dynamic data masking as a detective control for PII, PHI and PCI. As a detective control, anonymisation techniques help in addressing the disclosure risks. The various techniques Pseudonymisation, Aggregation, Replacement, Data Reduction, Data Suppression, Data Shuffling, Masking, etc. These security control measures need to be considered before the release of data for research or when transferring the data.

Thirdly, Corrective control, the hospital can document its policies and procedures, enforcing them by means of warnings when appropriate. When managers wisely back up data they can restore a functioning system in the event of a crash. If a disaster strikes, business recovery can take place when an effective continuity and disaster management plan is in place and followed. Full back up strategy is deployed within the hospital as well as offsite backup.

6.2 Access Control

Apart from the above three key security controls, Access control is also important security control. Role based access level is an approach to restricting system access to authorized users. In the hospital, we provide the role-based access level with least privilege that used to controls assigned to each person based strictly on their role assignment. With tight adherence to access requirements established for each role, access management becomes much easier.

For example, health information system in the hospital will not have access to customer credit card information while only medical staffs have access to patient medical records. Restricting access will be done in Ranger. As PR team is to promote the hospital products and treatments, they will not have any assess to PII, PHI and PCI. For example, Management team will have full access to every data in the hospital while CPF provider will not have any access data resided in the hospital big data environment (Table 1).

Table 1 Details of access controls

List of possible users	Roles and responsibilities	Access control (level with examples)	Ways to restrict the access
Management Team	Plays a wide range of roles and responsibilities including Planning, Organising, Leading, Controlling, Decision Making, etc.	Full Access	Access Control with rights, privileges and role-based access with restricting systems
Accounts Team	Key focus is on revenue and expenditure including financing, budgeting, etc.	Full Access: Accounts, CPF, HR, Patient Partial Access: Big Data matters	Access Control with rights, privileges and role-based access with restricting systems
Physician Team	Key focus in the Patient Management Matters including health records	Full Access: Patient, HR Partial Access: Accounts	Access Control with rights, privileges and role-based access with restricting systems
PR Team	Focus is on the promotion of the hospital	Full Access: HR Partial Access: Patient, Physician, Big Data matters	Access Control with rights, privileges and role-based access with restricting systems
Patients Relationship Team	Focus is on Patients Relationship matters	Full Access: CPF, HR, Patient Partial Access: Accounts	Access Control with rights, privileges and role-based access with restricting systems
HR and Lean Management Team	Key focus is on HR functions like staffing, training, development, learning, appraisal, etc	Full Access: HR, Patient, Accounts Partial Access: Big Data matters	Access Control with rights, privileges and role-based access with restricting systems
Big Data Project Team	Provide clear and effective data architecture platform for superior organisational performance in strategic directions	Full Access (where applicable)	Access Control with rights, privileges and role-based access with restricting systems

6.3 Model Developed

With the business initiative to improve the quality of healthcare services and customer satisfaction via proactive analysis, the following Processing Design and Dataflow Diagram is provided in Fig. 1.

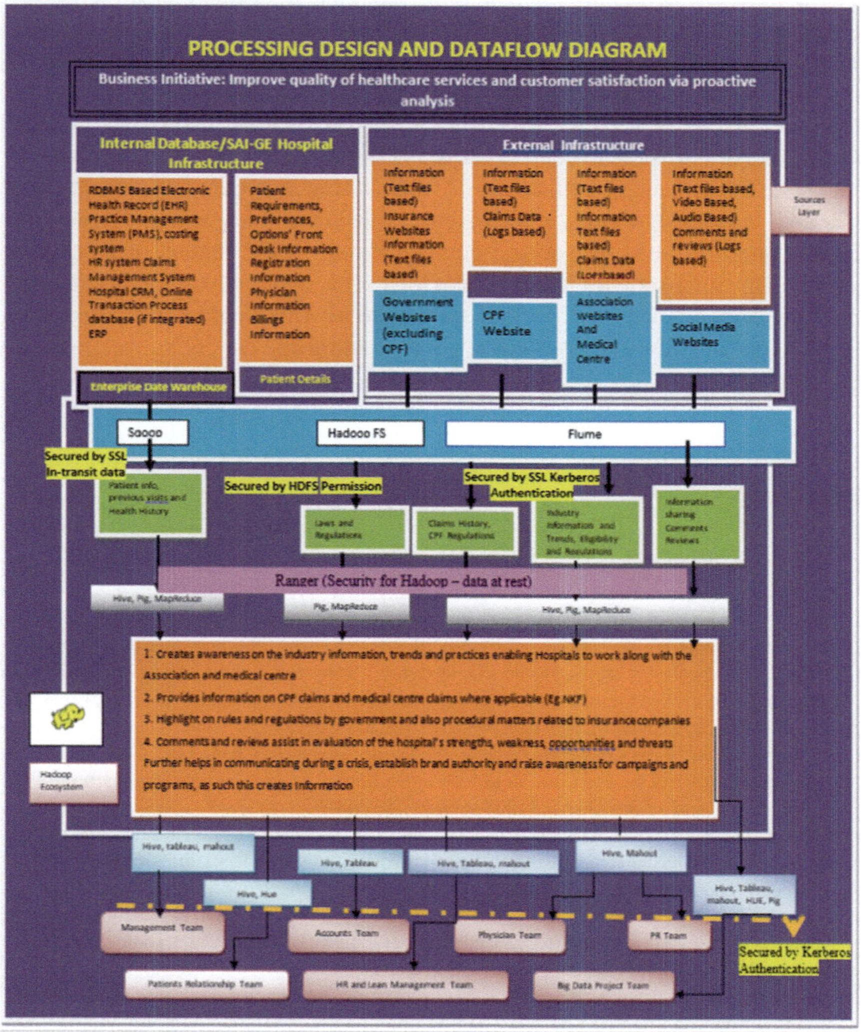

Fig. 1 Processing design and dataflow diagram

7 Conclusion

The present paper aims to address the issues related to the Big Data Systems Architecture and Data Security Fundamentals with a Case Study Approach for Hospital Project. High Level Design Diagram along with Processing Design and Dataflow Diagram has been proposed for a hospital in Singapore, which would be replicated in other hospitals across with world with minor customizations to the regulatory framework and infrastructural capabilities. Furthermore, technical challenges and recent advances needs to be constantly reviewed as the Data Systems Architecture and Data Security has been evolving. Other implications of the study is to have organizations to realize on need for what data and what type of data that is to be collected and processed in helping the decision makers. Further the study also indicates that data clutter has to managed effectively. Also, it is crucial that the key data should be separated, accessible, storable and also retrievable in the timely manner. Also, future trends would also need to work on possible layout in the context of blockchain. Further, in the future studies it is crucial to look at the data analytics in the context of the Artificial intelligence and Machine learning tools. This would be also approached from the general purposed technology, which would shape our digital worlds in very profound ways.

References

1. Mahbub M (2022) IoT ecosystem: functioning framework, hierarchy of knowledge, and intelligence. Internet Things, 47–76. https://doi.org/10.1007/978-3-030-87059-1_2
2. Tariq N et al (2019) The security of big data in fog-enabled IoT applications including blockchain: a survey. Sensors (Switzerland) 19(8):1–33. https://doi.org/10.3390/s19081788
3. Yang W, Aghasian E, Garg S, Herbert D, Disiuta L, Kang B (2019) A survey on blockchain-based internet service architecture: requirements, challenges, trends, and future. IEEE Access 7:75845–75872. https://doi.org/10.1109/ACCESS.2019.2917562
4. Akshay Kumaar M, Samiayya D, Vincent PMDR, Srinivasan K, Chang C-Y, Ganesh H (2022) A hybrid framework for intrusion detection in healthcare systems using deep learning. Front Public Heal 9. https://doi.org/10.3389/fpubh.2021.824898
5. Bahalul Haque AKM, Bhushan B, Nawar A, Talha KR, Ayesha SJ (2022) Attacks and countermeasures in IoT based smart healthcare applications. Intell Syst Ref Libr 215:67–90. https://doi.org/10.1007/978-3-030-90119-6_6
6. Dhanalakshmi G, George VS (2022) Security threats and approaches in E-Health cloud architecture system with big data strategy using cryptographic algorithms. Mater Today Proc. https://doi.org/10.1016/j.matpr.2022.03.254
7. EL Azzaoui A, Sharma PK, Park JH (2022) Blockchain-based delegated quantum cloud architecture for medical big data security. J Netw Comput Appl 198. https://doi.org/10.1016/j.jnca.2021.103304
8. Ibrahim M, Jamil F, Lee Y, Kim D (2022) Blockchain based secured load balanced task scheduling approach for fitness service. Comput Mater Contin 71(2):2600–2616. https://doi.org/10.32604/cmc.2022.019534
9. Zakharov I, Anderson J, Parsons G, Henschel MD, Ewenson B, Papanagiotou C (2022) Digest of blockchain technologies to design system for big image data provenance and security. In:

Communications in computer and information science, vol 1562. CCIS. pp 33–47. https://doi.org/10.1007/978-3-030-98883-8_3
10. Musleh Al-Sartawi AMA (ed) (2022) Artificial intelligence for sustainable finance and sustainable technology. In: ICGER 2021. Lecture notes in networks and systems, vol 423. Springer, Cham
11. Awotunde JB, Jimoh RG, Folorunso SO, Adeniyi EA, Abiodun KM, Banjo OO (2021) Privacy and security concerns in IoT-based healthcare systems. Internet Things, 105–134. https://doi.org/10.1007/978-3-030-75220-0_6
12. Zalloum M, Alamleh H (2020) Privacy preserving architecture for healthcare information systems. In: 2020 IEEE international conference on communication, networks and satellite, COMNETSAT 2020 - proceedings, pp 429–432. https://doi.org/10.1109/Comnetsat50391.2020.9328985
13. Yang C-T, Chen S-T, Yan Y-Z (2017) The implementation of a cloud city traffic state assessment system using a novel big data architecture. Cluster Comput 20(2):1101–1121. https://doi.org/10.1007/s10586-017-0846-z
14. Urmila MS, Hariharan B, Prabha R (2020) Enabling internet of things (IoT) security via blockchain framework. In: Lecture notes in networks and systems, vol 103, pp 575–582. https://doi.org/10.1007/978-981-15-2043-3_62
15. Veeramakali T, Siva R, Sivakumar B, Senthil Mahesh PC, Krishnaraj N (2021) An intelligent internet of things-based secure healthcare framework using blockchain technology with an optimal deep learning model. J Supercomput 77(9):9576–9596. https://doi.org/10.1007/s11227-021-03637-3
16. Jain S, Gupta S, Sreelakshmi KK, Rodrigues JJPC (2022) Fog computing in enabling 5G-driven emerging technologies for development of sustainable smart city infrastructures. Cluster Comput 25(2):1111–1154. https://doi.org/10.1007/s10586-021-03496-w
17. Shahid A, Nguyen T-AN, Kechadi M-T (2021) Big data warehouse for healthcare-sensitive data applications. Sensors 21(7). https://doi.org/10.3390/s21072353
18. Lee J (2018) DBS HK staff arrested over illegal use of customer data. Business Times. https://www.businesstimes.com.sg/banking-finance/update-dbs-hk-staff-arrested-over-illegal-use-of-customer-data. Accessed 22 Jan 2022
19. Tham I (2022) Personal data of 850 national servicemen and Mindef staff stolen in targeted cyber attack. The Straits Times, 1
20. Brandom R (2017) UK hospitals hit with massive ransomware attack. The Verge. https://www.theverge.com/2017/5/12/15630354/nhs-hospitals-ransomware-hack-wannacry-bitcoin. Accessed 15 Feb 2022
21. Glaser A (2017) U.S. hospitals have been hit by the global ransomware attack. Vox-Recode. https://www.recode.net/2017/6/27/15881666/global-eu-cyber-attack-us-hackers-nsa-hospitals. Accessed 15 Feb 2022
22. Patel P (2013) Bad data, bad decisions. Perficient 2013. https://blogs.perficient.com/2013/01/07/bad-data-bad-decisions/. Accessed 18 Feb 2022
23. Pdpc (2017) Advisory guidelines for the healthcare sector. In: Personal data protection commission. Singapore. https://www.pdpc.gov.sg/-/media/Files/PDPC/PDF-Files/Sector-Specific-Advisory/advisoryguidelinesforthehealthcaresector28mar2017.pdf. Accessed 18 Feb 2022
24. Justinia T (2019) Blockchain technologies: opportunities for solving real-world problems in healthcare and biomedical sciences. Acta Inform Medica 27(4):284–291. https://doi.org/10.5455/aim.2019.27.284-291
25. Sanad Z, Al-Sartawi A (2021) Financial statements fraud and data mining: a review. In: Musleh Al-Sartawi AM, Razzaque A, Kamal MM (eds) Artificial intelligence systems and the internet of things in the digital era. EAMMIS 2021. Lecture notes in networks and systems, vol 239. Springer, Cham

The Usage of Technology in Mass Communication and Its Applications

Abdulsadek Hassan, Sumaya Asgher Ali, and Nader Mohammed Sediq Abdulkhaleq

Abstract The study aims to identify the role of modern technology in influencing the media fields by applying to the press, radio and television sectors. Which strengthens the broadcast, which made it possible to transfer information from the most remote places, as it was relied on satellites to transmit information, and it became possible to transmit and receive information from within the event, due to the development of means of communication and the presence of the Internet, which was able to link the regions of the world together. Direct interaction options were also introduced by viewers and listeners, and the results revealed that the entry of newspapers into the era of electronic publishing made it imperative to find advanced, fast and unconventional means to store information that is received by the newspaper on an ongoing basis, and leaving this huge amount of information undocumented or stored represents a waste A real wealth of information for press publications and publishing houses that are now working in the most important global trade, which is the information trade.

Keywords Technology · Mass communication · Journalism · Radio & television

1 Introduction

Media technology plays a major role in making the presented media messages clearer and more transparent, which contributes to documenting facts, contents, data, and media information documented in audio and video, which takes place within the event sites. Thus, the use of technology in the media fields is through the use of many special and multiple techniques, which are represented in the direct broadcast technology, which in turn made the recipient live the moments of the event itself, and in addition to that, the media helped and supported modern technology in accessing news sources, where it is considered, these sources serve as neutral and new sources [7]. The media also contribute to the use of social networking sites, which are considered one of the most important research methods used to obtain information, as they contributed to

A. Hassan (✉) · S. A. Ali · N. M. S. Abdulkhaleq
Ahlia University, Manama, Kingdom of Bahrain
e-mail: dr.sadek1964@yahoo.com

A. M. A. Musleh Al-Sartawi et al. (eds.), *From the Internet of Things to the Internet of Ideas: The Role of Artificial Intelligence*, Lecture Notes in Networks and Systems 557,
https://doi.org/10.1007/978-3-031-17746-0_24

facilitating complex operations in social networking sites and made communication methods easy with all parties around the world, and modern media technology played a major role [12]. In getting to know all the cultures around the world, it also helped to facilitate communication with peoples and their cultures, and then learn multiple languages [15, 27]. It must also be emphasized that there is a strong relationship between technology and media, and therefore modern technologies used in media outlets have created many concepts related to information and media contents. It is also necessary to focus on the fact that the media sector in all its divisions is affected by the technical means that are developed from time to time and have also contributed to a qualitative leap in the ways and methods that deal with media news in all its forms [18]. Thus, the development of modern communication technology is linked to the frequent use of the media, as the computer played a major role in documenting, studying and storing information, facts and data related to the government sector; This is on the grounds that media technology serves as a mediator in the transmission of media messages to various press and media institutions and their transmission among individuals. Many press institutions are also tending to establish a special section in them related to the development and attraction of modern technology, which is used in order to transmit media messages to all means, whether audio, visual, written, or new electronic means [3].

The Concept of Information and Communication Technology: What is meant by information and communication technology is how to use modern technology to communicate different information and link it with different societies, using digital means of transportation and communication, using modern communication devices such as computers and smartphones [15]. And through multiple communication applications, those so-called social media, which have become the mediator that connects people, as well as one of the most important means of knowledge in the modern era, and even become an integral part of different areas of life such as e-commerce and industry [12].

The History of Media Technology. The first media revolution erupted in conjunction with World War II, and it became one of the most dominant features of international societies, the feature of the emergence of the media as an active force for international relations, and the time interval between the two world wars marked a turning point in the history of the emergence and development of media technology [15].

Media Revolution. The media revolution in the second half of the nineteenth century was represented by the invention of a number of devices; such as the telephone, the telegraph, fixed and mobile cameras, and by the beginning of the twentieth century the revolution was completed with the advent of radio and television, and technologies related to media technology played an important role in shortening distances, and left direct and indirect effects on societies, as they became psychological and logistical weapons with which countries fight each other, It also controls the international relations between them [1].

2 Positive Effects of Information and Communication Technology

There are many positive effects of media and communication technology on various aspects of life, the most important of which are the following: [3] Ease of access to information technology has made it easier for many to access the required information, as well as obtain various services such as facilitating trade, in addition to providing opportunities for entertainment and leisure [6]. Improving access to education ICT provides an opportunity for online learning, in addition to providing new job opportunities, such as flexible and mobile work, and many jobs in the field of communications [17]. Ease of access to new opportunities and tools Information and communication technology has opened up new horizons for people who did not have this type of technology at their fingertips, such as the following: photography, which previously required a professional studio and photographer, in addition to software and photo editing, and high-quality printers [12]. Integrating people with special needs into society through the use of this type of technology, such as the blind and visually impaired, through the Braille method of writing and reading on computers that interact with voice, and the screen magnification feature for the visually impaired [13].

Social Networks: There are many social networking sites available such as Facebook, Twitter, And LinkedIn, which allow users to keep in touch with each other all over the world [2].

Shopping and Commerce: Through information and communication technology, it is possible to shop online with ease, and this feature falls under the category of e-commerce that relies on technologies such as electronic transfer of funds, online transaction processing, and automated data collection systems.

3 Negative Effects of Information and Communication Technology

Despite the many positive effects of information and communication technology, it has some negative aspects, the most prominent of which are the following: Layoffs in favor of technology, i.e., replacing manual processes with automation, for example: the robot replaces people in some places, and the consequent economic consequences such as loss of income, and some social consequences such as loss of social status, and self-esteem [14].

- **Limit Face-to-Face Interaction**: Nowadays everyone prefers to communicate online through social networking sites; And e-mail instead of real face-to-face

conversations, which makes them more individual and introverted, and communication via the Internet makes individuals unable to express how they feel clearly [6].
- **Decreased physical activity and some health problems:** Prolonged use of technology causes health problems, such as: headaches, neck pain, eyestrain, and some psychological problems such as social isolation [12].
- Because of the high costs, many ICT tools are expensive, they are expensive to purchase or maintain, and accordingly some individuals with low income may not be able to benefit from e-learning services [5].
- **Misleading information**: Not all websites are reliable sources of information, and some may publish misleading or false information and news [12].
- **Loss of privacy**: Some data on the Internet is vulnerable to virus and malware attacks, for example, using social networking websites may lead to loss of personal data, as well as in the case of shopping and paying online, there is a risk of losing money [17].

4 Information Technology Jobs

The modern means of communication that technology has produced are characterized by several functions that cast a shadow and imposed their impact on human communication through its modern means, the most prominent of which are:

1. **Interactive**: one of its most prominent characteristics is the exchange of roles between the sender and the receiver, that is, there are common roles between them in the communicative process, and the communicators are called participants instead of sources, and from this we find the use of new terms in the communication process such as binary practice, exchange, control, and an example of This is the use of the (video texts) system, which allows a clear interaction between the sender and the receiver, and this system is one of the teletext systems [7].
2. **Determine the beneficiary:** This feature means that the information exchanged will be of a specific purpose, meaning that there is a degree of control over knowing the real beneficiary of certain information and not others. There is a wide scope to control the quantity and quality of the desired information, and this service is performed by a person called (the coordinator) who arranges this process by knowing the desires of the beneficiaries and their information needs and equipping them with it through the (electronic mailboxes) of each subscriber in return for a monthly or annual subscription paid for these services [4].
3. **Asynchrony:** The importance of this feature is that it allows the ability to communicate information between the parties to the communicative process without the condition that it be present at the time of its transmission, and this means that there is a possibility to store the information sent when it is received in the device and use it when needed. Any time [1].
4. **Mobility:** This feature allows the transmission and reception of information from anywhere else during the movement of the product and the receiver of the

information, using a number of different devices such as the mobile TV, the car phone, the TV built into the wristwatch, the fax machine that can be used in the car, as well as the portable electronic computer [11].

5. **Transferability:** It is the possibility of transferring information from one container to another using techniques that allow converting paper containers into film miniatures and vice versa, as well as the possibility of converting information recorded on film miniatures into magnetic or laser containers, as well as the possibility of converting texts from one language to another or the so-called machine translation system [8].
6. **Connectivity:** This feature is represented in the possibility of using devices manufactured by different companies that are governed by certain standards in unifying the manufacture of the different parts of these devices, which allows the possibility of information transfer between the beneficiaries, regardless of the manufacturers of the different devices [2].
7. **Prevalence and spread:** It means the systematic spread of means of communication around the world and in the different layers of society. Whenever a means of transmitting information appears, it is initially considered a luxury, but in the end, it becomes traditional after a while that can be used by different groups and classes in society such as the use of telephones, fax machines and other technologies [9].
8. **Global or cosmic:** It means the possibility of transferring information between beneficiaries worldwide, due to the availability of quantities and types of technologies that allow this. This feature of the capacity for information transfer between humans gives many advantages to scientific and technical communication and to the transfer of experiences between them and thus communication is global [17].

5 Fields of Use of Information and Communication Technology

Modern technology has become an integral part of our lives as individuals and societies. In fact, all areas of life now depend primarily on technology, including:

6 Education Field

Since the world began to know technology, the interest in the first place is to develop the field of education, and this is because education is the main field that works on the renaissance of the state and the development of society, and modern technology has contributed greatly to the development of education and the dissemination of knowledge, starting with the invention of the paper and pen [10].

Then the stage of inventing the means of printing, and even the invention of computers and the invention of the Internet. The computer has now become the main means of education. We find that schools and universities have become dependent on display screens to present lessons and lectures to students [16]. As well as the entry of the world now into the stage of distance education (online study), after the spread of the Corona pandemic, which caused total isolation of people in their homes, and here the importance of technology appeared, as universities and schools continued their activities without stopping, relying on computers, the Internet and social networking applications [13].

E-commerce Field: Modern technology and its technologies have greatly affected the field of commerce, with the emergence of the so-called e-commerce, which is the design of a virtual store on the Internet that includes different products, goods and merchandise, the consumer can browse in the contents of this store and choose the products he wants. Then he asks to buy it, and you can pay electronically via prepaid cards, or pay in cash when the required product arrives, and this method of buying and selling has spread a lot, and there are thousands of electronic stores on the Internet [1]. One of the advantages of electronic commerce is that it achieves great research for merchants and companies by presenting their products to an unlimited number of customers, and it also helps the customer to save time and effort in the shopping process [14].

Health Field: The medical field was able to benefit greatly from information and communication technology, by employing technological means and techniques to raise the health level of medical units and centers in rural areas for the local population, and to provide transportation expenses to capitals and major cities for treatment [10]. Modern technology and its multiple techniques have also helped to exchange medical knowledge and expertise, hold major medical conferences, and invent the latest medical devices and equipment that helped diagnose and treat many diseases that were difficult to treat in the past [7].

Communications: The field of communications has witnessed a very big technological revolution, as it has become possible to communicate with anyone anywhere in the world with ease, as well as the possibility of free voice and image communication via the Internet, as well as written text conversations, all via a cell phone or computer in addition to the ability to hold meetings and conferences remotely, via video conference technology [6].

The Field of Media and Culture: Modern technology and automated machine techniques have made great contributions to the media field, starting with printed newspapers, books and magazines, then the invention of radio and television, to the invention of the satellite, which made it possible to see all global events anywhere in the world as they occur, and with live broadcasting technology as well [16]. Then to the Internet and Social Networking Sites that monitor news everywhere as they happen and publish it instantly [15].

Industrial Field: Modern technology has constituted a huge revolution in the industrial field. All machines and mechanization worldwide have been modernized after the invention of the computer and communication technology. The computer is used in designing and checking complex machine models such as cars and planes. The industrial field has developed a lot after the invention, use and employment of robot technology to perform tasks. Dangerous that a person is unable to perform [17].

Administrative Field: Administrative institutions in our time have become completely dependent on modern technology and means of communication, as each institution has a computer network and internal and external communications linking its various departments and branches, and there is no room for manual paperwork as in the past [14]. All services, papers and administrative tasks in the various state institutions are now done electronically, such as obtaining personal cards, passports, licenses, birth certificates, and others [6].

Military Field: Man has been able to adapt the means of modern technology and exploit it in the military and war fields, and the arms industry has witnessed a great development than it was in the past, and a good section has also been attached to the army sections, which is the information war section [13].

As information is considered the basis for winning wars, whether ancient or modern, computers and robots have also become more widely used in the field of weapons industry, as there are computer-guided missiles, modern and secret communication systems between army departments and some of them [4].

Entertainment Field: As for the entertainment field, it is the field that has witnessed a great development due to the means of modern technology. The field of three-dimensional (3D) games has become the most prevalent among children and youth, and the demand for it has become very large [15]. As well as using communication technology and media in the production of modern films that depend on special audio and visual effects, and these films and games are spread on the Internet in huge numbers [11].

Technology Manifestations in Radio and Television: The media is the fourth authority in countries due to its prominent role in transmitting information and influencing the opinion of the local and external community [7]. The media varies between written, such as the press, audio, such as radio, and visual, such as television, but radio and television are among the old means that people relied on to obtain on the news and information, and with the development of technology and the entry of the world into the age of the Internet, it has evolved greatly, and interest in it has become increasing in our time [15]. There is no home that does not have a television and a means to receive radio [12]? What are the manifestations of the entry of technology to radio and television in the modern era? [5]. There are many changes that have occurred on radio and television due to the introduction of technology. The structure of communication processes has changed, and broadcasting has become stronger in all regions of the world. Devices that strengthen broadcasting have been used, making it possible to transmit information from faraway places [5]. Satellites have also been relied upon to transmit information, and it has become possible the transmission and reception of information from within the event, due to the development of means of communication and the presence of the Internet, which was able to link the regions of the world together [11]. Direct interaction options have also been introduced by viewers and listeners, whereby the viewer or listener feels in some programs that he is the one who chose the material that is being shown, and a sophisticated computer has been used to store and process information as required and bring it out to the viewer or listener in an attractive and likable way [2]. Communicating with them to get feedback on what they want and what they prefer from programs, and this led to the development of radio and television, and the combination of wired and wireless means of communication with computer technology has led to the creation of programs that allow the ability to print words on the television screen [18]. The emergence of Interactive Television services through (cable), which allows the ability to communicate directly between the receiver and the sender at the same time and obtain many services such as: banking services and medical services, and this cable provides more than a hundred television channels that the viewer can get it with advanced technology [7]. Technological development was not limited to improving the quality of services provided in radio and television, but rather updates and developments took place on the radio and television itself, so we find thin and small screens instead of large-sized television, and a person can choose the appropriate size of these screens commensurate with his desire, and it is possible to watch television stations on the mobile phone, as well as the radio, the service became available on all cell phones and various electronic devices [14].

7 Technology and Methods of Journalistic Work

In the last quarter century, there have been dramatic technological developments in the production, processing and dissemination of images, in a way that has transformed methods in journalism, entertainment and advertising, and in the visual environment itself as well. In this mass media age, the news industry has never been in a better state than it is today. With the click of a button, we get the news wherever we want [4]. With the great change in the environment and media technology, the question urgently asks about the nature of news journalism, in our definition of the concerns of the public today [9, 28]. The issue is very important, not only in terms of the future of news, but also in terms of the practice of democracy as well. The basic concept is that the traditional press was very effective in highlighting the problems and developments of that day [16]. But, for various reasons, she often fails to put the story in a context that makes it fully understandable. One of the most important reasons, is technology. Moreover, the traditional press generally works in one direction, from the journalist to the public. This conception of the receiving audience set limits to the reader's involvement in journalism and public work [13]. Technological innovations have reshaped the scene and made the 2014 "journalism" something that many could not imagine. Therefore, the new media work presented a set of professional challenges. Every day, new devices come into the field, with surprising speed, and come up with a variety of options, which journalists and their audience may not be able to anticipate [3]. These open borders have enriched news and media sources and rearranged what used to be the property of the traditional journalist, news anchor, and news agency. The challenge here is to extract the maximum benefit from the Internet and digital media, without doing any damage to civil liberties [14, 27].

8 Special Working Methods

There are countless ways in which digital technology has been able to influence the practice of journalism, starting with the way reporters gather information and deliver news, through how news organizations create their own structures, down to the speed of development of media technologies, in a way that makes them penetrate the social fabric in various forms and capabilities [12]. New to the world of communication. Hypertext, multimedia, and hypermedia have become part of the routine of the majority of professionals and are calling for a change in teaching and learning methods in institutions. Looking to the future, we can expect a media landscape dominated by a very fragmented, but vibrant audience. We can also expect intense media competition, and a lack of advertising budgets. The professional media, if it adopts the new technology, can retain its role as a vital source of media, continuing to serve as a tool for a successful democracy [17]. At the dawn of the information age (and electronic journalism), the tasks of journalists changed from transmitting information to processing information. Marshall McLuhan (the Canadian pioneer in the study of media theory, 1911–1980 CE) says that each media outlet is concerned with a different perceptual experience [1, 25]. Therefore, it is not surprising that the institution of the press begins to change, with the emergence of the new medium - the Internet - because the functions of the interactive Internet and multimedia media, impose an important development in the culture of the press. As the primary task of the informant, it evolved into the digital medium [22, 26], and the technology of the international network provided scope for a wide resource and audience participating in the new production. Researchers have begun calling journalist informants "gatekeepers" and "observers" of information, and they insist that they voluntarily exercise a certain authority [12]. Phrases such as "network journalism" and "community media" indicate that some people understand the technology of the international network, as an opportunity for a journalistic revolution, where the citizen has to make his say, participate in the content, and oppose the dominant trend in hypothetical places [9]. This creates a new web-based news production model that takes into account the dynamics of making, disseminating and consuming information on the web [13].

9 No Geographical Barriers

The world continues today in an uninterrupted manner, without borders or geographical barriers [8]. It is the reader who decides what, when and how to read, in addition to having a voice at a low cost, free from outside interference [2]. The digital revolution in media and communication has established a platform for the free flow of information, ideas and knowledge in the globe, and many ways through which changes and challenges reach news providers and consumers. We believe that the journalist is not an endangered species, but that its tasks and habits are changing dramatically [15]. Now, the website can be a platform for citizens to say their opinions and ask their questions about press releases that interest them. Dialogic journalism is a more liberal form of news production [12]. The content in the new media outlet is handled by journalists who interact with citizens [16, 24]. The viewer, or receiver, plays an active role in the press [4, 19]. In essence, these new models pose profound changes for journalists, in their role and in their ultimate control of the news [11]. Digital technology requires more effort from the detective, as it often requires him to learn a number of operating techniques, and at the same time work as a producer, editor and implementer [13]. In the past, all these tasks were divided into different sectors [8, 23].

10 What are the Benefits of Technology for Newspaper Owners and Communicators?

10.1 Benefits of Technology for Newspaper Owners

Abbreviating Some Stages of Newspaper Production: It is intended to dispense with large numbers of highly paid employees, as technology has contributed to dispensing with the services of typesetting, collecting, or light queuing technicians, whose numbers were huge in daily paper newspapers [12]. Thus, she contributed to shortening the two stages of montage or paper execution, as she had previously been implementing the newspaper page printed on a paper machete, which is filmed on a light-sensitive plastic film [9].

Abbreviating the Photographic Stages: This means that electronic newspapers no longer need to use negative films and chemicals, which were used in the stages of paper and photography, as technology contributed to the end of the stages of developing, printing and showing [16, 22, 25]. Thus, it also contributed to saving the time that the journalist was taking for a long time; To complete the photocopying and printing operations and then deliver them [18].

Integration of the Production and Implementation Stages: In the past, the stages of newspaper production depended on the director taking out the page, and then designing it on a pencil case [3], in addition to using the ruler to determine the places and spaces in which press materials, photos and typographical elements would be placed [15]. Thus, it worked on merging the two stages of output and implementation on the grounds that they are one stage, and thus the director is the one who implements the pages and outputs them together [12].

Distance Reduction: It means that some countries do not provide press cadres that are trained and qualified to carry out journalistic work [19, 20]. Therefore, here the elements of coordination and understanding are used between workers in adjacent rooms and offices on the Internet, so that advanced press production programs are used, which contributed to providing budgets on the Internet. press institutions [7, 18].

Benefits of Technology for Communicators: This means journalists, directors, photographers and illustrators, who no longer have to change their daily program, which is to go to the press editorial office to deliver the editorial material, but become through technology to write everything they have, either on a website The juvenile himself, at home [8], or while using the means of transportation, through the use of e-mail so that he sends it to the newspaper or newspaper in which he works [14, 21], and then it is possible to obtain a wealth of information about the topics he wants to write about by browsing the websites Internet [5].

11 The Dangers of Modern Information and Communication Technology

The merger between information technology and the media endows knowledge with unlimited capabilities and possibilities, and in return, it has negative effects that the user can be exposed to [11]. The occurrence of the knowledge gap between the countries that own this technology and the countries importing it, as is happening today between European countries and Arab countries, which can lead to local and regional conflicts [6]. The integration of information and communication technology into one system is one of the tools of the current globalization [4]. The danger of communication and media technology is embodied through the dismantling of cultures, cultural invasion, and the corruption of national culture and cultural identity [2]. Modern communication technology contributed to the decline of the standard Arabic language and planted many strange terms [16]. Communication and media technology contributed to perpetuating and spreading the values of Western consumption, imposing the Euro-American cultural model, and eliminating the cultural diversity of society [4]. The stereotyping of the world in a manner similar to the pattern of Western societies, particularly the American society, and the promotion of Western intellectual ideologies [11, 17]. Lack of privacy and there is no longer room for one's private life [9]. The health effects of communication and information technology on the biological, physiological and psychological aspects of individuals. Many diseases were caused by the excessive use of this technology [15].

12 Conclusion

The results revealed that the means of communication and modern technology helped to develop media performance at all levels. Among the most prominent manifestations of reliance on modern technological means in the media are the following: The ability to store a huge number of important information and news on CDs without having to write them down on paper. The ability to communicate with any person and obtain any important information or data, whether for people, companies or others, through the use of the Internet, biographical sites, social networking sites, etc., and this would help to obtain the information that the masses need and provide it to them easily and super-fast. The ability to support the news that is published or announced with some pictures, videos and other multimedia files, and this would enhance the credibility of the news. Electronic journalism conducted through various websites has managed to get rid of some of the limitations that surround traditional journalism; As it publishes local and international news, it can also cross the limits of law and censorship, although this is a double-edged sword, but it has helped to spread many important information and news among the masses. The use of various computer and Internet technologies in preparing media programs makes them more attractive, more distinguished, and more accurate, and helps deliver them quickly, easily and smoothly to the masses.

References

1. Wilding D, Fray P, Molitorisz S, McKewon E (2018) The Impact of Digital Platforms on News and Journalistic Content. University of Technology Sydney, NSW
2. Guanah SJ, Ojo GO (2018) The utilization of new media technologies in journalism practice in Delta State, Nigeria. In: Ozohu-Suleiman Y, Mohammed S (eds) Mass Communication Education and Practice in Nigeria. Ahmadu Bello University Press Ltd., Zaria, pp 174–195
3. Musleh Al-Sartawi AMA (eds) (2022) Artificial intelligence for sustainable finance and sustainable technology. In: ICGER 2021. Lecture notes in networks and systems, vol 423. Springer, Cham
4. Pradhan P (2018) A study on Journalistic use of social media. Amity J Media Commun Stud 8(1):49–59
5. Michael X, Carpini D (ed) (2019) Digital media and democratic futures. (Book ReviewPhiladelphia: University of Pennsylvania Press, 341, $69.95 USD (hbk.), ISBN: 9780812251166
6. Dowling DO (2021) The gamification of digital journalism: innovation in journalistic storytelling. Routledg, New York, 193 p, $44.95/£34.99 (Pbk), ISBN: 9780367076252
7. Bureš V, Mikuleckâl J, Ponce D (2017) Digital television as a usable platform for enhancement of learning possibilities for the elderly, SAGE Open, April–June 1–9

8. May SJ, Middleton C (2019) Jean-Pierre Blais' "Magic Items": over-the-air digital television delivery as Canadian regulatory revelation. Can J Commun 44(2):13–22. https://doi.org/10.22230/cjc.2019v44n2a3118
9. Al-Sartawi A (2019) Assessing the relationship between information transparency through social media disclosure and firm value. Manag Account Rev 18(2):1–20
10. Adeyanju A (2015) Comparative study of social media, television and newspapers' news credibility. In: [Paper Presentation] International conference on communication, media, technology and design, Dubai – United Arab Emirates, May 16–18
11. Al-Qudah AA (2022) Artificial intelligence in practice: implications for information systems research, case study UAE companies. In: Musleh Al-Sartawi AMA (eds) Artificial intelligence for sustainable finance and sustainable technology. ICGER 2021. Lecture notes in networks and systems, vol 423. Springer, Cham. https://doi.org/10.1007/978-3-030-93464-4_23
12. Guo Z (2018) Media credibility in the public mind: a critique. Commun Soc 46:121–182
13. Udende P, Mahmud A, Oyewo OO (2014) Public trust in the news: comparative study of NTA, AIT and channels. Centrepoint J (Hum Ed) 17(2):179–197
14. Al-Okaily M, Al-Okaily A (2022) An empirical assessment of enterprise information systems success in a developing country: the Jordanian experience. TQM J. ahead-of-print. https://doi.org/10.1108/TQM-09-2021-0267
15. Karolak M, Razzaque A, Al-Sartawi A (2021) E-services and M-services Using IoT: an assessment of the Kingdom of Bahrain. In: Musleh Al-Sartawi AM, Razzaque A, Kamal MM (eds) Artificial intelligence systems and the internet of things in the digital era. EAMMIS 2021. Lecture notes in networks and systems, vol 239. Springer, Cham
16. Thomas H (2017) Technological determinism and new media. Int J English Lit Soc Sci (IJELS) 2(2)
17. Kümpel AS, Karnowski V, Keyling T (2015) News sharing in social media: a review of current research on news sharing users, content, and networks, social media + society, pp 1–14, July–December 2015
18. Al-Qassim M, Chen W, Al-Sartawi A (2022) Blockchain in environmental compliance and enforcement. In: Musleh Al-Sartawi AMA (eds) Artificial intelligence for sustainable finance and sustainable technology. In: ICGER 2021. Lecture notes in networks and systems, vol 423. Springer, Cham
19. Ardèvol-Abreu A, Gil de Zúñiga H (2016) Effects of editorial media bias perception and media trust on the use of traditional, citizen, and social media news. J Mass Commun Q. https://doi.org/10.1177/1077699016654684
20. Al-Okaily M (2021) Assessing the effectiveness of accounting information systems in the era of COVID-19 pandemic. VINE J Inf Knowl Manag Syst. ahead-of-print. https://doi.org/10.1108/VJIKMS-08-2021-0148
21. Yohanna A (2020) The influence of social media on social interactions among students. Indonesian J Soc Sci 12(2):34–48
22. Al-Sartawi A (2020) Social media disclosure of intellectual capital and firm value. Int J Learn Intellect Cap 17(4):312–323
23. Razzaque A, Al-Sartawi AMM, Karolak M (2021) Rethinking tourism entrepreneurship amidst COVID-19 through the lens of artificial intelligence and big data analytics. In: Big data for entrepreneurship and sustainable development. CRC Press, pp 175–188
24. Al-Qudah AA, Al-Okaily M, Alqudah A, Ghazlat A (2022) Mobile payment adoption in the time of the COVID-19 pandemic. Electron Commer Res. https://doi.org/10.1007/s10660-022-09577-1
25. Almaqableh L, Reddy K, Pereira V, Ramiah V, Wallace D, Veron JF (2022) An investigative study of links between terrorist attacks and cryptocurrency markets. J Bus Res 147:177–188
26. Al-Okaily A, Al-Okaily M, Teoh AP (2021) Evaluating ERP systems success: evidence from Jordanian firms in the age of the digital business. VINE J Inf Knowl Manag Syst. ahead-of-print. https://doi.org/10.1108/VJIKMS-04-2021-0061

27. Almaqableh L et al (2022) Is it possible to establish the link between drug busts and the cryptocurrency market? Yes, we can. Int J Inf Manag 102488
28. Sanad Z, Shiwakoti R, Al-Sartawi A (2022) Gender diversity on corporate boards and earnings management: a review. In: Musleh Al-Sartawi AMA (eds) Artificial intelligence for sustainable finance and sustainable technology. ICGER 2021. Lecture notes in networks and systems, vol 423. Springer, Cham

The Need for Environmental Governance: A Literature Review Paper

Nabaa Qarooni and Abdalmuttaleb Al-Sartawi

Abstract This paper was accomplished to show the need for and importance of Environmental Governance in the current period. Additionally, it showed the main aim behind having environmental governance inside organizations and within each country. Also, addressed the impact of the environmental governance on governments, sustainability, and performance of different companies. Based on the literature conducted, it was proven that having a good environmental governance will improve economic growth for countries, enhance the financial performance of companies and will boost the sustainability programs to reach to higher levels of environmental protection and natural resource protection.

Keywords Environmental governance · Corporate governance · Sustainability · EG

1 Introduction

Environmental Governance (EG) concept is rapidly increasing now. Many different researchers defined governance and EG in order to show the importance of having such concepts within the organizations. Governance is defined as the processes and rules where society make decisions and affect the environment. Another definition is that governance is the application and implementation of the rules or a game, also it can be defined as the whole range of actions of people, and public expertise as they employ and produce public policy in societies [15]. Governance is considered as one of the most important and needed factor in order to enable the effectiveness of environmental management and preservation [6, 11, 13, 14, 30].

N. Qarooni—Freelancer.

N. Qarooni
Manama, Bahrain

A. Al-Sartawi (✉)
Ahlia University, Manama, Bahrain
e-mail: amasartawi@hotmail.com

A. M. A. Musleh Al-Sartawi et al. (eds.), *From the Internet of Things to the Internet of Ideas: The Role of Artificial Intelligence*, Lecture Notes in Networks and Systems 557,
https://doi.org/10.1007/978-3-031-17746-0_25

Moreover, when it comes to EG many concepts are there such as, it is the set of regulation where are people can influence through environmental activities and results. Also, it can be defined environmental regulations that supports sustainability [25, 27, 28].

Some researchers found that the aim behind having EG was summarized in three points, first, allow managers and scientists to contribute more effectively to the real-world protection programs. Second, promote expression on the expectations and values that set up their and others [1] responsibility in protection programs. Third, more recognition how protection occurs in a challenged and super-power social context [3, 6, 23].

[15, 29] mentioned that the main aim behind EG is that to manage individual actions in undertaking of community environmental benefits and related social outcomes. Moreover, objectives of environmental governance have varied among different researchers. Additionally, four main objectives were addressed by [10], it was mentioned that the EG is aiming to sustain and improve the capability of environmental methods in order to introduce ecology system facilities through the perseverance of classes, biodiversity or habitats. Additionally, it aims to be socially reasonable through the engagement in the decision-making process and generate socioeconomic results that may be characterized as sharing, comprehensive, honest, and right.

Based on the introduction a literature review has been conducted to explain more about the environmental governance based on the point of view of different researchers.

2 Literature Review

For several years, many countries have entered an environmental protection agreement and followed regulations which protects the natural resources and save the environment in order to reduce pollution. Therefore, firms as well started to implement environmental strategies within their operations in order to be in line with the country in protecting the environment and play a major role in applying the governmental regulations which are related to the environment [2, 14, 31].

A move in the past few years have been noticed in the way the environment is regulated. Researchers have mentioned that the concept behind EG have occurred from different actors (states, industries, public citizens, governments, and nongovernmental companies) activities and strategies in order to govern society in order to achieve to environmental protection objectives [10]. Previously governments and states used to put rules and regulations in order to control the environment but recently nongovernmental agencies and parties including organizations and public citizens have become more directly concerned about the environment and governing the environment [7, 30].

In line with the higher environmental problems and pollution in the recent years as this is threatening human and their countries and societies. These harms will continue unless the government interact and sets environmental governance and polices [9, 10]. It considers being the government responsibility to protect the environment while at the same time; there should be public support to ensure the full application to polices set by the governments. It was proven that higher quality governments will have higher and better implications for the EG and polices. Moreover, higher quality governments are those where hold good public health systems, environmental sustainability, and economic wealth. Furthermore, the quality of governments has different impacts depending on their sustainability goals whether it aims to enhance the local environmental situations or reduces global carbon emissions. As a result, the study believes that governmental quality will be linked to the good environmental results for example when having multiple displays of sustainability through the full public supports to the governance and polices set by the government as well as the spending on the environmental protection programs and systems [4, 12].

Higher awareness of citizen on the environmentally friendly programs and protection of natural resources lead to high demand for green products and environmentally friendly products [9]. Moreover, assimilation of environmental problems into new goods leads create new opportunities for companies such as, entering new markets, new and differentiated products as well as new technology. Firms which are involved in producing green products and having environmental policies in producing their products tend to carry higher financial performance when comparing with those with less green products and environmental policy focus [5, 28].

[16, 27] mentioned that companies will less waste and carbon emission will have the ability to get more improved technologies and hold the capability to prevent pollution (PP). Preventing pollution could be done through establishing environmental governance such as control and prevention methods. Control is done when the company dispose their waste and emission through the PP equipment's and prevention methods are done to reduce the waste and emission through recycling, material replacement or modernization technologies. Besides that, higher funds needed to have the control approach in order to provide the reduction equipment's while on the other hand the prevention methods will focus on minimizing waste and emission of current production therefore reducing cost for raw materials. Drastically, lowering pollution and applying approaches to reach to the minimal emission will positively impact the manufacturing company performance [19, 26].

A study in East Asia have been done and showed that the environmental pollution has caused many problems on the health, society, and natural resources as well as it threatens the economic growth of the country. As a response for the issue occurred in East Asia, the countries established environmental rules, law, and regulation in order to monitor and control individuals and companies' activities toward the environment. Also, they have established new technological systems in order to prevent environmental issues. Moreover, East Asia countries believe that having well EG and environmental law as well as developed systems to prevent environmental issues and pollution will lead to higher economic growth among the coming years [17, 24].

Many companies adopted systems to manage the environmental practices as part of the EG. Different researchers believe that applying environmental management system will enhance the environmental performance of organizations. Additionally, organizations that follow an EG and EMS will reduce their impact on the environment and carry better financial performance offered through higher return on assets and growing sales. Also, it was mentioned that the higher the adoption of environmental management systems will improve the country economic level and more growth in the economic [16, 22, 25].

Due to the high level of usage to the internet, mobile, and digital media, and communication role of citizens toward the environmental practices has increased in all over the world. Collaboration between governmental, non-governmental (citizens), and private entity with their different stakeholder parties should be reached to higher levels in order to build advanced environmental governance systems and transfer it from traditional to new and modern systems. Moreover, new systems will have the ability to promote the environmental practices which are related to the recent environmental issues and reduce it [18, 21].

Through the technology development and advancement, the ICT allowed citizen to easily access environmental information also spread and produce such information. Consequently, people tend to share their knowledge and information regarding environmental protection procedures in the country and increase the awareness of such thing [20].

A study has been done in China showed that good environmental management will result in higher protection to the natural system as well as encourage environmental sustainability. It also showed that humans will have higher quality life among the long run [8, 23].

[6] believe that governments now a days are not the only important part to set environmental decision, but also other non-governmental agencies can play major roles. Cooperation between governmental agencies and private entities will increase the efficiency of EG in the country therefore many citizens will have the responsibility to work toward protecting the environment and follow the EG.

3 Conclusion

Environmental Governance (EG) concept is trending in the recent business life. Various researchers have defined the environmental governance and governance (G) in order to show the importance of applying such strategies inside organizations and countries in general. Governance was defined as the rules and regulations where the public must make decisions and impact on the environment. While other researchers like [1, 3, 15] mentioned that governance is one of the most important and wanted element in order to insure environmental managements and preservation effectiveness. Furthermore, EG is the set of regulations in which citizens follow to act environmentally friendly. Also, it is the set of environmental protection regulations which support and enhance sustainability.

Environmental governance considered as an important system in order to govern environmental practices of different actors [7]. Good EG is important to reach to sustainable expansion and safeguard the environment [13]. A research paper done by [15] proved that governance is counted as one of the highly important and needed element in order to support the effectiveness of environmental management and preservation.

[10] addressed that the main objective behind environmental governance is to control human activities in undertaking of society environmental benefits and related social results. Additionally, it aims to it was mentioned that the EG is aiming to maintain and enhance the capability of environmental methods in order to introduce natural system.

In general, environmentally friendly programs are important to save the natural resources and work toward reducing pollution and emission as well as waste. Thus, researchers found that EG development will require higher funds in order to shift into more healthy and environmentally friendly operations therefore costs will be increased [1, 11]. While other investors found that the implication of EG will have no impact on the company performance [21]. On the other hand, other researchers like, found that EG is helpful for the economic growth and sustainability programs as well as to enhances the organizations performance including the country economic situations [2, 16].

Different organizations in different industries introduced systems in order to manage and control the environmental practices as it is considered as part of the EG. Various researchers suppose that using environmental management system (EMS) will improve the environmental performance of companies. As well, companies that go along with an EG and EMS will decrease their effect on the environment and hold better financial performance presented through greater return on assets (ROA) and sales growth. In addition, it was stated that the more the implementation of environmental management systems the better the country economic situations and higher economic growth will occur [16].

To conclude, this research paper founded those countries with higher implication of EG will have better environmental performance and more attraction for investors therefore better economic conditions. Also, it was proven that the higher the awareness about the environment and application of environmental governance and laws will lead to more sustainability development and better quality of governments.

4 Recommendations

As a recommendation for further research in future, this study could be applied on wider industries and countries. It could be applied on governmental agencies and nongovernmental companies and as a suggestion to be applied in GCC and listed companies as well as non-listed companies in GCC in order to address the level of environmental governance implication and their impact on the performance and economic of countries.

References

1. Al-Qassim M, Chen W, Al-Sartawi A (2022) Blockchain in environmental compliance and enforcement. In: Musleh Al-Sartawi AMA (eds) Artificial intelligence for sustainable finance and sustainable technology. In: ICGER 2021. Lecture notes in networks and systems, vol 423. Springer, Cham
2. Al-Qudah AA, Al-Okaily M, Alqudah H (2022a) The relationship between social entrepreneurship and sustainable development from economic growth perspective: 15 'RCEP' countries. J Sustain Financ Investment 12(1):44–61
3. Al-Okaily M, Alqudah H, Al-Qudah AA, Alkhwaldi AF (2022) Examining the critical factors of computer-assisted audit tools and techniques adoption in the post-COVID-19 period: internal auditors perspective. VINE J Inf Knowl Manag Syst. ahead-of-print. https://doi.org/10.1108/VJIKMS-12-2021-0311
4. Al-Okaily M, Alalwan AA, Al-Fraihat D, Rehman SU, Alkhwaldi AF, Al-Okaily A (2022) Investigating antecedents of mobile payment systems decision making: a mediated model. global knowledge, memory and communication. ahead-of-print. https://doi.org/10.1108/GKMC-10-2021-0171
5. Al-Okaily M, Alghazzawi R, Alkhwaldi AF, Al-Okaily A (2022) The effect of digital accounting systems on the decision-making quality in the banking industry sector: a mediated-moderated model. Global Knowl Memory Commun. ahead-of-print. https://doi.org/10.1108/GKMC-01-2022-0015
6. Al-Sartawi A (2015) The effect of corporate governance on the performance of the listed companies in the gulf cooperation council countries. Jordan J Bus Adm 11(3):705–725
7. Al-Sartawi A (2017) The effect of the electronic financial reporting on the market value added of the Islamic banks in Gulf Cooperation Council Countries. In: 8th global Islamic marketing conference, 4–6 May. International Islamic Marketing Association, Turkey
8. Al-Sartawi A (2020) Social media disclosure of intellectual capital and firm value. Int J Learn Intellect Capital 17(4), 312–323
9. Al-Sartawi A, Sanad Z (2019) Institutional ownership and corporate governance: evidence from Bahrain. Afro-Asian J Finan Account 9(1):101–115
10. Al-Sartawi A, Reyad S, Madbouly A (2021) Shariah, presentation and content dimensions of Web 2.0 applications and the firm value of Islamic financial institutions in the GCC countries. J Islamic Market. Accepted article
11. Al-Sartawi A (2020) Information technology governance and cybersecurity at the board level. Int J Crit Infrastruct 16(2), 150–161
12. Al-Sartawi A (2020) Does it pay to be socially responsible? Empirical evidence from the GCC countries. Int J Law Manag 62(5):381–394
13. Al-Sartawi A (2019) Assessing the relationship between information transparency through social media disclosure and firm value. Manag Account Rev 18(2):1–20
14. Armitage D, Loe RD, Plummer R (2012) Environmental governance and its implications for conservation practice. Wiley Periodicals 5:245–255
15. Benson D, Jordan A (2017) Environmental governance. Int Encyclopedia Geogr:1–9
16. Cheng R, Li W (2019) Evaluating environmental sustainability of an urban industrial plan under the three-line environmental governance policy in China. J Envrion Manag 251:1–11
17. Gogia J, Chakraborty D (2022) Open banking: a revolution in the Tech-Fin industry. Int J Electron Bank 3(2):100–120
18. Gupta M, Sikarwar TS (2020) Modelling credit risk management and bank's profitability. Int J Electron Bank 2(2):170–183
19. Hart S, Ahuja G (1996) Does it pay to be green? An empirical examination of the relationship between emission reduction and firm performance. Bus Strategy Environ 5:30–37
20. Kulin J, Seva I (2019) The role of government in protecting the environment: quality of government and the translation of normative views about government responsibility into spending preferences. Int J Soc:110–129

21. Lemos A (2013) Environmental governance
22. Liu H, Jiang J, Xue R, Meng X, Hu S (2022) Corporate environmental governance scheme and investment efficiency over the course of COVID-19:1544–6123
23. Lockwood M, Davidson J, Curtis A, Stratford E, Griffith R (2010) Governance principles for natural resource management. Soc Nat Res. 23:986–1001
24. Miroshnychenko I, Barontini R, Testa F (2017) Green practices and financial performance: a global outlook. J Clean Prod 147:340–351
25. Musleh Al-Sartawi AM, Hussainey K, Razzaque A (2022) The role of artificial intelligence in sustainable finance. J Sustain Finan Investment:1–6
26. Mori A (2013) Environmental governance for sustainable development: east Asian perspectives. United Nations University Press, Tokyo, New York, Paris
27. Musleh Al-Sartawi AM, Razzaque A, Kamal MM (eds) (2021) Artificial intelligence systems and the internet of things in the digital era. In: EAMMIS 2021. Lecture notes in networks and systems, vol 239. Springer, Cham
28. Musleh Al-Sartawi AMA (eds) (2022) Artificial intelligence for sustainable finance and sustainable technology. In: ICGER 2021. Lecture notes in networks and systems, vol 423. Springer, Cham
29. Soma K, Onwezen MC, Salverda IE, Dam RI (2016) Roles of citizens in environmental governance in the information age - four theoretical perspectives. Current Opinion Environ Sustain 18:122–130
30. Sanad Z, Al-Sartawi A (2016) Investigating the relationship between corporate governance and internet financial reporting (IFR): evidence from Bahrain bourse. Jordan J Bus Adm 12(1):239–269
31. Watson K, Klingenberg B, Polito T, Geirts T (2004) Impact of environmental management system implementation on financial performance: a comparison of two corporate strategies. Manag Environ Qual 15:622–628

Adoption Model of Mobile Technologies for Bahrain Bankers

Mahmood Saeed Mustafa Alalawi and Abdulla Al-Qallaf

Abstract Mobile banking is popular in financial matters since several factors contribute to its adaptation. Considering that Bahrain is the gulf's financial hub of the region it is important for this study to investigate what influential factors can enhance the adoption of mobile banking; especially for the bankers in Bahrain. A literature review gets conducted via journal papers and conference articles. A conceptual model is derived through this review of the literature. Five proposition get proposed the role of convenience, need for social acceptance, young people, attitude toward using technology and trust on the level of adoption of mobile banking.

Keywords Mobile banking adoption · Convenience · Social acceptance · Youth and attitude towards technology · TAM

1 Introduction

Technological innovation has influenced all facets of life, i.e. business and personal life. In modern days, banking services are an essential part of individual and business affairs. Banks are adopting new technologies to create superior customer value. Technology usage allows banking organisations to deliver quality customer services in cheaper and more convenient ways [46]. One emergent technological way to provide banking services to customers is through mobile connectivity. The provision of banking services through mobile phones get coined "*mobile banking*". Like many other new technologies, mobile banking gets influenced by various factors in gaining customer acceptance. Mobile banking adoption challenges are typical examples of the diffusion of any new technology to customers. An increasing number of customers are using Mobile banking applications on their smartphones to avail of

A. Al-Qallaf—Freelancer.

M. S. M. Alalawi (✉)
Ahlia University, Manama, Bahrain
e-mail: mahmood.alawi@gmail.com

A. Al-Qallaf
Manama, Bahrain

A. M. A. Musleh Al-Sartawi et al. (eds.), *From the Internet of Things to the Internet of Ideas: The Role of Artificial Intelligence*, Lecture Notes in Networks and Systems 557,
https://doi.org/10.1007/978-3-031-17746-0_26

mobile banking facilities [40]. There are two-fold challenges for banks to get the maximum benefit from Mobile banking. On one end, banks have to overcome technological challenges by designing software and network support to transfer manual banking services to the touch of mobile screens [55]. Secondly, banking clients' adoption of mobile banking services is full of challenges. Banking customers have many idiosyncrasies and fears about the use of Mobile banking [42, 50]. There is a long list of factors that hinder the adoption of Mobile banking worldwide. For example, mobile banking adoption is low because customers have less trust in using mobile for financial transactions. In addition, many customers are fearful regarding the security of their transactions through Mobile banking [17]. Banks are encouraging mobile banking applications for real-time financial transactions and financial information [24]. Banks consider the highest level of potential for mobile banking for both personal and business purposes [22]. With the advent of smartphones, mobile banking applications are get made available widely to increase the diffusion of Mobile banking. Mobile banking services enable the customer to manage and organize financial transactions without holding cash amounts. Banks are increasingly grasping the idea of mobile banking because it is a form of branchless banking for banks [23]. Previous research has investigated the adoption of Mobile banking in terms of factors that baulk and those that facilitate the use of Mobile banking. However, most of the evidence is accumulated from research in developed countries of Europe and America. In this regard, [53] called for more research exploring Mobile banking in a developing country. Therefore, this research responds to their call by examining Bahrain's mobile banking adoption. Unfortunately, any empirical evidence from developing countries in this regard is relatively nonexistent.

Additionally, the previous research uses only the convenience factor on the part of customers to investigate the adaptability of Mobile banking. The research should also include factors other than convenience that add to the adoption of mobile banking, as is the focus in this research. As [13] identified, behavioural intentions play a great role in adopting Mobile banking. Therefore, this study includes a mix of behavioural factors influencing Mobile banking adoption. This research fills in the gap by carrying out an empirical investigation of the adoption of mobile banking in a developing country, thus adding to the existing evidence for factors influencing Mobile banking adoption [42, 49, 50]. Until now, it is not clear about the perception of the adoption of mobile banking among banking customers in the Kingdom of Bahrain. Their positive and negative experience with mobile banking is relatively unchecked. Consequently, it is unknown about favourable factors inducing mobile banking, not about the factors hindering the use of mobile banking. Only with a sound understanding of factors that support Mobile banking adoption and the factors that baulk it, can banks device financial products to suit the expectation of customers. Because of a relatively limited understanding of factors influencing Mobile banking adaption in Bahrain, the banking organizations cannot devise a strategy to stimulate Mobile banking usage. Hence this study is justified in examining the behaviour of factors influencing Mobile banking. This research investigates the role of factors influencing customers' perception of mobile banking adoption. In other words, how mobile banking is gaining acceptance in Bahraini society, below are the main objectives: (1)

Identify key factors influencing the adoption & acceptance of mobile banking in the Kingdom of Bahrain by drawing a theoretical model and empirically testing it. (2) To examine how consumer characteristics (i.e. convenience seeking, life cycle stage, attitude towards technology) influence mobile banking adoption. (3) To examine how the social environment of the customer (i.e. social acceptance and trust) influences mobile banking adoption. (4) To examine the direction and magnitude of various factors influencing mobile banking in Bahrain. And (5) to suggest recommendations to banks for leveraging mobile banking profits.

2 Literature Review

Technology is influencing all aspects of business these days, and banking is no exception. Providing financial services through the internet and mobile phones is on the rise in all countries [1]. Mobile banking has been defined as "a channel whereby the consumer interacts with a bank via a mobile device, such as a mobile phone or personal digital assistant. It can be seen as a subset of electronic banking and an extension of internet banking with its own unique characteristics" [32–34]. Mobile Phones are ubiquitous in modern times, with more mobile phones than personal computers. Mobile banking provides an option of access to banking 24 h a day and seven days a week [51]. The banking services customers now have a great opportunity to get banking services without visiting their banks through mobile phones. Researchers have attempted to decode the technology adaptation process through various studies. An important technology adoption explanation comes from the work of [16] as they presented Technology Acceptance Model (TAM), delineating user's acceptance of information technology systems. In the context of mobile banking, [37] amended this model by adding perceived credibility, perceived self-efficacy and perceived trust. The TAM is then continuously tested in relation to other variables to explicate the mobile banking adaption rate [11]. For example, Akturan & Tezcan (2012) analyzed mobile banking adoption in combination with perceived benefits and perceived risks. Literature related to the main factors influencing Mobile banking adoption is summarized in the following section of the report. According to TAM, mobile banking adoption is influenced by a series of factors, i.e. Perceived Usefulness, Perceived Ease of Use, Attitude Toward Using, Behavioral Intentions and Actual System Use. (Fig. 1).

The following literature review is accumulated in light of all these important factors influencing Mobile banking Adoption. Mobile banking facilitates managing financial transactions without handling of cash. Mobile banking is also favourable to banks as these services are considered to be a form of branchless banking [23]. Banks are encouraging the use of mobile banking applications for real-time financial transactions and financial information [24]. Compatibility is considered to be an important indicator of mobile services adoption. For example, [26] reported that compatibility significantly influences the ease of use of mobile banking and the perceived usefulness of mobile banking services. Another important element in determining the

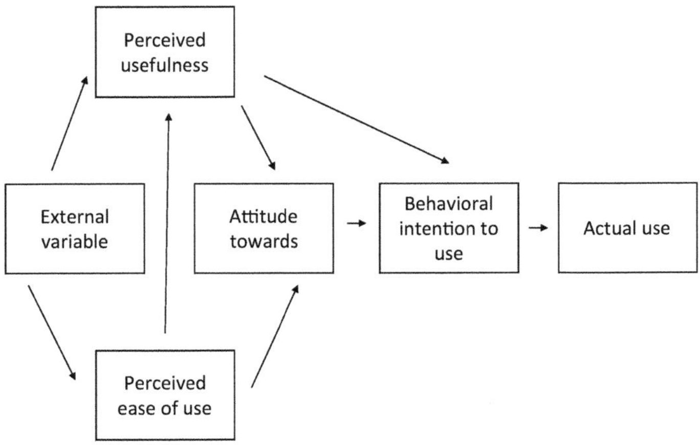

Fig. 1 Research model adopted from technology adoption model

adoption of mobile banking services is the ease of use. Ease of use indicates an increased likelihood of adopting mobile banking. Ease of use represents the degree to which customers feel that mobile banking is user friendly. Similarly, [35] indicated that ease-of-use means that mobile banking is understandable and offers operational ease. Technology is usually considered a complex phenomenon; however, in mobile banking, complexity will reduce adoption and user-friendliness (ease) will enhance adaptability. Therefore ease of use is considered to be a determinant of mobile banking adoption [3, 4, 7]. The mobile banking industry is characterized to have a low international adaption rate and hence is in infancy [16]. Consequently, all attempts to understand the antecedents and consequences of mobile banking adaption and the process of mobile banking adoption are justified for research. Mobile banking is also called, M-Banking, and SMS banking refers to internet-based banking using mobile devices. Customers consider m-banking a useful technology because they can perform various activities, i.e. make payments, check transaction history, make bill payments, and fund transfers [10]. Hence, enjoying the convenience of services through Mobile banking. Convenience is the main variable in Technology Adoption Model with strong implications for Mobile banking diffusion among customers. The convenience factor in Mobile banking adoption is examined in detail by researchers. All banking organizations are positioning themselves as a provider of convenient banking services through Mobile banking.

Regarding social acceptance, using technology products has a social image and status associated with it. People who use technologically advanced equipment and procedure are viewed as more socially acceptable. Social norms and social settings also shape the behaviour for the use of certain products and services. Mobile banking is a relatively new service and is considered an innovation in banking services. Therefore, the social norms account for the acceptability of Mobile banking. Social norms represent various combinations of factors such as family, friends, relatives and colleagues influencing the decision to adopt new technology. Social norms have been observed to influence email usage and internet usage [29, 30]; hence this factor also spells an important contribution to the adoption of mobile banking. [26, 27] researched social acceptance influence on Mobile banking adoption. The results indicate that social acceptance has a positive impact on the adoption of mobile banking adoption. Social norms are important determinants of behavioural intentions, i.e. customer behaviour is derived through social norms. Moreover, the nature of social norms determines the criteria for what is appropriate or not and what is acceptable or not. Social norms shape people's belief systems, hence determining what seems to be important for them or not unimportant. Social norms have also implied implications for social utility. Therefore, using Mobile banking represents social norms and social utility. In this line of research, (a) it is found that social norms positively influence behavioural intentions for technology adoption, (b) attitude towards mobile banking services, (c) and mobile banking adaptability [11]. Social norms have also been examined for influence on technology adoption and social blogging. [4] found that social and cultural factors harmonize and act as conditioning factors for the adoption of mobile banking services. The social context of the decision-maker demands that the person should weigh the cost and benefits associated with the adoption of new technology products and services. Customers only adopt those services and products that are in line with their social norms and cultural preferences. Therefore, acceptance of Mobile banking is reflected in the social and cultural context of a customer. In this research, the proposition also attempts to examine this relationship in mobile banking adoption in Bahrain. Similarly, social pressure from family, peers, friends and members of the social circle may influence the adoption of mobile banking by customers. For example, mobile banking and other online services are assumed to be adopted under the influence of social pressure by various customers [1] An investigation in line with the adoption of mobile banking indicates that social identity is an important social factor that influences the adoption of mobile banking. In this regard, it is found that mobile banking adoption is strongly associated with social norms, social identity and social pressure experienced by a customer [56].

Regarding trust, trust is an important variable that determines the adaptability of the electronic format of services. For example, trust influences the adoption of online news service subscriptions. Similarly, trust has been found to significantly influence the adoption of internet banking [18]. Additionally, empirical research has proved that trust has strong implications for the adoption of online health services [44–48]. Furthermore, trust has also been witnessed to significantly positively impact mobile shopping [36]. Based on these studies, this research assumes a strong conviction toward the feeling of trust in service providers and the use of mobile banking services. According to [38], mobile banking services are inherently effort saving independent of time and space constraints, therefore creating superior value for customers. On the other hand, many customers are reluctant to adopt mobile banking because they perceive it as riskier. However, if mobile banking organizations can establish a relationship of trust with their customers, this situation can be ameliorated. In this regard, previous research indicates that trust has an important role in adopting mobile banking. For example, trust in a service provider reduces perceived risk and expedites uncertain business transactions [14]. More specifically, the researchers have found that knowledge-based trust representing service providers' integrity, competence, and benevolence is instrumental in adopting online transactions [39, 43]. An important stream of research also exemplifies significant organizational level barriers that limit the applicability of mobile banking. For example, lack of significant information technology support, management's negative attitude and inefficient internal systems are responsible for not achieving mobile banking related objectives [52]. In the mobile banking context, it is to be noted that a customer's knowledge-based trust indicates that the mobile banking service provider has the necessary competence and integrity to offer mobile banking services [19, 35]. Trust binds a customer with a business or service provider. Initial trust in Mobile banking services creates a long-term relationship between the organization and the customer. Establishing a relationship of "initial trust" between the service provider and customer is critical for the long-term usage of Mobile banking [56]. Trust in service providers engenders information required to make a purchase decision or decision for the use of services of an organization. The information quality is directly associated with trust on the service provider and leads to the decision about the use of services provided. Higher trust means higher information quality, whereas lower trust means lower information quality [19, 35]. Hence, the level of information quality influences the adoption of technology such as mobile banking. Regarding Acceptance of Technology among Youngsters. The research by [1] indicates that non-users of mobile financial services require high mental effort to use mobile banking. All do not equally prefer mobile banking, i.e. young people are more oriented toward mobile banking than older people [20, 21]. Young people value convenience and time saving; therefore, they tend to use mobile banking.

In a study about Mobile banking adoption in Kuwait, it is found that gender moderates the adoption of mobile banking services. The findings of this study indicate that the Ease-of-Use feature influences females more than male users, and social norms have a higher influence on females than males [27, 28]. Mobile banking requires the users to interact through mobile phones with banks. Those users, who have a relatively higher understanding of mobile phone operations, are more prone to use Mobile banking. Banks are strategically positioning themselves to provide Mobile banking services to young clients who have got efficient skills in using and operating mobile phones. Young people regularly use various applications on smartphones and can subscribe or unsubscribe to various mobile phone services [8, 9]. Therefore, their inclination toward mobile banking is relatively higher than older customers. According to the research of [54], the very nature of financial services has been changed through the advent of mobile banking. Scholars have examined the diffusion pattern of mobile banking and have discovered that Mobile banking adoption varies from segment to segment. Moreover, their results indicate that young consumers possess the necessary characteristics to adopt Mobile banking at an earlier stage than that mature users of banking services and mobile services. n terms of the influence of Generation on Mobile banking adoption, it is generally noted that youth has a strong inclination towards the adoption of mobile banking. The behavioural intention of Generation Y towards the use of mobile banking has been examined recently [13]. The research findings indicate that trustworthiness and perceived risk correlate with mobile banking adoption. The research further indicates that mobile banking is rated high on innovation factors, i.e. trialability, compatibility, observability, and complexity. The above-mentioned literature regarding youth and the use of mobile banking indicates that mobile banking is largely acceptable to young customers [5–7]. This group of consumers is technology-driven and possesses the necessary skills and resources to use mobile banking technology. Therefore, current research has incorporated this variable as an important determinant for the adoption of mobile banking services. Regarding attitude toward Smart Phones (Technology), the list of factors hindering the adoption of mobile banking also includes behavioural and attitudinal issues related to individuals. A customer's demographic status, behaviour and motivation towards technology-enabled services and attitude towards technology acceptance have significant importance for technology adoption [31]. These individual traits influence how a customer adopts new technology and hence his motivation to adopt mobile technology. Once these characteristics are found in a number of individuals, then the adoption of mobile banking decreases. An attitude develops favourable or unfavourable evaluation of Mobile banking usage for individuals. The attitude towards Mobile banking is representative of technology readiness, computer skills and mobile technology readiness [27, 28].

Similar research from Turkey has provided important insight into Mobile banking adoption. The results indicate that mobile customers favour mobile marketing tools in Turkey. The survey indicates that the attitude toward mobile marketing is positive for customers, i.e. mobile advertising, mobile discounts, location-based mobile services, mobile coups, mobile internet and Mobile banking [10]. The stronger customer attitude towards mobile services indicates a favourable condition for the adoption and diffusion of mobile banking services. Moreover, it can be assumed that a favourable attitude towards mobile Technology and wireless Technology can get harnessed to increase mobile banking. On the other hand, customers' unfavourable attitude towards technology usage baulks the rapid adoption of Mobile banking services. Customers' attitudes towards mobile services also influence the adoption rate of mobile banking services. Customers who are more comfortable and frequently use mobile phone applications are more motivated to use mobile banking [41]. Empirical studies indicate that a technology feature that creates a relative advantage compared to predecessors makes adoption more rapid. Relative advantage means the degree to which a technological feature is more beneficial to the previous one [25]. Since mobile banking is new technology and an innovative feature in the banking sector, its perceived relative advantage has important implications for the adoption of mobile banking. A comparative study of internet banking and Mobile banking by [12] reveals that mobile banking substitutes for internet banking and other forms of banking by customers who have a high degree of preference for technology adoption a low degree of preference. Other researchers further support these findings, indicating that young customers are more likely to use mobile banking services as they are experts in using smartphones and various applications. The attitude toward technology is different from a different set of customers. Nonetheless, the attitude towards technology strongly impacts mobile banking adoption. Research work indicated that attitude towards computers [26–28] and attitude toward technology have positive implications for mobile banking adoption. The technology attitude towards the adoption of mobile banking includes factors like customers' previous technology experience, personal banking experience, and reference group influence [26, 27]. The benefits of relative advantage include enhanced status, efficiency and economic benefits. Through mobile banking, a customer can easily harvest these benefits. Therefore perceived benefits influence the mobile banking adoption rate. Moreover, the mobile banking industry gets characterized by offering convenience, affordability and immediacy [34]. The interplay of these variables has an important impact on mobile banking adaptation. Rapidly increasing mobile devices and the proliferation of smartphones have an exponential impact on the adaption of mobile banking. The extant research indicates that customer inclination towards the adaption of mobile banking is on the rise because of ease of use, perceived advantage, compatibility, competence and integrity [35].

3 Research Methodology

In line with the research mentioned above objectives, research question, literature review, and proposition, a theoretical model gets proposed in Fig. 2. Even though it shows five proposition, establishing the relationship between dependent and independent variables (Fig. 2).

For this deductive research, this study proposes that data get collected using a questionnaire with items adopted from studies that reflect items pertaining to the depicted variables. The suggested method of the sampling for future research is combination or purposeful mixed sampling. The reason for choose such a method is since it merges different sampling strategies, thus increasing the respondent participation rate. Through simple random sampling, the collected data reflects the influential factors for the adaption of mobile banking, a subset of a statistical population in which each subset member has an equal probability of being chosen. A simple random sample is meant to represent a group's unbiased representation.

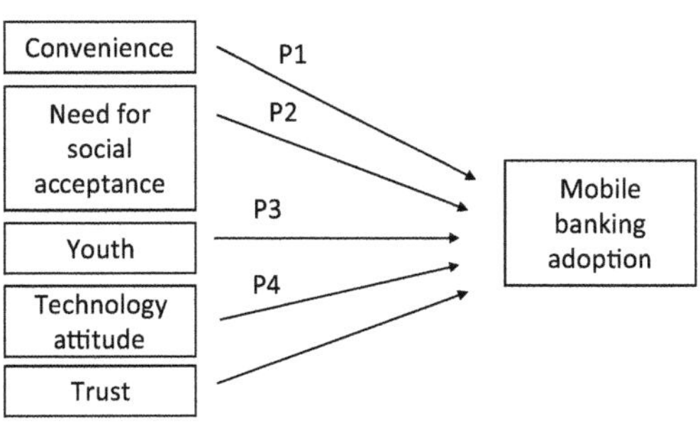

Fig. 2 Research model

4 Conclusion

The first proposition established the relationship between Convenience and Mobile banking Adoption. The proposition demonstrates a positive effect of Convenience on Mobile banking adoption. The diffusion of Mobile banking among mass-market becomes rapid if customers believe that Mobile banking is a convenient way to get banking services. Since Mobile banking allows the customer to manage financial affairs without cash, the element of convenience becomes prominent. Moreover, the customer does not have to travel to the bank's premises to avail of financial services, hence adding to their convenience. In line with such convenient features, the banks encourage mobile banking applications for real-time financial transactions and financial information [24]. In previous research, the Convenience factor has been given great importance for adopting Mobile banking. Convenience in Mobile banking adoption is a subset of ease of use and compatibility. Ease-of-use represents the degree to which customers feel that mobile banking is user friendly. With the increase in customers' perception of the ease-of-use of Mobile banking, the diffusion and adoption rate of Mobile banking increases. Previous research in the line of ease of use indicates the increasing likelihood of adopting Mobile banking. [35] indicated that ease of use means that mobile banking is understandable and offers operational ease. Technology is usually considered a complex phenomenon; however, in mobile banking, complexity will reduce adoption and user-friendliness (ease) will enhance adaptability. Therefore ease of use is considered to be a determinant of mobile banking adoption [34]. In addition to ease-of-use, compatibility features, i.e. the way mobile banking applications and software synchronizes with smart, also determine convenience for end-users. The higher the compatibility, the greater the convenience, and more mobile banking will be used. As proved by this thesis, mobile banking provides convenience to the users in managing their banking transactions. Hence, it is strongly recommended that banks attempt to provide banking services through mobile banking. Through such services, banks can capture customers seeking convenience in sending, receiving, and depositing money with their banks. The second proposition examines the influence of Social Acceptance needs on mobile banking Adoption. The results of this study along with the interviews conducted, indicate that Social Acceptance needs statistically significantly influence mobile banking Adoption. Customers have social contacts with several individuals around them, which influence their choices and buying behaviour. Social norms represent various combinations of factors such as family, friends, relatives and colleagues influencing the decision to adopt new technology. Therefore, a customer's decision to adopt mobile banking is strongly influenced by his need for Social Acceptance. Furthermore, social norms have been observed to influence email usage and internet usage [3] Hence this factor also spells an important contribution to the adoption of mobile banking. Moreover, it is noted that social influence shapes preferences towards product and service usage and consumption. Since customers have to show the kind of products and services they are using to various members of customer groups, therefore social norms play a critical role in the decision toward mobile banking adoption.

Social norms have also been examined for influence on technology adoption and social blogging. [17] found that social and cultural factors harmonize and act as conditioning factors for the adoption of mobile banking services. The social context of the decision-maker demands that the person should weigh the cost and benefits associated with the adoption of new technology products and services.

When friends, relatives and colleagues recommend the use of a particular brand, technology or service, then an individual has to comply with that pressure. Similarly, social pressure from family, peers, friends and social circle members may influence the adoption of mobile banking by customers. For example, Mobile banking and other online services are assumed to be adopted under social pressure by various customers [56] An investigation in line with the adoption of mobile banking indicates that social identity is an important social factor that influences the adoption of mobile banking. In this regard, it is found that mobile banking adoption is strongly associated with social norms, social identity and social pressure experienced by a customer [56]. In the current research, it is found that the need for Social Approval has a positive influence on mobile banking Adoption. The higher the social approval needs, the higher the use of mobile banking will be. Banking organizations can stimulate mobile banking adoption through advertisements highlighting social acceptance. By targeting the need for social approval through word of mouth and advertisements, the banks can increase the usage of mobile banking adopting. The recommendation by friends and family can strongly influence buying behaviour and service usage. The banks can leverage the recommendation features from friends and family. The third proposition examines the impact of technology attitude on mobile banking adoption. The results are consistent with previous studies, indicating that a positive attitude toward technology statistically significantly influences Technology Attitude. Customer having a preferable attitude towards the use of technology is more inclined to adopt new technologies of communication and services. An attitude develops un/favourable evaluation of mobile banking usage for individuals. The attitude towards mobile banking is representative of technology readiness, computer skills and mobile technology readiness [27, 28]. When customers, in general, have a positive attitude towards technology usage, they are not hesitant to adopt newer forms of technology. Consequently, these customers easily adapt themselves to mobile banking Technology. Customers' attitudes toward mobile services also influence the adoption rate of mobile banking services. Customers who are more comfortable and frequently use mobile phone applications are more motivated to use mobile banking [41]. Empirical studies indicate that a technology feature that creates a relative advantage compared to predecessors makes adoption more rapid. Relative advantage means the degree to which a technological feature is more beneficial to the previous one [25]. Since mobile banking is new technology and an innovative feature in the banking sector, its perceived relative advantage has important implications for the adoption of mobile banking.

A positive attitude towards new technology means that a customer understands the technical aspects of technologically sophisticated products. The attitude towards technology is different for a different sets of customers. Nonetheless, the attitude towards technology strongly impacts mobile banking adoption. Research work indicated that attitude toward computers [26–28] and attitude toward technology positively affect mobile banking adoption. The technology attitude towards the adoption of mobile banking includes factors like customers' previous technology experience, personal banking experience, and reference group influence [26–28]. Since attitude toward adopting new technology is becoming more and more acceptable, banks can encourage users to use mobile banking like they are willing to use technology means in other areas. In this regard, the banks can design a mobile application that customers can download on their smartphones. These applications can get used for fund transfer, bank account information, payment of bills etc. The banks can enable their customers to benefit from mobile banking services by using these applications. The fourth proposition in this study explored the nexus between trust and mobile banking Adoption. The findings and interviews of this study indicate that trust significantly influences mobile banking adoption. The customer perceives financial transactions through mobile phones and the internet as risk. With a higher level of perceived risk, the propensity to adopt Mobile banking decreases. Similarly, lesser risk and more trust engender a higher propensity to use mobile banking. Trust in mobile banking indicates that customers can make banking transactions without fear and with full confidence. The higher level of trust means that customers will rapidly adopt mobile banking. [38] indicate that mobile banking services are inherently effort saving, independent of time and space constraints, therefore creating superior value for customers. However, many customers are reluctant to adopt Mobile banking because they perceive it as riskier.

Trust in mobile banking reduces several risks like financial risk and social risk. With reduced risk, customers can have peace of mind and will manage financial transactions without any fear. However, if mobile banking organizations can establish a relationship of trust with their customers, this situation can be ameliorated. In this regard, previous research indicates that trust has an important role in adopting mobile banking. For example, trust in a service provider reduces perceived risk and expedites uncertain business transactions [14]. More specifically, the researchers have found that knowledge-based trust representing service providers' integrity, competence, and benevolence is instrumental in adopting online transactions [39]. Furthermore, trust binds a customer with a business or service provider. Initial trust in mobile banking services creates a long-term relationship between the organization and the customer. Establishing a relationship of "initial trust" between the service provider and customer is critical for the long-term usage of mobile banking [56]. Trust in service providers engenders information required to make a purchase decision or decision for the use of services of an organization. The information quality is directly associated with trust in the service provider and leads to the decision about the use of services provided. Higher trust means higher information quality, whereas lower trust means lower information quality. Hence, the level of information quality influences the adoption of technology such as mobile banking. Since trust has a strong influence in mobile banking, banks have to position themselves as providers of banking services in a trustful manner. As a result, the customers can see that using a mobile banking forum is free of risk. This position can be achieved through testimonials from customers who have successfully used mobile banking services. Similarly, the banks can provide testimonials about the safety of using banking services, through mobile banking, by showing certificates declaring that mobile banking is risk-free [2, 3]. The last proposition of this research examined the impact of youth on mobile banking Adoption. The results indicate that youth has positive impact on adoption of mobile banking. Previous research also reports similar kind of results for preference of youth towards technology adoption. Young customers are more accustomed to the use of various technology based products [2, 3].

They are also dexterous in using smartphones and the internet; consequently, the rate of adoption of mobile banking is related to young customers. As reported in the literature review that a study about mobile banking adoption in Kuwait found that gender moderates the adoption of mobile banking services. The findings of this study indicate that the ease-of-use feature more than male users influences females, and social norms have a high influence on females than on males [26–28]. Mobile banking requires the users to interact through mobile phones with banks. Those users, who have a relatively higher understanding of mobile phone operations, are more prone to use mobile banking. Banks are strategically positioning themselves to provide mobile banking services to young clients who have efficient skills in using and operating mobile phones. Young people regularly use various applications on smartphones and can subscribe or unsubscribe to various mobile phone services [9]. Therefore, their inclination towards mobile banking is relatively higher than older customers. In terms of the influence of Generation on mobile banking adoption, it is generally noted that youth has a strong inclination towards the adoption of mobile banking. The behavioural intention of the Generation toward mobile banking has been examined recently [13]. The research findings indicate that trustworthiness and perceived risk correlate with mobile banking adoption. This study reveals that mobile banking gets considered highly on various factors related to innovation, i.e., trialability, compatibility, observability, and complexity. The literature mentioned above regarding youth and mobile banking indicates that mobile banking is largely acceptable to young customers. This group of consumers is technology-driven and possesses the necessary skills and resources to use mobile banking technology. Therefore, current research has incorporated this variable as an important determinant for the adoption of mobile banking services.

References

1. Ahmed M, Almotairi MA, Ullah S, Alam A (2014) Mobile banking adoption: a qualitative approach towards the assessment of TAM model in an emerging economy. Acad Res Int 5(6):248–259
2. Al-Qudah AA, Al-Okaily M, Alqudah G et al (2022) Mobile payment adoption in the time of the COVID-19 pandemic. Electron Commer Res. https://doi.org/10.1007/s10660-022-09577-1

3. Alalawi M, Razzaque A (2016) Peak-end rule to promote social capital in a virtual community environment. In: International conference on emerging research for sustainable economic development (ERSED), Dubai, UAE
4. Al Mubarak H, Al Alawi M, Razzaque A (2017) The importance of national culture dimensions influence on quality management performance. Int J Bus Res 17(3):7–20
5. Al-Sartawi A (2020) Social media disclosure of intellectual capital and firm value. Int J Learn Intellect Cap 17(4):312–323
6. Alshaikh IY, Razzaque A, Alalawi M (2017) Positive emotion and social capital affect knowledge sharing: case of the public sector of the Kingdom of Bahrain. In: European, Mediterranean and Middle Eastern conference on information systems, EMCIS 2017, 7–9 September 2017. Lecture Notes in Business Information Processing. Springer, Coimbra
7. Al-Sartawi A (2020) Information technology governance and cybersecurity at the board level. Int J Crit Infrastruct 16(2):150–161
8. Alansari Y, Al-Sartawi A (2021). IT governance and E-banking in GCC listed banks. Procedia Comput Sci 183:844–848
9. Barati S, Mohammadi S (2009) An efficient model to improve customer acceptance of mobile banking. Paper presented at the world congress on engineering and computer science
10. Barnes SJ, Corbitt B (2003) Mobile banking: concept and potential. Int J Mobile Commun 1(3):273–288
11. Barutçu S (2007) Attitudes towards mobile marketing tools: a study of Turkish consumers. J Target Meas Anal Mark 16(1):26–38
12. Bauer HH, Barnes SJ, Reichardt T, Neumann MM (2005) Driving consumer acceptance of mobile marketing: a theoretical framework and empirical study. J Electron Commer Res 6(3):181–192
13. Calisir F, Gumussoy CA (2008) Internet banking versus other banking channels: young consumers' view. Int J Inf Manage 28(3):215–221
14. Chung K-C, Holdsworth DK (2012) Culture and behavioural intent to adopt mobile commerce among the Y generation: comparative analyses between Kazakhstan, Morocco and Singapore. Young Consum 13(3):224–241
15. Corritore CL, Kracher B, Wiedenbeck S (2003) On-line trust: concepts, evolving themes, a model. Int J Hum Comput Stud 58(6):737–758
16. Datamonitor (2009) Mobile banking: over-hyped gimmick or credible 21st century banking channel?
17. Davis FD, Bagozzi RP, Warshaw PR (1989) User acceptance of computer technology: a comparison of two theoretical models. Manage Sci 35(8):982–1003
18. Ellison K (2015) Mobile banking security still a barrier for consumers Retrieved 25 May. http://www.welivesecurity.com/2015/05/27/mobile-banking-security-still-barrier-consumers/
19. Flavian C, Guinaliu M, Torres E (2005) The influence of corporate image on consumer trust: a comparative analysis in traditional versus internet banking. Internet Res 15(4):447–470
20. Howard Chen Y-H, Corkindale D (2008) Towards an understanding of the behavioral intention to use online news services: an exploratory study. Internet Res 18(3):286–312
21. Howcroft B, Hamilton R, Hewer P (2002) Consumer attitude and the usage and adoption of home-based banking in the United Kingdom. Int J Bank Market 20(3):111–121
22. Hannoon A, Al-Sartawi A, Khalid A (2021) Relationship between financial technology and financial performance. In: Musleh Al-Sartawi AMA (eds) The big data-driven digital economy: artificial and computational intelligence. Studies in computational intelligence, vol 974. Springer, Cham
23. Ivatury G, Mas I (2008) The early experience with branchless banking. CGAP Focus Note (46)
24. Jacob K (2007) Are mobile payments the smart cards of the Aughts? Chicago Fed Letter 240:1–4
25. Karayanni DA (2003) Web-shoppers and non-shoppers: compatibility, relative advantage and demographics. Eur Bus Rev 15(3):141–152
26. Karjaluoto H, Koenig-Lewis N, Palmer A, Moll A (2010) Predicting young consumers' take up of mobile banking services. Int J Bank Market 28(5):410–432

27. Karjaluoto H, Mattila M, Pento T (2002) Factors underlying attitude formation towards online banking in Finland. Int J Bank Market 20(6):261–272
28. Karjaluoto H, Riquelme HE, Rios RE (2010) The moderating effect of gender in the adoption of mobile banking. Int J Bank Market 28(5):328–341
29. Kim G, Shin B, Lee HG (2009) Understanding dynamics between initial trust and usage intentions of mobile banking. Inf Syst J 19(3):283–311
30. Karolak M, Razzaque A, Al-Sartawi A (2021) E-services and M-services using IoT: an assessment of the Kingdom of Bahrain. In: Musleh Al-Sartawi AM, Razzaque A, Kamal MM (eds) Artificial intelligence systems and the internet of things in the digital era. EAMMIS 2021. Lecture notes in networks and systems, vol 239. Springer, Cham
31. Laforet S, Li X (2005) Consumers' attitudes towards online and mobile banking in China. Int J Bank Market 23(5):362–380
32. Laukkanen T (2007) Internet vs mobile banking comparing customer value perceptions. Bus Process Manag J 13(6):788–797
33. Laukkanen T (2007) Measuring mobile banking customers' channel attribute preferences in service consumption. Int J Mobile Commun 5(2):123–138
34. Laukkanen T, Pasanen M (2008) Mobile banking innovators and early adopters: how they differ from other online users & quest. J Finan Serv Market 13(2):86–94
35. Lin H-F (2011) An empirical investigation of mobile banking adoption: the effect of innovation attributes and knowledge-based trust. Int J Inf Manag:252–260
36. Lu H-P, Yu-Jen Su P (2009) Factors affecting purchase intention on mobile shopping web sites. Internet Res 19(4):442–458
37. Luarn P, Lin H-H (2005) Toward an understanding of the behavioral intention to use mobile banking. Comput Hum Behav 21(6):873–891
38. Musleh Al-Sartawi AM, Hussainey K, Razzaque A (2022) The role of artificial intelligence in sustainable finance. J Sustain Financ. Investment:1–6
39. Musleh Al-Sartawi AMA (eds) (2022) Artificial intelligence for sustainable finance and sustainable technology. ICGER 2021. In: Lecture notes in networks and systems, vol 423. Springer, Cham
40. O'Hear S (2015) Mobile banking Startup Monese picks up backing from Seedcamp, SmartCap, and Spotify's Shakil Khan. Accessed 25 May 2015. http://techcrunch.com/2015/05/27/monese/
41. Park Y, Kim H, Lee J (2010) An empirical analysis on consumer adoption of mobile phone and mobile content in Korea. Int J Mobile Commun 8(6):667–688
42. Prakash A, Mahajan Y, Gadekar A (2022) Adoption of mobile money among internal migrant workers during the corona pandemic in India: a study focused on moderation by mode of payments. Int J Electron Bank 3(2):144–162
43. Razzaque A, Alalawi M (2015) Knowledge sharing and positive emotions in virtual communities: the case of the healthcare sector. In: KS global research international conference on emerging trends in multidisciplinary research (ETMR-2015). KS Global Research, Istanbul
44. Razzaque A, Eldabi T (2018) Tackling obesity by diet quality, technology and social networks. Int J Food Eng 4(1):88–92
45. Razzaque A, Eldabi T (2020) Physicians social capital aids their medical decisions when they virtually share knowledge. Int J Knowl Manag Stud 11(3):229–257
46. Razzaque A, Eldabi T, Chen W (2020) Quality decisions from physicians' shared knowledge in virtual communities. Knowl Manag Res Pract, 1–13. https://doi.org/10.1080/14778238.2020.1788428
47. Razzaque A, Eldabi T, Jalal-Karim A, Karolak M (2013) Culture sustained knowledge management infrastructure and architecture facilitate medical decision making. In: 2013 IIAS-IASIA joint congress. IIAS-IASIA, Manama
48. Razzaque A, Eldabi T, Jalal-Karim A (2013) Physician virtual community and medical decision-making: mediating role of knowledge sharing. J Enterp Inf Manag 26(1):500–515
49. Razzaque A, Al-Sartawi AMM, Karolak M (2021) Rethinking tourism entrepreneurship amidst COVID-19 through the lens of artificial intelligence and big data analytics. In: Big data for entrepreneurship and sustainable development. CRC Press, pp 175–188

50. Rotchanakitumnuai S, Speece M (2003) Barriers to internet banking adoption: a qualitative study among corporate customers in Thailand. Int J Bank Market 21(6/7):312–323
51. Sakhnini Z, Elali W, Al-Sartawi A (2022) Behavioral and non-behavioral factors and the level of adapting and implementing Fintech and e-banking in Bahrain: suggested model. In: Musleh Al-Sartawi AMA (eds) Artificial intelligence for sustainable finance and sustainable technology. In: ICGER 2021. Lecture notes in networks and systems, vol 423. Springer, Cham
52. Sanad Z, Shiwakoti R, Al-Sartawi A (2022) Gender diversity on corporate boards and earnings management: a review. In: Musleh Al-Sartawi AMA (eds) Artificial intelligence for sustainable finance and sustainable technology. ICGER 2021. Lecture notes in networks and systems, vol 423. Springer, Cham
53. Sohail MS, Al-Jabri IM (2014) Attitudes towards mobile banking: are there any differences between users and non-users? Behav Inf Technol 33(4):335–344
54. Suoranta M, Mattila M (2004) Mobile banking and consumer behaviour: new insights into the diffusion pattern. J Financ Serv Market 8(4):354–366
55. Wood A (2015) http://www.thememo.com/2015/05/27/ing-brings-mobile-banking-to-your-apple-watch/. Accessed 27 May 2015
56. Zhou T (2011) An empirical examination of initial trust in mobile banking. Internet Res 21(5):527–540

Cybersecurity, Artificial Intelligence, and Internet of Things

Is Artificial Intelligence the Ideal Partner for Blockchain and Crypto Currencies?

Mohamed Bechir Chenguel

Abstract For several years everyone, companies and individuals have been more and more interested in new technologies concerning the blockchain, crypto currencies and also in the development of artificial intelligence. Blockchain is a technology that should transform our societies in depth, by providing a decentralized and ultra-secure system, and constitute the crypto currencies basis. Artificial intelligence is a new technology created to facilitate daily lives. But its application remains very technical and little known to the public. Both artificial intelligence and blockchain are technologies that are helping to change the world, but their applications are currently very technical and little known to the public. The present article tries to study the combination between blockchain technology and Artificial Intelligence and how this combination, can be so beneficial for the optimization of crypto-currency trading. The contribution of our work is to try to improve the notion of the use of these two new technologies, while trying to reveal the impact that can be generated on the financial, economic and social level. Our work will be organized as follows: In the first section we will define the blockchain concepts, and the artificial intelligence. In the second section, we will present the implication and effects between artificial intelligence and blockchain in the first place and in a second part, we will present the impact on cryptocurrencies. And finally we will present our conclusion. Certainly our work has limits, which can be surpassed by empirical work to affirm or not our basic hypothesis.

Keywords Blockchain technology · Artificial intelligence · Crypto-money · Trading

1 Introduction

New technologies are increasingly used by companies and individuals, and their development continues to grow. These new technologies such as blockchain, new

M. B. Chenguel (✉)
Kairouan University, Kairouan, Tunisia
e-mail: bechir.chenguel@gmail.com

A. M. A. Musleh Al-Sartawi et al. (eds.), *From the Internet of Things to the Internet of Ideas: The Role of Artificial Intelligence*, Lecture Notes in Networks and Systems 557,
https://doi.org/10.1007/978-3-031-17746-0_27

means of crypto-currency payment and artificial intelligence, are designed to facilitate professional life, especially in a world increasingly surrounded by constraints. Blockchain is a technology that should transform our societies in depth, and this, by providing a decentralized and ultra-secure system to authenticate any financial transaction or personal document (diplomas, contracts, etc.) [23]. Blockchain is currently known to follow the emergence of cryptocurrencies, since blockchain is the basic technology of using cryptocurrencies. In fact, this technology allows to decentralize the interventions and develop a world where the users are masters of their data. The absence of an intermediary in this process raises doubts about the use of this new technology [19]. Both artificial intelligence and blockchain are technologies that are helping to change the world, but their applications are currently very technical and little known to the public, even if artificial intelligence is becoming more and more known. Yet very distinct, these technologies could work together to offer a wider range of applications, from the economic, social, and financial levels. Indeed, the combination of blockchain technology with artificial intelligence can offer companies new growth opportunities. A new stage in the digital revolution is underway, likely to increase the challenge for SME managers in order to seize these new opportunities and quickly recreate value; [9] shows how the combination of blockchain technologies and artificial intelligence will disrupt certain established business models. Both artificial intelligence and blockchain are technologies that are helping to change the world, but their applications are currently very technical and little known to the general public, even if artificial intelligence is becoming more and more known. Yet very distinct, these two technologies could work together to offer a wider range of applications. The question is: can Artificial intelligence and blockchain combine together in order to bypass certain constraints?

The contribution of our work is to improve the use of these two new technologies, while trying to reveal the impact that can be generated on the financial, economic and social level. Certainly our work has limits, which can be surpassed by empirical work to affirm or not our basic hypothesis.

Our work will be organized as follows, in a first section we will define the two basic concepts, namely blockchain technology and artificial intelligence. In the next part, we will present the common fields of action between these two technologies. The third part will be devoted to the relationship between artificial intelligence and cryptocurrencies. And finally the conclusion.

2 Concepts and Definition

2.1 Blockchain Definition

The blockchain is a decentralized database, where each participant has information that is distributed to all users of the network. In the form of blocks linked to each other, successive transactions are stored and authenticated; [3]. Blockchain is a technological system based on a block chain for storing and transmitting information. This information chain constitutes a global database accessible to all users. The blockchain works without any regulatory body. To guarantee the authenticity of the information, each user validates it using cryptography [32]. As soon as a user performs a transaction by implementing information through the blockchain, it is grouped with other related information within the same block. It will be verified and approved by other members of the blockchain using cryptographic techniques. Members who ensure the technical structure of the blockchain implementation by checking and validating consistency with the other information in the block are compensated. As soon as it is validated, the block is time stamped and added to the blockchain [37]. The information will then be visible and accessible to everyone, but it can no longer be changed. In the event of an error, the correction is made with a new transaction [19]. A blockchain is a register, a large database which has the particularity of being shared simultaneously with all its users, all of whom are also holders of this register, and who also all have the capacity to enter data according to specific rules.

It is about a "ledger", that is to say of a ledger, or better of a large register, where is kept the trace of all the exchanges, of all the transactions which have already carried out. between the members of the network since its creation [21]. The advantage is that it is tamper-proof and safe from hackers: since the registry is decentralized, stored on users' servers, it cannot be hacked. Everything is constantly checked by all users. There is therefore not "one" blockchain, but blockchains, all different, created by anonymous people as well as by large institutions, for a wide variety of uses. Within modern blockchains, there are programs called "Smart Contract" which aim to control part of the blockchain by organizing the relations between the different parts, by establishing actions to be implemented in order to improve the existing one in the aim to correspond to new market expectations such as transparency of information [21]. Cryptocurrency and bitcoin in particular is indeed the most telling and well-known example of the use of blockchain; [30]. But spinoffs are expected very quickly and in many areas: banking, insurance, medical data, patents. This is what leads us to see in this technology. [3] Explain that blockchains, like any multi-agent system, create a competitive environment that can test the effectiveness of a artificial intelligence. In turn, this needs algorithms that are equally interpreted by all members of a system, regardless of their differing interests.

2.1.1 Blockchain: The Technology of Tomorrow

Blockchain is one of the most promising new disruptive technologies. It can be defined as a transparent and secure information storage and transmission technology that operates without a central control body [3]. Concretely, the exchange is done peer to peer, that is to say without intermediary, the register which lists all the transactions allows traceability and the conservation of registers by all users, constant verification of transactions is called a distributed [8]. The first use case of the technology emerged in 2008 in a context of crisis, by Satoshi Nakamoto to exchange goods and services in

Bitcoin (a virtual currency).In 2009, this technology enabled users to transfer virtual currency in a secure, traceable and unmediated manner. Blockchain should not be confused with cryptocurrency, which is a use case of Blockchain technology. [11] the main use cases are:

– Secure transfer of assets between two players: transfer of crypto-currencies, financial assets, ICO ...
– Smart contracts which are contracts that are self-executing automatically as soon as the condition is met: insurance indemnities, the example of drones in agriculture making it possible to raise the level of rain to trigger insurance indemnities.
– The keeping of a register that allows the sharing of authenticated and immutable data: For example, in some countries there are problems linked to the absence or irregularity of certain registers and cadastres.

Some countries have registered in the Blockchain a kind of cadastre (Honduras, Ghana ...) to ensure the authenticity of documents and information like diplomas for example, and it is for this reason several Councils of the Order of Chartered Accountants have set up a diploma registration prototype using Blockchain technology [36].

2.1.2 Blockchain Types

Blockchain can be public or private. A public Blockchain can be illustrated as an open ledger, which anyone can write on, anyone can read, but which is impossible to erase and indestructible. For example, Bitcoin. The private blockchain allows access only to those authorized by a private company or a consortium. Example: Ripple is a private (consortium) blockchain; [19].

2.1.3 Blockchain and Smart Contracts

Blockchain technology allows the implementation of smart contracts, contracts that automatically self-execute once predefined conditions have been met Enrolling smart contracts in a blockchain guarantees that the terms of the contract cannot be changed [8]. For example, smart insurance Fizzy, (a subsidiary of Axa group), compensates its customers for delays as soon as their plane arrives. In fact, the principle is that compensation is triggered automatically in the event of a flight delay of 2 h or more, without having to make a claim.

2.1.4 Blockchain and ICO (Initial Coin Offering)

It is a method of raising funds in cryptocurrencies. It works by issuing digital assets called "tokens" or tokens, in exchange for cryptocurrency (ether or bitcoin).Indeed, the issuing company issues tokens in pre-sale, against a virtual currency, accepted as a means of payment [7]. Tokens, which are at the center of the operating mechanism of ICOs, are subsequently admitted to buying and selling on exchange platforms whose value depends on supply and demand. Ethereum is the source of one of the first notable ICOs in history: In 2014, Ethereum raised more than $ 18 million (issuing $ 60 million of ether redeemable for bitcoins).Professionals are brought to meet this technology among clients who use cryptocurrencies to carry out transactions, those who are fundraising in ICOs and those whose application is based on Blockchain technology.

2.1.5 Blockchain the Future Key

Blockchain technology is a real revolution. The impact of blockchain can be compared to that of the Internet, as this technology has changed the way things are done since its establishment in the 1990s. [33] Consider that the financial world is upset by blockchain, in fact everyone can become their own bank. But this revolution extends far beyond the financial sector. Indeed, some professions will disappear, and others will categorically change, such as the profession of computer engineer. In fact, the profession of computer engineering has changed very quickly in recent years, as have several other professions that will adapt to these new realities: [21]. Companies also have to adapt and adjust to the impact of blockchain on their environment and their business models [8]. The blockchain will lead to a decentralization of platforms, where users will find greater simplicity and freedom. The implementation of these new technologies is not so simple, since the businesses and societies of tomorrow will not be similar to those of today.

2.2 Definition of Artificial Intelligence

Artificial intelligence is a scientific discipline aimed at making a machine reproduce human cognitive abilities (reasoning, language, perception, etc.).With learning methods, artificial intelligence analyzes data in order to organize information and learn how to solve problems, but in systems based on rules or logic, problem solving is programmed by humans [23]. In order to avoid any errors, as with Microsoft's "Nazi chatbot" which was designed to learn to adapt its language to the behavior of Internet users through the latter's exchanges. The exchanges of the chatbot (conversational robot: an automatic conversation software capable of responding to a human interlocutor in the most natural way possible) of Microsoft becomes Nazi by learning provocative messages from Internet users; [35]. So artificial intelligence needs a lot of reliable data, but that depends on the data repository to which the artificial intelligence has access. Nowadays the notion of Artificial Intelligence is everywhere, in fact this notion is used. To set up miracle solutions integrating artificial intelligence to facilitate the work of the number professional. Artificial intelligence can even herald the end of several professions and trades because of its simplicity and its facilitating role between economic agents [9]. The ultimate goal of Artificial Intelligence is to mimic the workings of the human brain. Artificial intelligence uses very advanced predictive analytics algorithms, using a quantity of data provided every day by the users themselves [36]. This is called self-learning. To define artificial intelligence, we must define the different levels of artificial intelligence:

- First and second level Artificial Intelligence: IBM's 1997 Deep Blue computer beat the world chess champion. It is a hyper-specialized Artificial Intelligence in a specific context. This is Artificial Intelligence that we encounter on a daily basis, mainly in the calculation of data.
- 3rd level Artificial Intelligence: In addition to performing the calculations, Artificial Intelligence level 3 anticipates the outcome. Based on the results of the (past) operations it performs, it can predict what will happen next.
- Fourth level Artificial Intelligence: this is self-learning Artificial Intelligence capable of reacting with the world around it and making decisions accordingly; [35].

2.2.1 Algorithm and Artificial Intelligence

We hear a lot about artificial intelligence integrated into certain tools such as algorithms. Let us first define the notion of algorithm, which is defined as being a finite and unambiguous sequence of operations or instructions allowing to solve a class of problems [25]. A recipe, for example, can be considered as an algorithm because it makes it possible to obtain the requested result while scrupulously following the recipe. Artificial intelligence makes it possible to accomplish tasks that classical algorithms cannot, such as the ability of the machine to learn itself from the data injected into it 18. There are two concepts that are important in artificial intelligence: machine learning and deep learning.

2.2.2 Machine Learning

Machine Learning is the field of study that gives computers the ability to learn without being explicitly programmed to learn. Machine learning or machine learning of the computer, it is the ability to inject data into the machine so that it can use to learn and detect [5]. For example in radiology medicine, the machine is taught to locate images of cancer by injecting it with a large quantity of data and images. In some cases, she has been able to overtake a man. We can even design and imagine an algorithm for detecting VAT fraud; This is based on the ability of this algorithm to be able to learn from its data and be able to correct it [24].

2.2.3 Le Deep Learning

This is deep learning, which was born in 2010 with the idea of being inspired by the functioning of the human brain, namely the connections of neurons, which will allow the machine to have an in-depth analysis and to be able to extract the data itself; [18]. Facebook already uses deep learning technology to recognize facial recognition.

3 Combination of Blockchain and Artificial Intelligence

By giving the possibility of networking data within the same ecosystem of actors, the Blockchain makes it possible to massify access to data in a secure manner and to accelerate the machine learning curve for intelligence systems Artificial; [13]. In perspective: the democratization of Artificial Intelligence algorithms and the possibility for business consortia to raise their computing power to the level of Facebook or Google [29].

3.1 Artificial Intelligence, a Massive Need for Data

Artificial intelligence could transform the world more than any other technical advance since the Industrial Revolution and fundamentally reinvent the way companies operate and grow [12]. To be effective, many artificial intelligence algorithms, and in particular machine learning, require large volumes of data, the value of which depends on its quality. Big Data, whose vocation is to collect, organize, and analyze large volumes of data, is an essential component of artificial intelligence solutions that are deployed in organizations today; [36]. It is in this sense that "data" is often presented as the new precious metal of the twenty-first century because it is the essential fuel of Artificial Intelligence models which mainly aim to operate classification or prediction. In fact, the two main challenges facing organizations that want to implement an artificial intelligence solution are access to data and the quality of that data for training Artificial Intelligence models. This lack of quality data means that some artificial intelligence solutions do not have a sufficient training dataset and therefore perform poorly or not. This is also a brake on the deployment of artificial intelligence, also creating disparities between companies (depending on their size and resources).Only a few large companies, which can afford to invest capital and skilled resources to build machine learning models and algorithms in-house [12]. However, if a few large companies have the means to build large databases and invest in the development of artificial intelligence systems, this represents a significant financial investment, which is even more difficult to assume in times of crisis. In addition, their ability to compile this data stops at their scope of activity and the operational cases they encounter on a daily basis. It should be noted that it is rare for a single company to be able to respond to a given problem, or to present a solution to a problem to which the Artificial Intelligence model claims to respond [33]. The challenge is to collect a volume of data large enough and representative of the different management cases, to introduce them into the models of Artificial Intelligence [36]. In this context, how to constitute and qualify large volumes of data, sufficiently representative, and to encourage data holders to share them? How to speed up the training of Artificial Intelligence models and reduce the resulting financial cost? Blockchain may provide the answer to these questions.

3.2 The Blockchain, a Factor in the Reliability of Artificial Intelligence

Blockchain is a technology that allows information to be stored and transmitted transparently, securely and without a central control body. It looks like a large database which contains the history of all the exchanges made and whose users contribute to the system by providing computing power and/or storage capacity [17]. Some of the essential components of the Blockchain are the token and the smart contract. The token is a digital asset issued and traded on a blockchain that can be sold and bought at any time, in particular on exchange platforms at a price set in real time by supply and demand. Smart contracts are irrevocable computer programs deployed on a blockchain which execute a set of predefined instructions and make it possible to secure an agreement between two or more parties, to automate payments, to eliminate the risk of non-payment, to drastically reduce the intermediate costs in the development, monitoring and signing of a contract.

In view of these elements, and to respond to the problem of access to quality data, these are used to train an Artificial Intelligence model, it is quite possible to imagine a smart contract that would offer a data holder to share his data within an ecosystem via a blockchain; [31]. Indeed, many data owners refuse to sell their data to intermediaries or platforms but who would not be opposed to sharing it with other organizations. But this on condition that this data sharing is done in a secure manner and for a fair remuneration linked to the use and quality of their data.

This is enabled by the Blockchain and its smart contracts by guaranteeing that each of the contributors to a blockchain is remunerated according to their actual contribution, within a secure contractual framework.

Data holders whose data would be validated would be automatically remunerated in tokens. The controllers who verify that the data is valid and of quality would also be paid in tokens. Organizations that wish to use this data to train their artificial intelligence algorithms would pay in tokens for each use, thus fueling the system. With a drastic effect on the ability of these organizations to accelerate the deployment of their Artificial Intelligence systems. With Blockchain technology, it would also be possible to manage the confidentiality issues of transactions and shared data, especially in very competitive contexts. Through the encryption of data that can only be decoded by the organization having the corresponding private key or allowing the establishment of exchange channels in a blockchain such as Hyperledger [34]. The Blockchain thus makes it possible to share data while preserving the confidentiality of transactions and data. This new economy based on data sharing is already finding concrete applications with several start-ups that can develop decentralized Blockchain platforms for sharing data sets and computing power. By thus facilitating the constitution of large and high-quality databases, the Blockchain can therefore help to accelerate the training of artificial intelligence algorithms and improve their performance. Together, Artificial Intelligence and Blockchain can empower organizations to exceed their current limits and access significant amounts of data and by extension value. In addition, the union of these two new technologies can allow companies to increase their computing power by combining their respective data assets and together they can reach the levels of large groups like Google or Facebook. The merger between artificial intelligence and blockchain makes it easier to access artificial intelligence in businesses. They allow small businesses to access a lot of data. The blockchain could also make it possible to restore confidence in the decisions of Artificial Intelligence, because it will be easier to carry out an audit of the decision-making process of artificial intelligence by the recovery of all the data collected by the latter in its taking. decision-making.

Finally, the data collected will be used by artificial intelligence is secured within the blockchain. Thus, the combination of blockchain and artificial intelligence will increase the possibilities of application. They can create a direct link between producers and consumers. The information will then be stored, secured and used in full transparency. To sum up, Artificial Intelligence could well improve the quality of information transmitted and simplify the execution of tasks, while Blockchain can provide good visibility on the execution of artificial intelligence and thus make it possible to secure information and connections.

4 Artificial Intelligence and Crypto Currencies

Advances in artificial intelligence can greatly benefit blockchain and cryptocurrencies, both in terms of data security and trading. Generally, Artificial Intelligence mainly contributes to simplify daily life by eliminating or simplifying certain tasks [4]. This technology has also helped in recent years to strengthen the appeal of cryptocurrencies. Emerging after the financial crisis of the late 2008, blockchain has steadily captured the attention of investors of all stripes ever since. The latter going on specialized platforms. While the market has had its ups and downs, the development of this industry is currently very promising [2]. Technology can also be the best asset for optimizing Bitcoin and cryptocurrency trading strategy. If the trading algorithms on the financial markets saw the light of day in the 1990s, the cryptocurrency market lends itself perfectly to the democratization of automated trading bots intended for a wide audience ranging from the amateur trader to the pro trader [5].

4.1 Artificial Intelligence and Trading

There are different classic trading strategies in the markets already implemented by any professional person. The first distinction to be made when talking about a trading strategy is of a temporal nature; we distinguish short-term strategies generally qualified as speculative. These strategies aim to analyze and anticipate price variations over short periods. Moreover, there are long-term strategies that aim to invest in increasing the value of an asset over a long period [2]. Thus, the use of trading algorithms may be more relevant when opting for the option of a short-term strategy, in a context where technology can be much more efficient than the human brain. The second distinction is made between "Swing Trading" and "Position Trading" strategies. With a "Position Trading" strategy as the name suggests, you will position yourself in an asset or in this case in a crypto with the objective of benefiting from an upward trend over several weeks or even months. Conversely, when you opt for "Swing Trading", the crypto holding period is much shorter, here we will no longer count in weeks or months but in days, or even in hours. Once again, you will understand, bots and algorithms bring all their intelligence in a context of "Swing Trading" rather than "Position Trading". While the development of high-frequency trading on the traditional financial markets has greatly developed in recent years to benefit from price variations over a short or very short period of time, taking transaction costs into account requires reviewing the timing of Swing Trading. so as not to be overwhelmed by unforeseen costs [14]. Trading strategies augmented by the power of algorithms and artificial intelligence. One thing is certain, that technological innovation is a trader's best asset.

These classic trading strategies (not using technological innovation) are revolutionized by super robot traders using the best artificial intelligence algorithms [27]. This is why when looking for a robot trader for your crypto, the majority of traders will look for the robots that are dedicated to unearthing medium and long term cryptocurrency trends; these robot traders thus minimize the potential risks. The second technological innovation in the financial markets is algorithmic "Quant Trading". Financial engineers have, thus, developed algorithms packed with artificial intelligence which constantly analyze the financial markets by focusing on two main metrics which are the evolution of prices of course and the volumes traded [20]. Classic financial markets have greeted technological developments with mixed eyes, with established traders for years looking down on the prospect of being replaced by an army of robot traders. The crypto-currency markets, for their part, offer many more possibilities for deploying technological solutions to optimize your trading strategy. Thus, at first, traders on the crypto-currency markets mostly went back to classic strategies [28]. However, today some companies intend to democratize the use of trading bots and more broadly the deployment of artificial intelligence in the crypto-currency markets.

4.2 Artificial Intelligence and Transaction Security

As investors bet on the blockchain, artificial intelligence has an increasingly important role in cryptocurrency trading. Indeed, traders can have access to more precise analyzes and make well-informed decisions thanks to the robots specially designed for trading. Thus resulting in a better return on investment [16]. Among the problems that the blockchain industry still faces are hacking operations and cyber attacks. Indeed, every day, millions of dollars in cryptocurrencies change hands fraudulently. To remedy this, Artificial Intelligence makes it possible to reinforce security and it is cybersecurity systems using artificial intelligence that prove to be the most effective in protecting the blockchain [22]. Indeed, they can instantly identify the nature of a threat. Then, they put it on a blacklist to prevent another intrusion. Thus, Artificial Intelligence evolves over time, increasing its effectiveness with each threat, and learning from previous operations [2]. To ensure the security and integrity of the blockchain, each transaction is verified before being added by miners. The latter will thus receive a reward in the form of cryptocurrency. But the only problem is that this process is too energy-consuming and has an impact because of the energy required by the processors and graphics cards that run it; [10]. Thus, many mining companies have turned to graphic cards equipped with Artificial Intelligence in order to reduce their carbon footprint. And some firms have established artificial intelligence-based ecosystems where miners can share the required energy load. Thus, these algorithms help make bitcoin mining faster, more efficient and more profitable. In conclusion, the technological sector as a whole has started to turn towards machine learning. Also, blockchain and cryptocurrencies should also benefit from artificial intelligence.

5 Conclusion

With the simultaneous development of two-key technologies: blockchain and artificial intelligence; which are technologies that can be used together, in order to put a direct link between the customer and the producer or the service provider. Thus, thanks to a blockchain, a company will not only be able to conclude a transaction directly with a customer in a transparent and secure manner and thus, saving the commission of a platform. It can also conclude a "smart contracts which are contracts with automatic execution, backed by traditional legal contracts that allow the automation of certain execution processes and guarantee all the conditions to be respected between the two parties for the payment to be made. They can also provide for compensation, avoiding, thus, to the customers the long and tedious stages of complaints [4]. Consequently, blockchain brings security and predictability; as for artificial intelligence, it contributes with simplicity in interfacing. Indeed, Artificial Intelligence is the missing link in the blockchain to allow it to become widespread and multiply throughout the economy. The combination of these two new technologies opens up promising possibilities for developing all the business models currently existing. Once companies have metabolized the security and predictability that blockchain guarantees them, when they have experienced the simplicity offered by artificial intelligence. This is a very important financial and economic step, where the business model will change and evolve [3]. Without delving into the apparent complexity of these technologies, SME managers should therefore seize upon them now and think about a promising strategy. It is in the blockchain that we must focus on primarily, and we must think carefully about the uses that could accelerate the growth of companies. Large groups have already understood this, and have already developed, for the most part, blockchain-based projects with a view to business between companies. But the articulation of these two technologies opens up a range of use cases that will soon also be perceptible by consumers [30]. These mutations that lead to disruption are presented as inevitable, without being sufficiently concerned with their social repercussions. Indeed, these new technologies, with the programs and the robotic combination with artificial intelligence, will allow the massive use of robots, which will adapt and react to their environment, thus replacing a large number of jobs and, ultimately, will transform society. This will result in widening the income inequalities that already exist between the creative class that created these novelties and the immense mass of employees in sectors working in the classic way.

References

1. Akkerman SWH (2021) Automatically trading small market capitalization cryptocurrencies using reinforcement learning, Bachelor's thesis)
2. Al-Ameer A, Fouad AS (2021) A methodology for securities and cryptocurrency trading using exploratory data analysis and artificial intelligence. In: 2021 1st international conference on artificial intelligence and data analytics (CAIDA). IEEE, pp 54–61
3. Alansari Y, Al-Sartawi A (2021) IT governance and E-banking in GCC listed banks. Procedia Comput Sci 183:844–848
4. Al-Sartawi A, Al-Okaily M, Hannoon A, Khalid AA (2022) Financial technology: literature review paper. In: Musleh Al-Sartawi AMA (eds) Artificial intelligence for sustainable finance and sustainable technology. ICGER 2021. Lecture notes in networks and systems, vol 423. Springer, Cham
5. An YJ, Choi PMS, Huang, SH (2021) Blockchain, cryptocurrency, and artificial intelligence in finance. In: Fintech with artificial intelligence, big data, and blockchain, vol 1
6. Awotunde JB, Ogundokun RO, Jimoh RG, Misra S, Aro TO (2021) Machine learning algorithm for cryptocurrencies price prediction. In: Artificial intelligence for cyber security: methods, issues and possible horizons or opportunities. Springer, Cham, pp 421–447
7. Badr Y (2021) On the integration of artificial intelligence and blockchains 3.0: prospects and challenges. In: 2021 IEEE 18th international conference on software architecture companion (ICSA-C). IEEE, pp 120–120
8. Bamakan SMH, Moghaddam SG, Manshadi SD (2021) Blockchain-enabled pharmaceutical cold chain: applications, key challenges, and future trends. J Cleaner Prod, 127021
9. Basu S, Majumdar B, Mukherjee K, Munjal S, Palaksha C (2022) The role of artificial intelligence in HRM: a systematic review and future research direction. Hum Resour Manag Rev, 100893
10. Belitski M, Boreiko D (2021) Success factors of initial coin offerings. J Technol Transf, 1–17
11. Brune P (2020) Towards an enterprise-ready implementation of artificial intelligence-enabled, blockchain-based smart contracts. arXiv preprint arXiv:2003.09744

12. Burgess N (2022) Int J Artif Intell Mach Learn
13. ChenguelM B (2021) Blockchain and ecological impact: between reality and accusation? Lecture notes in networks and systems. ISSN: 2367-3370 SCOPUS Q4 / Indexed proceeding published by Springer 2021. https://www.springer.com/series/15179
14. Deebak BD, Fadi AT (2021) Privacy-preserving in smart contracts using blockchain and artificial intelligence for cyber risk measurements. J Inf Secur Appl 58:102749
15. Ekramifard A, Amintoosi H, Seno AH, Dehghantanha A, Parizi RM (2020) A systematic literature review of integration of blockchain and artificial intelligence
16. Finlay S (2021) Artificial intelligence and machine learning for business: a no-nonsense guide to data driven technologies (No. 4th ed) Relativistic
17. Ganapathy A, Redwanuzzaman M, Rahaman MM, Khan W (2020) Artificial intelligence driven crypto currencies. Global Discl Econ Bus 9(2):107–118
18. Ge H, Ge Z, Sun L, Wang Y (2022) Enhancing cooperation by cognition differences and consistent representation in multi-agent reinforcement learning. Appl Intell, 1–16
19. Ghimire A, Thapa S, Jha AK, Adhikari S, Kumar A (2020) Accelerating business growth with big data and artificial intelligence. In: 2020 fourth international conference on I-SMAC (IoT in social, mobile, analytics and cloud)(I-SMAC). IEEE, pp 441–448
20. Ghosh A, Gupta S, Dua A, Kumar N (2020) Security of Cryptocurrencies in blockchain technology: state-of-art, challenges and future prospects. J Netw Comput Appl 163:102635
21. Gupta J, Singh I, Arjun KP (2021) Artificial intelligence for blockchain I. In: Blockchain, internet of things, and artificial intelligence. Chapman and Hall/CRC, pp 109–140
22. Gupta R, Srivastava D, Sahu M, Tiwari S, Ambasta RK, Kumar P (2021) Artificial intelligence to deep learning: machine intelligence approach for drug discovery. Mol Diversity, 1–46
23. Hu S, Liang YC, Xiong Z, Niyato D (2021) Blockchain and artificial intelligence for dynamic resource sharing in 6G and beyond. IEEE Wirel Commun
24. Iqbal M, Iqbal MS, Jaskani FH, Iqbal K, Hassan A (2021) Time-series prediction of cryptocurrency market using machine learning techniques. EAI Endorsed Trans Creative Technol, e4
25. Jha N, Prashar D (2022) Using blockchain in resolving the challenges faced by IIoT. In: Industrial internet of things. CRC Press, pp 189–216
26. Kakavand H, Kost De Sevres N, Chilton B (2017) The blockchain revolution: an analysis of regulation and technology related to distributed ledger technologies. SSRN 2849251
27. Majeed U, Khan LU, Yaqoob I, Kazmi SA, Salah K, Hong CS (2021) Blockchain for IoT-based smart cities: recent advances, requirements, and future challenges. J Netw Comput Appl, 103007
28. Mattos DMF, Krief F, Rueda SJ (2020) Blockchain and artificial intelligence for network security. Network Security, 101–102
29. Mikalef P, Gupta M (2021) Artificial intelligence capability: conceptualization, measurement calibration, and empirical study on its impact on organizational creativity and firm performance. Inf Manag 58(3):103434
30. Mohanta BK, Jena D, Satapathy U, Patnaik S (2020) Survey on IoT security: challenges and solution using machine learning, artificial intelligence and blockchain technology. Internet Things 11:100227
31. Mollaret S (2021) Artificial intelligence algorithms in quantitative finance, Doctoral dissertation, Paris Est
32. Musleh Al-Sartawi AMA (eds) (2022) Artificial intelligence for sustainable finance and sustainable technology. In: ICGER 2021. Lecture notes in networks and systems, vol 423. Springer, Cham
33. Nam D, Chapiro J, Paradis V, Seraphin TP, Kather JN (2022) Artificial intelligence in liver diseases: improving diagnostics, prognostics and response prediction. JHEP Rep, 100443
34. Sabry F, Labda W, Erbad A, Malluhi Q (2020) Cryptocurrencies and artificial intelligence: challenges and opportunities. IEEE Access 8:175840–175858
35. Sebastião H, Godinho P (2021) Forecasting and trading cryptocurrencies with machine learning under changing market conditions. Financ Innov 7(1):1–30

36. Shen Y (2021) The application of artificial intelligence in computer network technology in the era of big data. In: 2021 international conference on computer technology and media convergence design (CTMCD). IEEE, pp 173–177
37. Singh S, Sharma PK, Yoon B, Shojafar M, Cho GH, Ra IH (2020) Convergence of blockchain and artificial intelligence in IoT network for the sustainable smart city. Sustain Cities Soc 63:102364
38. Singhal T, Bhargavi MS, Hemavathi P (2021) Coalescence of artificial intelligence with blockchain: a survey on analytics over blockchain data in different sectors. In: Emerging Technologies in data mining and information security. Springer, Singapore, pp 703–711
39. Sreedevi AG, Harshitha TN, Sugumaran V, Shankar P (2022) Application of cognitive computing in healthcare, cybersecurity, big data and IoT: a literature review. Inf Process Manage 59(2):102888
40. Starnberger G, Froihofer L, Göschka KM (2009) QR-TAN: secure mobile transaction authentication. In: 2009 international conference on availability, reliability and security. IEEE, pp 578–583
41. Tariq A, Haq HB, Ali ST (2019) Cerberus: a blockchain-based accreditation and degree verification system. arXiv preprint arXiv:1912.06812
42. Thiebes S, Lins S, Sunyaev A (2021) Trustworthy artificial intelligence. Electron Mark 31(2):447–464
43. Treiblmaier H, Swan M, De Filippi P, Lacity M, Hardjono T, Kim H (2021) What's next in blockchain research?-an identification of key topics using a multidisciplinary perspective. ACM SIGMIS Database DATABASE Adv Inf Syst 52(1):27–52
44. Witowski J et al (2021) MarkIt: a collaborative artificial intelligence annotation platform leveraging blockchain for medical imaging research. Blockchain Healthcare Today
45. Zemčík T (2021) Failure of chatbot Tay was evil, ugliness and uselessness in its nature or do we judge it through cognitive shortcuts and biases? AI Soc 36(1):361–367
46. Zhang R, Xue R, Liu L (2019) Security and privacy on blockchain. ACM Comput Surv (CSUR) 52(3):1–34
47. Zhang Z, Song X, Liu L, Yin J, Wang Y, Lan D (2021) Recent advances in blockchain and artificial intelligence integration: feasibility analysis, research issues, applications, challenges, and future work. Secur Commun Netw

Risk Management of Human Genome Editing: Ethical and Islamic Perspectives

Ahmed Rashad Amin Al-Huwari, Saad Darwish, and Khalid Mohamad Al-Dweri

Abstract The idea behind editing the human genome started as a possible futuristic choice for people who suffer from specific genetic disorders and wish to have a child with their genes without the inherited disorder. This process, however, went beyond its objectives. Tremendous scientific and technological advancements, information and knowledge on the human genome have widely spread. Some started using these technologies to create a new breed of healthier, stronger, and more intelligent humans, called "eugenics". Considering the risks of such methods on the future of humanity and the possibility of permanent damage to future generations, these technologies were criticised to the extent that they were banned in several international conventions and national legislation. Thus, in using the human genome editing process, there is still a substantial chance of serious injury, which may be irreversible and uncontrollable.

The goal is to provide the necessary understanding by exploring the implications and ethical concerns. It would be offered to give knowledge about how genomic studies and experiments may require particular risk management considerations. The research will give insight into the significance of genome research and its ramifications in terms of Islamic Sharia.

Keywords Genome · Ethics · Risk · Islam · Eugenics

A. R. A. Al-Huwari · S. Darwish (✉)
Kingdom University, Riffa, Bahrain
e-mail: Saad.darwish@ku.edu.bh

A. R. A. Al-Huwari
e-mail: a.amin@ku.edu.bh

K. M. Al-Dweri
Kuwait International Law School (KILAW), Doha City, Kuwait
e-mail: k.aldweri@kilaw.edu.kw

A. M. A. Musleh Al-Sartawi et al. (eds.), *From the Internet of Things to the Internet of Ideas: The Role of Artificial Intelligence*, Lecture Notes in Networks and Systems 557,
https://doi.org/10.1007/978-3-031-17746-0_28

1 Introduction

The world has recently witnessed significant advancements in science, technology, and medicine, which have been the primary concern and focus of all recent developments to find a cure for the diseases from which most people suffer [21, 27]. Experimenting on humans in medicine has been very notable, but human genome editing has been the most prominent type of experimentation. This technology was first intended to find an effective remedy for genetic diseases, which could not be conventionally cured, or in cases where no cures had been discovered [8]. However, human genome editing has become a means to improve the genetic structure by providing grandchildren with the best possible genetic traits to enable them to face the hardships of life [19].

Although this issue intrigues human passion, because of the aims it wishes to achieve, it still holds risks and possible genetic consequences unknown to future generations. It also started a global legal and ethical controversy on whether this technique should be used in cases other than treatment, such as gender swaying and eugenics.

2 Research Objectives

This research aims to identify and analyse the ethical and Islamic frames that rule human genome editing names in societies.

3 Research Problem

Human genome editing raises multiple questions that we will answer through this research. The most critical issues tackled here are the definition of the editing of the human genome, the presentation of ethical impact on human, society, or country levels, and the Islamic Sharia rule regarding this challenging technology.

4 Research Methodology

This research will depend on secondary data, mainly the descriptive and analytical approach. We will study the idea itself and the risks associated with the ethical and Islamic judgement on the matter.

5 Research Importance and Significance

According to Islamic Sharia [10], the study will shed light on the importance of carrying out genome research and its implications. Its purpose accomplishes the needed understanding by discussing the implications and ethical consequences. It would be presented to share information about how operations of genome studies and experiments may need special treatment for risk management.

6 Literature Review

6.1 Gene Editing and Risk Factors

There are many issues regarding gene editing because it is a new technology; there are only a few insurance claims to conclude the most effective methods of loss management and risk reduction. Thus, risk managers have hurdles since gene editing is a new technology. There is a shortage of claims data on which to base loss control decisions and risk mitigation tactics at this time. Currently, risk managers and underwriters must utilise typical risk management strategies to decrease the risk of gene editing claims until more information concerning the incidence and severity of such claims. Traditional risk management measures must be utilised to mitigate the risks to companies and practitioners. Risk managers consider both the probability and the severity of a potential incident when evaluating and responding to risks [17].

Risk information and actuarial tables are used by underwriters, who work closely with risk managers to describe risk and establish insurance prices. Actuaries and insurance underwriters use data from previous claims to predict the probability and severity of future catastrophes with incredible accuracy. However, such a vision is seldom accompanied by reliable forecasts as underwriters can plan for various possible outcomes. Many factors go into setting an insurance rate; an underwriter sets the rate and determines the premium. It will be challenging to set insurance prices without the availability of historical data. Managing Risk entails limiting the number of exposures that may be taken on. As a rule of thumb, risk managers have four primary methods for reducing their exposure to risk. The easiest method to avoid risk is to avoid taking it first. Transferring or outsourcing risk is another option. Thus, Financial liability may be transferred to a third party as part of a transfer. Insurers, for example, assume some of the financial risks. In addition to shifting financial risk, outsourcing may also be used to transfer an act or service to another party. It is possible to accept a risk and then take it further [28].

6.2 Risk Implications for Gene Editing

Any number of factors may be used to estimate the anticipated risk. Somatic (Somatic cell gene therapy would aim to cure a disease only in the patient, not in the patient's descendants) or germline mutations (Germline mutations are inherited from a parent; the mutation was present in the parent's egg or sperm cells) may be altered through gene editing, which has a predicted risk. Germline editing, unlike somatic editing, can affect both the person who receives the genetic dealing and the individual's descendants. Only somatic therapies are being tested in clinical trials; somatic gene editing may be less concerning than germline editing—at least from an ethical and risk management standpoint—because the absence of heritability risk means that risk exposure is presumably lower for somatic than germline gene editing. This is because only somatic therapies are currently being tested in clinical trials. Thus, gene editing is not always accurate, and off-target modifications may occur.

The hazards of somatic gene editing may be compared to those of other treatments, such as those that affect just one patient, in terms of both known and undiscovered concerns. Most gene editing procedures take place outside the body of a patient subject. The patient in a gene-editing experiment is likely to be exposed to more risk than the patient in a gene therapy trial. In the future, when somatic or germline editing is widely accessible, it will be critical to prove that a patient subjprove informed about the dangers and advantages of a gene-editing treatment and its alternates [13, 25].

6.3 Risk Management Strategies

Risk managers in health care organisations may use the four risk management methodologies discussed previously in gene editing in one or more of the following ways. The simplest way is to avoid risk in the first place. Risk managers and insurance practitioners can say, "We do not do that" Somatic gene editing's unknown, unexpected, unknowable, or unforeseeable dangers may make avoidance an attractive risk management strategy, but it is also an appealing strategy for germline gene editing. As gene editing has therapeutic potential, avoiding it may not be morally acceptable if it offers comfort to a patient. The long-term availability of treatment alternatives, which vary over time, will require health care organisations' risk managers to reassess the hazards of gene editing as they become known or predictable, even if a short-term risk management approach does notice gene editing medicines. Transferring or outsourcing risk is an option. For example, if gene editing service references were available, outsourcing or transferring gene editing risk may be possible [22]. It is possible to reduce the risk of gene editing may and minimised in two ways:

An a. An informed consent procedure communicates the risks and advantages of gene editing and alternatives to gene editing d to a patient.

b. Through a robust credentialing process in which gene editing risks are transferred to a third party through outsourcing and could mitigate residual risks through third-party indemnification, which reduces or removes financial risk exposures by third parties agreeing not to sue or committing to pay damages if an individual patient sues them.

It is possible to accept a risk and then take further steps. As of right now, it is probably not a good idea to blindly embrace the hazards of gene editing, as neither the present risks to patients nor the future risks to their descendants are known in the case of somatic gene editing or germline editing. Unquantifiable risks can still be insured, but the premiums would be prohibitively high. Although it is not "ethically objectionable in principle".

6.4 New Worlds and Long Tails

It is not evident to a risk manager for a healthcare organisation or an individual practitioner how to protect themselves against a big claim arising from using technology with unknown and unanticipated dangers. There is currently no mechanism to guarantee against the descendants of a germline gene-edited person suing an organisation or physician. There is not enough data about or experience with gene editing to imagine—much less know or foresee—claims risks that future issues may offer. Longtail future claims are those that risk managers and underwriters expect to be filed 10, 20, 40, or more years after the initial insurance policy was established. While it is unclear what future responsibility will be for these claims during policy drafting, risk managers and underwriters will be experiencing nightmares dealing with such uncertainties. Medical risk managers must mitigate long-term potential claim risks generated by somatic and germline gene editing to be able to make predictions.

6.5 Historical Evolution of Human Genome Editing

The attempts to edit the human genome started in April 2015 when the Chinese stem-cell researcher "Junjiu Huang" and his team published an article describing their attempts to modify human embryos genetically. In their study, they used nonliving embryos, but the experiment failed. They could not correct the defect in the target gene. This experiment raised great controversy since it was the first experiment on the genetic alteration of humans. [24, 29] The experiment went through a second stage when a group of Chinese and British research groups 2016 tried to modify the embryo's genes to make them HIV resistant. Studies were conducted in the EU, the USA, and, recently, the Russian Federation [19].

In April 2016, the world witnessed the birth of the first genetically modified baby in Mexico. However, John Zhang, a Chinese-American fertility doctor, residing in New York and responsible for the operation, did not use the CRISPR technique but used instead a procedure called "pronuclear transfer" (also called mitochondrial replacement therapy). This operation has much less genetic effect than germline

CRISPR (Clustered Regularly Interspaced Short Palindromic Repeats) alteration. The genome alteration using this PNT procedure only affects the mitochondrial DNA, which is less than 1% of a person's DNA, thus leaving the DNA unaffected. Still, this operation is considered a form of genetic engineering of the human germline because this procedure affects, first, the genetic composition of the germline cells; and, second, if the baby is a female, the genetic alterations will pass down to her offspring as inheritance generally happens along the maternal line [32].

In the summer of 2017, the first successful modification of CRISPR occurred when the Kazakh-American biologist "Shoukhrat Mitalipov" and his American, Chinese, and South Korean team managed to correct a genetic mutation that caused a severe heart condition.

A year later, in 2018, Jiankui conducted an experiment believed to be the first human genome editing conducted on embryos using the same method. On November 25, 2018, Dr Jiankiu, Associate Professor of the Department of Biology at the Southern University of Science and Technology in Shenzhen, China, released a series of videos on YouTube declaring the birth of the first genetically modified baby. He edited the genome of two female embryos and then modified their cells, known as germline cells, which may be passed down to the offspring in the future [11, 26, 32].

The Chinese biophysicist was prosecuted before Shenzhen Nanshan District People's Court and was sentenced to 3 years in prison and fined 3 million RMB for illegal practice of medicine. The court ruled that the medical team members' behaviour was an intent behaviour violation of the Chinese law and medical code of ethics. They used techniques that had not been proven safe nor effective for clinical trials for human-assisted reproduction, and their actions have not met the minimum standards of research and the medical code of ethics [1, 20, 31].

The genome-editing method is usually used for reproduction-assisted techniques. The modification occurs in the sequence of the embryo's DNA before it is implanted in the mother's womb or the sperm or egg cell sequence before fertilisation. This is done to change the baby's inherited traits or improve them.

Undoubtedly, this technique effectively protects against some inheritable diseases because it affects the entire body of the offspring, unlike the genetic diagnosis occurring before the implantation or fertilisation at labs where healthy embryos are examined, selected, and then implanted in the womb; and the defect is discarded. But the process of genetic diagnosis before implantation could be an unbeneficial surgical procedure in cases when both spouses suffer from the same genetic disorder they wish to avoid in their children (20, Nimiec and Howard, 2020).

6.6 *Definition, Concept of Human Genome Editing*

The human body is based on small units called cells containing a group of genes. These genes are the genetic factors that determine the traits of every human [7]. The term 'genome' refers to the entire DNA chain of an organism or the complete set of genetic information relating to the traits of an organism (Alwi et al., 2017). It includes sequences of DNA with specific functions involved in producing proteins and other molecules necessary for growth, energy production, and disease control (Nuffield Council on Bioethics, 2018)

Genome editing is germline editing, and genetic modifications are made to the person's germ cells to change the set of genes inherited by one's descendants [33]. This process is done by the CRISPR-Cas9 technique, widely spread in recent times. It is used to turn on or off the genes (gene regulation), identify their functions, identify the possible mutations in cells, and identify why a disorder might occur to genes that make cells, leading to specific diseases technique which is used to eliminate the mutations that cause diseases, adding the desired characteristics and traits in children [30]. It is also used in the non-human sphere, such as altering the themes of animals to make, for example, police dogs more robust and plants to make the crops drought-resistant [5].

6.7 How CRISPR-Cas9 System Works in Human Genome Editing

The CRISPR-Cas9 system consists of two components: direct the editing mechanism to the target location in the genome where the alteration will occur. The other component is to carry out the alteration. The genome is encoded as a series of chemical bases aligned along the DNA molecule, a long molecule with a helix structure with two interlaced strands. This system works by causing a double-strand break (DSB) in the DNA at the specified location [23].

The genome-editing technique is one of the most efficient genetic engineering technologies used indirectly to alter the gene inside the genome in different organisms. It allows the addition, deletion, or substitution of the ACGT nucleotides where the DNA coding happens [12, 15, 23]. Hence, some believe that this technology is the most capable of treating inheritable diseases and ending human suffering resulting from such diseases by altering the patient's DNA [14].

Two possible applications of CRISPR on humans can be identified here; first, the type in which the genome alteration occurs in the human somatic line, i.e. in the cells of the affected person himself. Hence, the effect of modifications on line genes ends as soon as the person passes away without being passed down to his descendants. However, the other type has the alteration carried out on the germ cells, so the DNA of the embryo or gametes is altered, which is called human genome editing. The second type can affect all body cells of the intended humans in the future; it affects their brains, organs, vessels, and skin; and can be passed down to their descendants. Therefore, editing the human genome (germline) means editing the genomes of future generations [15] and Nuffield Council on Bioethics, 2018).

The first type does not raise any legal questions. It aims to modify the genome in the human somatic line for preventive, diagnostic, or curative purposes, provided that the alteration is intended for these reasons and not for mere alteration. The second application, which is editing the human genome, was banned in several international conventions, such as Oviedo Convention on Human Rights and Biomedicine in 1997 and UNESCO Declaration on the Human Genome in 1997. The bans considered the seriousness of this modification and the future side effects of such alteration on the human race, mainly if it aims to improve progeny or decide on gender. Following are the main reasons behind carrying out human genome editing:

a. **Editing Human Genome for Eugenics:**

Eugenics and gender swaying mean "intervention to determine the genetic characteristics of the embryo using assistive medical techniques for artificial insemination and fertilisation, to select favoured traits before implanting them into the uterine cavity" (Article (1) of Bahrain Law no. (26) of the year 2017 Regarding the Use of Assistive Medical Techniques for Artificial Insemination and Fertilization).

Humans have long imagined improving progeny and having designed children [30] because people wanted to live long and enjoy eternal youth. Hence, they sought cures for all diseases, ensuring a healthy environment and prolonging one's lifespan. This can be achieved through human genome editing, through which the genetic message can be mod, hence allowing for enhanced helical life [6].

In the earlier years of the past century, this idea was turned into practice as 'undesirable' individuals such as prisoners and people with mental disabilities were sterilised as a first step to eliminate such categories and maintain straight people. The first eugenics law was issued in the USA in Indiana in 1907. By the year 1931, most states had issued eugenics laws. By these laws, over 60,000 people were sterilised and deprived of the right to have children forever [18].

As a result of this notorious practice of eugenics, the idea of human genome editing was heavily criticised under the pretext of going against human rights. Nevertheless, its supporters see that earlier trait such as elitism, class discrimination, racism, and coercion, which were proposed last century as a social practice under eugenics, were indeed inconsistent with human rights; yet this can be avoided by requesting the full and informed consent of the persons involved as a condition before applying the editing procedure [26].

b. **Genome Editing for Selective Procreation:**

Selective Procreation, or the selection of the sex of the fetus, means "Human intervention to cause pregnancy with selected gender" [7]. It is close to what was defined by Article of Bahraini Law no. (26) of the year 2017 that fetal gender selection is an "intervention to determine the gender of the fetus using assistive medical techniques for artificial insemination and fertilisation".

The gender selection process can be done following different methods. Some depend on food types, intercourse timing, or assisted reproductive technologies. An example of the third type is selecting and separating the sperm, fertilising the eels with the selected sperm and separating the selected embryo before implanting it in the womb. This is done by examining the egg's fertilised sperms. The ones with the desired gender are then added implanted in the womb w, while the rest are discarded or damaged [3, 7].

c. **Editing Genome for Treatment:**

Editing the genome for treatment is one of the essential aims of human genome editing. This technique is used to examine the embryo before being implanted in the womb, identify the inheritable diseases that may be passed down to the embryo and genetically treat them. Genetic therapy involves replacing dysfunctional genes with other normal ones or preventing the production of unnecessary enzymes and proteins that may cause illnesses [9, 15].

It also helps in discovering which gender this inheritable genetic disease affects. Geneticists have proven that certain genetic diseases are affected by gender, as they appear in both sexes but are more common among one gender than the other. For example, gout and cleft palate are more common among males, and anencephaly and spina bifida among females. One of the most severe gender-related diseases, Hemophilia, is the inability of blood to clot properly, and this disease affects males more than females [3, 7]. It is worth noting that the Bahraini legislation banned the use of any reproduction assisted technologies for eugenics or gender selection by Law 26, Article 7, Clause D of the year 2017 unless the selection of gender prevents inheritable gender-based diseases.

7 Ethical Perspective on Editing Human Genome

7.1 Concept of Ethical Values and Whether Binding

Ethical values represent the ideal of what should be the behaviour of individuals and the role model people should seek to follow. These values are not binding unless stipulated in laws or international conventions; an example is the medical code of ethics. Although these ethics are non-binding, they are accompanied by a moral, which the person inflicts on himself, such as remorse of conscience, or the society inflicts on him, such as the society's disapproval of his behaviour [9].

7.2 Ethical Justifications for Disapproving Editing Human Genome

First of all, Dr Jiankiu's experiment, 2018, on editing the human genome and modifying of embryo's genes raised significant criticism in the scientific community worldwide.

a. Atat present is too soon to conduct a clinical experiment in editing embryo genes. There is a desperate need to conduct further studies that address the effectiveness of altering the genome by buying the disease-carrying gene without causing severe side effects [33]. The current techniques are not safe enough to be tested on humans since identical genome sequences outside the range to which modification is desired may be changed [19, 30].

b. The absence of ensured safety is one of the main reasons the experiment faced objections. Once carried out on a live human body, such experiments could lead to serious side effects for the person alone and his entire descendants. Since such a result cannot be avoided, the person whose genome was modified would need to be under observation and experimentation not until reaching the legal age only but their entire lives and the lives of their successors. This contradicts one's right to privacy and free self-growth. In addition, this procedure takes place without the consent of the intended person. Receiving parental consent is not enough in these cases because this is not similar to other routine parental decisions. Once the decision is made, it cannot be reversible. Admitting parents' authority to carry out such methods could extend their authority of decision-making power over their children to unacceptable levels [14, 26].
c. There is an imperative need for a more comprehensive understanding of the development of human organ systems such as the immune system and the nervous system to determine the effectiveness and potential risks of transferring and modifying embryos' genes. In addition, before conducting any modification to a human embryo's genes, there must be a pathophysiological understanding of the potential disease [33] and accurate identification of the effect of the various environmental factors on the emergence of the genetic disease in the offspring of the inflicted persons (Nuffield Council on Bioethics, 2018).
d. Before approving any medical procedure, it must undergo strict scientific evaluation by the experts. This has not yet happened with the medical procedure of the embryos' genome modification carried out by Jiankiu, who directly announced his experiment without any prior evaluation. Therefore, his work lacked the minimum level of good practice standards [19].
e. Such experiments contradict the fundamental values of dignity and equality, which are stressed by the modern international charters and constitutions such as the Convention on Human Rights and Biomedicine of 1997, the Universal Declaration on the Human Genome [21], the Universal Declaration on Human Rights, the National Action Charter of Bahrain of 2001 (Chapter 1: Basic Principles of the Society, Second Clause: Protection of Personal freedom and Equality).

8 Criticism of Human Genome Editing

In this context, criticism of human genome editing has significantly eased in recent years. Some now believe that clinical experiments of human germline genome editing (including the addition, removal, or replacement of pairs of DNA bases in the gametes or early embryos) may be permitted in the future, but only in critical conditions and under strict supervision and evaluation. Moreover, germline human genome editing may be permitted in the future if the related risks are appropriately resolved, and some standards are met. Such standards include strict and independent supervision, desperate medical need, absence of reasonable alternatives, development of a long-term follow-up plan, and attention to societal consequences. Accordingly, this technique may be put into effect if the tools applied are safe, the advantages outweigh the risks—or at least there are no reasonable alternatives- and no harm is caused to any third parties [19, 26].

The technique can also be used for purposes other than treating diseases, such as genetic modification for eugenic purposes. Proponents of this concept justify using this technique with the allegation that adding certain qualities to some people does benefit them alone. Instead, it would make their countries more robust and more prosperous, especially if those countries need such qualities. They believe that eugenics creates the most significant possible number of effects that can reasonably be employed to make the active groups of society contribute to the next generation more than their actual proportion (Galton, D.C.L.; Sc.D.; F.R.S).

9 Sharia Position on Human Genome Editing

Islamic Sharia seeks to achieve two main goals; the first is to achieve the interests of Man in the hereafter, and the second is to preserve and protect human life in this worldly life [2, 10, 16, 20]. Therefore, maintaining human life and protecting human dignity are the main Islamic Sharia objectives. Therefore, Islamic teachings permit everything that benefits humans, judging it as obligatory, recommendable, or allowed; and pro, it everything that might cause harm to their physical, mental, spiritual, or intellectual health—judging it as prohibited or discouraged. Moreover, Islamic Sharia urged the advancement of the human soul to the highest achievable degrees [1, 3]. Therefore, it can be established that Islamic Sharia has set a general protocol that regulates the profession of medicine, the treatment of diseases, and society's health as a whole. This is known in modern science as "preventive medicine", which establishes an ideal healthy society [4].

10 Conclusion

Upon completing this paper on the topic of human genome editing, we made the following conclusions and recommendations; all praise be to Allah.

First: Research results:

1- There is no sufficient scientific certainty on the consequences of applying this technique to humans, as it still lacks the safety and security criteria that must be ensured in any medical procedure. The risk of causing severe harm after applying the human genome editing technique is still high and may be irreversible and uncontrollable.

2- It must be differentiated between changing a person's genome and changing the human genome. The first can be permitted under the discussed conditions, especially for purposes. However, the second is prohibited from the perspective of sharia, especially if it is for eugenic purposes to provide the future generations with genes that add to their physical or mental capabilities exaggeratedly or to select a particular species (or sex) over another. Such acts may bring an end to the human race.

Second: Recommendations:
Before applying human genome editing technology, the following must be accomplished:

1. Expanding research and experimentation on this topic while keeping the research outside the human body and abiding by the rules and regulations stipulated in the international charters, and ensuring human rights are not violated.
2. Conducting societal dialogue at the local and international levels on human genome editing concluded with a binding international convention. The possible consequences of applying this technique are not exclusive to the persons involved alone but affects of the entire human community.
3. Do not use any medical techniques unless globally approved, and keep this use limited for treatment purposes and scientific experiments on nonliving human bodies.

References

First: Arabic References:

1. Abdul Rahim O (2007) The effect of necessity, need, and generality of distress on the prohibited and permissible professions. In: The fifth annual conference of Sharia Fiqh council of America held in Bahrain, Titled "And He Permits to them What is Good and Forbids What Evil is"
2. Al-Arabi B (2012) Rules of medical experiments on humans in light of islamic sharia and medical laws, comparative study. Dar Al-Thaqafa for Publishing and Distribution, Jordan
3. Al-Bar MA (1995) Rulings on medications and desperate cases and the issue of merciful death, 1st edn. Dar Al-Manar Publisher
4. Al-Fanjari AS (1991) Preventive medicine in Islam, Islam's medical teachings in the light of modern science, 3 edn. General Egyptian Book organisation
5. Al-Jamal A (1997) Secrets of genetics, general egyptian book organization
6. Al-Nasser I (2019) Genome and life and its ethical impact on muslim societies (brill is collaborating with JSTOR to digitise, preserve and extend access to Islamic Ethics and the Genome QuestionK)
7. Isma'il M, Ramadan RM (2009) Choosing the sex of a fetus to prevent hereditary disease. IUG J 17(1):27–48
8. Lutfi AM (2015) Medical experiments on the human body between the need to further medical advancement and the inviolability of the human body, a contemporary medical jurisprudence perspective, the winning research of Rashid bin Humaid Cultural and Science Award, UAE
9. Mansour A, Fayed A (2011) Introduction to legal sciences in Bahraini and comparative laws. Kingdom University

Second: Non-Arabic References:

10. Al-Sartawi A (2020) Shariah disclosure and the performance of islamic financial institutions. Asian J Bus Account 13(1):133–160
11. Annas GJ (2020) Ethics and human rights in germline editing in humans and gene drives in mosquitoes. Am J Law Med 46:143–165
12. Araki M, Ishii T (2014) International regulatory landscape and integration of corrective genome editing into in vitro fertilisation, Araki and Ishii Reproductive Biology and Endocrinology. http://www.rbej.com/content/12/1/108. Accessed 22 Feb 2022
13. Bizzarri M (2012) The new alchemists: the risks of genetic modification, Kindle edn. WIT Press
14. Bolton CC (2017) The consequences of human genome editing on human rights. https://ssrn.com/abstract=3444264. Accessed 24 Feb 2022
15. Christiane W, Christina G, Emmanuelle C, Hans S, (n.d.) Discussion paper focusing on the scientific relevance of genome editing and the ethical, legal, and societal issues potentially involved. Issues by the Ethics Council of the
16. Crisper FD (2019) Cas-9 genome editing and Islam: a religious perspective. Bang J Med Sci 18(1):7–13
17. Davies B (2019) The technical risks of human gene editing Human Reproduction, vol 34, no 11, pp 2104–2111. Welcome Centre for Human Genetics, University of Oxford.
18. Dohn MR (2018) Preventing an era of "New Englenics": an argument for Federal funding and regulation of gene-editing research in human embryos. Richmond J Law Technol XXV(2)
19. Ellacuria EP (2021) Scientific, ethical and legal issues of germline gene editing: light and shadow in the declaration of the Spanish bioethics committee. Ramon Llull J Appl Ethics (12). https://www.researchgate.net/publication/353463979. Accessed 20 Feb 2022
20. Ghaly, M. (2019) Islamic Ethical Perspectives on Human Genome Editing, spring.
21. Harmon SH (2005) The significance of UNESCO's universal declaration on the human genome & human rights, vol 2, no 1
22. Hopkin P (2018) Fundamentals of risk management: understanding, evaluating and implementing effective risk management, 5th edn. Kogan Publishing
23. Karen Y (2018) Genome editing and human reproduction: social and ethical issues. Nuffield Council on Bioethics, London
24. Krishan K, Singh B, Baruah N (2015) Genome editing of the human embryo - a question on editorial outlook and responsibilities, pp 661–663. https://www.researchgate.net/publication/281203150
25. National Human Genome Research Center (n.d.). https://www.genome.gov/Funded-Programs-Projects/eMERGE-Genomics-Risk-Assessment-and-Management-Network
26. Neri D (1999) On the concept of eugenics: preliminaries to a critical appraisal Sobre o conceito de eugenia: preliminares à uma avaliação crítica Cad. Saúde Pública, Rio de Janeiro 15(Sup. 1):27–34
27. Al-Sartawi AMAM (eds) (2022) Artificial intelligence for sustainable finance and sustainable technology. ICGER 2021. Lecture Notes in Networks and Systems, vol 423. Springer, Cham
28. Powell R (2015) In genes we trust: germline engineering, eugenics, and the future of the human genome. J Med Philos 40:669–695. https://doi.org/10.1093/jmp/jhv025
29. Singh AN (2021) Ethical controversies and challenges in human genome editing. Int Med J 28(2):134–137
30. Shinwari ZK, Tanveer F, Khalil AT (2017) ethical issues regarding CRISPR-mediated genome editing. Curr Issues Mol Biol 26
31. Townsend BA (2020) Human genome editing: how to prevent rogue actors, Townsend BMC Medical. https://doi.org/10.1186/s12910-020-00527-w. Accessed 1 Mar 2022
32. Van Beers BC (2008) Rewriting the human genome, rewriting human rights law? human rights, human dignity, and human germline modification in the CRISPR era. J Law Biosci, 1–36
33. Yotova R (2017) The regulation of genome editing and human reproduction under international law, eu law, and comparative law

The Use of Artificial Intelligence in Combating Crimes in the UAE: Critical Review

Maha Mohammed Yusr Othman and Maryam Mohammed Hassan Al Hammadi

Abstract The current study discusses artificial intelligence as one of the most important outcomes of the modern technological revolution, as it has multiple uses in all areas of security, economics, military and civil matters. According to the literature, research neglects to review studies regarding the detection of crimes in the security field. Nevertheless, Artificial intelligence enables police agencies to use resources effectively and achieve societal security and the speed of crime detection in the UAE. Furthermore, the synthesis of the research problems, research methods, and findings of the surveyed studies will assist the scholars in pursuing their future studies. Further discussions and implications are also presented in the study.

Keywords Artificial intelligence (AI) · Predictive police · UAE · Crimes · Future foresight · Police

1 Introduction

Crime is a social phenomenon that accompanies all societies, advanced and lagging behind the paths of progress. International studies in various countries indicate a rise in crime rates and a tendency to increase year after year. Due to the tremendous development in communication and transportation, crimes and the methods of its commission have evolved [1, 2]. In the beginning, artificial intelligence techniques were a field of academic research. Today, technology companies are now increasingly resorting to them to monitor the behaviour of their users on the Internet and predicting their next steps [3]. The large size of the Internet sites and social media means that the scope of the security responsibility or tracking information using traditional methods was nearly impossible [4]. Human intervention is no longer just an appropriate solution in tracking and preventing crimes. Artificial intelligence has

M. M. Y. Othman · M. M. H. Al Hammadi (✉)
School of Business Management, UUM College of Business, Changlun, Malaysia
e-mail: Mariam19766@hotmail.com

M. M. Y. Othman
e-mail: mmyusr@uum.edu.my

A. M. A. Musleh Al-Sartawi et al. (eds.), *From the Internet of Things to the Internet of Ideas: The Role of Artificial Intelligence*, Lecture Notes in Networks and Systems 557,
https://doi.org/10.1007/978-3-031-17746-0_29

tremendously changed as applications, tools, and software platforms are developed to process large amounts of information, financial data, transactions and crimes [5]. These tools work systematically to find, track and flag suspicious activities, whether it is a social media conversation or individual actions [6]. Electronic attacks, fraud, and forgery are also identifiable due to Artificial Intelligence; even funds transfer between banks and other financial institutions. Thus, the revolution in crime prevention made the concept of predictive police or machine learning programs to predict future crimes [4]. The stage of collecting inferences about crimes and what results from them in the investigation and investigation procedures affects the progress of the investigation process, tracks crimes, and leads to mistakes. Collecting information and evidence in a distinctive and highly efficient manner, using various technical and smart means [7]. The United Arab Emirates is one of the first countries in the region and the world to employ artificial intelligence in police work and prevent and track crimes. It operates the role of artificial intelligence technology in enhancing security and public safety in society, criminal induction [8, 9]. The use of artificial intelligence techniques in forensic evidence and criminology contributes to providing compelling evidence to the judicial authorities about crimes and providing information and evidence to the police agencies to decipher complex crimes [10]. As the UAE seeks, through these technologies, they help to access the maximum levels of community security by adopting smart technologies and by defining evaluation criteria, collecting criminal evidence, including criminal history, the severity of existing crimes, and using the computational risk assessment tool for many critical decisions, and Crime prediction [11]. It can specify whether an individual must be released on bail before or after a charge is issued or whether the suspect should be kept in custody for a longer period [12]. Where the current study attempts to identify the role of artificial intelligence in preventing and detecting crime and protecting society from it by the UAE law authorities, and a research gap in the most important techniques used in the law authorities in the United Arab Emirates and their impact on preventing and detecting crimes before they occur.

2 Literature Review

2.1 Artificial Intelligence as Security

Artificial intelligence (AI) is being used to monitor and prevent crimes in many countries. AI's involvement in crime management dates back to the early 2000s. AI is used in bomb detection and defused, surveillance, prediction, social media scanning, and suspect interviews [3]. However, there is room for growth in its role in crime management. Currently, some issues are proving problematic [10]. AI is not uniformly involved in all countries in crime management [13]. There is a fierce debate about the ethical limits of AI, forcing law enforcement authorities to tread carefully. Defining the scope and limits of AI, including collecting personal data, is a complex task [14]. AI promises a new paradigm in crime management, a solid case to pursue [8] Artificial intelligence techniques have been employed in security work in many innovative ways and methods. Many techniques used in security services and the inference process, such as facial recognition, can be found. and predictive analysis of crimes and the process of predictive analysis of crime [6]. It is a complex process that uses large amounts of data to predict and formulate possible results in light of data provided by police response devices such as traffic and movements of suspicious people, etc [3]. It also includes criminal crimes, and this function is mainly on the responsibility of the police and security authorities in employing all techniques to monitor and prevent crimes [4, 15]. Also, available information about the law and legal precedence, social information, media, identification of criminal institutions, prediction of their criminal activity and detection of people at risk through the activity of criminal institutions can be used [16]. Artificial intelligence systems provide computational methods for interpretation and data analysis that can increase the speed and quality of the security interpretation and reasoning process sought by security professionals, judges, lawyers, prosecutors, administrative staff, and other professionals. Artificial intelligence can also analyze large amounts of records related to potential future crimes [18]. Artificial intelligence provides investigators and security personnel with algorithms to analyze data sets that contain millions of records and orders [19]. These data algorithms are transformed into applications based on the security system so that security practitioners can determine the extent of crime and intervene quickly and prevent its occurrence Artificial intelligence constitutes many other policing areas, such as accommodation systems, traffic control, identifying congestion points, and discovering solutions before future accidents occur [12] It also includes detecting fraud and forgery cases through record keeping and similar issues. Nowadays, all security agencies and agencies worldwide are working towards more approaches based on artificial intelligence and data analysis to solve crimes [20] AI is exceptionally skilled at identifying patterns and can be very useful when determining how crimes were committed by the perpetrator [21].

2.2 Artificial Intelligence in the UAE

The world is witnessing an advanced technological revolution at various levels [22]. This was accompanied by significant security challenges imposed by that technology on multiple aspects of life, institutions, wills, and even countries, threatening national security or societal peace. Crime as a social phenomenon was not immune to this development, but instead had a share Clear and influential on the human community with a family [23]. Increased crimes demand the police and security services develop and advance the capacity that allows them, in the inference stage of the crime, to discover crimes, identify criminals and protect society from them [24]. Combatting cybercrimes through modern technologies in the various stages of corruption and an attempt to nip it in the bud [25]. The UAE, within its local and federal administrations and its security system, has adopted artificial intelligence in the inference stage of crime and crime in general, and to protect society and its economy from these crimes, by following advanced tracking and sensing systems in tracking criminal offences, and detecting them by employing the feature of artificial intelligence and smart applications [24]. According to [12, 13], the vision of its wise leadership is to become the first country in the field of global security. The state has tended to develop specialized scientific competencies and local capabilities in artificial intelligence. The aim is to organize technical training courses for government employees in data science and to create a culture of artificial intelligence among society groups to facilitate the spread of the use of applications [23]. Efforts by government, educational and media institutions to raise community awareness of the basics of this field, with the establishment of research centers that contribute to the development of the various sectors in the country and qualify them to receive the necessities of artificial intelligence [20] The United Arab Emirates has been keen on progress and success in development and competition for future technologies and challenges [23]. This explains the country's orientation towards the fourth generation technologies of the industrial revolution represented in artificial intelligence, the future language. As well as that is in the Artificial Intelligence report issued by the Ministry of Economics, the industrial sector is one of the main critical economic sectors at the level of the UAE [26]. Therefore, this sector affects the state's financial position, and this requires supporting this vital sector with the outputs of the Fourth Industrial Revolution, most notably artificial intelligence techniques. The need for intelligence will also serve other vital industries such as transport by drones, autonomous cars, flying taxis, the metro, and all means of land and sea transportation [25]. Artificial intelligence is rapidly developing in all areas of life, such as health, for its ability to assist the medical staff in diagnosing and treating diseases, prescribing medications, performing surgeries, and better accessing patient files by voice commands, and through which patients can book appointments [5, 27]. The possibility of teaching and self-development through artificial intelligence programs such as learning machines, logic, self-correction and self-programming [28]. Artificial intelligence provides legal advice and achieves interactive education, used in the security and military fields. Artificial intelligence relieves people of hardships and dangerous work such as exploration and rescue operations that require muscle strength [29]. It is possible to use artificial intelligence in expert systems based on knowledge bases and facial recognition technology [30]. The education system will change as a concept and as a result of applying the principles of the Fourth Industrial Revolution [31]. It will focus on sensory education to meet the requirements of the industrial sector and artificial intelligence. Artificial intelligence can provide new job opportunities, provide services at a cheap cost, contribute to maintaining security, and provide mechanisms and solutions to meet challenges, including cybercrime [32]. In the short term, the UAE AI strategy aims to increase government performance, productivity, and efficiency at all levels and sectors (such as saving 50% of annual costs using AI) [33].

2.3 Artificial Intelligence Security in the UAE

In line with the UAE's vision for artificial intelligence, the UAE Ministry of Interior has adopted artificial intelligence in its police operations and services through expert systems, which are programs that simulate the performance of a human expert in a specific field of expertise [34]. It simulates artificial intelligence and employs it in police work to maintain security and stability. Besides, AI also enhances the values of positivity and happiness in the Emirate and consolidates the concept of the strategy of employing artificial intelligence in security, data analysis and security information collection [34] The UAE Ministry of Interior has adopted, through artificial intelligence technology, to provide the best innovative services to the public and customers through a group of smart applications and platforms, as it includes a package of services of interest to the public such as: inquiring about reports, requesting criminal status, paying violations and reporting a traffic accident, as a safe service and a virtual assistant in the Ministry of Interior application To provide all the services in the application using voice, writing and other channels, such as the call center, and there is a service dedicated to people of determination and providing services to them [35–37]. Dubai has launched a new smart system that can predict crime before it occurs in the context of facilitating the work of the police and other investigative agencies [24]. Space Imaging Middle East "SIME" has announced that Dubai Police has developed special crime prediction software, in which patterns from police databases are analyzed in an attempt to discover when and where crime is likely to occur [25, 29] According to SIME, the software uses "sophisticated algorithms" to make predictions about crime, and the resulting data is "highly accurate". Modern smart technology can alert patrol teams in neighbourhoods that need more police attention to prevent crime [26]. For example, an intelligent system can be inserted into CCTV to send alerts to the police if a person is in a dark alley and behaving abnormally, indicating the possibility of a crime [38]. "This intelligent software is unique in terms of being able to discern complex patterns of criminal behaviour accurately [35]. Along with the predictions of experienced police officers, will create a powerful crime deterrent [38].

2.4 UAE Predictive Police

The matter is more straightforward for computers that can use data analysis techniques working with algorithms capable of helping police officers know where and when the crime is likely to occur through a set of rules [35]. Where, for example, the system can analyze the data of several years of crimes in a specific location to find out the crime patterns that occur in it, to match the most recent crime pattern with the previous patterns to know where the next similar crime is likely to occur, which can also help in understanding the extent to which a person can He returns to the world of crime after his release. Predictive police have recently made significant progress with artificial intelligence technologies. Many countries and companies are trying to exploit these technologies to improve the ability to predict crimes [38]. Dubai, in particular, has shown its determination to create the smartest police force in the world, which has been accelerating thanks to artificial intelligence [29] recently. The UAE, and specifically the Emirate of Dubai, recently announced that its innovative system based on artificial intelligence has led to the successful arrest of more than 100 wanted criminals since the beginning of the last year, 2021 The department's project includes advanced smart devices that help reduce the crime rate to maintain the security and safety of citizens [13, 39] This project comes from Dubai Police as part of its strategy for artificial intelligence to provide all possible police services to fight crime in the city. Many programs and projects are currently being modified to rely on data analysis using artificial intelligence by 2031 entirely [32] This strategy aims to create a work environment that balances artificial intelligence and the human element, where artificial intelligence is used to predict crimes and investigations in the field of forensic medicine and police operations. In addition to improving the rate of traffic and road safety, the United Arab Emirates is the first in the world in terms of investing in artificial intelligence in various fields [25] In addition to focusing on achieving goals that include the happiness of society by providing police services in the future with ease for all citizens to make Dubai a safe city by using artificial intelligence also in crisis and disaster management, which contributes to crowd management in the event of an enormous problem [35]. Artificial intelligence techniques also provide a quick and accurate response when responding to customers in various languages and improve the efficiency of police officers and several other benefits [12].

3 Discussion

Artificial intelligence systems have been employed in many innovative ways and methods in security work. With artificial intelligence, it is possible to use the available information about the law and legal precedence, social information, the media, identify criminal institutions, predict criminal activity and detect people at risk [3]. Artificial intelligence systems are algorithmic methods for the interpretation and analysis of data that can increase the speed and quality of the security interpretation and inference process sought by security professionals, judges, lawyers, prosecutors, administrative staff, and other professionals [4]. According to [4, 8, 34, 38], researchers published a report on this technology, which will have a positive impact on crime prevention, but warned that it could threaten millions of crime-related jobs. The researchers' predictions came from a research group called "Artificial Intelligence and Life in 2030", conducted by Stanford University, where they said that the "crime prediction system" would be highly adopted by 2030. AI is currently being used to combat financial crime [40]. In addition to assisting the law in uncovering unusual plans and activities on social sites. Despite this, researchers stated that there are legitimate concerns at the same time about the possibility of crossing borders and using smart systems to violate people's privacy [41]. They said that society's acceptance of this modern technology would determine its success and the scope of its use [6]. Nowadays, all security agencies and agencies worldwide are working towards more approaches based on artificial intelligence and data analysis to solve crimes [42]. AI is exceptionally skilled at identifying patterns and can be very useful when figuring out how the perpetrator committed crimes [7]. Thus, artificial intelligence tools can accelerate this work that may take longer to disclose when using human intelligence, which is what security systems seek in the future to take advantage of artificial intelligence of various kinds to detect crimes committed by people, organizations, and even countries [9, 36, 43].

4 Conclusion and Future Research

This The most critical smart technologies used by the UAE law authorities in criminal inference are the robot technology in detecting and dealing with crimes, smart traffic techniques in monitoring traffic accidents and accidents, and artificial intelligence systems that rely on algorithmic methods for interpretation [35, 38, 44] As well as the inference process, the police also use the artificial neural network to identify and detect crimes and rely on monitoring, intelligent control and inference of the police in the United Arab Emirates [37, 45, 46]. In addition, artificial intelligence systems detect fraud and forgery cases through record-keeping and similar cases. The use of artificial intelligence in the criminal inference stage also contributes to easing safe, economical and human requirements and in the complexity of criminal cases, the invention of new methods, the detection of the circumstances of the crimes, the speed of access to and processing of information, the identification of the method of committing crimes by the offender, and the diversity of experiences used in the detection of crimes [9, 13, 44]. There are some limitations to the study that it used a literary review [42]. Perhaps it will be the use of the qualitative approach and the best interview method to reach more accurate and diverse results about artificial intelligence and a course in combating crime in the UAE, other limitations that artificial intelligence systems in the security field are multiple and cannot be limited or disclosed from Before the authorities of the law for the sensitivity of its use in the security field. It would be good to foresee the future of police and security work and detect crime in all sectors through the application of artificial intelligence and the development of its systems to improve services and quality of life and through the enactment of legislation and laws that would employ these techniques in criminal inference processes and facilitate search and investigation of crimes.

References

1. Al-Skaf S, Youssef E, Habes M, Alhumaid K, Salloum SA (2021) The acceptance of social media sites: an empirical study using PLS-SEM and ML approaches. In: Advanced machine learning technologies and applications: proceedings of AMLTA 2021, pp 548–558
2. Salloum SA, AlAhbabi NMN, Habes M, Aburayya A, Akour I (2021) Predicting the intention to use social media sites: a hybrid SEM-machine learning approach. In: Advanced machine learning technologies and applications: proceedings of AMLTA, pp 324–334
3. Sukhodolov AP, Bychkov AV, Bychkova AM (2020) Criminal policy for crimes committed using artificial intelligence technologies: state, problems, prospects
4. Radulov N (2019) Artificial intelligence and security. Security 4.0. Secur Futur 3(1):3–5
5. Habes M, Ali S, Elareshi M, Tahat KM, Ziani A (2021) Understanding users' social TV content using artificial intelligence approach: a survey. In: International conference of modern trends in information and communication technology industry (MTICTI), pp 1–7
6. Sushina T, Sobenin A (2020) Artificial intelligence in the criminal justice system: leading trends and possibilities
7. Yeoh P (2019) Artificial intelligence: accelerator or panacea for financial crime? J Financ Crime
8. Harris H (2022) Artificial intelligence and policing of financial crime: a legal analysis of the state of the field. In: Financial technology and the law. Springer, pp 281–299
9. Younies H, Na T (2020) Effect of cybercrime laws on protecting citizens and businesses in the United Arab Emirates (UAE). J Financ Crime
10. Al Shibli M (2020) Hybrid artificially intelligent multi-layer blockchain and bitcoin cryptology (AI-MLBBC): anti-crime-theft smart wall defense. In: Encyclopedia of criminal activities and the deep web. IGI Global, pp 1089–1111
11. Habes M, Ali S, Salloum SA, Elareshi M, Ziani A-K, Manama B (2020) Digital media and students' AP improvement: an empirical investigation of social TV. In: International conference on innovation and intelligence for informatics, computing, and technologies
12. Alhumaid K, Habes M, Salloum SA (2021) Examining the factors influencing the mobile learning usage during COVID-19 Pandemic: an Integrated SEM-ANN Method. IEEE Access 9(1):02567–102578
13. Al Yazeedi AM (2021) An analysis of the forces influencing the implementation of artificial intelligence in the intellectual property sector in UAE. British University in Dubai (BUiD)
14. Horowitz MC (2018) Artificial intelligence, international competition, and the balance of power (May 2018). Texas Natl. Secur. Rev.
15. Salloum SA, Al-Emran M, Habes M, Alghizzawi M, Ghani MA, Shaalan K (2021) What impacts the acceptance of e-learning through social media? An empirical study. In: Recent Adv. Technol. Accept. Model. Theor., pp 419–431
16. Vaishya R, Javaid M, Khan IH, Haleem A (2020) Artificial Intelligence (AI) applications for COVID-19 pandemic. Diabetes Metab Syndr Clin Res Rev 14(4):337–339
17. S Plan (2016) The national artificial intelligence research and development strategic plan
18. Kumar S, Duhan M, Haleem A (2016) Evaluation of factors important to enhance productivity. Cogent Eng 3(1). https://doi.org/10.1080/23311916.2016.1145043
19. Tyugu E (2011) Artificial intelligence in cyber defense. In: 2011 3rd international conference on cyber conflict (ICCC), pp 1–11
20. Keding C (2021) Understanding the interplay of artificial intelligence and strategic management: four decades of research in review. Manag Rev Q. https://doi.org/10.1007/s11301-020-00181-x
21. Musleh Al-Sartawi AMA (eds) (2022) Artificial intelligence for sustainable finance and sustainable technology. In: ICGER 2021. Lecture notes in networks and systems, vol 423. Springer, Cham

22. Ziani A, Elareshi M, Habes M, Tahat K (2021) Digital media usage among Arab journalists during Covid-19 outbreak. In: Al-Sartawi MAM, Razzaque A, Kamal M (eds) Artificial intelligence systems and the internet of things in the digital era, EAMMIS 2021. Lecture note in network and systems, vol 239. Springer, Cham, pp 116–129. https://doi.org/10.1007/978-3-030-77246-8
23. Government UAE (2020) UAE strategy for artificial intelligence. UAE. https://government.ae/en/about-the-uae/strategies-initiatives-and-awards/federal-governments-strategies-and-plans/uae-strategy-for-artificial-intelligence
24. Alhashmi S, Salloum S, Abdallah S (2019) Critical success factors for implementing artificial intelligence (AI) projects in Dubai government United Arab Emirates (UAE) health sector: applying the extended technology acceptance model (TAM). In: International conference on advanced intelligent systems and informatics, pp 393–405
25. Almarzooqi A (2019) Towards an artificial intelligence (AI)-driven government in the United Arab Emirates (UAE): a framework for transforming and augmenting leadership capabilities. ProQuest Diss. Theses, p 204
26. Habes M (2019) The influence of personal motivation on using social TV: a uses and gratifications approach. Int J Inf Technol Lang Stud 3(1)
27. Elareshi M, Habes M, Ali S, Ziani A (2021) Using online platforms for political communication in Bahrain election campaigns. Soc Sci Humanit 29(3):2013–2031
28. Zhang T, Liu Y, Rao Y, Li X, Zhao Q (2020) Optimal design of building environment with hybrid genetic algorithm, artificial neural network, multivariate regression analysis and fuzzy logic controller. Build Environ 175:106810
29. Antwi-Boateng O, Alhashmi AA (2021) The emergence of the United Arab Emirates as a global soft power: current strategies and future challenges. Econ Polit Stud, pp 1–20
30. Clarke J (2019) Challenges of VAT implementation in the UAE. https://www.meed.com/vat-regime-uae-pinsent-masons/
31. Al-Harthei H, Garib AM, Hassan Y, El Halim A, Abd O (2010) Safety evaluation algorithm for signalized intersections in Abu Dhabi (UAE) using artificial intelligence. Adv Transp Stud 2010(20):29–38
32. Halaweh M (2018) Artificial intelligence government (Gov. 3.0): the UAE leading model. J Artif Intell Res 62:269–272
33. Al-Sartawi A (2020) Information technology governance and cybersecurity at the board level. Int J Crit Infrastruct 16(2):150–161
34. Qasim A, El Refae G, Eletter S (2021) Embracing emerging technologies and artificial intelligence into the undergraduate accounting curriculum: reflections from the UAE. J Emerg Technol Account
35. Zhongming Z, Linong L, Xiaona Y, Wangqiang Z, Wei L (2019) US-UAE joint statement on artificial intelligence cooperation
36. Ali S, Habes M, Youssef E, Alodwan M (2021) A cross-sectional analysis of digital library acceptance, & dependency during Covid-19. Int J Comput Digit Syst
37. Alhelli L, Al-Yafeai M, Daoud MS, Elamsy T (2021) An intelligent route finder For UAE desert driver based on A* algorithm. In: 2021 8th international conference on internet of things: systems, management and security (IOTSMS), pp 1–5
38. Almarashda HAHA, Bin Baba I, Ramli AA, Memon AH, Rahman IA (2021) Human resource management and technology development in artificial intelligence adoption in the UAE energy sector. J Appl Eng Sci 11(2):69–76
39. Elareshi M, Habes M, Youssef E, Salloum SA, Alfaisal R, Ziani A (2022) SEM-ANN-based approach to understanding students' academic-performance adoption of YouTube for learning during Covid. Heliyon 8(4):e09236
40. Wang H, Ma S (2021) Preventing crimes against public health with artificial intelligence and machine learning capabilities. Socioecon Plann Sci 101043
41. Russell SJ, Norvig P (2016) Artificial intelligence: a modern approach. Pearson Education Limited, Malaysia

42. Nemitz P (2018) Constitutional democracy and technology in the age of artificial intelligence. Philos Trans R Soc A Math Phys Eng Sci 376(2133):20180089
43. Khalifa SKH, Al-Kumaim NHS (2021) A conceptual model for prevention of efinancial crimes in UAE: a review paper. AcadStrateg Manag J 20:1–10
44. Mosteanu NR (2020) Artificial intelligence and cyber security–a shield against cyberattack as a risk business management tool–case of European countries. Qual Success 21(175)
45. Allen GC (2019) Understanding China's AI strategy: clues to Chinese strategic thinking on artificial intelligence and national security. Center for a New American Security Washington, DC
46. Matheson E (2020) UAE adoption of digital authoritarianism weakens US security and portends soft power shift

The Impact of Internet of Things on the Quality of Financial Reporting

Ebrahim Mansour, Rana Taha, and Noor Taha

Abstract In view of the many challenges that have emerged recently with regard to electronic disclosure, developing accounting systems, and information technology. Corporations aim to face these challenges to ensure the confidence and credibility of financial statements and reports and improve the quality of accounting information. Therefore, this study aims to investigate the impact of the Internet of Things (IoT) and big data on the quality of financial reporting. This paper found a significant positive relationship between the IoT and big data on the quality of financial reporting. The recommendation of this study is to apply these techniques in corporations to enhance financial reporting quality.

Keywords Internet of Things (IoT) · Big data · Quality of financial reporting

1 Introduction

These days, urbanization has significantly increased. Decision-makers must keep pace with this increase and provide a better life through a new future vision that is seen when converting traditional things into smart things. Using various technologies improves the quality of life, such as increasing security, reducing pollution, improving health care, increasing education quality, and developing industrial and economic sectors [1].

Advances in technology have emerged according to the needs of people. Technological developments that entered our lives slowly at first spread rapidly over time. With technology in businesses, organizational structures have also entered a change

E. Mansour · R. Taha (✉) · N. Taha
Department of Accounting, Faculty of Economics and Administrative Sciences, The Hashemite University, P.O. Box 330127, Zarqa 13133, Jordan
e-mail: 1970120@std.hu.edu.jo

E. Mansour
e-mail: ebrahim_mansour@hu.edu.jo

N. Taha
e-mail: 2070387@std.hu.edu.jo

A. M. A. Musleh Al-Sartawi et al. (eds.), *From the Internet of Things to the Internet of Ideas: The Role of Artificial Intelligence*, Lecture Notes in Networks and Systems 557,
https://doi.org/10.1007/978-3-031-17746-0_30

process. Although this change is mainly felt in the production process, it affects all departments of the organization [2]. We are now witnessing technologies that communicate with each other, learn, generate information, and analyze this information [3]. These developments will radically change the enterprises' departments and affect the financial reporting processes.

Financial scandals in several countries have served as justification for new legislation to regulate corporate governance practices. For instance, the USA passed the Sarbanes–Oxley Act in 2002. In 2005 the Financial Reporting Council (FRC) in the U.K. updated the Turnbull Guidance on internal control to be consistent with internal control reporting requirements as set out in section 404 of the Sarbanes–Oxley Act and the related SEC rules [4]. The Asian financial crisis started in 1997, which deteriorated many of the East Asian companies' financial performance, caused mainly by the failure of these companies to govern their activities. In Jordan, the Securities and Exchange Commission issued a code about corporate governance practices in 2007 to ensure the importance of applying corporate governance by Jordanian companies and promote international best practices in corporate governance [5–8]. The current study investigates the impact of IoT and big data on the quality of financial reporting.

The rest of this paper is organized as follows. The next section reviews the literature on the impact of IoT on financial reporting. The third section shows the research problem and hypotheses. The fourth section illustrates the research methodology. Section five concludes the paper.

2 Literature Review

Many researchers have investigated the real benefits of the quality of financial reporting. For example, Lin et al. [9] indicated that inadequate financial reporting might negatively influence business performance and decision-making. Meanwhile, there is varying evidence supporting the association between a firm's profitability and its financial reporting quality. The basic measures of performance are the profitability indicators.

Corporate governance prevents companies from making mistakes, frauds, and inaccuracies in financial reports by adapting management activities, especially International Financial Reporting Standards and International Audit Standards [10, 11]. In case of failure of corporate governance, it is seen that the reliability of financial reports is shaken and market efficiency is decreased, stock price fluctuations increase, and accounting scandals emerge [12]. Companies that implement corporate governance practices are more reliable for investors [4, 5, 13], so they are considered to implement fewer earnings management practices [14]. Ensuring the independence of the board of directors as a part of corporate governance practices prevents managers from making adjustments in financial reports in line with their interests and prevents financial reporting frauds and earnings management [15].

The corporate governance rules in the instructions have been updated in line with the new corporate governance principles issued by the Organization for Economic

Cooperation and Development (OECD) for the year 2015. The required reforms contained in the business report published by the World Bank Group contribute to raising the rating ranks in the report within the focus of protecting minority investors, which will enhance confidence in the investment climate in the market and the national economy because of the importance of the report at the global level.

In view of the many challenges that have emerged recently with regard to electronic disclosure, developing accounting systems and information technology is a must [16]. This has imposed on the authorities responsible for the accounting profession to reconsider international financial reporting standards to face these challenges, ensure the continuity of achieving confidence and credibility of financial statements and reports, and improve the quality of accounting information [14].

The information system function is to produce accounting information with a set of characteristics tò suit the needs of its users. The risks impact of electronic accounting information systems derives their importance as an element worth studying and analyzing from the importance of the information system for the firm [17]. Any defect in the information system or risks will undermine the chances of obtaining accounting information of adequate quality [15].

The quality of the accounting information system outputs of accounting information directly affects decision-making processes [18]. Different categories of financial statements users, whether internal or external parties, make their decisions based on accounting information [19]. The management use data to complete deals, distribute profits, request loans, or issue shares. The investors use accounting information to decide on the continued investment or knowing the firm's profitability. Banks use accounting information to decide on rejecting the loan or rescheduling it, as well as government institutions and regulatory agencies use financial statements to make decisions [20]. All these decisions depend entirely on accounting information of appropriate quality.

In the light of the significant progress that the world is witnessing in the technology and telecommunications sector in the era of the Internet of Things (IoT), remote sensing devices, digital machines, and smartphones have increased. They are present in many places. Thus, they contain software that can collect and store a considerable amount of data over networks, which must be processed and analyzed to enable decision-makers to make their decisions correctly. The essential things in these regards are the Internet systems that contribute to various applications that improve the quality of life.

Big data and IoT improve the quality of life, as they contribute to learning about the latest developments in all fields, which depends on the availability of different types of data, which is essential for researchers, companies, and other stakeholders [1, 21].

Big data is of great importance in all areas of life. Many studies have been conducted on Big data and IoT. IoT impacts the quality of Financial Reporting using Big Data techniques [22, 23]. However, IoT is the inter-networking of vehicles, physical devices, buildings, and other items embedded with electronics, software, sensors, actuators, and network connectivity, enabling these objects to collect and exchange data. According to Kumar and Santhosh Kumar [21], Big data techniques can handle many challenges that face data analysis and have the ability to extract valuable information. In the big data era, none of the existing storage systems works well while considering the three "V" features of big data (Velocity, Variety, and Volume). According to Wang [24], molecular profiling data size increases sharply due to new biological techniques. A new Key-Value data store like Apache Google Bigtablem and HBase can provide high-speed queries by the Key.

3 Research Problem and Hypotheses

There is growing attention to the significance of an effective information system providing timely and accurate information. Effective information systems benefit users of accounting information. One of the critical problems facing organizations nowadays is the enormous amount of data that passes to them and the pluralism of its forms and sources. And since the organizations are exposed to more data than others, the decisions should not be based on random events but rather be clear, correct, understandable, and ready to add value to users. Thus, the need for business intelligence to enhance the quality of financial reporting by making practical decisions builds on a continuous flow of quality information that is continuously updated. Therefore, the problem of the study is the role of electronic disclosure in light of the significant developments in the field of communication technology and information exchange, as one of the most critical challenges facing accounting presentation and disclosure at present. This prompted the researcher to shed light on the role of electronic disclosure in improving the quality of accounting information.

This study aims to enrich accounting literature related to IoT and business intelligence and underline areas for further research in this area. More specifically, this study seeks to investigate the potential impact of IoT on financial report quality. Therefore, this study assumes the following null hypotheses:

H0.1: IoT negatively affects the quality of financial reporting.
H0.2: Big data negatively affects the quality of financial reporting.

4 Methodology

A sample of data was obtained from the Jordanian Securities Commission (JSC). Databases were used to obtain the data and then analyze it using SPSS to test the research hypotheses.

Table 1 VIF of privacy dimensions

Variables	Tolerance	VIF
Internet of things (IoT)	0.556	1.799
Big data	0.556	1.799

4.1 *Collinearity Test*

Collinearity is a condition in which some of the independent variables are highly correlated. Collinearity tends to inflate the variance of at least one estimated regression coefficient. To ensure that there is no collinearity problem in the relationship between the independent variables, the variance inflation factor (VIF) was calculated. The variance inflation factor (VIF) statistically quantifies the severity of Multicollinearity in an ordinary least square regression analysis. It shows the measures of how much the variance of an estimated regression coefficient is increased because of collinearity.

According to Myers [20], the Value of VIF = (1/Tolerance) must be less than 10 (this means tolerance must be greater than (0.1), as shown in Table 1, the tolerance is (0.556) greater than (0.1) and the (VIF) is (1.799) less than 5. then there is no high relationship between dimensions of the independent variable. There is no collinearity problem in these variables.

4.2 *Hypotheses Testing*

To test the study hypotheses, the researchers used the SPSS program to find the effect of the Impact of IoT on the quality of financial reporting.

H.01: IoT Negatively Affects the Quality of Financial Reporting
To determine the relationship between IoT and the quality of financial reporting, the results in Table 2 show that the values of the correlation coefficient determination coefficient (R2) were (0.385) the standard for a good R-Squared reading can be much higher, more than (0.5) showing a high level of correlation. These results indicated a positive relationship between IoT and the quality of financial reporting. In contrast, the Beta (slope parameter in the regression model) for IoT was (0.511). When the beta coefficient is positive, the interpretation is that for every 1-unit increase in the predictor variable. The values of t-calculated (11.685) were significant at level (p-value. 0.05), then the null hypotheses are rejected to become, IoT positively affects the quality of financial reporting.

H0.2: Big Data Negatively Affects the Quality of Financial Reporting
To determine the relationship between dimensions of the big data and the quality of financial reporting, the results in Table 3 show that the values of the correlation coefficient determination coefficient (R2) were (0.351) less than (0.5). At the same

Table 2 Effect of the internet of things (IoT) on quality of financial reporting

	R	R^2	F	Sig*(F)	B	β	T	Sig*(T)
Constant	0.621[a]	0.385	136.528	0.000	1.981		11.564	0.000
The internet of things (IoT)					0.511	0.621	11.685	0.000
Independent variable: quality of financial reporting								

Table 3 Effect of big data on quality of financial reporting

	R	R^2	F	Sig*(F)	B	β	T	Sig*(T)
Constant	0.592	0.351	117. 866	0.000	2.291		14.685	0.000
					0.427	0.592	10.857	0.0
Independent variable: quality of financial reporting								

time, the value of Beta (slope parameter in the regression model) for big data was positive (0.427). The values of T calculated (10.857) were significant at level (p-value. 0.05), then the null hypotheses are rejected to become, big data positively affects the quality of financial reporting.

The study results are consistent with the study by Zraqat [23], who aimed to identify the impact of big data in terms of its dimensions (diversity, speed, volume, and honesty) on the quality of financial reports on business intelligence. The study results confirmed a statistically significant impact of using big data in improving the quality of financial reports.

5 Conclusion

Electronic disclosure is considered a new development in improving the quality of accounting information. Nevertheless, the problem of the study is the role of electronic disclosure in light of the significant developments in the field of communication technology and information exchange, as one of the most critical challenges facing accounting presentation and disclosure. In this study, a sample of big data was obtained and issued by the Jordanian Securities Commission (JSC). Databases were used to obtain the data and then analyze it using SPSS to test the research hypotheses. The analysis indicated a positive relationship between the internet of things (IoT) and the quality of financial reporting. In addition, this study found that big data significantly affects the quality of financial reporting. Consequently, the results of this study benefit accountants, auditors, and company managers in relying on the IoT and big data to improve the quality of financial reports, which has a significant impact on the company's continuity. The study recommends the importance of IoT and big data because of their significant role in enhancing the quality of financial reports.

References

1. Strohbach M et al (2015) Towards a big data analytics framework for IoT and smart city applications. In: Modeling and processing for next-generation big data technologies. Springer, pp 257–282
2. Peasnell KV, Pope PFF, Young S (2005) Board monitoring and earnings management: do outside directors influence abnormal accruals? J Bus Finance Account 32(7–8):1311–1346
3. Elseraiti AR (2011) Earnings management in the Libyan corporations (Doctoral dissertation, University of Gloucestershire)
4. Haloush HA, Alshurafat H, Alhusban AAA (2021) Auditors' civil liability towards clients under the Jordanian law: legal and auditing perspectives. J Gov Regul/Vol 10(1)
5. Al-Sartawi A, Reyad S, Madbouly A (2021) Shariah, presentation and content dimensions of Web 2.0 applications and the firm value of Islamic financial institutions in the GCC countries. J Islam Mark. Accepted article
6. Ananzeh H, Alshurafat H, Hussainey K (2021) do firm characteristics and ownership structure affect corporate philanthropic contributions in Jordan? J Financ Report Account. Forthcoming
7. Dahmash F et al (2021) The effect of a firm's internal factors on its profitability: evidence from Jordan
8. Olson D, Zoubi TA (2011) Efficiency and bank profitability in MENA countries. Emerg Mark Rev 12:94–110
9. Latif K, Bhatti AA, Raheman A (2017) Earnings quality: a missing link between corporate governance and firm value. Bus Econ Rev 9(2):255–280
10. Barghathi Y, Collison D, Crawford L (2017) Earnings management in Libyan commercial banks: perceptions of stakeholders. Int J Account Audit Perform Eval 13(2):123–149
11. Zaman M, Hudaib M, Haniffa R (2011) Corporate governance quality, audit fees and non-audit services fees. J Bus Financ Acc 38(1–2):165–197
12. Clark WW, Demirag I (2002) Enron: the failure of corporate governance. J Corp Citizsh 8:105–122
13. Ananzeh H et al (2022) The impact of corporate governance on forward-looking CSR disclosure. J Financ Report Account
14. Roy C, Debnath P (2015) Earnings management practices in financial reporting of public enterprises in India: an empirical test with M-Score. Available at SSRN 2551713
15. Sanad Z, Shiwakoti R, Al-Sartawi A (2022) Gender diversity on corporate boards and earnings management: a review. In: Musleh Al-Sartawi AMA (ed) Artificial intelligence for sustainable finance and sustainable technology, ICGER 2021. Lecture notes in networks and systems, vol 423. Springer, Cham
16. Sarea AM, Musleh Al-Sartawi A, Khalid AA (2018) Electronic financial disclosure: Islamic banking vs conventional banking in GCC. In: Conference on e-business, e-Services and e-Society. Springer
17. Abu-Musa AA (2006) Investigating the perceived threats of computerized accounting information systems in developing countries: an empirical study on Saudi organizations. J King Saud Univ-Comput Inf Sci 18:1–30
18. Sujud H, Hachem B (2019) Effect of the quality of the accounting information system outputs on customer satisfaction in Lebanese commercial banks. Int Res J Financ Econ 176:26–40
19. Nicolaou AI (2000) A contingency model of perceived effectiveness in accounting information systems: organizational coordination and control effects. Int J Account Inf Syst 1(2):91105
20. Susanto A (2015) What factors influence the quality of accounting information. Int J Appl Bus Econ Res 13(6):3995–4014
21. Musleh Al-Sartawi AM, Razzaque A, Kamal MM (eds) (2021) Artificial intelligence systems and the internet of things in the digital era, EAMMIS 2021. Lecture notes in networks and systems, vol 239. Springer, Cham

22. Rezaee Z, Wang J (2019) Relevance of big data to forensic accounting practice and education. Manag Audit J 34(3):268288
23. Zraqat OM (2020) The moderating role of business intelligence in the impact of big data on financial reports quality in Jordanian Telecom Companies. Mod Appl Sci 14(2):71–85
24. Wang X, Sun Y, Nanda S, Wang X (2019) Looking from the mirror: evaluating IoT device security through mobile companion apps. In: 28th USENIX security symposium (USENIX security 19), pp 1151–1167

The Effect of Organizational Virtuousness on the Organizational Citizenship Behavior: Evidence from Jordanian Health Sector.

Mohammad M. Taamneh, Lana M. Mhilan, and Manaf Al-Okaily

Abstract The purpose of this study was to determine the influence of organizational virtuousness, as defined by optimism, trust, empathy, integrity, and forgiveness on employees' organizational citizenship behaviors in Jordan's health sector. We test our proposed models in two big hospitals in Jordan's capital Amman. We analyze data using structural equation modeling (SEM) methods. The research sampled 338 personnel from a variety of medical, nursing, and ancillary medical specialties. The findings indicate that organizational virtuousness has a favorable effect on the health sector's organizational citizenship behavior.

Keywords Organizational virtuousness · Employees organizational citizenship behavior · Hospitals · Health sector · SEM · Jordan

1 Introduction

The health sector is beset by a slew of issues, including technology advancements and unique and non-traditional demands on the global health care system, all of which make it an appealing field for research. Many health institutions are facing greater pressure from globalization, changing funding mechanisms in the public health sector [1]. All of this necessitates attention to the human resources on the basis that healthcare institutions that depend on workers to perform official responsibilities (in-role) are frail and incapable of sustaining themselves for an extended period. Employees must do activities beyond their formal position (extra-role), which is referred to as organizational citizenship behavior, or OCB [2]. The factors that affect organizational citizenship behavior in health institutions always represent the vital core of interest in the field of organizational behavior studies. From this point of view and based on recent studies in this context, the organizational virtuousness variable plays a prominent role in achieving the characteristics of organizational citizenship behavior.

M. M. Taamneh · L. M. Mhilan · M. Al-Okaily (✉)
School of Business, Jadara University, 733, Irbid, Jordan
e-mail: m.alokaily@jadara.edu.jo

A. M. A. Musleh Al-Sartawi et al. (eds.), *From the Internet of Things to the Internet of Ideas: The Role of Artificial Intelligence*, Lecture Notes in Networks and Systems 557,
https://doi.org/10.1007/978-3-031-17746-0_31

The topic of organizational virtuousness is relatively recent, as interest in the concept of virtuousness emerged because of the moral and financial collapse that appeared in recent years in the business community which was discovered at the individual and organizational levels [3]. Most studies related to the concept of organizational virtuousness indicated that it is linked to the revival of the tradition of administrative ethics and virtue-based behavior. Moreover, it improves the lives of the organization's people and stakeholders, increases organizational efficiency and encourages employees to prepare to support the organization in times of crisis [4, 5]. One of the most important pillars of organizations are building human relations and organizational behaviors inside and outside the organization, which is called organizational citizenship behavior, as this concept reflects the organization's ability in its environmental and social interaction, and in its support for the values of compatibility within goals and interests, to raise the levels of efficiency and effectiveness in organizational performance. The concept of organizational citizenship behavior is one of the relatively recent philosophical concepts in administrative thought, especially in the Arab context in general and Jordan in particular. Many researchers agree that organizational citizenship behavior is a positive, voluntary behavior that distinguishes between the formal role versus the optional additional role of behavior [6].

More exactly, this conduct involves employees defending the company when it is criticized and assuming extra tasks that are not part of the employee's job description, as well as employees foregoing personal interests and perks. OCB adds to the organization's good ideals and aims through involvement, loyalty, and collaboration that are unrelated to the organization's formal rewards and incentive structure [7]. According to most of the research, organizational citizenship behavior is characterized by five distinct dimensions: altruism; politeness; honesty, loyalty, and belonging; and civic virtue and conscientiousness [8]. Most of the previous studies agree on the importance of the role of organizational virtuousness as one of the decisive factors that lead to the success of organizations and positively affect the job performance of employees, which in turn helps to increase the level of organizational citizenship behavior [6]. Previous studies expanded our knowledge of the factors that affect organizational citizenship behavior in organizations of all kinds.

Moreover, this conduct involves employees defending the company when it is criticized and assuming extra tasks that are not part of the employee's job description, as well as employees foregoing personal interests and perks. OCB adds to the organization's good ideals and aims through involvement, loyalty, and collaboration that are unrelated to the organization's formal rewards and incentive structure [7]. According to most of the research, organizational citizenship conduct is characterized by five distinct dimensions: altruism; politeness; honesty, loyalty, and belonging; and civic virtue and conscientiousness [8]. Recently, organizational virtuousness has received considerable attention among organizational behavior scholars who claim that this approach has a positive relationship to professional happiness, organizational citizenship behaviors and commitment among employees. To the best of the researchers' knowledge, no empirical research on these links in the Jordanian public health sector has been conducted. We contribute to the current literature in this research article by filling in the gaps stated above. Thus, the overall objective of this study is to evaluate the effect of organizational virtuousness on organizational citizenship behavior of employees in public health sector in Jordan.

2 Literature Review and Hypotheses Development

2.1 Organizational Virtuousness

The rise in moral and financial scandals that organizations have been exposed to in recent years is one of the key reasons for reassessing the function of virtuousness and pay more attention to it in an organizational context [9]. The concept of organizational virtuousness was first mentioned in by Cameron, a research leader at Michigan Business University. Organizational virtuousness is one of the most important concepts related to positive psychology, which is referred to through individual and group activities in the organization as well as organizational processes, cultural characteristics, and work ethics, which are related to the characteristics of organizational virtuousness. Organizational virtuousness has emerged as the most important indicators that indicate desirable behavioral patterns in institutions [10]. There are several definitions of the concept of organizational virtuousness according to a review of the theoretical literature. Zamahani et al. [11] define it as the pursuit of the highest possible status that an individual may attain or attain, with an emphasis on the strength or distinctiveness of moral virtue in social or personal interactions. Abedi [12] continued by defining organizational integrity as a collection of behaviors, motivations, and attitudes that promote personal and societal well-being.

As for the dimensions of organizational virtuousness, the views of researchers often vary in determining the dimensions of measurement for many variables through their contributions and research studies. This is compatible with the concept's recent discovery and the initiation of a study program into its substance in the last decade of the twenty-first century. Thus, relatively restricted perspectives emerged in departing from these dimensions, as Cameron et al. [13] referred to the most well-known model for evaluating organizational virtue as including five fundamental characteristics. (a) Optimism: It expresses the degree to which the organization's personnel believe they will be successful in completing their tasks when confronted with significant obstacles. Optimism is also seen as a positive idea in organizational behavior because it is a variable by which firms grapple with environmental changes to stay competitive [14], (b) trust; refers to the overall atmosphere within a business that is based on civility and mutual respect amongst its members; employees rely on one another, and senior management relies on them to accomplish the organization's goals [15]. Positive expectations about the functional efficiency of employees can be used to describe the concept of organizational trust; (c) forgiveness: Organizational forgiveness manifests itself in a regulated environment by allowing for the tolerance of errors and their utilization as learning opportunities. It is a common practice to substitute positivity for negative, which has several benefits, including increased morale, happiness, and learning [14]. Watanabe, [14] corroborates this pattern. In a nutshell, organizational forgiveness refers to the ability to forgive others inside an organization; (d) empathy; organizational empathy is defined as the ability of employees and management to cope with compassion and sympathy while doing the activities and obligations that are demanded of them, hence limiting the nature of the case and the amount of anxiety among workers. Organizational empathy [15] suggests that employees in a company care about one another; (e) integrity: refers to the shared values of honesty, good manners, and respect that exist in organizations as a result of employees' interactions with one another. Celik and Esen [16] described it as the responsibility to act in accordance with a set of moral beliefs and values. On a personal level, it is standing up for what is right, just, and fair.

2.2 *Organizational Citizenship Behavior*

The role of organizational citizenship behavior (OCB) is important. In an era of rapid change in public health institutions and work environments, the behavior of employees plays a critical role in their success by activating different practices in the workplace, strengthening social ties in organizations, and influencing employee behaviors related to organizational functions and outcomes [17, 18]. Because organizational citizenship is based on a bond of understanding and cooperation rather than a contractual relationship of interest, it is based on trust, honesty, and cooperation between employees and the organization. It also depends on the extent of the employee's attachment to the organization. To undertake unpaid volunteer work and activities, such as promoting and supporting others and initiating work without an assignment or the expectation of receiving incentives or benefits. Organizational citizenship behavior is defined by Kabir and Ali [19] as voluntary behavior that stems from an individual's will, exceeds official work requirements, is not included in the organization's incentive system, leaves a lasting impression on superiors and coworkers, and contributes to the organization's efficient performance. The value of organizational citizenship behavior stems from its contributions to attracting, investing, and directing human and material resources to achieve goals efficiently and effectively. It also contributes to boosting the overall performance of the organization by minimizing resistance to change and promptly adjusting to changes in the surrounding environment.

Although opinions on the elements of organizational citizenship behavior ranged, past research and academic literature in this area point in two ways. The first is the individual dimension, which expresses an individual's action aimed at assisting another individual or group of individuals rather than the organization. The second component, referred to as the organizational dimension, refers to the conduct that an individual engages in in order to benefit the organization as a whole such as making constructive suggestions that contribute to the development of performance, no absenteeism, and self-development [20]. According to most previous studies, organizational citizenship conduct has five main dimensions [21]: (a) altruism, which refers to an employee's willingness to assist colleagues, superiors, and subordinates in completing their tasks; (b) civility, which refers to an employee's effort to avoid causing problems and disputes at work, as well as providing necessary advice and consultation with Others when necessary; (c) sportsmanship, which refers to an individual's readiness to labor in inhumane working conditions, forgo gains, be patient with the organization's crises, and resist competing organizations' temptations. (d) civic virtue, which includes effective employee engagement, adherence to rules and regulations, and environmental stewardship;(e) Consciousness, it refers to voluntary conduct and willingness to take on additional responsibilities, as well as making recommendations that aid in the development and improvement of work.

2.3 Organizational Virtuousness and Organizational Citizenship Behavior

One of the administrative techniques that supports the management profession's ethics is organizational virtuousness. Organizational virtuousness fosters optimism among leaders and employees in today's workplaces, as well as increased organizational citizenship behaviors, high job satisfaction, and a sense of happiness and well-being [14]. It also inspires individuals to put forth more effort to do their professional duties and obligations with distinction, as well as to be more resolute and courageous when confronted with challenges, issues, and pressures [22]. According to Singh and Singh's [23] findings, there is a positive association between organizational virtuousness and professional happiness in Indian educational institutions, which boosts employees' attachment to and adherence to their jobs, and job involvement, as demonstrated by Marzouk et al. [9] study of Egyptian health sector employees. According to a study conducted by Gali and Amana [24], organizational virtuousness modifies and enhances the influence relationship between perceived organizational support and organizational citizenship behavior in Iraqi cement factories. To examine the relationship between organizational virtuousness and organizational citizenship behavior directly via the mediating role of organizational support in the Pakistani banking sector; the study discovered a significant positive relationship between organizational virtuousness and organizational citizenship behavior [25].

Employees' knowledge of organizational virtuousness encourages them to adopt greater organizational citizenship behavior practices. This can be explained by the fact that individuals generally operate in accordance with the organization's value system. As a result, we may anticipate the employee assisting his or her colleague and supervisor if the employee believes his or her job is valued by the business. It's worth emphasizing that, within this conceptual framework, the employee will be expected to do similarly if he believes his firm is virtuous. Then, the employees develop a favorable picture of their organization, strengthening their connection with it and fostering mutual trust among the organization's members. Numerous research has established that organizational virtuousness has an effect on organizational citizenship behaviors, either directly or indirectly [6]. On the basis of the foregoing debate, we propose the following hypothesis:

Main Hypothesis: Organizational virtuousness positively affects organizational citizenship behaviour.

3 Methodology

The study used a hypothetic-deductive technique with a structured design and a survey as an instrument to assess self-reported organizational virtuousness and organizational citizenship behavior among staff working in Jordanian public hospitals. Researchers can investigate pre-formulated hypothesis and test it using quantitative methods utilizing the hypothetic-deductive research strategy [26–29]. Because this is a quantitative study on the impact of organizational virtuousness on organizational citizenship behavior, the data has been collected using a closed-ended survey instrument., using a 5-point Likert scale with values ranging from 1 to 5. We employ well -developed items from earlier research to measure proposed factors in this study (the sources are described below). The researchers have selected the convenience sampling technique, participants being chosen based on their availability and accessibility. A group of medical, nursing, and auxiliary medical specializations consist of 400 employees were selected as a sample from the population that include two major hospitals in Amman's capital city. 350 questionnaires has been retrieved; 12 questionnaires were found to be not valid for statistical analysis, thus the study sample consisting of 338 participants.

The survey used in this study has three sections. the first section asked for demographic information. The second section of the survey (28 items) was designed to assess the participants' level of organizational virtuousness. The scale was comprised of five subscales measuring respect for optimism, trust, empathy, integrity, and forgiveness all of which were adopted from various studies (24–30) The survey's third component (39 items) includes four subscales that examine organizational citizenship behavior, including altruism, conscience, sportsmanship, courteousness, and civility, adopted from Malik, Zehra, [25]; Asad et al., [31] and changed to fit the setting of research, medical heath institutions. The survey's second and third portions each asked participants to score their agreement with specific assertions on a five-point Likert scale, with responses ranging from 5 to 1, with 5 indicating strong agreement and 1 indicating strong disagreement. The initial draft of the scale utilized in this study was examined by a panel of professors from several Jordanian universities, and their comments were incorporated into the development process. With a Cronbach alpha coefficient of 0.813 for organizational virtuousness and 0.802 for OCB, the final version of the survey was deemed to be reliable.

4 Data Analysis

Table 1 summarizes the demographic factors. According to the table, about 55% of participants were male, and 38% held administrative jobs and 34% held technical positions.

According to the table, 61% of respondents possess a bachelor's degree. Additionally, the data indicates that about 88 percent of participants had less than 15 years' experience, while others with more than 15 years' experience make up only 12% of the sample.

Prior to completing the linear regression analysis used to verify the study's hypothesis, several criteria and requirements must be satisfied to assure the regression analysis's safety and validity. The table below displays the normal distribution's findings. The Table 2 demonstrates that, based on the Shapiro–Wilk test results, all research variables had a normal distribution.

Following a discussion of the normal distribution, it is necessary to consider the possibility of outliers and the necessity of excluding them. The researchers used the (Mahalanops) test to identify and exclude anomalies from their statistical analysis. Following the test, it was discovered that twenty-five questionnaires were not

Table 1 Distribution of study sample members according to personal data

Variable	Category	Frequency	Percentage
Gender	Male	186	55.0
	Female	152	45.0
	Total	338	100.0
Job title	General Director/Deputy Director/Assistant Director	12	3.6
	Department Manager	30	8.8
	Head of Department	54	15.4
	Employee Administrative	129	38.1
	Technical employee	115	34.0
	Total	338	100.0
Education	Diploma (community college)	33	9.8
	Bachelor's	207	61.2
	Master's	95	28.1
	PhD	3	9
	Total	338	100.0
No. of years of service	Less than 5 years	118	34.9
	5 years - less than 10 years	96	28.4
	10 Less than 15 years	84	24.9
	More than 15 years	40	11.8
	Total	338	100.0

Table 2 Testing the normal distribution using Shapiro–Wilk test

Dimension	Value test	Significance level
Organizational optimism	0.91	0.16
Organizational trust	0.96	0.22
Organizational empathy	0.89	0.36
Organizational virtuousness	0.94	0.512
Organizational forgiveness	0.91	0.36

distributed normally. Because the remaining questionnaires have a normal distribution and the percentage of outliers is less than 10%, there is no reason to exclude them from the statistical analysis because they have no bearing on the accuracy of the results. The Variance Inflation Factor (VIF) and the allowable variance values (Tolerance) were used to ensure that the data were free of the problem of linear coupling between the independent variables, a problem that exists when an independent variable is a function of another independent variable, rising with its height and falling with its decline. The findings are summarized in Table 3.

The Pearson correlation coefficient was extracted between the study variables to identify the nature of the correlation between the variables. The results are shown in Table 4. Pearson's correlation coefficient was utilized to ensure that there was no significant association between the research variables. According to the accompanying table, the independent variables are moderately linked, with correlation coefficients ranging from 0. (0.048 - 0.604). It should not surpass 85 percent to guarantee that no meaningful relationship exists between the research variables [32].

To verify our hypothesis, we build a structural model in which a direct relationship between organizational virtue and organizational citizenship behavior is established. AMOS, a software for estimating and verifying structural equation models, was used to build the proposed model (SEMs). Five goodness of fit indices were used to determine the model's quality, as recommended by Anderson and Gerbing [33], namely the Chi-square/degrees of freedom (CMIN/DF), the goodness of fit index (GFI), the adjusted goodness of fit indices (AGFI), the comparative fit index (CFI), and the root mean square error of approximation (RMSEA). The findings indicate that the model fits are plausible and acceptable, suggesting the model's validity for hypothesis test (Tables 5 and 6).

Table 3 Linear duplications Test results between variables study

Variables	Inflation variation coefficient	Variation allowed
Organizational optimism	0.518	1.931
Organizational trust	0.552	1.813
Organizational empathy	0.569	1.757
Virtuousness	0.742	1.348
Organizational forgiveness	0.798	1.253

Table 4 Correlation matrix between study variables

No	Variables variable	1	2	3	4	5
1	Organizational optimism	1				
2	Regulatory trust	0.604**	1			
3	Organizational empathy	0.575**	0.565**	1		
4	Organizational virtuousness	−0.323**	−0.124*	−0.259**	1	
5	Organizational forgiveness	0.144*	0.048	0.221**	−0.425**	1

*Correlation coefficient is statistically significant at a significance level of 0.05
**Correlation coefficient is statistically significant at a significance level of 0.01

Table 5 Quality indicators of conformity of the study model

Indicator	Acceptance criterion	Value	Result
Chi-square ratio (CMIN/DF)	Less than 2	1.82	**Acceptance**
The quality of conformance index (GFI)	Greater than 0.90	0.951	**Acceptance**
The corrected quality of conformity index (AGFI)	Greater than 0.90	0.935	**Acceptance**
CFI	Greater than 0.90	0.921	**Acceptance**
Root squared error (RMSEA)	Less than 0.08	0.04	**Acceptance**
CMIN/DF: Chi-square/degrees of freedom, GFI: Goodness of Fit Index, AGFI: Adjusted Goodness of Fit Index, CFI: The comparative fit index, RMSEA: root mean square error of approximation			

Table 6 The result of testing the effect of dimensions of organizational virtuousness on organizational citizenship behavior

Path			Beta value (ß)	Value (T)	Significance level	Beta value (ß)
Organizational citizenship behavior	→	Organizational optimism	0.14	2.25	0.03	0.14
Citizenship behavior	→	Organizational trust	0.45	7.60	0.00	0.45
Citizenship behavior	→	Organizational empathy	−0.05	−0.79	0.43	−0.05
Citizenship behavior	→	Organizational integrity	0.09	1.76	0.08	0.09
Organizational citizenship behavior	→	Organizational forgiveness	0.29	5.90	0.00	0.29
Interpretation Factor (R^2): 0.589, value (P): 35.181, significance level: 0.000						

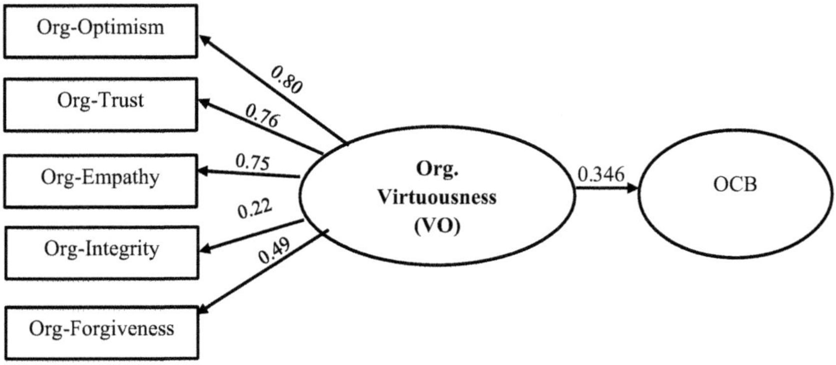

Fig. 1 Research model

To test the study's central hypothesis, the AMOS software was utilized to determine the influence of organizational virtuousness on organizational citizenship behavior among employees in Jordan's health sector. This is explained in the table 6

As indicated in the preceding Table 6 and Fig. 1, there is a statistically significant effect of organizational integrity dimensions (optimism, trust, empathy, integrity, and forgiveness) on the organizational citizenship behavior of health sector employees in Jordan, as measured by the value of the regression coefficient (B), which equaled (0 0.346) with a significance level less than (0.05). The coefficient of determination has the value R (0.589). Additionally, the correlation coefficients between the dimensions of organizational integrity and the field appear to vary between (0.22–0,80), indicating that the independent variables are connected to the dependent variable of organizational virtuousness as a whole.

5 Discussion

The purpose of this study is to determine the effect of organizational virtuousness, as defined by optimism, trust, empathy, integrity, and forgiveness on the organizational citizenship behavior of employees in the health sector in Jordan. Our study focused on staff at health sector institutions in Amman, Jordan's capital, specifically on two large hospitals (AL Basheer hospital and Jordan university hospital), which represent two distinct subsectors: the first is affiliated with the Ministry of Health, while the second is affiliated with university teaching hospitals. Regression analysis was performed on 338 valid data. We examined the offered hypothesis using PLS-SEM. Our main effect model indicates that organizational virtuousness is favorably associated with OCB in Jordan's public health institutions. The study's findings indicated that the characteristics of organizational virtuousness (optimism, trust, empathy, integrity, and forgiveness) had a statistically significant influence on the organizational citizenship behavior of employees in Jordan's health sector. The regression coefficient is equal to (0.589), indicating that the organizational virtuousness variable can account for 58.9 percent of the variation in organizational citizenship behavior, and the correlation coefficient is equal to (0.34), indicating that there is a direct relationship between organizational virtuousness and organizational citizenship behavior.The organizational virtuousness variable showed a positive and significant influence on all dimensions of organizational citizenship behavior, with the exception of the empathy component, which had a negative impact on organizational citizenship behavior. This can be explained by the fact that empathic behavior may send a negative message to subordinate staff, resulting in relaxation and potentially ignorance. However, organizational virtuousness is often associated with a favorable effect on organizational citizenship behavior, which corroborates the findings of several earlier studies [22, 30, 34, 35]. However, several research examined the effect of organizational virtuousness on organizational citizenship behavior via intermediary factors such as job satisfaction and well-being, demonstrating that organizational virtuousness has an effect on employee behavior when these variables are present [6]. While several prior studies in various contexts corroborated the findings of this investigation, the researchers were unable to locate any comparable study in the health sector in a country bordering Jordan, except for one that was most pertinent to the Jordanian setting, particularly Iraq [24].

To describe the conditions surrounding the link and the influence of organizational virtuousness on organizational citizenship behavior, as well as to comprehend the reasons for this effect, a theoretical framework of organizational citizenship behavior causes may be used. There is little question that several variables contribute to organizational citizenship behavior, some of which are personal, work-related, and others organizational. According to several studies, religious influences are one of the most significant [36], if we accept that the components of organizational virtue, which include optimism, empathy, honesty, trust, and forgiveness, are a true translation of the concept of piety [37]. Khalid et al. [37] even stated that the relationship between organizational virtuousness, or what can be boiled down to the concept of piety, and OCB transcends correlation and approaches the level of causality. Organizational culture is a collection of the ideas, values, and standards adhered to by the members of the company. This culture dictates how employees behave within businesses and how they conform to established ideals, conventions, and rules [38]. It is essential to analyze the religious beliefs of the population in which Jordanian hospitals operate at this stage. Religion and culture influence people's value systems, directing their conduct and how they treat one another [39, 40]. According to Dana [40], business behavior is intimately tied to cultural convictions and behavior norms imposed by culture and religion laws in some regions of the globe where religious and social viewpoints are strong, such as Japan and the Middle East - including Jordan. Finally, our study corroborates the findings of refers [41, 42], which found that administrators shown a high level of commitment to integrity in their HRM practices in Jordanian universities.

6 Managerial Issues

The study's findings indicated that the characteristics of organizational virtuousness (optimism, trust, empathy, integrity, and forgiveness) had a statistically significant influence on the organizational citizenship behavior of employees in Jordan's health sector. Given the substantial influence of organizational integrity on the organizational citizenship behaviors of employees in Jordan's health sector, managers at all levels should endeavor to maintain an optimistic attitude and to build trust between supervisors and their subordinates. It is laudable to substitute intimidation for trust-based management. Improved integrity (transparency) in relationships between superiors and subordinates is one of the concerns that must be addressed. Additionally, providing monetary and moral incentives to employees based on their moral values and views, as well as establishing a deterrent policy for those who violate ethical norms in interacting with subordinates, is advised. Promoting and cultivating a culture of tolerance among workers helps to the development of a spirit of innovation and experimentation as corporate citizenship behaviors. While establishing the conditions and requirements for organizational citizenship behavior is an administrative objective in and of itself, the consequences of its prevalence contribute to the enhancement of numerous positive administrative phenomena such as team building, decreased employee turnover, increased product or service quality, and increased client satisfaction.

7 Limitation and Future Research

The first research limitation is that it focused exclusively on the public health sector, ignoring the private health sector. This may be a partial instance for Jordan's health industry in this setting. As a result, it is advised that future research extend the study to health institutions in both the public and private sectors to increase the credibility of the generalization process. The current research improved our understanding of organizational virtuousness, its dimensions, and its relationship to and influence on organizational citizenship behavior. According to the result of this study, researchers recommend that the study be used in other sectors such as education and industry. While the current research focuses on the influence of organizational virtuousness on organizational citizenship behavior, future research may examine the effect of organizational virtuousness on other variables such as organizational commitment, organizational justice, and social capital. Finally, it is necessary to create an organizational environment that enhances the capabilities of employees and to ensure an increase in the level of organizational citizenship behavior among employees.

References

1. Alsaad A, Al-Okaily M (2022) Acceptance of protection technology in a time of fear: the case of Covid-19 exposure detection apps. Inf Technol People 35:1116–1135. https://doi.org/10.1108/ITP-10-2020-0719
2. Nikandrou I, Tsachouridi I (2015) Towards a better understanding of the "'buffering effects"' of organizational Virtuousness Perceptions on employee outccomes. Manag Decis 53(8):1823–1842
3. Lilius JM, Worline MC, Maitlis S, Kanov J, Dutton JE, Frost P (2008) The contours and consequences of compassion at work. J Organ Behav 29:193–218
4. Al-Okaily M (2021) Assessing the effectiveness of accounting information systems in the era of COVID-19 pandemic. VINE J Inf Knowl Manag Syst. https://doi.org/10.1108/VJIKMS-08-2021-0148
5. Al-Okaily A, Al-Okaily M, Teoh AP (2021) Evaluating ERP systems success: evidence from Jordanian firms in the age of the digital business. VINE J Inf Knowl Manag Syst. https://doi.org/10.1108/VJIKMS-04-2021-0061
6. Rego A, Ribeiro CM (2010) Perception of organizational virtuousness and happiness as predictor of OCB. J Bus Ethics 47(6):790–2004
7. Marashdeh A (2015) Employees performance appraisal and its impact on organizational citizenship behavior in irbid electric power company. Unpublished Master Thesis, Yarmouk University, Irbid -Jordan.
8. Tayeh A, Kareem B (2012) The effect of organizational justice on organizational citizenship behavior at central ministries in Jordan. J Islamic Univ Econ Adm Stud 20(2):145–186
9. Marzouk A-A, Al-Sindiouni MM, Al-Amawi BF (2019) The effect of organizational virtuousness on job engagement: an applied study on the nursing staff in government hospitals in Kafr El-Sheikh Governorate. J Contemp Bus Stud 1(6):185–222
10. Magnier- Watanabe R, Uchida T, Orsini P, Benton C (2017) Organizational virtuousness and job performance in japan: does happiness matter. Int J Organ Anal 4(25):628–646
11. Zamahani M, Ali A, Mohammad A, Hamideh S (2012) Moving towards virtuousness in organization: an analysis of factors affecting virtuous organizations. Manag Sci Lett 2(8):2889–2902
12. Abedi S, Hoseyn E, Nahid A (2014) The role of organizational virtuousness on the organizational commitment of employees (case study: Islamic Azad University). Interdisc J Contemp Res Bus 5(10):118–130
13. Cameron K, Bright D, Caza A (2004) Exploring the relationships between organizational virtuousness and performance. Am Behav Scientist 47(6):1–24
14. Watanabe R, Uchida T, Orsini P, Benton C (2016) Organizational Virtuousness and job performance in Japan: does happiness matter? Int J Organ Anal
15. Hurt YS, Rhee S-Y, Kim H (2017) Organizational virtuousness perceptions and task crafting: the Chinese airline industry. Gend Manag Int J Manag Paycholohy 27(4):330–346
16. Celik A, Esen E (2017) The moderating role psychological wellbeing: organizational virtuousness and work engagement. J Int Social Res 10(50):668–674
17. Al-Okaily A, Al-Okaily M, Teoh AP, Al-Debei M (2022) An empirical study on data warehouse systems effectiveness: the case of Jordanian banks in the business intelligence era. EuroMed J Bus. https://doi.org/10.1108/EMJB-01-2022-0011
18. Al-Okaily M, Al-Okaily A (2022) An empirical assessment of enterprise information systems success in a developing country: the Jordanian experience. TQM J. https://doi.org/10.1108/TQM-09-2021-0267
19. Kabir O, Ali A-R (2019) The effect of the organizational climate on the behavior of organizational citizenship among the employees of: Saidal Foundation - Annaba. Econ Visions J 9(2):226–242
20. Al-Sartawi A (2017) The effect of the electronic financial reporting on the market value added of the islamic banks in gulf cooperation council countries. In: 8th global islamic marketing conference, 4–6 May. International Islamic Marketing Association, Turkey

21. Polat S (2009) Organizational citizenship behavior display levels of the teachers at secondary schools according to the perception of the school administrators. Procedia Social Behav Sci 1(1):1591–1596
22. Sun H, Yoon H (2020) Linking organizational virtuousness, engagement, and OCB: the moderating role of individual and organizational factors. J Hosp Tour Res 46:879–904
23. Singh S, Singh A (2018) Interplay of organizational justice, psychological empowerment, m OCB and job satisfaction in the cotext of circular economy. Manag Decis 57(3):937–952
24. Hannoon A, Abdalla Y, Al-Sartawi A, Khalid A (2021) Board of directors composition and social media financial disclosure: the case of the United Arab Emirates. In: Musleh Al-Sartawi AMA (ed) The big data-driven digital economy: artificial and computational intelligence. Studies in computational intelligence, vol 974. Springer, Cham
25. Malik S, Zahra N (2016) Organizational virtuousness, perceived organizational support and organizational citizenship behavior: a mediation framework. J Behav Sci 26(1):113–129
26. Al-Okaily M, Alghazzawi R, Alkhwaldi AF, Al-Okaily A (2022) The effect of digital accounting systems on the decision-making quality in the banking industry sector: a mediated-moderated model. Glob Knowl Memory Commun. https://doi.org/10.1108/GKMC-01-2022-0015
27. Al-Bashayreh M, Almajali D, Altamimi A, Masa'deh RE, Al-Okaily M (2022) An EMPIRICAL INVESTIGATION OF REASONS INFLUENCING STUDENT ACCEPTANCE AND REJECTION OF MOBILE LEARNING APPS USAGE. Sustainability 14(7):4325
28. Al-Okaily A, Ai Ping T, Al-Okaily M (2021) Towards business intelligence success measurement in an organization: a conceptual study. J Syst Manag Sci 11(2):155–170. https://doi.org/10.33168/JSMS.2021.0210
29. Taamneh AM, Taamneh M, Alsaad A, Al-Okaily M (2021) Talent management and academic context: a comparative study of public and private universities. EuroMed J Bus. https://doi.org/10.1108/EMJB-08-2020-0088
30. Aslan S, Guzel S (2020) Job crafting, work engagement, organizational virtuousness, emotional self- efficacy and emotional intelligence differences of variables in socio-demographic. Ekev Acad Rev 24(82):435–462
31. Asad H, Naseem R, Faiz R (2017) Mediating effect of ethical climate between organizational virtuousness and job satisfaction. J Commer Social Sci 11(1):35–48
32. Al-Sartawi A (2021) The big data-driven digital economy: artificial and computational intelligence. In: Janusz K (ed) Studies in computational intelligence, vol 974. Springer, Heidelberg
33. Anderson JC, Gerbing DW (1988) Structural Equation modeling in practice: a review and recommended two-step approach. Sychol Bull 103(3):411–4213
34. Sania ZM, Ramsha N (2016) Organizational virtuousness perceived organizational support and organizational citizenship behavior: a mediation framework. J Behav Sci 26
35. Kooshki S, Zeinabadi H (2016) Organizational virtuousness in schools: a review of factor structure of a tool in secondary schools. J Adm Manag Educ Train (JAMET) 12(2):60–69
36. Abbasi AS, Rana AH (2012) Impact of islamic work ethics, reward system and Organizational environment on citizenship behavior of Employees. Sci Int (Lahore) 24(4):513–519
37. Khalid SA, Abdul Rahman N, Sintha Madar AR, Ismail M (2013) Undergraduates' organizational citizenship behavior: the role of religiosity. Int J Acad Res Bus Social Sci 39(7):572–584
38. Caputo A, Lombardi R, Akeel F, Almallah H, Dakkak B, Qubbaj N (2016) Youth employment in startup ventures in Jordan: an exploratory study. Int J Entrep Small Bus 28(4):468–491
39. Danley J, Harrick E, Stricklands D, Sullivan G (1991) HR situations. HRM 26(1):1–12
40. Musleh Al-Sartawi AMA (eds) (2022) Artificial intelligence for sustainable finance and sustainable technology. ICGER 2021. Lecture Notes in Networks and Systems, vol 423. Springer, Cham

41. Taamneh M, Athamneh S, Nsairat B (2017) The effect of administrators' work-related ethics on practicing human resource management functions at Jordanian universities. Int J Public Sector Perf Manag 3(3):337–354
42. Al-Okaily M, Alqudah H, Al-Qudah AA, Alkhwaldi AF (2022) Examining the Critical factors of computer-assisted audit tools and techniques adoption in the post-COVID-19 period: internal auditors perspective. VINE J Inf Knowl Manag Syst. https://doi.org/10.1108/VJIKMS-12-2021-0311.

The Effect of Information Sensitivity and Usage on Consumer Privacy: The Case of Electronic Payment Systems

Malak Mohammad Ghaith, Mohammad Abdalkarim Alzuod, and Manaf Al-Okaily

Abstract Electronic payment (e-payment) systems are considered a good advancement in the fields of business banking and commerce. In Jordan eFWATEERcom is the country's e-payment system. The study emphasis on examining how the factors affected consumer privacy (CP) by applying Information Sensitivity (IS), Information Usage (IU), also in using smart electronic devices for e-payment through eFWATEERcom website in Jordan. A questionnaire-based methodology was utilized to asses and clarify Jordanian's privacy on using eFWATEERcom. In means of examining such effects, a questionnaire of 33 items was constructed to measures study's variables; and the valid questionnaires for analysis are 324. The collected data was used through Statistical Package for Social Sciences (SPSS). The findings indicated that (IS) and (IU), are key predictors of CP to adopt the use of eFWATEERcom, at $\alpha \leq 0.05$, the analytical results of the study showed that factors; IS significantly impact on CP while IU had little negligible effect. On the basis of the results, it is recommended that future studies and researches could focus on other factors that could have an influence on the CP such as marketing efforts and new strategies that helps the online billing services serve more people that still using the traditional way of paying bills.

Keywords Electronic payment · Information sensitivity · Information usage · Privacy · Jordan

M. M. Ghaith
Luminus Technical University College (LTUC), Amman, Jordan

M. A. Alzuod
Faculty of Business, Zarqa University, Zarqa, Jordan

M. Al-Okaily (✉)
School of Business, Jadara University, Irbid 733, Jordan
e-mail: m.alokaily@jadara.edu.jo

A. M. A. Musleh Al-Sartawi et al. (eds.), *From the Internet of Things to the Internet of Ideas: The Role of Artificial Intelligence*, Lecture Notes in Networks and Systems 557,
https://doi.org/10.1007/978-3-031-17746-0_32

1 Introduction

The Internet emerged to the public by year 2001; such a change has produced many considerations and opportunities for the businesses. Dotcom and internet players have been grappling for survival as a majority of them continues to incur losses. Thus, such dilemma caused in huge academic debates regarding these new technologies: the debates have continued to fret about whether the Internet can cause some radical changes in the ways that people conduct their business or if it is just an evolutionary tool to offer a new communication and distribution channels [1].

For Jordan, [2] reported that about 24% of Internet users in Jordan use the e-commerce systems to pay bills and buy products in 2011 with an increase of 15.4% (approximately, 514,000 additional e-commerce users) from the preceding year. The penetration rate in this business to consumer (B2C) trading also has a noticeable rate of about 8.2% and revenues of $ 370 million in the same year. The e-commerce businesses still immature and infancy since the retail chains in Jordan are inadequately opened to the online channels. From the consumer perspective, though, the adoption of e-commerce is still relatively meagre, many studies have demonstrated that various people are expected to try it soon. The anticipated continued growth of online shopping via the Internet will allow limitless opportunities to market as well [1].

[3] define privacy as taking controls of information about ourselves. Before that, [4] define privacy as "legal right in the modern sense". Despite the definition, no one can contradict that the privacy concept is now part of information and communication technologies as it is applied in all electronic services and social media [3]. In other words, E-services and social media network usually require their users to use actual names and share personal and accurate information as sharing is the core of such systems and networks [3].

According [5] the ICT sector has grown significantly in the last five years in Jordan, as the number of Internet users in Jordan is 8.7 million, of whom 3.9 million used e-services [5].

The researcher is going to choose one of the most e-services companies in Jordan is eFWATEERcom that implement a leading system in financial services technology-based in Amman, Jordan. Since 2011, their solutions enable the consumers to inquire and pay their bills ease, comfort and fast manner regardless of the geographical location through a variety of payment channels that include ATMs, Tellers, Mobile and online banking services.

2 Literature Review

2.1 Information Sensitivity

[6] contribute by a new issue associated with privacy concerns which are the information sensitivity queried by the marketers, which may shrink the trust and increase the risks.

Nevertheless, these scholars do not discuss whether any request for extra sensitive data results in heightened consumers concerns about criminal information practices. For example, consumers are likely to be more fretful when requested to provide their names and income than their brand preferences. Hence, the more sensitive information, the more privacy concerns, which is moderating the effect of information sensitivity on the influence of privacy concerns on the trust and the perceived risks in the context of mobile advertising businesses [6].

[7] found that internet is dictated by free websites which are depending on the advertising as a source of revenues to survive; these websites would desire to magnify the advertising effectiveness and relevance rationally.

Targeted online advertising defined as "any form of online advertising that is based on information the advertiser has about the advertising recipient. Common examples are; demographics, current or past browsing or purchase behavior, information from preference surveys, and geographic information". Because of its decisive impact on revenues, targeted online advertising has appeared as a significant marketing trend [7].

However, it is still negotiated that the website must promise their user to publish more material advertisement based on more personal data. Such a practical argument "benefit appeal" is less effective as the practical benefits have limited impacts on the users' willingness to disclose their personal information if they are sensitive information [7].

2.2 Information Usage

[8] endorse the application of collaborative filtering techniques to personalize the quality of service (QoS) prediction. These approaches achieve high precision of QoS forecasts on the unobserved ones. Still, the obligation to get users' QoS data possibly put the users' privacy at danger and making them resistant to provide their data to a Web-based system. So, privacy becomes a crucial challenge in developing effective Web service systems. He proposed simple yet powerful privacy data obfuscation techniques from based on a publicly-available QoS dataset [8].

Generally, consumers are aware that all personal information is being gathered. In case, a Web site may ask that guest to access the location, and they might give such permission or decline to access the requested information. In all cases, they can't prevent the usage of such information after the approval [9]. If the information is only used for the transaction, consumers will not worry about their privacy. Nevertheless, if the companies use the information beyond the fundamental purpose, consumers will be concerned with privacy [9].

[9] addressed the feelings of lack of authority over how businesses manage their personal information and discover that the marketers ought to sell data as another vital problem. He perceived that inadequate knowledge of usage of information will cause severe complaints among consumers. Internet users typically had designated that they will be further willing to consider providing data when the sites explicitly notified them how such information is going to be used or shared with other organizations or by online disclosure [9].

2.3 Consumer Privacy

[10] shed light on how reliable personal identification is a critical determinant for business activities. Personal information means that "marketers are better able to identify the best scenarios, create promotions and reward programs that build customer loyalty, customize advertising and promotion strategies, and evaluate the effectiveness and cost-efficiency of advertising and promotions", also, technologies facilitate automatic and reliable personal recognition to enhance opportunities for the marketers by linking individuals with all personal activities. However, it causes many concerns about violating consumers' privacy and rights. Accordingly, [10] define the privacy concept as "a range from the popular conception of the right to be let alone, to the right to control information of oneself". Thus, consumer privacy is another face of control over all personal data.

The social contract theory stated that all marketers must observe consumers' exchange of personal data as a social contract. So, not reporting the consumer about information gathering or selling consumers' private information to any third party without authorization is a breach to this social contract. Conclusively, this violation of privacy may harm the association between marketers and their consumers [10].

[11] assume that consumer privacy concerns mobile advertising is also a mattering point for policymakers, and consumer advocates. For instance, the United States spam and unsolicited communications rose by 38% from the years 2006 till 2007, and they increased by 50% (1.5 million messages) in the year of 2008. Some of these spam messages ask to provide private information or try to infiltrate mobile devices with virus applications.

The mobile industry has implemented numerous self-regulations. For instance, the Mobile Marketing Association prevented the promotional contents for children unless there is a clarified free guideline [11]. Also, The U.S. Federal Trade Commission advocates the need for exploring issues of mobile privacy and consumer protection [11].

Japan, as well as assess the suitability of different methods to improve the extents of regulatory controls as this influence both government agencies and mobile service providers [11].

2.4 The Relationship Between Information Sensitivity and Consumer Privacy

[12] contributes to the knowledge about privacy as a basic to human rights. Also, [12] and [13] stated that the privacy is developed over hundreds of countries legal systems [14–17].

Further insight can be obtained by examining the privacy expectations of customers or nations to their real struggles when businesses process their own information. This change would provide for the classification of gaps, which would assist the improvements in the safeguarding of private information to build up a trustful relationship. It would be useful if the privacy concepts measured in this design that is aligned with best practice principles of privacy like those practices introduced in the Fair Information Practice Principles [12].

2.5 The Relationship Between Information Usage and Consumer Privacy

[18] found that the problem of information privacy has been appealing with 25% of Americans who consider themselves as victims due to their data privacy. Also, a total of 33.6 million Americans' data has been used for deceptive purposes in the last 30 years, about 75% of customers' report matters about the healthcare firms that sharing private medical data with third parties [18].

However, the literature is filled with studies on data privacy in medical and business transactional contexts. However, insufficient attention is devoted to examining information privacy interests with social networking when users are allowed options to reveal personal information. Besides, in social media networks, users have the options to use any privacy measures to guard their information privacy [18].

Based on the previous discussion of findings in literature, the present study proposes to test the following hypothesis:

H_1: There is statistically significant effect of information sensitivity on consumer privacy at selected company in Jordan at level ($\alpha \leq 0.05$).

H_2: There is statistically significant effect of information usage on consumer privacy at selected company in Jordan at level ($\alpha \leq 0.05$).

2.6 Research Model

In the development of consumer privacy, information sensitivity and information usage play a key role, particularly at the Selected Company (eFWATEERcom) in Jordan. The above explanation and discussion was made the basis of the development of the proposed research model in Fig. 1.

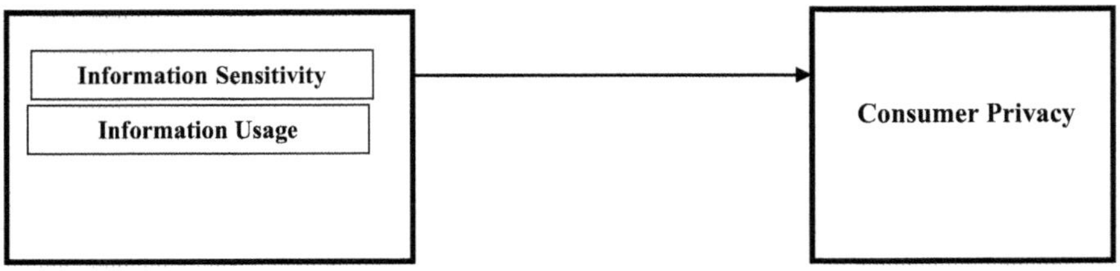

Fig. 1 Research model

3 Research Methodology

3.1 Study Population and Sampling

Based on the above, the total population size consists of all customers that use eFWA-TEERcom electronic payment services, with the total of (776, 000) [29]. Due to the population size, we will use purposive sampling method; depend on the sampling size of [19]. According to the population size (776,000) and depending on Sekaran's table [19], the sample size was determined based on the consumers' number. And this result aligned with the outcome of equation of [20]. Out of the (384) questionnaires distributed randomly to; public and private employees, universities employees and students, and others in Jordan, only (324) were usable questionnaires, (60) questionnaires invalid to the analysis and were excluded, so the total response rate approximately was 84%.

3.2 Measurements

In the light of reviews literature review from standing points, a self-administration questionnaire was designed. The questionnaire has three sections; the covering letter, the participants' demographic variables, and the items related to the factors affecting consumer privacy in using smart electronic devices. The cover letter of the questionnaire cleared the intent of the whole questionnaire and highlighted the matter of confidentiality where it will be used for research purposes only. The items in the questionnaire were measured along a 5-point Likert scale that ranged from 1 (strongly disagree) to 5 (strongly agree). Consumer privacy had five measuring items adopted from [21]. Meanwhile, information usage was measured by five items adopted from [22] and information sensitivity had five measuring items adopted from [9].

Table 1 Descriptive analysis/dimensions and variables

Item	Mean	Standard deviation
Information sensitivity	3.707	0.7877
Information usage	3.724	0.8128
Consumer privacy	3.429	0.5823

4 Data Analysis and Results

All data will collect by the questionnaire, and analyze by using Analysis of Statistical Package of the Social Sciences (SPSS). Descriptive statistics are used to obtain frequencies, means and standard deviations for sample's responses. Furthermore, multiple regression, conformity factor analysis was used to examined the study variables.

4.1 Descriptive Analysis (Statistics)

The results of the demographic data analysis (frequencies and percentages) show that 40.1% of the. Accordingly, for the age groups, the highest percentage is 34.6% which is assigned for the 21–30 years' group, followed by the age group 41–50 years with around 22.5%. Also, the most of the respondents hold a bachelor's degree and then a master's degree with percentages of 41.4% and 35.2% respectively; these percentages are aligned with the majority of Jordanian citizens. Nevertheless, the employment specifications indicate that 47.5% of the sample works for the private sector, 18.2% works for the public sector. While, 26.9% of the users are not employed, this means not limiting the eFWATEERcom usage with hired users. From Table 1 illustrates the mean value to reflect out the representative presentation of the gathered answers [23]. Though, the mean values for the affecting factors of the consumer privacy are 3.707 and 3.724, which show that the values are in the agreement areas according to the Likert scale. The information sensitivity item shows that the usage of the application requires providing sensitive and personal information like the actual username, account number and others. However, the payment system asks for the needed permissions to use such data.

4.2 Reliability Tests

The results of Cronbach Alpha coefficient (the reliability test); Table 2 shows up all dimensions are in the accepted ranges [more than 0.5] suggested by [24]. Thus, this confirms the ability to generalize the results of the study sample to the whole population.

Table 2 The reliability test

Item	Cronbach Alpha	Number of items
Information sensitivity	0.748	5
Information usage	0.754	5
Consumer privacy	0.598	5

4.3 Validity Tests

This study deployed the face validity by asking **a** panel of subject matter experts and academics to evaluate the questionnaire and to provide their comments and feedback about it. The Exploratory Factor Analysis (EFA) is commonly implemented to ensure the validity. EFA tests the "degree of correlations" between the latent (dimensions) and the observed variables (i.e., items or factors). Therefore, EFA sets the factor structure across grouping these factors according to the strength of correlations between these factors [25]. Factor loading tests the association between the (dimensions) latent variables and the survey items (factors) to evaluate the total variances. In order to accept the factor loading, its value must be more than 0.3. Or else, the factor (item) will be deleted. All items of this study are loaded in a value above 0.3.

4.4 The Results of the Regression and Hypothesis Testing

The hypothesis testing uses the "correlation coefficient" in defining the strength and the direction of the relationships between the variables, the values of it (r) ranges between -1 (full and negative linear relationship) and the value $+1$ (full and positive linear relationship). However, the value of zero means a non-existence of linear relationships [26].

The first hypothesis H1 shows that the R-value is 57.1% which indicates that the relationship between the information usage and consumer privacy are positive and high. Hence, the R^2 value is 32.7%, which means that the independent variable can justify around 32% of the dependent variable. The adjusted R^2 value also is 0.324, which means that the differences between the results (of justification) between the sample and the population are 0.003 as a minor value. The hypothesis H1 will be accepted as their significant values are less than 5%. Also, the results in table show that there the second hypothesis H2 will be rejected as the dimensions of information sensitivity is not impacting the consumer privacy as all the significant values are higher than 5%. Thus, H2 will be rejected, and their null hypothesis will be accepted (Table 3).

These factors are worth to be analysed as the beta values are positive and high, which means that such a development will enhance the situation in terms of privacy concerns across application users.

Table 3 Regression test for hypotheses

Hypothesis	R	R^2	Adj. R^2	F-value	Sig.	Beta (β)	Results
H1	0.571	0.327	0.324	100.859	0.000^b	0.456	Accepted
H2	0.006	−0.193	0.006	0.873	0.873^b	−0.143	Rejected

5 Discussion

The study results indicate the information sensitivity impact the consumer privacy. However, information usage does not impact the consumer privacy across the sampled respondents at confidence level of 95%. These results are harmonized with [27, 28, 6, 8] and others.

The results of the demographic data analysis (frequencies and percentages) show that 40.1% of the samples are male's vs 59.9% who are females; this reflects the diversity in the study sample. Though, it shows slight indications that the females tend more to use eFWATEERcom as they are most familiar with the online payment systems in general based on their willingness to shop using online platforms and channels. For the age groups, the highest percentage is 34.6% which is assigned for the 21–30 years' group, followed by the age group 41 – 50 years with around 22.5%. Unexpectedly, the elderly age group "61 years or more" is 4% of the eFWATEERcom users, which is a promising percentage towards the system easiness of use, and flexibility of eFWATEERcom application even for those who may not be experts in the applications and technological payment systems. Likewise, it indicates the extents of trustfulness in the application even that the paid amounts may be dull.

The study contributes to the literature that merges the technology with marketing and management aspects by providing sheds of evidence which stresses on the significant impact of factors and determinants that worry the consumers in terms of their privacy. Hence, the results provide pioneering contributions and a comprehensive view of the top payment application in Jordan that is managed and supported by the central bank and has a sound reputation in terms of low error tolerance.

Regarding to all mentioned above, this study deems the obligation to shape out key recommendations to reduce the worries about privacy concerns and then improve the usage of online payment systems. The following are the main recommendations for future research; firstly, that need to divide the study sample based on regions in Jordan, as there are normal variances between the users in Amman the capital and those who live in the north or south cites. Secondly, it should be to analyse the role of campaigns across the media channel to reduce the fears of using online payment systems. In addition, it is suggested to compare the e-FWATEERcom with other payment gateways such as the banks' payment and the brokers. Lastly, it is proposed to use the mixed research method in the future research to cover the technical and litigation aspects that protect the users from any data or monetary fraud of e-FWATEERcom.

6 Conclusions

This study aims to examining the impact of information sensitivity, information usage across the e-FWATEER.com users across Jordanian society [30]. For the methodology, the study data are obtained from the primary sources using a questionnaire instrument that was adapted from past literature and adjusted to be harmonized and aligned with the Jordanian environment. The study sample is 324 respondents. In the collected data analysis, the simple and then multiple regression model used for analysing on main hypothesis by used through Statistical Package for Social Sciences (SPSS). Lastly, the study results indicate the information sensitivity impact the consumer privacy. However, information usage does not impact the consumer privacy across the sampled respondents at confidence level of 95%. These results are harmonized with [6, 8, 27, 28] and others. The study contributes to the literature that merges the technology with marketing and management aspects by providing sheds of evidence which stresses on the significant impact of factors and determinants that worry the consumers in terms of their privacy. Hence, the results provide pioneering contributions and a comprehensive view of the top payment application in Jordan that is managed and supported by the central bank and has a sound reputation in terms of low error tolerance.

References

1. Al-Okaily M, Alalwan AA, Al-Fraihat D, Rehman SU, Alkhwaldi AF, Al-Okaily A (2022) Investigating antecedents of mobile payment systems decision making: a mediated model. Glob Knowl Memory Commun. https://doi.org/10.1108/GKMC-10-2021-0171
2. Arab Advisors Group (2012) The Consulting Survey-Based Project Report-2012. https://arabadvisors.com/uploads/2020/07/b41450ed-8dde-5f041ee2bb44.pdf
3. Al-Okaily M, Alqudah H, Al-Qudah AA, Alkhwaldi AF (2022) Examining the critical factors of computer-assisted audit tools and techniques adoption in the post-COVID-19 period: internal auditors perspective. VINE J Inf Knowl Manag Syst. https://doi.org/10.1108/VJIKMS-12-2021-0311
4. Warren SD, Brandeis LD (1890) Right to privacy. Harv. L. Rev. 4:193
5. The Ministry of Digital Economy and Entrepreneurship: Annual Report-2017 (2017) https://www.modee.gov.jo/ebv4.0/root_storage/ar/eb_list_page/%D8%A7%D9%84%D8%AA%D9%82%D8%B1%D9%8A%D8%B1_%D8%A7%D9%84%D8%B3%D9%86%D9%88%D9%8A2017.pdf
6. Musleh Al-Sartawi A.M.A. (eds), (2022), Artificial Intelligence for Sustainable Finance and Sustainable Technology. ICGER 2021. Lecture Notes in Networks and Systems, vol 423. Springer, Cham..
7. Al-Okaily M, Al-Okaily A (2022) An empirical assessment of enterprise information systems success in a developing country: the Jordanian experience. TQM J. https://doi.org/10.1108/TQM-09-2021-0267
8. Zhu J, He P, Zheng Z, Lyu MR (2015) A privacy-preserving QoS prediction framework for web service recommendation. In: 2015 IEEE international conference on web services, pp 241–248

9. Al-Okaily M, Alghazzawi R, Alkhwaldi AF, Al-Okaily A (2022) The effect of digital accounting systems on the decision-making quality in the banking industry sector: a mediated-moderated model. Glob Knowl Memory Commun. https://doi.org/10.1108/GKMC-01-2022-0015
10. Miltgen CL, Henseler J, Gelhard C, Popovič A (2016) Introducing new products that affect consumer privacy: a mediation model. J Bus Res 69(10):4659–4666
11. Al-Qudah AA, Hamdan A, Al-Okaily M, Alhaddad L (2022) The impact of green lending on credit risk: evidence from UAE's banks. Environ Sci Pollut Res. https://doi.org/10.1007/s11356-021-18224-5
12. Al-Sartawi A (2020) Information technology governance and cybersecurity at the board level. Int J Crit Infrastruct 16(2):150–161
13. Al-Okaily A, Al-Okaily M, Teoh AP, Al-Debei M (2022) an empirical study on data warehouse systems effectiveness: the case of Jordanian banks in the business intelligence era. EuroMed J Bus. https://doi.org/10.1108/EMJB-01-2022-0011
14. Smith HJ, Milberg SJ, Burke SJ (1996) Information privacy: measuring individuals' concerns about organizational practices. MIS Q 167–196
15. Al-Sartawi A, Sanad Z (2019) Institutional ownership and corporate governance: evidence from Bahrain. Afro-Asian J Finan Account 9(1):101–115
16. Malhotra NK, Kim SS, Agarwal J (2004) Internet users' information privacy concerns (IUIPC): the construct, the scale, and a causal model. Inf Syst Res 15(4):336–355
17. Goodall W, Dovey T, Bornstein J, Bonthron B (2017) The rise of mobility as a service. Deloitte Rev 20(1):112–129
18. Mohamed N, Ahmad IH (2012) Information privacy concerns, antecedents and privacy measure use in social networking sites: evidence from Malaysia. Comput Hum Behav 28(6):2366–2375
19. Al-Okaily A, Al-Okaily M, Teoh AP (2021) Evaluating ERP systems success: evidence from Jordanian firms in the age of the digital business. VINE J Inf Knowl Manag Syst. https://doi.org/10.1108/VJIKMS-04-2021-0061.
20. Al-Sartawi A (2019) Assessing the relationship between information transparency through social media disclosure and firm value. Manag Account Rev 18(2):1–20
21. Lwin MO, Wirtz J, Stanaland AJ (2016) The privacy dyad: antecedents of promotion-and prevention-focused online privacy behaviors and the mediating role of trust and privacy concern. Internet Res 26(4):919–941
22. Al-Okaily M (2021) Assessing the effectiveness of accounting information systems in the era of COVID-19 pandemic. VINE J Inf Knowl Manag Syst. https://doi.org/10.1108/VJIKMS-08-2021-0148
23. Salkind NJ (2016) Statistics for people who (think they) hate statistics. Sage Publications
24. Pallant J (2020) SPSS survival manual: a step by step guide to data analysis using IBM SPSS. Routledge
25. Al-Sartawi A (2018) Corporate governance and intellectual capital: evidence from gulf cooperation council countries. Acad Account Finan Stud J 22(1):1–12
26. Hair JF, Ringle CM, Sarstedt M (2011) PLS-SEM: indeed a silver bullet. J Mark Theory Pract 19(2):139–152
27. Al-Okaily A, Ai Ping T, Al-Okaily M (2021) Towards business intelligence success measurement in an organization: a conceptual study. J Syst Manag Sci 11(2):155–170. https://doi.org/10.33168/JSMS.2021.0210
28. Lee S, Lee Y, Lee JI, Park J (2015) Personalized e-services: consumer privacy concern and information sharing. Soc Behav Pers Int J 43(5):729–740
29. Central Bank of Jordan: Annual Report-2019 (2019). https://www.cbj.gov.jo/EchoBusV3.0/SystemAssets/d7dec196-6888-44b9-a189-652a1ee8a881.pdf
30. Al-Okaily M, Natour ARA, Shishan F, Al-Dmour A, Alghazzawi R, Alsharairi M (2021) Sustainable FinTech innovation orientation: a moderated Model. Sustainability 13(24):1–12

Leveraging Artificial Intelligence for Enforcing Competition in the Digital Economy

Syed Mohammad Yawar and Rahul Shaw

Abstract Artificial Intelligence (AI) is a transformative technology which is rapidly changing the nature of consumers, markets, and even the economy. The digital economy is rife with various anti-competitive practices and the current competition law regimes around the world struggle to keep up with it. The competition law enforcers and regulators need to closely monitor and study the advancements in AI as it keeps changing the dynamics of the digital economy. This research paper analyzes the interface of AI and competition laws and evaluates the possibility of using AI to enforce better competition in the digital economy.

Keywords AI · Competition · Digital economy · Antitrust · Monopoly

1 Introduction

The global economy is in a precarious position. The COVID19 global pandemic and various other geopolitical factors have plunged economies across the world into an unforeseen recession. Technology can indisputably be said to be one of the factors keeping society afloat. Technology is the very thing guiding the fourth industrial revolution [1]. Technological advancements in artificial intelligence (AI), machine learning (ML), augmented reality (AR), blockchain, distributed ledger technology (DLT), smart contracts, and decentralized autonomous organizations (DAO), etc. have the potential to alter the shape of economy and society as we know. These technologies while offering innumerable new possibilities, also pose some intense threats which need the immediate attention of law and policymakers.

The fact that technology is good for both the economy and society is undisputed. Financial technology and financial performance have also been determined to have a beneficial association [2]. The global lockdown protocols enforced due to the COVID19 pandemic has led to an increase in the volatility of the global stock markets and further hampered the economy [3]. A very faint bright side to the pandemic is

S. M. Yawar (✉) · R. Shaw
Department of Law, Aliah University, Kolkata, India
e-mail: smyawar@gmail.com

A. M. A. Musleh Al-Sartawi et al. (eds.), *From the Internet of Things to the Internet of Ideas: The Role of Artificial Intelligence*, Lecture Notes in Networks and Systems 557,
https://doi.org/10.1007/978-3-031-17746-0_33

that it has accelerated the assimilation of digital technology in our daily lives and made almost everyone cognizant with its benefits [4].

The increased dependance on the digital ecosystem has also lent urgency to the legal and regulatory aspects of technologies. It is also an established fact that the principles of competition law and policy are not sufficiently enforced on the digital platform [5]. The need to enforce better competition in the digital ecosystem while also promoting innovation is a quest that cannot remain unfulfilled for long.

1.1 Research Gap

The interface of AI and competition laws is not extensively explored. This research aims to fill this gap and stimulate further study in this area.

1.2 Research Objective

This research aims to analyze the threats and possibilities offered by AI for competition law enforcement.

1.3 Research Methodology

This research conducts qualitative analyses of the recent studies exploring the interface between competition law and AI technology. This is a doctrinal research paper that conceptualizes the problems faced by competition laws due to AI and ways to remedy them.

2 Artificial Intelligence and the Law

Artificial Intelligence (AI) is at the frontier of contemporary technological advancements and is a foundational block of the digital ecosystem. In this research paper, we use the broad definition used by the EU which considers AI to be an umbrella term and defines it as a range of software-based technologies that encompass machine learning, logic and knowledge-based systems, and statistical approaches, that can, for a given set of human-defined objectives, generate outputs such as content, predictions, recommendations, or decisions influencing the environments they interact with [6].

Although AI is still in its nascent stages, it does not mean that it has no practical application. In the United States, AI has been used for making a plethora of important decisions including university admissions selections, loan disbursements and determinations, and even for the distribution of COVID-19 vaccines [7]. Other examples of AI being used for government services include facial recognition, criminal risk assessment, public benefits administrations, health insurance applicability, and even autonomous weapons [8]. The integration of AI in society has been so rapid and extravagant that it has attracted some grave accusations which allege that AI has tainted democracy and spoiled political processed [9]. AI has sparked serious discussion regarding whether it would reinforce democracy or simplify authoritarianism [9]. The influence of AI in several decision making roles has also led to a newfound assertion for the right to a human decision [10].

The most interactive aspect of AI can be claimed to be a digital assistant. It is no wonder that the most sophisticated and advanced digital assistants are offered by the technological enterprises who dominate their respective spheres in the digital economy and have a vast amount of user data fueling their algorithms. These digital assistants like Apple's Siri, Google Assistant, and Amazon's Alexa, have already become widely popular. The combination of network effects, big data, big analytics, machine learning, and an unending supply of user data, will likely enable these digital assistants to circumvent regulatory and competition law policies, and threaten our economy, privacy, and democracy itself [11].

The numerous issues revolving around the legal status of AI has sparked much discourse. While some believe that AI and machines should be strictly regulated in their operations, others even go as far as claiming that AI and machines operating independently and at a cogent level, should be held directly responsible for their actions, at least in the eyes of law [12]. AI aided robots, such as automated machines and self-driving cars, also pose a serious regulatory challenge because the technology behind them keeps evolving while law struggles to keep up with them [13].

AI enabled decision-making and governance has also brought forth a host of regulatory concerns, specifically in the realm of political operations. The governance-by-design norms of engagement establish a framework for avoiding the dysfunction that has followed the shift to technology-based regulation. Policymaking is completely detached from its conventional foundation in processes and institutions guided by public standards of rationality, discussion, expertise, accountability, openness, and participation with every conflict over which values should be embedded in the coercive and enduring command of code. Instead of suppressing rights and privatizing political decision-making, the governance-by-design battle has resulted in negativity, overbroad remedies, and unforeseen effects, such as the overlooking of rights and the privatization of political decision-making [14]. AI itself can help lawmakers and enforcement agencies in reigning in the negative effects of AI enabled digital platforms but for that, the law needs to supplement AI with proper regulation and scrutiny, such that it promotes innovation, not constrains it [15]. The approach towards law making and enforcement should harmonize between considering the accumulated wisdom of the past and evaluating the evolution of society and technology [16].

The current state of AI legal regulation is very obscure and uncertain. The approach towards this issue varies from jurisdiction to jurisdiction and such an attitude does not help the cause in any way.

3 Network Effects in the Digital Economy- Bane or Boon?

A relatively simple business strategy has allowed digital platforms to firmly establish themselves in our lives. They offer their services to customers for free in exchange for showing them advertisements and collecting user data, which they then utilize and sell to other industry players. This straightforward business approach has allowed them to control nearly every area of the digital ecosystem over time. The prevalence of network effects only adds to their market dominance.

The enterprises dominating the digital ecosystem have all amassed an exorbitant amount of data from the users who use their platforms for free. Feeding this data to their algorithms has allowed them to use the power of ML and AI to manipulate their platforms according to their own needs. Their algorithms also analyze the data and formulate new strategies to keep the users hooked on their platform, creating a loop where they constantly keep the users engaged and generate new data to feed upon, further entrenching their hold on the market due to network effects.

It can satisfactorily be claimed that AI by design thrives on exclusionary principles and holding a monopoly on its resources. While the operational nature of AI cannot be changed marginally without frustrating its purpose, a regulatory framework where all players on the digital ecosystem have a level playing field due to mitigated network effects, might be optimal for both the involved players and consumers.

4 Competition in an AI-Powered Economy

The role of competition in a sustainable economy cannot be exaggerated. Competition is a natural phenomenon, which is widely known but poorly understood.

4.1 The Need for Competition

Charles Darwin claimed that the struggle for existence leads to the 'survival of the fittest' [17]. This statement is perhaps the best example of the natural occurrence and merit of competition. In the economic sphere, Hayek famously claimed that the force of competition is the best tool to minimize the exercise of power by man over man [18]. Adam Smith's seminal analogy of the 'invisible hands' shaping the markets [19], is also a widely cited and often quoted example of competition.

Simply said, competition is the impulse that motivates people (and organizations) to work for their own advantage and compete for superiority or influence over others. The goals of competition might be widely debated, but the end is understood to be maximizing consumer welfare [20]. The financial crisis of 2008–2009 led to an overhaul in the global regulation of markets [21] and the current economic scenario needs an even bigger reform, preferably on the back of contemporary technological advancements.

Data has become the most valuable commodity in this digital age and has consecutively altered the ideologies regarding the use and regulation of data. Such ideological revisions are necessary in light of the different sorts of power that can be gained through new organizational models, the use of AI, data, and analytics that platform firms rely on, and the long-term economic stability that can result [22]. The issue of data privacy, consumer rights, and competition laws have also destabilized the modern competition law jurisprudence [23]. Revisiting the efficacy of competition laws is necessary.

4.2 The Threat to Competition from AI

The application of AI programs in the digital ecosystem is a necessary evil. AI needs an endless supply of data to reach its innovative capabilities.

The major social media platforms using AI based content moderation and governance policies have time and time again prioritized their own user-engagement and monetary profit at the expense of their users' data, privacy, and freedom of speech [24]. The effects of data-based governance in the digital platforms have already started posing several threats to legal enforcement [25].

Exclusionary practices by enterprises in the digital platforms, even if they do not hold a monopoly, are hugely rewarding for them and grossly harm the consumers [26]. Contemporary computational methods particularly promote gaining and maintaining a monopoly in the digital ecosystem [27].

A supposedly new branch of law known as 'Computational Law', which is simply understood to be a branch of legal informatics concerned with the mechanization of legal analysis, has paved the way for a new wave of competition law utilizing computational tools from economic theory, business and management science, computer science, statistics, behavioral insights, known as 'Computational Antitrust' [28]. Machine learning has enabled researchers to glean valuable insights from competition law cases decided by competition law authorities [29]. Computational analysis has also enabled researchers to ascertain the difference in understanding of different terms by various stakeholders [30].

The limits of competition laws keep changing with society [31] and AI has further exacerbated the process. AI not only makes it difficult to prosecute big tech enterprises for indulging in anti-competitive behavior, it also makes it difficult to sanction them and enforce court-mandated remedies [32]. The digital ecosystem already makes it difficult to define several aspects which are necessary for competition law enforcement [33]. The evolution of AI does not help this even a little bit. The only hope of reigning in AI is through the use of computational antitrust and even that is contested [34]. Several scholars stress on the importance of introducing new reforms to revive competition law in the digital ecosystem [5]. Several other scholars even argue about the role of competition laws in regulating platform monopolies [35].

The future of economy is powered by AI and its derivative technologies, and they pose considerable challenge to competition law policy and enforcement.

5 Conclusion and Suggestions

AI is the torchbearer of the future economy. While AI offers possibilities previously unimagined by many, it also poses unforeseen challenges. So far, since AI is still in its early stages, the common people have only been reaping its benefits, giving little thought to the threats posed by it. In such an event, it becomes even more important for the policymakers and lawmakers to be vigilant in protecting the interests of the consumers. The best tool to fight the threats posed by AI, is AI itself. Perhaps the best argument in favor of that proposition is evident from an article written completely by an AI program which answers the question of 'How should antitrust agencies approach AI?' [36].

The most alarming part of this entire predicament is that competition agencies have been suspiciously uninterested in exploring the possibilities offered by AI to combat AI. So far, the competition enforcers and regulators have been focusing on regulating the digital platforms and failing abysmally. Researchers and scholars are seriously contemplating computational methods for addressing AI and its derivative technologies, but the competition authorities have shown no such initiative.

Competition enforcers and regulators need to start conducting more research and familiarize themselves with the various nuances of AI and its interactions with competition laws if they ever hope to regulate and reinforce the digital ecosystem. The competition authorities can also start cooperating with various government enterprises already dealing with AI to help them in this endeavor. Stimulating further research in this area would also go a long way in aiding the competition law authorities gather better data and gaining a better understanding of AI.

References

1. Adhikari R (2020) Fourth industrial revolution: from least developed countries to knowledge societies
2. Hannoon A, Al-Sartawi AMAM, Khalid AA (2021) Relationship between financial technology and financial performance. In: Al-Sartawi AMAM (ed) The big data-driven digital economy: artificial and computational intelligence. Springer, Cham, pp 337–344. https://doi.org/10.1007/978-3-030-73057-4_26
3. Kusumahadi TA, Permana FC (2021) Impact of covid-19 on global stock market volatility. J Econ Integr 36(1):20–45. https://doi.org/10.11130/jei.2021.36.1.20
4. Shi C, Xinhong G (2021) Outlook on the 2021 global digital economy — a silver lining amid the crisis. China Econ Transit 4(2):84–90. https://doi.org/10.3868/s060-012-021-0026-2
5. Mahari RZ, Lera SC, Pentland A (2021) Time for a new antitrust era: refocusing antitrust law to invigorate competition in the 21st century. Stanford Comput. Antitrust 1:52–63. Accessed 18 Dec 2021. https://www.ftc.gov/public-statements/1996/08/evolution-us-merger-law

6. European Parliamentary Research Service (2022) Artificial intelligence act, no. January. European Parliament. https://www.europarl.europa.eu/RegData/etudes/BRIE/2021/698792/EPRS_BRI(2021)698792_EN.pdf
7. Kaminski ME, Urban JM (2021) The right to contest AI. Columbia Law Rev 121(7):1957–2048. https://doi.org/10.2307/27083420
8. Crawford K, Schultz J (2019) AI systems as state actors. Columbia Law Rev 119(7):1941–1972. https://doi.org/10.2307/26810855
9. Ünver AH (2018) Artificial intelligence, authoritarianism and the future of political systems. EDAM Res Rep 9:1–20
10. Al-Sartawi A (2020) Does it pay to be socially responsible? empirical evidence from the GCC countries. Int J Law Manag 62(5):381–394
11. Stucke ME, Ezrachi A (2017) How digital assistants can harm our economy, privacy, and democracy. Berkeley Technol Law J 32(3):1239–1300. https://doi.org/10.15779/Z383B5W79M
12. Chatterjee M, Fromer JC (2019) Minds, machines, and the law: the case of volition in copyright law. Columbia Law Rev 119(7):1887–1916. https://doi.org/10.2307/26810853
13. Lemley MA, Casey B (2019) Remedies for robots. Univ Chicago Law Rev 86(5):1311–1396. https://doi.org/10.2307/26747441
14. Mulligan DK, Bamberger KA (2018) Saving governance-by-design. Calif Law Rev 106(3):697–784. https://doi.org/10.2307/26577731
15. Pasquale F (2019) Data-informed duties in AI development. Columbia Law Rev 119(7):1917–1940. https://doi.org/10.2307/26810854
16. Al-Sartawi AMAM (ed) (2022) Artificial intelligence for sustainable finance and sustainable technology. ICGER 2021. Lecture Notes in Networks and Systems, vol 423. Springer, Cham
17. Darwin C (2021) The Origin of Species, 2021 repri. New Delhi: Fingerprint! Classics
18. Hayek FA (2006) The road to serfdom. Routledge, New York. https://ctheory.sitehost.iu.edu/img/Hayek_The_Road_to_Serfdom.pdf
19. Smith A (2018) The Wealth of Nations. 2018 repri. New Delhi: Fingerprint! Publishing
20. Gundlach GT, Moss D (2015) The role of efficiencies in antitrust law: introduction and overview. Antitrust Bull 60(2):91–102. https://doi.org/10.1177/0003603X15591991
21. (2012) Market and regulatory balance. J Bank Regul 13(1):1–3. https://doi.org/10.1057/jbr.2011.24
22. Bamberger KA, Lobel O (2017) Platform market power. Berkeley Technol Law J 32(3):1051–1092. https://doi.org/10.2307/26488977
23. Ohlhausen MK, Okuliar AP (2015) Competition, consumer protection, and the right [approach] to privacy. Antitrust Law J 80(1):121–156. https://doi.org/10.2307/26411522
24. Sylvain O (2019) Recovering tech's humanity. Columbia Law Rev 119(7):252–282. https://doi.org/10.2307/26960743
25. Elkin-Koren N, Gal MS (2019) The chilling effect of governance-by-data on data markets. Univ Chicago Law Rev 86(2):403–432. https://doi.org/10.2307/26590560
26. Khan LM (2019) The separation of platforms and commerce. Columbia Law Rev 119(4):973–1098. https://doi.org/10.2307/26632275
27. Chayjan MR, Bagheri T, Kianian A, Someh NG (2020) Using data mining for prediction of retail banking customer's churn behaviour. Int J Electron Bank 2(4):303–320
28. Schrepel T (2021) Computational antitrust: an introduction and research agenda. Stanford Comput. Antitrust 1:1–15. Accessed 18 Dec 2021. https://github.com/Liquid-Legal-Institute/Legal-Text-
29. Massarotto G, Ittoo A (2021) Gleaning insight from antitrust cases using machine learning. Stanford Comput Antitrust 1:16–37. Accessed 18 Dec 2021. https://www.justice.gov/opa/speech/assistant-
30. Di Porto F, Grote T, Volpi G, Invernizzi R (2021) I see something you don't see: a computational analysis of the digital services act and the digital markets act. Stanford Comput Antitrust 1:84–116. https://doi.org/10.2763/407537
31. Petit N (2021) A theory of antitrust limits. Georg Mason Law Rev 28(4):1399–1460

32. Himes JL, Nieh J, Schnell R (2021) Antitrust enforcement and big tech: after the remedy is ordered. Stanford Comput Antitrust 1:64–83. Accessed 18 Dec 2021. https://perma.cc/MH4K-U4CD
33. Robertson VHSE (2021) Antitrust market definition for digital ecosystems. Concurrences 2:3–9
34. Lim D (2021) Can computational antitrust succeed? Stanford Comput Antitrust 1:38–51. Accessed 18 Dec 2021. https://law.stanford.edu/codex-the-
35. Hovenkamp H (2021) Antitrust and platform monopoly. Yale Law J 130:1952–2050. https://ssrn.com/abstract=3639142
36. GPT-3 (2021) (under the supervision of Thibault Schrepel), "How should antitrust agencies approach AI? Concurrentialiste. https://leconcurrentialiste.com/antitrust-ai/. Accessed 06 Mar 2022

The Impact of Using Smart Algorithms and Blockchain Technology on the Profits' Quality in Saudi Financial Market

Abeer Mahmoud M. Abdelhalim and Nahla Mohamad El Sayed Ibrahim

Abstract The study aimed to demonstrate the impact of using smart algorithms and blockchain technology on the profits' quality in Saudi financial market. As these forms of the recent technological revolution lead to great changes in business models and affect consequently the accounting profession, study was applied on Saudi Money (Tadawul), the shares of which were traded during the time period (2018/2021), amounted to (223) companies divided into (21) sectors, financial institutions were excluded due to the nature of the study, where a sample was taken consisting of (60) companies. The banking and insurance companies sectors were also excluded due to the different nature of the activity (10 banks + 32 insurance companies), and thus the number of observations was (240) for the years covered by the study, then the proposed model was applied using data extracted from the annual reports and financial statements of companies listed in the Saudi capital market, and the study concluded that the use of smart algorithms contributes significantly to improving the quality of profits in companies listed on the Saudi capital market as it has the ability by using its forecasting models to detect the frauds in the financial statements, also the use of blockchain technology contributes significantly to improving the quality of profits in the companies listed in the Saudi stock market as the blockchain is a system that ensures against hacking and the ability to alter records, study conclude that Saudi financial market may adopt blockchain technology in accounting information systems in the future to take advantage of the advantages and opportunities that it achieves, and for future research implications. Researchers may work to shed light on smart algorithms and block chain technology

A. M. M. Abdelhalim (✉)
Department of Accounting, College of Applied Studies, King Faisal University, Al-Hassa 31982, Saudi Arabia
e-mail: aabdelhalim@kfu.edu.sa

A. M. M. Abdelhalim · N. M. E. S. Ibrahim
Department of Business Information Systems, College of Management and Information Systems, Port Said university, Port Said, Egypt
e-mail: netman@kfu.edu.sa

N. M. E. S. Ibrahim
Department of Accounting, College of Business Administration, King Faisal University, 31982, Al-Hassa, Saudi Arabia

A. M. A. Musleh Al-Sartawi et al. (eds.), *From the Internet of Things to the Internet of Ideas: The Role of Artificial Intelligence*, Lecture Notes in Networks and Systems 557,
https://doi.org/10.1007/978-3-031-17746-0_34

effects and changes in the accounting systems, and the need for global and Saudi regulatory authorities to issue standards and guidelines governing the application of smart algorithms and block chains in accounting information systems.

Keywords Smart algorithms · Blockchain technology · Artificial intelligence · Quality of profits

1 Introduction

Information and communication technology plays a strategic role in increasing economic growth rates and reforming economic, commercial and financial mechanisms. In particular, information technology and its tools change the patterns of economic performance in money, business, trade and investment from the traditional form to the immediate form in order to improve competitive positions and thus profitability [41].Where the impact of the tremendous technological development and the resulting modern applications in the business models, as these technologies helped to provide modern digital solutions that enabled companies to attract and retain customers more efficiently than traditional methods [65] which helped them to gain a competitive advantage, and the impact of reliance on information and communication technology In our current era, to reach the so-called second age of machines to describe the strong impact of technology on the way business is performed, digital transformation refers to the use of new technology and developments accompanying the fourth industrial revolution such as social media, mobile, big data analytics, artificial intelligence, clouds, and the Internet of computers To achieve major breakthroughs in the way of doing business by improving relations with customers and achieving good operational efficiency [67].

The world has witnessed many rapid changes in various fields in conjunction with the emergence of the scientific and industrial revolution, and one of these areas is the technological and scientific development, which has had a great impact on the lives of individuals and society positively. This revolution has contributed to the production of artificial intelligence, which has become an integral part of our lives. And life flourished in all its fields [69], not only in science, but also in the fields of humanities, social and economic sciences [45]. Blockchain technology is considered a pioneering innovation in information technology, and although this technology is still in its early stages and has not yet reached adoption by companies, the blockchain technology is in a state of great progress [38], and with the succession of generations of this technology has multiplied Its uses in various fields, especially in the accounting field, and its exchange in a safe, reliable and transparent manner that enables easy and quick access to it.

2 The Problem of the Study

It is noted that the need to switch to the application of artificial intelligence programs has become one of the necessities in accounting work, which is characterized by information acceleration, technological progress and the information revolution, due to what this sector requires in providing quality requirements in the provision of services and ensuring the safety and security of information and response and adaptation with the requirements of the surrounding environment in accordance with the controls Legal and administrative approved by the senior management in order to improve and raise the efficiency of accounting performance and achieve continuous communication and communication between data and information to improve the quality of profits. Blockchain technology is also considered the technology that is likely to change the way companies work. This emerging technology and its multiple capabilities are able to overcome the problems in the accounting system [1], and it is characterized by advantages such as that the data stored on the chains cannot be modified, and the transparency of transactions. On the chains, and the possibility of verifying transactions on the chains, as well as the possibility of reviewing all the data and transactions that took place on the chains at the same moment they occur, these features can have an impact on improving the quality of profits.

Blockchain technology lowers the cost structure by reducing transaction costs such as negotiation costs, search costs, and eliminating intermediaries' costs. 2019). Also, accountants' dealing with smart programs and technologies determines the full responsibility of each of them, as a result of considerations of privacy and confidentiality, as each accounting employee has his own identifier, and a clear separation of responsibilities to a certain level, and this reduces the possibilities of financial fraud, which improves the quality of profits [43]. Accordingly, the study raises the following question: What is the impact of the use of smart algorithms and blockchain technology on the quality of profits in the Saudi financial market.

3 Reasons and Motivations of the Research

Rapid developments in technology have introduced dramatic changes to the practice of the accounting systems which may lead to great change in the shape and content of the financial statements, we assume that the modern technologies may affect positively the outcome of the accounting systems, especially its quality features because of the smart models and techniques included in these technologies, we chose the most influencing technologies on business models and therefore accounting systems to study their impact on the quality of profits.

4 Literature Review and Hypothesis Development

4.1 *Smart Algorithms*

The application of artificial intelligence is manifested in the correct prediction of demand and supply of different products, as artificial intelligence techniques can contribute to improving the accuracy of forecasting sales volume and consumer trends more accurately, and achieve a greater opportunity to manage warehousing and logistics operations more efficiently if smart robots are used [18], in addition to the fact that artificial intelligence helps determine the vision and future direction of the company through the use of advanced technologies such as text analysis technology and machine learning algorithms in many sectors by analyzing customers and identifying the customers who must be targeted accurately [31].

And artificial intelligence can work on a very large amount of data that is difficult to be handled by humans, as the artificial intelligence system is designed to monitor the environment and interact with it according to the reaction [35]. It is the behavior and certain characteristics of computer programs that make them simulate human mental abilities and patterns of work, and among the most important of these characteristics is the ability to learn, infer and react to situations that have not been previously programmed in the machine, and artificial intelligence is classified into two types, the first type is intelligence Weak artificial intelligence, which focuses on a set of specific and narrow tasks, and the other type is strong artificial intelligence, which is known as artificial general intelligence, and this type is capable of performing most of the cognitive functions that a person may possess, in addition to applying intelligence to more than one problem [26], and among the main techniques in artificial intelligence are expert systems, neural networks, robotics, and fuzzy logic.

4.2 *Blockchain Technology*

Blockchain technology may be a new boom for the use of the Internet and will change the way in which all business is practiced in various fields [25], including the field of accounting, as it is considered an important tool for automated control of data and information and ensuring its reliability [15], and its use leads to Lower costs in the long run, avoid human errors and control manipulation and fraud by promptly controlling information and enhancing its integrity to complete all operations [13, 37]. Blockchain technology is defined as a distributed database characterized by its ability to manage an ever-increasing number of records called blocks so that each block contains its own time and a link to the previous block [12, 24]. It is stored in it and preventing it from being modified, meaning that when a piece of information is stored in the block chain, it is not possible to modify this information later [1, 64]. One of the advantages of using blockchain technology in accounting systems is that the use of blockchain technology depends on its decentralized, distributed nature, as in the absence of a central authority, the entire network must agree on all operations recorded in the chain [70], and is characterized by providing complete security that records What is registered on it is fixed and cannot be changed or modified, as once it is registered on the blockchain technology, it cannot be removed or changed [55].

The effective application of blockchain technology in accounting systems contributes to reducing costs, making it superior to traditional databases by reducing the costs of recording information related to commercial transactions [54], as well as the possibility of controlling it in a timely manner and reducing verification costs [48], and is characterized by great speed in processing transactions and saving time and effort, as well as speed in accessing that information and records, as well as speed in disclosing and controlling information, tracking operations and auditing [56, 71].

4.3 *Quality of Profits*

The concept of profit quality is one of the basic concepts in accounting thought, but so far there is no specific and accurate definition of profit quality. It is a multifaceted concept [36]. There are many definitions provided by researchers in accounting thought and this difference came for several reasons, including Their different point of view regarding the characteristics of profits, where some used the continuity of profits as a measure of their quality, where the continuity of profits refers to the extent to which current profits are related to future profits [68], and this means that the cash flows are greater than the receivables and thus increase the quality of profits [53], and others believe that the quality of profits is achieved without profits from profit management practices, and this indicates that the lower the percentage of voluntary accruals in profits, the higher their quality. The quality of profits was also defined in terms of being free from error that profits are of high quality if they reflect (accurately economic reality, generally accepted accounting principles, include some long-term estimates, are subject to high-quality audits, have a strong governance structure) [50], and the credibility of management is the main determinant. For the quality of profits [3].

The quality of profits derives its importance from the importance of profits themselves, in that they are considered one of the important inputs in the process of making financial and investment decisions [49], and it is also considered one of the most important figures disclosed in the financial statements due to the fact that it reflects the application of many accounting principles and rules that The company adopts it [11], and the stakeholders depend on it when making their decisions related to the company [4], and profits are of high quality whenever they are able to reflect the real financial performance of the company [29, 30].

4.4 Analysis of Previous Studies and Derivation of Hypotheses

Among the studies that dealt with smart algorithms: [26] study, the results of which indicated seeking to improve their knowledge of artificial intelligence, as this will help improve their performance in various accounting functions, and [27] study [23] and their results indicated that The level of performance of management systems based on artificial intelligence was effective and qualified in performing various administrative tasks.

The studies [8], the study, the study [63] and the study [52] indicated that the application of artificial intelligence positively affects the performance of Accounting jobs With the expectation of the positive impact of artificial intelligence in the coming years. In a study conducted by the regression analysis that was conducted showed that the perceived performance of artificial intelligence, success and failure of task performance are important factors for predicting perceived confidence as perceived confidence depends on user expectations, in addition, the analysis conducted shows that The growth of task complexity impairs the perceived confidence and expected performance of AI. [28, 40, 61] also emphasized that reliance on modern artificial intelligence techniques contributed to directing the customer and making decisions on his behalf, and that he could fully automate business activities as it facilitates Analyzing data and giving indications of consumer behavior through the algorithms it records, as indicated [16] to the importance of relying on artificial intelligence techniques in processing big data, as the availability and abundance of unstructured and unorganized data on the Internet makes it vulnerable to loss and non-exploitation.

With regard to blockchain technology, many studies have emerged that attempt to study the impact of the application of blockchain technology on the accounting system in an attempt to draw attention to the importance of this modern technology and the consequences of its application in the business world. The use of this technology in the accounting field [9], where the study of [20] and the study [42] aimed to identify the impact of using digital trust chain technology and distributed ledger technology, which is one of its applications on the future of The accounting profession, and a study [47], which confirmed that the use of trust chain technology does not completely eliminate the role of accountants, but changes how they perform their work. Therefore, it is clear that the use of blockchain technology may add more accuracy and transparency in accounting operations compared to the previous traditional methods [51], where the use of blockchain technology in the field of accounting, auditing and transactions related to it online differs from the use of traditional accounting and the traditional methods represented in In recording accounting operations in many records, preparing invoices, processing data and inventory, whether by manual method or accounting programs [14].

This is consistent with the results of the study of [5], which indicated that block chains are a decentralized database capable of ensuring data integrity and preserving it from tampering, manipulation and fraud, and a study [7], which results indicated that the use of block chain technology Digital will lead to an increase in the auditor's responsibility towards all parties, and the study [19, 21, 44], which emphasized the need to take advantage of this modern technology and integrate it into accounting information systems. to take advantage of its advantages. The study [46] also addressed the existence of some new roles for auditors in light of the accounting systems based on block chains, and indicated that there are many skills that must be available in order to adapt to the new environment of block chains, as indicated [2]. That the technology of block chains represents the best technical solutions available at the present time, in order to reduce the material costs of transactions while ensuring the achievement of the element of security. Several studies such as [22, 34, 57, 66] provided evidence of the importance of adopting the concept of blockchain by companies that led to lowering the cost of the product. Increasing the market share in the current competitive situation, and that blockchain technology is seen as an opportunity to exploit the existing resources and competencies that derive from its ability to provide stable data.

With regard to the quality of profits, the study (Hamid, 2022) dealt with theoretical indications, including the adoption of a wide range of models that dealt with the quality of profit on the components and changes of working capital, and the study [6], which indicated that there is a significant statistically significant effect between Profit continuity (accounting profit quality) and bank continuity, as well as a significant statistically significant effect between the predictive ability of accounting profits and bank continuity. The study [10] confirmed that there is a positive and moral effect of voluntary benefits on the cash retained, and this means that there is a negative and moral effect of the quality of profits on the cash retained, and the study [58] indicated that the sustainability of companies is negatively related The quality of discretionary profits in mitigating administrative risks (profit manipulation and opportunistic unethical reporting behaviour).The results of the study [33] also demonstrated that high-quality profits reflect consistent financial reporting, avoiding long-term estimates, and that high-quality profits are characterized by continuity, and [59, 60] study. The results indicated an indirect relationship between the level of The quality of profits and the risks of financial failure, as it was found that there is a positive correlation between the quality of profits and informatics.

Study Hypothesis

In light of the literature review and in order to achieve its objectives and questions, and based on the results of previous studies, we can develop research hypotheses in tow main hypotheses as follows:

H1: The use of smart algorithms affects the quality of profits in the Saudi financial market.

H2: The use of blockchain technology affects the quality of profits in the Saudi financial market.

5 Study Methodology

The current study relies on the use of the descriptive analytical approach because it is the appropriate method for the study, as through this approach, we can compare, interpret and evaluate to reach meaningful generalizations that increase the balance of knowledge about the subject of the study.

5.1 Study Population and Sample

The study population consists of joint stock companies listed on the Saudi Stock Exchange (Tadawul), whose shares were traded during the time period (2018AD/2021AD), which numbered (223) companies divided into (21) sectors,. A sample was taken consisting of (60) companies, and the banking and insurance companies' were excluded due to the different nature of activity (10 banks + 32 insurance companies).

5.2 Description and Measurement of the Study Variables

5.2.1 Independent Variables

5.2.1.1 Intelligent Algorithms (IA)

In measuring the extent to which intelligent algorithms are used in the study sample companies, the two researchers relied on a questionnaire list consisting of (28) phrases distributed on (6) fields,Where the level of using Intelligent Algorithms in the study sample companies is measured by giving a dummy value = 1 in the case of achieving each statement in each axis indicating the use of Intelligent Algorithms and a dummy value = zero in the case of non-verification of the statement in the light of the opinions of the study sample, and after The number of statements indicating the achievement of each field indicating the use of Intelligent Algorithms is divided by the total sum of the axis as a whole.

5.2.1.2 Blockchain Technology

In measuring the extent of the use of Blockchain technology in the study sample companies, we relied on a questionnaire list consisting of (20) phrases distributed on (4) axes, Where the level of use of Blockchain technology in the study sample companies is measured by giving a dummy value = 1 in the case of achieving each statement in each axis indicating the use of Blockchain technology, and a dummy value = zero in the event that the statement is not achieved in the light of the views of the study sample, and then it is Divide the number of statements indicating the achievement of each axis indicative of the use of Blockchain technology by the total number of the axis as a whole.

5.2.2 Dependent Variable

Quality of Earnings (QE)

There are many indicators by which to measure the quality of earnings, but the two researchers relied on (3) indicators, which are as follows:

- Total benefits and symbolized by the symbol (TAC): It can be measured through the following equation:

$$\mathrm{TAC} = \mathrm{IBX} - \mathrm{CFO}$$

- Voluntary benefits and symbolized by the symbol (DAC): It can be measured through the following equation:

$$\mathrm{DAC} / \mathrm{A} = \mathrm{B} + \mathrm{B1}(1/\mathrm{A}) + \mathrm{B2}\,(\mathrm{REV} - \mathrm{AR})/\mathrm{A}$$

- Profit continuity, symbolized by the symbol (EARN): It can be measured through the following equation:

$$\mathrm{EARN} / \mathrm{A} = \mathrm{B} + \mathrm{B1}(\mathrm{EARN} / \mathrm{A}) + \mathrm{E}$$

5.2.3 Control Variables

- **The size of the company**, symbolized (SIZE): It is measured by the natural logarithm of total assets.
- **The degree of financial leverage** and symbolized (LEV): It is measured by dividing the total liabilities by the total assets.
- **Cash flows from operating activities** and symbolized (CFO): They are measured by dividing the net operating cash flows by the total assets.

  Liquidity and symbolized (LIQUID): It is measured by dividing current assets by current liabilities.

  Profitability and symbolized (PROF): It is measured by dividing the net profit for the year before tax by the total assets.
- **Report on losses** and symbolized (LR): It refers to the report on negative profits, and it is measured using a binary criterion that takes the value (0) if the net income is less than zero, and takes the value (1) if the net income is greater than zero.

6 Hypothesis Tests and Analysis of Results

6.1 Descriptive Statistics

Table 3 presents the descriptive statistics of the study variables classified into three groups: the independent variables (Intelligent Algorithms and Blockchain technology), the dependent variable (Quality of Earnings), and the control variables, in order to show the characteristics of these variables at the level of the study sample.

It is evident from Table 1 that:

- With regard to the independent variables, the arithmetic mean of smart algorithms (IA) was (0.629) with a minimum of (0.379) and a maximum of (0.843), while the arithmetic mean of the Blockchain technology reached (0.471) with a minimum of (0.214) and a maximum of (0.691).
- With regard to the dependent variable (quality of earnings) measured by three measures (total accruals TAC, optional benefits DAC, EARN):

  * The arithmetic mean (total benefits TAC) reached (0.520) with a minimum of (0.009) and a maximum of (0.819), while the arithmetic mean (optional benefits DAC) reached (0.462) with a minimum of (0.000) and a maximum of (0.781), also the arithmetic mean (EARN) (0.521), with a minimum of (0.003) and a maximum of (0.928).
- With regard to the control variables: for (company size) the arithmetic average reached (14,289) with a minimum of (7.641) and a maximum of (20,192), and for (the LEV) the arithmetic average reached (6.201) with a minimum of (2.661)

Table 1 Descriptive statistics for the study variables

Variables	Views	Minimum	Maximum	Average	Average
Independent variables					
Intelligent algorithms (IA)	240	0.379	0.843	0.629	0.926
Blockchain	240	0.214	0.691	0.471	0.823
Dependent variables					
Total accruals (TAC)	240	0.009	0.819	0.52	0.616
Optional accruals (DAC)	240	0	0.781	0.462	0.599
Earning Continuity (EARN)	240	0.003	0.928	0.521	1.001
Control variables					
Company SIZE (SIZE)	240	7.641	20.192	14.289	1.92
Leverage (LEV)	240	2.661	10.184	6.201	0.927
cash flow from operating activities (CFO)	240	0.629	0.755	0.687	0.002
Liquidity (LIQUID)	240	0.091	2.833	1.436	0.871
Profitability (PROF)	240	469	0.943	0.794	0.756
Loss Report (LR)	240	−0.003	0.825	0.57	0.801

Table 2 Pearson's correlation matrix for the relationships between the study variables

variables	IA	Blockchain	TAC	DAC	EARN	SIZE	LEV	CFO	LIQUID	PROF	LR
IA											
Blockchain	0.47										
TAC	0.41	0.47									
DAC	0.38	0.29	0.66								
EARN	0.22	0.25	0.59	0.52							
SIZE	0.05	0.08	0.42	0.47	0.41						
LEV	0.11	0.18	0.28	0.22	0.25	-0.1					
CFO	0.23	0.19	0.43	0.39	0.32	0.38	0.08				
LIQUID	-0.25	-0.22	0.39	0.31	0.35	0.35	-0.11	0.4			
PROF	0.3	0.32	0.33	0.28	0.25	0.22	0.22	0.38	0.5		
LR	0.31	0.28	0.19	0.17	0.2	0.13	0.24	-0.21	-0.15	-0.22	

and a maximum of (10,184)), for (cash flows from operating activities CFO) the arithmetic average reached (0.687) with a minimum (0.629) and a maximum (0.755), and for (LIQUID) the arithmetic average reached (1.436) with a minimum (0.091) and a maximum (2.833), For (ProF) the arithmetic mean reached (0.794) with a minimum (0.469) and a maximum (0.943), and for (Reporting on losses LR) the arithmetic average reached (0.570) with a minimum (-0.003) and a maximum (0.825).

6.2 *Analysis of the Correlation Between the Variables of the Study*

(See Table 2).

6.3 *Results of the Study Model's Validity Tests*

Three regression models were used in the current study in order to test the study hypotheses through the use of multiple linear regression equations to test the study models using the method of least squares "OLS" Ordinary Least Squares to estimate the parameters of these models and to ensure the validity of those models and study the problems that Multiple linear regression analysis was based on the following methods:

A- Autocorrelation between errors:

The Durbin-Watson test was applied to identify the existence of a self-correlation problem between errors in the models. The results of this test ranged between (1.502:

Table 3 The results of the multiple regression model to study the impact of smart algorithms on the quality of profits in the Saudi financial market in terms of total accruals TAC

Independent variables and control variables	Dependent variables	B	T	SIG
Intelligent algorithms (IA)	Total accruals (TAC)	9.401	5.021	9.401
Company size (SIZE)		0.79	3.127	0.54
Leverage (LEV)		1.278	3.902	0.987
cash flow from operating activities (CFO)		0.67	5.591	1.631
Liquidity (LIQUID)		0.271	1.23	0.163
Profitability (PROF)		0.71	0.971	0.332
Loss report (LR)		−0.602	0.98	0.327
Constant = 11.101				
R = 0.892 R^2 = 0.796				
F = 21.502 sig = 0.00				

2.922), which indicates the absence of self-correlation between errors for each of the regression models used in the study.

b- Normality of random errors:

The extent of the moderation of random errors, which is one of the important conditions for using the method of least squares "OLS", to ensure that the errors follow a normal distribution, was revealed by the Kolomogrov-Siminrov test, where the (Sig) values of the test ranged between (0.229: 0.631). It is greater than the level of significance (0.05), which indicates that the random errors follow a normal distribution.

C- Multi-collnearity:

To verify that the estimated model did not fall into the problem of linear multi-collinearity for the independent and control variables, the Multicollinearity Test was conducted by finding a variable (Variance Inflation Factor-VIF) for each variable, where the strength of the regression model depends on the hypothesis of the independence of the explanatory variables.

It was found that all the values of (VIF) for the variables of the regression models are greater than (1) and less than (10), where the values of (VIF) ranged between (2.007: 3.801), which indicates that the models used in interpreting the effect on the dependent variables do not suffer from the presence of The problem of linear interference, and therefore there are no problems in measuring regression models that affect their results, and this indicates the strength of regression models in explaining the effect of the independent variable on the dependent variable.

From the above it is clear that the necessary conditions for applying the multiple regression model are met without any standard problems that could affect the accuracy of the results.

Table 4 The results of the multiple regression model to study the impact of smart algorithms on the quality of profits in the Saudi financial market in terms of optional accruals (DAC)

Independent variables and control variables	Dependent variables	B	T	SIG
Intelligent algorithms (IA)	Optional accruals (DAC)	9.401	4.65	0.00
Company size (SIZE)		0.54	3.523	0.00
Leverage (LEV)		0.987	4.12	0.00
cash flow from operating activities (CFO)		1.631	6.45	0.00
Liquidity (LIQUID)		1.509	0.85	0.349
Profitability (PROF)		2.102	1.12	0.205
Loss report (LR)		-0.23	0.73	0.496
Constant = 9.560				
R = 0.795 R^2 = 0.632				
F = 18.734 sig = 0.00				

6.4 *Results of the Study Hypotheses Test*

The study hypothesis states the following:

H1: The use of smart algorithms affects the quality of profits in the Saudi financial market.
H2: The use of blockchain technology affects the quality of profits in the Saudi financial market.

To test the hypotheses of the study, multiple regression analysis was used using the Enter method to explain the relationship between the independent variables (smart algorithms, blockchain technology) and the dependent variable (profit quality) and the control variables (company size, degree of financial leverage)., cash flows from operating activities, liquidity, profitability, reports on losses), and Tables 3, 4, 5, 6, 7 and 8 explain the most important results that have been reached:

6.4.1 The First Hypothesis in the Study

H1: The use of smart algorithms affects the quality of profits in the Saudi financial market.

The hypothesis can be formulated to be tested statistically in the following regression models:

$$\mathbf{TAC} = B + B1(AI) + B2(SIZE) + B3(LEV) + B4(CFO)$$
$$+B5(LIQUID) + B6(PROF) + B7(LR) + E$$

Table 5 The results of the multiple regression model to study the impact of smart algorithms on the quality of profits in the Saudi financial market in terms of profit continuity EARN

Independent variables and control variables	Dependent variables	B	T	SIG
Intelligent algorithms (IA)	Profit continuity EARN	5.22	3.79	0.00
Company size (SIZE)		1.27	4.23	0.00
Leverage (LEV)		1.60	5.50	0.00
cash flow from operating activities (CFO)		0.56	3.78	0.00
Liquidity (LIQUID)		0.29	1.22	0.21
Profitability (PROF)		0.68	1.78	0.09
Loss report (LR)		−0.12	0.87	0.42
Constant = 10.702				
R = 0.881 R^2 = 0.776				
F = 19.501 sig = 0.00				

Table 6 The results of the multiple regression model to study the impact of the blockchain on the quality of profits in the Saudi financial market in terms of total accruals TAC

Independent variables and control variables	Dependent variables	B	T	SIG
Blockchain	Total accruals (TAC)	6.74	3.18	0.00
Company size (SIZE)		2.03	4.48	0.00
Leverage (LEV)		1.37	4.92	0.00
cash flow from operating activities (CFO)		1.69	3.29	0.00
Liquidity (LIQUID)		0.79	0.54	0.64
Profitability (PROF)		0.48	0.68	0.56
Loss report (LR)		-0.47	0.49	0.71
Constant = 8.915				
R = 0.917 R^2 = 0.841				
F = 24.780 sig = 0.00				

$$\mathbf{DAC} = B + B1(AI) + B2(SIZE) + B3(LEV) + B4(CFO) + B5(LIQUID) + B6(PROF) + B7(LR) + E$$

$$\mathbf{EARN} = B + B1(AI) + B2(SIZE) + B3(LEV) + B4(CFO) + B5(LIQUID) + B6(PROF) + B7(LR) + E$$

Analysis of the results of the first model for the first hypothesis of the study

Table 7 The results of the multiple regression model to study the impact of the blockchain on the quality of profits in the Saudi financial market in terms of voluntary accruals (DAC)

Independent variables and control variables	Dependent variables	B	T	SIG
Blockchain	Optional accruals (DAC)	5.76	5.26	0.00
Company size (SIZE)		1.23	4.20	0.00
Leverage (LEV)		2.45	3.48	0.00
cash flow from operating activities (CFO)		0.58	3.48	0.00
Liquidity (LIQUID)		0.69	0.76	0.47
Profitability (PROF)		0.52	0.82	0.46
Loss report (LR)		−0.44	0.54	0.64
Constant = 13.210				
R = 0.901 R^2 = 0.812				
F = 23.302 sig = 0.00				

Table 8 The results of the multiple regression model to study the impact of the blockchain on the quality of profits in the Saudi financial market in terms of profit continuity EARN

Independent variables and control variables	Dependent variables	B	T	SIG
Blockchain	Profit continuity EARN	7.52	4.81	0.00
Company size (SIZE)		0.60	5.38	0.00
Leverage (LEV)		0.83	3.69	0.00
cash flow from operating activities (CFO)		0.77	3.56	0.00
Liquidity (LIQUID)		0.37	0.24	0.96
Profitability (PROF)		0.72	0.60	0.60
Loss report (LR)		−0.37	0.77	0.47
Constant = 11.950				
R = 0.888 R^2 = 0.808				
F = 27.119 sig = 0.00				

$$\mathbf{TAC} = \mathrm{B} + \mathrm{B1(AI)} + \mathrm{B2(SIZE)} + \mathrm{B3(LEV)} + \mathrm{B4(CFO)} + \mathrm{B5(LIQUID)} + \mathrm{B6(PROF)} + \mathrm{B7(LR)} + \mathrm{E}$$

From the data of Table 3, the regression equation that can explain the first applied model of the first hypothesis of the study can be formulated as follows:

$$\mathbf{TAC} = 11.101 + 9.401\mathrm{(AI)} + 0.790\mathrm{(SIZE)} + 1.278\mathrm{(LEV)} + 0.670\mathrm{(CFO)} + 0.271\mathrm{(LIQUID)} + 0.710\mathrm{(PROF)} - 0.602\mathrm{(LR)} + \mathrm{E}$$

The results of Table 3 indicate that the regression model is statistically significant based on the (F) test, where the result of (SIG) for the test is (0.000), which is less than the level of significance (0.05), and the results of the (T) test indicate that (smart AI algorithms), company size (SIZE, LEV, cash flow from operating activities CFO) have a positive, statistically significant effect on the quality of earnings in terms of total accruals TAC at a significant level (0.05), as for (LIQUID, PROF, loss report LR) was not statistically significant.

The coefficient of determination reached (R2) (0.796), which means that the explanatory variables that show the significance of the estimated regression model from the study sample explain an amount of (79.6%) of the total changes that occur in the quality of earnings in terms of TAC, which means The importance of AI smart algorithms to improve the quality of earnings in companies listed in the Saudi capital market.

Analysis of the results of the second model for the first hypothesis of the study:

$$\mathbf{DAC} = B + B1(AI) + B2(SIZE) + B3(LEV) + B4(CFO) + B5(LIQUID) + B6(PROF) + B7(LR) + E$$

From the data in Table 4, the regression equation that can explain the second applied model of the first hypothesis of the study can be formulated as follows:

$$\mathbf{DAC} = 9.560 + 9.401(AI) + 0.540(SIZE) + 0.987(LEV) + 1.631(CFO) + 1.509(LIQUID) + 2.102(PROF) - 0.230(LR) + E$$

The results of Table 4 indicate that the regression model is statistically significant based on the (F) test, where the result of (SIG) for the test is (0.000), which is less than the level of significance (0.05), and the results of the (T) test indicate that (smart AI algorithms), company size (SIZE, LEV, cash flow from operating activities CFO) have a positive, statistically significant effect on the quality of earnings in terms of optional accruals DAC at a significant level (0.05), as for (LIQUID, PROF, PROF, report on losses LR) was not statistically significant.

The coefficient of determination reached (R2) (0.632), which means that the explanatory variables that show the significance of the estimated regression model from the study sample explain the amount of (63.2%) of the total changes that occur in the quality of earnings in terms of the optional accruals DAC, which means The importance of AI smart algorithms to improve the quality of earnings in companies listed in the Saudi capital market.

Analysis of the Results of the Third Model for the First Hypothesis of the Study

$$\mathbf{EARN} = B + B1(AI) + B2(SIZE) + B3(LEV) + B4(CFO) + B5(LIQUID) + B6(PROF) + B7(LR) + E$$

From the data of Table 5, the regression equation that can explain the third applied model of the first hypothesis of the study can be formulated as follows:

$$\textbf{EARN} = 10.702 + 5.218(\text{AI}) + 1.274(\text{SIZE}) + 1.601(\text{LEV}) + 0.560(\text{CFO}) + 0.294(\text{LIQUID}) + 0.681(\text{PROF}) - 0.118(\text{LR}) + \text{E}$$

The results of Table 5 indicate that the regression model is statistically significant based on the (F) test, where the result of (SIG) for the test is (0.000), which is less than the level of significance (0.05), and the results of the (T) test indicate that (smart AI algorithms), company size (SIZE, LEV, cash flow from operating activities CFO) have a positive, statistically significant effect on the quality of earnings in terms of profit continuity EAEN at a significant level (0.05), as for (LIQUID, PROF, report on losses LR) was not statistically significant.

The coefficient of determination reached (R2) (0.776), which means that the explanatory variables that show the significance of the estimated regression model from the study sample explain an amount of (76.6%) of the total changes that occur in the quality of earnings in terms of EAEN, which means The importance of AI smart algorithms to improve the quality of earnings in companies listed in the Saudi capital market.

6.4.2 The Second Hypothesis in the Study

H2: The use of blockchain technology affects the quality of profits in the Saudi financial market.

The hypothesis can be formulated to be tested statistically in the following regression models:

$$\text{TAC} = \text{B} + \text{B1}(\text{Blockchain}) + \text{B2}(\text{SIZE}) + \text{B3}(\text{LEV}) + \text{B4}(\text{CFO}) + \text{B5}(\text{LIQUID}) + \text{B6}(\text{PROF}) + \text{B7}(\text{LR}) +$$

$$\text{DAC} = \text{B} + \text{B1}(\text{Blockchain}) + \text{B2}(\text{SIZE}) + \text{B3}(\text{LEV}) + \text{B4}(\text{CFO}) + \text{B5}(\text{LIQUID}) + \text{B6}(\text{PROF}) + \text{B7}(\text{LR}) + \text{E}$$

$$\text{EARN} = \text{B} + \text{B1}(\text{Blockchain}) + \text{B2}(\text{SIZE}) + \text{B3}(\text{LEV}) + \text{B4}(\text{CFO}) + \text{B5}(\text{LIQUID}) + \text{B6}(\text{PROF}) + \text{B7}(\text{LR}) + \text{E}$$

From the data of Table 6, the regression equation can be formulated that can explain the first applied model of the second hypothesis of the study as follows:

$$\text{TAC} = 8.915 + 6.741(\text{Blockchain}) + 2.034(\text{SIZE}) + 1.370(\text{LEV}) + 1.689(\text{CFO}) + 0.791(\text{LIQUID}) + 0.481(\text{PROF}) - 0.470(\text{LR}) + \text{E}$$

The results of Table 6 indicate that the regression model is statistically significant based on the (F) test, where the result of (SIG) for the test is (0.000), which is less than the level of significance (0.05), and the results of the (T) test indicate that (Blockchain, The size of the company, the financial leverage LEV, the cash flow from operating activities (CFO) have a positive, statistically significant effect on the quality of earnings in terms of the total accruals TAC at a significant level (0.05), as for (LIQUID, PROF, LR) was not statistically significant.

The coefficient of determination reached (R2) (0.841), which means that the explanatory variables that show the significance of the estimated regression model from the study sample explain the amount of (84.1%) of the total changes that occur in the quality of earnings in terms of TAC, which means The importance of the Blockchain to improve the quality of earnings in companies listed in the Saudi capital market.

Analysis of the Results of the Second Model for the Second Hypothesis of the Study

$$\mathbf{DAC} = B + B1(\text{Blockchain}) + B2(\text{SIZE}) + B3(\text{LEV}) + B4(\text{CFO}) \\ + B5(\text{LIQUID}) + B6(\text{PROF}) + B7(\text{LR}) + E$$

From the data in Table 7, the regression equation that can explain the second applied model of the second hypothesis of the study can be formulated as follows:

$$\text{DAC} = 13.210 + 5.761(\text{Blockchain}) + 1.231(\text{SIZE}) + 2.450(\text{LEV}) \\ + 0.581(\text{CFO}) + 0.691(\text{LIQUID}) + 0.519(\text{PROF}) - 0.436(\text{LR}) + E$$

The results of Table 7 indicate that the regression model is statistically significant based on the (F) test, where the result of (SIG) for the test is (0.000), which is less than the level of significance (0.05), and the results of the (T) test indicate that (Blockchain, The size of the company, the financial leverage LEV, the cash flow from operating activities (CFO) have a positive and statistically significant effect on the quality of earnings in terms of the voluntary accruals DAC at the level of significance (0.05), as for (LIQUID, PROF, LR) was not statistically significant.

The coefficient of determination reached (R2) (0.812), which means that the explanatory variables that show the significance of the estimated regression model from the study sample explain the amount of (81.2%) of the total changes that occur in the quality of earnings in terms of the optional accruals DAC, which means The importance of the Blockchain to improve the quality of earnings in companies listed in the Saudi capital market.

Analysis of the Results of the Third Model for the Second Hypothesis of the Study

$$\text{EARN} = B + B1(\text{Blockchain}) + B2(\text{SIZE}) + B3(\text{LEV}) + B4(\text{CFO}) \\ + B5(\text{LIQUID}) + B6(\text{PROF}) + B7(\text{LR}) + E$$

From the data of Table 8 it is possible to formulate the regression equation that can explain the third applied model of the second hypothesis of the study as follows:

$$\text{EARN} = 10.702 + 5.218(\text{Blockchain}) + 1.274(\text{SIZE}) + 1.601(\text{LEV}) + 0.560(\text{CFO}) + 0.294(\text{LIQUID}) + 0.681(\text{PROF}) - 0.118(\text{LR}) + \text{E}$$

The results of Table 8 indicate that the regression model is statistically significant based on the (F) test, where the (SIG) result of the test is (0.000), which is less than the level of significance (0.05), and the results of the (T) test indicate that (Blockchain, The size of the company, the financial leverage LEV, the cash flow from operating activities (CFO) have a positive, statistically significant effect on the quality of earnings in terms of profit continuity EAEN at a significant level (0.05), as for (LIQUID, PROF, LR) was not statistically significant.

The coefficient of determination reached (R2) (0.808), which means that the explanatory variables that show the significance of the estimated regression model from the study sample explain an amount of (80.8%) of the total changes that occur in the quality of earnings in terms of earnings continuity (EAEN), which means The importance of the Blockchain to improve the quality of earnings in companies listed in the Saudi capital market.

7 Results of the Study

The current study reached several results, the most important of which are:

The use of smart algorithms contributes to improving the quality of profits in companies listed on the Saudi capital market.

The use of blockchain technology contributes to improving the quality of profits in companies listed on the Saudi capital market.

- The high morale of the proposed models and their validity to achieve the goal of the study, where all the calculated (F) values were statistically significant at the level of significance (0.000).

8 Recommendations

- Emphasis on the use of smart algorithms and blockchain technology in companies listed on the Saudi stock market due to their importance in improving the quality of profits.
- Emphasis on the use of the applied models proposed in the study, which will reflect positively on the quality of profits in the companies listed in the Saudi stock market.

- Paying attention to the preparation, qualification and refinement of those managing companies listed in the Saudi capital market on how to deal and use smart algorithms and blockchain technology.
- Emphasis on the use of more than one indicator to verify the quality of profits.
- Conducting more studies to measure the impact of using other digital transformation techniques on the quality of profits for companies listed on the Saudi capital market.
- The need for many studies that address the challenges facing the accounting and auditing profession in light of modern technology.
- Conducting future research in the field of the relationship of trust chain technology to corporate governance and in the field of the impact of the integration between trust chain technology and big data analytics on the performance of supply chains.
- The need to seek to increase the adoption of blockchain technology in accounting information systems to take advantage of the advantages and opportunities that it achieves.
- The need for academics to go to more research and conferences to identify and shed light on smart algorithms and block chain technology, and the effects of their application on the rest of the accounting environment elements.
- The need for global and Saudi regulatory authorities to issue standards and guidelines governing the application of block chains in accounting information systems.

References

1. Al-Maati A, Hassan M (2019) An analytical study of the effect of the effectiveness of trust chains technology in the accounting environment and its reflection on different business sectors. J Account Thought 23(1):58–132
2. Al-Sayed BA (2021) A proposed model to reduce transaction costs in the real estate sector using blockchain technology (case study). J Finan Commer Res 22(4):380–414
3. Saeed BDA (2020) Studying the impact of the quality of accounting profits and dividends on stock returns (a test study on Egyptian joint stock companies). Sci J Account Stud 2(3):112–157
4. Salem BH (2017) The relationship between the quality of profits and cash retained (a test study on Egyptian joint stock companies). J Account Fikr 21(4):200–248
5. El-Sayed HM, El-Ghannam SH, Murr NA (2020) The effect of using block chains on external audit. J Finan Commer Res 21(1):85–111
6. Hamid TK, Hassan WAA (2020) The effect of the quality of accounting profits using the sustainability model (profit continuity) and the predictive ability on the continuity of banks, applied research in a sample of commercial banks listed in the Iraqi Stock Exchange. J Account Finan Stud 15(52):209–226
7. Sabry AM (2020) The effect of using digital block chain technology (Block Chain) on the responsibility of the auditor. J Account Thought 24(1):122–165
8. Rahim A, Rahim NA (2021) The mediating role of the trend towards modern technologies in the relationship between the use of artificial intelligence and marketing innovation (an applied study on the electronic retail stores sector in Egypt). J Finan Commer Res 22(3):1079–1127
9. Abdel-Fattah SS (2019) The Impact of blockchain properties on improving digital financial reports (a field study). J Contemp Bus Stud 8:170–205

10. Kandil YS (2019) The effect of the quality of profits and family ownership on the cash retained by applying to companies listed on the Egyptian stock market. J Account Audit 1:286–355
11. Mohamed AS (2018) The relationship of joint auditing and double auditing to the quality of profits (applied study). J Account Thought 22(1):868–903
12. Mustafa NF (2020) A proposed approach to accounting and disclosure of cryptocurrencies according to the business model within the framework of blockchain technology and under the umbrella of International Financial Reporting Standards (IFRS), a survey study of the Egyptian market. Sci J Account Stud 2(2):258–302
13. Abreu PW, Aparicio M, Costa CJ (2018) Blockchain technology in the auditing environment. In: Paper presented at the 13th Iberian conference on information systems and technologies (CISTI), Caceres, Spain
14. Kwillinski A (2019) Implementation of blockchain technology in accounting sphere. Acad Account Finan Stud J
15. Hughes A, et al (2019) Beyond bitcoin. What blockchain and Distributed ledger Technologies Mean for Firms, Business Horizons
16. Balducci B, Detelina M (2018) Unstructured data in marketing. J Acad Mark Sci 46(4):557–590
17. Bargarai FA, Tiryaki V, Zeebaree D (2020) Management of wireless communication systems using artificial intelligence-based software defined radio
18. Bolton RN, Kennedy M, Janet R, Cheung L, Gallan A, Zaki M (2018) Customer experience challenges: bringing together digital, physical and social realms. J Serv Manag 29(5):776–808
19. Bonson E, Bednarova M (2019) Blockchain and its implications for accounting and auditing. Meditari Account Res 27(5):725–740
20. Seng TB, Low KY (2019) Blockchain as the database engine in the accounting system. Aust Account Rev
21. Brender N, Gauthier M, Morin JH, Salihi A (2019). The potential impact of blockchain technology on audit practice. University of Applied Sciences and Arts
22. Cao S, Cong LW, Yang B (2018) Auditing and blockchains: pricing, misstatements, and regulation. Working paper, Georgia State University
23. Chang WY (2019) A data envelopment analysis on the performance of using artificial intelligence-based environmental management systems in the convention and exhibition industry. Ekoloji Dergisi
24. Chedrawi C, Howayeck P (2018). Audit in the Blockchain era within a principal-agent approach. In: Paper presented at the conference: information and communication technologies in organizations and society (ICTO 2018), University Paris Nanterre - Pole Léonard de Vinci, Paris-France
25. Christidis K, Devetsikiotis M (2016) Blockchains and smart contracts for the internet of things. IEEE Access 4:2292–2303
26. Chukwuani VN, Egiyi MA (2020). Automation of accounting processes "impact of artificial intelligence". Int J Res Innov Social Sci (IJRISS) IV(VIII):444–449
27. Davenport T, Guha A, Grewal D, Bressgott T (2020) How artificial intelligence will change the future of marketing. J Acad Mark Sci 48(3):24–42
28. de Bellis J, Venkataramani G (2020) Autonomous shopping systems: identifying and overcoming barriersto consumer adoption. J Retail 96(1):74–90
29. Dechow PM, Ge W, Schrand CM (2010) Understanding earnings quality: a review of the proxies, their determinants and their consequences. In: Accounting Papers, University of Pennsylvania, Wharton Faculty Research, pp 1–174
30. Wu D (2019) What are affecting earnings quality: a summarization. J Mod Account Audit 5(5):53–58
31. Dekimpe MG (2020) Retailing and retailing research in the age of big data analytics. Int J Res Mark 37(1):3–14
32. Derouiche I, Jemai K, Mansali H (2018) Accruals quality, financial constraints and corporate cash holdings. In: The 2018 international conference of the African federation of operational research societies, Tunisia

33. Dichev ID, Graham J, Harvey C, Rajgopal S (2013) Earnings quality: evidence from the field. J Account Econ 56(2–3)
34. Dnyaneshwar J et al (2020) Architecture to enhance transparency in supply chain management using blockchain technology. Procedia Manuf 51:815–826
35. Dzyabura D, Hauser JR (2019) Recommending products when consumers learn their preference weights. Mark Sci 38(3):417–441
36. Eliwa Y, Haslam J, Abraham S (2016) The association between earnings quality and the cost of equity capital: evidence from the UK. Int Rev Finan Anal 1–43
37. Enrique B, Michaela B (2019) Blockchain and its Implication for accounting and Auditing. Meditàri Account Res 27(5):725–740
38. Evelina P, Baur K (2018) Impacts of blockchain technology on supply chain collaboration. Master Thesis in Business Administration, Jonkoping University
39. Farinha J, Mateus C, Soares N (2018) Cash holdings and earnings quality: evidence from the main and alternative UK markets. Int Rev Financ Anal 56:238–252
40. Gacanin H, Wagner M (2019) Artificial intelligence paradigm for customer experience management in next-generation networks: challenges and perspectives. IEEE Netw 33(2):188–194
41. Greenman C (2017). Exploring the impact of artificial intelligence on the accounting profession. J Res Bus Econ Manag (JRBEM)
42. Schmitz J, Leoni G (2019) Accounting and auditing at the time of blockchain technology : aresearch agenda. Aust Account Rev
43. Jedrzejka D (2019) Robotic process automation and its impact on accounting. Zeszyty Teoretyczne Rachunkowościtom 137–166
44. Liu M, Wu K, Xu JJ (2019) How will blockchain technology impact auditing and accounting: permission less versus permissioned blockchain. Am Account Assoc Curr Issues Audit 31(2):19–29
45. Ma Y, Siau K (2018) Artificial intelligence impacts on higher education. In: Proceedings of the thirteenth midwest association for information systems conference, Saint Louis, Missouri, 17–18 May 2018
46. Liu M, Wu K, Xu J (2019) How will blockchain technology impact auditing and accounting? permissionless vs. permissioned blockchain. Am Account Assoc 13(2)
47. Canemi MP (2019) Blockchain auditing accounting the need for automated audits. EDPACS 59(4):1–11
48. Morkunas et al (2019) How blockchain technologies impact you business model. BUSHOR 12:1–10
49. Musa S (2019) Ethical accounting practices , board characteristics and financial reporting quality of listed firms in Nigeria. Dissertation submitted in fulfillment of the requirements for the degree of Master to Department of Accounting and Finance, College of Humanities Management and Social Sciences, University, Nigeria
50. Al-Sartawi AMAM (eds) (2022) Artificial intelligence for sustainable finance and sustainable technology. ICGER 2021. Lecture Notes in Networks and Systems, vol 423. Springer, Cham
51. Niranjanamurthy B, Nithya N (2018) Analysis of Blockchain technology: pros, cons and SWOT, JagannathaSpringer Science+Business Media, LLC, part of Springer Nature
52. Odoh LC, Echefu SC, Ugwuanyi UB, Chukwuani NV (2018) Artificial intelligence in accounting firms. Asian J Econ Bus Account 7(2):1–11
53. Persakis A, Iatridis GE (2016) Audit quality, investor protection and earnings management during the financial crisis of 2008: an international perspective. J Int Finan Mark Inst Money 41(C):73–101
54. Puthal D, Malik N, Mohanty SP, Kougianos E, Das G (2018) Everything you wanted to know about the blockchain: its promise, components, processes, and problems. IEEE Cons Electron Mag 7(4):6–14
55. Zhang Q, Liao B, Yang S (2020) Application of blockchain in the field of intelligent manufacturing: theoretical basis, realistic plights, and development suggestions. Front. Eng. Manag 7(4):578–591

56. Rebecca Y, et al (2020) Public and private blockchain in construction business process and information integration, Autom Constr 1–21
57. van Hoek R (2019) Exploring blockchain implementation in the supply chain. Int J Oper Prod Manag 39(6):829–859
58. Rezaee Z, Tuo L (2017) Are the quantity and quality of sustainability disclosures associated with the innate and discretionary earnings quality? J Bus Ethics 155:763–786
59. Ke R (2012) Predicting financial distress: "The role of earnings quality", A Dissertation of Ph.D. in Management, University of California, Los Angeles
60. Schroeder R, et al (2012) Financial accounting theory and analysis, 7th edn. John Wiley and Sons
61. Seranmadevia R, Senthil KA (2019) Experiencing the AI emergence in Indian retail early adopters approach. Manag Sci Lett 9:33–42
62. Shin M, Shi K, Lee J (2018) Earnings quality effect on corporate excess cash holdings and their marginal value. Emerg Mark Finan Trade 54(4):901–920
63. Shrestha YR, Menahem SM, Krogh G (2019) Organizational decision-making structures in the age of artificial intelligence. California Manag Rev 66–84
64. Simon AD, Kasale S, Manish PM (2017) Blockchain technology in accounting & audit. J Bus Manag 6:6–9
65. Suanpang P (2020) Factor analysis of using social media in tourism enterprises for competitiveness. Int J Innov Manag Technol
66. Kim S, Aseem K (2019) How the blockchain enables and constrains supply chain performance
67. Verma S, Siddharth T (2021) Social media, an emerging platform for relationship building: a study of engagement with nongovernment organizations in India. Int J Nonprofit Volunt Sector Marketing 23(2)
68. Wells K (2020) Who manages the firm matters: the incremental effect of individual managers on accounting quality. Account Rev 95(2):365–384
69. Wirth N (2018) Hello marketing, what can artificial intelligence help youwith?. Int J Mark Res 6(5):435–438
70. Wust K, Gervais A (2018) Do you need a blockchain? In: Crypto valley conference on blockchain technology (CVCBT), zug, pp 45–54
71. Zhang R, Xue R (2019) Security and privacy on blockchain. ACM Comput Surv 1(1)

A Broader Perspective on eXtensible Business Reporting Language (XBRL) in Earnings Management Research

Zakeya Sanad and Abdalmuttaleb Al-Sartawi

Abstract The literature on earnings management is vast. Surprisingly, far too little attention has been paid to investigating the impact of adopting eXtensible Business Reporting Language (XBRL) on earnings management practices. The current study reviews and discusses the previous studies that investigated the impact of adopting XBRL disclosure on earnings management practices. The study also highlights the gap in the literature and revealed that most of the studies agreed so far that XBRL could lead to a reduction in accrual-based earnings management, however, the real activities are more likely to be increased instead. The current findings ring a bell regarding the need to dig more in depth regarding the factors that could lead to managing real earnings management as well as the other types of earnings management (i.e., classification shifting) as these methods could lead to severe unpreferable consequences in the long run. This research will serve as a base for further studies as it suggests several future research topics. The results of this study can inform regulators, investors, and corporate management on how XBRL adoption is associated with corporate financial reporting.

1 Introduction

XBRL is considered the latest technology in non-structure information and was rated as one of the top ten technologies in the accounting field [56, 79]. Researchers called for accelerating theadoption of XBRL interactive data format for financial reporting purposes in all countries[13, 68] Unlike the traditional disclosure, the XBRL has a distinct advantage in providing an interactive data disclosure [9, 14]. Study revealed that firms perceived XBRL disclosure to be a more transparent than HTML disclosure because it limits the errors in stakeholder data input.

The accounting information system literature documented that the usage of XBRL disclosure could lead managers to provide updated and transparent financial reports

Z. Sanad (✉) · A. Al-Sartawi
Ahlia University, Manam, Bahrain
e-mail: zsanad@ahlia.edu.bh

A. M. A. Musleh Al-Sartawi et al. (eds.), *From the Internet of Things to the Internet of Ideas: The Role of Artificial Intelligence*, Lecture Notes in Networks and Systems 557,
https://doi.org/10.1007/978-3-031-17746-0_35

[15, 20, 28, 52] Furthermore, by providing interactive data, the financial information become more accessible and easier to be analyzed by different stakeholders like regulators, analysts, and investors. Also, the adoption of XBRL would lead to information asymmetry reduction [41, 53, 58], decrease credit default swap spreads [16, 21, 31], improve analyst forecast quality [27, 34, 43], reduce cost of capital [76, 78], increase market reactions to earnings surprises[18, 33], and increase market liquidity [61].

Besides, it could enhance the financial reports quality by boosting the reporting efficiency and providing timely and reliable data for firms' internal users such as the board of directors, this would help them in closely monitoring the managerial activities and making sure that shareholders' interests are met [11, 24, 63]. Thus, regulators and policy makers in different countries were actively promoting its adoptions in listed firms. For instance, the U.S. Securities and Exchange Commission (SEC) in 2009 [24, 39] issued a new rule that requires publicly traded firms to disclose their financial statements using interactive data (XBRL) to enhance the financial reporting. Also, the Chinese Ministry of Finance enforced its own XBRL regulation since 2010.

With regards to earnings management (EM), Surprisingly, far too little attention has been paid to investigating the impact of adopting XBRL on EM practices. Therefore, the current study reviews and discusses the previous studies that investigated the impact of adopting XBRL disclosure on earnings management practices. The study also highlights the gap in the literature and serves as a base for further studies as it suggests several future research topics. The results of this study can inform regulators, investors, and corporate management on how XBRL adoption is associated with corporate financial reporting.

2 Literature Review

2.1 eXtensible Business Reporting Language (XBRL)

XBRL has attracted the attention of the accounting scholars to better understand its impact on accounting and auditing professionals. [1, 38, 65] defined XBRL as "an open international standard for business reporting in the disclosure of financial statement". The idea of XBRL was first launched in 1998 by Hoffman which was called back then as XFRML (eXtensible Financial Reporting Markup Language) [6, 7]. Since then, XBRL has been widely used for financial reporting purposes by many public firms internationally [8, 51]

Along with a growing trend of XBRL usage around the world, prior research focused on discussing the implementation of XBRL in different countries. Generally, the researchers concurred that XBRL adoption has resulted in better outcomes as it provides comparable and reliable financial reports and enhances the efficiency of reporting proves by presenting data in a timely manner [10]. For example, using

European non-financial firms' sample, [37] study findings revealed that the information asymmetry has decreased after XBRL adoption. [2] found that XBRL adopters receive lower rate spreads because XBRL voluntary data available decrease the costs of information processing in Belgium. In China, [14–16] documented that after XBRL implementation, the information asymmetry is reduced.

Early studies were motivated to investigate the impact of XBRL adoption on the capital market. For instance, [20] added that the information transparency of capital markets has enhanced after the XBRL implementation and investors' speculation of future crash risk has decreased. [30]claimed that after the SEC requirement to provide detailed footnote information using XBRL, the financial analysts' information environment has enhanced significantly. [40] study revealed that XBRL adoption has increased the forecast accuracy of the analysts.

However, other studies documented an adverse impact of XBRL mandatory implementation. [12] stated that the consequences of XBRL mandatory implementation could vary internationally due to the countries' accounting values differences. [17] claimed that XBRL mandatory adoption could result in lowering the financial statement comparability. [19] added that XBRL implementation had enlarged information asymmetry between small and large investors.

Some studies were motivated to investigate XBRL more in depth by comparing its different strategies and their impact on firms' financial reporting. The first strategy, stand-alone solution (SAS), deals with generating firms' traditional financial statements, and then translate them into XBRL format reports. The other method, disclosure management solution (DMS), produces XBRL reports format in a more integrated and automatic manner as it needs firms to apply XBRL taxonomies to the financial data sources [29]. [22] stressed that DMS strategy would generate a more automated, transparent, and efficient financial reports and reduce the errors associated with the manual data entry. As suggested by the Financial Executives Research Foundation's 2013 survey findings, most of the financial executives preferred using DMS method over SAS [72]. [25, 26] found that DMS method could have a better impact on improving financial reporting process.

2.2 *Earnings Management Practices (EM)*

There are different factors that contribute on managers' EM method preference, such as the timing, risks and costs associated with them [3]. Previous studies concurred that managing earnings can possibly result in negative outcomes in the long run. For example, researchers agreed that REM could cause greater negative economic consequences since it directly alters firms' cashflows and affects firms' operating performance resulting in a reduction in earnings quality [30, 42, 47]. More precisely, [10, 36, 57] documented that managing R&D expenditure as part of REM could lead to a reduction in firms' future value. In addition, although CS was described in the literature as a soft form of EM ([46], it could cause negative impact on future operating performance and cashflows [44]

[48] stressed that having solid reporting regulatory environments did not eliminate EM practices, instead, it motivated the managers to seek for alternative methods rather than ABM. For example, Researchers concurred that ABM is more likely to be detected by regulators and auditors while REM is considered less susceptible which resulted in a significant increase of REM practices recently compare to ABM [31, 54].

[21, 22] conducted an interview with auditors and the findings showed that the majority of the interviewees agreed REM is not easy to be identified and it causes discomfort feelings because it shows that managers are trying to meet their opportunistic short-term goals. Researchers documented a number of reasons related to managers' preference to engaging in REM. For instance, researchers argued that REM is ambiguous and does not violate GAAP although it would negatively harm firm value in the long run [21, 22, 32, 35, 48] and there is no certain benchmark to know what is the right operational action to be made under any specific situation [53].

Studies regarding the litigation risks associated with three EM methods showed that ABM and REM are associated with high litigation risks, while CS is associated with low litigation risk. For instance, researchers argued that ABM could violate GAAP causing high level of litigation risks on managers [55]. [45, 49] stated that ABM is associated with higher potential litigation risk and reputation costs due to the extensive public and regulatory attention. [69] study showed that firms in the post-IFRS period [78] are less likely to manage ABM. Furthermore, [59] stressed REM is more probably to be associated with deceiving disclosures and financial reporting misrepresentations because managers try to hide their opportunistic REM practices. Consequently, if opportunistic REM were suspected or uncovered, it could attract high litigation risk and higher litigation risk is associated with higher costs which could discourage managers to manage REM.

Unlike ABM and REM, the extant literature agreed that CS method is associated with low litigation risks because it is associated with high managerial discretion; thus, regulators and auditors have limited ability to verify it [76–78] resulting in low litigation and reputation concerns than other EM practices [2]. Moreover, under the International Financial Reporting Standards (IFRS), regulations regarding the non-recurring items in the income statement are less rigid ([76–78] Researchers argued that no CEOs have been sued due to the engagement in CS to-date which gives an indication that the litigation risks associated with CS are low [76–78].

A number of studies were interested in discussing the costs associated with EM practices. Researchers argued that when ABM costs are relatively high, managers are more likely to substitute it with less costly EM methods [62]. [66] said that compared to ABM, REM is more costly and more detrimental to firms' operations. Studies suggested that CS is associated with the low costs compared to REM and ABM because it does not involve reversal of accruals, or future returns decline [76–78]. Similarly, [67] found that managers tend to use CS when the cost of within-GAAP EM is relatively high and when managers are restricted from using other EM methods.

Another important factor is timing. As stated by [60] and [62], timing is the most important aspect in deciding which EM method to apply. As a matter of fact, ABM and CS are more likely to be used at the end of the fiscal year, while REM is usually done during the fiscal year [70]. Accordingly, a number of researchers were motivated to test the relationship between these practices and see how managers are using them. [72] showed that managers manage ABM according to the level of REM. [76–78] agreed noted that REM and ABM can be used as substitutes. Managers generally depend on REM more than ABM since it is based on business rather than accounting decisions [20] and it is easier for managers to reduce R&D expenses than aggressive revenue recognition [43].

[4] argued that the managers have the capacity to apply ABM, REM and CS methods to meet earnings targets. [4] stated that CS practices are used when REM is constrained by different factors such as the presence of institutional investors, poor financial condition, and when the firm has low industry market share. The authors added that CS practices are increased when ABM practices are eliminated when having less flexible accounting system. Researchers confirmed that CS is used as a substitute for the ABM and REM [6]. [40] revealed that when firms are performing well and they can meet forecasts, managers tend to avoid the engagement in ABM, REM and CS practices. Based on the studies' findings mentioned earlier, an increasing number of studies have emphasized on the importance of investigating different types of EM rather than one method because managers use multiple EM methods at the same time and focusing on one EM method would provide only one part of the actual situation [76–78]

2.3 XBRL and Earnings Management

Regulators around the world are continuously seeking for practices that could eliminate earnings management practices. EM literature mainly classified EM methods into three categories: accruals-based management (ABM), real activities management (REM) and classification shifting (CS) [2, 76–78, 80] Regardless of the method used to manage firms' earnings, all practices increase the information asymmetry between managers and interested parties and hide firm's actual performance, thereby, diminishing financial reporting reliability and credibility [16, 80].

To reduce the probability of manipulation and misleading the stakeholders, high attention was given to developing effective financial reporting technological approaches such as XBRL. The previous literature documented that XBRL format could help different stakeholders, such shareholders, and policymakers by providing a unified format for exchanging and preparing data as well as producing uniform filings from various reports sources with minimum effort [14–17]. Also, XBRL can help the stakeholders in simultaneously view similarly tagged financial information which enable them in comparing deeper sets of information [10]. Previous studies claimed that XBRL implementation can decrease the costs associated with information processing for financial statement users and help them in effectively overseeing managers' behavior [38, 39] and most importantly, reducing the information asymmetry, especially in large sized firms [43, 50]

Most importantly, regulators can easily identify the issues using XBRL which automatically verifies the presented data [64], thus, the reliability and accuracy of the financial data are improved. To sum up, using XBRL software facilitates the communication for stakeholders by providing an accurate and user-friendly data.

With regards to earnings management practices, as declared by [20] information asymmetry between managers and stakeholders is an essential condition for earnings management. XBRL does not provide extra information about the firm, but it simply provides a user-friendly information environment for the financial reports interested users [13] at a lower information processing cost. The reduction of these costs could lead to increasing the detection risk by auditors, regulators or investors and eventually eliminating the opportunistic managerial practices through ABM [65]

Accordingly, some researchers argued that in order to avoid the detection risk, managers are avoiding the engagement in ABM practices when XBRL is adopted because it is considered a riskier EM method [11]. This was supported by a number of previous studies which indeed concurred that XBRL adoption could assist in lowering the magnitude of ABM practices [65]. On the other hand, there is a little research about the impact of XBRL on REM [25] Although REM is less likely to be uncovered compared to ABM as it is less subject to external auditor scrutiny (Ge and Kim 2014) and has the advantage of being less likely to violate accounting rules compared to ABM [71, 76, 78]. As suggested by the agency theory, managers are motivated to engage in opportunistic practices to achieve their own goals [14–17]. Therefore, when there is a fewer chance to engage in ABM, managers are more likely to search for other available EM options such as REM [22, 33]

A study by [65] tested if the adoption of XBRL by publicly traded firms on the Shanghai Stock Exchange and Shenzhen Stock Exchange is related to discretionary accrual-based EM practices that firms report in the pre-XBRL versus post-XBRL periods. The study findings revealed that the level of ABM in the post-XBRL period is lower compared to the pre-XBRL period. A more recent study by [62] investigated if the adoption of XBRL for financial reporting could mitigate EM practices during the period of XBRL implementation by the SEC. the findings showed that ABM was lowered significantly from the pre- to the post-XBRL period indicating that XBRL software eliminate ABM. In addition, further analysis revealed that XBRL adoption increases REM level. The authors justified the result by stating that managers may prefer the engagement in REM when the detection risk of ABM practices is higher in the post-XBRL environment [76]

Furthermore, a study conducted in Indonesia by [64] tested the impact of XBRL implementation on ABM. The study sample included the manufacturing firms listed on the Indonesia Stock Exchange before the adoption of XBRL (2012-2014) and the period after the adoption of XBRL (2015-2017). The results showed that the adoption of XBRL could mitigate ABM practices by managers and enhance the earnings quality. In addition, a study by [21] examined if XBRL disclosure eliminate ABM as well as REM practices. The authors found that there is a negative relationship between DMS strategy and ABM, while a positive relationship is found between DMS and REM measured using abnormal cash flows. Also, [14–17] focused mainly on the impact of XBRL on REM practices in China and found that REM level has increased after the adoption of XBRL, especially after the dual regulation (i.e., China Securities Regulation Committee and Ministry of Finance).

3 Discussion and Conclusion

As shown in the previous discussion, the focus of the previous research has been largely on the capital market consequences of XBRL implementation from the investors' point of view. However, although EM literature includes a massive number of studies compared to the other accounting literature, a limited studies undertook investigating the impact of XBRL adopting on managerial opportunistic reporting behavior [14, 22, 76, 77] This is a critical issue specially that an increasing number of countries are encouraging the listed firms to adopt XBRL interactive data for financial reporting purposes. Surprisingly, the few studies so far concurred that the adoption of XBRL would constrain the accrual earnings management method but at the same time, the real earnings management practices would increase. The studies justified the results by stating that mangers would prefer to shift to less risky earnings management method (i.e., REM). Besides, researchers argued that REM is ambiguous and does not violate GAAP [76–78], and there is no certain benchmark to know what the right operational action is [15–17]. However, the current findings imply that in the long run, severe consequences and huge damages compared to accrual-based literature may occur as documented in the literature. For example, researchers agreed that REM could cause greater negative economic consequences since it directly alters firms' cashflows and affects firms' operating performance [36], resulting in a reduction in earnings quality [20]. Moreover, REM would negatively harm firm value in the long run [20, 39, 46],

Also, the previous studies' findings should be carefully interpreted as although REM is associated with firms' actual activities and it is not easy for stakeholders to uncover it, sometimes engaging heavily in REM practices could result in manipulating disclosures. Hence, affecting XBRL as well. As argued by [18], REM is more probably to be associated with deceiving disclosures and financial reporting misrepresentations because managers try to hide their opportunistic REM practices. Consequently, if opportunistic REM were suspected or uncovered, it could attract high litigation risk and higher litigation risk is associated with higher costs which could discourage managers to manage REM. Therefore, using XBRL could possibly help in providing more detailed disclosure which eventually reduce REM misleading practices.

One possible reason could be that the proxies used to measure REM may not accurately capture REM practices as some researchers believed that this method is still not well understood due to its vagueness [21, 22]. Some researchers claimed that REM is associated with business decisions rather than accounting choices, and it directly affects firms' cashflow [75], thus, distinguishing between REM and regular business choices is not easy [21, 22]. Similarly, researchers raised concern regarding REM proxies and claimed that REM is complicated, and it cannot accurately differentiate between the opportunistic from the actual fluctuations in the business dynamics [37]. Hence, REM proxy might not provide an accurate picture of opportunistic REM behavior.

Previous studies used a comprehensive REM proxy which included the three REM practices as suggested [48, 52, 70] However, some researchers argued that some activities would lead to abnormally high production costs and at the same time would result in abnormal low cashflow from operations, hence, having one aggregate REM proxy would result in double counting [71, 76] Accordingly, another aggregate measure could be adopted as suggested by [78] and [76] which is the total of abnormal production costs proxy and abnormal reduction of discretionary expenditures only. [73] argued that the abnormal operating cashflow is not included in this aggregate proxy because the other REM methods could possibly have reverse effects on operating cashflow [72]

Also, an important concern that needs also to be addressed in the literature is the impact of XBRL on misclassification of income statement items as a third type of EM practices. [64] described CS as "the recent form of earnings management". Although CS might not sound as a substantial EM concern since it does not deal with managing the bottom-line earnings, it became an increasingly critical EM issue. Despite the fact that CS differs from the other two EM methods with regard to different aspects, what is in common between these methods is that they have the same purpose which is deceiving the interested parties. [5] argued that all three methods of EM (i.e., ABM, REM and CS) used by managers could mislead the investors in anticipating firms' future performance.

Besides, like ABM and REM, managing core earnings opportunistically can have a negative impact on future operating performance and cashflows [58]. Accordingly, regulators recently are paying more attention to the core earnings reporting because it became a popular performance metrics in the capital markets [37, 53]. Unlike ABM and REM, the extant literature agreed that CS method is associated with low litigation risks because it is associated with high managerial discretion; thus, regulators and auditors have limited ability to verify it [76–78] resulting in low litigation and reputation concerns than other EM practices [2]. Therefore, testing the association between XBRL adoption and CS is an interesting research topic that needs to be further investigated.

Future studies should also focus on the specific XBRL strategies that influence EM practices. Testing the relationship between XBRL and EM practices in different context would provide a broader explanation. For instance, investigating the role of XBRL during a crisis like COVID 19 pandemic and how the EM practices are affected. More importantly, some researchers argued that the complexity of XBRL adoption and the limited assurance for XBRL based financial reports may result in greater concerns regarding the quality of financial reporting [40, 41]. Therefore, qualitative studies are needed to deeply investigate the factors that affect the adoption of XBRL in different countries and maybe conducting interviews with board members and auditors to better understand their perception toward XBRL and if it could really help in eliminating opportunistic practices and providing transparent financial reports to the interested users.

References

1. Abbott LJ, Parker S, Presley TJ (2012) Female board presence and the likelihood of financial restatement. Account Horiz 26(4):607–629
2. Alfonso E, Cheng CSA, Pan S (2015) Income classification shifting and mispricing of core earnings. J Account Audit Finan. https://doi.org/10.1177/0148558X15571738
3. Al-Khadash H, Al-Sartawi A (2010) The capability of sarbanes-oxley act in enhancing the independence of the Jordanian public accountant & its impact on reducing the audit expectation gap. Jordan J Bus Adm 6(3):294–316
4. Al-Sartawi A (2016) Measuring the level of online financial disclosure in the Gulf Cooperation Council Countries. Corp Ownersh Control 14(1):547–558
5. Al-Sartawi A (2019) Performance of Islamic banks do the frequency of Shari'ah supervisory board meetings and independence matter? ISRA Int J Islamic Finan 11(2):303–321
6. Al-Sartawi A (ed) (2021) The big data-driven digital economy: artificial and computational intelligence. In: Kacprzyk J (ed) Studies in computational intelligence, vol 974. Springer, Heidelberg
7. Baker TA, Lopez TJ, Reitenga AL, Ruch GW (2019) The influence of CEO and CFO power on accruals and real earnings management. Rev Quant Financ Account 52:325–345. https://doi.org/10.1007/s11156-018-0711-z
8. Banker RD, Wattal S, Plehn-Dujowich JM (2011) R&D versus acquisitions: role of diversification in the choice of innovation strategy by information technology firms. J Manag Inf Syst 28(2):109–144
9. Berger P, Hann R (2007) Segment profitability and the proprietary and agency costs of disclosure. Account Rev 82(4):869–906
10. Black EL, Christensen TE, Taylor Joo T, Schmardebeck R (2017) The relation between earnings management and non-GAAP reporting. Contemp Account Res 34(2):750–782
11. Blankespoor E, Miller BP, White HD (2014) Initial evidence on the market impact of the XBRL mandate. Rev Acc Stud 19(4):1468–1503
12. Cai G, Li W, Tang Z (2020) Religion and the method of earnings management: evidence from China. J Bus Ethics 161(1):71–90
13. Carol C, Kalin K, Sarah M (2020) Detecting opportunistic special items. Manag Sci 66(5):2099–2119. https://doi.org/10.1287/mnsc.2019.3285
14. Chen S, Guo J, Tong X (2017) XBRL implementation and post-earnings-announcement drift: the impact of state ownership in China. J Inf Syst 31:1–19. https ://doi.org/https://doi.org/10.2308/isys-51374
15. Chen S, Harris LL, Li W, Wu D (2015) How does XBRL affect the cost of equity capital? Evidence from an emerging market. J Int Account Res 14:123–145. https://doi.org/10.2308/jiar-51211
16. Chen S, Guo J, Liu Q, Tong X (2021) The impact of XBRL on real earnings management: unexpected consequences of the XBRL implementation in China. Rev Quant Financ Acc 56(2):479–504
17. Cheng Q, Lee J, Shevlin T (2016) Internal governance and real earnings management. Account Rev 91:1051–1085. https://doi.org/10.2308/accr-51275

18. Chi W, Lisic LL, Pevzner M (2011) Is enhanced audit quality associated with greater real earnings management? Account Horiz 25(2):315–335
19. Cohen DA, Zarowin P (2010) Accrual-based and real earnings management activities around seasoned equity offerings. J Account Econ 50(1):2–19
20. Cohen DA, Dey A, Lys TZ (2008) Real and accrual-based earnings management in the pre- and post-Sarbanes-Oxley periods. Account Rev 83(3):757–787
21. Commerford BP, Hermanson DR, Houston RW, Peters MF (2016) Real earnings management: a threat to auditor comfort? Audit J Pract Theory 35(4):39–56
22. Commerford BP, Hermanson DR, Houston RW, Peters MF (2019) Auditor sensitivity to real earnings management: the importance of ambiguity and earnings context. Contemp Account Res 36(2):1055–1076
23. Cooper LA, Downes JF, Rao RP (2018) Short term real earnings management prior to stock repurchases. Rev Quant Financ Account 50:95–128. https://doi.org/10.1007/s11156-017-0624-2
24. Dhole S, Lobo GJ, Mishra S, Pal AM (2015) Effects of the SEC's XBRL mandate on financial reporting comparability. Int J Account Inf Syst 19:29–44
25. Dichev ID, Graham JR, Harvey CR, Rajopal S (2013) Earnings quality: evidence from the field. J Account Econ 56(2–3):1–33
26. Doyle JT, Jennings JN, Soliman MT (2013) Do managers define "Street" earnings to meet or beat analyst forecasts? J Account Econ 56(1):40–56
27. El Diri M, Lambrinoudakis C, Alhadab M (2020) Corporate governance and earnings management in concentrated markets. J Bus Res 108:291–306
28. Evans ME, Houston RW, Peters MF, Pratt JH (2015) Reporting regulatory environments and earnings management: US and non-US firms using US GAAP or IFRS. Account Rev 90(5):1969–1994
29. Fan Y, Barua A, Cready WM, Thomas W (2010) Managing earnings using classification shifting: evidence from quarterly special items. Account Rev 85:1303–1323
30. Felo AJ, Kim JW, Lim J-H (2018) Can XBRL detailed tagging of footnotes improve financial analysts' information environment? Int J Account Inf Syst 28:45–58. https://doi.org/10.1016/j.accinf.2017.12.003
31. Francis B, Hasan I, Li L (2016) A cross-country study of legal-system strength and real earnings management. J Account Public Policy 35(5):477–512
32. Galdi FC, Johnson ES, Myers JN, Myers LA (2019) Accounting for inventory costs and real earnings management behavior. SSRN 3480492
33. Gao J, Gao B, Wang X (2017) Trade-off between real activities earnings management and accrual- based manipulation-evidence from China. J Int Accounting, Audit Tax 29:66–80. https://doi.org/10.1016/j.intaccaudtax.2017.08.001
34. Ge W, Kim. (2014) Boards, takeover protection, and real earnings management. Rev Quant Financ Acc 43(4):651–682
35. Gettler L (2014) XBRL, making the world turn faster, institute of chartered accountants of Australia. www.charteredaccountants.com.au/News-Media/Charter/Charter-articles/Reporting/ 2011-11-XBRL-Making-the-World-Turn-Faster.aspx
36. Golden R (2017) Why the FASB Cares About Non-GAAP Performance Measures, FASB Outlook: from the Chairman's Desk, 1Q 2017. http://www.fasb.org/jsp/FASB/Page/SectionPage&cid=1176168752402
37. Graham JR, Harvey CR, Rajgopal S (2005) The economic implications of corporate financial reporting. J Account Econ 40:3–73
38. Gray GL, Miller DW (2009) XBRL: solving real-world problems. Int J Discl Gov 6(3):207–223
39. Griffin PA, Hong HA, Kim JB, Lim JH (2014) The SEC's XBRL mandate and credit risk: Evidence on a link between credit default swap pricing and XBRL disclosure. Working paper, University of California–Davis, University of Memphis, City University of Hong Kong, and University of Waterloo
40. Guragai B, Hunt N, Neri MP (2015) Accounting information systems and ethics research: review, syntheses, and the future. J Inf Syst 31(2):65–81

41. Hannoon A, Abdalla Y, Al-Sartawi A, Khalid A (2021) Board of directors composition and social media financial disclosure: the case of the United Arab Emirates. In: Al-Sartawi AMAM (ed) The big data-driven digital economy: artificial and computational intelligence. Studies in computational intelligence, vol 974. Springer, Cham
42. Haw IM, Ho SS, Li AY (2011) Corporate governance and earnings management by classification shifting. Contemp Account Res 28(2):517–553
43. Ho LCJ, Liao Q, Taylor M (2015) Real and accrual-based earnings management in the pre-and post-IFRS periods: evidence from China. J Int Financ Manag Acc 26(3):294–335
44. Hoffman C, Kurt C, Koreto RJ (1999) The XML files. J Account 187:71–77
45. Hope O, Thomas W (2008) Managerial empire building and firm disclosure. J Account Res 46(3):591–626
46. Hopkins J (2018) Do securities class actions deter misreporting? Contemp Account Res 35(4):2030–2057
47. Hsieh TS, Wang Z, Abdolmohammadi M (2019). Does XBRL disclosure management solution influence earnings release efficiency and earnings management? Int J Account Inf Manag
48. Huang S, Roychowdhury S, Sletten E (2020) Does litigation deter or encourage real earnings management? Account Rev 95(3):251–278
49. Karolak M, Razzaque A, Al-Sartawi A (2021) E-services and M-services using IoT: an assessment of the Kingdom of Bahrain. In: Al-Sartawi AMM, Razzaque A, Kamal MM (eds) Artificial intelligence systems and the internet of things in the digital era. EAMMIS 2021. Lecture notes in networks and systems, vol 239. Springer, Cham
50. Kim JB, Kim JW, Lim JH (2019) Does XBRL adoption constrain earnings management? early evidence from mandated US filers. Contemp Account Res 36(4):2610–2634
51. Kim JW, Lim JH, No WG (2012) The effect of first wave mandatory XBRL reporting across the financial information environment. J Inf Syst 26(1):127–153
52. Kothari SP, Lester R (2012) The role of accounting in the financial crisis: Lessons for the future. Account Horizons 26(2):335–351
53. Kothari SP, Mizik N, Roychowdhury S (2016) Managing for the moment: the role of earnings management via real activities versus accruals in SEO valuation. Account Rev 91(2):559–586
54. Krishnan GV, McDermott JB (2012) Earnings management and market liquidity. Rev Quant Financ Acc 38:257–274
55. Kuo JM, Ning L, Song X (2014) The real and accrual-based earnings management behaviors: evidence from the split share structure reform in China. Int J Account 49(1):101–136
56. Lai SC, Lin YS, Lin YH, Huang HW (2015) XBRL adoption and cost of debt. Int J Account Inf Manag 23(2):199–216
57. Li OZ, Lin Y, Ni C (2013) Does XBRL adoption reduce the cost of equity capital? Working paper, National University of Singapore
58. Li V (2019) The effect of real earnings management on the persistence and informativeness of earnings. Br Account Rev 51(4):402–423
59. Liu C, Luo X, Wang FL (2017) An empirical investigation on the impact of XBRL adoption on information asymmetry: evidence from Europe. Decis Support Syst 93:42–50. https://doi.org/10.1016/j.dss.2016.09.004
60. Liu C (2013) XBRL: a new. J Glob Inf Manag 21(3):60–80
61. Liu C, Wang T, Yao LJ (2014) XBRL's impact on analyst forecast behavior: an empirical study. J Account Public Policy 33(1):69–82
62. Mayapada AG, Afdhal M, Syafitri R (2020) Earnings management in the pre and post eXtensible business reporting language period in Indonesia. Indon J Account Res 23(1):29–48
63. Al-Sartawi AMAM (eds) (2022) Artificial intelligence for sustainable finance and sustainable technology. ICGER 2021. Lecture notes in networks and systems, vol 423. Springer, Cham
64. Al-Sartawi AMM, Razzaque A, Kamal MM (eds) (2021) Artificial intelligence systems and the internet of things in the digital era. EAMMIS 2021. Lecture notes in networks and systems, vol 239. Springer, Cham
65. Peng EY, Shon J, Tan CEL (2011) XBRL and accruals: empirical evidence from China. Account Perspect 10:109–138. https://doi.org/10.1111/j.1911-3838.2011.00021.x

66. PricewaterhouseCoopers (2012) Disclosure management: Streamlining the last mile. www.pwc.com/gx/en/xbrl/pdf/pwc-streamlining-last-mile-report.pdf
67. Rapoport M (2016) SEC probes whether companies are misusing adjusted earnings metrics. Wall Street J
68. Roohani S, Furusho Y, Koizumi M (2009) XBRL: improving transparency and monitoring functions of corporate governance. Int J Discl Gov 6(4):355–369
69. Roychowdhury S (2006) Earnings management through real activities manipulation. J Account Econ 42(3):335–370
70. SEC (2009) Interactive data to improve financial reporting. http://www.sec.gov/rules/final/2009/33-9002.pdf
71. Sanad Z, Al-Sartawi A (2021) Financial statements fraud and data mining: a review. In: Musleh Al-Sartawi AM, Razzaque A, Kamal MM (eds) Artificial intelligence systems and the internet of things in the digital era. EAMMIS 2021. Lecture notes in networks and systems, vol 239. Springer, Cham
72. Sanad Z, Al-Sartawi A (2016) Investigating the relationship between corporate governance and internet financial reporting (IFR): evidence from Bahrain bourse. Jordan J Bus Adm 12(1):239–269
73. Watts R, Zimmerman J (1978) Towards a positive theory of the determination of accounting standards. Account Rev 112–134
74. XBRL International (2007) Frequently asked questions. http://www.xbrl.org/faq.aspx
75. Yen JC, Wang T (2015) The association between XBRL adoption and market reactions to earnings surprises. J Inf Syst 29(3):51–71
76. Zalata A, Roberts C (2016) Internal corporate governance and classification shifting practices: an analysis of UK corporate behavior. J Acc Audit Financ 31(1):51–78
77. Zalata AM, Tauringana V, Tingbani I (2018) Audit committee financial expertise, gender, and earnings management: does gender of the financial expert matter? Int Rev Financ Anal 55:170–183
78. Zalata AM, Ntim C, Aboud A, Gyapong E (2019) Female CEOs and core earnings quality: new evidence on the ethics versus risk-aversion puzzle. J Bus Ethics 160(2):515–534
79. Zang A (2012) Evidence on the trade-Off between real activities manipulation and accrual-based earnings management. Account Rev 87(2):675–703
80. Zhang Y, Guan Y, Kim J-B (2019) XBRL adoption and expected crash risk. J Account Public Policy 38:31–52. https://doi.org/10.1016/j.jaccpubpol.2019.01.003

Internet of Ideas, Artificial Intelligence, and the Business Systems of the Future

E-commerce and Impact of COVID-19 on Consumer Behaviors Globally: A Review

Saad Darwish and Anjali Gomes

Abstract The background of e-commerce, shopping, online banking, and E-enterprise are only a few of the services offered by e-commerce. Consumers may buy, order, and watch goods and services online using their gadgets, no matter where they are in the world. This industry has emphasized the importance of doing business in rural and remote areas. This paper suggests that e-commerce platforms must ensure consumer privacy and security and protect data from violence and fraud. Consumers are afraid of online transactions and purchases because e-commerce platforms may not ensure the proper protection, trust, and trust. Companies can guarantee fast delivery nationwide to attract customers to e-commerce and shop online. Hence, consumers' desire for online purchases will increase, and e-commerce will become more used. The study also examined the leading theories of adopted behaviour and the theory of planned behaviour (TPB). A survey of selected countries will be conducted to examine literature challenging the Covid 19 issues and their effect on consumers' behaviour. The COVID-19 situation has had an immediate and widespread impact on consumer behaviour. Public health laws are contributing to increased Internet usage.

Keywords Pandemic · Shopping habits · Attitudes · Online shopping · Internet commercial transaction

1 Introduction

In many regions, stores and other public spaces are still undergoing additional operations, necessitating masks and social distancing behaviour. People worldwide recognize and respect the importance of these parameters; yet consumer behaviour has evolved as a result of COVID-19. At the same time, people who see such safeguards as infringing on their freedom or who have misgivings or objections to vaccinations may stop progress toward eliminating limitations and safety measures. When lockdowns were at their worst, conventional healthcare was put on hold—or delivered

S. Darwish (✉) · A. Gomes
Kingdom University, Riffa, Bahrain
e-mail: saad.darwish@ku.edu.bh

A. M. A. Musleh Al-Sartawi et al. (eds.), *From the Internet of Things to the Internet of Ideas: The Role of Artificial Intelligence*, Lecture Notes in Networks and Systems 557,
https://doi.org/10.1007/978-3-031-17746-0_36

online. Many individuals resorted to telemedicine services, such as the Patient Access system in the United Kingdom. However, it is unclear if telehealth will become a more permanent aspect of healthcare solutions. Americans are somewhat more likely to anticipate utilizing telemedicine after a pandemic than those who have previously used it (18% vs 16%). Although many people will return to doctors' offices for specific reasons, now is the moment for the healthcare business to reconsider the best channels for its services while not neglecting the intricacies and limitations of digital health offers.

One thing is clear: dependence on chemical sanitation and cleaning solutions in the near term has not diminished long-term greater attention on the environment. With heightened awareness of mounting environmental risks and new hazards arising from the pandemic, sustainability measures will remain a "must-have." As they reach maturity and gather greater buying power, Generation Z will urge the wider population to be more environmentally conscious [12].

Since work and school are most people's principal daily activities, continuing trials in remote employment and education might have a massive long-term influence. Some have fared better than others. The general assumption is that remote work is likely to become a permanent fixture for certain employees at some point. Still, online education is only advantageous to some sectors, such as college students and those with independent mindsets. COVID-19 consumer behaviour is changing at school and work.

Despite Zoom's weariness and the desire to meet people in person, there is no denying that technology has been an enormous aid in helping people relax over the last year. The popularity of video streaming, internet gaming, and social networking has surged. Although growth in certain services is already slowing, it is expected to level out higher than before the epidemic. Many people find or rediscover low-tech hobbies, such as board games and crafts. About half of all Americans reported taking up new hobbies, learning a new skill, cooking using new recipes, or working on home renovations in the previous year. People are social creatures, and they will begin going out, gathering, and travelling as soon as possible. This is already seen in areas where limitations have been relaxed or abolished. This is likely to result in a spurt of activity in the near term, should funds allow. Simultaneously, some of the virtual experiences generated during the crisis will become permanent features of the work, school, and leisure scene. For example, 29% of respondents who have increased their use of streaming services in the last year think they would continue to do so [13, 29].

Consumer research provides more specific information on a client's needs, preferences, expectations, and behaviour. By appropriately locating this information, tactics for enticing consumers may be enhanced, and businesses may profit from learning exactly what people desire during the pandemic's vital phase. This assessment provides a cross-section of how worldwide nations dealt with the crisis. It will provide researchers with the option to refer to situations in which they may expound on the changes in consumer behaviour.

2 Research Methodology and Conceptual Framework

This study employs an interpretive methodological approach due to its exploratory review nature. According to this approach, researchers form and communicate experiences based on their interpretations of the current knowledge of Covid19 research in certain countries. This knowledge and experiences existed outside of the researchers' comprehension in the literature, but it was collected through a comprehensive assessment of the existing material; it was then interpreted and developed by the researchers using their concise resources and personal knowledge. The conceptual framework is based on the discussion of current research trends examining the impact of consumer behaviour and adjustments to these trends based on the experiences of other nations in terms of culture and population size. The conversation gravitated towards the importance of combating Covid 19 after a critical examination of existing trends and the occurrence of fast changes. It explores the potential dependence of these sources on data collection using credible secondary data and literature. We used a systematic methodology to show case each country's standing; following that, we talked about the challenges and issues involved with consumer behaviour during the Covid era. We summarized and ended the paper in the final section using this method.

3 Research Objectives

The objective of consumer behaviour analysis is essentially consumer studies that are conducted to learn about the consumer's thoughts and emotions about a product. Their choices and dislikes influence the marketer's ongoing modernization of sales methods. During Covid19, experts noticed the shift. Thus, the researchers decided to target the situation in many countries by reviewing the research conducted.

4 Literature Review

4.1 ASIA: A Reflective Dimension

The coronavirus has an impact on e-commerce, technology, business travel, and the economy, among other things (Covid-19). It has already taken the lives of thousands of people and is still taking lives. Furthermore, as a precautionary step, all nations have implemented lockdown measures, and Malaysia has followed suit. China serves as a key industrial base for a large number of businesses across the globe. As a consequence, any disruption in China's manufacturing production will have repercussions. National governments, on the other hand, try to reintroduce their goods from countries other than China to recommence their economic activities. The presence of Chinese goods in the Malaysian market also has the effect of diminishing demand for goods by altering consumer buying behaviour in the country [16]. COVID-19 has a good influence on green consumption patterns in China, according to the organization. Fear, anxiety, and incapability, in particular. Individuals who exhibit COVID-19's negative emotion-cognition-behaviour instinctively need to respond to risk and pay more attention to their safety and interests to promote green consumption. Individuals who exhibit COVID-19's negative emotion-cognition-behaviour inherently need to acknowledge the risk and pay more significant consideration to their protection and interests to advertise green consumption [8].

Positive emotion-cognition-behaviour, on the other hand, is a powerful combination. COVID-19 stimulates higher-order thinking skills such as adoration, motivation, and optimism. It encourages people to focus on the environment, social networks, and green consumer behaviour. This would support the authorities to establish sensible guidelines for emotional orientation to combat the effects of COVID-19 and promote green consumer behaviour. Past research has examined the great possibility of emotion-cognitive behaviour in explaining various behaviours, including social behaviour, word of mouth, participation in collective action, and hostile workplace feedback. Therefore, emotional-cognitive-behaviour has become a new topic that has received much attention in personal behaviour. Past studies have shown that positive and negative emotions, cognitive behaviours have paths to individual behaviours (Sun et al., (2021). A study examined the factors influencing consumers' online shop behaviour during Bangladesh's coronavirus (COVID-19) pandemic. The variables highlighted were the product aspect, price aspect, time-saving aspect, payment aspect, security aspect, administrative aspect, and psychological aspect of consumers' online shopping behaviour during the coronavirus disease (COVID-19) pandemic. The study's outcome showed that all aspects, except the price and safety, were significantly and positively associated with consumer online shopping behaviour during Bangladesh's coronavirus disease pandemic (COVID-19). This has an association with e-marketing about the basic internet shopping strategy for Bangladesh's consumers during Bangladesh's coronavirus (COVID-19).

The characteristics of technology users, their awareness of technology and whether it will affect the acceptance of technology and the value of the use of measurement as an advocate; the attitude toward mobile payment technology was seen as most effective during the pandemic due to lockdown consumers were forced to shop online and considered, especially in this research area, Central Java Province, Indonesia SME customers. To reduce the risk of the spread of coronavirus [8]. The COVID-19 epidemic has radically altered the world. Another study was conducted to identify consumers' purchasing behaviour during Covid-19 in one of India's cities, Coimbatore. As a result of this study, it was observed that the consumer buying behaviour has altered, and consumers are spending to a greater extent; money on fruits and vegetables, epidemics have also brought changes in the choice of consumers; in choosing the brands and Covid 19 has changed consumer buying behaviour. Indian consumers said they were shopping more health-conscious and focused more on reducing food wastage. Personal care practices are also experiencing significant changes in purchasing products while consumers reduce their purchases of additional items (https://www.warc.com). Indian consumer goods businesses face many challenges. Enterprises address challenges in all three phases: permanent security measures in offices and manufacturing floors, revaluation of brand portfolios, increased relationships with third-party e-commerce platforms and shared warehouse research. It's essential to devise a strategy immediately and prepare to accommodate new habits consumers may develop after living during a pandemic. It has been observed that there is a significant effect on the behaviour of the Indian consumers; most of the consumers are spending on essential goods, and they favor using online payment for buying.

The required response to the epidemic has impacted the global economy and has changed consumer preferences, behaviour and buying behaviour. This has led to new challenges for supply chains, distribution, retail stores, and workers [9]. In this study, we investigate the effect of Covid 19 outbreaks on consumer behaviour by discussing how the epidemic affects consumer usage of products and services and shopping behaviour. Consumers, in particular, have seen a significant reduction in spending on traditional shopping malls, but there has been an increase in spending on e-commerce. The results are particularly significant because outbreaks are due to psychological factors, fear of infection, and budgetary restriction [27]. Consumer behaviour is a complex process. It depends on several factors, including socio-political status, financial status, and psychological profile of the buyer. Now, buyers' attention has shifted from general consumer goods to advanced essential products. The protests among retailers are just spreading. Since then, shopkeepers have been financially prudent and gradually think about hygiene and wellness items. Business opportunities are expanding in different new dimensions. Consumer needs are determined based on their fundamental needs. This will increase your interest in the items you need and reduce the number of non-essential classes.

4.2 GULF COUNTRIES (GCC): A Reflective Dimension

The COVID-19 pandemic is believed to have affected consumer behaviours worldwide, affecting the regional and international economies. Research has shown that consumer behaviours have altered in reaction to the pandemic. Thus, many companies innovate (innovate) to adapt to new situations. In particular, technology has become more effective in the daily lives of consumers and businesses. Research shows that COVID-19 has hurt the global economy, rattling supply chains and sales channels. This study shows that for two years, the number of online purchases among Iraqi consumers has grown along with the increase in the number of COVID-19 cases worldwide and Iraq. This may indicate a correlation that should be the subject of further investigation. More will be included in a rigorous reversion analysis of consistent variables both in Iraq and globally. With the spread of the internet, online shopping is omnipresent as payments and deliveries become easier. Additionally, Covid-19 has made online shopping even more attractive in limited conditions. A study was conducted to examine the potential relationship of COVID-19 with consumer buying behaviours in electronic goods in Iraq, with particular attention to understanding how consumers adapt to relevant constraints [2].

The current epidemiological institute government's significant adverse effects of certain constraints and measures require people to recognize and act accordingly. Thus, customer behaviours critically impact the demand for business activities. On the other hand, shopping behaviours are affected by the closure of businesses around the globe that lack products and services. Therefore, consumers rely primarily on online platforms to meet their needs which are essentials like food items, grocery etc. [24]). Another study shows different risk factors that changed online shopping behaviours in Kuwait during the Covid-19 epidemic. The results demonstrate that risk exposure, risk intensity, and the risk of legal penalties positively affect the online shopping behaviours of Kuwaiti consumers. In contrast, product, financial, and non-delivery risks show no particular effect (Alessa 2021). The risk of rest is the only determinant that negatively affects attitude. In addition, we are looking for a new way to impose formal penalties on violators of blocking rules that can directly and positively impact consumer behaviours towards online shopping during an epidemic. The research also observed that determinants affecting users' attitudes and behaviours towards online shopping in regular times outside of crisis differ from the situation of crisis [3, 11].

The influence of COVID-19 on consumers' buying behaviour is investigated in another research, which also predicts the future of online store logistics in Oman. The goal of this paper is to analyze the influence of COVID-19 on the behaviours of e-grocery businesses and e-shoppers, as well as to speculate on the future of online grocery buying in the United Kingdom. It was decided to undertake a survey of food consumers in Oman, both citizens and residents, to better understand their purchase habits and level of satisfaction with online grocery shopping in Oman. The findings revealed that the COVID-19 epidemic has contributed to shifts in consumer behaviour, with many people opting to switch from their conventional in-store purchasing habits to online shopping as a consequence. While the present status of COVID-19 and the lifestyle of Oman have influenced the future of electronic logistics in Omani food shops, this has not yet been established. However, the future of electrical items in Oman seems to be quite promising. Because of Covid-19, there has been a significant rise in the number of people who purchase online in Oman. The findings revealed that COVID-19 had an impact on the purchase habits of consumers. Respondents said that to attract more consumers, Oman's electronic grocery shops should enhance their overall efficiency. It was also asked to sell and add a diverse range of things that clients may need in all of Oman's cities, rather than just a few.

As per a discussion carried out related to Egypt, it has been understood that COVID-19 has affected consumer behaviours in five significant ways, some of which will have a continuing impact such as there will be a change or shift in the consumption behaviours about value products and essential items. Revenues of many consumers worldwide are still declining, and optimism about an economic recovery is not alive. Consumers are more cautious about their costs and trade as they expect the effect of Covid-19 to last more than four months. Customers want to defer. They spend more on essentials like food and household goods and reduce costs on arbitrary categories. Another effect was seen related to flight to digital and Omnichannel. Most segments have increased their online customer count by more than 10% during the epidemic - and a lot of consumers say they prefer to continue shopping online even when the malls and stores reopen in the market with high online conversion rates before the epidemic; as e-commerce segments are increasing. The next was a shock to loyalty; COVID-19 caused supply chain turmoil for some commodities and brands, resulting in consumers who could not get their favorite products at their favorite retailer changing their shopping behaviours, even across multiple brands and shops. Worldwide, the price was the main factor for consumers trying a new brand or place of purchase. The impact of loyalty although these changes in consumer behaviours occur comprehensively, there are differences in each country. Last but not least is the homebody economy; the majority of countries, with more than 70% of consumers, do not feel it convenient to start their normal outdoor activities. When many consumers think about buying groceries and socializing with friends, they stay away from travel and crowded places [5].

4.3 Europe: A Reflective Dimension

Another study was conducted in the United Kingdom to analyse and determine the long-term impact on sales and consumer behaviours. The long-term effect is expected to create a new standard of online consumption, higher than before COVID. It is unlikely to revert to old values as the change has already started, and the average UK consumer has realized the potential benefits of consumption. Internet leads customers to adapt their behaviours to a more user-friendly Internet, which is expected to become a trend in the coming years. The continuing shift and attitude shift brought about by the pandemic indicates an increase in the volume and value of online sales, as the analysis of the data available in this UK study has shown. The connection between the independent variable -COVID-19 infection rate and online retail sales- was identified by examining the time series of various retail markets. They are determined by the numbers shown after smoothing the data, their visual representation, and the implicit nature of these two interrelated variables. This indicates the chances of COVID-19 occurring due to UK usage changes, as the argument for the sizeable mandatory change in the proportion of customer behaviours and income designated to each retailer's sales channel. The research outcome shows that the broad question of the long-term impact of the outbreak on sales and behaviours is as follows. The long-term impact is anticipated to create a new moderate level of online usage, higher than pre-Covid and probably will not rebound to old values. The change begins, and an average of the United Kingdom consumer consumption capacity is known. Via the Internet, customers adapt their behaviours to a more online-friendlier trend that will likely be the trend for the next few years. Both the current change and the trends driven by the pandemic indicate an increase in the volume and value of online sales [28].

A study showed the UK adapted digital strategies to manage challenges due to Covid [1]. To limit the spread of the virus, Italy has adopted interim norms, which has resulted in a variety of psychological responses; It has mental pressure and stress, which causes people to alter their lives, particularly in the area of food familiarization. There is a sense of unease as a result of these developments, but it is well recognized that new opportunities may emerge in these types of circumstances. Regarding consumer understanding of the strong relationship between human health and ecosystems, the pandemic might serve as a watershed moment in promoting the "One Health" viewpoint and boosting the orientation toward environmentally-friendly consumer food goods. The psychological impact of the COVID-19 situation on consumers' purchases of sustainable food items, on the other hand, is yet little understood (20). The psychological effect of the COVID-19 catastrophe has also resulted in a shift in consumer views, with an increased interest in environmental and animal concerns, as well as human health. It seems that these considerations will drive future objectives in terms of purchasing wholesome food products. The main focus of this study is on how psychological responses to health emergencies have changed consumers' attitudes toward sustainability issues, leading them to adopt a more sustainable diet that is recognized as a means of preserving human health, preserving the environment, and preserving animal welfare for both the present and future generations, among other things [7].

As per an article by [19] that changing trend in consumer behaviours shows that the rapid spread of COVID-19 in Italy has had an instantaneous effect on consumer behaviours; the majority of Italians have been quarantined, Which means that all citizens are locked up in their houses. The startling circumstances have certainly affected everyone's lifestyle, including consumers' consumption behaviours, and have led to a change in business models. Consumers buy water, stationery products, and other essential items. The other product for which its sales are growing, though not as much as the original product, is books. However, the sales of makeup and perfumes fell by about −60%. In addition, an exciting occurrence was seen. Seemingly, in the first week of the lockdown, consumers were shopping casually in the supermarkets; however, from the second week onwards, consumers became more knowledgeable, and their shopping lists seemed more organized with less mandatory buying. In this regard, Jesse Garcia, a Los Angeles-based consumer psychologist, said: "COVID-19 has wholly changed consumer behaviours around the world. Consumers are scared, and they go into survival mode when they are scared." It is very striking to know that consumers are starting to use technology tools actively and mainly buy essentials during emergencies as we are experiencing. Thus, Italians mostly use technological tools despite people being unable to buy them. Continuing to buy essentials means that demand for the product is still there even though demand has changed during the crisis [4].

Since Covid 19 broke out around two years ago, Spanish consumers carry on adapting to a lot of elements of their lives in which they buy food. The anticipated rise in e-commerce, more mindful spending, a lower purchase frequency, and a more excellent choice for a close-by business establishment that sells goods are a few of the apparent and instantaneous results. Additionally, the consumer census will continue to affect consumer behaviours in the long run (20).

The proportion of total retail sales in the United Kingdom is represented by Fig. 1 above where estimated figures are given as monthly estimates and quarterly data is a simple average. In the case of the United States, the data estimates e-commerce as a proportion of total retail sales based on data from monthly retail and management surveys. A preliminary estimate of the data for the second quarter of 2020 has been provided. A percentage change in retail sales is shown in the data for the 27-member European Union (EU-27) when compared to the same time, the previous year. Automobiles are excluded from the total retail sales figure. Retail sales conducted via mail order or Internet sites comprise of retail sales in which a user selects based on adverts, catalogues, website information, models, or any other forms of advertising and then puts their purchase through the mail or an Internet site, "you may reach us by phone or online". By the Organization for Economic Cooperation and Development (OECD), only the most recent may be deemed electronic commerce. There is a disagreement on what constitutes e-commerce (21).

Both public-health and economic crises urge alteration in consumers' behaviours which might have enduring results. A survey was carried out to understand consumers to these changes better. It was observed in this survey that the customers are in a state of uncertainty. 49% of the customers avoid leaving their homes, 50% prefer to work from home either wholly or partly, 42% try to avoid using public transport, and 57% follow social distancing with others. The survey also reflects that the cost of this crisis affects homes 78% is the economic cost, 71% human cost such as death and

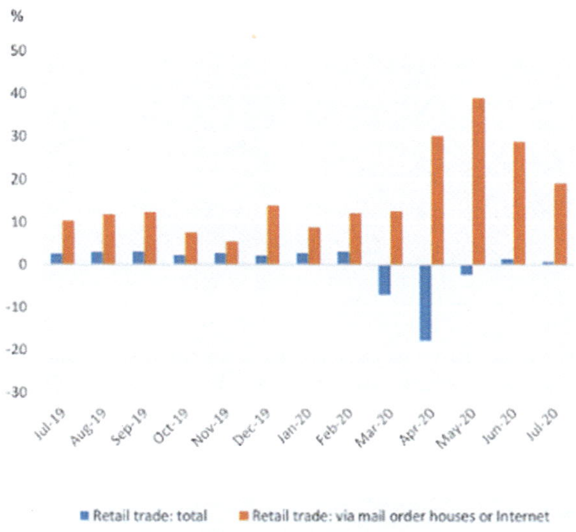

Fig. 1 The COVID-19 crisis has increased the share of e-commerce in total retail. *Source* OECD's elaboration based on data from the US Census Bureau, the Office for National Statistics in the United Kingdom and Eurostat. [23]

being sick, 51% personal health, 49% financial well-being and 48% spreading the virus to a family. As per the survey, the consumers are now purchasing more of the essential commodities; 27% more consumers are buying non-perishable groceries; 25% more consumers are buying household and cleaning items; 25% more consumers are buying frozen items; 12% more consumers are buying perishable groceries; 11% more consumers are buying; 11% more consumers are buying pet care; 11% more consumers are buying personal care and pharmacy; 8% more consumers are buying nonalcoholic drinks; 7% more consumers are buying baby and infant products; 6% more consumers are buying alcohol and tobacco [14]. As per a survey conducted by SRG (Sterling-Rice Group), Covid 19 Consumer Attitude tracker, 74% of the population in the US have made changes to their buying behaviours, mainly related to the grocery. The number of consumers who store food has risen by 26%, with 50% of consumers saying they buy longevity or frozen food. There has been a significant increase in consumers who pay minimal attention to the price.

The retail demand for food has grown, as seen in Fig. 2 below. Sales of frozen and packaged goods, in particular, saw a substantial increase. By late March, weekly sales of frozen foods in France had increased by 63% over the same period the previous year, while sales of packaged goods had increased by 56%. In Germany, the year is denoted by the letter "J." In other nations, there was a comparable need. The retail demand for fresh, frozen, and packaged items has been around 15–20% more than typical from the beginning of this initial flurry.

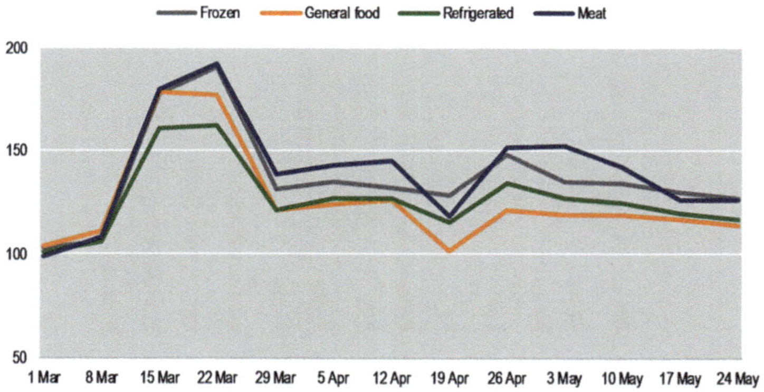

Fig. 2 Retail food demand soared as consumers stockpiled and have remained high. *Note* The IRI CPG Demand Index measures weekly changes in consumer purchases (in dollar terms) relative to the same period in the previous year. *Source* IRI, https://advantage.iriworldwide.com/Engineering/covid19/ [17]

5 Conclusion

The pandemic has had a profound impact on practically every facet of daily life. Some developments have happened unexpectedly and accidentally, such as social distance, the wearing of masks, travel limitations, and other factors, while others have occurred deliberately. Because of this, attitudes that have already gained holds, such as the digitization of buying, banking, and other services, have seen rapid growth in popularity. Finally, we look at five key behavioral shifts and their implications for risk and security in the workplace. As a consequence of the impacts of COVID-19, we have identified five important themes in behavioral change:

1. Increased adoption of digital platforms: People are more reliant on digital platforms for their daily requirements.
2. Modifications in movability patterns, such as reduced use of public transportation and increasing use of remote work, among other things.
3. Shifts in buying behaviour, including a shift toward value-based and online shopping.
4. improved health awareness, such as the use of masks, enhanced cleanliness, good diet, and so on.
5. Changes in interpersonal behaviours, such as a rise in divorce rates and an increase in pet adoption rates, among other things.

These tendencies are mutually reinforcing and overlapping. To remain connected in a physically disconnected global society, individuals have expanded their usage of digital gadgets in their personal lives and enterprises as a result of the epidemic. Because of the rising usage of digital devices, it is becoming more difficult to distinguish between work, personal life and social contact, mobility, health, and financial matters. We anticipate that this will continue in the post-COVID-19 environment [25]. Changes in consumer attitudes, purchasing behaviours, and purchases have occurred: Consumer goods companies are seeing several positive trends. Sales of groceries will climb by 160% from new or low-frequency users in the grocery category, which has the lowest penetration rate in the United States. In addition to their own houses, customers report feeling more satisfied when they visit the homes of relatives and friends during their homestay. As a result of this study, consumers are more worried about the safety of their employees than they were in the three prior rounds of research. 10 consumer trends that will influence CPGs in 2020, according to a research study on COVID-19.

Consumer behaviour has changed dramatically as a result of the COVID-19 disaster, which occurred all at once and throughout the country. Many individuals have become homeless as a result of public health laws, which has also increased Internet use. A large number of customers are familiar with high levels of anguish produced by direct interactions with the crisis or dread of the crisis's physical, economic, and social ramifications, and they are particularly susceptible to commercial exploitation in this situation. Traditional modes of trade are being restricted by the detention procedures, prompting enterprises to transfer their operations online.

While some companies have managed to do so effectively while retaining customer confidence and protecting workers, many others have failed to do so, resulting in a reduction in consumer options in certain circumstances. Additional factors such as border closures, travel restrictions, and social distance measures have lowered consumer demand for services such as housing, travel, and large-scale events, while demand for other items has risen significantly.

Out of fear and anxiety, consumers were making purchase decisions; effective strategies could be developed to predict this kind of behaviour [10]. Covid-19 has brought about a huge number of changes in every walk of life such as lifestyle; preferences; decision on online purchases. So also, the customer demand is changing [15]. The study focused on the factors that affected the behaviour of the consumers in an eCommerce business. The results reflected that social platforms and media advertising had no key effect on consumer behaviour [18]. A study showed how in Turkey the consumers adapted to the new normal- Ecommerce as they are trying to avoid physical stores [6]. We observe a shift in the online consumer purchasing behaviour on different factors at this time of Covid. It is interesting to note the fast pace of decision making by consumers while making purchases online [26].

6 Recommendations

Based on the previous study's findings, the researchers made a few critical suggestions. Examples include providing inexpensive and high-quality internet across the board to all persons without regard to their location, enhancing financial inclusion, and building trust and skills to engage in e-commerce without regard to their location or income. Reduce the digital gap and encourage the participation of the most vulnerable individuals in e-commerce by developing community-based distribution programs for seniors and limiting the number of distribution slots available. Defenseless customers are protected from illicit business activities and potentially dangerous items. It encourages the development of new e-commerce business models while also ensuring that the administrative structure stays flexible enough to accommodate a mix of online and offline company activities. Increasing openness and reducing administrative uncertainty may be accomplished via information exchange. For example, SMEs may participate in e-commerce by assigning laws, regulations, or financial incentives for sales diversification and establishing a level playing field for SMEs that depend on the services of online platforms to compete fairly. It will reduce environmental impediments to e-commerce, such as connectivity, trade, logistics, and postal services, all beneficial to consumers (22).

References

1. Alhomaid A (2021) How Covid-19 has impacted global consumer behaviours and the adoption of digital business strategies by companies in the UK. Turkish J Comput Math Educ. https://turcomat.org/index.php/turkbilmat/article/view/11641
2. Ali BJ (2020) Impact of Covid 19 on consumer buying behaviours towards online shopping in Iraq. Econ Stud J
3. Al-Sartawi A (2020) Social media disclosure of intellectual capital and firm value. Int J Learn Intellect Cap 17(4):312–323
4. Al-Sartawi A (2020) Information technology governance and cybersecurity at the board level. Int J Crit Infrastruct 16(2):150–161
5. Arora N et al (2020) A global view of how consumer behaviours are changing amid COVID-19. Discussion document. http://www.amcham-egypt.org/bic/pdf/corona1/a-global-view-of-how-consumer-behaviours-is-changing-mid-COVID-19.pdf
6. Belbağ AG (2021) Impacts of Covid-19 pandemic on consumer behaviour in Turkey: a qualitative study. J Consum Aff. https://doi.org/10.1111/joca.12423
7. Castellini G, Savarese M, Graffigna G (2021) The impact of COVID-19 outbreak in italy on the sustainable food consumption intention from a "One Health" perspective. https://doi.org/10.3389/nut.2021.622122. Frontiers in Nutrition | www.frontiersin.org 1 March 2021 | Volume 8 | Article 622122
8. Cannito L et al (2021) Temporal discounting of money and face masks during the COVID-19 pandemic: the role of hoarding level. Front Psychol 12:642102. PMID:34177697
9. Chohan UW (2020) A post-coronavirus world: 7 points of discussion for a new political economy. SSRN J
10. Crosta A, Ceccato I, Marchetti D (2021) Psychological factors and consumer behaviour during the COVID-19 pandemic
11. Darwish S, El Naggar N (2021) The effect of covid-19 on corporate social responsibility (CSR) practices reflections from the Kingdom of Bahrain. J Legal Ethical Regul Issues 24(Special Issue 1):1–10
12. Darwish S, Gomes AM, Ahmed U (2021) Risk management strategies and impact on sustainability: the disruptive effect of Covid 19. Acad Strateg Manag J 20(SpecialIssue2):1–19
13. Euromonitor (2020) How is COVID-19 affecting the top 10 global consumer trends in 2020? https://go.euromonitor.com/webinar-ec-2020-covid-19-impact-ongct.html?utm_source=blog&utm_medium=blog&utm_campaign=WB_20_04_28_REC_Coronavirus_Impact_on_GCT
14. Evolving priorities: COVID-19 rapidly reshapes consumer behaviours. https://www.pwc.com/us/en/industries/consumer-markets/library/covid-19-consumer-behaviours-survey.html
15. Firmandani W, Sya'bania AF, Abdani F, Madani E (2020) Analysis of Indonesian consumer online shopping behavior during the Covid-19 pandemic: a Shopee case study. Int J Bus Rev (Jobs Rev) 4(2)
16. Hasanat MW, Hoque A, Shikha FA, Anwar M, Hamid ABA, Tat HA (2020) The impact of coronavirus (Covid-19) on E-business in Malaysia. Asian J Multi Stud 3(1)
17. IRI (n.d). https://advantage.iriworldwide.com/Engineering/covid19/
18. Jawaid MH, Karim E (2021) Factors affecting consumer buying behavior in E-commerce business during outbreak of Covid-19: a case study on top E-commerce websites
19. Mazzeschi Legal Counsels (2021) The effects of Covid 19 on the Italian consumer behaviours/Articles 2020. https://www.mazzeschi.it/the-effects-of-covid-19-on-italian-consumer-behaviour/. Voluntary Report – Voluntary - Public Distribution
20. Musleh Al-Sartawi AMA (eds) (2022) Artificial intelligence for sustainable finance and sustainable technology. In: ICGER 2021. Lecture notes in networks and systems, vol 423. Springer, Cham
21. OECD (2020) Food supply chains and COVID-19: impacts and policy lessons. OECD Policy Responses to Coronavirus (COVID-19). http://www.oecd.org/coronavirus/policy-responses/food-supply-chains-and-covid-19-impacts-and-policy-lessons-71b57aea/

22. OECD Policy Responses to Coronavirus (COVID-19) E-commerce in the time of COVID-19 (2020). https://www.oecd.org/coronavirus/policy-responses/e-commerce-in-the-time-of-covid-19-3a2b78e8/
23. OECD (2020) Protecting online consumers during the COVID-19 crisis. https://www.oecd.org/coronavirus/policy-responses/protecting-online-consumers-during-the-covid-19-crisis-2ce7353c/
24. Tyagi P, Pabalkar V (2021) Impact of COVID-19 over purchasing behaviour of the consumers. Eur J Mol Clin Med 08(02):1–11
25. Puttaiah MH, Raverkar AK (2020) All change: how COVID-19 is transforming consumer behaviours. https://www.swissre.com/institute/research/topics-and-risk-dialogues/health-and-longevity/covid-19-and-consumer-behaviour.html
26. Shengyu G, Slusarczyk B, Hajizada S, Kovalyova I, Sakhbieva A (2021) Impact of the COVID-19 pandemic on online consumer purchasing behavior. J Theor Appl Electron Commer Res
27. Sood S (2020) Impact of COVID-19 on consumer behaviours in India. In: Conference paper: sustainable management practices and economic slowdown in India at Delhi, India. https://www.researchgate.net/publication/343153169
28. Volchkov F (2021) Long-term effect of COVID-19 outbreak on consumer behaviours and online retail in the United Kingdom. Bachelor thesis, Modul University, Vienna. https://www.modul.ac.at/uploads/files/Theses/Bachelor/Undergrad_2021/BSC_2021/1821059_VOLCHOV_Fedor_Thesis_no_sig.pdf
29. Zwanka RJ, Buff C (2020). COVID-19 generation: a conceptual framework of the consumer behavioural shifts to be caused by the COVID-19 pandemic. J Int Consum Market. https://www.tandfonline.com/doi/abs/10.1080/08961530.

Toward a Digital Sustainability Reporting Framework in Organizations in the Industry 5.0 Era: An Accounting Perspective

Carmen Olsen

Abstract Organizing digital sustainability reporting in organizations is a crucial in organizations facing the upcoming tsunami of sustainability reporting requirements. The aim of this paper is to apply the accounting perspective in the implementation of digital sustainability reporting in the current scene of Industry 5.0 era. Under Industry 4.0, the accounting perspective guided external financial reporting which is produced by Accounting Information System (AIS) and financial accounting modules in Enterprise Resource Planning (ERP) systems. This paper applies this accounting perspective of digital financial reporting to organize digital sustainability reporting in the same fashion. The novelty of this paper is excelling upon the prior accounting perspective of digital financial reporting to create a conceptual framework for the implementation of digital sustainability reporting. The framework for digital sustainability reporting is a first step in helping the accounting profession and sustainability management level in organizations in the implementation of digital sustainability reporting. The framework also opens the door for new research in accounting research.

Keywords Accounting profession · Accounting Information system (AIS) · Digital sustainability reporting · Enterprise Resource Planning (ERP) · Industry 5.0 · Sustainability reporting systems

1 Introduction

In the sustainability era of industry 5.0, research [1–4], sustainability accounting standard setters [5, 6], such as Sustainability Accounting Standards Board (SASB) and European Financial Reporting Advisory Group (EFRAG), and information system providers [7, 8], such as Accounting Information Systems (AIS) and financial accounting modules integrated in an Enterprise Resource Planning (ERP) system,

---

C. Olsen (✉)
Western Norway University of Applied Sciences, Bergen, Norway
e-mail: carmen.olsen@hvl.no

Inland Norway University of Applied Sciences, Hamar, Norway

A. M. A. Musleh Al-Sartawi et al. (eds.), *From the Internet of Things to the Internet of Ideas: The Role of Artificial Intelligence*, Lecture Notes in Networks and Systems 557,
https://doi.org/10.1007/978-3-031-17746-0_37

work on developing the foundations of digital sustainability reporting in organizations. Today, the accounting profession provides services similar to digital sustainability reporting services called accounting services to facilitate financial reporting. Still, to the best of my knowledge, neither research nor the profession seem to use their prior knowledge from financial reporting to guide client organizations in the implementation of sustainability reporting. I propose using the financial reporting rational in organizing sustainability information in organizations. This paper aims to create a digital sustainability reporting framework based on the financial reporting framework. The digital sustainability reporting framework can be relevant for the accounting profession and management when implementing digital sustainability reporting in organizations.

In this paper, digital sustainability reporting refers to reporting activities in organizations that incorporate the applicable sustainability reporting framework in their digital systems to provide stakeholders with high quality sustainability reporting information. This definition is inspired by definitions from prior research [e.g. 9] and the definition of Industry 5.0 era [10]. In general, sustainability is defined as a concept and a process that aim to improve environmental, economic, and social aspects for life preservation on our planet for our and future generations [11]. Industry 5.0 embraces both digitalization and sustainability and is defined by the EU by a "re-found and widened purposefulness, going be-yond producing goods and services for profit. This wider purpose constitutes three core elements: human-centricity, sustainability and resilience" [10].

Industry 5.0 embraces digitalization by inheriting digital transformation from the fourth industrial revolution [2, 12, 13]. Industry 4.0 has provided the technologies that helped accountants in the digitalisation of financial reporting. Similarly, Industry 5.0 provides the technological means and simultaneously enhances the importance of sustainability [10] to the accounting profession and digital system providers when implementing digital sustainability reporting in organizations. As such, digitalization is one of the building blocks of digital sustainability reporting, the other building blocks are provided by financial reporting research and theory.

External financial reporting is produced through software such as AIS or/and financial accounting modules in ERP systems [14]. Whether an organization relies on an AIS or an ERP system depends on the organization size. ERP systems are costly and often used by organizations that have more than 500 employees and 500 million dollars in sales [14]. A fully integrated ERP system in multinationals may take years to implement and millions of dollars to build. This explains why smaller organizations prefer an AIS instead of an integrated ERP system.

Conventionally, good quality financial reporting relies on AIS/financial accounting ERP module [14], compliance to the applicable accounting framework and good internal control [15]. Figure 1 illustrates the relationship between financial reporting, AIS, applicable framework and internal control as the four building blocks to produce good quality financial reporting. Along this paper, I will use the financial reporting framework in Fig. 1 to arrive to a digital sustainability reporting framework depicted in Fig. 3 in the end of the paper. The framework in Fig. 3 aims to help

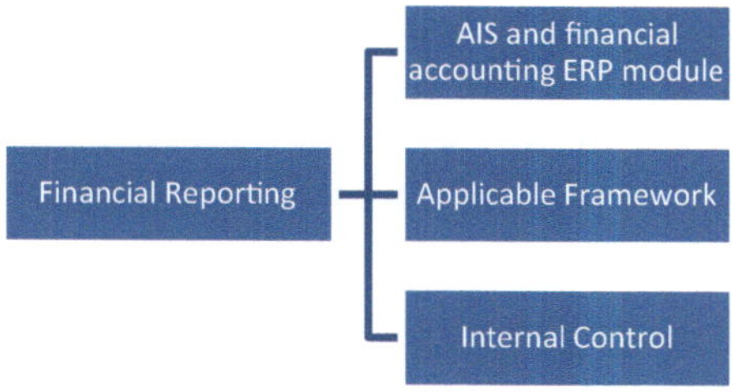

Fig. 1 Financial reporting framework

the implementation of digital sustainability reporting in organizations. Accounting research is warranted to further elaborate the framework through qualitative research.

The rest of the paper will proceed as follows. The four sections below, apply every building block from the financial reporting framework in Fig. 1 onto the domain of digital sustainability reporting in organizations. Finally, the paper ends with concluding remarks and unravels the digital sustainability reporting framework.

2 From Financial Reporting to Digital Sustainability Reporting

Digital sustainability reporting refers to reporting activities in organizations that incorporate the applicable sustainability reporting framework in their digital systems to provide stakeholders with high quality sustainability reporting information. For the sake of completeness, the research definitions that inspired the paper's definition of digital sustainability reporting are provided here.

Definitions of digital sustainability is to be found since the 2000s. The definitions are often referring to a technology or a digital resource. Technically, digital sustainability as a field could apply techniques from computer science, information science, operations research, and other fields interested in balancing environmental, economic and societal needs for sustainable development [16]. The concept of digital sustainability includes the digital resource and its community of people [17]. Other research have defined digital sustainability: "as the organizational activities that seek to advance the sustainable development goals through creative deployment of technologies that create, use, transmit, or source electronic data" [9]. Finally, research also considers digital sustainability as an interdisciplinary field intertwining digitalisation and sustainability [18]. Consequently, digital sustainability would easily adopt notions from the discipline of accounting and the AIS research.

3 From Accounting Information System (AIS) and Financial Accounting ERP Module to Digital Sustainability Reporting Systems and Modules

Traditionally, AIS or/and financial accounting modules in ERP systems produced external financial reporting [14]. To this end, digital system providers such as Microsoft Dynamics and SAP ERP systems, and providers of AIS systems, are adapting their systems to produce sustainability reporting adopting the same approaches from the digital financial reporting. Accounting Information System (AIS) is defined as a software producing financial reports that provides decision relevant information to shareholders and external stakeholders [19]. The AIS is either on premise or in the cloud and is often integrated or communicating with other ERP systems.

Similar to AIS, Enterprise Resource Planning (ERP) system is a software (but bigger than the AIS) that integrates more than the financial accounting function such as control management, production, logistics, sales management, marketing, and human resources [20]. One of the objectives of ERP system is to use databases (on premise or in the cloud) fast and effectively. Consequently, the ERP systems could merge many types of data to produce sustainability data for reporting purposes.

Researchers have been working on integrating sustainability data and transitioning the ERP systems into Sustainable ERP (S ERP) systems. Chofreh, Goni, Shaharoun, Ismail, and Klemeš [21] described S ERP as a holistic solution to support sustainability data integrated into traditional ERP system [22]. In practice, SAP [23] developed several modules to integrate sustainability data and ease sustainability reporting [24]. For instance, the "SAP Product Footprint Management" module can collect data on product CO^2 footprint. The Fig. 2 proposes a depiction of S ERP system based on the SAP ERP system.

Within the SAP module, data analytics capabilities are embedded. Clearly, applications analyzing and managing sustainability data provides opportunities to help organizations steer toward their sustainability targets [7, 12]. According to Ghobakhloo [12], sustainability data analytics are important in articulating and mitigating organizations negative impact on Environmental, Social and Governance (ESG) factors.

Another SAP module in the figure above, is called "SAP Sustainability Control Tower" [25]. This module establishes a central ledger for sustainability reporting and metrics. This module provides a warehouse to store the sustainability data, sustainability analytics, reporting based on the different sustainability standards offered today, and dashboards that provides sustainability Key Performance Indicators (KPIs).

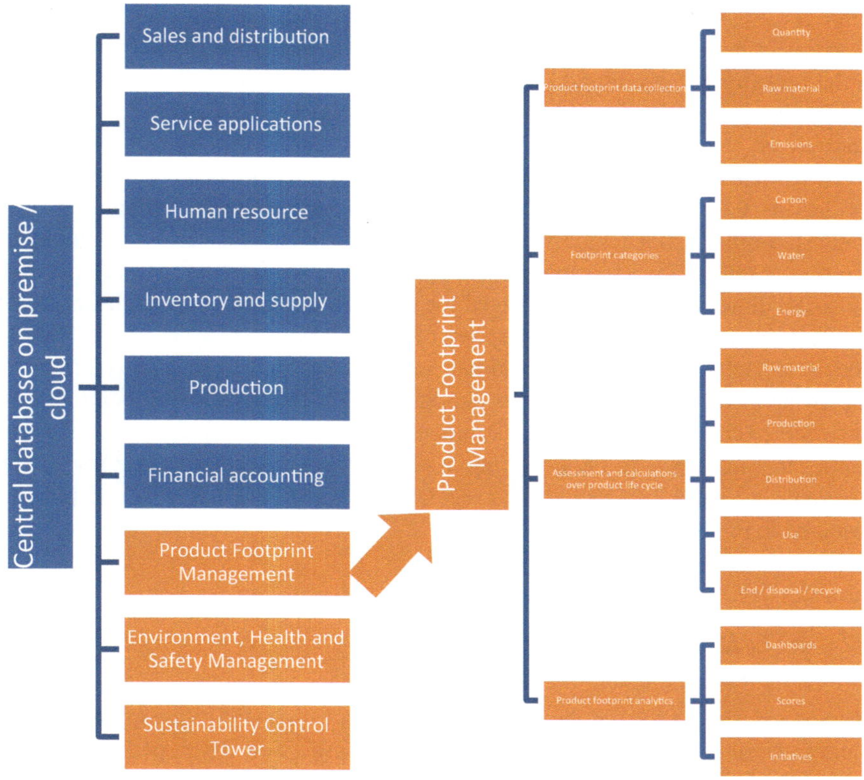

Fig. 2 S ERP model of sustainability modules based on SAP products

4 Current Applicable Framework for Sustainability Reporting

This section presents the ESG theory and philosophy, ESG regulations, and ESG standards that may provide organizations with the motivation behind implementing ESG based sustainability reporting [26].

4.1 Theory and Philosophy

Mainstream financial performance measurement had the unique ability to recognize, measure and report on financial performance. One traditional reported bottom line has been reflecting the financial performance of organizations. Now, organizations wish to add several bottom lines in their reporting to account for ESG factors [e.g. 27]. Clearly, the sustainability mindset may clash with traditional financial performance mindset [26]. Traditional financial performances in organizations followed a homo economicus mindset [28] where the human is selfish, rational, utility maximizer [29]. In the Industry 5.0 era, organizations are expected to act altruistic, ecological [10] and emphasize the ESG framework and are leaving the notion of a selfish profit utility maximizer.

The United Nations (UN) adopted the ESG dimensions [30] in many sustainability regulations. Chief among the ESG dimensions is the environmental dimension. This dimension reflects stakeholders' interest in how organizations' activities and operations may impact the environment (e.g., product CO^2emissions) [31, 32]. Second, the social dimension has been accepted for several decades, e.g., through Corporate Social Responsibility (CSR) and the work of Howard Bowen [33]. The social dimension is related to sustainability [27] and includes ensuring high quality products, improving customer satisfaction, increased focus on employees' wellbeing, health and safety [31]. Third, the governance dimension reflects governance risks and opportunities that influence decision-making in organizations. According to Rezaee [31], governance factors may include the regulations, rules, best practices and policies designed to improve corporate governance.

4.2 ESG Regulations

Leading in issuing ESG regulations is the European Union (EU). According to Singhania and Saini [34], EU lands are classified as having a well-developed ESG framework. USA has a medium to high developed ESG framework, while developed Asian countries have an ESG framework at developing stage [34]. Below an overview of major recent development in ESG regulations in the EU, Asia and the US.

The ESG framework of the EU is leading with its' European green deal in driving ahead several sustainability regulations.[1] Lately the EU adopted in 2021, a proposal for a Corporate Sustainability Reporting Directive (CSRD) [15]. The CSRD requires sustainability disclosures from all large companies and all listed companies [13]. A second regulation is called Taxonomy and is a classification system for environmentally sustainable activities in organizations. A third voluntary regulation is the EUropean Green Bond Standard (EUGBS). This proposed standard helps in raising green funds.

In Asia, recent years have seen an increase in ESG reporting [35, 47]. In 2018, the Chinese Securities Regulatory Commission launched the ESG information disclosure framework [36]. This framework developed a sustainability standard for listed companies' disclosures. In 2019, Hong Kong Exchanges and Clearing Limited (HKEX) issued guidelines for mandatory ESG disclosures for listed Companies

[1] Please consult the following paper for more details on the EU regulations: 24.Olsen, C., *An overview of key sustainability theories, regulations and S ERP for business education, business research and digital business practitioners*. 2022, Springer Nature: Lecture Notes in Networks and Systems. overview of recent key EUs regulations).

[36]. Similarly, India has established mandatory ESG requirements for 1000 listed companies. The Securities and Exchange Board of India (SEBI) introduced the format Business Responsibility and Sustainability Report (BRSR) that is mandatory from financial year 2022–23 [37].

Finally, in the United States (US), the Securities and Exchange Commission (SEC) recognizes that the US market is having trouble to keep pace with leading markets within sustainability regulations [38]. The first set of sustainability accounting standards for the US capital markets are designed by Sustainability Accounting Standards Board (SASB) and are voluntary to use in filings with the SEC [39].

4.3 Main Sustainability Standards

To translate the ESG regulations into sustainability disclosures in organizations, regulators and standard setters supported the creation of standards on ESG reporting. In fact, Truant [40] reports at least ten reporting standards in the world. Among the popular standards for large organizations, the Task Force on Climate-related Financial Disclosures (TCFD) [41] focuses on disclosures on climate-related metrics such as disclosures of Scope 1, Scope 2, and, if appropriate, Scope 3 greenhouse gas (GHG) emissions [42]. Another widespread used standards among large organizations is the Global Reporting Initiative (GRI) standards [40].

Other sustainability standards are forthcoming or widely used in other parts of the worlds. In Asia, the Hong Kong Exchanges and Clearing Limited (HKEX) issued standards on sustainability disclosures that are mandatory for listed companies. In India, the Business Responsibility and Sustainability Report (BRSR) format will be used for sustainability related disclosures and effective from FY 2022–23. In the US, the SASB [39] has developed industry standards identifying, managing and communicating financially-material sustainability information to their investors. These standards are voluntary standards on sustainability for US listed companies. In the Nordics, standards are being designed for SMEs called the Nordic Standard Sustainability Reporting Standards (NSRS) [1]. Finally, forthcoming standards developed by EFRAG are the new European Sustainability Reporting Standards (ESRS) [43].

In addition to the standards, sustainability reports include international metrics such as KPIs on sustainability developed by the World Economic Forum [44]. The KPIs are designed in collaboration with the accounting profession (Deloitte, EY, KPMG and PwC). Such international sustainability KPIs can be integrated into the digital sustainability reports.

5 Internal Control for Digital Sustainability Reporting

Best practices for internal control follow the Committee of Sponsoring Organizations of the Treadway Commission (COSO) framework [45]. COSO defines internal

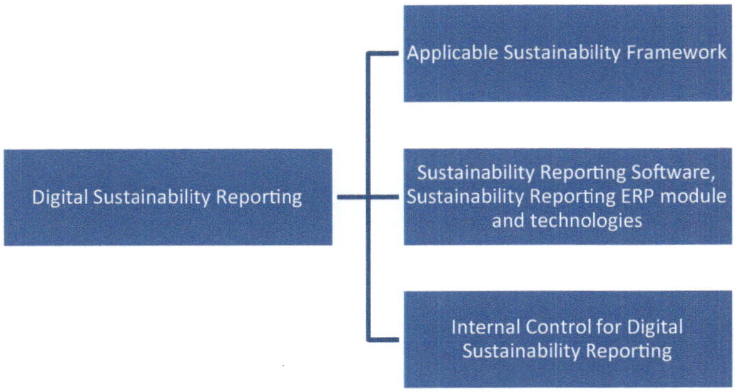

Fig. 3 Digital sustainability reporting framework

control as helping "entities achieve important objectives and sustain and improve performance" [45]. Such, internal control is crucial in providing timely relevant financial and sustainability reporting to the stakeholders [46]. Internal control includes five components: 1) control environment, 2) risk assessment, 3) control activities, 4) Information and communication, and 5) monitoring activities. Herz et al. [46] suggest applications for the five components to achieve good sustainability reporting.

The first component in COSO Framework is control environment. This component refers to the tone at the top set by top management. This means that top management and the board of directors have taken actions to set a strong tone of commitment to sustainability to prevent greenwashing. The second component of the Framework, risk assessment, it relates to the assessment of risks an entity faces either internally or externally that threaten the achievement of its sustainability objectives. For instance, an internal risk that may hamper digital sustainability reporting is the integration of new Information Technology (IT) systems. Internal changes and the lack of appropriate controls on the IT systems may put at risk the quality of data underlying the sustainability reporting. To prevent or mitigate such risks to the objective of timely relevant sustainability reporting, management can design control activities.

Control activities relate to activities in the organization that mitigates the risks that hinder the achievement of its objectives. Control activities may include manual and IT controls. Controls activities related to digital sustainability reporting would be regular performance reviews at all levels and across teams in the entity to check sustainability indicators. The information and communication component of internal control focuses on that timely, accurate and reliable information is disseminated internally and externally in accordance with regulations and stakeholders' expectations. Perhaps, this component ensures that ultimately the underlying information conveyed to the digital sustainability reporting is accurate. Finally, the component monitoring activities keeps track of whether the five COSO components are present and functioning effectively to detect and correct eventual errors in the internal control system.

6 Concluding Remarques and Digital Sustainability Reporting Framework

The framework in Fig. 3 provides a roadmap for the implementation of digital sustainability reporting in organizations. Accounting researchers may use the framework in their research on digital sustainability reporting. In practice, the framework for digital sustainability reporting helps the accounting profession and sustainability management level in in the implementation of digital sustainability reporting. First, I recommend identifying the objectives (e.g., which stakeholders the reporting is designed for) of digital sustainability reporting. Then, to identify the applicable sustainability framework of regulations and standards that the organization must or voluntary follows. Then, the best sustainability reporting software should be chosen. Finally, the internal control activities should be designed to match the identified digital sustainability reporting objectives.

References

1. Innovation N (2021) NSRS - nordic standard sustainability reporting for SMEs. Nordic Innovation, Oslo
2. Tiwari K, Khan MS (2020) Sustainability accounting and reporting in the industry 4.0. J Clean Prod 258:120783
3. Çalıyurt KT, Said R (2018) Sustainability and social responsibility of accountability reporting systems : a global approach. Springer, Singapore
4. Klymenko O, Halse LL, Jæger B (2021) The Enabling role of digital technologies in sustainability accounting: findings from norwegian manufacturing companies. Systems 9(2):33
5. SASB (2017) SASB Conceptual Framework. Sustainability Accounting Standards Board (SASB): San Francisco
6. EFRAG (2021) Conceptual Framework For Non-Financial Information Standard Setting. European Financial Reporting Advisory Group (EFRAG): Luxembourg
7. Chofreh AG et al (2020) Development of guidelines for the implementation of sustainable enterprise resource planning systems. J Clean Prod 244:118655
8. Chofreh AG, Goni FA, Klemeš JJ (2017) Development of a roadmap for sustainable enterprise resource planning systems implementation (part II). J Clean Prod 166:425–437
9. George G, Merrill RK, Schillebeeckx SJD (2020) Digital sustainability and entrepreneurship: how digital innovations are helping tackle climate change and sustainable development. Entrep Theory Pract 45(5):999–1027
10. Breque M, De Nul L, Petridis A (2021) Industry 5.0: towards a sustainable, human-centric and resilient European industry. European Commission
11. Brundtland GH (1987) Our common future: report of the world commission on environment and development. United Nations (UN), Geneva
12. Ghobakhloo M (2020) Industry 4.0, digitization, and opportunities for sustainability. J Clean Prod 252:119869
13. Yadav G et al (2020) A framework to achieve sustainability in manufacturing organisations of developing economies using industry 4.0 technologies' enablers. Computers in Industry 122:103280
14. Weygandt JJ, Kimmer PD, Kieso DE (2019) Accounting information systems, in accounting priciples IFRS version. Wiley, Singapore

15. Moen T, Havstein B, Kvalvik KME (2017) Accounting organization: business management and internal control, 7th edn. Cappelen Damm Akademisk, Norway
16. Institute for Computational Sustainability (2022) Computational Methods for a Sustainable Environment, Economy, and Society. https://www.computational-sustainability.org/
17. Stuermer M, Abu-Tayeh G (2016) Digital Preservation through Digital Sustainability. In: iPRES
18. Musleh Al-Sartawi AMA (eds) (2022) Artificial intelligence for sustainable finance and sustainable technology. ICGER 2021. Lecture Notes in Networks and Systems, vol 423. Springer, Cham
19. Belfo F, Trigo A (2013) Accounting information systems: tradition and future directions. Procedia Technol 9:536–546
20. Osnes KB et al (2018) ERP systems in multinational enterprises: a literature review of post-implementation challenges. Procedia Comput Sci 138:541–548
21. Chofreh AG et al (2014) Sustainable enterprise resource planning: imperatives and research directions. J Clean Prod 71:139–147
22. Chofreh A et al (2016) A master plan for the implementation of sustainable enterprise resource planning systems (part I): concept and methodology. J Clean Prod 136:176–182
23. SAP (2021) SAP Sustainability Summit - Virtual Live Experience over two days. SAP
24. Olsen C (2022) An overview of key sustainability theories, regulations and S ERP for business education, business research and digital business practitioners. Springer Nature: Lecture Notes in Networks and Systems
25. SAP (2022) Products: SAP Sustainability Control Tower. SAP
26. Scruton R (2013) Green philosophy: how to think seriously about the planet. Atlantic Books Ltd
27. Shim J et al (2021) The impact of CSR on corporate value of restaurant businesses using triple bottom line theory. Sustainability 13(4):2131
28. Becker C (2006) The human actor in ecological economics: philosophical approach and research perspectives. Ecol Econ 60(1):17–23
29. Simon HA (1990) Invariants of human-behavior. Annu Rev Psychol 41:1–19
30. UN and FDFA (2004) The Global Compact - Who Cares Wins: Connecting Financial Markets to a Changing World, in United Nations (UN) and Swiss Federal Department of Foreign Affairs (FDFA), UN, Editor
31. Rezaee Z (2015) Business sustainability: performance, compliance, accountability and integrated reporting. Greenleaf Publishing, UK
32. GSSB (2016) Global Reporting Initiative (GRI) 101: Foundation. Global Sustainability Standards Board (GSSB): Amsterdam
33. Bowen HR (1953) Social responsibilities of the businessman. University of Iowa Press
34. Singhania M, Saini N (2021) Quantification of ESG regulations: a cross-country benchmarking analysis. Vision 09722629211054173
35. Alsayegh MF, Rahman RA, Homayoun S (2020) Corporate economic, environmental, and social sustainability performance transformation through ESG disclosure. Sustainability 12(9):3910
36. PwC (2020) Environmental, Social and Governance (ESG) in Asia-Assets and Wealth Management. PricewaterhouseCoopers, PwC
37. Bhatia, A., *Environmental, Social, and Governance Compliance in India: New Reporting Requirements*. 2021, India Briefing.
38. Guillot J (2020) US Government Accountability Office: ESG Disclosure Is Prevalent but Problematic. Sustainability Accounting Standards Board (SASB)
39. SASB (2022) Standards Overview. 2022, Sustainability Accounting Standards Board: California
40. Truant E, Corazza L, Scagnelli SD (2017) Sustainability and risk disclosure: an exploratory study on sustainability reports. Sustainability 9(4):636
41. FSB (2017) Final Report-Recommendations of the Task Force on Climate-related Financial Disclosures. Financial Stability Board (FSB)

42. WBCSD and WRI (2004) World Resources Institute and World Business Council for Sustainable Development: USA.
43. Lemessiou L, Slomp S (2020) sustainability reporting standards roadmap. European Financial Reporting Advisoy Group (EFRAG)
44. WEF (2020) Measuring stakeholder capitalism towards common metrics and consistent reporting of sustainable value creation. World Economic Forum, Switzerland
45. COSO (2013) Internal control — integrated framework: executive summary, framework and appendices, and illustrative tools for assessing effectiveness of a system of internal control. The Committee of Sponsoring Organizations (COSO) USA
46. Herz RH, Monterio BJ, Thomson JC (2017) Leveraging the COSO internal control—integrated framework to improve confidence in sustainability performance data. In: COSO Framework and sustainability. Institute of Management Accountants
47. Al-Sartawi A (2019) Assessing the relationship between information transparency through social media disclosure and firm value. Manag Account Rev 18(2):1–20

The Impact of Hris Usage on Organizational Efficiency and Employee Performance: A Research in Higher Education Sector in Jordan

Mohd Bashir Ra'ed Bani Ismail, Mohd Saiful Izwaan Saadon, Jasem Taleb Al-Tarawneh, and Aya Naser Maqableh

Abstract Human Resources as a strategic resource one of an organization's human resources and primary strategic management techniques is management. Organizations must keep track of changes in their internal and external climates on a regular basis. Developing countries like Jordan, on the other hand, appeared to have difficulty implementing HRIS. The impact of HRIS on organizational performance has been investigated in previous research. The focus of this study is to investigate the influence of the use of HRIS in higher institutions in Jordan on organizational efficiency and employee performance. The research used SmartPLS 3.3.2 to model partial least square structural equations to examine 326 questionnaires completed by staff working at higher institutions in Jordan. The rate of response was 65.20 percent. The results showed that the use of HRIS had a substantial positive impact on both employee performance and organizational efficiency, while the latter served as a partial mediator in the higher institutions of Jordan between the usage of HRIS and employee performance. The results of the study are anticipated to be useful for human resource managers to further study the contribution of HRIS in implementing and recognizing the conductive effects of HRIS applications for academics as well as for organizations.

Keywords Employees' performance · HRIS usage · Organizational efficiency

1 Introduction

It has become generally accepted that within organizations, internet technology has a critical role. The success of an institution in today's knowledge-based economy ultimately determines the efficiency of its human resources. HRIS is described as 'a system used to obtain, preserve, manage, evaluate, retrieve and provides insights information about the human resources of an organizations' [7]. HRIS can still include a range of advantages for Organizational structure as well as for line managers

M. B. R. B. Ismail (✉) · M. S. I. Saadon · J. T. Al-Tarawneh · A. N. Maqableh
Faculty of Business, Universiti Malaysia Terengganu, Kuala Nerus, 21030 Terengganu, Malaysia
e-mail: Rayanababneh7@gmail.com

A. M. A. Musleh Al-Sartawi et al. (eds.), *From the Internet of Things to the Internet of Ideas: The Role of Artificial Intelligence*, Lecture Notes in Networks and Systems 557,
https://doi.org/10.1007/978-3-031-17746-0_38

and the institution as a whole. Unpreventable, competitive and dynamic business environments compel organizations to coordinate a rapid adaptation of the internal operational climate and reactions with new techniques. Incredibly, organizations use information systems as technology advances and globalization increases. HRM has undergone major changes in the recent past, shifting from organizational management activities to the role of a technology relationship. In all sectors, including the higher institutions, there are many advanced programmed that are useful for handling and arranging staffing. Recent rapid development in HRIS has emphasized its significance and prospective effects on the performance of an organization [6, 10] There is increasing concern regarding the effects on organizational performance and efficiency of HRIS usage [11] HR operations will gradually be forced to converge with computerization by the latest legal standards and the trend of digitization. HRM has faced the pressure since then to play a more strategic role in helping organizations meet their long-term organizational goals. An organization's greatest efficiency and performance are not based exclusively on its external funding or on the implementation of new technologies, but rather on the degree to which and how it uses its committed, empowered and high capital. Therefore, HRM has been one of the programmers or divisions that often use information systems for management [12]. HRIS can be classified as reducing permanent expenditures and lowering the cost of looking for work recruits, regulatory compliance, retirement education, role monitoring (staff stipulation), participation and analysis reports, disaster monitoring and preventing, promotion planning, and hazard reporting and prevention. There are, however, few researches on the effect of HRIS usage on the performance of HR workers in developing countries. No study has yet been performed in relation to our study context, Jordan. While institutions recently invested an enormous amount of money on adopting different information systems, their adoption and usage by end-users in developing countries especially in Jordanian higher institutions remains an essential problem. There are, however, few researches on the effect of HRIS usage on the performance of HR workers. But it continues poorly in Jordan, where, according to [4], there is still strong resistance to the adoption of technology. The primary aim of this study is, therefore, to examine the influence of HRIS usage on organizational efficiency and employee's performance in the Jordanian higher institutions.

2 Review of Literature and Theoretical Perspective

2.1 HRIS Usage

HRIS usage are the quantity of HR-related administrations accessible on an association's HRIS conveyance channels. Analysts have made different recommendations with respect to how HRIS usage may uphold hierarchical cycles [4]. The degree of vital HRIS use across associations shifts, and the greater part of the associations just keep on utilizing HRIS. For instance, [3] have revealed a few usages for HRISs, such as recruitment and determination; turnover following/work investigation; finance, advantages and pay (the executives); preparing and advancement; performance analysis; interior and outer correspondence and progression HR arranging.

2.2 *Employees Performance*

HRISs are important because they can enhance the performance of organizations and employees. Employee performance is a subject that is constantly studied by academic researchers and realistic managers. Employee commitment is the work force commitment to the business objectives. HRISs can likewise help the executives produce exceptionally successful performance surveys to additionally improve performance the board [1].

2.3 *Organizational Efficiency*

Organizational efficiency can be defined as the efficient use of human and material resources. It is about minimizing waste and optimizing the opportunities of resources in order to provide consumers with better services. Reducing costs is the best solution in the face of tough competition, as internal waste leads to higher costs [2]. In an unprecedented competitive climate that has never been matched, companies in the business world today have to work. Personnel efficiency is the commitment of personnel to company objectives. High-performing individuals are those who have effectively discharged their duties and obligations. The main achievement basis for HR is the evaluation of performance (achievement) [9]. HRIS use in HR would diminish the mechanization expenses of the information and the quantity of workers yet at the same time add to representatives' checking their own information. It would offer HRIS directors' simple admittance to the pertinent information and information, and permit them to settle on choices and speak with others without talking with HR experts [11]. It was exhibited in the study on HRIS Use and Impacts that there are some normal contrasts in HRIS use in SMBs and huge organizations as to works for deciding the size of HRIS' use and use on key and operational level. Moreover, they demonstrated that despite the fact that HR experts are furnished with uncommon HRIS use because of the expert vital association, they had lower experience contrasted with people from different businesses. It showed the effect of individual and e-HR framework qualities on four significant performance factors (information stream, social cooperation's, seen control and framework acknowledgment) in their study on adequacy and acknowledgment of Human Resources Systems and variables influencing these, and they offered a progression of models.

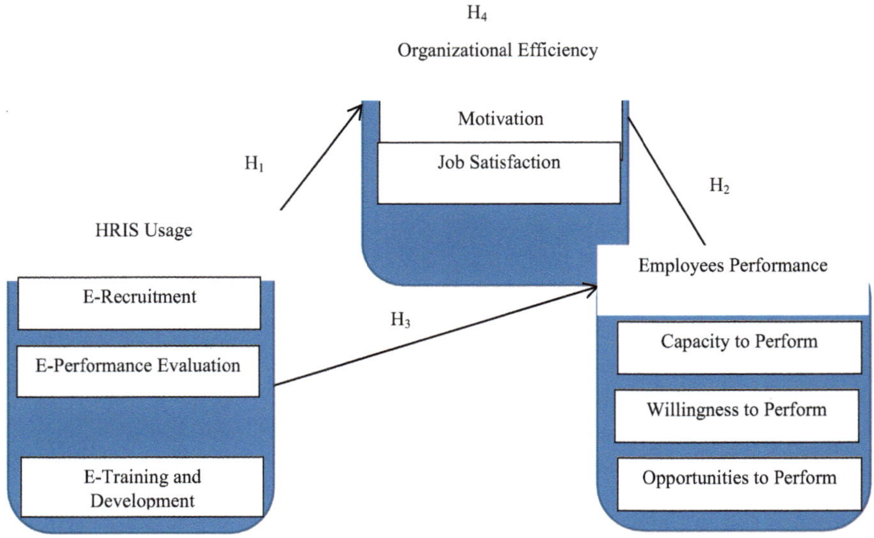

Fig. 1 Model of the research

2.4 Model of the Research

The model described here was proposed to investigate the effect of e-recruitment, e-performance and evaluation on employee performance by mediating influence of organizational efficacy. Three main classes of variables compose the conceptual framework. These are: HRIS usage as the independent variable; organizational efficiency (motivation and job satisfaction) as the mediating variable; and employees' performance as the dependent variable (Fig. 1).

2.5 Research Methodology

About 3300 h workers employed in higher institutions in Jordan were the aim respondents. 326 questionnaires were obtained by the researchers, which represented a 65.20 percent response rate. The content of the questionnaire (structures and measures) was chiefly selected and adapted from previous related studies.

3 Results of the Study

3.1 Measurement Model: Reliability and Validity

The reliability and validity for each construct was assessed, Cronbach's alpha results for the overall measurement model and convergent validity are presented in Table 1. AVE was put to use for the assessment of Convergent Validity showing a degree measuring positive correlates with measures that were alternative but of the same construct.

3.2 Discriminant Validity

The acceptable value for discriminant validity is 0.850 or less. The Fornell-Larcker criterion has been questioned (Table 2).

3.3 Structural Model and Hypotheses Testing

A bootstrapping method with a sample of 5,000 was analyzed the structure model by computing beta (β), R2, and the corresponding t-values. The relationship between HRIS use, organizational efficiency and employee performance was assessed within the structural model. Figure 2 and Tables 3 and 4 present standard regression weights together with the hypothesized direct effects.

At a p-value of less than 0.05, hypotheses H1, H2 and H3 are important. The CR and p-values of HRIS use in predicting employee performance were 2.222 and 0.026, respectively. This implies that employee performance rises by 0.119 standard deviations when HRIS usage rises by one standard deviation. Employee performance improves by 0.618 standard deviations when organizational efficiency is one standard deviation above or below 0.001. This implies that the mediating role of organisational efficiency between the use of HRIS and the performance of employees in the higher institutions of Jordan was mediating.

3.4 Discussions

The aim of this study was to analyze the influence of the use of HRIS in higher institutions in Jordan on organizational HR workers are required to be attentive, respond faster to questions, and provide more reliable information. HRIS can also allow these staff to be more autonomous. As indicated in hypothesis H2, the study presented empirical evidence concerning the influence of organizational efficiency

Table 1 The measurement model

Variables	Loading	CA	CR	AVE
Employees performance		**0.948**	**0.953**	**0.506**
Capacity to perform		**0.926**	**0.941**	**0.695**
CP1	0.846			
CP2	0.833			
CP3	0.736			
CP4	0.820			
CP5	0.871			
CP6	0.856			
CP7	0.867			
CP8	Deleted			
Opportunities to perform		**0.875**	**0.906**	**0.619**
OP1	Deleted			
OP2	Deleted			
OP3	0.825			
OP4	0.812			
OP5	0.791			
OP6	0.837			
OP7	0.777			
OP8	0.665			
Willingness to perform		**0.929**	**0.942**	**0.673**
WP1	0.654			
WP2	0.854			
WP3	0.825			
WP4	0.811			
WP5	0.857			
WP6	0.849			
WP7	0.871			
WP8	0.820			
Organizational efficiency		**0.954**	**0.959**	**0.628**
Job satisfaction		**0.934**	**0.947**	**0.718**
JS1	0.853			
JS2	0.799			
JS3	0.846			
JS4	0.871			
JS5	0.875			
JS6	0.806			

(continued)

Table 1 (continued)

Variables	Loading	CA	CR	AVE
JS7	0.879			
Motivation		**0.922**	**0.939**	**0.721**
M1	0.772			
M2	Deleted			
M3	0.849			
M4	0.888			
M5	0.884			
M6	0.843			
M7	0.854			
HRIS usage		**0.941**	**0.948**	**0.503**
E-recruitment		**0.884**	**0.916**	**0.686**
ER1	0.778			
ER2	0.887			
ER3	0.879			
ER4	0.819			
ER5	Deleted			
ER6	Deleted			
ER7	0.771			
E-performance and evaluation		**0.895**	**0.918**	**0.617**
EPE1	0.821			
EPE2	0.866			
EPE3	0.868			
EPE4	0.815			
EPE5	0.761			
EPE6	0.688			
EPE7	0.653			
E-training and development		**0.831**	**0.877**	**0.544**
ETD1	0.658			
ETD2	0.698			
ETD3	0.690			
ETD4	0.757			
ETD5	0.803			
ETD6	0.807			

Table 2 Fornell-Larcker criterion analysis discriminant validity

	Employees performance	HRIS usage
Employees performance	**0.711**	
HRIS usage	0.664	**0.709**

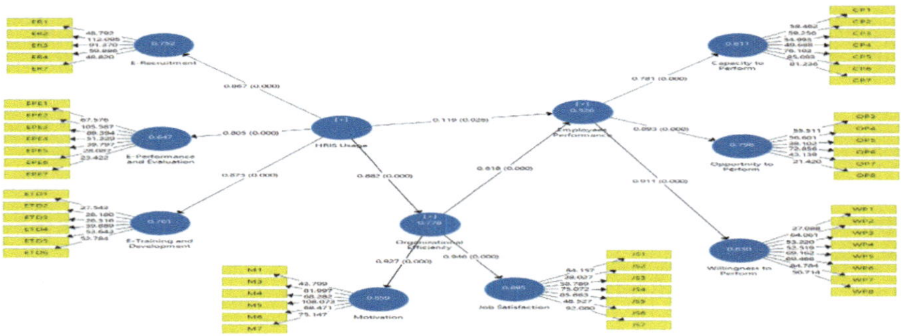

Fig. 2 The standardized result

Table 3 The path coefficients

Hypothesis		Coefficient	St. dev.	T statistics	P values	Decision
H_1	HRIS usage –> Employees performance	0.119	0.054	2.222	0.026	Supported
H_2	HRIS Usage –> Organizational efficiency	0.882	0.009	95.972	0.000	Supported
H_3	Organizational efficiency –> Employees performance	0.618	0.051	12.097	0.000	Supported

Table 4 The mediating effect

Hypothesis		Coefficient	St. dev.	T statistics	P values	Decision
H_4	HRIS usage –> Organizational efficiency –> Employees performance	0.545	0.046	11.792	0.000	Supported

on employee performance. The findings showed that HRIS had an important and direct positive influence on the output of employees. The use of HRIS also had a direct impact on organizational efficiency, and this impact was also significant. No previous empirical research studies examined the mediating impact of organizational efficiency on the relationship between HRIS and employee efficiency.

References

1. Alansari Y, Al-Sartawi A (2021) IT governance and E-banking in GCC listed banks. Procedia Comput Sci 183:844–848
2. Al-Sartawi A (2021) The big data-driven digital economy: artificial and computational intelligence. In: Kacprzyk J (ed) Studies in computational intelligence, vol 974. Springer, Heidelberg
3. Al-Sartawi A (2020) Social media disclosure of intellectual capital and firm value. Int J Learn Intellect Cap 17(4):312–323
4. Balogun MA, Egharevba OJ (2020) Assessment of organisations incentives in enhancing employees'performance in Nigeria: a study of obafemi awolowo university, ile-ife, osun state. Fulafia J Social Sci 3(1):13–26
5. Barišić AF, Poór J, Pejić Bach M (2019) The intensity of human resources information systems usage and organizational performance. Interdisc Desc Complex Syst INDECS 17(3-B):586–597
6. Hosain MS, Arefin AHMM, Hossin MA (2020) The role of human resource information system on operational efficiency: evidence from MNCs operating in Bangladesh. Asian J Econ Bus Account 18:29–47
7. Jayabalan N, Makhbul ZKM, Selvanathan M, Subramaniam M, Nair S, Perumal I (2020) HRIS
8. Mahadik R, Ayarekar S (2020) Impact of size of organization on usage of HRIS. Asian J Multidisc Stud 8:5
9. Manley SC, Hair JF, Williams RI, McDowell WC (2020) Essential new PLS-SEM analysis methods for your entrepreneurship analytical toolbox. Int Entrep Manag J 17:1–21
10. Al-Sartawi AMAM (ed) (2022) Artificial intelligence for sustainable finance and sustainable technology. ICGER 2021. Lecture Notes in Networks and Systems, vol 423. Springer, Cham
11. Nixon C (2019) Global human resource management. Scientific e-Resources
12. Reina R, Scarozza D (2020) Human resource management in the public administration. In: Organizational development in public administration. Palgrave Macmillan, Cham, pp 61–101
13. Shihadeh F (2020) Online payment services and individuals' behaviour: new evidence from the MENAP. Int J Electron Bank 2(4):275–282

Impact of Firm Characteristics on Related Party Transactions: Evidence from Listed Real Estate Companies of GCC

Raj Bahadur Sharma, Gagan Kukreja, Atul Bansal, and Mariam Waleed Alhamar

Abstract The aim of this study is to explore the impact of firm characteristics on Related Party Transactions (RPTs) of listed real estate companies of GCC. The firm characteristics consist of board independence, board size, profitability (ROA), and financial leverage. A correlation and multiple regression analysis conducted on data collected from thirty listed real estate companies in the GCC region. The correlation analysis shows that all four independent variables have a weak positive relationship with the RPTs. The regression analysis results partially support the correlation results as they shows that the board size and leverage have a significant impact on the RPTs. The results of the data analysis imply that companies should focus on having a small-sized board of directors and companies with a high level of leverage may increase their RPTs, to overcome these situations. The study suggests that the regulations regarding RPTs should increase and that companies should disclose the information regarding RPTs should be disclosed in one uniform structure.

Keywords Related party transactions · Board size · Board independence · Profitability · Leverage

1 Introduction

Society has recognized the importance of the development of the accounting field after it had faced the scandals that affected it. Scandals like Enron and Parmalat showed that people in power often use tools to benefit themselves and modify the

R. B. Sharma (✉) · G. Kukreja · A. Bansal · M. W. Alhamar
College of Business Administration, University of Bahrain, Zallaq, Kingdom of Bahrain
e-mail: rsharma@uob.edu.bh

G. Kukreja
e-mail: gkukreja@uob.edu.bh

A. Bansal
e-mail: aradhey@uob.edu.bh

M. W. Alhamar
e-mail: mariam.w.alhamar99@gmail.com

A. M. A. Musleh Al-Sartawi et al. (eds.), *From the Internet of Things to the Internet of Ideas: The Role of Artificial Intelligence*, Lecture Notes in Networks and Systems 557,
https://doi.org/10.1007/978-3-031-17746-0_39

accounting information presented in the financial statements to serve their interests [10, 14]. Thus, the financial statements lose their purpose in helping decision-makers by being biased and incorrect. Accounting regulators and standard setters have been investigating special transactions that would serve as means for controlling interests to exercise opportunistic behaviour that harms minority interests for their gain [5, 8]. Related Party Transactions (RPTs) grabbed the attention of many accounting regulators following the recent scandals. RPTs actions are transactions conducted between the company and a person or entity related to the reporting company. Companies normally perform these transactions to lower the transactions and operational cost and also to support their growth facing the challenges of the incompetence of local markets by creating an efficient market within the companies themselves [2, 14]. However, despite its benefits, RPTs have attracted some scepticism due to their special nature, many regulators view RPTs as an indicator of potential fraud [1].Controlling interests can use these transactions in earning management and transfer companies' resources to themselves. RPTs were identified as one of the most significant tools that managers can abuse to manipulate the financial statements as many of the regulators have found evidence linking these transactions to those scandals [4, 7].

In order to prevent these scandals from reoccurring accounting, regulators have enforced regulations on RPTs in order to restrict managers from using these transactions in opportunistic behaviour. The accounting standards have identified the transactions and the related party that needs to be disclosed in the financial statement.

The International Accounting standard board (IASB) (2010) has identified RPTs in its International Accounting Standards No. 24 as "A transfer of resources, services or obligations between reporting entity and a related party, regardless of whether a price is charged". Several previous studies have used this definition while discussing related party transactions [8, 13]. The IASB has also identified the individuals and entities that may be considered as related parties. As for individuals, IAS 24 has specified them as a person, or a close relative to that person, that either has control or significant influence over the company that is considered as the reporting entity or they are a key management member of the company or its parent. Moreover, there are many circumstances that make an entity be considered as a related party to the company. [19] among those circumstances are the entity being connected to the company by being members of the same group, whether it is a parent and subsidiary relationship, or them being subsidiaries to the same parent. In addition, the entity can be a joint venture of the company. Furthermore, an entity can be considered as a related party if it is being controlled or significantly influenced by an individual who is a related party to the company.

Accounting Theories Related to the Related Party Transactions

RPTs can be used as a tool that allows controlling interests to commit fraud for their benefit at the expense of the minority interest. On the other hand, RPTs are utilized to benefit the whole company. This divided the researchers into adopting two main viewpoints of these transactions. Each group have used an accounting theory to support their view of the related party transactions. These accounting theories are conflict of interest theory and efficient transaction theory.

Conflict of interest theory suggests that controlling shareholders would use RPTs to benefit themselves at the expense of minority shareholders. Researchers like [5] and [11] have called this act tunnelling or self-dealing. This theory suggests that the nature of related party transactions allows controlling parties to transfer assets and wealth to themselves. There are many forms that controlling parties can abuse their control using these transactions. First, companies can enter into a sale or purchase transaction with related parties at higher or lower than normal prices that benefit these parties but harm the companies' performance. In addition, companies may grant loans to these parties with no regard to their negative impact on the companies. Moreover, a company's performance may be harmed if managers were paid a large amount of compensation that disrupts its resources [1, 5, 11]. RPTs can also present an opportunity to managers to manage their companies' earnings to avoid the risk of delisting and facilitatethe issuing of new shares [11].

On the other hand, Efficient transaction theory or can be called propping views the RPTs as tools that benefit the entire company [4]. These transactions offer the companies some benefits compared to its arm-length transactions [16]. First, these transactions save companies' time and effort by conducting transactions with related parties that require less negotiation. Furthermore, companies can overcome difficult financial situations through the support of related parties. Moreover, companies can lower their transactions, operating costs and increase the effectiveness of the resource's distribution. Finally, companies conduct these types of transactions to create an internal market facing the lack of local markets' efficiency [6, 7, 12].

1.1 Research Objective

The objective of the study is to examine the firm characteristics on RPTs of listed real estate companies of GCC. The problem and the objectives of this study achieved through reviewing previous literature related to the related party transactions. In addition, analyzing data collected from 30 GCC listed companies from the real estate sector through the period of five years (from 2016 to 2020).

2 Literature Review

Many previous studies studied different aspects concerning the RPTs such as their effects, potential risks and determinants. These studies were conducted in different areas and durations. These previous studies are summarized as follows:

Desoky, A. et al. [5] are one of the few studies that have examined what can determine the RPTs in the companies. It focused on how governance aspects of the companies can impact their related party transactions. Their study consisted of four main GCC countries which are Bahrain, Saudi Arabia, Kuwait and the United Arab Emirates. It included listed companies from all sectors except the financial sector, the sample reached up to 460 companies for the years between 2016 and 2018. They conducted a correlation analysis between different types of RPTs and governance characteristics in companies such as the number of board members, audit quality and ownership concentration. They reached the conclusion that only one type of related party transaction, management compensation, have been significantly impacted by companies' governance characteristics. Another study that was conducted in an Arabian country is a study done by [1]. It studied the impact of both financial and governance characteristics of companies, mainly the profitability, ownership concentration, leverage, cashflows, board independence and separation between CEO and chairman positions, in relation to their related party transactions.

On the other hand, the study found that the separation between CEO and chairman position and the level of independence of the board are negatively associated with related party transactions. A very similar study to the previous one is conducted by [10] which have had similar objectives and variables. It was conducted on listed companies in Egypt. The study provided a few additional findings that differ from the previous one. It resulted in deducing that RPTs have a positive association with the size of the company, while it is negatively associated with the government's influence over the companies. The study found that most types of RPTs lead to an increase in companies' audit risk. [3] have stated that RPTs often lead to damaging companies' overall value. They used this revelation as evidence of the conflict- of-interest theory that specifies how RPTs serve as tools that controlling interests use to benefit themselves at the expense of minority interests. This study has reached these results by conducting a descriptive analysis over a sample gathered from listed companies in Spain from 2004 to 2012. [7] is a study that aimed at reviewing previous studies regarding related party transactions. It focused on several matters such as the adverse effects generated from these transactions as mentioned in the conflict-of-interest theory. In addition, it studied the efforts done to limit opportunistic use of RPTs through enforcing strict regulations over these transactions.

Moreover, the study looked into other matters such as related party transactions' role in increasing the audit risk and the various RPTs variable measures used in previous studies. The study concluded that weak regulations and governance structure lead to utilizing RPTs in an opportunistic manner which increases the audit risk of these companies. [14] is done in a similar approach as [7] as it analysed the work done by previous researchers of the related party transactions. But it is also worth noting that this study has focused on mainly one topic concerning RPTs which was examining the two main theories associated with related party transactions, the efficient transaction theory and the conflict- of-interest theory. It deduced that RPTs cannot be viewed fully as harmful or beneficial. It suggested a new framework that combines both views to improve the monitoring of these transactions. [3] have examined a very distinct matter regarding the related party transactions. Their study was mainly a comparison between the two major regulations of the disclosure of related party transactions, which are the International Accounting Standards regulations issued by the International Financial Regulations Standards (IFRS) and the regulations issued by the Security Exchange Community (SEC). [8] the study observed thirty companies selected from the New York Stock Exchange (NYSE) to examine how the companies comply with both of these regulations. The study has found that there is no significant difference between the requirement of the two regulations and most companies combine the RPTs required by the two regulations. [9] in Korea took a sample of 2712 companies from the period between 2011 to 2015, to examine how their board characteristics affect their related party transactions [15].

One of the few studies that have provided evidence that support the transaction efficiency theory is the one conducted by [12]. The study observed how RPTs affect the performance of 300 Malaysian companies taken as a sample for the study. They took the characteristics of the companies' board directors into consideration. The study has found that companies that engage in transactions with their subsidiaries and related entities have higher overall performance. In addition, it was concluded that having independent directors does not impact the effects of related party transactions.

Another study that has explored the governance factors that determine the related party transactions conducted and these transactions' effect on the companies' performance is performed by [16, 17]. The study has taken the listed in the Indian markets after eliminating the financial and government-owned entities, they reached 246 companies between the years 2008 and 2011. The most significant finding of the study was consistent with the similar studies discussed earlier, but it also found results that are distinct from other studies. For instance, the study has not found any association between ownership concentration and related party transactions. Moreover, the study did not find a significant relationship between board independence and related party transactions. However, evidence has proved that transactions performance market share declines when companies' conduct transactions with related parties. [2] have concentrated on the correlation between revenues generated by RPTs and the profitability of the companies. they have collected data from 100 Italian companies in 2011 to test this situation. They have reached the conclusion that companies that struggle in having high profitability resort to conducting RPTs to generate profit.

3 Research Methodology

This section discusses the process of establishing the sample used to test the study's hypothesis. In addition, the variables are exploring the hypothesis and the model of the study constructed in order to test the suggested hypothesis.

3.1 Sample Selection and Scope

To achieve the objectives of the study, a sample of 30 real estate listed companies were selected for the study's analysis. The sample is derived from the GCC region, it includes two companies from the Kingdom of Bahrain, one company from Oman, four companies from the Kingdom of Saudi Arabia, two companies from the United Arab Emirates, two companies from Qatar and nineteen companies from Kuwait. The companies were derived from a single industry in order to improve the comparability of the sample in aspects of industry conditions and regulations. The data is collected for the duration of five years from 2016 to 2020.

3.2 *Variables of the Study*

3.2.1 Dependent Variable

The objective of this study is to examine the characteristics that the companies possess that would positively or negatively affect the RPTs. Thus, the RPTs are set as the dependent variable of the study. As discussed in the conceptual framework part in the introduction section, the RPTs are identified as the transactions that companies conduct with parties that are identified as a related party.

As the topic of RPTs only has recently gained the interest of the accounting field, different measures and classifications were used among the studies. For instance, some researchers used the total amount of all types of RPTs or certain types of these transactions [10]. Others used the number of RPTs types used for the measurement of the variable [5, 18]. In this study, the log of one type of RPTs is used to represent the dependent variable. This method of measurement is chosen because it signifies a major amount among the related party transactions conducted by the sampled companies. This method also helps avoid misstatement in the classification of these transactions as most companies classify these transactions differently.

3.2.2 Independent Variables

The determinants of the RPTs form the independent variables that the study aims to examine their effect on the dependent variable. The study has four independent variables including two governance variables, the board size and independence. The study would also include two financial variables, profitability and leverage. Each of the independent variables would be discussed as follows:

Board Size

The board of directors are crucial to ensure that the company they are operating within is working effectively to accomplish the objectives and strategies set to maintain its success and serve the interest of the shareholders. Therefore, companies take great care in selecting their board members. There are conflicting opinions regarding the proper size of the board that may serve the company best.

Many conflicting opinions exist regarding the impact of a large-sized board of directors on the company and the level of RPTs it conducts. Some researchers viewed that companies that have large board sizes have more effective monitoring systems and lower the risk of engaging through opportunistic behavior through RPTs [10]. Other researchers state that having large boards in the companies might disrupt its operations as it would cause disputesin the interests of the members of the board of directors. They have also stated that companies with a large board of directors have damaged communications that present opportunities for controlling interests to abuse the RPTs for their benefits. Thus, they have concluded that a large board of directors positively affects the RPTs conducted by the companies [1, 5]. Most of the researchers found evidence that supports this assumption in their studies. We hypotheses the following hypothesis:

H01: There is no relationship between board size and RPTs.

Board Independence

As discussed in the previous sections, RPTs have gained scepticism due to the difficulty in distinguishing between harmful related party transactions from normal transactions. However, the accounting field has found that companies with good governance structures have less risk of committing fraudulent related party transactions. One of the main indicators of good governance structure in companies is the percentage of independent directors present on the board of directors. These directors are viewed as unbiased and have no interest and relation to the company. Researchers suggest that the presence of independent directors prevents controlling interest from abusing the company's resources through related party transactions. Most previous studies agreed on the fact that the level of independence negatively impacts RPTs [9]. We hypotheses the following hypothesis:

H02: There is no relationship between board independence and RPTs.

Leverage

Many researchers overlooked the financial characteristics of companies as aspects that affect the RPTs or only utilized them as control variables. However, studies like [1] and [10] examined these financial characteristics in a new light. These studies suggested that companies with high obligations and debts may resort to RPTs to improve their company's condition. In addition, the study by Abdullatif et al. (2019) indicated that auditing regulations perceive high leverage as an indicator of the risk of committing fraud. Therefore, the leverage is assumed to negatively impact the RPTs of the company. There are many ratios that previous studies use to measure companies' leverage such as debt to equity ratio and debt to assets ratio. The debt to assets ratio is used in this study to measure the leverage variable. We hypotheses the following hypothesis:

H03: There is no relationship between Leverage and RPTs.

Profitability

Another financial characteristic that that can affect the conducted RPTs is the profitability of the companies. Most previous researchers stated the possibility of having a negative relationship between the profitability and related party transactions, as companies with high profitability have the incentive in conducting fraudulent RPTs

Table 1 Variables definitions and measurement

Name	Measurement
Dependent variable	
Related party transactions (RPTs)	The log of monetary amounts of one type of related party transactions utilized to represent the dependant variable RPTs
Independent variables	
Board size (BDS)	The number of members in the board of directors
Board independence (BDI)	The percentage of independent members in the board of directors. Calculated by dividing the number of independent members over the number of members in the board of directors
Leverage (LEV)	Total debts divided by the total assets
Profitability (ROA)	Measured by return on assets (ROA), which is measured by dividing net income over total assets

[1, 2]. Various ratios measure the profitability of the companies including the return on assets, return on equity and earnings per share. The study divided the net income of companies over their total assets to measure profitability by the return on assets ratio. We hypotheses the following hypothesis:

H04: There is no relationship between financial Profitabilityand RPTs (Table 1).

3.3 *Study Model*

To assess the hypothesis suggested by the study, the following model is constructed to showcase the dependent and independent variables:

$$\mathbf{RPTs} = \boldsymbol{\beta 0} + \boldsymbol{\beta 1}\,\mathbf{BDS} + \boldsymbol{\beta 2}\,\mathbf{BDI} + \boldsymbol{\beta 3}\,\mathbf{LEV} + \boldsymbol{\beta 4}\,\mathbf{ROA} + \boldsymbol{\varepsilon}$$

where:
β0 is a constant;
RPTs signifies the dependant variable related party transactions;
β1 BDS signifies the independent variable Board Size;
β2 BDI signifies the independent variable Board Independence;
β3 LEV signifies the independent variable leverage of the companies;
β4 ROA signifies the independent variable profitability of the companies;
ε is an error term.

3.4 *Data Sources and Analysis Software*

The data collection process was started by identifying the listed companies of each GCC country through their Bourse websites. These websites provided companies with basic information; some also provided the financial statements of the companies. Information that was not found in the bourse websites can be found in the companies' websites. The dependent variable, RPTs can be found in the notes of the financial statements. Furthermore, the two financial characteristics, profitability and leverage are calculated using numbers found in the financial statements of the companies. Moreover, the governance characteristics, board size and board independence are found in either the companies' Governance reports or the annual reports. The data of the last five years were collected and organized in the Microsoft Excel program.

In order to achieve the purpose of determining the associations between the RPTs and companies' governance and financial characteristics, a correlation analysis is conducted through Microsoft Excel to show the nature and strength of the suggested association between the variables. In addition, a multiple regression analysis is conducted to examine the effect of the company characteristics over its related party transactions.

4 Data Analysis and Interpretation

To examine the relationship between the variables of the study, a correlation analysis is conducted with the Microsoft Excel program. Table 2 shows the results of the correlation analysis.

First, Table 2 shows that the dependent variable related party transactions (RPTs) has weak positive relationships with all the four independent variables. Thus, the hypotheses H01, H02, H03, H04 that state there is no relationship between the RPTs and the other variables are rejected. At 0.458 the board size (BDS) has the strongest positive relationship to RPTs compared to other independent variables. The leverage has a weak positive relationship at 0.274. finally, the independence of the board of directors (BDI) and the profitability (ROA) have the weakest positive relationship to the RPTs at 0.142 and 0.198 respectively.

Table 2 Correlation analysis

	RPTs	BDS	BDI	ROA	LEV	
RPTs	1					
BDS	0.458478	1				
BDI	0.141986	0.251111	1			
ROA	0.198111	0.418972	0.296164	1		
LEV	0.273808	−0.29818	−0.27397	−0.41195		1

Table 3 Testing of hypotheses

Hypotheses (null)	Result	Significant value
H01: There is no relationship between RPTs and board size	Reject	0.005199099
H02: There is no relationship between RPTs and board independence	Accept	0.515676255
H03: There is no relationship between related party transactions and profitability	Accept	0.361962064
H04: There is no relationship between RPTs and financial leverage	Reject	0.004744597

The following Table 3 summarizes the hypotheses testing done in this study.

4.1 Model Testing

In order to test the model constructed in the methodology section, a multiple regression is conducted through Microsoft Excel program to examine the impact of each independent variable over the related party transactions. The analysis resulted in constructing three tables that represent the significance of the model and each independent variable. The regression analysis is conducted at a 95% confidence level.

First, Table 4 shows the percentage in which the independent variables can explain the dependent variable related party transactions. It can be concluded from the value shown in the table that only 43% of the dependent variable RPTs can be explained by the independent variables as concluded by the R Square value. The Adjusted R Square 43% may be considered; this value takes the multiple independent variables into consideration. The table also shows the size of the sample which consists of 30 companies.

The following Table 5 shows the significance of the model constructed in the study. The F value (4, 29) is equal to 4.72, which is higher than one, which indicates the constructed model is effective. The p-value of the model is shown in the table as Significance F with the value of 0.0057, thus it is concluded that the model is significant as the p-value is less than the $\alpha = 0.05$.

After ensuring the significance of the model through the ANOVA table, Table 5 is examined to assess each independent variable's impact on related party transactions.

Table 4 Regression statistics

Multiple R	0.655774126
R square	0.430039704
Adjusted R square	0.338846057
Standard error	0.455131548
Observations	30

Table 5 ANOVA table

	Df	*SS*	*MS*	*F*	*Significance F*
Regression	4	3.90730974	0.976827	4.715676109	0.005656809
Residual	25	5.178618147	0.207145		
	29	9.085927888			

Table 6 shows that the board size (BDS) significantly impacts the related party transactions, as its related p-value of 0.0052 is less than the alpha value of 0.05. It has a positive Beta of 0.16, meaning that an increase of one unit in board size leads to a 0.16 increase in the value of related party transactions. This means that companies that have a high number of members in the board of directors are likely to conduct more related party transactions.

The board independence (BDI) is an insignificant variable in the model with a p-value of 0.516 which is much higher than the alpha value 0.05. This can be due to the fact that most companies included in the sample have only appointed few members in its board of directors to comply with the regulations. In addition, most independent members do not have enough power to prevent conducting related party transactions.

Another insignificant variable is the profitability (ROA) of the company, as it has a p-value of 0.362 which also exceeds the alpha value of 0.05. This means that companies may engage in RPTs regardless of their level of profitability.

The last variable is the financial leverage of the companies (LEV), it has the p- value of 0.0047, which indicates that it has significant effect on the dependent variable RPTs as it is less than the value of alpha 0.05. the variable has the beta value of 1.546 which means that in each increase of one unit in the leverage leads to an increase in the RPTs in the amount of 1.546. This may be due to the fact that companies with high level of leverage may resort in conducting RPTs to overcome this situation.

Interpretation

Through the correlation and regression analysis, the evidence was provided for the determinants of related party transactions. The correlation analysis showed that all independent variables have weak positive relationship with the dependent variable related party transactions.

Table 6 Statistics related to the variables of the model

Coefficients	Standard error	t stat.		P-value	Lower 95%	Upper 95%
Intercept	4.3519	0.4612	9.4361	0.0000	3.4020	5.3017
BDS	0.1606	0.0525	3.0621	0.0052	0.0526	0.2686
BDI	0.3027	0.4590	0.6594	0.5157	−0.6427	1.2480
ROA	2.0410	2.1979	0.9286	0.3620	−2.4856	6.5677
LEV	1.5465	0.4989	3.0998	0.0047	0.5190	2.5740

The results of the analysis of the correlation are partially consistent with the results of the regression analysis. The board size and the leverage have significant positive impact on the related party transactions. Moreover, the board independence and the profitability have insignificant impact on the related party transactions. these results indicate that companies should have smaller-sized board of directors and that companies with high leverage may engage in RPTs to improve their situation.

The results are partially consistent with previous studies' findings. For instance, [1, 10] and [5] have also found that the size of the board has positive significant impact on the related party transactions. however, [9] found that the board size has insignificant impact on related party transactions. In addition, [10] and [5] were also consistent with the results of this study in terms of finding insignificant impact of board independence on related party transactions, while [9] and [1] found that this independent variable has significant negative impact on related party transactions. Regarding the leverage, the results that present the significant positive impact on RPTs were in line with the results of [1] study. Finally, the insignificant impact of profitability on RPTs agrees with the results of [10] studies.

5 Findings, Suggestions and Conclusions

The study examined previous studies to gain insights regarding the topic of related party transactions. This topic is not yet thoroughly explored by previous studies, the determinants of the RPTs have few previous studies, especially in the Arab region. This study focused on how both the governance and financial factors may impact the related party transactions.

There are few aspects that affect the findings of this study, some companies do not fully disclose the information required for the data collected for the study. Moreover, the companies in some countries have just started following the international accounting standards, thus many of those companies constantly changed its accounting methods and reporting structure.

There are suggestions that were thought of during the process conducted in the study. First, more regulations should be enforced to protect the minority interest from harmful related party transactions. in addition, companies should follow a uniform structure while reporting the related party transactions. In addition, companies must have agreed upon classifications of the RPTs to improve the comparability of these transactions among the companies.

Due to the recent attention to the topic of related party transactions, there are many areas that have not been discussed and supplied with evidence. Further studies can be conducted to examine the regulations and their effect on the RPTs performed by the companies listed in the GCC region. In addition, studies can be conducted to examine the effect of the RPTs on firm value of listed companies in GCC. Throughout the process of this study, there were some limitations and challenges that faced the study. The inability to access many research papers related to the topic of the study has been a major limitation. Another major limitation was the time constraint.

References

1. Abdullatif M, Alhadab M, Mansour I (2019) Determinants of related party transactions in Jordan: financial and governance factors. Aust Account Bus Finan J 13(1):44–75
2. Bava F, Trana MGD (2017) The influence of profitability on related party revenues. Int J Bus Gov Ethics 12(2):107–126
3. Beerbaum D, Piechocki M (2017) Related party transactions–empirical study based on IFRS and SEC disclosures., Maciej
4. Bona-Sánchez C, Fernández-Senra CL, Pérez-Alemán J (2017) Related-party transactions, dominant owners and firm value. BRQ Bus Res Q 20(1):4–17
5. Desoky AM, Mousa GA, Yassin M, AbuRaya R (2020) Ownership concentration and board features as determinants of related party transactions-evidence from GCC countries. Int J Manag Financ Account 12(3–4):304–327
6. Diab AA, Aboud A, Hamdy A (2019) The impact of related party transactions on firm value: Evidence from a developing country. J Finan Report Account 17:571–588
7. El-Helaly M (2018) Related-party transactions: a review of the regulation, governance and auditing literature. Manag Audit J 33:779–806
8. International Accounting Standards Board (IASB) (2010) Related Party Disclosures, IAS 24. https://www.iasplus.com/en/standards/ias/ias24
9. Jeon K (2019) The characteristics of board of directors and related party transactions. Acad Account Finan Stud J 23(3):1–11
10. Madbouly D (2020) The determinants of the related party transactions' occurrence: an empirical study on firms listed in the Egyptian stock exchange. Alex J Account Res Second Issue 4
11. Al-Sartawi AMAM (ed) (2022) Artificial intelligence for sustainable finance and sustainable technology. ICGER 2021. Lecture notes in networks and systems, vol 423. Springer, Cham
12. Nor HM, Ismail KNIK (2017) The moderating effects of independent directors' human capital on the relationship between related party transactions and firm performance: evidence from Malaysia. Jurnal Pengurusan (UKM J Manag) 51:37–52
13. Peng WQ, Wei KJ, Yang Z (2011) Tunneling or propping: Evidence from connected transactions in China. J Corp Finan 17(2):306–325
14. Pizzo M (2013) Related party transactions under a contingency perspective. J Manag Gov 17(2):309–330
15. Rasheed PCA, Hawaldar IT (2021) Related party transactions and audit risk. Cogent Bus Manag 8(1):1888669
16. Ryngaert M, Thomas S (2012) Not all related party transactions (RPTs) are the same: ex ante versus ex post RPTs. J Account Res 50(3):845–882
17. Sanad Z, Shiwakoti R, Al-Sartawi A (2022) Gender diversity on corporate boards and earnings management: a review. In: Al-Sartawi AMAM (eds) Artificial intelligence for sustainable finance and sustainable technology. ICGER 2021. Lecture Notes in Networks and Systems, vol 423. Springer, Cham
18. Wahab EAA, Haron H, Lok CL, Yahya S (2011) Does corporate governance matter? evidence from related party transactions in Malaysia. In: International corporate governance. Emerald Group Publishing Limited
19. Zimon G, Appolloni A, Tarighi H, Shahmohammadi S, Daneshpou E (2021) Earnings management, related party transactions and corporate performance: the moderating role of internal control. Risks 9(8):146

The Impact of Intellectual Capital on Performance of Service and Industrial Firms in Middle East

Tariq Nasir Maqableh, Aya Naser Maqableh, and Hamzeh Naser Maqableh

Abstract Intellectual capital (IC) has become an essential contributor to the performance of whole economies and individual companies. However, this contribution is relatively small in the countries of Middle East compared with other regions. In addition, previous literature focuses on examining the effect of IC on the performance of banks while other sectors received less attention. This research aims to examine the effect of value added intellectual capital (VAIC) and its components (Human capital efficiency (HCE), Structural capital efficiency (SCE), and capital employed efficiency (CEE)) on the performance of industrial and service sector of five countries in the Middle East. Data was collected from listed companies in the five selected countries in the Middle East. This research finds that VAIC has a significant effect on financial performance and CEE is the most important component of VAIC and has strong positive effect on ROA and MTBV. The effect of HCE, SCE, on both ROA, and MTBV was found to be insignificant. The findings are varied based on countries and sectors. Policy makers in the five selected countries need to pay more attention to the contribution that VAIC can make to the performance of their economies. The heavy dependence on financial capital and physical capital (observed in the CEE) that is common in these economies, should be redirected to the institutional and human intellectual categories (SCE and HCE).

Keywords Intellectual capital · Performance of service · Industrial firms · Middle east

T. N. Maqableh (✉)
School of Business and Finance, University of Westren Sydney, Parrmatta, Westren Sydney 2150, Australia
e-mail: tariqmagableh@hotmail.com

A. N. Maqableh
Faculty of Business, Universiti Malaysia Terengganu, 21030 Kuala Nerus, Terengganu, Malaysia

H. N. Maqableh
Faculty of Business, Amman Arab University, Amman 2234, Jordan

A. M. A. Musleh Al-Sartawi et al. (eds.), *From the Internet of Things to the Internet of Ideas: The Role of Artificial Intelligence*, Lecture Notes in Networks and Systems 557,
https://doi.org/10.1007/978-3-031-17746-0_40

1 Introduction

Knowledge based economy has become the most after sought goal for all economies around the world. In this process, intellectual capital (IC) is increasingly recognised as an essential contributor to the economic growth and performance of countries and organizations. At a national level, a nation's IC can account for 70% of the variation in the economic growth and performance of the country [44]. At an organisational level, IC can account for over 75% of a company's value [34]. Companies with high levels of IC can create new products and services and implement innovative and creative solutions. This, ultimately, gains them a competitive advantage [20, 39].

In a highly competitive environment, companies seek to gain a competitive advantage over other companies by increasing customer value. One way of achieving this is by exploiting the capacities of individuals [4]. IC management is essential for making intangible assets a competitive advantage by strengthening human potential to improve performance [45]. Consequently, organisational structure, strategic planning, capital management, and investments in human resources affect the utilisation of IC. They also potentially impact the innovation, efficiency, and profitability of an organisation [14, 34, 39, 63].

Previous studies on IC focus on developed economies. Meanwhile, studies on emerging and developing economies are limited [44]. In many developing countries, the importance of IC for companies is largely unrecognised [15, 64, 65]. In Middle East Countries, IC studies are in their infancy stage, and IC's effect on the performance of companies remains largely unknown. IC's contribution to economic growth and organizational performance and value accounts for over 70% globally. However, findings indicate that the contribution of IC is only 20% in Middle Eastern countries [17].

Previous findings are mixed regarding the effect of IC and its components on the financial performance of companies. One study found that the effect of IC is negative and insignificant [8] Meanwhile, other studies found a positive and significant effect [38] This contradiction also occurs in relation to the effect of IC's components like human capital, structural capital, and relational capital on the financial performance of companies [3, 7, 16, 44, 56, 58]. Other issues in the literature are related to the measurements of financial performance. Financial performance has been measured differently across previous studies. Some studies applied only accounting-based performance measurements [16], while others analysed market-based performance [33] Few have used both indicators to measure financial performance [24].

2 Literature Review

2.1 Intellectual Capital

During the 1980s, it was observed that the market value of companies began to grow beyond their respective book values. This increase in the prices of the shares was linked to the experience, knowledge, skills, patents and copyrights of the companies. These different organisational capabilities were described by a group of Swedish researchers in 1986 as "intangible assets" [37]. This use of the term intangible assets was the first indirect reference to this new concept of IC. The first time the specific term IC was mentioned in an article was by [59], and published in Fortune Magazine. This seminal article highlighted the importance of IC for the wealth and prosperity of nations [31]. Several studies divided IC to include three components. They were used on the organizational and the national level [44]. Studies on IC value the three components differently [58] Some found the effect of IC and its components on financial performance as inconclusive [3]. Others identify relational capital as the most critical component for financial performance in companies [7, 44]. Still, others identify human capital as the most critical component of a company's financial performance [16, 56]. Certainly, IC can offer competitive advantages [24]. Still, any company attempting to devise an IC strategy should balance its three components to ensure stability and positively influence performance [16, 21, 56].

2.2 Theoretical Framework

Among the first researchers who investigated the concept of IC is [62]. This author divided IC into three distinct components that include external structure, the internal structure, and the employee competence of companies. At the same time that [62] introduced the three components of IC, another early study by [18] categorised the elements of IC in a slightly different way to that provided by [62]. They identified IC as consisting of human capital and structural capital. Stewart (1998) followed the same approach as [18], but used some slightly different terminology and added some additional components. [60] divided IC into three components. However, he added the concept of customer capital in his approach, which includes structural capital, human capital, and now the new component of customer capital [59].

A new and more comprehensive is the method known as Value-Added Intellectual Coefficient (VAIC). VAIC was developed by the Austrian Intellectual Capital Research Centre (AICRS), under [40], to measure the efficient of key resources in firms. The method of VAIC is a model to measure the efficiency of resources used in organizations, regions, and nations, in that sense it is a very flexible model that can be applied at several different levels of analysis [28]. As a measurement of IC, it is a comprehensive measure [57]. VAIC outperforms other measurement models because it offers a standard and reliable framework for calculating the IC value and the performance of companies.

[40, 41] indicates that VAIC can highlights the sectors of the economy that might need intervention, which can lead to increases in output and wealth. Some of the additional advantages of VAIC include that it is objective, supports the statistical analysis of large sample sizes, and provides useful indicators and information for stakeholders. VAIC has also been widely used commercially [22, 42]. This study contributes to our understanding of the VAIC measurement and applies to the calculation of VAIC in two different sectors in order to measure the IC in five selected Middle Eastern countries.

2.2.1 An Appraisal of the VAIC Model

Researchers have deployed the VAIC and referred to some strengths and weakness of the model. VAIC has become an established model for measuring the value of IC in different locations. In terms of the accepted components of the VAIC approach, it can be seen that majority of the previous studies have used structural capital, human capital, and a measurement of capital employed efficiency. VAIC provides a relatively simple method to calculate the IC, which is based on data available in the annual reports of companies [25]. VAIC also helps in identifying ways to improve the internal capabilities of organisations, and then to provide them with a competitive edge [61].

[11] examined the relationship between VAIC and firm performance across countries. The author criticized VAIC and pointed out that the model is largely influenced by exogenous technological factors that are unrelated to IC. However, [11] also acknowledged the wide usage of VAIC for measuring the IC of companies. The author modified VAIC to include human capital, physical capital, and internet rate. [32] pointed out that researchers overcome the criticism of VAIC by modifying the measurement of the variables. Researcher also acknowledged the criticism of VAIC. However, they have used it to measure the IC.

Despite the existence of criticism regarding the measurement of the variables [35] and inconsistencies in the model as well as the variation in the results of testing the effect of IC on the firm performance [32]. The VAIC model has been widely accepted by academics and practitioners [12]. The biggest advantage of the VAIC model is its simplicity and comparability [3, 27, 65]. Further, VAIC relies heavily on the audited financial data (Chen et al., 2005). In addition, the use of VAIC by UK's Department for Business, Innovation and Skills to measure IC of firms in the UK further strengthened the validity of the model [64].

Overall, VAIC is considered a useful and practical measurement of IC, and its use of data from financial reports provides a point of strength in offering a consistent basis through which to compare IC in different locations (and potentially across industries). However, the use of it across different locations has also highlighted the fact that different regions present different results. Furthermore, in terms of the specific components, a range of (sometimes contradictory) results have been observed. In this study, the VAIC model is adopted to measure the IC of the listed companies in two sectors (industrial and service) in the five selected countries of the Middle East (Qatar, Kuwait, UAE, Saudi Arabia, and Jordan).

2.3 *Conceptual Model and Hypothesis Development*

As outlined above, this study deploys the VAIC model originally developed by [41]. It widely accepted that the components of IC are divided into three primary elements: Human capital, Structural capital and Relational or Customer [30]. However, in the VAIC model of [41] the author considered these components in terms of their efficiency. Pulic transformed the assessment of these three components into efficiency measurements through emphasis on the economic value added (VA) by the individual components, thereby attempting to consider the effectiveness of these three IC components in generating value and contributing to output, i.e., capital employed (CE), human capital (HC) and structural capital (SC). This extension of the VAIC method enabled the formation of a metric to gauge IC efficiency, referred as intellectual capital efficiency (ICE). The three VAIC components therefore become Human Capital Efficiency (HCE), The Capital Employed Efficiency (CEE), and the Structural Capital Efficiency (SCE). The study proposes that the VAIC and its components i.e., HCE, SCE, and CEE will have a positive effect on a financial performance of companies in the Middle East. The study proposes also that firm size, leverage and asset turnover, are the appropriate control variables. Figure 1 shows the conceptual framework of this study.

2.3.1 VAIC and Financial Performance

Previous studies investigated the link between IC and financial performance and found a positive relationship. For instance, [13] found empirically that investors

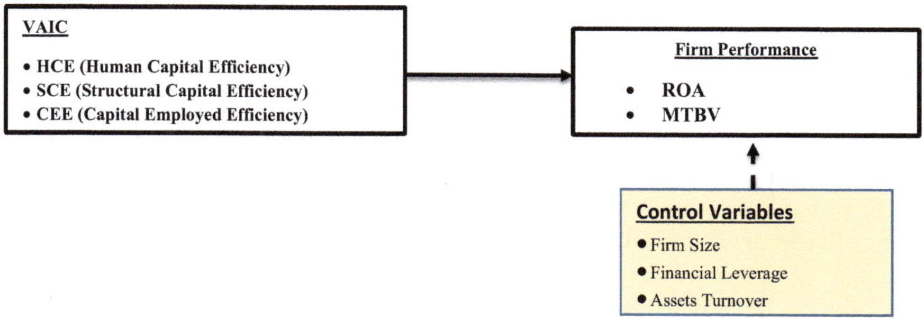

Fig. 1 Conceptual framework

place greater value on companies with better IC because they achieve greater profitability and revenue. [55] determined that VAIC has a positive correlation with profitability and market valuation. However, interestingly, it has a negative correlation with productivity. The findings of [43] indicate that there is a positive relationship between VAIC and financial performance measurements, such as ROA, ROE and return on investment (ROI). [14] noted that the individual IC components commonly have a positive influence, either directly or indirectly, on financial performance. Similarly, [26] showed that IC is significantly positively correlated with the operational performance of companies, as measured through a productivity index.

The different regional based studies of the impact of IC (through the VAIC model) on firm performance demonstrate a broad consistency that supports a positive relationship between the two such as in the study of [16] in Australia, [7] in Pakistan, and [24] in Saudi Arabia (IC affects positively ROA but does not affect Tobin's Q). In this study, it is expected that the VAIC as a composite variable is expected to have a positive effect on the financial performance (both accounting and market-based measurements) of companies in the Middle East. Accordingly, it is hypothesized:

H1: VAIC is positively associated with financial performance (ROA and MTBV) across sectors and countries.

2.3.2 Human Capital Efficiency and Financial Performance

Various studies have examined the effect of HCE on financial performance. [56] considered HCE to be the most important component of VAIC among Indian companies and identified a positive and a direct effect on financial performance. The findings of the study of [16] showed that HCE is an important component, and it positively affects the performance of companies in Australia while [21] derived similar findings in Kenya. While the above studies identify a positive relationship, other studies have obtained mixed results in assessing the effect of HCE on financial performance. In the studies of [9] and [36], the findings showed that HCE has an insignificant effect on the financial performance of companies. On the other hand, the findings of the study of [11] showed that the effect of HCE on financial performance is positive and significant. In this study, it is expected that the HCE is expected to have a positive effect on the financial performance of companies across the two sectors in the five selected countries in the Middle East. Accordingly, it is hypothesized:

H2: HCE is positively associated with the financial performance (ROA and MTBV) of companies across sectors and countries.

2.3.3 Structural Capital Efficiency and Financial Performance

Structural capital efficiency (SCE) is important for the performance of companies. Previous research has found a positive link between SCE and financial performance. For instance, in the study of [56], SCE has a significant effect on the financial performance of companies. [58] also found that the effect of SCE on financial performance is positive and significant. [10] found that SCE is critical for the performance of companies and has positive effect on financial performance. Similar findings were derived from the study of [13]. In this study, SCE is expected to have a positive effect on the financial performance of companies in the two sectors of the five selected countries in the Middle East. Therefore, the following hypotheses are proposed.

H3: SCE is positively associated with the financial performance (ROA and MTBV) of companies across sectors and countries.

2.3.4 Capital Employed Efficiency and Financial Performance

Capital Employed Efficiency (CEE) has been investigated by only a few studies in the literature. Existing studies demonstrate that the effect of CEE on financial performance is positive and significant. [16] found that the effect of CEE on the financial performance of companies in Australia is positive and significant. Similarly, the study of [56] found that CEE has an important and positive effect on the financial performance of companies in India. Nadeem et al. (2016) in the UK found also that CEE has a significant effect on financial performance of companies in UK. Accordingly, this study predicts that the effect of CEE on financial performance, of companies in the two sectors of the five selected countries in the Middle East, is positive and significant. Thus, it is hypothesized:

H4: CEE is positively associated with the financial performance (ROA and MTBV) of companies across sectors and countries.

3 Research Methodology

3.1 Population

This study investigates the VAIC and its effect on financial performance of five countries in the Middle East. The countries include the Kingdom of Saudi Arabia, Jordan, Qatar, Kuwait, and United Arab Emirates (UAE). While the IC of some of these countries have been examined previously, they have not been comprehensively analyzed as a group, and not using sector specific measurements within the VAIC framework. This study considers companies from two sectors namely, the industrial sector and the services sector. The choice of these two sectors is due to the notion that prior literature focused on the financial sector and in particular the banking sector. This is may well be related to the fact that the VAIC model was developed with original application to the banking sector by [41]. For this reason, previous studies do not often pursue a multiple sector analysis, and instead commonly restrict focus to a single sector or two. Accordingly, companies in industrial and service sectors are included in this study in the five selected countries in the Middle East.

Table 1 Sample of the study

		Jordan	Kuwait	UAE	Qatar	SA	Total
Service	# Of Observation	92	16	12	27	40	187
Industrial	# Of Observation	91	23	5	18	74	211
	Total Observation	183	39	17	45	114	398

Number of observation = 398

These companies are each listed on the stock exchanges of their respective countries. However, due to the fact that not all companies declare information about the IC and its components, only companies that have declared the information are selected from each individual country stock exchange. From Jordan, a total of 183 companies were included in this study, 114 companies from KSA, 45 companies from Qatar, 17 companies from UAE, and 39 companies from Kuwait. This makes the total sample population of this study, after excluding the companies that do not declare or have IC, 109 companies. These companies are the sample of this study. The overall number of observations that is collected across the time period under consideration is 398. Table 1 presents a summary of the sample of this study, including the distribution of the sample based on country and sectors.

3.2 *Variables Measurement*

3.2.1 Financial Performance

Financial performance in this study is measured using ROA and MTBV. This measurement includes accounting based performance such as ROA, and market based performance such as MTBV. The equations of both variables are given below. Understanding the calculation of these two measurements of performance, reveals the potential to compare them in terms of market and accounting measurements of the return's companies earn on their balance sheet assets.

- Y1 = Price to Book Value = Market Capital/Total shareholders define equity [36].
- Y2 = Return on Assets (ROA) = Net income/Total assets [41]

3.2.2 VAIC Variable Measurement

The VAIC model estimated the degree to which a company creates benefit based on IC efficiency. There are criticisms of VAIC in term of measurement of variables. [32] pointed out that researchers overcome these criticism by modifying the measurement of VAIC. This study also specified the measurement of VAIC's components based on sectors. Individual VAIC calculations depend on:

A. **Human Capital (HC)**, which is essentially defined as employee expertise, knowledge and skills.
B. **Structural Capital (SC)**, which is described as the difference in produced benefit (VA) and HC

$$SC = VA - HC$$

III. **Capital Employed (CE)**, which is represented as financial capital, e.g., book value [32]

$$VAIC = HCE + SCE + CEE$$

To calculate the aggregate and individual components of the VAIC, and in particular the efficiency metric for each, we must first measure the value added (VA). In this study the approach to this initial calculation follows previous studies. However, the measurement of VA needs to differ according to the individual sector being measured, as is now explained. This study employs sector measurements that are individually tailored for each sector. For the industrial and services firms, the variable calculation follows the approach of [8]:

$$X11(VA) = \text{operating revenue} - \text{operating expenses.}$$

This formula differs from the original formula developed by [41] which measured the VA as (VA) = Interest Income - Interest Expense.

Following the measurement of value added in this sector specific approach, the measurements of the individual components of VAIC can then be pursued. Here the analysis begins with the HC efficiency measurement. In terms of the industrial and service firms, HC is obtained from the general and administrative expense. The Industrial and Services Sectors - General and Administrative Expenses [8]. Following the calculation of the sector specific value of HC, the overall measurement of HCE can be formed for each sector using the sector specific VA measurement obtained above. The HCE measurement follows [41] and is given as:

- HCE = VA/HC

In terms of the SCE, most previous studies use the measurement of SCE based on [41]. This component has not been modified, as the conventional approach developed by [41] can be consistently applied to all sectors. As SC is simply the residual of the VA once HC has been subtracted. This can be easily summarized, is applied in a consistent manner in this study across each industry.

- SCE = SC/VA - Where, SC = VA – HC

Finally, the third individual component of VAIC, the CEE, is again measured based on the original specification of [41], which has become the standard and conventional approach to this efficiency measurement.

- CEE = VA/CE

However, while the basic structure of the calculation is the same, the source of information is once again sector specific. As with the calculation of the HCE values above, the estimation needs to reflect the sector differences in terms of financial statement records. Accordingly, in this study the sector specific sources of information follow the previous work on this topic. The Industrial and Services Sectors: Working Capital [8].

3.2.3 Control Variables

Three control variables are used in this study. The size of the firm (SIZE) and the degree of financial leverage (LEV), and assets turnover. These three variables may influence the dependent variables, the effects of which should be excluded in the study of panel regression. The inclusion of these variables is in line with previous studies [13, 55].

1. Leverage: debt used for financing of the assets. The leverage in financial terms offers other benefits to businesses including tax cover, lower costs than equity investments and exposure and repayment versatility.

$$\text{LEV} = \text{Financial Debt/Shareholder's Equity}$$

2. Size (The company size): This is measured by total revenues. Size serves as an indicator of the company's size and is computed as a natural log of the book value of equity.

$$\text{Size} = \text{Natural log of the book value of equity (total assets).}$$

3. Asset Turnover (ATO): The literature uses asset turnover (ATO) often to explain company efficiency, and several research trials have been used to determine the relevance of IC to the efficient use of assets by companies [17] Asset turnover is measured by dividing net sales or revenue by the average total assets.

3.3 Data and Data Sources

This section describes the data collection and the descriptive statistic of the panel data.

3.3.1 Data Sources and Time Period

In this study, most of the data are collected from annual reports. This type of data is called secondary data. A secondary archival database, Research Insight (formerly known as Compustat), is used to collect data of the variables of this study. The Compustat database is used extensively in the accounting literature and includes data of active and inactive firms. All companies that had available annual reports in the Compustat database are selected and are included in the analysis. The study covered the years 2007 to 2013. The time period of the data set was chosen due to two important reasons that are relevant to this region. First the Global Economic Crisis of 2008–2009 had a significant impact on the world economy, and this extended to the economies of the countries selected in this study [19, 36]. Interestingly, while this crisis impacted financial markets in the Middle East, the "real economies" of these countries demonstrated some resilience [1, 5]. The time period of this study can contribute to our understanding if whether the accumulation of IC was part of the reason for some of the observed resilience.

The second reason to select this time period is the outbreak of the Arab spring, which created an unstable environment that affected the performance of economies, sectors, and companies [6]. The major effect of the Arab spring lasted from 2011 to 2013 with the collapse of several regimes such as Libya, Egypt, Yemen and Tunisia which are linked economically to the economies of the selected countries. Overall, therefore, this is a relatively tumultuous period in the history of the MENA region, and studies dedicated to this time period are particularly valuable to document the social, political, and in this case economic, events of the time.

4 Findings

Results of the descriptive information are given in Table 2. In the case of the MTBV, the mean score of MTBV is 1.517 while for ROA, the mean is 2.031 with standard deviation of 7.012, indicating differences in the ROA of the sample. The mean of VAIC is 3.411, HCE is 0.117, SCE is 1.468, the mean score of the CEE is 1.824 with large variability among countries as demonstrated by the high standard deviation of 10.543. This relatively homogeneity in the values of HCE could be due to the fact that the levels of education background are relatively similar across the countries included in the sample. In contrast, the countries differ in term of CEE and SCE as each country (and sectors within each country) has different resource and capabilities. The mean score of firm size is 19.47 with standard deviation of 3.316 implying that the size variable incorporates moderate differences among the sample of this study. The mean of ATO is -1.433 and the standard deviation is 1.336. The LEV data showed that the mean is 0.181 indicating an average in the sample of this study is using 18.1% as LEV.

Table 2 Descriptive statistics of variables

Variable	N	Mean	Std. dev.	Min	Max
MTBV	398	1.517	1.255	−2.239	10.045
ROA	398	2.031	7.012	−62.08	54.482
VAIC	398	3.411	9.868	0.04	198.975
HCE	398	0.117	1.001	−18.624	1.867
SCE	398	1.468	2.335	−6.261	44.617
CEE	398	1.824	10.543	−10.927	217.547
Size	398	19.47	3.316	9.801	27.756
ATO	398	−1.433	1.336	−7.382	1.489
LEV	398	0.181	0.183	0	0.739

5 Discussion

The findings showed that the effect of VAIC on financial performance (ROA and MTBV) is positive and significant across all countries and sectors. The positive effect of VAIC can be explained by the fact that companies in the Middle East has started to deploy and utilize IC to enhance financial performance. Several economies in this sample are emerging economies and have established strategies to move their economies toward knowledge-based economies. In agreement with the findings of this study, many previous studies indicate that the effect of VAIC on the financial performance is positive and significant [55, 24].

The study found that the effect of HCE on financial performance (ROA and MTBV) is not significant. A possible explanation of this finding is the notion that human development in the countries of this study is still emerging, and the contribution of this human capital is still not utilized.

The effect of SCE on ROA and MTBV is not significant. In terms of the variable itself, this result is surprising. Unlike the value of HCE, which has a relatively limited variance, the SCE value exhibits significant variation, and this heterogeneity across companies was expected to exhibit an impact on measures of financial performance. A possible explanation of this insignificant effect is the nature of the SCE component itself.

The effect of SCE on ROA and MTBV is not significant. In terms of the variable itself, this result is surprising. Unlike the value of HCE, which has a relatively limited variance, the SCE value exhibits significant variation, and this heterogeneity across companies was expected to exhibit an impact on measures of financial performance. A possible explanation of this insignificant effect is the nature of the SCE component itself.

In agreement with the findings of this study, [16] in Australia, [56] in India and [36] in the UK, each found that the effect of CEE on financial performance of companies in positive and significant.

The results, obtained in the analysis of individual components in each country, do not provide strong support for the hypotheses motivating this aspect of the study. However, these mixed results, limited in terms of components and their impact on ROA and MTBV in the five countries, are in line with the previous study of [3] who also found that there is diversity and mixed results. Results related to the components of international cooperation in the Arab Gulf states.

6 Conclusion

The results of the study showed that there is a significant association between VAIC and the dependent variables; ROA and MTBV. However, the association is greater with ROA. In addition, the findings showed that among the components, CEE is the most important and has substantial effect on ROA and MTBV. Among countries, the highest association of VAIC with MTBV is found in Kuwait and Qatar. However, in terms of the ROA, the best performing country is UAE. The associations differ to large degree based on sector. This study is conducted based on secondary data collected from two sectors in five countries between 2007 and 2013. Therefore, the study is limited to the two sectors in the five selected countries. The study also focused on the effect of VAIC on financial performance represented by ROA and MTBV. Thus, the findings of this study should be interpreted based on these characteristics of sectors, countries, and measurements. As a way forward, further studies are needed to examine the VAIC and its implications on financial performance in developing countries. Along with VAIC, there are other important variables that can be included. This includes the corporate governance and ownership structure and concentration. A specific sector-based study is a direction for future work.

References

1. Abdelbaki HH (2010) Assessing the impact of the global financial crisis on GCC countries. J Bus Econ Res 8(2):139–152
2. Al-Sartawi A (2018) Ownership structure and intellectual capital: evidence from the GCC countries. Int J Learn Intellect Cap 15(3):277–291
3. Altaee HHA, Talo IMA, Adam MHM (2013) Testing the financial stability of banks in GCC countries: pre and post financial crisis. Int J Bus Social Res (IJBSR) 3(4):93–105
4. Bayraktaroglu AE, Calisir F, Baskak M (2019) Intellectual capital and firm performance: an extended VAIC model. J Intellect Cap 20(3):406–425
5. Bontis N (2004) National intellectual capital index. J Intell Cap 5:1–13
6. Bontis N, Keow WCC, Richardson S (2000) Intellectual capital and business performance in Malaysian industries. J Intell Cap 1:1–85
7. Borenstein M, Hedges LV, Higgins JPT, Rothstein HR (2010) A basic introduction to fixed-effect and random-effects models for meta-analysis. Res Synth Methods 1(2):97–111
8. Chen H, Manning AK, Dupuis J (2012) A method of moments estimator for random effect multivariate meta-analysis. Biometrics 68(4):1278–1284

9. Choong KK (2008) Intellectual capital: definitions, categorization and reporting models. J Intell Capital
10. Chowdhury LAM, Rana T, Azim MI (2019) Intellectual capital efficiency and organisational performance. J Intell Cap 20:784–806
11. Clarke M, Seng D, Whiting RH (2011) Intellectual capital and firm performance in Australia. J Intellect Cap 12(4):505–530
12. Dženopoljac V, Janoševic S, Bontis N (2016) Intellectual capital and financial performance in the Serbian ICT industry. J Intell Cap 17:373–396. https://doi.org/10.1108/JIC-07-2015-0068
13. Dzenopoljac V, Yaacoub C, Elkanj N, Bontis N (2017) Impact of intellectual capital on corporate performance: evidence from the Arab region. J Intell Cap 18:884–903
14. Edvinsson L, Malone MS (1997) Intellectual capital: the proven way to establish your company's real value by finding its hidden brainpower. Piatkus
15. El-Chaarani H (2019) The impact of oil prices on stocks markets: new evidence during and after the Arab spring in Gulf Cooperation Council Economies. J Energy Econ Policy 9(5):214–223
16. Elamer AA, Ntim CG, Abdou HA, Zalata AM, Elmagrhi M (2019) The impact of multi-layer governance on bank risk disclosure in emerging markets: the case of Middle East and North Africa. Account Forum 43(2):246–281
17. Felin T, Zenger TR, Tomsik J (2009) The knowledge economy: emerging organizational forms, missing microfoundations, and key considerations for managing human capital. Human Res Manag Publ Cooper School Bus Adm Univ Michigan Alliance Soc Human Res Manag 48(4):555–570
18. Gerramosca S, Chio A (2016) Leveraging intellectual capital in developing countries: evidence from Kenya. J Intellect Cap 17(2):397–412
19. Hamdan A (2018) Intellectual capital and firm performance: Differentiating between accounting-based and market-based performance. Int J Islam Middle East Financ Manag 11(1):139–151
20. Hatane SE, Djajadi AT, Tarigan J (2017) The impact of corporate governance on intellectual capital and firm value: Evidence from Indonesia and Malaysia consumer goods. Petra Christian University
21. Iazzolino G, Laise D (2013) Value added intellectual coefficient (VAIC): a methodological and critical review. J Intell Cap 14:547–563. https://doi.org/10.1108/JIC-12-2012-0107
22. Javornik S, Tekavcic M, Marc M (2012) The efficiency of intellectual capital investments as a potential leading indicator. Int Bus Econ Res J (IBER) 11(5):535–558
23. Karolak M, Razzaque A, Al-Sartawi A (2021) E-services and M-services using IoT: an assessment of the Kingdom of Bahrain. In: Al-Sartawi AMM, Razzaque A, Kamal MM (eds) Artificial intelligence systems and the internet of things in the digital era. EAMMIS 2021. Lecture Notes in Networks and Systems, vol 239. Springer, Cham
24. Krušinskas R, Bruneckienė J (2015) Measurement of intellectual capital of Lithuanian cities by a composite index. J Bus Econ Manag 16(3):529–541
25. Martín-de-Castro G, Delgado-Verde M, López-Sáez P, Navas-López JE (2011) Towards 'an intellectual capital-based view of the firm': origins and nature. J Bus Ethics 98:649–662. https://doi.org/10.1007/s10551-010-0644-5
26. Martin-de Castro G, Diez-Vial I, Delgado-Verde M (2019) Intellectual capital and the firm: evolution and research trends. J Intell Cap 20:555–580
27. Massaro M, Dumay J, Garlatti A, Dal Mas F (2018) Practitioners' views on intellectual capital and sustainability: from a performance-based to a worth-based perspective. J Intellect Cap 19(2):367–386
28. Nawaz T, Haniffa R (2017) Determinants of financial performance of Islamic banks: an intellectual capital perspective. J Islamic Account Bus Res
29. Al-Sartawi AMAM (ed) (2022) Artificial intelligence for sustainable finance and sustainable technology. ICGER 2021. Lecture Notes in Networks and Systems, vol 423. Springer, Cham
30. Ousama AA, Hammami H, Abdulkarim M (2019) The association between intellectual capital and financial performance in the Islamic banking industry. Int J Islamic Middle East Finan Manag 13:1–75

31. Popescu GH, Comanescu M, Sabie OM (2016) The role of human capital in the knowledge-networked economy. Psychosociol Issues Human Res Manag 4(1):168
32. Pulic A (2000) VAIC^{TM}–an accounting tool for IC management. Int J Technol Manag 20(5–8):702–714
33. Pulic A (2004) Intellectual capital–does it create or destroy value? Measuring Business Excellence, Emerald Group Publishing Limited
34. Sakhnini Z, Elali W, Al-Sartawi A (2022) Behavioral and non-behavioral factors and the level of adapting and implementing fintech and E-banking in Bahrain: suggested model. In: Al-Sartawi AMAM (ed) Artificial intelligence for sustainable finance and sustainable technology. ICGER 2021. Lecture Notes in Networks and Systems, vol 423. Springer, Cham
35. Serenko A, Bontis N (2013) Investigating the current state and impact of the intellectual capital academic discipline. J Intell Cap 14:476–500. https://doi.org/10.1108/JIC-11-2012-0099
36. Shiu H-J (2006) The application of the value added intellectual coefficient to measure corporate performance: evidence from technological firms. Int J Manag
37. Smriti N, Das N (2018) The impact of intellectual capital on firm performance: a study of Indian firms listed in COSPI. J Intellect Cap 19(5):935–964
38. Soewarno N, Tjahjadi B (2020) Measures that matter: an empirical investigation of intellectual capital and financial performance of banking firms in Indonesia. J Intellect Cap 21(6):1085–1106
39. Ståhle P, Bounfour A (2008) Understanding dynamics of intellectual capital of nations. J Intellect Cap 9(2):164–177
40. Stewart (1998) Intellectual capital: the new wealth of organization, Currency
41. Stewart TA (1991) Brainpower: how intellectual capital is becoming america's most valuable asse. 3 June 2019. https://archive.fortune.com/magazines/fortune/fortune_archive/1991/06/03/75096/index.htm. 27 Srpnja 2019
42. Pulic A, Bornemann M (1997) The physical and intellectual capital of Austrian banks
43. Tayles M, Pike RH, Sofian S (2007) Intellectual capital, management accounting practices and corporate performance. Account Audit Account J
44. Xu J, Li J (2019) The impact of intellectual capital on SMEs' performance in China. J Intell Cap 20:488–509
45. Yusliza MY, Yong JY, Tanveer MI, Ramayah T, Faezah JN, Muhammad Z (2020) A structural model of the impact of green intellectual capital on sustainable performance. J Clean Prod 249:119334

Corporate Governance Characteristics and Firm Performance: Evidence from Bahraini Listed Financial Sector Companies

Gagan Kukreja, Raj Bahadur Sharma, Atul Bansal, and Sara A. Husain Moh'd Alqassab

Abstract The purpose of this study is to examine the impact of corporate governance practices on firm performance of the listed financial sector companies in Bahrain. The study used the dependent variable (Return on Assets) as firm performance, and three independent variables as Board size, Board independence and Board meetings. The study also considered two control variables Firm size and Firm leverage. The period of the study is from 2016 to 2020 with a sample of 18 financial sector companies listed on the Bahrain Bourse. The results of the study found that there is a positive relationship between return on assets and board meetings, while a negative relationship between return on assets and board size, board independence, firm size, and firm leverage. The multiple regression analysis revealed a negative relationship between the dependent variable (return on assets) with the independent variables, the board size, board meeting, board independence, and the control variables, firm size, and firm leverage. The results of this study recommend for future studies extend the period to additional five years and also extend it to other GCC countries to get more accurate results and impact.

Keywords Return on assets · Board size · Board meeting · Board independence · Firm size · Corporate governance

G. Kukreja (✉) · R. B. Sharma · A. Bansal · S. A. H. Moh'd Alqassab
College of Business Administration, University of Bahrain, Shakir, Bahrain
e-mail: gkukreja@uob.edu.bh

R. B. Sharma
e-mail: rsharma@uob.edu.bh

A. Bansal
e-mail: aradhey@uob.edu.bh

S. A. H. Moh'd Alqassab
e-mail: sarah_alqassab@hotmail.com

A. M. A. Musleh Al-Sartawi et al. (eds.), *From the Internet of Things to the Internet of Ideas: The Role of Artificial Intelligence*, Lecture Notes in Networks and Systems 557,
https://doi.org/10.1007/978-3-031-17746-0_41

1 Introduction

Corporate governance is the system that controls the practices of the companies to perform efficiently. The composition of the board of directors (board size, board independence and board meetings) is important for the firm performance. The corporate governance practices have a positive impact on the firm's performance and are going to attract potential shareholders. The issue of corporate governance and its impact on firm performance has been a major subject over the past several years. Another crisis that has affected the economy worldwide is Covid-19, as it highly impacted the organizational performance worldwide. Covid-19 has resulted in a decline in the stock markets and many investors withdrew their investments as the value of their shares fell, resulting in poor firm performance in many organizations [27]. The precautionary measures taken by various nations to control the spread of the pandemic, such as company shutdowns and lockdowns, indicated a negative impact on stock values and firm performance. Because the firm's primary goal is to have an effective corporate governance practice, to attract and maintain long-term relationships with existing and potential investors, and to raise the future performance, profit, and long-term activities.

Corporate governance assists organizations in attaining their goals, making decisions, controlling risk, and meeting compliance standards. Generally, corporate governance is defined by a set of rules that govern stakeholder relationships, management, and the board of directors [21].

In response to the financial crises, organizations assume that corporate governance practices differ in normal situations and abnormal situations [19]. In the events of the crisis, the mechanisms became more effective as they have learned from previous financial collapses. This is related to the agency theory, and the role of the board of directors with the management. The role of the management is to ensure good corporate governance to sustain the firm's performance. To ensure an effective corporate governance practice, important variables should be measured to determine the effectiveness of the governance, which include (board meetings, board independence, board size, firm size, and firm Leverage).

Measuring a firm's performance is also challenging as there are two models of measurements used in previous research. The first model is, accounting-based measures, which are return on assets (ROA), return on equity (ROE), return on investment (ROI), and return on sale (ROS). The second model is the market-based measurements which are Tobin's Q and market return. In this study, we're going to measure a firm's performance by using ROA as it has been used in a majority of the research.

Many studies have been conducted to examine the impact of corporate governance on firm performance [4, 6, 7, 11]. Based on the previous study, agency problems arise because the management of a firm has to be cautious when dealing with the firm's funds as dealing with their own money [19]. While [25] discussed that by aligning management and director interests with shareholders' interests, corporate governance mechanisms reduced the agency problems. The firms with weaker governance structures face more agency problems and managers of such firms gain more private benefits [10]. The problem of "principal-agent" or "agency" has long been linked to corporate governance. The principal and agent relationship is formed when the owner and manager of a company are not the same people. Managing a business on behalf of principals (investors or financiers) is a common practice. Moreover, management's specialized human capital is required for investors to earn returns on their investments.

1.1 Research Objectives

The study focuses on the 'Corporate Governance Characteristics' and 'Firm Performance'. The main objective of the study is to examine corporate governance practices' impact on the firm performance in financial sector companies listed on Bahrain Bourse. "What is the influence of corporate governance procedures on business performance of the financial sectors listed in Bahrain Bourse between 2016 and 2020?" is the research question in this study.

A few studies were conducted in Bahrain to study the effectiveness of the corporate governance practices in Bahrain yet the financial sector was not covered in many. Bahrain is keen to promote good firm performance by issuing the corporate governance code re-issued on the 19th of March 2018. The objective of a corporate governance code is to provide firms with policies and standards for managing their performances in the most efficient way possible. Corporate governance aims to establish a process that guides and regulates the businesses and practices of firms to achieve efficient organizations.

2 Literature Review

[4] study examined the impact of corporate governance practices and institutional ownership on firm performance. This study examined the impact of corporate governance on the companies listed on the Bahrain Bourse. This study found that there is a significant relationship between institutional ownership and the corporate governance of the firms listed in the Bahrain Bourse.

The impact of corporate governance on Sharia corporate governance was investigated by [12]. It investigates the connections between the audit committee, the board of commissioners, corporate governance, corporate performance, as well as their implications for businesses. The study collected data for 63 firms registered on the Indonesian stock exchange and the Sharia stock list, the study covers the years 2014 and 2015. The data were analyzed using multiple regression, the researchers discovered a link between board commissioners, audit committee effectiveness, and sharia company performance. [20] investigated the relationship between corporate governance and business performance in GCC countries. It investigated if board size, audit method, government shareholding, corporate social responsibility, and leverage have an impact on GCC companies' financial performance [1]. The data was collected for the years 2005 to 2012, and it included 349 firms in the financial and non-financial sectors that were listed on GCC stock exchanges. The authors discovered that the variables have an impact on financial performance in the majority of GCC countries; as a result, future-proofing GCC firms will require unique regulatory and management implications.

The purpose of [2] research is to determine whether there is a link between corporate governance and financial success in Bahraini financial firms. It investigates how board size, ownership concentration, and auditor reputation affect financial performance. The study's sample includes 13 banks and 2 insurance businesses that are publicly traded on the Bahrain Stock Exchange, and it spans the years 2011 to 2016. According to this study, the yearly board of directors' meetings, the firm's return on equity, and the firm's size all had a negative impact on the firm's success. [27] studied the impact of corporate governance on the business performance of Vietnam's publicly traded companies. The size of the board, board independence, ownership concentration, and dual form of the CEO are the variables that will impact the outcome of this study, which will be measured using four different methods: ROA, ROE, Z-score, and Tobin'sQ. This study's database focuses on 177 firms that have been listed in Vietnam during the last five years, from 2008 to 2012. The dual role has a positive correlation with firm performance, the ownership structure has significant changes with firm performance, and board independence has a negative relationship with firm performance. Finally, there is no evidence to prove the relationship between board size and business performance in this study.

According to [22], a study was undertaken in Bangladesh to examine corporate governance practices and business value. It aims to establish a link between corporate governance practices such as board composition, audit committee composition, and business performance variables ROA and ROE. The study's sample for 2006 includes 93 Dhaka Stock Exchange-listed companies. The findings show that, according to the OLS estimate, the independent director and chief executive officer duality has a strong relationship with ROA and ROE. However, no link was found between the firm's measures (ROA and ROE) and the size of its board of directors or audit committee. [28] investigated the link between corporate governance and company performance in Malaysian publicly traded companies. To test the hypothesis against company performance, the ROA and ROE proxies were used, which were the dependent variables. The independent variables of corporate governance were board size and board independence. The hypotheses were evaluated against the top 100 public firms listed on Bursa Malaysia from 2008 to 2012 using descriptive and correlation analysis. The tests revealed that board size had a negative relationship with ROA but a positive relationship with ROE. Another finding was that board independence had no impact on business performance. [11] the study investigated the impact of corporate governance on financial institution performance in Sri Lanka. According to the study, return on asset and return on equity are important factors to consider when evaluating a company's performance. In addition, corporate governance is evaluated using board features such as board size, board meetings, and the audit committee.

From 2008 to 2012, a sample of 25 Sri Lankan publicly traded firms was employed in this study, using data collected quantitatively. The findings of the study revealed that corporate governance variables like the size of the board of directors and the audit committee have a positive impact on a company's success. Board meetings, on the other hand, have a negative impact on corporate governance. [7] examined the corporate governance practices and firm performance in Saudi Arabia. The research was conducted on 171 companies listed in the Saudi Stock exchange 'Tadawul' and collected data for the period 2012–2014. The dependent variable in this study is Firm performance and used three measurements 'ROA, ROE, and Tobin's Q. [3]'. Additionally, the researchers used five control variables in their measurements. [13, 16] The findings of the research concluded that corporate governance adoption has no substantial impact on the firm performance of companies. And by using Tobin's Q model, the ownership and Board of director's independence have no impact on the firm's performance. However, the size of the Board of directors and ownership had a significant impact on the firm's performance [8, 9].

The purpose of [5] research is to see how the characteristics of the board of directors and audit committee, as well as corporate governance, influence the financial outcomes and performance of Thailand-listed companies. On 452 firms listed on Thailand Stock Exchange between 2000 and 2016, the study used the generalized method of moments (GMM) methodology to estimate data and used OLS estimate and fixed effects to verify robustness. The outcomes of the study reveal that family ownership and ownership concentration have a negative link with market-based firm performance. Managerial ownership, on the other hand, is associated with improved corporate performance. Furthermore, board structure characteristics such as independence, size, meetings, dual tasks, and audit committee meetings have a significant impact on company performance.

According to [2], the purpose of their study is to investigate the internal corporate governance monitoring in Pakistani businesses. The dependent variable is board size, board independence, CEO duality, firm size, and management ownership. [14] The dependent variables were used to measure their impact on the dependent variable (firm performance). There are four metrics used to measure firm performance (ROA, ROA, EPS, and market to book ratio). [15] The research method is quantitative, using pooled ordinary regression utilized to estimate the results with data from 2004 to 2008. The findings revealed that a firm's size is positively related to its performance, and a positive relationship exists between CEO duality and EPS. However, leverage, board independence, and managerial ownership are negatively related to ROA, ROE, and EPS. Internal governance mechanisms are also important. [24] conducted a study to obtain scientific data on the influence of corporate governance on tax evasion. The study looks at the impact of effective corporate governance on firm success, using data from 120 manufacturing firms on the Indonesian Stock Exchange between 2010 and 2013. As a dependent variable, the researchers looked at tax evasion. Executive salary, executive character, concentrated ownership, size, board of commissioners' ratio, audit committee, and quality are the independent factors in this study. [17, 18] For analyzing the dependent variable, effective tax rates (ETR) were utilized as a proxy. (ETR) is a method of calculating how large corporations avoid paying taxes.

A multiple regression model was used to examine the data collected. Based on the result, audit committees have a positive effect on tax avoidance in part. However, the other independent variables, executive compensation, executive character, company size, institutional ownership, board proportion, audit committee, and audit quality have negatively impacted tax avoidance simultaneously.

According to [23], the study examined the quality of corporate governance procedures in an Egyptian stock exchange setting as well as their impact on firm behavior and financial distress. This study explored the variables of the board of directors' independence, shareholder rights, investor interaction, and ownership and control structure through 86 non-financial companies, using Tobin's Q and Altman Z-score as indicators; the higher the Z-score, the higher the risk [26]. Results of this study indicate that corporate governance does not impact business performance and financial distress. Additionally, the study indicates that firm-specific characteristics can serve as a first-pass screening tool for forecasting firm performance.

H1: There is a relationship between Independent Variables (board size, board independence, board meetings, firm size) and the dependent variable (ROA).

3 Methodology

The purpose of the research is to examine the impact of corporate governance on the firm performance of the financial companies listed in the Bahrain Bourse. The study analyzed a sample of 18 financial companies out of a total of 42 listed companies. The data was collected using the quantitative method (secondary method) and covered the period 2016–2020. The study included information about the financial sectors because they are subject to different regulations and have more restrictions than industrial and service companies. The measurements in this study will include one dependent variable, firm performance using the ROA measurement, and three independent variables board size, board independence, and board meetings. Furthermore, the study used two control variables, firm size calculated using (the logarithm of total assets), and firm leverage calculated using (Debt to Equity ratio), to better understand the relationship. The data was collected manually by reviewing the annual reports of the companies. Using this sampling and period will allow us to ensure more reliable and accurate results, as well as the most up-to-date results of our research. The data collected would be analyzed using the multiple regression model.

Measurement of Variables (Dependent, Independent, Control) and Definitions

Dependent Variable

Firm performance

The proxy that will be used in this study to measure the firm performance is the ROA. **ROA** is one of the most used accounting ratios when measuring a firm's performance. It is measured by calculating net income divided by the total assets of the company.

Table 1 Variables abbreviation

$\beta 0$	Constant
Firm performance	Firm performance Return on assets (ROA)
$\beta 1$ BSIZE	Board size
$\beta 2$ BIND	Board independence
$\beta 3$ BMTING	Board meetings
$\beta 4$SIZE	Firm size
$\beta 5$ LEV	Firm leverage
ε	Error term

It indicates how effectively a company uses its resources to generate a profit. The higher ratio indicates the more efficient the company's management.

Independent Variables

When assessing a firm's performance, the research measured the independent factors that are linked to it, as the variables affected the firm's performance (ROA), and determined how significant the relationship is with corporate governance processes (Board size (BSIZE), board independence (BIND), board meetings (BMTING)) for the financial companies listed in Bahrain. The study also measured the correlation between the firm performance using return on assets proxy (ROA) with the independent variables, is there a negative or positive correlation and how it affects the firm's performance. After measuring the corporate governance variables (Board of directors' size, meeting. Independence, the following model would be hypothesized and tested (Table 1):

$$\text{ROA} = \beta 0 + \beta 1\ \text{BSIZE} + \beta 2\ \text{BIND} + \beta 3\ \text{BMTING} + \beta 4\text{SIZE} + \beta 5\ \text{LEV} + \varepsilon$$

Control Variables

The two control variables studied in this research are Firm size (SIZE), Firm leverage (LEV).

The definitions and measurements of all variables indicated above will be gathered from annual public reports, as shown in the Table 2 below:

The data was analyzed using an excel sheet. To test and present the research relationship, the model was analyzed via data analysis in excel. It has one dependent variable, the Firm's performance, which is measured by Return on assets (ROA), three independent variables, Board Size (BSIZE), Board Independence (BIND), Board Meetings (BMTING), and two control variables Firm size (SIZE), Firm leverage (LEV).

Data Sources

To collect data related to the research and evaluate research hypotheses, the study employed a variety of resources:

Table 2 Description and measurement of variables

Variables' symbol	Variable name	Description and measurement
Dependent variable		
Firm performance return on assets	Return on assets	Net Income to Total assets
Independent variables		
BSIZE	Board size	Total number of board of directors in the firm
BIND	Board independence	Total independent directors to total directors' ratio
BMTING	Board meeting	The frequency of board meetings in a particular year
Control variables		
SIZE	Firm size	Logarithm of Total Assets
LEV	Firm leverage	The firm's total debt to total equity ratio

- Numerical reports of all selected firms, which are available on their websites.
- Thomson Reuters, google scholar, UOB library, and Bahrain Bourse website for gathering programs that can cut down the time it takes to manually collect data.
- Previous research in the field by looking at previous pieces of literature.

4 Data Analysis and Interpretation

Table 3 shows the relationship between dependent variables and independent variables. The dependent variable of this study is the return on assets (ROA), the independent variables are board size, board Independence, and board meetings, and the control variables are firm size and firm leverage. The table shows a positive correlation between return on assets and board meetings. On the other hand, there is a negative correlation between board size, board independence, firm size, and firm leverage with return on assets.

Table 4 using the R squared formula, you can calculate the percentage of change in the dependent variables explained by changes in the independent variables. In the summary output, the first table shows the R square with a 41.3% change in firm performance (ROA) and is driven by the independent variables which are Board size, Board Independence, and Board meetings. The Adjusted R square (0.169) it adjusts the R square whenever the independent variables increases and there is an increase in sample size. The higher the number indicates a better model. In this model it is 16.9% which represents the model is insignificant.

Table 5 is the ANOVA it represents the significance level at which the model fits perfectly with the results. However, the table shows a significance level of 21% which is statistically insignificant.

Table 3 Correlation matrix

	Return on assets (ROA) dependent variable	Board size	Board independence	Board meetings	firm size	firm leverage
Return on Assets (ROA) Dependent Variable	1					
Board size	−0.306	1				
Board Independence	−0.259	0.646	1			
Board meetings	0.195	0.332	−0.02	1		
Firm size	−0.331	0.153	0.432	0.087	1	
Firm leverage	−0.358	0.284	−0.054	0.116	-0.121	1

Table 4 Regression statistics

Regression statistics	
Multiple R	0.643
R square	0.413
Adjusted R square	0.169
Standard error	0.015
Observations	18

Table 5 ANOVA table

ANOVA					
	df	SS	MS	F	Significance F
Regression	5	0.002	0	1.69	0.211
Residual	12	0.003	0		
Total	17	0.005			

When the regression coefficient intercept value $\beta 0 = 0.041$, this means if the board size, board meeting, board independence, as well as firm size, and firm average will equal zero, the firm performance (ROA) will equal = 0.041. We can see in Table 4 the coefficient value of board size is -0.003, this means as the board size increases, the ROA decreases by 0.003. Alongside, the board independence's coefficient value is equal to positive 0.002, it means that an increase in the board independence will increase the ROA by 0.002. Additionally, the board meetings coefficient value is positive 0.006, it means that an increase in the board meetings will increase the ROA increases by 0.006. Furthermore, the coefficient value of firm size is −0.004,

Table 6 Regression analysis output

	Coefficients	Standard Error	t Stat	P-value	Lower 95%	Upper 95%	Lower 95.0%	Upper 95.0%
Intercept	0.041	0.025	1.663	0.122	−0.013	0.094	−0.013	0.094
Board size	−0.003	0.003	−1.037	0.320	−0.010	0.004	−0.010	0.004
Board independence	0.002	0.004	0.427	0.677	−0.007	0.010	−0.007	0.010
Board meetings	0.006	0.004	1.568	0.143	−0.002	0.013	−0.002	0.013
Firm size	−0.004	0.002	−1.630	0.129	−0.009	0.001	−0.009	0.001
Firm leverage	−0.003	0.002	−1.386	0.191	−0.008	0.002	−0.008	0.002

to interpretate, as the firm size increases the ROA decreases by 0.004. Moreover, the firm leverage coefficient is -0.003, this means that as the firm leverage increases the firm's performance decreases by 0.003 (Table 6).

$$\text{ROA} = \beta 0 + \beta 1\ \text{BSIZE} + \beta 2\ \text{BIND} + \beta 3\ \text{BMTING} + \beta 4\text{SIZE} + \beta 5\ \text{LEV}$$

$$\text{ROA} = \beta 00.042 - \beta 10.003 + \beta 20.002 + \beta 30.006 - \beta 40.006 - \beta 50.003$$

The results show that corporate governance negatively correlates with the firm performance measurement ROA. Furthermore, the independent variables and control variables P−value is equal to; (board size = 0.320, board independence = 0.677, and board meetings = 0.143, the control variables, firm size = 0.129 and firm leverage = 0.191). The independent and control variables have a relationship with ROA at P-value > 0.05, which means the relationship is statistically insignificant.

5 Conclusion

This study aims to examine the influence of corporate governance mechanisms (Board size, Board independence, Board meetings, and control variables firm size and leverage) on the performance of companies listed in the financial sector of the Bahrain Bourse. The study employed one dependent variable, firm performance which is measured by return on assets (ROA). Three independent variables, are governance mechanisms, such as board size, board independence, and board meetings. Finally, two control variables: are firm size (logarithm of total assets) and firm leverage (debt-to-equity).In this study, it has been found that over the period 2016 to 2020, most companies' board meetings and firm leverage (debt to equity) increased. On the other hand, the collected data indicated that board size, board independence, and firm size were not affected over time. Due to Pandemic Covid-19, the loss of capital assets in most companies adversely affected the firm's performance and led to a decline in share prices as investors withdrew their funds, negatively affecting the overall profitability of Bahraini financial firms. Moreover, board meetings did not have a significant impact on the firm's performance (ROA). Nevertheless, the hypothesis tested contradicts the regression model, and it seems there is a weak positive impact on a firm's performance. In addition, firm size and firm leverage showed a negative relationship with firm performance. This supports the hypothesis that the variables are negatively correlated with the firm's performance.

6 Recommendations

As a proxy for measuring firm performance in this study, our study used return on assets (ROA), however, we recommend future studies to employ additional proxies for measuring firm performance, such as return on equity (ROE), earning per share (EPS), Tobin's, return on investment (ROI), and Z-Score to obtain more accurate results and more understanding on the relationships. Furthermore, the study recommends that the study period should be extended by over five years to consider the dynamic nature of governance, which would lead to more conclusive analysis through time-series or panel data. Since the current study used five years that showed no relationship between independent variables and the dependent variable, the significance level was statistically insignificant. Also by examining more aspects of board characteristics for corporate governance mechanisms, such as board of directors' frequency, board of directors' composition, dual role of CEO, government shareholding, corporate social responsibility, and ownership concentration. In addition, corporations around the world have been affected differently by the pandemic. Accordingly, this study examines the financial sectors in Bahrain. While it's better to focus on just one sector, and not consider the other sectors because it would interrupt the statistical analysis. Therefore, this study recommends comparing data by including the financial sectors in GCC countries or comparing it between two different countries rather than focusing on just Bahrain, because default information more likely to occur when using a smaller size sample. As a final recommendation, the study encourages future work to explore how COVID-19 affects organizational behavior to obtain more accurate results and a deeper understanding of the impacts and connections.

Implication: The study would be able to gain a better understanding of the relationship between corporate governance and firm financial performance in GCC. This study is useful to board directors and chairmen of listed companies, as it can help them to better understand and manage firm performance.

Future Research Perspective: This study covers only the financial sectors listed in the Bahrain Bourse. Future research can consider all companies from GCC, the middle east, or worldwide. More dependent variables can be used such as ROE, Tobin Q, EPS, and so on. More independent variables related to CG can be added. Pre- and Post-Covid-19studies can be conducted to see how Covid impacted the relationship.

References

1. Abdalkrim G (2019) Chief executive officer compensation, corporate governance and performance: evidence from KSA firms. Corp Gov Int J Bus Soc 19(6):1216–1235
2. Ahmed Sheikh N, Wang Z, Khan S (2013) The impact of internal attributes of corporate governance on firm performance: evidence from Pakistan. Int J Commer Manag 23(1):38–55
3. Aktan B, Turen S, Tvaronavičienė M, Celik S, Alsadeh HA (2018) Corporate governance and performance of the financial firms in Bahrain. Pol J Manag Stud 17:39–58
4. Al-Sartawi A, Sanad Z (2019) Institutional ownership and corporate governance: evidence from Bahrain. Afro-Asian J Finan Account 9(1):101–115
5. Al-Sartawi A (2017) The effect of the electronic financial reporting on the market value added of the islamic banks in gulf cooperation council countries. In: 8th global islamic marketing conference, 4–6 May 2017. International Islamic Marketing Association, Turkey
6. Bhambra RK (2022) Adoption of online banking in Goa amidst the pandemic. Int J Electron Bank 3(2):163–175
7. Buallay AM, Abuwadi RM, Kukreja G, Hassan A (2020) Evaluating ESG disclosures of Islamic Banks: evidence from the organization of Islamic Cooperation Members. Int J Innov Sustain Dev 14(3):266–287
8. Buallay A, Abuhommous AA, Kukreja G (2020) The relationship between intellectual capital and employees' productivity: evidence from the Gulf Cooperation Council. J Manag Dev (NA). https://doi.org/10.1108/JMD-05-2019-0210
9. Buallay A, Kukreja G, Aldhaen ES, Almubarak MS, Hamdan AM (2020) Corporate social responsibility disclosure and firms' performance in mediterranean countries: a stakeholders' perspective. EuroMed J Bus 15(2):1–16
10. Chauhan N, Tekta P (2020) Fraud detection and verification system for online transactions: a brief overview. Int J Electron Bank 2(4):267–274
11. Danoshana S, Ravivathani T (2019) The impact of the corporate governance on firm performance: a study on financial institutions in Sri Lanka. SAARJ J Bank Ins Res 8(1):62–67
12. Faradisi RJ, Setyaningrum D (2017) The effect of corporate governance on sharia corporate performance in Indonesia. In: 6th international accounting conference (IAC 2017), pp 273–277. Atlantis Press
13. Hermuningsih S, Kusuma H, Cahyarifida RA (2020) Corporate governance and firm performance: an empirical study from indonesian manufacturing firms. J Asian Finan Econ Bus 7(11):827–834
14. Jarboui S, Guetat H, Boujelbene Y (2015) Evaluation of hotels performance and corporate governance mechanisms: empirical evidence from the Tunisian context. J Hosp Tour Manag 25:30–37
15. Jebran K, Chen S (2021) Can we learn lessons from the past? COVID-19 crisis and corporate governance responses. Int J Finan Econ 1–9
16. Jensen M, Meckling W (1976) Theory of the firm: managerial behavior, agency costs and ownership structure. J Financ Econ 3(4):305–360
17. Khatib SFA, Nour ANI (2021) The impact of corporate governance on firm performance during the COVID-19 pandemic: evidence from Malaysia. J Asian Finan Econ Bus 8(2):0943–0952
18. Letza S, Sun X, Kirkbride J (2004) Shareholding versus stakeholding: a critical review of corporate governance. Corp Gov Int Rev 12(3):242–262

19. Lin Z, Zhao X, Ismail KM, Carley KM (2006) Organizational design and restructuring in response to crises: lessons from computational modelling and real-world cases. Organ Sci 17(5):598–618
20. Al-Sartawi AMA (eds) (2022) Artificial Intelligence for Sustainable Finance and Sustainable Technology. ICGER 2021. Lecture Notes in Networks and Systems, vol 423. Springer, Cham
21. Rathod L (2018) Why is corporate governance important?. https://www.diligent.com/en-gb/blog/why-is-corporate-governance-important/
22. Abdur RM (2011) The relationship between corporate governance and value of the firm in developing countries: evidence from Bangladesh. Int J Appl Econ Finan 5(3):237–244
23. Sanad Z, Al-Sartawi A (2016) Investigating the relationship between corporate governance and internet financial reporting (IFR): evidence from Bahrain Bourse. Jordan J Bus Adm 12(1):239–269
24. Tandean VA, Winnie W (2016) The effect of good corporate governance on tax avoidance: an empirical study on manufacturing companies listed in IDX period 2010–2013. Asian J Account Res 1(1):28–38
25. Vintila G, Gherghina SC (2012) An empirical investigation of the relationship between corporate governance mechanisms, CEO characteristics and listed companies' performance. Int Bus Res 5(10):175
26. Vo DH, Nguyen TM (2014) The impact of corporate governance on firm performance: empirical study in Vietnam. Int J Econ Financ 6(6):1–13
27. Zabri SM, Ahmad K, Wah KK (2016) Corporate governance practices and firm performance: evidence from top 100 public listed companies in Malaysia. Procedia Econ Finan 35:287–296
28. Zhang D, Hu M, Ji Q (2020) Financial markets under the global pandemic of COVID-19. Financ Res Lett 36:101528

The Impact of Financial Performance on Firm's Value During Covid-19 Pandemic for Companies Listed in the Palestine Exchange (2019–2020)

Israa Aref Abu Asa'd, Abdulnaser Nour, and Sameh Atout

Abstract The purpose of this study is to examines whether Covid19 pandemic has an impact on the relationship between the financial performance of companies in return to assets, equity, and EPS and the company value with regard to stock return, Tobin Q ratio, P/E ratio for Companies Listed in the Palestine Exchange (2019–2020). Based on 216 observations, this study utilizes panel data is used to analyze the relationship between financial performance (ROA, ROE, and EPS) and company's value (stock returns, Tobin's q and price to earnings ratio). Correlation matrix and linear regression analysis are considered a tool to analyze the research hypotheses. The results show that found no effect of the Covid19 pandemic on the relationship between the financial performance (ROA, ROE, EPS) and the company's value (stock return), but it was turned out that the Covid19 pandemic had significant effect weakening the relationship between financial performance (ROA, ROE, EPS) and the company value (Tobin's Q), and it also had a significant impact on the relationship between the financial performance (ROA, ROE) and the company's value (P/E ratio) as the results showed that the relationship is negative. The novel Coronavirus (COVID-19) has out broken a health and economic catastrophe all over the world. Palestine is not excluded from being affected by this virus, this research is that it is one of the first studies in this field—within the limits of the researcher's knowledge—that is interested in studying the impact of this virus on the economic situation in Palestine and the extent to which the investor's decision is affected in light of the existence of an ambiguous pandemic globally and locally.

Keywords PEX · (COVID-19) · (Tobin's Q) · Financial performance

I. A. A. Asa'd · A. Nour (✉) · S. Atout
Accounting Department, An-Najah National University, Nablus, Palestine
e-mail: a.nour@najah.edu

I. A. A. Asa'd
e-mail: israa.abuasad@ptuk.edu.ps

S. Atout
e-mail: sameh.atout@najah.edu

A. M. A. Musleh Al-Sartawi et al. (eds.), *From the Internet of Things to the Internet of Ideas: The Role of Artificial Intelligence*, Lecture Notes in Networks and Systems 557,
https://doi.org/10.1007/978-3-031-17746-0_42

1 Introduction

The world of economy is experiencing important shifts due to unprecedented situation that represented in a serious health phenomenon threatening the global economy. This phenomenon launched at the Republic of China in late December of 2019, which was diagnosed as a global epidemic at the beginning of March 2020 by the World Health Organization (WHO). It was termed the Covid-19 (Corona) virus pandemic and classified as a rapid epidemic and required special medical attention [29].

The outbreak of the virus and with the global opening up of countries, and after disease cases were diagnosed with the virus in the State of Palestine, Palestinian government had promptly implemented the instructions of the World Health Organization and took preventive governmental measures such as declaring the state of emergency in Palestine on 3/5/2020. These Procedures included restrictions of citizens' movement and closure of borders and companies, which led to halting the supply and production chains for the majority of economic sectors while health and food sectors were on the run as they are essential for citizens [20].

The pandemic has been addressed in all sectors with the inability to predict future scenarios for the pandemic at the present time which highlights the importance of successful financial management of the pandemic as it is the basic element of future plans. The complexity of the pandemic has resulted most notably the lack of liquidity and debt default. The key for companies is to maintain market continuity and ensure that debt is not defaulted and entered into real financial hardship as it may take a long time to remedy the consequences over the market value of companies' shares because the world faces an unspecified virus [10].

There is no doubt that, after the end of the pandemic, the world will accept political, economic and social changes, and the longer the pandemic lasts, the more it will have a greater impact on humanity, which is an important juncture and watershed in the history of the world as it is an event with a fundamental impact whose consequences are difficult to predict [24].

Accordingly, the State of Palestine isn't of course immune to the global pandemic, which causes disruption in sectors such as the tourism sector and aviation. A state that calls for adapting some facilities and measures to mitigate the economic and financial consequences of the pandemic.

The remained of the study is organized as follows; the second section presents a review of literature, while the third section describes the data collection and methodology. The fourth section presents the parametric regression outcomes. In the fifth section, the results are discussed while the sixth section contains the conclusions and recommendations for future studies.

The study's significance emerges through highlighting the economic implications of governmental measures and the consequent stopping of productivity for the majority of economic sectors and the decrease in cash liquidity due to the closure. Many of the effects, that will be addressed during the study, are focused on the impact of the Covid 19 pandemic with regard to the performance according to the financial statements of the companies listed on the Palestine Exchange and their implications for the stock returns of these companies.

Considering the negative repercussions of the pandemic and the contradictions between the economic and health sectors in light of the governmental restriction measures in Palestine, the main question the study's problem arises: **What is the effect of financial performance on firm value during the new Covid 19 pandemic—a study on companies listed on the Stock Palestine for the period 2019–2020?** Other questions also arose:

1. **Do the financial performances influence the companies' values?**
2. **Does Coronavirus pandemic affect the relationship between companies' financial performances and the stock returns?**
   2.a. Does the return on assets affect the stock return during the Coronavirus pandemic?
   2.b. Does the return on equity affect the stock return during the Coronavirus pandemic?
   2.c. Does the earning per share affect stock return during the Coronavirus pandemic?
3. **Does Coronavirus pandemic affect the relationship between the companies' financial performances and Tobin's Q?**
   3.a. Does the Return on assets affect the Tobin's Q during the Coronavirus pandemic?
   3.b. Does the Return on equity affect the Tobin's Q during the Coronavirus pandemic?
   3.c. Does the Earning per share affect Tobin's Q during the Coronavirus pandemic?
4. **Does Coronavirus pandemic affect the relationship between the companies' financial performances and P/E ratio?**
   4.a. Does the Return on assets affect P/E ratio during the Coronavirus pandemic?
   4.b. Does the Return on equity affect the P/E ratio during the Coronavirus pandemic?
   4.c. Does the Earning per share affect the P/E ratio during the Coronavirus pandemic?

The study mainly seeks to achieve the companies' financial performances listed in the Palestine Exchange and their impacts on the stock returns, Tobin's Q, and the P/E ratio during the coronavirus pandemic, in addition to the following objectives:

1. Studying the association between the companies' financial performances and their value.
2. Determining the effect of Coronavirus on the association between the companies' financial performances and their values.
3. Studying the effect of the Covid-19 pandemic on the profitability of companies as well as its effect on the stock returns on the Palestine Exchange.

4. Studying the effect of the Covid-19 pandemic on the profitability of companies and its effect on the Tobin's Q (TQ) on the Palestine Exchange.
5. Studying the effect of the Covid-19 pandemic on the profitability of companies and its effect on price to earnings ratio (P/E) on the Palestine Exchange.

2 Literature Review Hypotheses Development

Coronavirus Pandemic (Covid19)
Coronavirus (Covid-19) Pandemic it is a large group of viruses that cause disease in animals and humans and causes respiratory infections, which range from common colds to more severe diseases such as Middle East Respiratory Syndrome and acute respiratory syndrome (SARS) and recently appeared Covid-19. [22] reported by: "the World Health Organization (WHO) on February 11, 2020.

The World Health Organization renamed the new variable of this virus as Covid-19, and identified it as an infectious disease, and was classified as a global pandemic [10].

The world is experiencing many successive economic and financial crises, the last one was the 2008 mortgage crisis, as will also be listed later from pandemics that disrupted the global economic system, and with China announcing the first case of Coronavirus (Covid 19) in Wuhan in late December 2019, where The virus has prevalence to most of the world countries shortly, which led the governments of the countries to take precautionary measures, such as closing airports, borders, and preventing movement of citizens, creating a state of alarm and fear of sudden and rapid virus outbreaks, as the Corona virus differs from the rest of the epidemics that struck the world, which were limited Its spread in certain geographical regions such as the SARS virus, which spread exclusively in the country of China. [26] The virus began to spread quickly in most countries of the world, amounting to 199 countries until the end of March 2020, and the World Health Organization recorded 800,000 cases, including 39,000 deaths [10].

Accordingly, the entire world and the Arab region are experiencing an exceptional situation in light of the spread of the Corona virus and its wide-ranging economic effects, including many economic sectors, especially the tourism and aviation sectors, in light of the countries tendency to impose the policy of preventing movement, which affected the global supply chains, the international trade movement, Consumption activities, investment and manufacturing, high levels of uncertainty, and low consumer confidence [2].

Economic Effects of Coronavirus Pandemic (Covid19) on the Palestinian Economy

The Palestinian economy is greatly affected by the crises and shocks to which it is exposed, especially with the succession of these shocks and crises as the outbreak of the Corona virus may be more lethal to the economy, especially with the persistence of previous crises, led by the occupation of the occupation of the Palestinian territories, and the current political situation and the resulting significant decrease in The amount of international support provided by donors, and accordingly, the economic effects of Corona virus on the Palestinian economy are listed as follows, where these effects are linked in the period of pandemic and emergency in the country [21].

1. Increase in public spending: Given the overlap between the health and economic aspect of the pandemic, and adapting preventive measures to confront coronavirus has to increase government spending on several levels. **2**. Decreased government revenues: The declaration of a state of emergency, the closure of Palestinian crossings and borders, and the stopping of the foreign trade movement will affect the clearing revenues. As indicated by the Palestinian Prime Minister Dr. Muhammad Shtayyeh, the cost of the new coronavirus may reach from 300–400 million shekels during the slowdown in the global economy, which is reflected on Palestinian treasury revenues, with a decrease of more than 50% [23]. High government debt: The Palestinian public budget suffers from a deficit before the pandemic, and with the outbreak of the virus and the consequent rise in public expenditures, the government will resort to external and domestic borrowing to cover these expenses. **4**. Economic loss from preventing workers from working in 48 lands and in settlements: The implementation of preventive measures, including preventing workers from working in 48 lands and in settlements. **5**. Stoppage of import, export and the productive sectors: The procedures followed within the decree declaring the state of emergency and the preventive measures included closing all crossings and borders and restricting them to the necessary and medical consumer goods, i.e. **6**. The investment disturbance: investment needs a stable environment, but the spread of creating a state of panic and fear for the consumers and the investors together reflected negatively on the prices of assets and aggregate demand, accompanying by high debt and the loss of a large number of investors because of the low ability to predict the future giving rise to taking a decision to stop future projects.**7**. Decrease in foreign investments: This includes investments of institutions and individuals residing in Palestine and they have investments abroad (foreign investment) as well as currency and deposits investments [21]. 0.8. The economic sectors, especially the tourism sector, are affected: stopping the Palestinian life completely and the consequent cancellation of all reservations for tourists and the closure of tourist areas to domestic tourism [45].

Global Financial Crises

The financial crisis is characterized by its sudden and unexpected occurrence and requires immediate interventions and treatments at all levels, whether for individuals, companies or governments, in order to save the total economy from collapse and bankruptcy as a result of this serious event [8]. The financial crisis is a sharp and sudden turbulence in economic balances followed by a collapse in a number of financial institutions and their performance indicators and their effects extending to other economic sectors, then followed by a severe turmoil in financial markets (stocks and bonds) accompanied by a rapid development of events, which destabilizes the basic system of countries and financial markets [16].

Major Global Financial Crises

1. The Great Depression 1929

It is considered one of the most severe financial crises witnessed by the global economy, as a result of which shares in the US financial market fell by 13% as the manifestations of the crisis were represented by a decrease in total consumption and a decrease in investments with high unemployment rates [27].

2. The Mortgage Crisis 2008

The pandemic began in the United States of America due to the global risk of bank loans to the real estate sector as real estate loans were granted to large segments with high interest rates for American banking markets creating a state of panic in the global markets. The collapse and losses of Lehman Brothers began in succession drifting all over the world. The most important of which was the aggravation of the budget deficit, the high debt and the slowing economic growth. [4] Therefore, the immediate causes of the pandemic were the sudden drop in interest rates, the lowest quality mortgage i.e. the mortgage resulting from the mortgage of the property a second time, and finally the financial derivatives [5].

3. Emergence of the Coronavirus Pandemic (Covid-19): In view of the researcher's opinion)

What distinguishes Coronavirus (Covid-19) pandemic from the previous crises is that it is a pandemic caused mainly by a health pandemic represented in the spread of a dangerous virus that has resulted in the closure of many sectors, especially the financial markets. It is also a health pandemic with economic and financial repercussions. Hence, in the opinion of the researcher, the contradiction has emerged between the health and the economic aspect. On one hand, the importance of the human element and the necessity of preserving it, and minimizing the negative effects on the economy, on the other.

Therefore, the difference of this pandemic from previous crises is that its source is not the financial markets or the monetary or the economic policies of countries, which makes the methods of dealing with them different from the methods used in the previous crises. In this instance, the monetary policies and the financial measures are to reduce the severity of the pandemic and its impact on the economy and not to address it because its treatment is represented in a vaccine. It makes governments to proceed with the procedures for opening markets, airports and other sectors affected by the closure and declaring a state of emergency in most countries of the world [36].

Financial Performance

Stock markets occupy an important role in most of the global economies and have the attention of all developed and developing countries alike and by varying degrees, according to the existing economic philosophy, due to the tools and mechanisms provided by these markets to evaluate projects as well as their primary tasks in financing various economic activities [37].

The financial markets provide the liquidity necessary to achieve sustainable development and the provision of medium and long-term financing for various economic units [1, 38]. The financial evaluation is considered one of the important issues for companies that can examine the financial policy of the company during a certain period of time, by studying and analyzing the financial statements of the company to understand its implications and know the reasons that lead to its appearance of the quantities and how it is, which helps the company's management to identify weaknesses. And the strength in financial policy to follow financial measures address weaknesses and maintain the strength of the company's financial performance [6]. Financial Performance it is the ability and efficiency of the company to manage the activity in various administrative, productive, technical, marketing aspects. And others during a specific time period as well as its ability to convert the inputs to outputs with the required quality and quantity. It is a measure to judge the soundness of the financial position of companies during a specific time period [3]. The globalization of markets, the global economy, and the resultant work in highly competitive markets show the importance of financial performance in Monitoring and identifying the activity of the economic institution and its nature, following up and identifying the financial and economic conditions surrounding it, [39]. and assisting in conducting the process of analysis, comparison and evaluation of financial statements, and assisting in understanding the interactive understanding between the financial statements.. [3] Stock returns are defined as profits or losses resulting from investment during a certain period, which are measured by cash distributed over the time period, which is important for the analysis of investment in the company's shares in the sense that the higher the return on shares, the more profitable the investment resulting from the difference between the selling price and the purchase price per share [19]. This return is used to measure the performance of the shares by reflecting the change in stock prices on the company's financial objective of maximizing the returns of the stock, where the return per share is used to measure the financial performance of the company [6], as stipulated in the study [28] by having a positive relationship with many factors affecting the return of shares, namely financial performance, capital structure and equity, as the return on shares is a strategic objective of the investor to be achieved from investment [40, 41].

1. *Stock markets*

1.1: The Rate of Return on Assets (ROA)

It is defined as an "indicator that measures the profitability of a company relative to its total assets, and gives the return on assets an idea of how efficient the management is in using its assets to achieve profits, and is the ratio of net profit to total assets (i.e. total current assets and long-term assets") [15]. The study of [17] which was titled "Determinant analysis of financial ratio on stock returns in constructions companies registered at Indonesia stock exchange 2025–2019″ dealt with a positive relationship between the rate of return on assets and the market return of shares of companies listed on the Palestine Stock Exchange, and accordingly the hypothesis of the study is formulated as follows:

H1a: Return on assets has a positive relationship on the company's stock return.

H1b: Coronavirus pandemic moderates the relationship between return on assets and company's stock return.

1.2: Equity Return Rate (ROE)

The ROE rate is used to reveal the company's ability to invest its resources in a manner that achieves a great return on property rights, meaning that in the event of a high rate, the company can distribute greater profits than other companies, which affects shareholders' expectations by providing profits from investment in the company's shares, which has The positive effect on the rise in stock prices. This is what the study [7] concluded that increasing the ROE rate has a positive effect on stock prices, and therefore the study hypothesis is formulated as follows:

H2a: Return on equity has a positive impact on stock returns.

H2b: Coronavirus pandemic moderates the relationship between return on equity and stock return.

1.3: The earnings per share ratio (EPS)

The earnings per share ratio is considered one of the important factors affecting the share price as the earnings per share indicates the unit profit on the ordinary share by dividing the profits after taxes and interest on the number of issued or existing shares in which the number of ordinary shares purchased from the shareholders is intended. The ratio plays an important role in influencing the market value of the stock and its sleep in the market, according to the shareholders' preferences and investment objectives [14]. Earnings per share is used as an indicator in seeing the attractiveness of the stock through the effect of the earnings per share on investor confidence. The shares are bought and the buyer attaches importance to the earnings per share affecting the share price, [18]. The study showed that the earnings per share greatly affects the price of the shares, i.e. there is a positive relationship between earnings per share and stock prices. On the other hand, the results showed that the earnings per share do not affect the market ratio as a relationship between two variables. Accordingly, the earnings per share affects one variable, which is the stock market prices, but does not affect the ratio of price and price profit to the book value [43].

The study showed that the stock returns do not mainly reflect the financial performance of the company as there are many other factors that affect the financial performance, including macroeconomic conditions, political positions, government policies, and the technical aspects of the company. They are factors that affect the returns of shares other than the financial performance of the company, and therefore, it formulates the study hypothesis as follows:

H3a: Earning per share ratio has a positive effect on stock returns.

H3b: Coronavirus pandemic moderates the relationship between earning per share and stock return.

2. *Tobin's Q*

Current studies are moving towards new tools in order to reach a high forecast of stock returns in order to make a better decision with the result that the TQ ratio provides a high forecasting capability for stocks returns. The definition of TQ by researchers is "the TQ ratio is defined as the market value of a firm divided by the replacement cost of the firm's assets [13] it is confirmed by the researchers' findings [9] in their study on the existence of a relationship between TQ and returns on stocks where it had the highest interpretive power among other variables indicating that TQ has an important role and a high predictability of the company's future performance when the ratio Low TQ leads to high stock returns and vice versa. Tobin's Q is one of the most important indicators of the company's value, showing the market value of total capital assets divided by the replacement cost of those assets, and the first to propose the Concept of Tobin's Q is global economist James Tobin [11]. James Tobin's assumption of Q concept was that the combined market value of all companies on the stock market should be equal to the costs of their replacement, and companies with a Q value of 0–1 would be attractive to potential buyers rather than setting up a similar company and then increase interest in them and increase their value, unlike companies with a higher than 1% of the top 1 where they are experiencing great competition because they earn a higher rate of return than the cost of replacement. [30]. Among the studies that examined the relationship between ROA and Tobin's Q study [11]" Effect of Return on Assets (ROA) Against Tobin's Q: Studies in Food and Beverage Company in Indonesia Stock Exchange Years 2007–2011″, where among the results of the study is the existence of a direct relationship between ROA and Tobin's Q, and therefore, the hypothesis of the study is as follows:

H4a: Return on asset has a positive impact on Tobin's Q.

H4b: Covid-19 moderates the effect of ROA on Tobin's Q.

One of the studies that took the Q index as one of the most important indicators to measure the value of the company and predict its future performance, including the study [30] "Tobin's Q and its Determinants: A study on Huawei Technologies Co., Ltd.," where the study community took Huawei, and the study aimed at studying the impact of Tobin's Qi with the company's factors from 2011 to 2015, using the ROE factor, and the result was The study had a significant effect of ROE on Tobin's Kew. From here, we derive the hypothesis of our study, which is:

H5a: Return on equity ROE has a positive impact on Tobin's Q.

H5b: Covid-19 moderates the effect of ROE on Tobin's Q.

In addition to previous studies, there are many researchers who have studied the impact of profitability variables on the Tobin's Que Index, and among these variables is EPS, which is one of the most important profitability ratios because it helps investors make decisions at first sight, and among these studies [12]," Factors influencing firm value as measured by the Tobin's Q: Empirical evidence from the Saudi Stock Exchange (TADAWUL) "which They took the study sample consisting of 40 companies from 6 sectors listed on the Saudi Stock Exchange and it was among the factors that affect Tobin's Q and had a positive impact on the value of the company, and from here we derive the hypothesis of our study, which is as follows:

H6a: Earning per share (EPS)ratio has a positive effect on Tobin's Q.

H6b: Covid-19 moderates the relationship of EPS on Tobin's Q.

3. Price of Earnings Ratio

The decision to invest in any company is based on the expectation of the future performance of the company and the evaluation of the company's performance by the investor by relying heavily on the relative valuation, especially the price-to-earnings ratio P/E. It is measured by dividing the market value of the share by the profit earned per share [43]. This percentage indicates the number of times the share price exceeded to profit for shares and therefore shows the investor's desire to pay each unit of the company's revenue. [25] study checks the company's profitability performance indicators (ROA and ROE) as child variables. The independent variables are DY, PE, PB and EPS. The sample was from 51 companies in BIST over ten years from 2006 to 2015. Among the results were a significant negative relationship between ROA and the P/E variant, and a significant negative relationship between ROE and the P/E variable. Hence the following hypotheses:

H7a: Return on assets (ROA) has a positive impact on Price to earnings ratio (P-E).

H7b: Covid-19 moderates the effect of ROA on Price to earnings ratio (P-E).

[15] study examines the relation between the forward price-to-earnings (P/E) ratio and profitability. This paper finds a U-shaped relation between the forward P/E ratio and return on equity (ROE). Besides, firms with high P/E ratios tend to have lower ROE in the subsequent years, and their ROE is very volatile and wide-distributed. Hence the following hypotheses is formed:

H8a: Return on equity ROE has a positive impact on Price to earnings ratio (P-E).

H8b: Covid-19 moderates the effect of ROE on Price to earnings ratio (P-E).

In [4] study, researchers have has found that the initial differences in the portfolio continue to be studied until 14 years. The P/E ratio constitutes a negative relationship with profit growth (EPS) in the early years of the portfolio and a positive relationship in subsequent years, indicating that investors only predict in short-term periods. The negative and positive P/E correlation is commensurate with market risk. Hence the following hypotheses is formulated:

H9a: Earning per share (EPS) ratio has a positive effect on Price to earnings ratio (P-E).

H9b: Covid-19 moderates the effect of EPS on Price to earnings ratio (P-E).

4. Stock Trading by Volume

The volume of trading indicates the share prices of listed companies as an important indicator of the performance of the financial market. The whole volume of trading of shares is high indicating the preoccupation of a large number of shares, and one of the ratios used to measure the volume of trading, which is (Trade Volume activity) because of its impact on the revenue of shares. This ratio can be calculated by dividing the number of shares traded by the number of shares traded on the number of shares listed. The fact that the ratio is close to one is true indicator that the volume of trading is normal and vice versa. The result showed a positive relationship between stock returns and trading volume [48].

5. Leverage ratio

This ratio means the company's ability to face the obligations and continuity in the market for the longest possible period as it indicates the company's ability to face adverse working conditions and resilience without exposure to losses or financial insolvency. Therefore, this ratio is concerned with measuring the company's ability to pay debts by calculating the ratio of debt to rights ownership, which is an important measure that indicates the margin of safety for creditors, shareholders and investors in the company, as well as an indication of the company's dependence on debt and borrowing or shareholders' equity (equity) to finance its activities and operations [5].

A study by [46] showed that the solvency ratio that measures the debt ratio has no effect on stock returns, which is the increase and decrease in profitability ratios that does not affect the returns of shares after the capital structure based on the fact that the use of debt will be more alive than the use of capital of the private sector, especially through borrowing from banks, where the company's use of debt does not affect the returns of the shares when the company's financial conditions are strong.

6. The Company Size

The growth rate reflects the company's ability to maintain its economic position, meaning that a high rate of growth means that the company has an opportunity to earn extra profits, which is measured through the growth rate of sales, that is, the change in sales during a specific time period. The higher the rate of sales grow, the higher the company's ability to cover obligations. In addition to encouraging the shareholders to invest in the company's shares, the market value of the company's shares is increased. The [47] study showed that the growth rate has a positive yet insignificant effect on stock returns for companies listed on the ZSE for the period June 2009 to July 2013.

3 Research Methodology

3.1 Data and Data Methods

As it provides a coherent framework for analyzing and collecting data [31], this study applied quantitative approached through using secondary data collected from DataStream. This study utilizes panel data method to investigate the implication of coronavirus pandemic into the relationship between company performance and its value. This study restricts its sample to the 36 publicly listed companies on the Palestine Exchange (PES) with a period of first three quarters of two years (2019 & 2020). Firms that were excluded from this study due to not having financial data for the year 2019 and 2020 by the time of conducting.

Regression model is showing the relationship between financial performance (ROA, ROE, and EPS) and company value (stock returns, Tobin's q and price to earnings ratio). The framework of relationship between each variable can be seen in the model below (Fig. 1 and Table 1):

$$Rit = \alpha + \beta_1 \mathbf{ROA}it + +\beta_2 \mathbf{ROE}it + \beta_3 \mathbf{EPS}it + \beta_4 \mathbf{Lev}it + \beta_5 \mathbf{TBV}it + \beta_6 \mathbf{Size}\, it + \beta_7 \mathbf{year}\, it + \varepsilon\, it$$

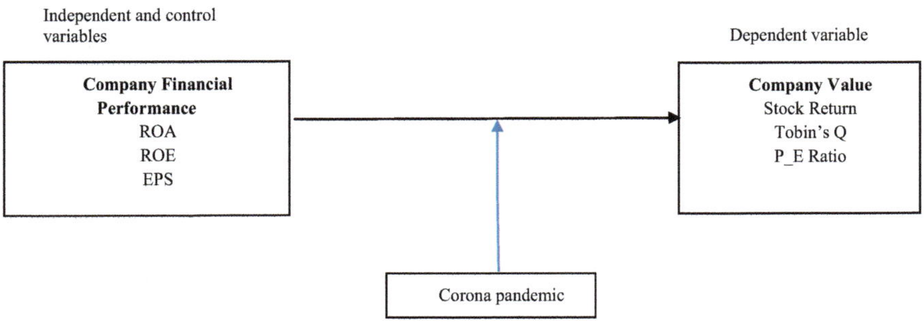

Fig. 1 Research Model. *Source* Developed by researcher

Table 1 Definition of variables used in the study

Name	Definition
R	Quarterly Stock return, extracted from the DataStream data base
TQ	Equals the quarterly market value of a company divided by its assets' replacement cost, extracted from the DataStream data base
P_E	The price-to-earnings ratio (P/E ratio) is the quarterly ratio for valuing a company that measures its current share price relative to its per-share earnings (EPS), extracted from the DataStream data base
ROA	Quarterly Return on Assets, extracted from the DataStream data base
ROE	Quarterly Return on Equity, extracted from the DataStream data base
EPS	Quarterly earnings per share, extracted from the DataStream data base
Lev	Quarterly leverage ratio, the ratio of total liabilities to total assets extracted from the DataStream data base (Control Variable)
TBV	Quarterly Turnover by Volume, divide the total number of shares traded during the measurement period by the average number of shares available for sale, extracted from the DataStream data base (Control Variable)
Size	firm size is quarterly logarithm of total assets (Control Variable)
Year	The impact of the CORONA virus in 2020 compared to 2019(Year Dummy (0,1))

Note(s) i refers to the company, t refers to year and ε Error term.

Model 2:

$$TQ\,it = \boldsymbol{\alpha} + \beta_1\,\mathbf{ROA}\,it + \beta_2\,\mathbf{ROE}\,it + \beta_3\,\mathbf{EPS}it + \beta_4\,\mathbf{Lev}it + \beta_5\,\mathbf{TBV}it + \beta_6\,\mathbf{Size}\,it + \beta_7\,\mathbf{year}\,it + \varepsilon\ it$$

Model 3:

$$PE\,it = \alpha + \beta_1\mathbf{ROA}it + \beta_2\,\mathbf{ROE}\,it\,\beta_3\,\mathbf{EPS}\,it + \beta_4\,\mathbf{Lev}it + \beta_5\,\mathbf{TBV}it + \beta_6\,\mathbf{Size}\,it + \beta_7\,\mathbf{year}\,it + \varepsilon\ it$$

Table 2 Pearson's correlation between financial performance (ROA, ROE, EPS) and firm value (Stock return, TQ, P/E)

Variables	(1)	(2)	(3)	(4)	(5)	(6)	(7)	(8)	(9)
(1) Stock Return	1.000								
(2) P_E Ratio	−0.155	1.000							
(3) lnTQ	0.336*	−0.042	1.000						
(4) ROE	0.497*	−0.030	0.110	1.000					
(5) ROA	0.529*	−0.048	0.370*	0.809*	1.000				
(6) EPS	0.224*	−0.374*	0.051	−0.117	0.123	1.000			
(7) TA	−0.049	−0.070	−0.401*	0.064	−0.164	−0.222*	1.000		
(8) Leverage	−0.169	−0.013	−0.371*	−0.161	−0.332*	−0.100	0.726*	1.000	
(9) Turnover	−0.178*	0.019	−0.167	0.025	−0.166	−0.217*	0.424*	0.416*	1.000
(10) corona	0.006	0.073	−0.020	−0.102	−0.076	−0.041	0.032	0.023	−0.200*

Note *** $p < 0.01$, ** $p < 0.05$, * $p < 0.1$.

4 Results and Recommendations

4.1 Multi-collinearity and Correlation Analysis

The correlation test explains the correlation score between the study variables. Multi-collinearity phenomena occurs when the correlation analysis shows high score between two or more variables in the same model [35]. Based on [32, 34] they argued that the multi-collinearity problem occurred when the correlation score among two variables is higher than 0.8 or 0.9, whereas [33] proposed the correlation score is higher than 0.7. Table 2 presents Pearson's correlation between independent variables (ROA, ROE, EPS) and dependent variables (Stock return, TQ, P/E). Moreover, the Variance Inflation Factor (VIF) results in Table (3, 4) show that there is not a multi-collinearity problem in the model. In other words, the VIF statistic values for all the independent variables are smaller than 5, indicating that the multi-collinearity is insignificant.

4.2 Regression Analysis

This study employed panel data to match with this study's objectives and research questions. Therefore, this study tested additional statistical tests related to panel data before testing the linear regression. Firstly, Breusch and Pagan Lagrangian multiplier

Table 3 Financial performance and stock return, Tobin's Q, Price to Earnings

Variables	Stock return	Stock return	Stock return	lnTQ	lnTQ	lnTQ	P/E	P/E	P/E
ROA	0.0193***			0.0205**			−0.254		
	−2.19			−2.06			(−0.50)		
ROE		0.0039			0.0029			0.0669	
		−1.08			−0.58			−0.26	
EPS			0.0861			0.1501*			−28.46***
			−1.07			−1.73			(−5.66)
TA	0.0002	0	0	−0.0087***	−0.0086***	−0.0081***	0	0	0
	−0.62	−0.12	−0.74	(−3.77)	(−3.66)	(−3.85)	(−0.59)	(−0.68)	(−1.48)
Leverage	0.0017*	0.00125	0.0017	0.0033***	0.0029***	0.0031***	0.0376	0.0504	0.0498
	−1.7	−1.19	−1.35	−2.73	−2.46	−2.75	−0.63	−0.85	−0.96
Turnover_by_volume	0.0048***	0.0044**	0.0005**	0.0031***	0.0029**	0.0031***	−0.0012	−0.0012	−0.0014
	−3.32	−2.91	−3.32	−2.16	−2.03	−2.19	(−0.95)	(−0.97)	(−1.21)
Cons	1.706***	1.829***	1.757***	−7.004***	−6.930***	−6.977***	18.20***	16.40***	27.48***
	−15.99	−8.18	−16.55	(−19.36)	(−19.14)	(−18.56)	−4.38	−4.28	−7.86
N	216	216	216	216	216	216	216	216	216
VIF Test	1.76	1.78	1.72	1.76	1.78	1.72	1.76	1.78	1.72
Breusch and Pagan Lagrangian Test	486.20***	450.93***	502.39***	528.38***	532.64***	534.68***	146.81***	146.74 ***	133.38***
Hausman Test	26.66***	4.48	61.50***	0.2134	0.9651	0.9238	0.48	2.85	2.31
Best Model	Fixed Effect	Random Effect	Fixed Effect	Random Effect	Random Effect	Random Effect	Random Effect	Random Effect	Random Effect
R-sq	10%	8%	8%	37%	34%	36%	1.20%	1.30%	20%

(continued)

Table 3 (continued)

Variables	Stock return	Stock return	Stock return	lnTQ	lnTQ	lnTQ	P/E	P/E	P/E
Prob > F	0.0000***	0.0107**	0.0039***	0.0000***	0.0000***	0.0000***	0.7922	0.8251	0.0000***

* $p < 10\%$, ** $p < 5\%$, *** $p < 1\%$. Note: Stock return: firm's stock return (dependent variable), ROA: Return on Assets, ROE: Return on equity, EPS: earning per share. Model 1 examines the relationship between ROA and Stock Return, Model 2 examines the relationship between ROE and Stock Return, Model 3 examines the relationship between EPS and Stock Return, t statistics in parentheses. Note: lnTQ: firm's Tobins' Q (dependent variable), ROA: Return on Assets (independent variable), ROE: Return on equity (independent variable), EPS: earning per share (independent variable). Model 1 examines the relationship between ROA and Stock Return, Model 2 examines the relationship between ROE and Stock Return, Model 3 examines the relationship between EPS and Stock Return, t statistics in parentheses, Note: P/E: firm's price to earnings (dependent variable), ROA: Return on Assets, ROE: Return on equity, EPS: earning per share. Model 1 examines the relationship between ROA and P/E, Model 2 examines the relationship between ROE and P/E, Model 3 examines the relationship between EPS and P/E, t statistics in parentheses.

Table 4 Moderating role of corona on the relationship between financial performance and Stock Return, Tobin's Q, Price to Earnings

Variables	(1)	(2)	(3)	(1)	(2)	(3)	(1)	(2)	(3)
	Stock Return	Stock Return	Stock Return	lnTQ	lnTQ	lnTQ	P/E Ratio	P/E Ratio	P/E Ratio
ROA	0.0233**			0.0125*			0.379		
	−1.89			−1.93			−0.74		
ROA*Corona	−0.0023			−0.0073**			−1.518***		
	(−0.45)			(−2.46)			(−3.79)		
ROE		0.0041			0.0026			0.608**	
		−0.93			−0.92			−2.01	
ROE*Corona		0.0008			−0.0047**			−0.729***	
		−0.27			(−2.55)			(−2.92)	
EPS			0.1058			0.127**			−25.95***
			−1.27			−2.36			(−4.81)
EPS*Corona			−0.0397			0.0502			−6.323
			(−0.69)			−1.34			(−1.30)
Corona	0.0308	0.0062	0.0182	−0.0192	−0.0294	−0.0783***	8.329***	7.791***	3.714
	−0.8	−0.2	−0.61	(−0.99)	(−1.51)	(−4.01)	−3.39	−2.99	−1.56
TA	0	0	0	−0.0001***	−0.0001***	−0.0001***	0	0	−0.0000*
	−0.15	(−0.13)	−0.07	(−6.96)	(−5.66)	(−5.80)	(−0.51)	(−0.62)	(−1.70)
Leverage	0.0016	0.0014	0.0013	0.0027***	0.0023***	0.0030***	−0.0044	0.02	0.0546
	−1.2	−1.22	−1.23	−3.8	−3.29	−4.55	(−0.08)	−0.34	−1.03
Turnover_by_volume	0.0001***	0.0001***	0.0001***	0.0001**	0.0001**	0.0001	0.0004	−0.000212	−0.000736
	−3.2	−2.84	−3.1	−2.46	−2.14	−1.45	−0.3	(−0.16)	(−0.58)

(continued)

Table 4 (continued)

Variables	(1)	(2)	(3)	(1)	(2)	(3)	(1)	(2)	(3)
	Stock Return	Stock Return	Stock Return	lnTQ	lnTQ	lnTQ	P/E Ratio	P/E Ratio	P/E Ratio
Cons	1.725***	1.839***	1.819***	−6.997***	−7.020***	−7.060***	15.11***	11.73**	25.71***
	−14.28	−8.25	−7.56	(−21.32)	(−20.49)	(−20.00)	−3.58	−2.9	−6.86
N	216	216	216	216	216	216	216	216	216
VIF Test	2.08	2.28	2.09	2.08	2.28	2.09	2.08	2.28	2.09
Breusch and Pagan Lagrangian Test	487.55***	448.34***	503.34***	525.97***	527.60***	534.55***	164.12***	156.17***	135.17***
Hausman Test	64.08***	4.31	5.28	1.42	2.16	1.01	1.32	3.24	3.01
Best Model	Fixed Effect	Random Effect	Random Effect	Random Effect	Random Effect	Random Effect	Random Effect	Random Effect	Random Effect
R-sq	11%	8%	8%	43%	43%	43%	R-sq	9%	8%
Prob > F	0.0027***	0.0393**	0.0251**	0.0000***	0.0000***	0.0000***	0.0075 ***	0.0661*	0.0000***

*P < 10%, **P < 5%, ***P < 1%.

test was conducted in STATA software version 14.2 to select the appropriate model for regression linear analysis. This test chose the best model between Ordinary Least Squares (OLS) or Random Effect Model (REM). More specifically, the null hypothesis (Ho) claims that if the p-value (the prob chi-square) is more than α level (0.05). In this case, the null hypothesis is accepted, then the OLS is the suitable model. Whereas the p-value of this test is less than 0.05. In this case, the null hypothesis is rejected, then the REM is the suitable model. This study has 18 models which were carried out to test the research hypotheses. Table 3 provides the summary of random and fixed effect model for the direct hypothesis. Table 4 present the results for Corona pandemic as a moderating factor for the three dependent variables.

Financial Performance and Stock Return, Tobin's Q, Price to Earnings

Table 3 shows that there is a positive significant relationship between return on asset (ROA) and stock return. Result shows that there is a positive and significant value (t-value 0.193) and significant level at 1%. This result indicates that when company has a good financial performance through higher return on assets, the company would have a higher value through higher stock return. This result is supported with prior research [17]. Model 1 shows that the first hypothesis which says ROA has a positive and significant relationship on company financial performance. However, ROE and EPS does not have relationship with stock return. Hence, hypotheses two and three are rejected.

The Fourth hypothesis is stated that there is a positive significant relationship between return on asset (ROA) and Tobin's q. Table 3, model 1 presents result that indicate a positive and significant value (t-value 0.0205) and significant level at 5%. This result is supported with prior study e.g., [11]. This result indicates that when company has a high performance in terms of ROA, the company has higher Tobin's q. Hence, hypothesis H4a is accepted. In addition to, model 3 presents the relationship between earning per share (EPS) and Tobin's q. Hypothesis H6a stated that there is a positive significant relationship between EPS and Tobin's q. Result shows that there is a positive and significant value (t-value 1.501) and significant level at 10%. This result is supported with past study e.g. [12]. This result indicates that when firm has a high performance in terms of EPS, the firm have higher Tobin's q. Hence, (H6a) hypothesis is accepted.

The ninth hypothesis is stated that there is a significant relationship between earning per share (EPS) and price to earnings (P/E ratio). Model 3 shows that there is a negative value (t-value -28.46) and significant level at 1% between EPS and P/E Ratio. This result indicates that when firm has a high performance in terms of EPS, the firm have lower P_E ratio. This result is supported with [4] study. Thus, the ninth hypothesis is accepted.

Moderating Role of Corona on the Relationship Between Financial Performance and Stock Return, Tobin's Q, Price to Earnings

The (the first (b), second(b), and third (b)) hypothesis is stated that coronavirus pandemic is moderating the relationships among the financial performance (ROA, ROE, EPS) and the stock return. Table 4 shows that that there the moderating factor (ROA *Corona), (ROE *Corona), (EPS *Corona) has no moderating effect on stock return. This result indicates that coronavirus does not have effect on the relationship between ROA, ROE, EPS and stock return. Thus, the hypothesis (H1b, H2b & H3b) is rejected.

This study analyzes the role of the corona pandemic on the association between the financial performance and Tobin's' Q. Examining the moderating hypothesis, the interaction term is the key factor of interest here. (i.e., ROA*Corona, ROE*Corona, and EPS*Corona). The interaction term represents the disparity between the periods before and after coronavirus pandemic. The following hypothesis is that coronavirus pandemic is moderating the relationships between the financial performance (ROA, ROE, EPS) and Tobin's q. Table 4, model 1 shows a negative coefficient of ROA*Corona (p value significant at the 5% level) states that when controlling for other variables, the average increase in Tobin's q led by the ROA is higher for firms in which corona virus is affected. It also indicates that coronavirus has negative effect towards the company's performance and its value. Hence, the hypothesis (H4b) is accepted. Moreover, the interaction coefficients of the interaction terms ROE* coronavirus. (Table 4, model 2) has a negative coefficient with (p value significant at the 5% level) states that when controlling for other variables. The average increase in Tobin's Q led by the ROE is higher for firms in which coronavirus is affected. This result indicates that coronavirus weakens the association between ROE and Tobin's q. It also indicates that coronavirus has negative effect towards the company's performance (ROE) and its value (Tobin's Q). Hence, the hypothesis (H5b) is accepted. The result indicates that when value of ROE decreased the value of TQ and indicates the low performance of the company, the financial performance is one of the considerations in determining their investment decisions. Hence, the hypothesis (H5b) is accepted.

The following hypothesis is stated that coronavirus pandemic is moderating the relationship between EPS and Tobin's q. The result shows that there is the moderating factor (EPS*Corona) that has a significant value (t-value -0.0783) and significant level at 1%. This result indicates that coronavirus weakens the relationship between EPS and Tobin's Q. Hence, the hypothesis (H6b) is accepted. This study also analyses the moderating impact of coronavirus on the association between the components of financial performance indicators and the firm's value indicators of sample publicly listed firms in Palestine. Table 4, model 1 shows a negative coefficient of ROA*Corona (p value significant at the 1% level) states that when controlling for other variables, the average increase in P/E ratio led by the ROA is higher for firms in which corona virus is affected. Seventh hypothesis is stated that corona pandemic (Corona) is moderating the relationship between ROA and P-E. Result shows that there the moderating factor (ROA*Corona) have a significant value (t-value -1058) and significant level at 1%. This result indicates that Coronavirus weakens the relationship between ROA and P-E. Hence, the hypothesis (H7b) is accepted. The following hypothesis is stated that coronavirus pandemic is moderating the relationship between ROE and P-E. The result shows that there the moderating factor (ROE*Corona) has a significant value (t-value -0.0047) and significant level at 1%. This result indicates that the coronavirus weakens the relationship between ROE and P-E. That increased the value of ROE decreased the value of P-E and indicates the low performance of the company. Hence, hypothesis (H8b) is accepted.

5 Conclusion

This study examined The Impact of Financial Performance on Firm's value during Covid-19 Pandemic for Companies Listed in the Palestine Exchange (2019–2020). In summary, this paper provides evidence that coronavirus pandemic had an impact on the relationship between ROA, ROE, EPS and TQ as it was found that the relationship was negative, i.e., the coronavirus pandemic had an effect and a mitigation of the relationship between the dependent and independent variables. in addition to The coronavirus pandemic has an impact on the relationship between P/E ratio and ROA, and ROE and P/E ratio where the relationship was negative. In contrast, the coronavirus pandemic has no effect on the relationship between the financial performance of companies listed on the Palestine Exchange during the period 2019–2020 and the company's value. Accordingly, the Corona virus (Covid-19,) is considered one of the most serious pandemics that have affected the global economy due to its difference from previous pandemics that are faced through economic policies, whereas the Corona pandemic directly threatens the human element, requiring preparing different policies to prevent further spread, especially in the absence of Vaccine for the virus [21, 42].

The findings of such an examination can Seeking future solutions to face such crises eventually as the pandemic revealed the inability of the regimes to meet the requirements of the communities and institutions, and filling the gap of their lack of readiness to face the general and comprehensive closures, especially the electronic services. 2. Supporting national banks to implement procedures for postponing customers' debt payments, and restructuring granted credit facilities. 3. Conducting additional studies to examine the impact of the Coved-19 pandemic and its impact on the economy and the share prices of companies on the Palestine Exchange, and to deal with future studies on the dimensions of unemployment and inflation, which can be entered as a result of the pandemic. 4. Researchers should take into consideration in their future research a larger sample size and a long-term period of time for the impact of the coronavirus pandemic, and encourage researchers to study additional mechanisms and indicators other than indicators of profitability for financial performance such as indicators of debt, liquidity and others.

References

1. Al-Sartawi A (2019) Performance of Islamic banks Do the frequency of Shari'ah supervisory board meetings and independence matter? ISRA Int J Islamic Finan 11(2):303–321
2. Al-Sartawi A (2020) Shariah disclosure and the performance of Islamic financial institutions. Asian J Bus Account 13(1):133–160
3. Awadallah A (2015) Impact of return on assets and financial leverage multiplier on financial performance of by applying application in the global financial crisis case study of Jordan steel company and its subsidiaries. Sudan University of Science and Technology and the College of Graduate Studies

4. Salman Abbas S (2009) The financial pandemic (concept, causes, effects) from the point of view of different economic schools 2007–2008 pandemic as a model, p 13
5. Wijaya M, Yustina AI (2016) The impact of financial ratio toward stock price: evidence from banking companies. Accountancy, 1–36. http://e-journal.president.ac.id/presunivojs/index.php/JAAF/article/view/174
6. Ngunjiri N (2012) Relationship between firms' financial performance and stock return for firm listed at NAIROBI SECURITIES EXCHANGE, School of Business, University of Nairobi
7. Christina O, Robiyanto R (2018) The effect of financial performance and firm size on stock prices of manufacturing company in 2013–2016. Prosiding SENDI, 559–563
8. Dalal B (2018) Measuring the stability of the financial performance of Islamic banks in light of the global financial crisis of 2008, an applied study for a group of Islamic banks for some countries. Mohamed Boudiaf University, M'sila, p 76
9. Davidson I, Leledakis GN (2002) Tobin' s q and the cross-sectional variation of stock returns: evidence from the London stock exchange Tobin's Q and the cross sectional variation of stock returns: evidence from the London stock exchange by Department of Accounting and Finance Athen, May 2014
10. Arab Center for Research and Policy Studies (2020), p 1/p 30
11. Alghifari ES, Triharjono S, Juhaeni YS (2013) Effect of return on assets (ROA) against Tobin's Q: studies in food and beverage company in Indonesia stock exchange years 2007–2011. Int J Sci Res (IJSR), India Online ISSN: 2319-7064
12. Gharaibeh AMO, Qader AAAA (2017) Factors influencing firm value as measured by the Tobin's Q: empirical evidence from the Saudi stock exchange (TADAWUL). Int J Appl Bus Econ Res 15(6):333–358
13. Fu L, Parkash M, Singhal R (2017) Tobin's Q ratio and firm performance. Int Res J Appl Finan. https://doi.org/10.0704/article-2
14. Jain N, Bajaj K (2017) Impact of earnings per share on market price of share with special reference to selected companies listed on NSE. Int J Eng Manag Res 7(3):1–9. www.ijemr.net
15. Liu H, Manzoor A, Wang C, Zhang L, Manzoor Z (2020) The COVID-19 outbreak and affected countries stock markets response. Int J Environ Res Public Health 17(8). https://doi.org/10.3390/ijerph17082800
16. BasoulM M, Abdel Ghanim YH (2018) The financial performance of banks in light of the global economic crisis (A comparative study between Islamic banks and Jordanian conventional banks). J Econ Adm Legal Sci 2(15) AD. ISSN: 2522-3372
17. Siregar AMR, Sihombing P (2020) Determinant analysis of financial ratio on stock returns in construction companies registered at Indonesia stock exchange 2015–2019. 2(1). E-ISSN: 2686-6331. P-ISSN: 2686-6358
18. Ngunjiri N (2016) Relationship between firms' financial performance and stock return for firms listed at Nairobi securities exchange Ndirangu Ngunjiri
19. Nurfadila N (2020) Does CSRD and GCG moderate the effect of financial performance on stock return? ATESTASI: Jurnal Ilmiah Akuntansi 3(2):133–141. https://www.jurnal.fe.umi.ac.id/index.php/ATESTASI/article/view/570
20. Palestinian Council of Ministers, decision No. 1, 7 of 2020
21. PMA (2020) The expected impact of Covid-19 pandemic on the Palestinian economy in 2020. https://www.pma.ps/Portals/0/Users/002/02/2/Publications/English/WorkingPapers/2-PositionPaperCovid19impactonPalestinianeconomy.pdf
22. Saber B (2020) Corona COVID-19 crisis management by promoting mental health under domestic quarantine. University of Mohamed Lamine Debaghin Setif 2 / Algeria, p 15
23. Sadaqah J (2020) An unprecedented financial pandemic on the horizon and the required tools to mitigate its repercussions required. https://bit.ly/39o0I1O
24. Mujib M (2020) The repercussions of the pandemic—analytical and critical insights for the repercussions of the corona pandemic for 2020. Bibliotheca Alexandrina, Center for Strategic Studies, Egypt, p 22
25. Şamiloglu F, Oztop AO, Kahraman YE (2017) The determinants of firm financial performance: evidence from Istanbul stock exchange (BIST). IOSR J Econ Finan (IOSR-JEF) 8(6):62–67. https://doi.org/10.9790/5933-0806016267

26. Talha EWA (2020) The implications of the new corona virus outbreak on the banking and financial sector in the Arab countries. J King Abdulaziz University Islamic Econ 33(3):131–140. https://doi.org/10.4197/Islec.33-3.7
27. Al-Khazraji T (2009) The current global financial crisis and its impact on Arab economies: challenges and ways to confront. J Baghdad College Econ Sci Univ 1(20):297–340
28. Ghi TN (2015) The impact of capital structure and financial performance on stock returns of the firms in hose. Int J Inf Res Rev Ba Ria Vung Tau Univ Vietnam xx(xxx):xxx-xxx
29. WHO (2020). https://www.who.int/ar/news-room/detail/04-09-1441-who-timeline---covid-19
30. Fen Y, P'ng Y (2019) Tobin's Q and its determinants: a study on Huawei Technologies Co., Ltd., Universiti Utara Malaysia 14 May 2019. https://mpra.ub.uni-muenchen.de/93894/MPRA. Paper No. 93894, posted 15 May 2019 13:40 UTC
31. Bryman A, Bell E (2013) Business Research Methods. Oxford University Press, USA
32. Gujarati DN, Porter D (2009) Basic Econometrics. Internatio. Mc Graw-Hill, New York
33. Kennedy P (2008) A guide to econometrics. 6th Editio. Wiley-Blackwell, New Jersey
34. Pallant J (2011) SPSS survival manual: a step by step guide to data analysis using SPSS. Open University Press
35. Sekaran U, Bougie R (2010) Research methods for business a skill building approach. Wiley
36. Nour A, Alia MA, Balout M (2022) The impact of corporate social responsibility disclosure on the financial performance of banks listed on the PEX and the ASE. In: Musleh Al-Sartawi AMA (eds) Artificial intelligence for sustainable finance and sustainable technology. In: ICGER 2021. Lecture notes in networks and systems, vol 238. Springer, Cham. https://doi.org/10.1007/978-3-030-93464-4_5
37. Mahamid FA, Bdier D, Nour AI (2022) Mental health, E-learning, and future of education in Palestine after the COVID-19 pandemic. In: Hamdan A, Hassanien AE, Mescon T, Alareeni B (eds) Technologies, artificial intelligence and the future of learning post-COVID-19. Studies in computational intelligence, vol 1019. Springer, Cham. https://doi.org/10.1007/978-3-030-93921-2_15
38. Musleh Al-Sartawi AM, Razzaque A, Kamal MM (eds) (2021) Artificial intelligence systems and the internet of things in the digital era. In: EAMMIS 2021. Lecture notes in networks and systems, vol 239. Springer, Cham
39. Sharabati A-AA, Shamari NS, Nour A-NI, Durra A-BI, Moghrabi KM (2016) The impact of intellectual capital on business performance in Kuwaiti telecommunication industry. Int J Bus Performance Manag 17(4):428–446. https://doi.org/10.1504/IJBPM.2016.079278
40. Khatib SFA, Nour A-NI (2021) The impact of corporate governance on firm performance during the COVID-19 pandemic: evidence from Malaysia (2021). J Asian Financ Econ Bus 8(2):0943–0952. Print ISSN: 2288-4637 / Online ISSN 2288-4645. https://doi.org/10.13106/jafeb.2021.vol8.no2.0943
41. Najjar M, Alsurakji IH, El-Qanni A, Nour AI (2022) The role of blockchain technology in the integration of sustainability practices across multi-tier supply networks: implications and potential complexities. J Sustain Finan Investment. https://doi.org/10.1080/20430795.2022.2030663
42. Al Momani K, Nour AN, Jamaludin N, Zanani Wan Abdullah WZW (2021) Fourth industrial revolution, artificial intelligence, intellectual capital, and COVID-19 pandemic. In: Hamdan A, Hassanien AE, Khamis R, Alareeni B, Razzaque A, Awwad B (eds) Applications of artificial intelligence in business, education and healthcare. studies in computational intelligence, vol 954. Springer, Cham. https://doi.org/10.1007/978-3-030-72080-3_5
43. Saleh MWA, Shurafa R, Shukeri SN, Nour AI, Maigosh ZS (2020) The effect of board multiple directorships and CEO characteristics on firm performance: evidence from Palestine. J Acc Emerging Econ 10(4):1–18. https://doi.org/10.1108/JAEE-12-2019-0231
44. Al Momani KM'K, Jamaludin N, Abdullah WZWZW, Nour A-NI-H (2021) The influence of relational capital on the relationship between intellectual capital and earnings per share in the digital economy in the Jordanian industrial sector. Stud Comput Intell 974:59–76
45. Alqaraleh MH, Thuneibat NSM, Nour AI (2020) The factors affecting the adherence of Islamic accounting standards AAOIFI in Jordan. J Gov Regul 9(4):69–75

46. Sucipto A, Chasanah N (2019) Liquidity ratio, profitability, and solvency on stock returns with capital structure as an intervening variable (study on food and beverage sub sector listed in Indonesia stock exchange (Idx) period 2013–2017. Ekspektra Jurnal Bisnis Dan Manajemen 3(1):52. https://doi.org/10.25139/ekt.v3i1.1476
47. Mazviona BW, Nyangara D (2014) Does firm size affect stock returns? Evidence from the Zimbabwe stock exchange. Int J Bus Econ Dev 2(3):13–17. http://search.ebscohost.com/login.aspx?direct=true&db=bth&AN=100271396&site=eds-live&authtype=uid
48. Nugraha NM, Nugraha DNS (2020) The influence of stock prices and trading volume on abnormal return of LQ45 shares pre and post 2019 election. Solid State Technol 63(3):2523–2534

The Essential Concept of Audit Quality and Earnings Management: A Conceptual Literature Review

Sawsan Ismail, Merwiey Alaqrabawi, Safaa Ahmad, Ezzat Ghaidan, and Mahmoud Mahmoud

Abstract Audit quality and earnings management have become pressing issues in today's capital markets. This paper aims to demonstrate many aspects of earnings management and motivations for controlling earnings, as well as investigate the relationship between audit quality and earnings management behavior. The significance of this research stems from the link between accounting earnings and the quality of audit procedures that can regulate various earnings management approaches. This research provides a comprehensive view of the audit quality and earnings management concepts. A review of relevant academic studies has yielded a range of related concepts. Therefore, this paper contributes to the ongoing debate about audit quality and earnings management and highlights the need for a new approach to these issues.

Keywords Audit quality · Earnings management · Conceptual framework

1 Introduction

In today's capital markets, earnings management has become a critical concern. It is a tool used by a company's management to impact its earnings so that the numbers meet a predetermined objective. This approach is used for various reasons, one of which is income smoothing. As a result, instead than have years of unusually good or negative earnings; corporations aim to keep the results reasonably consistent by employing innovative accounting tactics.

A financial statement audit's primary goal is to provide users confidence that the statements are prepared in accordance with International Financial Reporting Standards (IFRS) Financial reporting should include representationally faithful information, according to IFRS (i.e., financial statements should correct-ly portray the information they purport to represent). It stands to reason that audit quality is inversely related to the occurrence of this type of earnings manipulation.

S. Ismail (✉) · M. Alaqrabawi · S. Ahmad · E. Ghaidan · M. Mahmoud
Accounting Department, Faculty of Economics and Administrative Sciences, The Hashemite University, Zarqa, Jordan
e-mail: sawsani@hu.edu.jo

A. M. A. Musleh Al-Sartawi et al. (eds.), *From the Internet of Things to the Internet of Ideas: The Role of Artificial Intelligence*, Lecture Notes in Networks and Systems 557,
https://doi.org/10.1007/978-3-031-17746-0_43

Earnings management and audit quality are two critical interrelated topics that have been studied in literature and have been linked together in many ways [1, 2]. The two topics are also linked to each other in the covered terms and concepts. The basic concepts contained in each topic are repetitive and interrelated, so a conceptual framework that summarizes the most important basic concepts related to profit management and audit quality appears to be necessary.

The aim of this paper is to cover the main concepts related to earnings management as well as audit quality. These topics share many standard financial and accounting concepts.

A significant number of prior studies link the impact of earnings management to audit quality issues [2]. The significance of this research stems from the it attempts to bridge the gap in finding a conceptual literature review about the essential concept of audit quality and earnings management. This research provides implications for researchers, educators, students, and policymakers on the relevant terms and concepts in the earning management and audit quality areas.

2 Literature Review and Theoretical Background

2.1 Underpinning Theory

This paper is supported by the agency theory basis, which states that the firm is involved with the nexus of contracts among the owners of the economic resource (principles) and managers (agents) [3]. These two parties are responsible for the overall utilization and control of founded resources. This theory has been adopted by several previous auditing-related research in many issues such as resolving conflicts resulting in agency relationships [2, 4]. The unaligned information arises conflicts between these two parties, which agency theory states asymmetry in communication with information to avoid conflicts of interest. Linking this paper's purpose with agency theory, the board of directors will have to support the internal audit function to manage and evaluate the management procedures that will result in firm performance and stakeholder expectations.

2.2 Audit Quality

Auditing is a monitoring tool aimed at reducing information asymmetry among agents and stakeholders. [5]. The quality of an audit is determined by two factors: (1) the auditor's competence and (2) the auditor's independence [6]. The literature frequently argues whether big auditing firms produce higher-quality audits than small auditing firms. In most empirical research, a dummy variable (i.e., Big4 versus non-Big4 audit firms) is used to proxy for audit quality differences between large and small audit firms [6].

Large audit firms may be able to deliver higher-quality audits than small audit firms because the former set aside, more significant resources for staff training and establishing industry knowledge. Compared to small audit firms, large audit companies are in a stronger position to question or negotiate with customers who try to utilize aggressive accounting tactics because of their size..In comparison to small audit firms, large audit firms can more easily absorb the financial loss associated with losing a single customer, giving them more independence. Blokdijk, Drieenhuizen [7] found that there is a difference in audit quality between Big N and non-Big N auditors, not because of the length of the engagement but because of differences in audit technologies and how audits are conducted [8].

There is also an assumption that large audit companies perform higher-quality audits than small audit firms since large audit firms have more extensive customer bases and thus have more to lose in reputational damage connected with a poorly done audit [4].

Previous literature has investigated audit quality from different perspectives [9]. DeAngelo [10] evident that the chance that the auditor will both detect and report a violation in the client's accounting system is referred to as audit quality. Al-Mousawi and Al-Thuneibat [11] defined audit quality as an external technique demonstrating how well an auditor may perform his professional duties. Deis Jr and Giroux [12] stated that audit quality as the possibility that the audit company will discover and identify weaknesses in the client's accounting system. In addition, Teoh and Wong [13] defined Audit quality as the procedure that produces high-quality financial reporting. Therefore, as a result of the preceding definitions, the main idea is that audit quality indicates that the audit process is a false-free procedure that produces false-free financial reports.

Audit services provided by larger audit firms are of greater quality than those provided by smaller ones. This is since more prominent auditors are less able to rely on their specific customers [14]. As a result, they are less inclined than small audit firms to compromise with their client's force in order to report misstatements. They also have a stronger motivation to provide high-quality services in order to protect their brand's reputation. According to Rezaei and Shabani [14] larger audit businesses may suffer greater reputational damage as a result of audit failures than smaller firms [9, 10]. Furthermore, large audit companies have more technological expertise and resources, also have more qualified and skilled auditors and more experience than small audit firms. Therefore, from an agency theory lens, the high-quality auditor is less willing to adopt high-quality methods that are more likely to be free from errors [15].

2.3 Earnings Management

Earnings management is the deliberate manipulation of earnings to achieve a specific goal. That target could be one set by management, a forecast provided by analysts, or a figure that aligns with a more consistent, long-term earnings stream. Earnings management usually means taking steps to reduce and "store" profits during good years for use in weaker years.

External stakeholders want the independent auditor to keep earnings management under control and ensure accurate financial reporting. As a result, if stakeholders believe there has been a failure in financial reporting, they are more likely to sue the auditor [16]. The agency theory supports this statement by which it approves that the stakeholders seek high-quality auditing services from external auditors as they present the reliability of financial reports [15, 17].

Furthermore, controlling accruals allow managers to affect accounting reports opportunistically. However, according to Kaplan [18], "normal" accruals occurring during normal business periods are unlikely to reflect managerial manipulation. Any accounting numbers that have been tampered with will most likely show up in "abnormal" accruals.

Companies' executives have two types of earnings management tools: (1) genuine operating decisions (such as whether to sell an asset) and (2) pure financial reporting decisions (e.g., determining an asset impairment). It should be obvious that auditors can only control financial reporting choices intended to manage reported earnings. [6].

Earnings management has gotten much attention, with many articles looking into different possibilities. However, there is still no agreement on detecting and measuring earnings management effectively. Nonetheless, most authors employ the accrual-based technique, employing complex models to divide total accruals into discretionary and non-discretionary components. Alternative models for measuring earnings management may be used in a variety of ways. The Jones [19] model, the modified Jones model Dechow, Sloan [20], Teoh, Welch [21] model, Kasznik [22] model, and Kothari, Leone [23] model were all employed by the researchers. However, it is verified that the company's operating environment has an impact on earnings management [24].

2.4 Discretionary Accruals (DA)

Accrual accounting is a type of accounting that measures a company's performance and status by recording economic events regardless of when cash transactions take place. The central premise is that economic events are recorded when revenues and expenses are matched (the matching principle) at the time of the transaction instead of when payment is made (or received). This strategy combines current cash inflows and outflows with predicted future cash inflows and outflows to provide a more accurate view of a company's present financial condition.

Normal or expected accruals (known non-discretionary accruals). Moreover, abnormal, or unexpected accruals (referred to as discretionary accruals) are the two accruals. Because discretionary accruals are unobservable, estimating them necessitates using models to help in the estimation process, one of which is the modified Jones model. The difference between total and non-discretionary accruals is referred to as discretionary accruals. Previous authors proposed discretionary accruals as a proxy variable for earnings management. According to the literature review, Mihret [17] evident that when management measures the earnings management by discretionary accruals, they correlated negatively with audit quality. While Yasser and Soliman [4] found that audit firm size negatively correlates with discretionary accruals, big 4 auditors generate lower discretionary accruals than non-big 4 did.

In the agency theory premise, firms' managers use DA for their interest to influence the content of the overall financial statements. Despite that, DA is easily to be found by auditors and supervisory agencies [25].

2.5 *Cash Flow from Operations*

Receivables from the sale of goods or services, receipts for the sale of loans, debt or equity instruments in a trading portfolio, interest earned on loans, dividends earned on equity securities, expenditures to suppliers for goods and services, expenses to employees or on behalf of employees, tax amounts, and interest payments are all examples of operating cash flows.

Depreciation, deferred tax, amortization, and any profits or losses associated with an asset sale are all added back to the net income to arrive at cash flows from operations (unrealized gains/losses are also added back from the income statement). Cash flow from operations is used to calculate the total accruals required to estimate discretionary accruals using the Modified Jones Model to see if the results have been manipulated.

Literature evidence that operating cash flow influenced the earnings management of the listed firm. According to Boujelben, Khemakhem-Feki [26], the study revealed that operating cash flow is a factor that impacts the level of earnings management.

In other words, operating cash flow provides an insightful overview of the firm's ability to generate operating cash flow to pay their overall debt and equity or assets acquisitions. Based on Yoon and Miller [27] case study found that when firms have a lower performance, managers make some initiative plan to increase the level of earnings by lowering the revenue, which tends to implement discretionary accruals. These findings align with Jang and Kim [28], where this paper reported that low operating cash flows would increase discretionary accruals accordingly.

2.6 Net Income and Stock Price

Because creating profit is the most significant responsibility that for profit firms have to their shareholders, net income is the most crucial metric of a company's profitability. It is the reason why so many businesses exist. The primary purpose of earnings management is to meet a predetermined target, which is reflected in a predetermined income at the end of each year, demonstrating the relationship between net income and earnings management, as manipulating earnings necessitates manipulating net income components.

Stocks are a financial commodity with a high risk, high return, and flexible trading that many investors prefer [29]. By correctly predicting stock price patterns, investors can earn much money. However, numerous factors influence stock prices, including the macroeconomic environment, market conditions, critical social and economic events, investor preferences, and company management decisions. As a result, stock price forecasting has always been a priority and a challenging research area [30].

The stock price is the amount a buyer pays to buy a stock or the amount a seller receives when they sell a stock. Market forces, such as demand (buyers' force) and supply (sellers' force), determine the stock price in a stock market. Environmental influences and individuals' future expectations/assumptions drive demand and supply. The price of a stock (market) differs from its par value and book value. From the agency theory perspective, several studies found that the firm's financial performance is motivated by the stakeholder's demand [31] to increase revenues. Therefore, based on the principle and agent relationship, management is motivated to occur earnings management in presenting financial statements, including the net income. Despite that, it is found that audit quality contributes to better financial performance and reduces earning management's presence [32].

2.7 Par Value and Book Value

"When a company is formed, it is given the authority to issue up to a certain number of shares of common stock, each with a specific par value. A business may be legally barred from making payments to common stockholders if doing so would lower the balance sheet value of stockholders' equity below the amount represented by the outstanding stock's par value. As a result, the par value is usually low compared to the price at which the stock is first offered. Some businesses offer stock with no par value. In this scenario, a stated value must be recorded in the same location as the par value; if stocks are issued at a premium or discount, the original offering price of a share may differ from the par value.

"A corporation generates income over time, much of which is distributed to creditors (as interest) and stockholders" (as a dividend). Any remaining amount is added to the amount indicated on the company's accounts as cumulative retained earnings. The book value of stockholder's equity is the sum of cumulative retained earnings and other entries (such as "common stocks" and "capital contributed more than par value") under stockholder's equity:

$$\begin{aligned}\text{Book Value of Equity} &= \text{Cumulative Retained Earnings} \\ &\quad + \text{Capital Contributed Over Par} + \text{Common Stock}\end{aligned}$$

Divide the book value of the equity by the number of shares outstanding to get the book value per share.

2.8 Earnings Per Share and Dividend

The earnings per share of a company is often of importance to current and potential stockholders, as well as management. The amount earned on each outstanding share of common stock during the accounting period is computed by dividing the total earnings available to the firm's common stockholders during the period by the number of outstanding shares of common stock.

According to, Arsal [33] a dividend is a portion of a company's earnings that it gives to its shareholders in cash. Of course, the dividend reduces the amount of earnings maintained in the company and has an impact on the total amount of internal financing. Dividends to stockholders are the most important factor. They buy a stock in the company with the expectation of receiving a portion of the profits. Stockholders' primary motivation is to make a profit on their investment; nothing satisfies them more than knowing the company's profitability, and more earnings imply higher dividends.

Cash dividends are payments made in cash to stockholders. This necessitates enough cash in a company's bank account. When a cash dividend is declared, the cash amount and reserves account of the company are reduced, lowering the total assets and net capital of the company.

3 Conclusion

This paper investigated the essential concept of audit quality and earnings management. In addition, this paper adopted the literature review findings to conclude findings in previous studies. This paper aimed to evaluate several aspects related to earnings management and motivations for controlling earnings and investigate the relationship between audit quality and earnings management behavior. According to the literature, there is a link between accounting earnings and the quality of audit procedures that can be used to control the earnings management strategies. This implies that the audit process adds value to accounting data and is recognized to enhance financial data quality. A limitation of this study is that the paper is limited to literature points of view in specific contexts that might differ from one to another. In addition, this paper contributes to the existing literature review on the relationship between earnings management and the quality of audits. This research provides a comprehensive view of the audit quality and earnings management concepts. A review of relevant academic studies has yielded a range of related concepts. Therefore, this paper contributes to the ongoing debate about audit quality and earnings management and highlights the need for a new approach to these issues. It is suggested to further research to expand the scope of this research and narrow the geographical or sector contexts. In addition, it is recommended to design a quantitative method to test the proposed perspectives.

References

1. Haloush HA, Alshurafat H, Alhusban AAA (2021) Auditors'civil liability towards clients under the Jordanian law: legal and auditing perspectives. J Gov Regul 10(1):42–48
2. Chowdhury SN, Eliwa Y (2021) The impact of audit quality on real earnings management in the UK context. Int J Account Inf Manag 29(3):368–391
3. Singh KSD et al (2021) Antecedents and internal audit quality implications of internal audit effectiveness. Int J Bus Sci Appl Manag 16(2):1–21
4. Yasser S, Soliman M (2018) The effect of audit quality on earnings management in developing countries: the case of Egypt. Int Res J Appl Finan 9(4):216–231
5. Herrbach O (2001) Audit quality, auditor behaviour and the psychological contract. Eur Account Rev 10(4):787–802
6. Van Caneghem T (2004) The impact of audit quality on earnings rounding-up behaviour: some UK evidence. Eur Account Rev 13(4):771–786
7. Blokdijk H et al (2006) An analysis of cross-sectional differences in big and non-big public accounting firms' audit programs. Audit J Pract Theory 25(1):27–48
8. Jordan CE, Clark SJ, Hames CC (2010) The impact of audit quality on earnings management to achieve user reference points in EPS. J Appl Bus Res (JABR) 26(1):19–30
9. Chen JZ, Elemes A, Lobo GJ (2021) David versus goliath: the relation between auditor size and audit quality for UK private firms. Eur Account Rev 1–34
10. DeAngelo LE (1981) Auditor size and audit quality. J Account Econ 3(3):183–199
11. Al-Mousawi RJ, Al-Thuneibat A (2011) The effect of audit quality on the earnings management activities. Dirasat Adm Sci 38(2):614–628
12. Deis Jr DR, Giroux GA (1992) Determinants of audit quality in the public sector. Account Rev 67:462–479
13. Teoh SH, Wong TJ (1993) Perceived auditor quality and the earnings response coefficient. Account Rev 68(2):346–366
14. Rezaei F, Shabani S (2014) The effect of audit firm size and age on the quality of audit work. Eur Online J Nat Soc Sci 3(1):56–64
15. Raimo N et al (2021) Do audit committee attributes influence integrated reporting quality? an agency theory viewpoint. Bus Strateg Environ 30(1):522–534
16. Alkebsee RH et al (2022) Audit committee independence and financial expertise and earnings management: evidence from China. Int J Bus Gov Ethics 16(2):176–194
17. Mihret DG (2014) How can we explain internal auditing? the inadequacy of agency theory and a labor process alternative. Crit Perspect Account 25(8):771–782
18. Sanad Z, Shiwakoti R, Al-Sartawi A (2022) Gender Diversity on corporate boards and earnings management: a review. In: Al-Sartawi AMAM (ed) Artificial intelligence for sustainable finance and sustainable technology. ICGER 2021. Lecture notes in networks and systems, vol 423. Springer, Cham

19. Jones JJ (1991) Earnings management during import relief investigations. J Account Res 29(2):193–228
20. Dechow PM, Sloan RG, Sweeney AP (1995) Detecting earnings management. Account Rev 70:193–225
21. Teoh SH, Welch I, Wong TJ (1998) Earnings management and the long-run market performance of initial public offerings. J Financ 53(6):1935–1974
22. Al-Sartawi AMAM (ed) (2022) Artificial intelligence for sustainable finance and sustainable technology. ICGER 2021. Lecture Notes in Networks and Systems, vol 423. Springer, Cham
23. Kothari SP, Leone AJ, Wasley CE (2005) Performance matched discretionary accrual measures. J Account Econ 39(1):163–197
24. Gastón SC, Jarne JIJ, Wróblewski D (2017) Detecting earnings management investigation on different models measuring earnings management for emerging Eastern European countries
25. Chen RC, Hung SW (2021) Exploring the impact of corporate social responsibility on real earning management and discretionary accruals. Corp Soc Responsib Environ Manag 28(1):333–351
26. Boujelben S, Khemakhem-Feki H, Alqatan A (2020) Real earnings management and the relevance of operating cash flows: a study of French listed firms. Int J Discl Gov 17(4):218–229
27. Karolak M, Razzaque A, Al-Sartawi A (2021) E-services and M-services using IoT: an assessment of the Kingdom of Bahrain. In: Al-Sartawi AMM, Razzaque A, Kamal MM (eds) Artificial intelligence systems and the internet of things in the digital era. EAMMIS 2021. Lecture Notes in Networks and Systems, vol 239. Springer, Cham
28. Jang GB, Kim W-J (2017) Effects of key financial indicators on earnings management in Korea's ready mixed concrete industry. J Appl Bus Res (JABR) 33(2):329–342
29. Dahmash F et al (2021) The effect of a firm's internal factors on its profitability: evidence from Jordan. Invest Manag Finan Innov 18(2):130–143
30. Ji X, Wang J, Yan Z (2021) A stock price prediction method based on deep learning technology. Int J Crowd Sci 5:55–72
31. Mahrani M, Soewarno N (2018) The effect of good corporate governance mechanism and corporate social responsibility on financial performance with earnings management as mediating variable. Asian J Account Res
32. Ching CP et al (2015) The relationship among audit quality, earnings management, and financial performance of malaysian public listed companies. Int J Econ Manag 9(1):211–231
33. Arsal M (2021) Impact of earnings per share and dividend per share on firm value. ATESTASI Jurnal Ilmiah Akuntansi 4(1):11–18

The Effect of Cost Efficiency on Profitability and Value: Evidence from Jordanian Banks Sector

Omar K. Gharaibeh, Mohammad Abdel Mohsen Al-Afeef, and Manaf Al-Okaily

Abstract the aim of the current research is to examine the impact of cost efficiency on the profitability and value of Jordanian commercial banks for the period 2012–2019. Using time-series cross-sectional data and depending on the pooled data model and random effects, the results of this study showed strong evidence that other operating income and net interest income have a positive and statistically significant impact on profitability in Jordanian commercial banks, while other operating expenses have a negative and significant impact on profitability. For the bank value model, other operating expenses has a negative effect, but it is weak statistically significant, while the net interest income has a positive effect and statistical significance. The net interest income in the two models, whether profitability or value, achieved the highest coefficient and the largest statistically significant effect. Bank size plays a positive and always statistically significant role in the profitability and value of banks. For the inflation and growth variables, they do not have any statistically significant effect, except for inflation which has a weak, statistically significant and positive effect on the profitability model. The study recommends that the managers of Jordanian commercial banks attempt to reduce other operating expenses because of their negative impact on the profitability and value of the bank.

Keywords Bank profitability · Cost efficiency · Value · Interest · Expenses · Jordanian banks

O. K. Gharaibeh
Finance and Banking, School of Business, Al-alBayt University, Al-Mafraq, Jordan

M. A. M. Al-Afeef
Faculty of Business, Jerash University, Jerash, Jordan

M. Al-Okaily (✉)
School of Business, Jadara University, Irbid 733, Jordan
e-mail: m.alokaily@jadara.edu.jo

A. M. A. Musleh Al-Sartawi et al. (eds.), *From the Internet of Things to the Internet of Ideas: The Role of Artificial Intelligence*, Lecture Notes in Networks and Systems 557,
https://doi.org/10.1007/978-3-031-17746-0_44

1 Introduction

During the COVID-19 pandemic, several financial institutions have faced major challenges related to their profitability and value [1, 2] In recent years, the liberalization of financial markets, the globalization of finance and the use of technology have increased, especially after the global financial crisis of 2009 [3, 4]. This led to an increase in competition and a rise in credit and operational risks in the banking environment in all countries of the world [3]. This development directly contributed to the growth of global financial markets, as well as prompted banks to adjust their strategic directions and motivate them towards reducing costs, increasing revenue generation and improving performance rates for continuity and growth [5, 6]. The Jordanian banking sector has been affected by globalization and market liberalization like other countries. For example, the Jordanian banks are exposed to a high level of risks, including credit risks, which constituted 87% of the total risks in 2020, operational risks of 11.7%, and market risks that constituted about 1.3% of the total risks. For operational efficiency, which is measured by the ratio of non-interest expenses to total income, it rose in 2020 to 74.4%, compared to 57.1% in 2019 due to the repercussions of the COVID-19 pandemic [7].

Because of the change in cost efficiency related to operating expenses and operating income resulting from the liberalization of financial markets, the globalization of finance and the use of technology. Also, the lack of studies that research this subject at the level of Jordanian commercial banks as an emerging country, this study came to fill the gap and examine the impact of cost efficiency on the profits and value of Jordanian commercial banks. By answering the following questions: Is there an impact of cost efficiency related to other operating expenses (OE) on the profitability and value of Jordanian commercial banks? Is there an impact of cost efficiency related to other operating income (OI) on the profitability and value of Jordanian commercial banks? Is there an impact of cost-efficiency related to net interest income (NIN) on the profitability and value of Jordanian commercial banks? Is there an effect of the controlling variables, namely the size (SIZE) of the bank, inflation (INF) and growth domestic product (GDP) on the profitability and value of Jordanian commercial banks?

The purpose of this paper is to examine the cost efficiency in Jordanian commercial banks and its impact on profitability and value. This paper contributes to the cost efficiency literature in several ways. Firstly, it provides evidence of the impact of cost efficiency on the profitability and value of Jordanian commercial banks. Secondly, the results in this study also indicate that Jordanian commercial banks are not isolated from the results found in other commercial banks, whether in developing or developed countries. Finally, although there are many studies that discuss the impact of cost efficiency on the profitability, there are few studies that discuss the impact of cost efficiency on the bank's value. This study bridges this gap and examines not only the effect of cost efficiency on profitability, but also on the value of the Jordanian bank using the Tobin's q.

2 Literature Review

2.1 Cost Efficiency

The concept of efficiency is usually associated with the concept of performance of financial institutions [8]. Efficiency has different types, namely cost efficiency, revenue efficiency and profit efficiency [9, 10]. Technical efficiency is fundamental to all efficiency measures, which measures how efficient a management is in obtaining maximum output from the available inputs. Efficiency has been divided into three types: the first is economic efficiency or what is known as total cost efficiency, which is the sum of technical efficiency and allocated efficiency [11, 12]. The second type is the efficiency of economies of scale resulting from the expansion of the volume of activity, which is important in determining the volume of investment and pricing products [13–15]. The third type is the efficiency of diversification in products, which is an advantage of banks [16] the indicate that determining the efficiency of banks is done through a set of inputs and outputs. They point out that previous studies related to banking efficiency focused on the role of banks as financial intermediaries and that these inputs and outputs affect the efficiency of the bank [17–19]. They divide the inputs into capital, labor and the cost of money, while the outputs were divided into different types of deposits and loans. The cost efficiency of an organization is measured by its performance relative to the estimated performance of banks that are characterized by minimal costs or use of inputs in the industry, while keeping the scale and output mix constant [17]. In this vein, Berger and Humphrey [20] confirmed that cost inefficiencies in the range of (20–30%) dominate the effect of lack of economies of scale and scope, which were generally found in the range of (5–10%) [20].

The relationship can be negative between the operating cost and the profitability of banks if the banks do not utilize operational resources in an optimal manner and do not respond to the requirements of keeping pace with technical and informational developments [21]. The Islamic banks, relatively, are less efficient in costs compared to their traditional commercial banks, if there is no regulatory environment supportive of their operations, but it is effective in making a profit [22]. However, Zuhroh, Ismail and Maskie [23] show that Islamic banks in countries where the banking system prevails according to the rules of Islamic law will be more efficient in terms of allocation and that all efficiency measures are closely related to return on assets and return on equity [23]. The cost efficiency that leads to increased profitability can be enhanced by controlling operating costs, using optimal production plans, and using new technologies [24].

Otero, Razia, Cunill and Mulet-Forteza [25] examine the determinants of cost efficiency in the Middle East and North Africa (MENA) countries during the period 2005–2012. Show that cost efficiency is positively related to capital level, and economic performance, while cost efficiency is negatively related to concentration level [25]. However, Fungáčová, Klein and Weill (2020) find that the five largest state-owned banks in China suffer from low cost efficiency. This is due to persistent low cost efficiency, and structural problems [26]. Haralayya and Aithal [3] point out that mismanagement is a reason for inefficiency, and banks need to strengthen their tasks and control different types of costs as they are the main factor for financial inefficiency [3].

In the banking sector in the euro area, Huljak, Martin and Moccero [27] find that technological progress has the largest role in enhancing efficiency, followed by technical efficiency and there is a need to enhance productivity and enhance profitability [27]. In an emerging market, Duho [28] shows that the difference in efficiency between banks converges over time, and that banks are effective in leveraging intellectual capital resources to achieve a competitive advantage [28]. Therefore, his study recommend focusing on intangible assets because of their role in enhancing efficiency, and that competition has a significant negative impact on technical efficiency [28]. Based on the above discussion, the following hypotheses are proposed:

H1: Cost Efficiency has a statistically significant effect on return on assets of Jordanian Commercial Banks.

H2: Cost Efficiency has a statistically significant effect on value of Jordanian Commercial Banks.

2.2 Size

Hasan and Marton [29] find that there is a positive relationship between bank size and profits [29]. This means that banks will be more cost-effective and profitable if they are larger. Hassan (2006) confirmed that the larger the size of the bank is associated with higher efficiency, and most Islamic banks are smaller in size compared to the conventional banks [22]. Therefore, he recommends that the merger between Islamic banks have a positive impact on efficiency in terms of technology and competition by making use of modern technology and expanding the degree and volume of operations [22]. Tyas and Rusydiana (2021) confirmed the merger and point out that the merger process between banks in Indonesia will have a high performance efficiency [30]. Otero, et al., [25] support this positive effect in MENA countries [25]. In this vein, Duho [28] show that the size of the largest bank is more efficient [28]. Gharaibeh and Jaradat [1] found that size is positively and statistically significant effect on stock prices of Jordanian commercial banks [31].

On the other hand, in both domestic and foreign banks in Thailand, Leightner and Lovell [32] has found a negative relationship between bank size and efficiency [32]. Examining the determinants efficiency among the Nigeria's commercial banks, Ojeyinka and Akinlo [33] demonstrate that big banks do not have a cost advantage over their smaller counterparts [33]. Gharaibeh, Alown and Al-Eitan [34] showed that the relationship between the size and profitability of the Jordanian banking sector is negative and statistically significant. Also, the size of the bank does not affect the cost efficiency [34]. The main factors determine the efficiency is the capitalization ratio, loan-to-deposit ratio, operating expenses and loans to total assets [34]. Based on the above discussion, the following hypotheses are proposed:

H3: Size has a statistically significant effect on return on assets of Jordanian Commercial Banks.

H4: Size has a statistically significant effect on value of Jordanian Commercial Banks.

2.3 Inflation

In Latin American countries, Hassan and Sanchez (2007) find that most of the sources of inefficiency are organizational rather than technical (administrative) in most of these countries [35]. Inefficiency drives most of the technical shortcomings of banking industries. Their results show that the net interest margin is negatively related to cost efficiency, which is consistent with the assertion that higher net national income associated with lower efficiency because wider margins indicate lower competition. Their result also show that a high return on equity is positively correlated with high efficiency, while high inflation rates or real interest rates are detrimental to banking efficiencies [35]. Otero, et al., [25] show that inflation are positively related to the total cost function in MENA countries [25]. Based on the above discussion, the following hypotheses are proposed:

H5: Inflation has a statistically significant effect on return on assets of Jordanian Commercial Banks.

H6: Inflation has a statistically significant effect on value of Jordanian Commercial Banks.

2.4 Growth Domestic Product

At the level of the banking systems in the European Union, Maudos, Pastor, Perez and Quesada (2002) show that there is a greater discrepancy in the performance of banks in terms of profit efficiency than in terms of cost efficiency. They indicate that achieving higher levels of efficiency can depend on markets that achieve high rates of economic growth, and banks that operate in markets that have a higher network density and have higher structural expenditures are less cost-effective [36]. In the countries of the southern Mediterranean, Ayadi, Arbak, Naceur and Groen (2015), confirmed that credit to the private sector and bank deposits are negatively related to growth, and there are problems in the allocation of credit. In addition, they indicated the weakness of financial regulation and financial supervision in these countries. They point out that stock market size and liquidity play a role in growth, especially when the quality of institutions is low, while investment had a role in economic growth [37].

Niţoi and Spulbar (2015) find that the high ratio of bank credit to GDP has a positive impact on inefficiency. They also show that commercial banks that focus on traditional loan-granting activity are more efficient as compared to banks that have a lower share of loans to total assets [38]. Furthermore, Niţoi and Spulbar (2015) demonstrate that banks that take on more risks are more ineffective. Therefore, they conclude that banks with less liquidity, lower solvency and higher credit risk are more ineffective than more cautious credit institutions. Otero, et al., [25] demonstrate that GDP are positively related to the total cost function in MENA countries [25]. Based on the above discussion, the following hypotheses are proposed:

H7: Growth domestic product has a statistically significant effect on return on assets of Jordanian Commercial Banks.

H8:Growth domestic product has a statistically significant effect on value of Jordanian Commercial Banks.

3 Methodology

3.1 Data Collection

Banking data is downloaded from the Amman Stock Exchange website,[1] while macroeconomic data is downloaded from the Central Bank of Jordan website.[2] The data covers 8 years between 2012 and 2019 for all Jordanian commercial banks. The time period for the study was chosen from 2011 to 2018 to avoid the impact of the global financial crisis in 2009 and its repercussions and before the impact of the COVID-19 or Corona pandemic in 2020. The data consist of two dependent and three independent variables, as well as three control variables.

Dependent Variables
The first dependent variable is profitability as measured by return on assets (ROA) as measured as follows [39]:

$$\text{Return on Assets} = \frac{\text{Net Income}}{\text{Total Assets}}$$

This ratio shows the bank's ability to generate profits from total assets.

The second dependent variable is the bank's value (Tobin's q) as measured as follows:

$$\text{Tobin's q} = \frac{(\text{Market Capitalization} + \text{Total Liabilities})}{\text{Total Assets}}$$

This ratio indicates the bank's value.

Independent Variables
The first independent variable of cost efficiency is the cost of operating efficiency measured as follows:

[1] https://www.ase.com.jo/ar.

[2] https://www.cbj.gov.jo/.

$$\text{Cost of Other Operating Efficieny} = \frac{\text{Other Non - Interset Operating Expenses}}{\text{Total Assets}}.$$

This ratio measures the cost of other operating expenses such as depreciation, amortization, employee expenses and other expenses divided by total assets.

The second independent variable of cost efficiency is the profit of other operating measured as follows:

$$\text{Profit of Other Operating Efficieny} = \frac{\text{Other Non - Interset Operating Income}}{\text{Total Assets}}$$

This ratio measures the profit of other operating income such as gains from foreign currencies, gains from financial assets and Instruments, net commissions and others generated from total assets.

The third independent variable of cost efficiency is the profit of net interest operating efficiency measured as follows:

$$\text{Profit of Operating Efficieny} = \frac{\text{Net Interset Operating Revenue}}{\text{Total Assets}}$$

This ratio measures the profit of net interest operating income, which is interest income minus interest expenses generated from total assets.

Control Variables

The first control variable is capitalization as measured by logarithm of total assets (TA).

$$\text{Size} = \log(\text{TA})$$

To control the business cycle in the model used in this study, the two macroeconomic variables are employed in this study, namely growth and inflation, used as follows:

$$\text{Gross National Production (GDP)} = \frac{\text{GDP}_1 - \text{GDP}_0}{\text{GDP}_0}$$

$$\text{Inflation (INF)} = \frac{\text{CPI}_1 - \text{CPI}_0}{\text{CPI}_0}$$

3.2 Method

Banking transactions depend mainly on interest. In addition, there are other income and operating expenses that are not based on interest such as transaction fees, commissions, employee salaries and others. The current study focuses on cost efficiency through attention to other operating expenses, and profit efficiency by other operating income [24, 40, 41] indicate that non-interest-based transactions reduce risks and increase profits. Following the [24] methodology, the study also focuses on efficiency in profit through net interest income. Therefore, this study examines not only the impact of cost efficiency on the profitability, but also on the value of Jordanian commercial banks. This study uses the panel data set to achieve the objectives of the study. The first and second model is used in this study as follows:

$$ROA_{it} = \alpha + \beta_1 OE_{it} + \beta_2 OI_{it} + \beta_3 NIN_{it} + \beta_4 SIZE_{it} + \beta_5 GDP_t + \beta_6 INF_t + e_{it} \quad (1)$$

$$Tobin's\, q_{it} = \alpha + \beta_1 OE_{it} + \beta_2 OI_{it} + \beta_3 NIN_{it} + \beta_4 SIZE_{it} + \beta_5 GDP_t + \beta_6 INF_t + e_{it} \quad (2)$$

where:
ROA_{it} = The return on assets of bank *i* in time *t*.
$Tobin's$ q = The value of bank *i* in time *t*.
OE_{it} = The other operating expenses of bank *i* in time *t*.
OI_{it} = The other operating income of bank *i* in time *t*.
NIN_{it} = The net interest income of bank *i* in time *t*.
$ZISE_{it}$ = log of total assets of bank *i* in time *t*.
GDP_t = The percentage of growth domestic product in Jordan in time *t*.
INF_t = The inflation rate in Jordan in time *t*.
$\beta_1, \beta_2, \beta_3, \beta_4, \beta_5, \beta_6$ = Coefficients of independent variables and α is intercept, while e_{it} is the errors of bank *i* in time *t*.

Commercial banks more efficient than Islamic banks in terms of cost. Cost efficiency also has a direct impact on profitability performance and increased growth in banking industries. In this regard, Jensen (1986) suggests the cost control hypothesis which supports debt financing rather than equity [42]. However, control hypothesis will only reduce the tendency to bankruptcy but not increase performance in returns.

4 Results

This section of the study is devoted to presenting descriptive statistics and pre-tests for the dependent and independent variables used in this study, as well as discussing the results of the study.

Table 1 shows in the first column the names of Jordanian commercial banks. The second and third columns detail the dependent variables used in the current study over the period of 2011–2018, which is the average profitability measured by return on assets (ROA) and average bank's value measured by Tobin's q, respectively. The last three columns report the independent variables used in this study, namely, average other operating expenses (OE), average other operating income (OI), and average net interest income (NIN), respectively.

Table 1 Profitability of Jordanian commercial banks over the period of 2011–2018

Bank's name	ROA %	Tobin's q	OE %	OI %	NIN %
BOJ	1.77	1.04	3.09	1.55	4.12
CAB	1.56	1.01	3.11	1.38	3.97
JKB	1.54	0.98	2.50	1.19	3.47
INVB	1.53	0.97	2.81	1.67	3.27
HBTF	1.49	1.16	2.38	1.09	3.35
ABC	1.26	0.97	2.41	0.90	3.37
AJIB	1.23	1.02	1.86	1.02	2.62
CBJ	1.12	0.94	2.79	1.77	2.61
ARB	1.12	1.01	2.28	1.38	2.47
BETH	1.03	0.96	2.29	0.98	2.95
SGDB	0.85	0.97	1.31	0.84	1.82
JAB	0.78	0.97	3.30	1.28	3.11
JCB	0.46	0.99	3.11	1.21	2.61
Average	**1.21**	**1.00**	**2.56**	**1.25**	**3.06**
St. dev.	**0.005**	**0.056**	**0.56**	**0.29**	**0.63**

The first note taken from Table 1 that profitability of Jordanian commercial banks varies among banks during the years of study. For example, the Bank of Jordan had the highest return on assets of 1.77%, while the Jordan Commercial Bank achieved the lowest an average return on assets of 0.46%. The difference between the Jordanian commercial banks is relatively large 1.31% compared to the difference in the value of the bank, which is close to 1. This means that the difference between the values of these banks is relatively small.

For the independent variable that measures other operating expenses (OE), Table 1 shows that the difference in the average of other operating expenses differs among Jordanian commercial banks about 2%. For example, the average other operating expenses in JAP is large 3.30% while the percentage of other operating expenses for SGDB is 3.30%. For average other operating income, the difference among Jordanian commercial bank is small around 0.5%. Looking at the average net interest income, we note that the difference is relatively large and exceeds 2%. These differences in average other operating expenses and average net interest income could have an impact on the profitability and value of Jordanian commercial banks.

Table 2 shows that the first dependent variable used in this study, which is profitability, achieves an average return of 1%, while the second dependent variable, the value, achieves an average equal to 1. Also, the volatility in the average profitability is very small, equal to 0.005, while the volatility in the value of banks increases to 0.056. This indicates that the profitability of Jordanian banks is relatively stable, while the value of Jordanian banks changes. With regard to the cost efficiency, which is the independent variables, Table 2 shows that the average of other operating income (OI) is very small 1% compared to average of other operating expenses (OE) that are

Table 2 Descriptive statistics

Variables	Mean	Std. dev.	Max.	Min.	Skew.	Kurt.	Obs.
ROA	0.0121	0.005	0.02	−0.002	−0.61	2.91	104
Tobin's q	1.00	0.056	1.19	0.90	1.82	6.35	104
OI	0.012	0.005	0.03	0.002	0.88	4.36	104
OE	0.03	0.01	0.04	0.01	−0.23	3.18	104
NIN	0.03	0.01	0.04	0.01	−0.18	3.01	104
SIZE	21.53	0.93	23.98	19.7	1.16	4.26	104
GDP	0.03	0.01	0.03	0.02	0.21	1.79	104
INF	0.03	0.02	0.05	−0.01	−0.92	2.12	104

equal to 3%. However, the average of net interest income (NIN) is 3%. On the other hand, the volatility of these cost efficiency variable OI, OE, and NIN is relatively small 0.005, 0.01, and 0.01, respectively. Given the Skewness results, all the variables values are around zero. In addition, the Kurtosis results also indicate that most variables values are less than 3, except of Tobin's q, OI, and SIZE. This means that the data does not suffer from the problem of skewness and the presence of outliers.

Unit Root Test

This study uses test Levin, Lin and Chu test, which is suitable for examining cross-sectional time-series data. This test is based on two main hypotheses. The null hypothesis indicates that the time series is not stationary, while the alternative hypothesis indicates that the time series is stationary. If the significance value is less than 5%, the alternative hypothesis will be accepted [43]. Based on the results of Table 3, the study variables are stable at the level because all variables are statistically significant and less than 5%. This means rejecting the null hypothesis and concluding that the study variables are stationary.

Table 3 The results of levin, lin, and chu tests

Variables	Level	P-value
ROA	−2.34785	0.0094
Tobin's q	−9.29550	0.0000
OE	−9.87825	0.0000
OI	−13.4349	0.0000
NIN	−2.55261	0.0053
SIZE	−3.17629	0.0007
INF	−9.12612	0.0000
GDP	−4.04768	0.0000

Table 4 Correlation matrix

Independent variables	OI	OE	NIN	SIZE	INF	GDP
OI	1					
OE	0.48***	1				
NIN	0.24**	0.58***	1			
S	−0.06157	−0.04	0.01	1		
INF	0.19**	0.04	−0.05	−0.06	1	
GDP	0.39***	0.03	0.07	−0.13	0.15	1

*** **1%, ** 5%, * 10%**

The Correlation Matrix Test

Table 4 shows the correlation matrix between the independent and control variables. Gujarati (2009) indicates that the correlation between the independent variables should not be more than 80%, and if the correlation value is greater than 80%, one of the two variables should be deleted. Looking at the results in Table 4, the correlation values between the variables are less than 80%, with the highest correlation value reaching 58% between the two variables of other operating expenses (OE) and net interest income (NIN) [44]. This indicates that the variables used in this devoid of multicolinearity problem.

Breusch-Pagan (LM) Test

The current study uses the Breusch-Pagan test to discover the problem of heterogeneity of the random error variance in the two models. The residual errors are considered sound and homogeneous if the value of the test significance is greater than 5%. Table 5 shows that the probability value is greater than 5% in the bank's profitability (ROA) model, while it is less than 5% in the bank's value (Tobin's q) model. This indicates that the first model (ROA) is free from heterogeneity problems, while the second model (Tobin's q) suffers from a problem of heterogeneity. Thus, the Pooled regression model will be used in the first model, while in the second model, the appropriate model depends on Huesman test to select the fixed or random effects model.

Cost Efficiency Effect and Bank's Profitability

Table 6 shows that the independent variables used in this model explain 72.6% of the changes in the bank's profitability measured by ROA. Table 6 also shows that the other operating incomes (OI) have a positive statistical significance on bank's profitability. The coefficient of OI is 0.317596, which means that the more OI increase

Table 5 Breusch-Pagan (LM) test for panel data

Breusch-Pagan (LM) test	ROA	Tobin's q
Breusch-Pagan	2.447925	241.296
P-value	(0.1177)	(0.0000)

Table 6 Cost efficiency effect and profitability of Jordanian commercial banks

Dependent variable: ROA	Pooled test			
Independent variables	Coefficient	Std. error	t-statistic	Prob.
C	−0.00861	0.005017	(−1.72)	0.0895
OI	0.317596	0.092581	(3.43)	0.0009
OE	−0.5899	0.046146	(−12.78)	0.0000
NIN	0.709707	0.041455	(17.12)	0.0000
SIZE	0.000446	0.000212	(2.10)	0.0384
INF	0.013651	0.007309	(1.87)	0.0648
GDP	0.00559	0.073181	(0.08)	0.9393
R-squared	0.742582			
Adjusted R-squared	0.726659			
Cross-sections included	13			
Observations:	104			

by 0.317596, the bank's profitability increases by 1%. This result is consistent with result of [21, 24]. This can be attributed to the fact that operating income from gains from financial assets and instruments, foreign exchange, net commission income and other operating income increased in Jordanian commercial banks.

Other operating expenses (OE) have a statistically significant negative impact on bank's profitability. The coefficient of OE is equal to −0.5899, which means that the increase in OE leads to a decrease in profitability by 1%. This result is consistent with result of [24]. Although the percentage of expenses to total income reached about 57% at the end of 2019 (Financial Stability Report, 2019), expenses for banks decreased in general due to the decline in provisions for credit losses and the decrease in the risk of clients' default. However, this indicates that operating expenses are still relatively high and must be reduced in Jordanian commercial banks by adopting and expanding the use of technology.

Net interest factor (NIN) plays an important role in profitability because it has the highest coefficient in the analysis. It equals to 0.709707 and is statistically significant (*t*-stat 17.12). This result is in line with result of [21, 24]. The net interest income is one of the most important determinants of the operational efficiency of banks due to its importance and being the mediator in managing and allocating loans, which in turn affects profitability.

The bank size (SIZE) has a positive and statistically significant effect at the level of 5%. This indicate that the larger the bank, the greater its ability to generate profits. This result supports [22, 29] result who confirmed that the banks of the largest size are the ones that achieve higher efficiency. On the other hand, this result contradicts the result of [34] because their study investigated the bank sector as a whole based on time series analysis, while the current study investigates all Jordanian commercial banks based on panel data analysis. Thus, the methodology and time period for both studies are different. For inflation (INF), it has a positive and statistically significant

Table 7 Hausman test

Random effect	Tobin's q
Chi-Sq. statistic	0.561668
P-value	(0.9672)

effect at the %10 level, while growth domestic product (GDP) has no statistically significant effect. This can be attributed to the fact that the Jordanian commercial banks witnessed a relative increase in credit growth in the last period, while there was a decrease in the economic growth rates as measured by the GDP during the same period. The percentage of bank credit granted to the gross domestic product, according to the reports (CBJ), ranged between 85–83% during the period 2018–2019. This refers to an increase in the money supply that has led to an increase in inflation. This implicitly indicates that inflation has a positive effect with weak statistical significance, while GDP has no statistically significant effect [7].

Cost Efficiency Effect and Bank's Value

Table 7 shows the result of Hausman test, which determines the best model, should be used to test the effect of cost efficiency on the bank's value. Hausman test demonstrates that the probability value is greater than %5. This means that the random effects model is more suitable than the fixed effects model. Therefore, based on Hausman test result this study uses the random effects model to explain the effect of cost efficiency on bank's value.

Table 8 demonstrates that the independent variables employed in this model explain 26.3% of the changes in the bank's value measured by Tobin's q. In contrast to the statistically significant positive impact on profitability, Table 8 indicates that other operating income (OI) does not have a statistically significant effect on the bank's value. The coefficient of OI is −0.91108, which means that the more OI decrease by 0.91108, the bank's value increases by 1%.

Consistent with previous result related to effect of cost efficiency and profitability documented by Table 6, other expenses (OE) have negative effect but weak statistically significant impact on bank's value. The coefficient of OE is equal to −1.77708, which means that the increase in OE leads to a decrease in bank's value by 1%. This result is consistent with result of [24]. This means that operating expenses are still have a relatively negative impact on bank's value and must reduce other operating expense in Jordanian commercial banks by implementing and increasing the use of technology.

Table 8 also confirms previous result documented by Table 6 and show that net interest income (NIN) plays an important role in bank's value because it has the highest coefficient in the analysis. It equals to 3.244965 and is statistically significant (*t*-stat 5.16). This finding is in line with result of [21, 24]. The net interest income is one of the most important determinants that affect the bank's value because banks make the main profits from the difference between deposit and lending interests.

The bank size (SIZE) has a positive and statistically significant effect at the level of 5%. This means that the larger the bank, the greater its ability to generate profits.

Table 8 Cost efficiency effect and value of Jordanian commercial banks

Dependent variable: Tobin's q				
Independent variables	Random effects			
	Coefficient	Std. error	t-statistic	Prob.
C	0.364828	0.021361	17.07904	0.0000
OI	−0.91108	1.254589	−0.7262	0.4695
OE	−1.77708	0.971849	−1.82856	0.0705
NIN	3.244965	0.627848	5.16839	0.0000
SIZE	0.026807	0.001335	20.07638	0.0000
INF	0.024501	0.050799	0.482313	0.6307
GDP	0.539767	0.485403	1.111999	0.2689
R-squared	0.306777			
Adjusted R-squared	0.263897			
Cross-section included	13			
Observations:	104			

This finding is supported by [22, 29] who asserted that larger banks achieve higher efficiency and value. For inflation (INF) and growth (GDP), they have no significant effect.

5 Conclusion

This paper examines the impact of cost efficiency on the profitability and value of Jordanian commercial banks for the period 2012–2019. This study provides the first strong evidence at the level of Jordanian commercial banks that other operating income and net interest income have a positive and statistically significant impact on profitability, while other operating expenses have a negative and statistically significant impact on profitability. For the banks value model, over operating expenses have a negative impact with weak statistical significance, while net interest income has a positive and statistically significant effect. The net interest income in the two models, whether profitability or value, achieved the highest coefficient and the largest statistic significant impact.

The control variables used in this study showed that the size of the bank plays a positive and statistically significant role on the profitability and value of the banks. For inflation and growth variables, they have no statistically significant effect, except for inflation, which has a positive and weak statistically significant effect on the profitability model. The study recommends that the managers of Jordanian commercial banks reduce other operating expenses because of their negative impact on the profitability and value of the bank. The study also recommends a comparison between conventional banks and Jordanian Islamic banks. Finally, the study recommends conducting research on non-financial sectors.

References

1. Kharabsheh B, Gharaibeh O, Mahafza A (2022) Is There an impact of COVID-19 on the returns of the amman stock exchange? Invest Manag Finan Innov 19(2):24–36
2. Alsaad A, Al-Okaily M (2022) Acceptance of protection technology in a time of fear: the case of Covid-19 exposure detection apps. Inf Technol People 35:1116–1135. https://doi.org/10.1108/ITP-10-2020-0719
3. Haralayya D, Aithal P (2021) Study on productive efficiency of financial institutions. Int J Innov Res Technol (IJIRT) 8(1):116–159
4. Liu J, Lai MY, Liu ZS (2022) Trade liberalization, domestic reforms, and income inequality: evidence from Taiwan. Rev Dev Econ 26:1286–1309
5. Hamdan M, Gharaibeh O, Al-Quran A, Nusairat N (2021) The impact of information technology investment on the financial performance of the banks. Acad Strateg Manag J 20(6):1–12
6. Al-Okaily M, Al-Okaily A, (2022) An empirical assessment of enterprise information systems success in a developing country: the Jordanian experience. TQM J. https://doi.org/10.1108/TQM-09-2021-0267
7. Central bank of Jordan, financial stability report (2020) Financial Stability Department. https://www.cbj.gov.jo/EchoBusv3.0/SystemAssets/PDFs/EN/JFSRE2019Final_23-11-2020.pdf
8. Al-Okaily M (2021). Assessing the effectiveness of accounting information systems in the era of COVID-19 pandemic. VINE J Inf Knowl Manag Syst. https://doi.org/10.1108/VJIKMS-08-2021-0148
9. Al-Okaily M, Alghazzawi R, Alkhwaldi AF, Al-Okaily A (2022) The effect of digital accounting systems on the decision-making quality in the banking industry sector: a mediated-moderated model. Glob Knowl Memory Commun. https://doi.org/10.1108/GKMC-01-2022-0015
10. Al-Okaily A, Al-Okaily M, Teoh AP, Al-Debei M (2022) An empirical study on data warehouse systems effectiveness: the case of Jordanian banks in the business intelligence era. EuroMed J Bus. https://doi.org/10.1108/EMJB-01-2022-0011
11. Al-Okaily M, Natour ARA, Shishan F, Al-Dmour A, Alghazzawi R, Alsharairi M (2021) Sustainable FinTech innovation orientation: a moderated model. Sustainability 13(24):1–12
12. Gogia J, Chakraborty D (2022) Open banking: a revolution in the Tech-Fin industry. Int J Electron Bank 3(2):100–120
13. Al-Qudah AA, Al-Okaily M, Alqudah H (2022) The relationship between social entrepreneurship and sustainable development from economic growth perspective: 15 'RCEP' countries. J Sustain Finan Invest 12(1):44–61
14. Al-Qassim M, Chen W, Al-Sartawi A (2022) Blockchain in environmental compliance and enforcement. In: Al-Sartawi AMAM (ed) Artificial intelligence for sustainable finance and sustainable technology. ICGER 2021. Lecture Notes in Networks and Systems, vol 423. Springer, Cham
15. Anderson RI, Fok R, Springer T, Webb J (2002) Technical efficiency and economies of scale: a non-parametric analysis of REIT operating efficiency. Eur J Oper Res 139(3):598–612
16. Jovanovic B, Gilbert RJ (1993) The diversification of production. Brook Pap Econ Act Microecon 1993(1):197–247
17. Al-Sartawi A (2019) Performance of Islamic banks Do the frequency of Shari'ah supervisory board meetings and independence matter? ISRA Int J Islamic Finan 11(2):303–321
18. Al-Okaily A, Al-Okaily M, Teoh AP (2021) Evaluating ERP systems success: evidence from Jordanian firms in the age of the digital business. VINE J Inf Knowl Manag Syst. https://doi.org/10.1108/VJIKMS-04-2021-0061

19. Al-Okaily A, Al-Okaily M, Ai Ping T, Al-Mawali H, Zaidan H (2021) An empirical investigation of enterprise system user satisfaction antecedents in Jordanian commercial banks. Cogent Bus Manag 8(1):1918847
20. Berger AN, Humphrey DB (1991) The dominance of inefficiencies over scale and product mix economies in banking. J Monet Econ 28(1):117–148
21. Mohammed F, Belkacem M, Achouak B (2020) Determinants of costs-efficiency for commercial banks operating in Algeria econometric study during the period 2010–2016. Milev J Res Stud 6(1):382–403
22. Al-Sartawi AMAM (ed) (2022) Artificial intelligence for sustainable finance and sustainable technology. ICGER 2021. Lecture Notes in Networks and Systems, vol 423. Springer, Cham
23. Zuhroh I, Ismail M, Maskie G (2015) Cost efficiency of islamic banks in Indonesia–a Stochastic frontier analysis. Procedia Soc Behav Sci 211:1122–1131
24. Aliyu S, Yusof RM (2016) Profitability and cost efficiency of Islamic banks: a panel analysis of some selected countries. Int J Econ Financ Issues 6(4):1736–1743
25. Otero L, Razia A, Cunill OM, Mulet-Forteza C (2020) What determines efficiency in MENA banks? J Bus Res 112:331–341
26. Fungáčová Z, Klein P-O, Weill L (2020) Persistent and transient inefficiency: explaining the low efficiency of Chinese big banks. China Econ Rev 59:101368
27. Huljak I, Martin R, Moccero D (2019) The cost-efficiency and productivity growth of euro area banks
28. Duho KCT (2020) Intellectual capital and technical efficiency of banks in an emerging market: a slack-based measure. J Econ Stud 47(7):1711–1732
29. Hasan I, Marton K (2003) Development and efficiency of the banking sector in a transitional economy: Hungarian experience. J Bank Finan 27(12):2249–2271
30. Al-Okaily M, Alalwan AA, Al-Fraihat D, Rehman SU, Alkhwaldi AF, Al-Okaily A, (2022) Investigating antecedents of mobile payment systems decision making: a mediated model. Glob Knowl Memory Commun. https://doi.org/10.1108/GKMC-10-2021-0171
31. Gharaibeh OK, Jaradat MA (2021) Determinants of stock prices in jordanian banks: an empirical study of 2006–2018. J Asian Finan Econ Bus 8(7):349–356
32. Leightner JE, Lovell CK (1998) The impact of financial liberalization on the performance of Thai banks. J Econ Bus 50(2):115–131
33. Ojeyinka TA, Akinlo AE (2021) Does bank size affect efficiency? evidence from commercial banks in Nigeria. Ilorin J Econ Policy 8(1):79–100
34. Gharaibeh O, Alown B, Al-Eitan GN (2017) Factors affecting profitability of the Jordanian banking sector. Int J Econ Perspect 11(3):2059–2066
35. Hassan MK, Sanchez B (2007) Efficiency determinants and dynamic efficiency changes in Latin American banking industries. SSRN 3263102.
36. Maudos JN, Pastor JM, Perez F, Quesada J (2002) Cost and profit efficiency in European banks. J Int Finan Mark Inst Money 12(1):33–58
37. Ayadi R, Arbak E, Naceur SB, Groen WPD (2015). Financial development, bank efficiency, and economic growth across the Mediterranean Economic and social development of the Southern and Eastern Mediterranean countrie, pp 219–233
38. Niţoi M, Spulbar C (2015) An examination of banks' cost efficiency in Central and Eastern Europe. Procedia Econ Finan 22:544–551
39. Al-Qudah AA, Al-Afeef MAM (2015) The relationship between the investment in current assets and profitability & liquidity. J Finan Invest Anal 4(4):11–22
40. Stiroh KJ (2004)\ Diversification in banking: Is noninterest income the answer? J Money Credit Bank 853–882
41. Demirgüç-Kunt A, Huizinga H (2010) Bank activity and funding strategies: the impact on risk and returns. J Financ Econ 98(3):626–650
42. Jensen MC (1986) Agency costs of free cash flow, corporate finance, and takeovers. Am Econ Rev 76(2):323–329
43. Greene W (2000) Econometric analysis, 4th edn. Prentice Hall, Upper Saddle River

44. Gujarati D, Dan Porter DC (2009) Basic econometrics, 5th edn. McGraw-Hill International Edition, Boston
45. Al-Okaily M, Alqudah H, Al-Qudah AA, Alkhwaldi AF (2022) Examining the critical factors of computer-assisted audit tools and techniques adoption in the post-COVID-19 period: internal auditors perspective. VINE J Inf Knowl Manag Syst. https://doi.org/10.1108/VJIKMS-12-2021-0311

Investigating the Influence of COVID-19 Pandemic on Global Currency Exchange

Tareq A. M. Atiany, Jameel J. Alawneh, Osama Ala'yed, and Manaf Al-Okaily

Abstract The outbreak of coronavirus disease (COVID-19) as a pandemic has affected the economy globally. In this research, we use the correlation constancy test to explore the effect of COVID-19 on the construction of exchange rates among 15 currencies from European and US countries. Jennrich's control chart and its root sources are utilized to investigate the constancy of the correlation currency's structure. To filter the evidence from Minimum Spanning Tree, we employ a degree centrality measure. The analysis shows that the structures of correlation from January 2019 to December 2021 are not stable. To be detailed, based on October 2021, we determine that COP, HUF, and USD are unsteady currencies amongst others.

Keywords COVID-19 · Global currency exchange · Multivariate statistical · Jennrich's M · Stability

1 Introduction

In December 2019, coronavirus disease (COVID-19) caused a series of sharp atypical aerobic illnesses started from Wuhan in China [1–3]. COVID-19 pandemic is concern globally as a result of its high stretching and alarming fatality rate. Therefore, large number of countries lockdown and do gregarious distancing [4–6]. These measures have a great impact on the economy [7, 8] and especially on Foreign Exchange Market (FOREX) [9, 10].

T. A. M. Atiany
Department of Applied Mathematics, Palestine Technical University, Tulkarem, Palestine

J. J. Alawneh
Faculty of Administrative and Economic Sciences, Al-Quds Open University, Nablus, Palestine

O. Ala'yed
Faculty of Science and Information Technology, Jadara University, Irbid 733, Jordan

M. Al-Okaily (✉)
School of Business, Jadara University, Irbid 733, Jordan
e-mail: m.alokaily@jadara.edu.jo

A. M. A. Musleh Al-Sartawi et al. (eds.), *From the Internet of Things to the Internet of Ideas: The Role of Artificial Intelligence*, Lecture Notes in Networks and Systems 557,
https://doi.org/10.1007/978-3-031-17746-0_45

Numerous different events of near-meltdown in markets have happened. However, a wide range of political businesses to have been suggested failed, this circumstance, produced a lot of destruction to the rest of the world financial stability herewith caused the crisis to FOREX. With the global pandemic, the financial markets are witnessing a huge disruption with the revaluation and modification procedures in the world [9]. The pandemic has roughly limited the economic activity due to safeguarding measures. The sudden large shocks for the reason structural changes in the economy and financial markets have prospect impacts on efficiency markets. Since beginning the pandemic, the FOREX market has observed unparalleled activities, and therefore market is subject to close observation by investors and policymakers. Most central banks jostle to regulate the pecuniary framing to exchange rate movements in a bid to the monetary setbacks of the outbreak pandemic [9].

However, during this pandemic, the FOREX market has demonstrated volatility sharp. Whilst the current gross volatility levels are irregular, the effect of COVID-19 may differ between currencies based on their market stature and strategic response from governments. Regarding the COVID-19 influence on FOREX, a critical phase that needs to be taken into account is the efficiency of the FOREX, which is difficult to detected [6, 11]. However, the exchange efficiency rates changes with time, especially, over the crisis such conditions [12, 13]. This made the COVID-19 to be one of the hottest subjects of argumentation around the world, cause of its impact on the economy and the universal FOREX.

The crisis affects the showing and progress of countries everywhere the world and the economic events, such as natural disasters and the global monetary market [14]. Conversely, its effect on the FOREX has been much less argued [15]. During the crisis, the exchange rates of the currencies are affected in different ways. For instance, through the global financial crisis (2007–2009), the US dollar appreciated compare to some other currencies [16], which did not occur in the Asian crisis in (1997–1998) or the Russian crisis (1998–1999). Therefore, in this research, we aim to focus on the research question: *What is* the effect of COVID-19 pandemic *on* FOREX*?* Moreover, in order to detail the effect of COVID-19, European and US currencies which are the controlling currencies have been preferred in this research.

2 Case Study

Intending to analyze, we employed daily data from 15 main currencies involved in European and US countries. The data is recovered daily in 15 currencies from the 1st January, 2019, to 31st December, 2021. The data is retrieved from Sauder School of Business (2011) Pacific exchange rate facility (http://fx.sauder.ubc.ca/data.html*)* on a daily source. We presented the currencies symbols in detail. The Europe countries are the United Kingdom (GBP), Czech Rep (CZK), Hungary (HUF), Iceland (ISK), Poland (PLN), Russia (RUB), Sweden (SEK), Switzerland (CHF), Romanian (RON) and Norway (NOK), and then the America countries are Argentina (ARS), United States (USD), Brazil (BRL), Colombia (COP) and Chile (CLP).

Generally, there are 5 days trading in the week (20 days trading in the month). For base exchange, valuable metallic such as gold, platinum, and silver can be utilized [17–19]. The consequence could be impacted by the alterations of the values of these minerals. Atiany and Alawneh, [20] and Jang, Lee and Chang, [21] advised utilizing the Special Drawing Right (SDR) as a rule, which is practical freely on the exchanges of worldwide financial. Thus, here we use the SDR as the ruling exchange.

3 Methodology

Multivariate Statistical Process Control (MSPC) procedure is used to monitor the constancy test of numerous independent samples of correlation matrices at the same time. We construct the control chart to identify the stability or instability of currency exchange during the COVID-19 pandemic. In order to investigate the root sources of any out-of-control indicator, the network analysis is implemented [17] (Table 1).

Let m samples drown from p-varite normal distributions $N_k(\mu_1, \Sigma_1), N_k(\mu_2, \Sigma_2), \ldots, N_k(\mu_m, \Sigma_m)$ these samples are independent and the size $n_1, n_2, \ldots, n_m$, and Ω_i are denoted the correlation matrix of $i - th$ population where $i = 1, 2, \ldots, m$. Thus, to examine the hypothesis of two correlation matrices are equal $H_0 : \Omega_i = \Omega_0$ against $H_1 : \Omega_i \neq \Omega_0$. The Jennrich statistic [22], formula is come after:

Table 1 Countries and their symbols

Continent	Countries	Symbol
America	Argentine	ARS
	Brazil	BRL
	Colombia	COP
	Chile	CLP
	United States	USD
Europe	Switzerland	CHF
	Czech Rep	CZK
	Hungary	HUF
	United Kingdom	GBP
	Iceland	ISK
	Norway	NOK
	Poland	PLN
	Romania	RON
	Russia	RUB
	Sweden	SEK

$$\mathrm{J} = \sum_{i=1}^{m}\left\{\frac{1}{2}Tr(A_i{}^2) - (A_d)'U^{-1}A_d\right\} \quad (1)$$

where:

i. $A_i = n_i^{\frac{1}{2}} R_p^{-1}(R_i - R_p)$.
ii. R_i is the $i - th$ Sample correlation.
iii. R_p is the mean of all samples correlation matrices.
iv. $U = I + R_p * R_p^{-1}$, the * is Hadamard product of two matrices.
v. A_d is a diagonal of A_i.
vi. I_p is the identity matrix of size $(p \times p)$.

$Jennrich$ is distribute asymptotically as Chi-Square (χ^2) with degrees of freedom $df = \frac{1}{2}(m-1)p(p-1)$ where m is the number of dimensions. If J > $\chi^2_{\alpha,df}$ at the $(1-\alpha)^{th}$ quantile the hypothesis is rejected of Chi-square distribution.

4 Examining the Constancy of Correlation Constructing

In this section, we explore the constancy of correlation constructions, and next, we utilize the network analysis in followed part. To construct the compliant controls chart. We compute the Jennrich statistic every month from the 1st January 2019 to 31st December 2021 to realize the constancy (inconstancy) of the correlation structure.

To detect the correlation constancy, the J control chart was obtained and its rate UCL = 82.35374 in Fig. 1, was rejected if J since the statistic value falls external the UCL. This indication shows that there is an alteration in correlation constructions.

The horizontal axis is the month from January 2019 to December 2021 and the vertical axis is the value of the J test. From Fig. 1 there are signals of instability of correlation construction happening at the graph, we show that all J values fall outside of UCL, and this brings an indication about the history of method inconsistency. Since

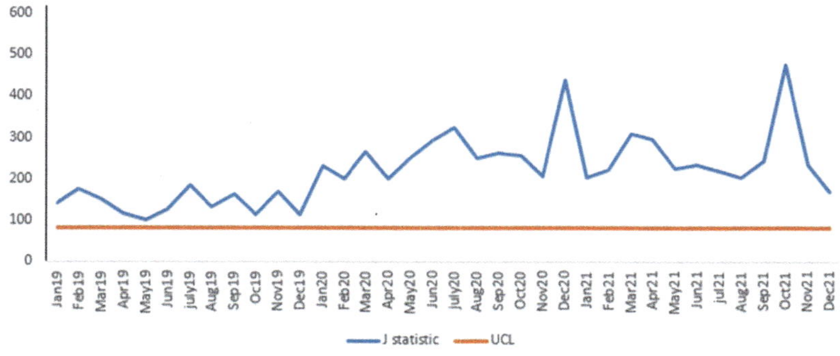

Fig. 1 J chart

Table 2 Calculation J statistic

Month	J statistic	Month	J statistic	Month	J statistic
Jan19	143.1546	Jan20	231.2202	Jan21	203.1926
Feb19	177.3670	Feb20	199.7874	Feb21	221.5407
Mar19	152.3504	Mar20	267.4439	Mar21	310.1795
Apr19	118.0698	Apr20	201.9094	Apr21	296.4476
May19	100.3608	May20	250.2481	May21	227.1382
Jun19	127.5618	Jun20	295.0379	Jun21	236.6150
july19	184.7934	july20	326.2193	jul21	219.0065
Aug19	133.4099	Aug20	252.0624	Aug21	203.7706
Sep19	162.5604	Sep20	264.6399	Sep21	245.1365
Oc19	115.1315	Oct20	256.2724	Oct21	478.1860
Nov19	170.2278	Nov20	206.1646	Nov21	235.5498
Dec19	115.0318	Dec20	439.8884	Dec21	171.3657

all the J values fall out of UCL, this means there is a special cause; the process is out of control (unsteady).

Next, we accomplished the correlation analysis to detect which is the most dominant currency (IES) out of the others by utilizing a minimum spanning tree. On behalf of that issue, the data structure of June 2021 compared with the out-of-control correlation matrix; the reference sample. In the following, the correlation diagnostics analysis was discussed.

5 Network Analysis

Primarily, network examination is a process of human social relations, established for assessing social connections to preferable research the connections among psychosomatic and gregarious structures [17]. This process is applied to resolve the proportional importance of nodes in the link [20]. Sharif, Yusoff, and Djauhari [23] recommended employing this approach as a basis causes examination in observing procedure inconsistency. For that objective, there are other processes that can be used principal component analysis and neural network. The Minimum Spanning Tree (MST) is a category of map currency that is applied in the pecuniary purse of overseas exchange average [17]. Furthermore, to filter the most vital evidence network analysis is an essential instrument for that issue [20]. For examining that network, originally the major evidence we strainer it via utilizing MST process, next, the details will be described by using the measure of centrality [17]. We explore the relationship of 15 exchanges by employing the network analysis, this procedure was created depending on matrix correlation; we computed the correlation matrix between all currencies, the nudle contains $\frac{15*(15-1)}{2} = 105$ links. We used Matlab version R2010a and Pajek software. To simplify the structure, let the currencies underneath study. After that, we convert the data into price returns logarithm exchange rate r_i:

$$r_i = \ln\hbar_i(t+1) - ln\hbar_i(t)$$

$r_i(t)$ is a daily exchange rate at the time $i = 1, 2, \ldots, 15$. Then, the correlation matrix is calculated from the formula:

$$\boldsymbol{\rho}_{ij} = \frac{\sum(\boldsymbol{r}_i\boldsymbol{r}_j) - \sum\boldsymbol{r}_i\sum\boldsymbol{r}_j}{\sqrt{(\sum\boldsymbol{r}_i^2 - (\sum\boldsymbol{r}_i)^2)(\sum\boldsymbol{r}_j^2 - (\sum\boldsymbol{r}_j)^2)}}$$

Next, convert matrix correlation to distance matrix to exploring the network by utilizing the following equation:

$$d_{ij} = \sqrt{2 \times (1 - \rho_{ij})}$$

where d_{ij} is an ultrametric distance among the ith and the jth characteristics, which fulfills the following properties.

(i) $d_{ij} \geq 0$ and $d_{ij} = 0 \Leftrightarrow i = j$ cause $\rho_{ij} = 1$
(ii) $d_{ij} = d_{ji}$ since $\rho_{ij} = \rho_{ji}$
(iii)$d_{ij} \leq d_{ik} + d_{kj}$, moreover $0 \leq d_{ij} \leq 2$

The correlation coefficient measures the degree of direct association among i^{th} and j^{th} stocks, where:

1- $\rho_{ij} = 1$, i.e. positive linear relationship
2-$\rho_{ij} = 0$, i.e. no relationship
3-$\rho_{ij} = -1$, i.e. negative relationship

To brief the foremost evidence we compute d_{ij} the distance matrix for 15×15 the concept of Sub-Dominant Ultrametric (SDU) was utilized for that issue which recommended by Mantegna and Stanley 1999. The SDU matrix 15×15 with δ_{ij} as the element in i^{th} and the jth column. Where $\delta_{ij} \leq d_{ij}$ for i and$j$, there are some things will be accomplished from SDU:

(i) $\delta_{ij} = \delta_{ji}$ for all i and j
(ii) $\delta_{ij} = 0 \Leftrightarrow i = j$
(iii) $\delta_{ij} \leq max\{\delta_{ik}, \delta_{kj}\}$ for all i and j.

We specified MST by utilizing an algorithm suggested by kruskal, [24] to deduce the network analysis and understand which is the importance and or effect of each currency to others the degree centrality was used [25]. The degree centrality can be denoted by way of a number of links that a specified currency has. Exactly, the degree of the nodule i is, $d_i = \sum_{i=1}^{p} a_{ij}$. The most dominant currency has a higher number of links [17].

6 Discussion and Result

TO clarify the evidence in Table 2, we present Fig. 2, with colour and size for significance. It shows in Fig. 2a, that COP and NOK have 4 links mean takes the maximum number of links (red colour) in the network. The following currency CZK has 3 links (orange colour), then ARS, PLN, RON, RUB, and SEK (green colour) have 2 links respectively. Finally, the following currencies (yellow colour) GBP, BRL, CLP, HUF, ISK, CHF, and USD are of 1 link. From Fig. 2a, we learn that NOK and COP take the maximum number of inks means are the most significant currencies where NOK controls RUB, SEK, CZK AND GBP and COP controls ARS, RUB, RBL, and CLP. In this investigation we determine the currencies excruciating into 2 big clusters where COP in America and NOK in Europe.

On another hand, Fig. 2b, October 2021 it shows that USD, CZK, HUF, and NOK has 3 links (orange colour) which are the highest number of links, and COP, ARS, PLN, and BRL has 3 links (green colour). While CLP, GBP, ISK, CHF, SEK, and RUB has 1 link (yellow colour). We can show from this figure there is one cluster controlled by CZK and RON serve as a key to the connectedness through RON and ISK. We make evaluation between reference sample and October 2021 to realize which currencies have alteration on their centrality degree it shows COP is very unstable since the links are changed from 4 to 2 (Table 3).

Based on Table 2, mean of weight centrality COP is the greatest important, next imperative (1.22–1.33) HUF, ARS, USD and CHF, the third imperative (0.95–1.1) GBP, CZK, BRL, CLP, SEK, RUB, NOK, and ISK. The last important currencies have a small average weight (0.71–0.86) PLN and RON. While, the most important currency for weight average in October PLN 1.80 the second important (1.22–1.27) CLP, ARS and COP, the third (1–1.07) CHF, USD and SEK, the fourth (0.90–0.96) ISK and BRL, lastly NOK, HUF, CZK and RON have the small average weights.

It shows PLN is the very unsteady currency in October 2021 since the mean is changed from 0.71 to 1.80 followed by GBP, HUF, and CZK.

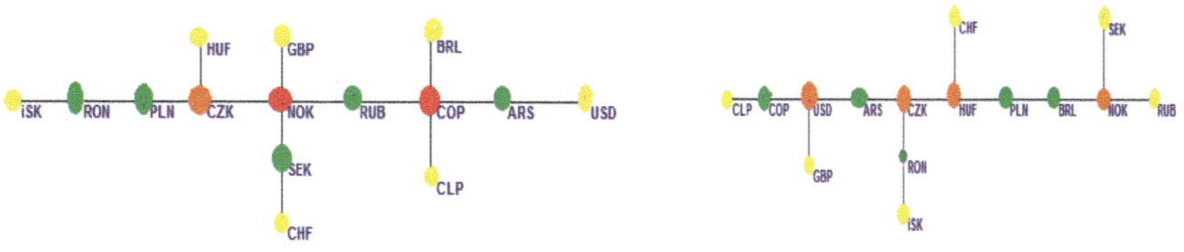

(1) (2)

Fig. 2 Degree of centrality **1** all data reference sample and **2** October 2021

Table 3 Degree and mean weight centrality

Label	All degree all data	Average	All degree october	Average
GBP	1	1.10	1	0.66
ARS	2	1.32	2	1.24
BRL	1	1.01	2	0.94
CLP	1	1.01	1	1.27
COP	4	1.90	2	1.22
CZK	3	1.03	3	0.76
HUF	1	1.33	3	0.83
ISK	1	0.98	1	0.96
NOK	4	0.99	3	0.86
PLN	2	0.71	2	1.80
RON	2	0.86	2	0.70
RUB	2	0.99	1	1.00
SEK	2	1.00	1	1.01
CHF	1	1.22	1	1.07
USD	1	1.31	3	1.02

7 Conclusion

In eras of the COVID-19 pandemic, there are significant changes happen to the procedures of the business and financial organizations around the world. In order to support the economy, central banks reduced their interest rates [26, 27]. Furthermore, the international trading has been seriously impacted by this pandemic. As a result, the exchange rate systems have significantly changed through the pandemic.

In this research, we display the MSPC process to apprehend the stability or instability of the currencies of US and European countries during the COVID-19 pandemic. To monitor the variability procedure, we construct a J control chart. If there is an out-of-control indication happens, network analysis can be performed which identifies the current procedure inconsistency to understand the status. However, in this research, we clarify the network analysis for October 2021 since the value of the J statistic is the largest value. To simplify the specifics, the degree of centrality and the mean weight was used. The biggest significance is the more impressive of the individual currency on others. From the discussion above, we can conclude that there are three currencies that are unstable these currencies are COP, HUF, and USD.

References

1. Alsaad A, Al-Okaily M (2022) Acceptance of protection technology in a time of fear: the case of Covid-19 exposure detection apps. Inf Technol People. https://doi.org/10.1108/ITP-10-2020-0719
2. Al-Bashayreh M, Almajali D, Altamimi A, Masa'deh RE, Al-Okaily M (2022) An empirical investigation of reasons influencing student acceptance and rejection of mobile learning apps usage. Sustainability 14(7):4325
3. Al-Okaily M (2021) Assessing the effectiveness of accounting information systems in the era of COVID-19 pandemic. VINE J Inf Knowl Manag Syst. https://doi.org/10.1108/VJIKMS-08-2021-0148
4. Al-Okaily M, Al-Okaily A (2022) An empirical assessment of enterprise information systems success in a developing country: the Jordanian experience. TQM J. https://doi.org/10.1108/TQM-09-2021-0267
5. Taamneh A, Alsaad A, Elrehail H, Al-Okaily M, Lutfi A, Sergio RP (2022) University lecturers acceptance of moodle platform in the context of the COVID-19 pandemic. Glob Knowl Memory Commun. https://doi.org/10.1108/GKMC-05-2021-0087
6. Al-Okaily A, Al-Okaily M, Teoh AP (2021). Evaluating ERP systems success: evidence from Jordanian firms in the age of the digital business. VINE J Inf Knowl Manag Syst. https://doi.org/10.1108/VJIKMS-04-2021-0061
7. Abdoul-Azize HT, El Gamil R (2021) Social protection as a key tool in crisis management: learnt lessons from the COVID-19 pandemic. Glob Soc Welfare 8(1):107–116
8. Qudah AA, Al-Okaily M, Alqudah H (2022) The relationship between social entrepreneurship and sustainable development from economic growth perspective: 15 'RCEP' countries. J Sustain Finan Invest 12(1):44–61
9. Aslam F, Aziz S, Nguyen DK, Mughal KS, Khan M (2020) On the efficiency of foreign exchange markets in times of the COVID-19 pandemic. Technol Forecast Soc Chang 161:120261
10. Al-Qudah AA, Hamdan A, Al-Okaily M, Alhaddad L (2022). The impact of green lending on credit risk: evidence from UAE's banks. Environ Sci Pollut Res. https://doi.org/10.1007/s11356-021-18224-5
11. Katusiime L, Shamsuddin A, Agbola FW (2015) Foreign exchange market efficiency and profitability of trading rules: Evidence from a developing country. Int Rev Econ Financ 35:315–332
12. Levich R, Conlon T, Potì V (2019) Measuring excess-predictability of asset returns and market efficiency over time. Econ Lett 175:92–96
13. Al-Okaily A, Ai Ping T, Al-Okaily M (2021) Towards business intelligence success measurement in an organization: a conceptual study. J Syst Manag Sci 11(2):155–170. https://doi.org/10.33168/JSMS.2021.0210
14. Longstaff FA (2010) The subprime credit crisis and contagion in financial markets. J Financ Econ 97(3):439–450
15. Melvin M, Taylor MP (2009) The crisis in the foreign exchange market. J Int Money Financ 28(8):1317–1330
16. Kohler, M. (2010). Exchange rates during financial crises. BIS Q Rev 39–50
17. Atiany TA, Sharif S (2015) Asia-Pacific currencies structure aftermath tohoku earthquake. Res J Appl Sci Eng Technol 11(2):215–220
18. Al-Okaily M, Alghazzawi R, Alkhwaldi AF, Al-Okaily A (2022). The effect of digital accounting systems on the decision-making quality in the banking industry sector: a mediated-moderated model. Global Knowl Memory Commun. https://doi.org/10.1108/GKMC-01-2022-0015
19. Mizuno T, Takayasu H, Takayasu M (2006) Correlation networks among currencies. Physica A 364:336–342
20. Atiany TA, Alawneh J (2019) Influence of war on currency exchange using jennrich's statistics. Int J Recent Technol Eng (IJRTE) 8(4):10939–10942. https://doi.org/10.35940/ijrte.D7298.118419

21. Jang W, Lee J, Chang W (2011) Currency crises and the evolution of foreign exchange market: evidence from minimum spanning tree. Physica A 390(4):707–718
22. Jennrich RI (1970) An asymptotic χ2 test for the equality of two correlation matrices. J Am Stat Assoc 65(330):904–912
23. Sharif S, Yusoff NS, Djauhari MA (2012) Network topology of foreign exchange rate. Mod Appl Sci 6(11):35
24. Hannoon A, Abdalla Y, Al-Sartawi A, Khalid A (2021) Board of directors composition and social media financial disclosure: the case of the United Arab Emirates. In: Al-Sartawi AMAM (eds) The big data-driven digital economy: artificial and computational intelligence. Studies in computational intelligence, vol 974. Springer, Cham
25. Abbasi A, Altmann J (2011) On the correlation between research performance and social network analysis measures applied to research collaboration networks. In: Proceeding of the 44th Hawaii international conference on system sciences (HICSS), pp 1–10
26. Al-Okaily A, Al-Okaily M, Teoh AP, Al-Debei M (2022). an empirical study on data warehouse systems effectiveness: the case of Jordanian banks in the business intelligence era. EuroMed J Bus. https://doi.org/10.1108/EMJB-01-2022-0011
27. Al-Okaily M, Alalwan AA, Al-Fraihat D, Rehman SU, Alkhwaldi AF, Al-Okaily A (2022). Investigating antecedents of mobile payment systems decision making: a mediated model. Glob Knowl Memory Commun. https://doi.org/10.1108/GKMC-10-2021-0171

Intrinsic Factors Influencing the Accounting Information Systems' Effectiveness in Jordanian Listed Companies

Hamza Alqudah, Manaf Al-Okaily, Abdalwali Lutfi, Malek Alshirah, Aws Al-Okaily, Mahmoud Al-Kofahi, and Hamzah Alqudah

Abstract The purpose of the current study is to investigate the effect of intrinsic factors (staff qualifications, data quality, internal control and company size) on the effectiveness of Accounting Information Systems (AISs) in the Jordanian Listed Companies (JLCs). This study used 141 responses from financial managers of JLCs to measure the studied variables, whereas the annual report of each company was used to measure the company size. By following a specific procedure for developing and distributing the questionnaire, the collected data was then analyzed using Statistical Package for the Social Sciences (SPSS). Using the descriptive analysis and multiple regression the study hypotheses have tested. The findings indicate that all factors have a positive effect on the effectiveness of AIS in JLCs except the company size which found insignificant in this study. These findings are necessary for the regulators and decision-makers in advancing regulation and legislation for the AISs in the JLCs. It is pointed out that the studied intrinsic factors in this study can evidently improve the AIS effectiveness to be able to achieve the demanded objectives.

Keywords İntrinsic factors · Staff qualifications · Accounting information systems · Company size · Jordanian listed companies

---

H. Alqudah · A. Al-Okaily
Faculty of Administrative and Financial Sciences, Irbid National University 2600, Irbid, Jordan

M. Al-Okaily (✉)
School of Business, Jadara University 733, Irbid, Jordan
e-mail: m.alokaily@jadara.edu.jo

A. Lutfi
Department of Accounting, School of Business, King Faisal University, Al-Ahsa 31982, Saudi Arabia

M. Alshirah
Department of Accounting, Al Al-Bayt University, AL-Mafraq, Jordan

M. Al-Kofahi
Graduate School of Business, Universiti Sains Malaysia, Penang, Malaysia

H. Alqudah
School of Engineering, Auckland University of Technology, Auckland 1010, New Zealand

A. M. A. Musleh Al-Sartawi et al. (eds.), *From the Internet of Things to the Internet of Ideas: The Role of Artificial Intelligence*, Lecture Notes in Networks and Systems 557,
https://doi.org/10.1007/978-3-031-17746-0_46

1 Introduction

Nowadays organizations are working and racing in the information age. The information has converted to the major resource of most economies, organizations and societies [1–6]. The information systems are broadly utilized by most organizations to automate their operation and to improve the efficiency performance of their activities [7–11]. Accounting Information System (AIS) is one of an essential part of the modern information systems [12]. AIS is a framework that can convert economic data into financial information which promotes the process of decision-making [13, 14]. The advancements in the fields of information systems, accounting, and information technology during the last decades of twentieth-century have extended the AISs' range and roles [15, 16].

Jordan, like other developing countries worldwide is affected by the technological development [14, 17, 67]. Despite, the differences in the degree of this development are relatively, but it noted that most companies and large enterprises are using computer technology in the implementation of their financial and administrative as well as operational processes [18]. Companies that listed on the Amman Stock Exchange are the most important parts in the Jordanian economy, due to their importance in providing jobs and supporting the Jordanian economy [1, 19]. Hence, this companies need a dramatically technology, including AISs to facilitate their works [20].

AIS has become one of the main elements of the Jordanian organizations, by assisting them in obtaining, classification, processing, analyzing, and reporting appropriate financial information to make proper decisions within the organization and for outside parties too [7, 21]. Moreover, AISs have helping managers by providing them with the information needed to perform the managerial functions. Several authors demonstrate that AIS effectiveness rely on multiple factors, such as intrinsic or extrinsic factors [22–24]. To the best of our knowledge, most of extrinsic factors have been addressed in the previous studies to measure the adoption or usage of AIS, such as technological, organizational and environmental factors [24–26]. However, few studies have focused on the intrinsic factors (e.g., staff qualifications, data quality, internal control and company size) that potentially play a vital role in supporting the effectiveness of AISs [17, 22, 23, 28, 29]. Therefore, research to understand the intrinsic factors that affecting the AIS effectiveness in JLCs is very much needed.

In the context of our study, the presence of qualified staff plays an important role in raising the level of AIS performance [18, 30]. Thus, the ability of AISs' characteristics to explain business performance rely on the accounting staff competency level. Wherein, the competency of accounting personnel is an imperative factor in the AISs success [31]. Besides, Nwokeji [29] confirmed that, the staff qualification is essential for the AISs success since the qualified accountant can assure the quality of the provided data (inputted data) to improve the company performance. Data quality is another important factor to the operational and decisional processes, whereby it enables companies to proactively react and provide desired services and products.

Thus, data quality influences organizations' ability to function and increase their performance. Executing data quality policies at all levels means that suitable information is available to the appropriate person at the proper time and in the right format [32].

Large companies have also a high possibility of utilizing AIS due to the huge transactions and complicated nature of their business compared with small ones [15, 33–35]. The size of the company refers to its capability to make the developments for the AISs [36, 37]. Thus, when the company has financial and technical resources that needed will contributes in raising the AIS effectiveness. Further, the need to protect unit assets, and to produce accurate statements for the purpose, have increased the importance to have a solid internal control system [7, 22], that comprised of policies and procedures developed for the purpose of ensuring satisfaction to the management that the objective of a company has been achieved [17, 38].

Based on aforementioned, the present study seeks to investigate the effect of the intrinsic factors (i.e., staff qualifications, data quality, internal control and company size) on the effectiveness of AIS among JLCs. Wherein, its dealing with a large and important sector in the economy of Jordan [45], as well as it addressed such significant issue. In addition, to the scarcity of studies pertaining to this subject in the Jordanian environment [17, 22], and with the tough competition in markets at all levels, (i.e., local regional, and international) in the application of modern technology [39], it is essential to consider the role of staff qualifications, data quality, internal control, and company size in enhancing AISs effectiveness in JLCs. The current study may present a new contribution to the literature in this subject.

2 Literature Review

2.1 Accounting Information Systems

AIS defines as a correlated elements that collect, processes, stores and distributes information to promote decision making within organizations [22]. While, Bodnar and Hopwood [40] stated that the AIS as a set of resources (i.e., people, equipment) provided to convert financial data and other raw data into accounting information. AISs contains a set of business applications employed in conducting broad business functions, e.g., human resources management, financial analysis, stock management and accounting [41]. The comprehensive basic nature of the AIS is to computerized business processes and most importantly for real-time data production [9, 42]. AIS use computer-based subsystems to link company departments, e.g., sharing information among purchases, control and accounting departments [15, 43].

Accordingly, AIS is responsible for elaborating the financial information, as well as refining the obtained information from the collection and handling of transactions' data [37, 44]. Furthermore, AIS represents a subsystem that processes financial/non-financial transactions that directly influence the financial transactions [17, 29]. In line with Nwokeji [29], Etim [21] stated that, the information systems play a key role in the survival continuous and successful of organizations. It produces reports that contain operational, financial and non-financial, and information about external operations, activities and the necessary conditions to enable decision-making process. Hence, possessing such information system could help companies' managements and the other authorities to accomplish the tasks and make decisions concerning the fulfilment of assigned goals [22]. For example, Susanto [46] stated that the AIS is a set of both physical/non-physical components that are interconnected and cooperation among each other, to treat the data of transactions that associated with financial problems. Whereas, Prihatni et al. [37] mentioned that, AIS is a critical system of reporting that produces quality accounting information for managers at all organizational levels that are utilized in creating useful decisions. AIS provide an opportunity for business to improve their efficiency and effectiveness in decision making thereby allowing firms to gain competitive advantage [9, 26, 47].

2.2 Accounting Information Systems' Effectiveness

The AISs effectiveness indicates to the observed satisfaction of information users concerning decisions and controls that have produced from AIS, through coordination and control over the produced information within the company [21, 47, 48]. An effective AIS can enhance the confidence of financial statements [6, 22, 49, 50]. In this vein, the effective AISs usually assist managers to understand their tasks more explicitly and overcome uncertainty while making their decisions [20, 64]. To estimate the effectiveness of AISs there are various dimensions, wherein according to DeLone and McLean [10] and Nicolaou [48], the systems' effectiveness can be demonstrated by varied viewpoints. In the present study, the decision-makers' satisfaction of the information that provided by the AISs will be used to measure the effectiveness of AIS, besides timeliness, information scope and aggregation that presented by the AISs.

2.3 Staff Qualifications

In Accordance with Public Accounting Act / Public Accounting Standards, the accountants are obliged to have sufficient education [51]. Indeed, the accountants require an appropriate level of education commensurate with their works [52]. An employee's education is a very important factor that needs to be considered before hiring an accountant or any employee in the accounting department, in such job that demanding high integrity, skilled and capability [31]. The accountancy education for the employees is essential for the correctness of company financial reporting presentation [37]. The knowledge obtained either by "formal education, previous experience, or informal training" are key elements in a successful venture [53]; they are parts of the educational processes which intended to enhance the abilities and distinct skills of the AISs users with a company [37].

According to a study by Al-Hanini [55], one of the obstacles facing the AISs in the Jordanian banks is the employees' lack of experience in managing the security system, wherein Al-Hanini indicated that, new employees are not trained on handling the protecting means of the accounting system were they begin their work, also there was no suitable employment system to hire the appropriate person in the proper place. In this vein, the Jordanian banks have recommended holding continuous training programs to both old and new employees. These programs could present to the employees the significance of their commitment to the control procedures concerning the safety and security of the AISs and train them to use protection procedures.

2.4 Data Quality

The growing dependence of business on AIS to fulfil their purpose in the current information era requires a strategic and proactive approach to data quality management [32]. Information and data are key ingredients of all activities [22, 56]. The data are the key driver of the material acquisition process, preparing estimates, billing, and budget process [29]. Illustrating the value of data, Al-Hakim [57] declared that, data and information have become a strategic requirement for an organization's prosperity and future success, such as oxygen to human life. Data quality is fundamental in information systems [8, 58]. The quality of AISs relies on the quality of inputs, processes, and outputs. The inputs of AISs are the data that associate to financial transactions that properly documented [32, 59]. If data are not correct and hold a lot of errors and discrepancies, hence the information quality declines and then becomes unmeaning [17].

Rahayu [60] measured data quality by applying the following dimensions: accuracy (ensuring correct procedures, consistency), validity (existence, recording, and integrity), management expression (expectation of management, satisfaction), and simplicity (not posing much of perception, no redundancy). Data accuracy refers to the valid or error free condition with respect to some known, designated, or measured value [60]. Validity is a concept like truth, which represents an ideal or desirable situation [58]. Management expression denotes that the inputted data are in accordance with the management's expectation and in line with the required information. Simplicity of the data indicates that data are not redundant and generate different perceptions on each information user [60]. Thus, this study focuses on the dimensions used by Rahayu. The objectives of this study will be attained if the above dimensions are perceived.

2.5 Internal Control

The system of internal control has defined as a collection of procedures designed, to supply the management with plausible assurance that the accounting of financial information that presents by AISs are credible and timely available [61]. The system of internal control can be as a security tool for management in all related company transactions. Further, internal controls encompass all procedures utilized by the organization managements to assist in attaining their goals. These goals including e.g., ensuring the efficiency of the business, compliance with management policy, prevention the error and cheating, the integrity and correctness of the accounting documents, asset preservation, and prepare reliable and timely financial reports [62]. An effective internal controls system can prove the suitability of the inputted data, storage techniques, processing methods, and finally the efficiency of produced information. In this vein, there is a necessary need to have a solid internal controls system in the electronic data entry units. Hence, unless specific rules and practices are installed for data entry, the discovering errors' chances will be affected since the electronic systems will not be able to recognize errors [61]. There are three main goals behind preferring managements to create a solid internal control system, these goals are: financial statements accuracy, financial operations effectiveness, and the acquiescence with the assigned regulations and laws [63]. However, if the internal control systems keep pace with technological development and have the ability to detect tampering and prevent its occurrence, then the accounting information will be easy to tamper with and accounting information and difficult to detect [61]. Thus, it important to have good internal controls and pay attention to information integrity.

Mndzebele [25] announced that there is a crucial association between the effectiveness of the AISs and the effective internal controls system. Whereby, AISs gives companies timely and accurate information and a clarity internal controls system. Meanwhile, AISs are also used in internal control to have more timeliness, accuracy, convenient and cost-effectiveness. Accordingly, it could be announced internal controls affect AISs effectiveness.

2.6 Company Size

Company size refers to its capability to make the developments for the AISs [22], there are many ways to measure company size such as by revenue size, capital size and board of director's size [41]. But in this study the authors will using capital size as tool to measure company size. The size of a company could provide the financial sources needed to obtain a new technology, activity, technique, or innovation and spending on them [36]. A large company is more likely to invest a big volume of capital, time and learning to adopt new technologies or systems, than a small firm [9, 30]. On the other hand, the small company does not possess funds to spend on the information technology and also does not possess the support to choose the fit technology [17, 65]. Specifically, several scholars found that the limited capital and resources have affected the training and the upgrading of the effectiveness of AISs in the small and medium enterprises [37, 65], as they claimed that in those firms the computers are employed only for the purpose of producing sales report and printing out invoices. Edison et al. [9] argued that AISs use is not linked with company size. But Prihatni et al. [37] affirmed that the company size determines its capability to make developments on AISs. Consequently, it is anticipated that larger companies that have large financial resources are more probable to reach advanced degrees of AISs effectiveness [36]. Thus, the capital size of the companies will contribute to the effectiveness of AIS.

Table 1 present the operational definitions and measurements of each variable has used in the present study:

Table 1 Variables operational definition and measurements

Types of Variables	Variables	Operational Definition	Measurements
Dependent Variables	AISs effectiveness	The satisfaction of decision-makers that the output information available to them meets their requirements, information scope, timeliness and aggregation of information that produced by AIS. The items in the questionnaire were adopted from Nicolaou (2000) and Sajady (2008)	1. The outputs of AIS are presented in a useful format 2. The outputs of AIS are presented in clear information 3. The usage of AIS provides more data accuracy 4. The AIS provide sufficient information 5. The AIS provide up-to-date Information 6. The AIS is easy to use 7. The information provided by AIS is available at any time 8. Decision-makers have satisfaction from the information content provided by AIS
Independent Variables	Staff qualifications	Academic background of staff (i.e., level of education, professional qualifications of staff). The items was dopted from Alsharayri (2012)	1. The staff who deals with AIS are fully equipped with knowledge and AIS application 2. The company is recruiting people with appropriate qualifications 3. Levels education of staff contributes to the effectiveness of AIS 4. The professional qualification of the staff plays an important role towards AIS effectiveness in the company 5. All the staffs in the company have specializations related with accounting systems such as (AIS)

(continued)

Table 1 (continued)

Types of Variables	Variables	Operational Definition	Measurements
	Data quality	The quality of data includs: accuracy (ensuring correct procedures, consistency), validity (existence, recording, and integrity), management expression (expectation of management, satisfaction), and simplicity (not posing much of perception, no redundancy). By the items in the questionnaire which adopted from Rahayu (2012)	1-All inputted data really exist and are in accordance with the facts 2-All data are tape recorded 3-All data are securely inputted 4-All data are inputted with correct procedures 5-All data inputted into the AIS are in line with the needed information 6-All data inputted into the AIS can satisfy the information users 7-All data inputted into the AIS are not redundant
	Company size*	The size of capital for each company using an annual report. The item was adopted from Prihatni et al., (2012)	the capital was used to measure company size using annual report for each company
	Internal Control	Consists of rules, policies and procedures that followed in the company. The items in the questionnaire were adopted from Neogy (2014)	1. The internal control quality will help to effectiveness of AIS 2. The development of an internal control in AIS will help the managements to have complete information 3. The goal of an internal control in companies is to assure that all transactions are recorded in accurate numbers, in appropriate accounts, and in proper time 4. Monitor of operations and performance can help users understand the requirements of AIS 5. The development of an internal control in AIS will help the companies to keep financial data accuracy 6. The internal control quality will help the companies to keep consistency of financial data 7. The internal control quality will help the companies to provide up-to-date Information

3 Research Model and Hypotheses

Using the variables presented in the previous section, the research model is depicted in Fig. 1. Hypothesis testing would be carried out to achieve the study's objective, i.e., to determine the effect of intrinsic factors (i.e., staff qualifications, data quality,

Fig. 1 Research model

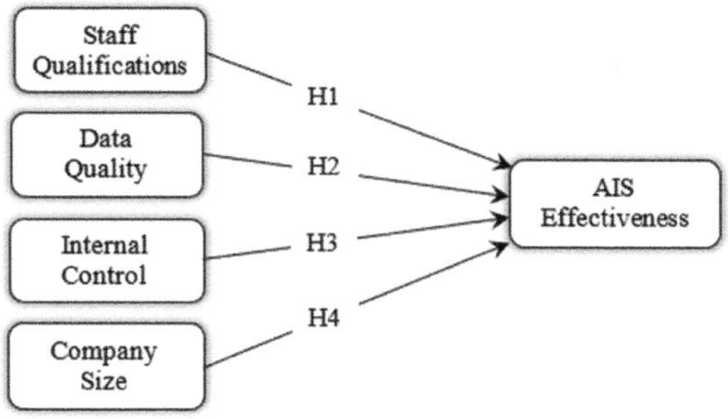

internal control and company size) on the effectiveness of AISs in the JLCs. Based on the literature review, the following hypotheses are proposed:

H1: *The AIS effectiveness is affected positively by staff qualifications in the JLCs.*
H2: *The AIS effectiveness is affected positively by data quality in the JLCs.*
H3: *The AIS effectiveness is affected positively by the internal control in the JLCs.*
H4: *The AIS effectiveness is affected positively by the company size in the JLCs.*

4 Data Analysis and Findings

The present study utilized a five-point Likert scale to measure the variables' items. 4.1The Validity and Reliability of the Measurements, and Descriptive statistics have conducted.

4.1 The Validity and Reliability of the Measurements

For Validate the instrument, the questionnaire was sent to a group of experts with specialization and academic experience in AIS in a number of Jordanian universities, their number is (6) experts, for the purpose of expressing their opinion on it and ensuring its suitability, in terms of the number of paragraphs, clarity, accuracy and coherence, and any other observations they consider it appropriate, by deleting, changing or adding, and considering this review process and the subsequent correction and modification for most of them as a test of the apparent validity of the instrument.

For reliability of the instrument, it is highly encouraged for the reliability coefficient to be closed to 1.0 as it will be better, and the values that are over 0.80 are considered good. Values equal to 0.70 is taken as acceptable, and those in the range of 0.7 are acceptable and those over 0.8 as good [14]. The reliability was established by testing for both consistency and stability [14]. So, based on the result of Cronbach's Alpha for variables items (independent variable, dependent variable), indicated that

Table 2 Summary of reliability

Ser	Variables	Cronbach's Alpha
1	AIS Effectiveness	0.889
2	Staff Qualification	0.786
3	Data Quality	0.827
4	Internal Control	0.774
5	Company size	-*

* The authors used annual report to measure the company size

all the items of variables are reliable (refer to Table 2), where all the items of variables have value more than 0.7, which supported by Cavana et al. [14], an alpha coefficient of 0.7 or higher is accepted.

4.2 Descriptive Statistics

Descriptive statistics shows the quality of the data utilized, which allows to continue with the proper methodology for valuation [27]. In this vein, the scores of the mean for less than 2.5 will be treated as low and for scores of the mean that between 2.5 to 5 as moderate, whereas for 5 as high, as recommended by Hair et al. [66].

As shown in Table 1, all variables are evaluated based on a 5-point scale. Where, the results show that the mean on AIS effectiveness (3.63), that is mean AIS provides accuracy information and sufficient. Therefore, decision-makers have satisfaction from the information content provided by AIS. While staff qualifications (3.34), that is meaning the staff who deals with AIS are fully equipped with knowledge and the companies are recruiting people with appropriate qualifications, also has possessed appropriate staff expertise especially those who deal with the AIS, so that contributes to the effectiveness of AIS. Internal control (3.48) that is meaning the procedures designed to provide management with reasonable assurance that the accounting information that produces by AIS is credible and available timely. Therefore, the internal control procedure plays an important role in improving the information that offered by AISs within the JLCs. Data quality (3.25) that is meaning the data entered in the system is accurate, usable, reflect management's expectations and satisfaction, and it is characterized by simplicity, thus contributes in increasing the effectiveness of AIS. Finally, Table 1 showed also the mean score of company size (capital size) are 33,244,468 JD with a maximum and minimum capital of 1,100,000,000 JD and 500,000 JD respectively.

4.3 Normality

Normality indicates that data have a symmetrical distribution exhibiting a shape that forms a bell curve (Pallant, 2007). Normality is a required assumption in multivariate analysis and in most statistical tests because any fundamental violation of this assumption may lead to unreliable findings (Hair et al., 2010). Skewness is a measure that illustrates to what range a data distribution has deviated from the centre of the mean (George & Mallery, 2006). The findings in Table 4 show that the skewness values of the study variables were within the range of + 3 and −3, indicating that the data have a normal distribution. Kline (2016) recommended that it should only be a problem when the absolute value for skewness exceeds three. Kurtosis is a measure that shows whether the data group is flat or peaked relative to a normal distribution (Hair et al., 2014). The values of kurtosis can be either positive or negative, and positive values indicate that distribution is flatter whereas negative values indicate a peaked distribution (George & Mallery, 2006). In the present study as demonstrated in Table 4, the kurtosis test was conducted for all variables and it fell within the ideal suggested range (+7 and −7) as Curran, West, and Finch (1996) recommended. Therefore, the data have no problem with kurtosis, indicating the data is normally distributed.

4.4 Multi-collinearity Test

The variance inflation factor (VIF) test and the tolerance test were used to ensure that there is no multi collinearity problem among study variables as shown in Table 5.

According to Hair et al. (2014a), multicollinearity is not an issue when the value of VIF is less than five and tolerance is above 0.20. The findings reported in Table 5 indicates that there was no multicollinearity among the latent variables as all tolerance values ranged between 1.402 and 1.657, which were all higher than 0.20, and VIF values ranged from 2.946 to 3.485, which were below five. Therefore, it can be concluded that there was no multicollinearity among the independent variables.

4.5 Hypotheses Testing

In this study, a simple regression analysis has been conducting to test the research hypotheses, regarding the effectiveness of AISs, in general, besides the decision-makers' satisfaction on the information that produces by AISs, as well as information aggregation, scope and timeliness that provided particularly by the AISs. Table 3 shows the significance level of the variables' effect on the AIS effectiveness. The model validity has evaluated by the value of R^2 and the structural paths [66, 68]. The

Table 3 Descriptive statistics

Construct	Mean	Standard deviations	Rank
AIS Effectiveness	3.631	0.75332	1
Staff Qualification	3.344	0.88771	3
Data Quality	3.247	0.79028	5
Internal Control	3.481	0.79028	2
Company Size	3,324,470	10,152,412	4

Table 4 Normality test

Variables	Skewness	Kurtosis
AIS effectiveness	0.220	0.385
Data quality	0.380	0.339
Staff qualifications	0.173	0.328
Internal control	0.265	0.238

Table 5 Multi-collinearity result

	Tolerance	VIF
(Constant)		
Data quality	1.514	2.946
Staff qualifications	1.519	2.928
Internal control	1.657	2.523

R^2 value shows that approximately %52 of the variance in the AISs effectiveness is considered by the anticipated framework, which displays moderately predicted by its predictors [66].

Table 6 displays the multiple-regression coefficient (R) of exogenous variables to the AISs effectiveness as the endogenous variable. As shown in Table 3, the factors i.e., staff qualification, data quality and internal control have a significant effect on the AIS effectiveness at $p < 0.05$. In regard to beta value, the highest beta coefficient was to the staff qualification, at Beta = 0.35. This finding implies that staff qualifications obtained the most effective contribution to explain the AIS effectiveness. Where, the regression analysis results (Beta = 0.35, $p < 0.05$) indicates a positive statistically significant effect between staff qualifications and AIS effectiveness. That is mean the staff qualifications in JLCs play a significant role in increasing the effectiveness of AIS. This result is also in line with studies by Prihatni et al. [37] and Alsharayri [54]. The presence of qualified staff plays an important role in raising the effectiveness of AIS. Furthermore, the educational levels of accountants are crucial in the comprehension and application of AIS, and thus to accuracy of corporate financial reporting presentation. Therefore, companies are required to recruit people with appropriate qualifications and education. Therefore, a significant relationship is observed between the academic background of staff and AIS, which is consistent

Table 6 Summary of multiple regression analysis

Ser.	Variables	Beta	Significance level
1	Staff Qualification	0.350	0.015
2	Data Quality	0.142	0.013
3	Internal Control	0.241	0.000
4	Company Size	0.019	0.638

with those reported in prior studies by Prihatni et al. [37] and Alsharayri [54]. Thus, **H1** is accepted in this study.

The second-highest beta coefficient was to the internal control, at Beta = 0.241. This result (Beta = 0.241. $p < 0.05$) means that internal control obtained a strong contribution to explain the AIS effectiveness. Besides, the third beta value is seen in data quality. Where the regression analysis results (Beta = 0.142, $p < 0.05$) indicates a positive statistically significant effect between data quality and AIS effectiveness. This result is in line with the study of AL Shbiel and Bani Ahmad [71]. The items in internal control (roles, policies, procedures in the company) would help the companies to increase the effectiveness of AIS in the companies listed in ASE.

This study also proved that internal control has a positive (0.241) effect on AIS effectiveness, with significance value of 0.013, which is less than 0.01. When a company has a good internal control system (i.e., the rules, policies, and procedures in the company), all of these components lead to the increase of AIS effectiveness, by the decision-makers' satisfaction on information produced by AIS, as well as scope, timeliness, and aggregation of information produced by AIS. Thus, contingency theory supports the relationship between internal control and AIS effectiveness [62]. This result is in line with the studies by Neogy [62], Mndzebele [25] and AL Shbiel and Bani Ahmad [71]. With good internal control in any organization, management can use the information with greater reliance to properly maintain their business activities and provide AIS. However, if internal control is weak, management cannot achieve its goal. Thus, a significant relationship between quality of internal control and AIS exists [25, 62, 71]. Thus, **H2** is accepted in this study.

This study also proved that data quality has a positive (0.142) effect on AIS effectiveness, with significance value of 0.000, which is less than 0.01. When a company has a good quality of data (i.e., accuracy (ensuring correct procedures, consistency), validity (existence, recording, and integrity), management expression (expectation of management, satisfaction), and simplicity (not posing much of perception, no redundancy) [60], all of these components lead to the increase of AIS effectiveness, by the decision-makers' satisfaction on information produced by AIS, as well as scope, timeliness, and aggregation of information produced by AIS. Thus, contingency theory supports the relationship between data quality and AIS effectiveness [62]. This result is in line with the studies by Rahayu [60], Al-Hakim [57] and Xu [32]. With good quality of data in any organization, management can use the information with greater reliance to properly maintain their business activities and provide AIS. However, if quality of data is weak, management cannot achieve its goal. Thus,

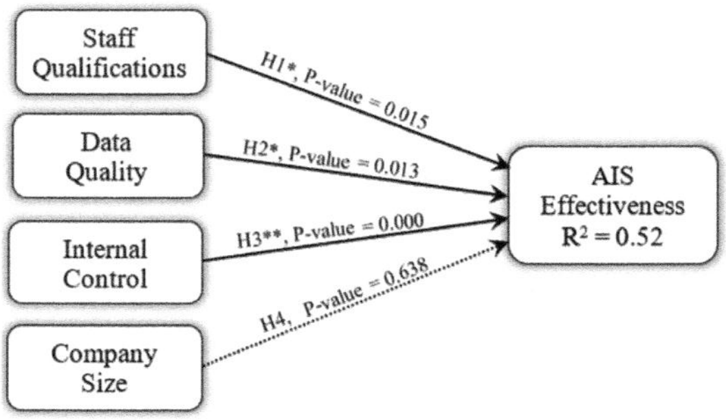

Fig. 2 Summary of hypotheses testing

a significant relationship between quality of data and AIS exists [25, 62, 71]. Thus, **H3** is accepted in this study.

While, the result shows that the company size has insignificant effect on AIS effectiveness. Where, the regression analysis results (Beta = 0.019, $p > 0.05$) indicates insignificant effect between company size and AIS effectiveness. Where, the company size determines the ability to adopt and develop AIS. The size of a company provides the financial resources needed for acquiring a new technology and develop it, and enables spending on innovative activities [9, 37]. However, the findings of this study have not supported the prior studies' results. That is meaning there is no effect of company size to increasing the effectiveness of AIS in the JLCs. Ismail and King [28] study and Edison et al. [9] study. Where, postulated Ismail and King [28] that larger firms would achieve alignment AIS better than smaller firms. Interestingly, found that companies that have achieved AIS alignment were smaller when compared to the companies that have not achieved AIS alignment. This finding contradicts the common belief that firm size would contribute to AIS alignment. Thus, **H4** is rejected in this study. One explanation of this results may be due to the different context and conditions of the current study population. Figure 2 present the research model with the P-values of each factor affecting AIS.

5 Theoretical and Practical Implications

The AISs effectiveness is intimately associated with the company's performance, whereby AISs gives information that is, accurate, usable, reflect management's expectations and satisfaction. The information provided AISs are also characterized by simplicity for the businesses, and usually are associated with the administrative decisions under processing to make the right decision [72, 73].

Some unique theoretical and practical implications are revealed in this study. For theoretical implication, in the previous studies on the effectiveness of AIS, scholars have given attention to the factors influencing the effectiveness of AIS in general from different perceptions. Instead of grouping such factors into several themes from financial managers to facilitate the understanding and the way of dealing with them in order to be clearer for the decision-makers, this study has grouped respective factors which are relevant to the AIS effectiveness, which in turn, improves the performance of JLCs. Hence, to enhance the JLCs performance, it is necessary to recruiting the staff who deals with AIS are fully equipped with knowledge and the companies are recruiting people with appropriate qualifications, also has possessed appropriate staff expertise especially those who deal with the AIS. Furthermore, data related to the inputs of AISs, the data are treated in the AISs to produce information that required decisions making. Accordingly, businesses should give the data great concern, in terms of testing and monitoring them, to ensure that it is free of errors. Further, accurate data can empower AIS to affords high-quality information, hence elevating the company's performance.

For practical implications, the current study provides information on how internal control, company size, and staff qualifications affect AIS effectiveness. The context of Jordanian listed companies, which are public shareholding companies, requires attention on the satisfaction of its decision-makers, such as financial managers, on information produced by AIS to improve the performance of those companies, then to achieve their strategic goals. The result of this study could lead to, if the companies are interested, continuous development in internal control systems, hiring of staff with appropriate expertise and academic background, and the proper allocation of funds, which will contribute in raising the effectiveness of AIS. Furthermore, the findings contribute to literature, particularly for the improvement and development of AIS effectiveness in companies that are listed on ASE and other countries. Based on these results, the study encourages companies to train new employees on the optimal use of accounting systems before starting work to have an understanding of all the system functions and to be able to adapt to the changes in controlling the function of the AIS. The study also encourages the adoption of a proper recruitment system to appoint the right person in the right place, so employees have specializations related to their works. Companies must develop internal control systems in AIS to help the management achieve complete information and to ensure that all transactions are recorded accurately in appropriate accounts and in a timely manner. Companies must also be able to provide the financial resources needed for acquiring and developing new technology, such as AIS, and enable spending on innovative activities.

6 Recommendations

The following are some recommendations that may be extremely important for the Jordanian listed companies for future improvement and modification purposes. The researcher recommends doing more studies on AIS effectiveness in unlisted Jordanian companies and government companies and similar studies in other sectors, such as banks, government, or small and medium enterprises. The researcher also recommends conducting a study on the effects of AIS on financial reporting quality in a Jordanian environment. This study provided only an idea of the effect of internal control, company size, and staff qualifications on AIS effectiveness in Jordanian listed companies. Examining these variables in another sector or comparing them with various sectors is recommended. The results can be used by the management to review the policy related with enhancement of AIS effectiveness. This kind of study would help explain any similarities and differences in the influence between internal control, qualifications of staff, and company size toward satisfaction of decision makers on information produced by AIS.

Future research could replicate this research in more specific sampling contexts and expand the theoretical model by studying the effects of a larger set of variables. For instance, there is a need to include more variables, such as management commitment, financial reporting quality, external organization, organizational structure, and manager knowledge. Expanding the study to include more companies to enhance the consistency and applicability of the results is also recommended.

References

1. Al-Halalmeh, Mohammad I, Sayah AM (2010) Impact of foreign direct investment on shares market value in Amman exchange market. Am J Econ Bus Adm 2:1–35
2. Alshirah M et al (2021) Influences of the environmental factors on the intention to adopt cloud based accounting information system among SMEs in Jordan. Accounting 7(3):645–654
3. Alsaad A, Al-Okaily M (2022) Acceptance of protection technology in a time of fear: the case of Covid-19 exposure detection apps. Inf Technol People. https://doi.org/10.1108/ITP-10-2020-0719
4. Daoud L et al (2021) Moderating the role of top management commitment in usage of computer-assisted auditing techniques. Accounting 7(2):457–468
5. Daoud H, Triki M (2013) Accounting information systems in an ERP environment and tunisian firm performance. Int J Digit Acc Res 13
6. Xu H (2009) Data quality issues for accounting information systems' implementation: systems, stakeholders, and organizational factors. J Technol Res 1:1
7. Al-Attar K (2021) The effect of accounting information system on corporate governance. Accounting 7(1):99–110
8. Al-Kofahi MK, Hassan H, Mohamad R, Intan TP, Com M (2020) Information systems success model: a review of literature. Int J Innov Creativity Change 12
9. Edison G, Manuere F, Joseph M, Gutu K (2012) Evaluation of factors influencing adoption of accounting information system by small to medium enterprises in Chinhoyi. Interdiscipl J Contemp Res Bus 4:1126–1141
10. DeLone WH, McLean ER (1992) Information systems success: the quest for the dependent variable. Inf Syst Res 3(1):60–95
11. Saracina, Tara H (2012) An examination of the use of accounting information systems and the success of small businesses in South Carolina. Nova Southeastern University
12. Avacheva, Gennadyevna T, Yablochnikov SL, Milovanova OA (2018) Expanding the capabilities of medical information systems to automate the document flow of health care institutions. Proceedings of the 21st International Conferenceon Information Technology for Practice 7–14
13. Sori ZM (2009) Accounting information systems (AIS) and knowledge management: a case study. Am J Sci Res 4(4):36–44
14. Al-Okaily A, Al-Okaily M, Teoh AP, Al-Debei M (2022b) An empirical study on data warehouse systems effectiveness: the case of Jordanian banks in the business ıntelligence era. EuroMed J Bus. https://doi.org/10.1108/EMJB-01-2022-0011.
15. Adenike, Akanbi Taibat, and Jonathan Oyerinde Adewoye (2018) Investment in accounting information system and sales growth: an investigation of Nigeria small and medium enterprise. J Acc Tax 10(6): 71–77

16. Habiba Y et al (2019) Computerized accounting information system adoption among small and medium enterprises in Addis Ababa, Ethiopia. Int J Acc Financ Bus (IJAFB) 4(19): 44–60
17. Alqudah, Ahmad HM (2015) Factors affecting accounting information system effectiveness among Jordanian companies. School of business innovation & Technopreneurship. Universiti Malaysia Perlis, Arau
18. Al-Okaily A, Ai Ping T, Al-Okaily M (2021) Towards business intelligence success measurement in an organization: a conceptual study. J Syst Manag Sci 11(2):155–170. https://doi.org/10.33168/JSMS.2021.0210
19. Saira K, Ashhari ZM, Nassir AM (2010) Information system and firms' performance: the case of Malaysian small medium enterprises. Int Bus Res3(4): 28
20. Mohammed Z, Alqudah H (2007) The ımpact of liquidity and financial leverage on profitability: the case of listed Jordanian ındustrial firm's
21. Jawabreh, Abdalla O, Alrabei AM (2012) The impact of accounting information system in planning, controlling and decision-making processes in Jodhpur hotels. Asian J Financ Acc 4(1): 173–188
22. Etim EO (2011) Enhancing the efficiency of accounting information system in organizations. Int J Econ Develop Res Invest 2(2):19–27
23. Alqudah H, Shukeri B (2014) The role of data quality and ınternal control ın raising the effectiveness of ais ın Jordan Companies. Int J Sci Technol Res 3(8): 298–303
24. Alqudah, Mohammad H, Amran NA, Hassan H (2019) Extrinsic factors ınfluencing ınternal auditors' effectiveness in Jordanian public sector. Rev Eur Stud 11: 67
25. Hamdan, Mousa MN (2013) Improving the performance of accounting information systems of commercial banks in Jordan by using the balanced scorecard approach. Curtin University
26. Mndzebele N (2013) The usage of accounting information systems for effective internal controls in the hotels. Int J Adv Comput Technol 2(5):1–3
27. Lutf, Alwali A, Mohamad R (2016) The influence of technological, organizational and environmental factors on accounting information system usage among Jordanian small and medium-sized enterprises. Int J Econ Financ Issue 6(7): 240–248
28. Adeosun OT, Owolabi KE (2021) Gender inequality: determinants and outcomes in Nigeria. J Bus Soc Econ Devel
29. Ismail, Azizi N, King M (2007) Factors influencing the alignment of accounting information systems in small and medium sized Malaysian manufacturing firms. J Inf Syst Small Bus 1(1–2): 1–20
30. Nwokeji NA (2012) Repositioning accounting information system through effective data quality management: a framework for reducing costs and improving performance. Int J Sci Technol Res 1(10):86–94
31. Abdalwali L et al (2020) The ımpact of AIS usage on AIS effectiveness among Jordanian SMEs: a multi-group analysis of the role of firm size. Global Bus Rev
32. Dewi EK, Himam F, Sobirin A (2020) Developing security climate in educational organizations in the cyber era. IOP Conf Ser: Ear Environ Sci 448 (1)
33. Xu H (2015) What are the most important factors for accounting information quality and their impact on AIS data quality outcomes? J Data Inf Qual (JDIQ) 5(4):1–22
34. AbdAlwali, Khassawneh L (2014) The influence of organizational factors on accounting information systems (AIS) effectiveness: a study of Jordanian SMEs. Int J Market Technol 4(10): 36
35. Alqudah, Mohammad H, Amran NA, Hassan H (2019) Factors affecting the internal auditors' effectiveness in the Jordanian public sector. Euro Med J Bus
36. Reid GC, Smith JA (2000) The impact of contingencies on management accounting system development. Manag Acc Res 11(4):427–450
37. Binh V et al (2020) Firm size, business sector and quality of accounting information systems: evidence from Vietnam. Acc 6(3):327–334
38. Rida P et al (2012) The comprehension and application of accounting ınformation system for the small and medium enterprise. J Global Entrepr 3(1): 26–41

39. Al-Qudah AA, Al-Okaily M, Alqudah H (2022) The relationship between social entrepreneurship and sustainable development from economic growth perspective: 15 'RCEP'countries. J Sustain Financ Invest 12(1):44–61
40. Al-Okaily M, Al Natour AR, Shishan F, Al-Dmour A, Alghazzawi R, Alsharairi M (2021) Sustainable FinTech innovation orientation: a moderated model. Sustainability 13(24):13591
41. Bodnar, George H, Hopwood WS (2004) Acconting ınformations systems, Ninth Editions
42. Alqudah, Hamzah Mohammad A, Osman A, Al-Qudah HEM (2014) The effect of human resources management practices on employee performance. Int J Sci Technol Res 3(9): 129–134
43. Nah, Fui-Hoon F, Lau J LS, Kuang J (2001) Critical factors for successful implementation of enterprise systems. Bus Process Manag J
44. Trabulsi, Rehab U (2018) The ımpact of accounting ınformation systems on organizational performance: the context of Saudiâ€™ s SMEs. Int Rev Manag Market 8(2): 69–73
45. Idris, Kamil Md, Mohamad R (2017) AIS usage factors and impact among Jordanian SMEs: the moderating effect of environmental uncertainty. J Adv Res Bus Manag Stud 6(1): 24–38
46. Al Kurdi OF (2021) A critical comparative review of emergency and disaster management in the Arab world. J Bus Soc-Econ Devel
47. Susanto A (2007) Accounting ınformation systems concept and development computer-based. Bandung: Lingga Jaya
48. Yuvaraj S, Assefa KB (2013) Evaluating the design of accounting information system and its implementation in Ethiopian manufacturing industries. Res J Sci IT Manag 2(7): 16–29
49. Nicolaou, Andreas I (2000) A contingency model of perceived effectiveness in accounting information systems: organizational coordination and control effects. Int J Acc Inf Syst 1(2): 91–105
50. Albashabsheh, Nahar AA (2018) An appraisal of accounting ınformation systems in minimizing risk in banking sector of Jordan
51. Salehi M, Rostami V, Mogadam A (2010) Usefulness of accounting information system in emerging economy: empirical evidence of Iran. Int J Econ Financ 2(2):186–195
52. Sumaryati A, Praptıka Novıtasarı E, Machmuddah Z (2020) Accounting ınformation system, ınternal control system, human resource competency and quality of local government financial statements in Indonesia. J Asian Financ Econ Bus 7(10): 795–802
53. Wisna N, Agung AAG, Yuniar I (2020) User competency and ıts effect of accounting ınformation system ımplementation". Int J Tourism, 5
54. Aldrich, Howard E, Martinez MA (2007) Many are called, but few are chosen: an evolutionary perspective for the study of entrepreneurship." Entrepreneurship, pp. 293–311 Springer, Berlin, Heidelberg. https://doi.org/10.1007/978-3-540-48543-8_14
55. Alsharayri M (2012) Evaluating the performance of accounting information systems in Jordanian private hospitals. J Soc Sci 8(1):74
56. Hanini E (2012) The risks of using computerized accounting information systems in the Jordanian banks; their reasons and ways of prevention. Euro J Bus Manag 4(20):53–63
57. Kofahe, Khaled M, Hassan H, Mohamad R (2019) Factors affecting successful ımplementation of government financial management ınformation System (GFMİS) in Jordan public sector: a proposed framework. Int J Acc 4(20): 32–44
58. Al-Hakim, Latif, (ed) (2007) Information quality management: theory and applications. IGI Global
59. Borsboom, Denny, Mellenbergh G J, Van Heerden J (2004) The concept of validity. Psychol Rev 111(4): 1061
60. Al-Okaily M (2021) Assessing the effectiveness of accounting information systems in the era of COVID-19 pandemic. VINE J Inf Knowl Manag Syst. https://doi.org/10.1108/VJIKMS-08-2021-0148
61. Rahayu SK (2012) The factors that support the implementation of accounting information system: a survey in Bandung and Jakarta's taxpayer offices. J Global Manag 4(1):69–73
62. Guan Y (2010) A study on the Internal Control of accounting information system. 2010 International conference on computer and communication technologies in agriculture engineering. Vol 2. IEEE

63. Al-Sartawi A (2020) Social media disclosure of intellectual capital and firm value. Int J Learn Intellect Cap 17(4):312–323
64. Messier WF, Glover SM, Prawitt DF (2008) Auditing & assurance services: a systematic approach. McGraw-Hill Irwin, Boston
65. Al-Okaily M, Al-Okaily A (2022) An empirical assessment of enterprise information systems success in a developing country: the Jordanian experience. TQM J. https://doi.org/10.1108/TQM-09-2021-0267
66. Al-Okaily M, Alqudah H, Al-Qudah AA, Alkhwaldi AF (2022) Examining the critical factors of computer-assisted audit tools and techniques adoption in the post-COVID-19 period: ınternal auditors perspective. VINE J Inf Knowl Manag Syst. https://doi.org/10.1108/VJIKMS-12-2021-0311
67. Alqudah H, Al Natour AR, Al-Kofahi M, Rahman MSA, Abutaber TA, Al-Okaily M (2022) Determinants of the cashless payment systems acceptance in developing countries: evidence from jordanian public sector employees. In: Musleh Al-Sartawi AMA (ed) Artificial Intelligence for Sustainable Finance and Sustainable Technology, vol 423. Lecture Notes in Networks and Systems. Springer, Cham, pp 593–601. https://doi.org/10.1007/978-3-030-93464-4_58
68. Musleh Al-Sartawi AM, Razzaque A, Kamal MM (eds) (2021) Artificial ıntelligence systems and the ınternet of things in the digital era. EAMMIS 2021. Lecture Notes in Networks and Systems, vol 239. Springer, Cham. https://doi.org/10.1007/978-3-030-77246-8
69. Al-Qudah A A, Hamdan A, Al-Okaily M, Alhaddad L (2022) The impact of green lending on credit risk: evidence from UAE's banks. Environ Sci Pollut Res 1–13. https://doi.org/10.1007/s11356-021-18224-5
70. Al-Okaily A, Al-Okaily M, Teoh AP (2021) Evaluating ERP systems success: Evidence from Jordanian firms in the age of the digital business. VINE J Inf Knowl Manag Syst. https://doi.org/10.1108/VJIKMS-04-2021-0061
71. Al-Sartawi A (2020) Information technology governance and cybersecurity at the board level. Int J Crit Infrastruct 16(2):150–161
72. Lutfi A et al (2022) Antecedents and ımpacts of enterprise resource planning system adoption among Jordanian SMEs. Sustainability 14(6): 3508
73. Al-Okaily M, Alalwan AA, Al-Fraihat D, Rehman SU,Alkhwaldi AF, Al-Okaily A (2022) Investigating antecedents of mobile payment systems decision making: a mediated model. Global Knowl Memory Commun. https://doi.org/10.1108/GKMC-10-2021-0171

The Capabilities of Using Artificial Intelligence for Value-Added

Noora Kaladari and Abdalmuttaleb Al-Sartawi

Abstract The financial service industry, particularly the tax function has undergone substantial transformations due to advanced technologies, the most important of which is Artificial Intelligence (AI). This paper provides a comprehensive account of Value Added Tax (VAT) and the role that AI has played, is playing, and will play in its implementation. The results reveal that because of the widespread use of digital technologies and huge improvements in algorithmic abilities, increased computer power, and access to richer data, AI adoption has risen dramatically in the previous years. Nevertheless, as deployment trends increase, advanced tax analytics – and their advantages – will become more widely used in the coming years. It's likely that AI may disrupt corporate plans, therefore it's critical for all players to grasp what's ahead as AI influences the tax function of the future.

Keywords Value added tax · VAT · Artificial intelligence · AI · Financial services · Businesses

1 Introduction

The Value-Added Tax (VAT) is a consumption-based tax that was, in 1954, initially implemented in France. In the late 1960s, it became more widespread. Each region appears to have a different explanation and main motive for implementing the VAT system. It is widely regarded as one of the most effective revenue-raising taxes. Since the mid-1980s, VAT has been the dominant consumption tax when it comes to both geographic reach and revenue. 170 nations and territories throughout the world have introduced a VAT as of November 1, 2020 [15]. Almost every country has general consumption taxes, which are levied on the sale of nearly all products and services. VAT constitutes the majority of countries' general consumption tax. VAT is collected

N. Kaladari
Manama, Bahrain

A. Al-Sartawi (✉)
Ahlia University, Manama, Bahrain
e-mail: amasartawi@hotmail.com

A. M. A. Musleh Al-Sartawi et al. (eds.), *From the Internet of Things to the Internet of Ideas: The Role of Artificial Intelligence*, Lecture Notes in Networks and Systems 557,
https://doi.org/10.1007/978-3-031-17746-0_47

throughout all the stages of production and distribution of goods and services, with avoiding tax accumulation throw permitting businesses the deduction of the tax paid on inputs from the tax paid on outputs. Except in the U.S. that imposes sales taxes at the state level, all other OECD nations apply VAT [15].

Artificial intelligence (AI) is a type of technology that, via self-learning, can duplicate and even surpass human performance. It is capable of comprehending complicated information and drawing its own judgments [9]. Overall, AI applications are assemblages of many technologies (software, cloud computing, big data, sensory interfaces, and algorithms) that learn, respond, observe, decide, and suggest in humanlike ways (Milner & Berg, 2017). As such, AI refers to a set of techniques for having software adjust its functionality automatically to varying input with no need for explicit instructions. For decades, university, government, and industrial research facilities have been developing AI. Recent developments, on the other hand, enable AI to make the leap to general use in everyday life, and sectors that involve high-data-processing like the tax function.

2 Literature Review

2.1 Artificial Intelligence Usage on Business

Because of the widespread use of digital technologies and huge improvements in algorithmic abilities, increased computer power, and access to richer data, AI adoption has risen dramatically in the previous years, according to a study conducted by [18] fraud analysis on transactional data, Chatbots, and workflow optimization are among the most prominent AI applications. Computer-assisted diagnostics, market segmentation, virtual assistants, opinion mining and emotion analysis, face identification and detection, and applications in human resources like screening resumes are all considered examples of emerging uses for AI [1]. Retail, insurance, software and IT services, and telecom, are the most popular industries to use these AI application. Since 2015, AI businesses have tended to get more funding rounds and possess greater valuations than startups that did not employ AI [5, 10] assert that AI is just about to cause a new wave of digital disruption. Thus, businesses should be prepared and ready for it by now. Tangible results have been demonstrated by a number of early AI adopters in business, which makes it even more crucial for other companies to accelerate their digital revolutions. According to the same report, tech giants like Baidu and Google are leading the way in AI investment. However, outside of the IT industry, AI adoption is still in its early stages and is frequently experimental. Data shows that AI might bring about considerable benefits to the early adopters as well as acting as a disruptive force [5]. According to the research, those who adopt AI earlier with active initiatives and outstanding digital capabilities have bigger profit margins. AI's ability to optimize and automate operations, improve forecasting and sourcing, enhance the user experience, and establish targeted marketing and pricing is shown in the research's case studies across many industries including education, retail, healthcare, electric utilities, and manufacturing.

In another research by [2] governments and companies across the Middle East are beginning to recognize the worldwide move toward AI and sophisticated technologies in the aftermath of the fourth industrial revolution. They must decide whether to participate in the technological upheaval or risk being left behind. When it comes to the region's economic influence, being left behind is not an option. In 2030, the Middle East is predicted to get 2 percent of the global overall AI advantages, according to the research estimates. This equates to around US$320 billion [17].

2.2 Previous Studies on VAT

In the Consumption Tax Trends research by the OECD [15], from 4.1% of GDP in 1975 to 7.1% of GDP in 2018, OECD nations' reliance on general consumer taxes (which includes VAT) has climbed by more than 70%. The implementation of VAT in most OECD nations is the primary reason for this. VAT, currently, is considered amongst important sources of consumer taxes. In 2018, it accounted for about. 6.8% of Gross Domestic Product GDP and about 20.4% of the total tax income in OECD nations. Consumption taxes can be further divided into main two categories which are: (1) general taxes on goods and services ("generic consumption taxes") and (2) particular taxes on certain commodities and services. VAT alongside its counterpart GST, sales taxes, and additional general taxes on goods and services are all examples of general consumption taxes. Customs and import tariffs, taxes on particular services, and excise taxes are the most common taxes on certain commodities and services like financial services and insurance [15].

The present VAT approach is intended to be a fee added to ultimate consumption. It stays business-neutral in international trade and during the manufacturing process, based on particular standards that must be met. One of those standards is the need for taxpayers to file VAT returns frequently in a manner that allows for the reduction of tax due as the tax receivable increases. Such a paradigm is vulnerable to errors in both reporting and computation, as well as fraudulent behavior [3]. Thus, it could result in revenue loss, which would lead to a VAT gap that has an influence on the government's economic policy objectives.

In a paper by [19], the most pressing issue is boosting VAT collection while also addressing the system's vulnerability to compliance and fraud. The European Commission indicated in 2010 that technology advancements might lead to new means of collecting VAT [8]. Despite the fact that governments all over the globe have established systems that allow and require businesses to submit their tax filings online, this is not true of all governments. Since the introduction of VAT, the collection model has remained mostly constant. The study indicates that the discrepancy between planned and actual VAT income in the European Union was EUR [11] 140 billion in 2018. It proposes digitization of the VAT collection system as a necessary measure which includes employing Artificial Intelligence AI and other technologies to eliminate these variables.

2.3 Artificial Intelligence Usage on VAT

[16] asserts that Artificial Intelligence (AI) might be utilized to make procedures easier. Machine learning, for example, is a type of AI that allows a computer to learn by detecting patterns in data using algorithms. Some organizations are currently using AI to calculate VAT/GST liabilities, and it's feasible that, in the future, this technology, along with robots, might constitute the foundation of an end-to-end automated tax procedure [13]. There are currently applications related to tax that are not restricted to rule-based solutions, such as ones that learn from users' previous tax decisions then analyze large databases quickly, reducing the danger of sample bias and human mistake. In the future, tax solutions that are enabled by AI might be able to take into account both relevant tax practice and legislation and the parent organization's own rules, allowing for the optimal blend of artificial and human intelligence [12].

According to [4], developing a tax intelligent system can aid in effectively monitoring tax risks and handling tax issues. It can also help in reducing administrative expenses, standardizing taxation practices, and reducing human-induced judgment standards. On some note, [6] mentions that taxpayers and their advisors are well-versed in the use of cloud, Big Data, and Robotic Process Automation technologies. Taxpayers nowadays are looking for increasingly more modern tax technologies and applications that are powered by Artificial Intelligence. Optical character recognition enabled by Artificial Intelligence, for example, has already become a standard function of printers scanning checks and VAT invoices, identifying data and incorporating it into accounting systems, and completing different tax forms. Another example is the employment of a device named Dexter to that can read thousands of tax forms optically without the need for human interaction or data entry. These examples demonstrate that Artificial Intelligence is presently being employed at a simple level, referred to as "self-service". Some tax experts go a step farther and create tax data visualizations.

To reap the benefits of AI, research by [7] mentions that it is required to construct a knowledge corpus that has to be entered into and absorbed by AI in order for AI to operate on the data. The second objective is to teach the AI the appropriate language, which in this situation is the context and language of taxation. It's a big investment to develop the capabilities needed to construct learning techniques and select, organize, and present knowledge for such a corpus. For instance, both the training and the corpus must be adapted to the appropriate case law precedents, jurisdiction's tax policies, and legislation. The third major responsibility is giving the AI numerous questions and teaching it the answers as well as correcting it when it produces incorrect responses. The training and feedback loop aids in the refinement of algorithms with the goal of increasing the degree of confidence in the AI responses.

3 Conclusion

Many useful tax applications are starting to develop from professional service organizations and research, indicating that AI has already arrived in the tax function. It may appear doubtful that artificial intelligence (AI) robots would ever act as human-like tax accountants. AI applications, on the other hand, are quite likely to play a part in the tax function in the future, if not now. They would help tax specialists classify transactions, detect errors, estimate audit risk, and propose favorable tax options within increasingly complicated worldwide regulatory frameworks. In summary, academics, taxation authorities, taxable organizations, and tax service providers are all using AI and machine learning prototype systems in tax. As deployment trends increase, advanced tax analytics—and their advantages—will become more widely used in the coming years [14]. It is likely that AI may disrupt corporate plans, therefore it's critical for all players to grasp what's ahead as AI influences the tax function of the future.

References

1. Al-Sartawi A (2015) The effect of corporate governance on the performance of the listed companies in the gulf cooperation council countries. Jordan J Bus Adm 11(3):705–725
2. Al-Sartawi A (2020) Social media disclosure of intellectual capital and firm value. Int J Learn Intellect Cap 17(4):312–323
3. Al-Sartawi A (2017) The effect of the electronic financial reporting on the market value added of the Islamic banks in gulf cooperation council countries, 8th Global Islamic Marketing Conference. 4–6 May. International Islamic Marketing Association, Turkey.
4. Al-Sartawi A, Sanad Z (2019) Institutional ownership and corporate governance: evidence from Bahrain. Afro-Asian Journal of Finance and Accounting 9(1):101–115
5. Al-Sartawi A (ed) (2021) The big data-driven digital economy: artificial and computational intelligence. In: Kacprzyk, Janusz (ed) Studies in Computational Intelligence, Vol. 974. Springer, Cham. https://doi.org/10.1007/978-3-030-73057-4
6. Bughin J et al (2017) Artificial intelligence: the next digital frontier? McKinsey & Company.
7. Burneikaitė I (2019) Impact of tax technologies on current and future tax compliance. Vilnius University.
8. Deloitte (2019) Artificial intelligence-entering the world of tax. https://www2.deloitte.com/content/dam/Deloitte/global/Documents/Tax/dttl-tax-artificial-intelligence-in-tax.pdf
9. Commission E (2010) Green paper on the future of VAT–Towards a simpler, more robust, and efficient VAT system. Brussels
10. Karolak M, Razzaque A, Al-Sartawi A (2021) E-services and M-services using IoT: an assessment of the kingdom of Bahrain. In: Musleh Al-Sartawi AM, Razzaque A, Kamal MM (eds) Artificial intelligence systems and the internet of things in the digital era. EAMMIS 2021. Lecture Notes in Networks and Systems, vol 239. Springer, Cham. https://doi.org/10.1007/978-3-030-77246-8_48
11. MMC (2019) The state of AI 2019: divergence. MMC ventures report in partnership with Barclays. https://www.stateofai2019.com/
12. Musleh Al-Sartawi AMA (2020) E-learning improves accounting education: case of the higher education sector of Bahrain. In: Themistocleous M, Papadaki M, Kamal MM (eds) Information Systems. EMCIS 2020. Lecture Notes in Business Information Processing, vol 402. Springer, Cham. https://doi.org/10.1007/978-3-030-63396-7_20

13. Musleh Al-Sartawi AM, Razzaque A, Kamal MM (eds) (2021) Artificial intelligence systems and the internet of things in the digital era. EAMMIS 2021. Lecture Notes in Networks and Systems, vol 239. Springer, Cham. https://doi.org/10.1007/978-3-030-77246-8
14. Musleh Al-Sartawi AMA (eds) (2022) Artificial intelligence for sustainable finance and sustainable technology. ICGER 2021. Lecture notes in networks and systems, vol 423. Springer, Cham. https://doi.org/10.1007/978-3-030-93464-4
15. Musleh Al-Sartawi AM, Hussainey K, Razzaque A (2022) The role of artificial intelligence in sustainable finance. J Sustain Financ Invest: 1–6
16. OECD (2021) Consumption Tax Trends 2020. OECD Publishing. https://doi.org/10.1787/19990979
17. Pritchard G, Hatherell D, Young L (2019) Industry 4.0's impact on VAT/GST reporting and compliance Harnessing disruption to enhance processes. https://www2.deloitte.com/content/dam/insights/us/articles/22675_industry-four-point-zero-s-impact-on-vat-gst-reporting-and compliance/Industry-four-point-zero-s-impact-on-VAT-GST.pdf
18. PWC (2021) The potential impact of AI in the Middle East. https://www.pwc.com/m1/en/publications/documents/economic-potentialai-middle-east.pdf
19. Strusani D, Houngbonon G (2019) The role of artificial intelligence in supporting development in emerging markets. Int Financ Corp
20. Szewczyk R (2021) COVID-19 and its Impact on VAT Gap in the EU: Lessons from and for Poland. European Research Studies Journal. Abraham, A (2007) Student-centred teaching of accounting to engineering students: comparing blended learning and traditional approaches. Faculty of Commerce-Papers, University of Wollongong 435:1–11

The Effect of Social Responsibility Disclosure on the Economic Value Added

Thaer Ahmad Abutaber and Sanaa Nazami Maswadeh

Abstract The current study aimed at analyzing the effect of disclosing social responsibility in its various dimensions (social, workers, and environment), on the Economic Value Added (EVA). The study population consisted of the Jordanian industrial companies listed in the Amman Stock Exchange (ASE), which numbered (43) companies until the end of the year (2019), so that the sample size of this study was (36) companies by the simple random sampling method. The financial reports published in the ASE were relied upon, and for the purpose of determining the elements of disclosure of the social responsibility items (society, workers, and environment), each of the social responsibility variables was measured as a dummy. The study relied on the model suggested by Khan et al., (Khan S, Chouhan V, Chandra B, Goswami S (2012) Measurement of value creation Vis-À-Vis EVA: analysis of select BSE companies. Pacific Bus Rev Int 5(3),114–131.)[1] to measure the EVA. In order to test the hypotheses of the study a multiple regression test was used. The study reached the disclosure of social responsibility towards the society came in the first order with a positive impact on the EVA, followed by the disclosure of social responsibility towards the environment. Among the most important recommendations the necessity of activating the laws and legislations that obligate companies to disclose social responsibility in the Jordanian annual financial reports.

Keywords Social responsibility · Economic value added · Jordanian industrial companies

1 Introduction

The concept of corporate social responsibility disclosure has evolved significantly in recent years, and corporates have become aware of its importance for the continuity of businesses and to obtain the intended benefits. Corporates know they should act in a "socially responsible" manner, which requires the disclosure of their environmental

T. A. Abutaber (✉) · S. N. Maswadeh
Accounting Department, School of Business, Jadara University, Irbid, Jordan
e-mail: thaer@jadara.edu.jo

A. M. A. Musleh Al-Sartawi et al. (eds.), *From the Internet of Things to the Internet of Ideas: The Role of Artificial Intelligence*, Lecture Notes in Networks and Systems 557,
https://doi.org/10.1007/978-3-031-17746-0_48

and social information [2]. The corporates have become aware that violating their social responsibilities may create an atmosphere of opposition in society, which will impede the achievement of their goals. Therefore, the interest in disclosing social responsibility by corporates should be one of the main objectives; not only to serve the community and the environment, but also because of the effects on the performance and value of the corporate and the achievement of benefits for it [3].

Previous studies had shown that there is a difference in the level of corporate social responsibility practices between developed and developing countries. The developed countries have used practical procedures to motivate corporates to disclose information related to their role in corporate social responsibility [3]. For instance, the UK government has appointed a Minister for Corporate Social Responsibility in the industry and commerce sectors. France has passed a mandatory law whereby companies with 300 or more employees are required to disclose corporate social responsibility reports. Developing countries, including Jordan, have proceed to increase their interest in social responsibility practices and to disclose these practices in different dimensions i.e., society, humans, and environment. The differences between developed and developing countries indicate that the practice of corporate social responsibility is greatly influenced by the institutional context in which corporates operate [4]. In Jordan, for example, Legal Article (14/e) of the Corporate Governance Regulations for Listed Shareholding Companies for the year 2017 indicated the need for companies to disclose their policies and programs towards the local community and the environment.

Corporates have encountered a great challenge because they perform in a complex work environment characterized by rapid change economically, politically, socially, and technically [5] The more increase awareness among society members, the more it becomes unacceptable to be satisfied with the economic goals of the corporate without achieving social goals [6]. Thus, it is necessary to clarify the advantages of disclosing corporate social responsibility, for corporates that are able to provide appropriate and complete information, both financial and non-financial, would be more capable of maintaining business continuity. On the other hand, corporate social responsibility is used as a tool for evaluating the social performance of these corporates [2]; as well as helping companies to grow and achieve their goals and competitive advantage [7, 37].

Disclosure of corporate social responsibility information increasing the number of monitors of managerial behavior, and creating implicit contracts with an expanded group of stakeholders, and improves the information environment and the quality of accounting information, which increases the coordination of information between managers and shareholders [8]. This leads to the improvement of the company's performance in the short term, which is reflected positively on the prices of its shares, as well as on the benefits and financial incentives for the management [9]. In the long term, it is supposed to reflect positively on the EVA of corporates.

This study derives its importance from the importance of disclosing social responsibility, which is one of the important tools that are used by corporate departments to communicate the societal role played by their corporates and as a reflection of the interaction with their stakeholders. Therefore, the disclosure of corporate social responsibility information not only transfers more internal company information to external parties, but also has an important value related to investor decisions. From the perspective of corporate information transparency, disclosure of social responsibility provides investors with more internal information, which is limited to the management of the corporate in the event that it is not disclosed. When companies do not disclose their fulfillment of social responsibility, this may have a negative impact on the value of the company and its financial performance due to the lack of support from the community in which it operates, and this has a negative impact on its ability to achieve competitive advantage and long-term continuity.

This study tries to bridge the research gap associated with measuring the impact of disclosure of social responsibility on financial performance, as measured in an unconventional financial performance indicator, which is the EVA, which differs from other financial performance measures by calculating the remaining profits after calculating the costs of the total invested capital. The costs of debt and equity are deducted from the operating profit, to show the real economic profit of the company and its ability to achieve output that exceed the cost of the invested capital as a whole.

2 Literature Review

The legitimacy theory and the signaling theory explain the relationship between the disclosure of corporate social responsibility and the value of the company. Based on the signaling theory, disclosure of corporate social responsibility is a signal that can influence investor decisions. According to the legitimacy theory, the disclosure of corporate social responsibility is an implementation of the social contract between the company and society, and it is a form of community support stimuli for the company's business operations in order to increase the value of the company. [10] identified the factors that affect the implementation of social responsibility, the level of awareness of the benefits of social responsibility, the participation of stakeholders, the characteristics of work, the social responsibility report, and the legal standards.

The dimensions of corporate social responsibility are widely discussed in both developing and developed countries due to the different work environments of the business community in both of them, and social and economic problems; such as, unemployment, poverty, debt, and beliefs, which affect the analysis and understanding of the concept of corporate social responsibility and its dimensions. The disclosure of corporate social responsibility is an implementation of the Global Reporting Initiative, which referred to the categories of corporate social responsibility disclosure in accordance with the Global Reporting Initiative 4.0 standards launched in May 2013 [11].

Chandler and Werther [12] believe that the community surrounding the corporate is the first factor influencing its success or failure, even if such community is not a direct market for its product(s). Therefore, the company must build positive relationships with the community so that the interests are mutual, as the company serves the people of the communities through employment, training, job creation, health care programs, and support for clubs and charities, in addition to other means used in serving the community surrounding the company. On the other hand, society plays an important role by its positivity towards the company and its products, as the positive image contributes to building the reputation of the company with all its positive consequences.

Teoh et al., [13] stated that social responsibility towards the economy is represented by the responsibility that corporates bear towards the economy in which they operate, as they contribute to solving unemployment problems, reducing inflation, and increasing GDP, in addition to their contribution to the payment of fees and taxes. Its tendency to export products contributes to creating a balance in the trade balance with all its implications for the national currency and the economy in general. Therefore, social responsibility towards the economy is one of the drivers of sustainable economic development.

On the other hand, the study of Nuswantara and Pramesti [14] was applied to mining companies in Indonesia by researching stakeholders' factors and agency theory in influencing disclosure on social responsibility, which was measured using the GRI index. The study found a positive impact of foreign ownership and company size on Indonesian corporate social responsibility practices. However, the study could not prove the existence of an effect of the financial leverage and the size of the board of directors on the disclosure of the social responsibility of the Indonesian companies.

Moreover, Zaky et al., [15] tried to identify the impact of the disclosure using the corporate social responsibility index on the value of Indonesian companies. The study covered the period from 2013 to 2017. The most important results of the study were that there is a positive relationship between disclosure of social responsibility and the value of the company, and that profitability has a modified effect on this relationship.

Faisal et al., [16] aimed at indicating the extent to which environmental disclosure has changed, whether government laws have affected the extent of this change, and its relationship to the value of the company. Multiple linear regression was used to test the study hypotheses. The study concluded that there was a change of 21.6% in environmental disclosure during the years 2010–2014, that government legislations affected this change, and that the value of the company had changed positively according to environmental disclosure.

The study of Luisa and Alwiya [17] sought to detect the motives for adopting the concept of corporate social responsibility by Saudi private companies; especially that social responsibility has become a vital topic for corporate management and researchers. The results showed that the main motivation for Saudi companies to adopt social responsibility is to improve the company's image and moral commitment, and that there are obstacles to the application of social responsibility practices; such as, the weak commitment of management and investors to such guidelines.

Gerda and Dalia [18] aimed at analyzing the benefits of adopting the concept of corporate social responsibility by reviewing the theoretical literature and developing a conceptual framework linking social responsibility and the financial performance of companies. The study concluded that most of the previous studies found a positive relationship between social responsibility and financial performance, and a few previous studies indicated a negative relationship between social responsibility and financial performance.

The Study of Awaysheh et al., [19] sought to retest the relationship between social responsibility and financial performance by dividing companies into the best and worst performing companies in the United States of America. The study concluded that the top performing companies – as measured by Tobins Q – are the best in the disclosure of social responsibility, which confirms that these companies had a higher market value than their counterparts in the industry.

In a study conducted by Bambang et al., [20] which aimed at testing the relationship between social responsibility and financial performance by selecting data from 1998–2017 and reviewing 55 articles within the same period, the results demonstrated that the nature of the relationship between social responsibility and financial performance is very complex and requires the use of other factors to reach generalized results due to the differences between countries, models, and dimensions of social responsibility and financial performance measures.

Wenbin et al., [21] examined the relationship between social responsibility and financial performance in China. The results showed that social responsibility positively affects financial performance and that government ownership weakens the relationship between social responsibility and financial performance. On the other hand, industrial competition strengthens this relationship for all companies owned by the government or private companies.

Relying on previous studies, this study depends on studying the effect of disclosing social responsibility in its various dimensions (social, workers, and environment), as independents variables on the Economic Value Added (EVA) as dependent variable, even though most of previous studies studied the effect of social responsibility disclosure on the financial performance measured by accounting and marketing performance indicators such as rate of return and market value of shares. While this study came and relied on a different financial indicator, which is the -economic value added- according to the Khan et al., [1] model.

3 Methodology

In order to achieve the objectives of the field study, the researchers collected data related to the study's variables based on the financial reports issued by industrial public shareholding companies published on the website of the Jordan Securities Commission during the period from 2015 to 2019. Using the statistical program (SPSS), the study data were analyzed to test the study's hypotheses and reach its results.

3.1 Study Population and Sample

The study population consisted of the Jordanian public shareholding industrial companies listed on the ASEe i.e., 43 companies until the end of the year (2019). The sample of the study was determined based on the table of Krejcie and Morgan [22], consequently, the sample size of the study was (36) companies using the simple random sampling method.

3.2 Measuring Study Variables

In order to measure the variables of the study, it was relied on the financial reports published in the ASEe, and for the purpose of determining the elements of disclosure of the items of social responsibility, which are represented in the items of disclosure of social responsibility activities towards: society, workers, and environment.

Each social responsibility variable was measured as a dummy, so it is given (0) if the item is not mentioned in the annual financial reports and (1) if the item is mentioned in the annual financial reports. The mean was calculated for each of the variables of disclosure of social responsibility (society, workers, and environment) for each of the study sample companies and during the study period from 2015 to 2019. To express the percentage of the company's implementation of the dimensions of social responsibility in its financial statements, the average percentage of disclosure of the dimensions of social responsibility was measured using the following equations:

$\text{DSC} = Mean\left(\sum_i^n DSC_i\right)$

$\text{DSW} = Mean\left(\sum_i^n DSW_i\right)$

$\text{DSE} = Mean\left(\sum_i^n DSE_i\right)$

DSC, DSW, DSE, disclosure towards: society, workers, and environment respectively

In order to measure Economic Value Added (EVA) the Khan et al., [1] model was used.

3.3 Statistical Methods

The descriptive analysis of the study data was used in order to describe the extent of disclosure of the items of social responsibility in the financial statements of industrial companies, in addition to describing the EVA. In order to verify the suitability of the study data to the regression test, the normal distribution of the study data was tested by checking the residuals of the regression analysis model (Kolmogorov–Smirnov). To test the Multicollinearity between the independent variables, the variance inflation factor was used. In order to test the hypotheses of the study related to measuring the effect of social responsibility in its three dimensions on the EVA, it was relied on the multiple regression test, which was represented by the following equation:

$$EVA_{i,t} = \mathbf{a_0} + \beta 1\, DSC_{i,t} + \beta 2 DSW_{i,t} + \beta 3\, DSE_{i,t} + \varepsilon$$

4 Statistical Analysis Results

After Applying different statistical methods, which started in descriptive statistics for the study variables appeared as shown in Table 1.

Presents the descriptive analysis of the study variables where the mean of the EVA of the study sample companies appeared negative, which is an indication that the operating profit after taxes on average was not sufficient to cover the cost of capital invested in Jordanian industrial companies. The mean of the disclosure of social responsibility appeared in its three studied dimensions: society, workers, environment (0.781), (0.896), and (0.765) respectively and these values indicate that the disclosure of social responsibility activities towards society, workers, and environment averaged 78.1, 89.6, and 76.5 %, respectively. It is an indication that the Jordanian industrial companies disclose their social responsibility activities, and it was evident that the mean of the disclosure of social responsibility towards workers increased compared to other dimensions.

Table 1 Descriptive statistics for the study variables

variables	mean	min	max	Standard deviation
EVA: EVA	−26,783,552	−590,055,230	97,005,752	84,647,457
DSC: Disclosure toward society	0.781	0	1	0.402
DSW: Disclosure toward workers	0.896	0	1	0.315
DSE: Disclosure toward environment	0.765	0	1	0.352

Source Result Output from SPPS Statics Analysis

Table 2 The results of the normal distribution of the regression analysis residuals model

	Kolmogorov-Smirnov[a]		
	Statistic	N	Sig
Standardized Residual	0.0310	180	0.28[a]0

Source Result Output from SPPS Statics Analysis

4.1 Data Validity Test

To ensure the validity of the data for the multiple regression analysis, a test of the normal distribution of the residuals of the regression analysis model was relied upon. The following are the results of the tests to verify the data of the study represented by the residual test in the regression analysis model through the (Kolmogorov–Smirnov). The results of the normal distribution of the regression analysis residuals shown in Table 2.

Based on Table 2, the residuals in the regression model follow the normal distribution because the significance (K-S) value is (0.28) which is greater than (0.05), this indicates that the residuals in the regression model follow the normal distribution and that it is valid for the multiple regression model [23, 24]. To test Multicollinearity to detect the autocorrelation of all independent variables and their inflation, the Variance Inflation Factor (VIF) was used, the results of which appeared in Table (3) as follows:

Referring to Table 3 indicates that there is no problem of self-correlation in the study model [25, 26]. The fact that VIF is less than 5, and according to the Tolerance coefficient, the value of Tolerance is greater than (20%), which means the ability of the independent variables are expected to remain within the study models and that they are suitable for the multiple regression models.

4.2 Regression Analysis Results

The study model was designed to measure the effect of disclosing social responsibility represented in disclosing the social responsibility towards society (DSC), towards workers (DSW), and towards the environment (DSE), on the dependent variable measured in the natural logarithm of EVA (EVA) for industrial companies listed in the ASE. Table 4 below shows the results of the multiple regression analysis of the study model in order to test the study hypotheses:

Table 4 shows that the calculated F-value is (35,321) at the significance level of ($\alpha \leq 0.05$), which indicates that the proposed model has the appropriateness and explanatory power in a statistically acceptable way. The results showed that the value of Adjusted R^2 is (0.402), which means that 40.2% of the fluctuations that occur in the

Table 3 The results of the multicollinearity test of all independent variables

	VIF	Tolerance
DSC: Disclosure towards society	1.09	0.917
DSW: Disclosure towards workers	1.824	0.548
DSE: Disclosure towards the environment	1.779	0.562

Source Result Output from SPPS Statics Analysis

Table 4 The results of the multiple regression model of the study

Variable	T-Statistic	Coefficient (β)	Prob. (Sig T)
Constant	10.345	----------	0.000
DSC	6.386	0.441	0.000
DSW	4.125	0.259	0.000
DSE	5.123	0.394	0.000
S.E. of regression	0.475		
Prob.(F-statistic)	0.000		
F-statistic	35.321		
R	0.589		
R^2	0.487		
Adjusted R^2	0.402		

Source Result Output from SPPS Statics Analysis

EVA of industrial companies can be explained by the changes that occur in the level of disclosure of social responsibility towards society, workers, and environment. This result is similar to the findings of many other studies, including [15, 18, 19, 21] these studies agreed on the existence of a statistically significant effect of corporate social responsibility on the financial performance of companies. However, such findings have contradicted by Bambang et al., [20] the results demonstrated that the nature of the relationship between social responsibility and financial performance is very complex, and there is a difference between countries, models, and dimensions of social responsibility and financial performance measures. Likewise, in the study of Al-Awdat [27] and Timmons [28] which concluded that there is no statistically significant impact of the disclosure of social responsibility on the market value of the company.

With reference to Table 4, the value of the Prob. (Sig T) of social responsibility towards the society, workers and environment were less than (0.05), where its values were (0.000). Therefore, disclosure of social responsibility towards society workers and environment affects the (EVA) of industrial companies listed on the ASE. The results of the regression analysis also showed that the value of the social responsibility coefficient towards the society, workers and environment were (B = 0.441), (B = 0.259), (B = 0.394) respectively, indicates the existence of a positive effect of the disclosure of social responsibility towards society, workers and environment on the EVA. These results agree with Gerda and Dalia [18]; Andayani and Atmini [29], these studies agreed on the existence of a significant effect of corporate social responsibility towards society on the performance of companies and agrees with Wenbin et al., [21]; his study agreed on the existence of a significant effect of corporate social responsibility towards workers on the value of the companies, and agrees with Faisal et al., [16] that concluded the company had changed positively according to environmental disclosure.

5 Findings and Recommendations

From reviewing the results of the statistical analysis, the following important conclusions can be drawn: The disclosure of industrial companies listed on the ASE about social responsibility towards society, workers, and the environment has a positive impact on the company's image in its local community, increases the confidence of the users of reports, and helps them in rationalizing their investment decisions and serving their purposes. It also leads to upgrading the company's products by increasing customers' confidence in its products, which in turn is positively reflected in the company's performance and value [30–36].

The disclosure of social responsibility towards society has the highest positive effect on the EVA, followed by disclosure of social responsibility towards the environment, which is an indication of the importance of disclosing any amounts of money associated with social responsibility towards society, including donating to charitable societies, and contributing to alleviating unemployment by preparing productive projects, providing educational grants to the community, and training university students. The interest of industrial companies in the activities of social responsibility towards the environment, by moving away from projects that harm the environment, and participating in programs to protect the environment and preserve natural resources, and disclosing all these environmental activities, contributes to improving the value of the company and its positive image in the community and the local environment in which it operates. The study recommended the necessity of activating the laws and legislations that oblige companies to disclose social responsibility in the annual financial reports, and increasing the awareness level of companies' administrations of the need to direct their attention towards disclosing all the items related to social responsibility, whether they are towards the society, the workers, or the environment, and to avoid focusing on financial items only, as this has a positive impact on the company's value, performance, then ability to achieve competitive advantage.

References

1. Khan S, Chouhan V, Chandra B, Goswami S (2012) Measurement of value creation Vis-À-Vis EVA: analysis of select BSE companies. Pacific Bus Rev Int 5(3):114–131
2. Wali AF, Uduma IA, Wright LT (2016) Customer relationship management (CRM) experiences of business-to-business (B2B) marketing firms: a qualitative study. Cogent Bus Manag 3(1):1183555
3. Elmagrhi MH, Ntim CG, Elamer AA, Zhang Q (2019) A study of environmental policies and regulations, governance structures, and environmental performance: the role of female directors. Bus Strateg Environ 28(1):206–220
4. Barakat FS, Lopez Perez M, Rodríguez Ariza L (2015) Corporate social responsibility disclosure (CSRD) determinants of listed companies in Palestine (PXE) and Jordan (ASE). Rev Manag Sci 9(4):681–702
5. Albuquerque R, Koskinen Y, Zhang C (2019) Corporate social responsibility and firm risk: theory and empirical evidence. Manage Sci 65(10):4451–4469

6. Hasan KK (2021) The relationship between intellectual capital and organizational trust and its impact on achieving the requirements of entrepreneurship strategy (The case of Korek telecom company, Iraq). Int J Multicultural Multireligious Underst 8(2):130–146
7. Al-Sartawi A (2020) Does it pay to be socially responsible? Empirical evidence from the GCC countries. Int J Law Manag 62(5):381–394
8. Nyeadi JD, Ibrahim M, Sare YA (2018) Corporate social responsibility and financial performance nexus: empirical evidence from South African listed firms. J Global Responsib
9. Hedblom D, Hickman BR, Listss JA (2019) Toward an understanding of corporate social responsibility: theory and field experimental evidence (No. w26222). Natl Bur Econ Res
10. Lin TL, Lai WH (2018) Analyzing the factors that affect the implementation of CSR in enterprises. Rev Integrative Bus Econ Res 7(3):66–92
11. Gutierrez-Huerter OG (2016) The adaptation of social responsibility reporting within a multinational corporation: a multi-level study (Doctoral dissertation, University of Nottingham)
12. Chandler D, Werther WB (2013) Strategic corporate social responsibility: stakeholders, globalization, and sustainable value creation, 3rd edn. SAGE Publications Inc., Thousand Oaks, pp 21–30
13. Al-Okaily A, Al-Okaily M, Teoh AP, Al-Debei M (2022) An empirical study on data warehouse systems effectiveness: the case of Jordanian banks in the business intelligence era. Euro Med J Bus. https://doi.org/10.1108/EMJB-01-2022-0011
14. Nuswantara DA, Pramesti DA (2020) Corporate social responsibility regulation in the Indonesian mining companies. J Asian Financ Econ Bus 7(10):161–169
15. Machmuddah Z, Sari DW (2020) Corporate social responsibility, profitability and firm value: evidence from Indonesia. J Asian Financ Econ Bus 7(9):631–638
16. Faisal F, Situmorang LS, Achmad T, Prastiwi A (2020) The role of government regulations in enhancing corporate social responsibility disclosure and firm value. J Asian Financ Econ Bus 7(8):509–518
17. Pinto L, Allui A (2020) Critical drivers and barriers of corporate social responsibility in Saudi Arabia organizations. J Asian Financ Econ Bus 7(11):259–268
18. Barauskaite G, Streimikiene D (2021) Corporate social responsibility and financial performance of companies: the puzzle of concepts, definitions and assessment methods. Corp Soc Responsib Environ Manag 28:278–287. Wileyonlinelibrary.Com/Journal/Csr
19. Awaysheh A, Klassen RD (2010) The impact of supply chain structure on the use of supplier socially responsible practices. Int J Oper Prod Manag
20. Subroto B, Sutrisno T, Saraswati E (2020) Explaining the complexity relationship of CSR and financial performance using neo-institutional theory. J Asian Bus Econ Stud 27(3):227–244
21. Musleh Al-Sartawi AMA (eds) (2022) Artificial intelligence for sustainable finance and sustainable technology ICGER 2021. Lecture Notes in Networks and Systems, vol 423. Springer, Cham. https://doi.org/10.1007/978-3-030-93464-4
22. Krejcie, Morgan (1970) Determining sample size for research activities educational and psychological measurement 1(30): 607–610
23. Babbie E, Wagner WE III (2018) Adventures in social research: data analysis using IBM® SPSS® statistics. Sage Publications, USA
24. Alqudah H, Al Natour AR, Al-Kofahi M, Abd Rahman MS, Abutaber TA, Al-Okaily M (2022) Determinants of the cashless payment systems acceptance in developing countries: evidence from Jordanian public sector employees; Lecture Notes in Networks and Systems. https://doi.org/10.1007/978-3-030-93464-4_58
25. Schreiber-Gregory DN, Jackson HM (2017) Multicollinearity: what is it, why should we care, and how can it be controlled? 1–12
26. Al-Okaily M, Al-Okaily A (2022) An empirical assessment of enterprise information systems success in a developing country: the Jordanian experience. TQM J. https://lnkd.in/gVYWFSqQ
27. Al-Okaily, M., Al-Sartawi, A., Hannoon, A., & Khalid, A. A. (2022) Information Technology Governance and Online Banking in Bahrain. In: Musleh Al-Sartawi A.M.A. (eds), Artificial Intelligence for Sustainable Finance and Sustainable Technology. ICGER 2021. Lecture Notes in Networks and Systems, vol 423. Springer, Cham. https://doi.org/10.1007/978-3-030-93464-4_60

28. Al-Awdat Anas (2015) The impact of disclosure of social responsibility on the market value of Jordanian commercial banks according to (Tobin's Q) model, unpublished master's thesis, business faculty, middle east university, Amman, Jordan
29. Timmons JA (2011) The relationship between corporate social responsibility and market performance, (Unpublished Doctoral Dissertation). University of Phoenix, USA
30. Andayani W, Atmini S (2015) Corporate social responsibility (CSR) and good corporate governance (GCG), and firm performance. J Mod Acc Audit 8(10):1484–1495
31. Abutaber T, Bzur A, Odeh M, Alathamneh M, Al-Okaily M, Afaneh M (2021) The effect of corporate governance indicators on enhancing the financial performance of industrial listed companies on the ASE. Accounting 7(2):415–422
32. Al-Okaily A, Al-Okaily M, Teoh AP (2021) Evaluating ERP systems success: evidence from Jordanian firms in the age of the digital business. VINE J Inf Knowl Manag Syst. https://doi.org/10.1108/VJIKMS-04-2021-0061
33. Al-Okaily M (2021) Assessing the effectiveness of accounting information systems in the era of COVID-19 pandemic. VINE J Inf Knowl Manag Syst. https://doi.org/10.1108/VJIKMS-08-2021-0148
34. Al-Qudah AA, Al-Okaily M, Alqudah H (2022) The relationship between social entrepreneurship and sustainable development from economic growth perspective: 15 'RCEP' countries. J Sustain Financ Invest 12(1):44–61
35. Al-Okaily M, Al-Okaily A (2022) An empirical assessment of enterprise information systems success in a developing country: the Jordanian experience. TQM J. https://doi.org/10.1108/TQM-09-2021-0267
36. Al-Qudah AA, Hamdan A, Al-Okaily M, Alhaddad L (2022) The impact of green lending on credit risk: evidence from UAE's banks. Environ Sci Pollut Res. https://doi.org/10.1007/s11356-021-18224-5
37. Al-Okaily M, Alghazzawi R, Alkhwaldi AF, Al-Okaily A (2022) The effect of digital accounting systems on the decision-making quality in the banking industry sector: a mediated-moderated model. Global Knowl Mem Commun. https://doi.org/10.1108/GKMC-01-2022-0015

What Do We Know About Forensic Accounting? A Literature Review

Seif Al Shbeil, Hashem Alshurafat, Noor Taha, and Mohannad Obeid Al Shbail

Abstract A review of relevant academic studies has yielded a range of insights on the forensic accounting profession. This review aims to provide a comprehensive view of what is forensic accounting. This paper reviews published forensic accounting studies to explore the reality of the profession in Australia mainly and in many international contexts. This paper notes that the forensic accounting profession reality is varied across contexts and well established in the Australian context. It is desired that this structured examination of the forensic accounting profession will provide a sufficient overview of the critical issues and concerns that must be addressed to comprehend and advance the critical application of forensic accounting worldwide. This article is expected to have an impact on future policies, which will help to improve the forensic accounting profession. This paper contributes to the ongoing debate about the forensic accounting services that lead to enhanced fraud investigation, litigation support, business valuation, and cybercrime detection.

Keywords Forensic accounting · Profession · Skills · Services

1 Introduction

According to Huber and DiGabriele [1], before determining the research purpose in forensic accounting, it is crucial to know the concept of forensic accounting, its fields, disciplines, and domains encapsulated by it. In this paper, forensic accounting and its critical aspects are viewed from the profession and discipline lens and are examined in the Australian and global context. The paper presents evidence of empirical studies concerning the services that forensic accountants provide, the professional's desired characteristics, accreditation methods, and educational pathways

S. Al Shbeil · M. O. Al Shbail
Department of Accounting, School of Business, Al Al-Bayt University, P.O. Box 130040, Mafraq 25113, Jordan

H. Alshurafat (✉) · N. Taha
Department of Accounting, Faculty of Economics and Administrative Sciences, The Hashemite University, P.O. Box 330127, Zarqa 13133, Jordan
e-mail: hashema@hu.edu.jo

A. M. A. Musleh Al-Sartawi et al. (eds.), *From the Internet of Things to the Internet of Ideas: The Role of Artificial Intelligence*, Lecture Notes in Networks and Systems 557,
https://doi.org/10.1007/978-3-031-17746-0_49

that forensic accountants can take. In addition, the skills and knowledge of practicing forensic accountants in Australia are also discussed through a review of literature demonstrating past studies relating to forensic accounting literature. The paper ends with a summarized version of the literature and confirmation of the gap related to the topic.

This paper is also dedicated to presenting a historical overview of the forensic accounting concept, taking into account two major forensic accounting perspectives, namely the academic perspective (viewpoints in literature) and professional perspective (forensic accounting standards and ethical code of conduct in Australia and the U.S.). Literature concerning forensic accounting services and the needed skills for the profession's effectiveness are presented and discussed.

This paper provides implication regarding the fundamental skills and the scope of services that is important to enhance the integration process of sustainable technologies, artificial intelligence, and the internet of ideas in fraud investigation and cybercrime detection services.

The paper proceeds as follows. The next section defines forensic accounting. The third section demonstrates the forensic accounting scope of services. The fourth section reports on the required forensic accounting skills. Section five concludes the paper.

2 What is Forensic Accounting?

Forensic accounting can be traced back to the work of Maurice Peloubet in 1946 in essay form entitled, "Forensic Accounting—Its Place in Today's Economy" [2]. However, studies claimed that intrinsic forensic accounting dates back even further to the time of the Pharaohs of Ancient Egypt, where specialists were designated to conduct inventories of grain and gold to protect them from fraudulent activities [3]. In fact, Crumbley [4] found evidence that in 1817, a Canadian accountant's investigative work landed him in court to testify as an expert witness.

Generally speaking, literature on forensic accounting leads to two branches, intern-courtroom direction and the more general services direction. The former direction defined forensic accounting in light of its legalistic role. On the other hand, the latter direction focused on emphasizing the concept's holistic role deviating from the narrow in-courtroom focus and taking into account the crucial role played by forensic accounting in detecting fraud, valuation of the business, and IT forensics. Such broad direction is defined by the Australian Accounting Professional and Ethical Standards Board (APESB), which is in line with Botes and Saadeh [2].

According to some past studies [2, 5–7], forensic accounting is a discipline that has no clear definition and this is manifested through the profession's general lack of public recognition, the complex scope of the concept, and the complexity involved in developing the forensic accounting education programs. In some studies, the concept's variation and inconsistency have been attributed to the minimal interest in the subject [8]. while others indicated that the majority of the academics need to understand and appreciate the subject of forensic accounting Heitger and Heitger [9]. In another related study, McMullen and Sanchez [10] contended that because of the lack of understanding and appreciation of the subject, studies on it are few and far between but that being said, academic literature, professional literature, and professional standards have proposed various definitions of forensic accounting.

3 Forensic Accounting Scope: What do Forensic Accountants do?

Forensic accounting involves supporting individuals, organizations, and governments by the provision of different professional services for their maintenance and sustainability [11]. Due to the extensiveness of the services offered by forensic accounting, it is consisting multiple disciplines [12]. Practitioners of forensic accounting can thus be employed in different organizational types, public institutes, insurance firms, banks, police forces, and government agencies [13]. Despite the overlapping nature of the forensic accounting services such as litigation services and business valuation, fraud detection, and cybercrimes detection, most studies in the literature have categorized them into four major types, namely litigation services [8, 14], computer forensic analysis [15], fraud investigation [16], and business valuation [17].

Added to the above, forensic accountants may take on the role of an advisor to provide professional assistance to lawyers or provide evidence in a lawsuit Huber, Domino [18]. Such types of litigation services may cover the following; resolution of disputes, expert witness engagements, consulting services for forensic accounting, business advisory, corporate misconduct, corporate governance, and assessment of contracts [19]. In this line of argument, forensic accountant's may also provide services in matrimonial litigation and take on the role of expert witnesses in the family court for family property evaluation, provision of the family income fiscal profile, and provision of lifestyle analysis. The above-mentioned services take up a few of the many litigation services that an expert witness may provide [9, 20]. Some cases may need a shadow expert who plays a role in facts interpretation, analyzing the report of experts, and discussing the pros and cons of the logic behind the employed method of valuation [21].

Moreover, forensic accounting is an extensive accounting field for the process of investigation, with forensic accountants considered qualified individuals to make improvements in the organization by mitigating the occurrences of fraud [22]. In this regard, a forensic accountant possesses the ability to investigate, a role that goes beyond that of a normal auditor [23]. In other words, a forensic accountant plays a key role in the detection, prevention, and prosecution of fraudulent individuals involved in financial misstatement, identity theft, money laundering, cybercrime, administration flaccid, and theft of government funds [15, 24, 25].

One of the major specialty areas of forensic accountants is business valuation [26]. According to Crumbley, Rezaee [27], forensic accountants are another term for experts' valuators of losses, while Smith [28] contended that they are capable of providing business valuations services in some cases like personal injury, risk analysis, bankruptcy, estate and gift, stock-based compensation, family law, matrimonial issue, and financial reporting fair value. Other authors promote the role of forensic accountants in this specific area and suggest adding business valuation to the accounting curricula [8, 9].

According to Cook and Clements [29], using computers for fraud detection is at an increase and forensic accountants are expected to handle computerized work environments and take part in different related works like information technology, data mining, and computer forensics [30]. In the same line, Pearson and Singleton [16] mentioned a range of cybercrimes like credit card fraud, identity theft, online narcotic sales, online/email extortion, telecommunications fraud, computer privacy theft, online gambling, and intellectual property rights as a part of the forensic accountants work. Such professionals have also been commended for their effective use of computerized tools in fraud detection [29]. Additionally, the majority of authors suggested that curricula of forensic accounting integrate the subject's technological aspects [16].

In the case of Australia, the field of studies in forensic accounting requires additional intensive research as recommended by several past authors [14, 31–34]. Specifically, Van Akkeren, Buckby [31] revealed that the perceptions of Australian forensic accountants as well as forensic accounting educators vary as to the best practices that can develop the forensic accounting profession among the practitioners in Australia. Nevertheless, when it comes to the offered services, Australian forensic accounting companies actively respond to the forensic accounting work complexity, providing various services that meet many of the needs of the clients [32]. However, accounting firms primarily concentrate on offering the following services; fraud investigation, dispute-solving, family law services, expert testimony services, opinion delivery & value estimations, and services for loss and damage claims.

4 Desired Skills of Forensic Accountants

It has been contended by several authors in literature [35, 36] that forensic accounting students need to develop distinguished skills (investigative, adversarial, analytical and technological) for their future work success. Following the introduction of forensic accounting, the required skills set have been constantly discussed with the majority of the studies [10, 36] exploring the importance of skills from different viewpoints (stakeholders, academic, and practitioners). Based on Rezaee, Lander [37] study, forensic accounting discipline is an integration of accounting knowledge and skills, and those of auditing, criminology, tax, law, sociology, and psychology. Such skills, knowledge, and attributes vary according to the services category offered by forensic accountants [9].

Studies in the same line have examined the nature of the skills, knowledge, and attributes needed by forensic accountants. For instance, DiGabriele [36] determined the presence of differences in perceptions concerning forensic accountants' relevant skills among practitioners, accounting academics, and forensic accounting services users. The author proposed guidance towards the enhancement of the subject curricula through the identification and integration of suitable skills, knowledge, and attributes. The study findings supported a consensus between practitioners and academics as to the required skills being critical thinking, investigative flexibility, unstructured problem-solving, and analytical proficiency.

In the Australian case, accounting students' required set of skills, which have been increasingly discussed in the literature [38]. Prior studies mainly contended that graduates need to be able to work in teams, possess good interpersonal skills, communication skills, presentation skills, problem-solving, technological competence, initiative, self-management, and enterprise planning and organizing.

Contrastingly, some other studies claimed that only a few studies focused on determining the required skills for effective forensic accounting engagements in Australia [14, 31, 32]. In this context, the required skills are similar to those on a global level, as evidenced by Van Akkeren, Buckby [31], who stated that forensic accounting firms in Australia expect their staff to provide excellent work-based skills involving technology, oral and written communication and analytical skills.

The role of education in ensuring that students are prepared for forensic accounting work has been addressed by many studies. To begin with, it was argued by Brickner, Mahoney [39] that applied learning programs should be used to enhance the students' skills in forensic accounting curricula as such skills are needed for organizing evidence, the process of investigation, interview methods, and communication. Additionally, Lee, Cefaratti [40] conducted an examination of the forensic accounting education effect on the creativity of students and found those students engaging in forensic accounting to be more creative compared to their counterparts that lack engagement. Moreover, a forensic accounting course was found by Carpenter, Durtschi [22] to boost the skepticism level of the students. Based on the results of their longitudinal experiment, teaching forensic accounting courses provides the students with the ability to perform optimally in fraud risk assessments. This finding supported McMullen and Sanchez [10] statement that skepticism should be integrated into classrooms along with surveillance methods, puzzle skills, interrogation, and investigation.

Kern and Weber [35] found that forensic accounting implementation could bring advantages to the students as it enables them to gain knowledge and skills that they can use for the efficient administration of interviewing methods, increasing their analytical, critical thinking, and collaboration skills, and developing their appreciation for internal control systems. In this regard, the specialization of forensic accounting with its multidisciplinary nature has formed expectations as to what additional capabilities and skills forensic accounting graduates should possess (technological and digital analysis skills) [41]. These views were brought up by Pearson and Singleton [16] who contended that students need to understand the tools and methods that IT forensic specialists use for digital evidence retrieval and forensic computer work.

5 Conclusion

To this end, it is crucial for forensic accounting firms to enhance their member qualification and competencies in order to offer domain services and maintain benchmarking and competencies. In fact, studies referred to it as a challenge to be a specialist that is capable of working in the profession's various domains [13]. Nevertheless, potential forensic accounting professionals could obtain an educational degree in order to comprehend the legal ramifications involved in the profession and the work environment although a professional accreditation will be useful for specialized practitioners in distinct services categories.

This research is associated with different implications for the literature and practice of forensic accounting. The paper highlights the definition, scope, and skills of forensic accounting. In addition to the legal support services, the field of forensic accounting encompasses fraud investigation and cybercrime detection services. This paper provides implication regarding the fundamental skills and the scope of services that is important to enhance the integration process of sustainable technologies, artificial intelligence, and the internet of ideas in fraud investigation and cybercrime detection services.

References

1. Huber W, DiGabriele JA (2015) What is the purpose of research in forensic accounting? J Theo Acc Res 10(1):30–49
2. Botes V, Saadeh A (2018) Exploring evidence to develop a nomenclature for forensic accounting. Pac Acc Rev 30(2):135–154
3. Wells JT (2000) So that's why it's called a pyramid scheme. J Acc 190(4):91
4. Crumbley DL (2001) Forensic accounting: older than you think. J Forensic Acc 11(2):181–202
5. Alshurafat H (2021) Forensic accounting as a profession in Australia? A sociological perspective. Meditari accountancy research
6. Alshurafat H, Al Shbail M, Almuiet M (2021) Factors affecting the intention to adopt IT forensic accounting tools to detect financial cybercrimes. Int J Bus Excell
7. Alshurafat H, Al Shbail MO, Mansour E (2021) Strengths and weaknesses of forensic accounting: an implication on the socio-economic development. J Bus Soc Econ Develop
8. Rezaee Z, Burton EJ (1997) Forensic accounting education: insights from academicians and certified fraud examiner practitioners. Manag Audit J 12(9):479–489
9. Heitger LE, Heitger DL (2008) Incorporating forensic accounting and litigation advisory services into the classroom. Issues Acc Educ 23(4):561–572
10. McMullen DA, Sanchez MH (2010) A preliminary investigation of the necessary skills, education requirements, and training requirements for forensic accountants. J Forensic Investig Acc 2(2):30–48
11. Howieson B (2018) What is the 'good'forensic accountant? A virtue ethics perspective. Pac Acc Rev 30(2):155–167
12. Hegazy S, Sangster A, Kotb A (2017) Mapping forensic accounting in the UK. J Int Acc Audit Tax 28:43–56
13. Tiwari RK, Debnath J (2017) Forensic accounting: a blend of knowledge. J Financ Regul Compliance 25(1):73–85

14. Van Akkeren J, Tarr J-A (2014) Regulation, compliance and the Australian forensic accounting profession. J Forensic Investig Acc 6(3):1–26
15. Lindberg DL, Seifert DL (2016) Microinsurance, fraud, and fraud controls. J Forensic Investigative Acc **8**(1): 29–35
16. Pearson TA, Singleton TW (2008) Fraud and forensic accounting in the digital environment. Issues Acc Educ 23(4):545–559
17. DiGabriele JA, Lohrey PL (2016) The valuation of economic damages: a case study for the forensic accountant. J Forensic Investig Acc 8(2):338–348
18. Huber W et al (2015) Factors which may bias judges' decisions to exclude accounting expert witnesses testimony. Acc Res J 28(1):59–77
19. Quirin JJ, O'Bryan D (2015) The marriage of Sharon and Henry sawbones: a forensic case illustrating the use of a tax return in a litigation advisory services context. Issues Acc Educ 31(3):347–354
20. Brennan NM (2005) Accounting expertise in litigation and dispute resolution. J Forensic Acc 2:13–35
21. DiGabriele JA (2008) Gender choices, business valuation, state specific variables in the division of marital assets. J Leg Econ 15(2):1–16
22. Carpenter TD, Durtschi C, Gaynor LM (2011) The incremental benefits of a forensic accounting course on skepticism and fraud-related judgments. Issues Acc Educ 26(1):1–21
23. Apostolou N, Crumbley DL (2008) Auditor's responsibilities with respect to fraud: a possible shift? CPA J 78(2):32
24. Alhusban AAA et al (2020) The regulatory structure and governance of forensic accountancy in the emerging market: challenges and opportunities. J Govern Regulation/Volume 9(4):149–161
25. Alshurafat H et al (2020) Perceptions of the usefulness of various teaching methods in forensic accounting education. Acc Educ 29(2):177–204
26. DiGabriele JA (2012) A case study on the determination of lost profits for the forensic accountant. Issues Acc Educ 27(3):751–759
27. Crumbley LD, Rezaee Z, Elmore RC (2004) Forensic accounting education. Adv Acc Educ Teach Curric Innov 6:193–231
28. Smith EP (2012) The basics of business valuation, fraud and forensic accounting, and dispute resolution services. The CPA J 82(6):6–11
29. Cook GJ, Clements LH (2009) Computer-based proactive fraud auditing tools Gary J. Cook. J Forensic Investigative Acc 1(2):1–23
30. Smith GS (2005) Computer forensics: helping to achieve the auditor's fraud mission. J Forensic Acc 6(1):119–134
31. Van Akkeren J, Buckby S, MacKenzie K (2013) A metamorphosis of the traditional accountant: An insight into forensic accounting services in Australia. Pac Acc Rev 25(2):188–216
32. Musleh Al-Sartawi AMA (2020) E-Learning improves accounting education: case of the higher education sector of Bahrain. In: Themistocleous M, Papadaki M, Kamal MM (eds) Information Systems, vol 402. Lecture Notes in Business Information Processing. Springer, Cham, pp 301–315. https://doi.org/10.1007/978-3-030-63396-7_20
33. Alshurafat H et al (2019) Forensic accounting core and interdisciplinary curricula components in Australian universities: Analysis of websites. J Forensic Investig Acc 11(2):353–365
34. Alshurafat H et al (2019) The domain of forensic accounting services: evidence from Australia. 19th Asian Academic Accounting Association (FourA) Annual Conference
35. Sanad Z, Al-Sartawi A (2021) Financial statements fraud and data mining: a review. In: Musleh Al-Sartawi AMA, Razzaque A, Kamal MM (eds) Artificial Intelligence Systems and the Internet of Things in the Digital Era, vol 239. Lecture Notes in Networks and Systems. Springer, Cham, pp 407–414. https://doi.org/10.1007/978-3-030-77246-8_38
36. Musleh Al-Sartawi AMA (ed) (2022) Artificial Intelligence for Sustainable Finance and Sustainable Technology, vol 423. Lecture Notes in Networks and Systems. Springer, Cham. https://doi.org/10.1007/978-3-030-93464-4
37. Rezaee Z, Lander GH, Gavin TA (1992) Forensic accounting in the curriculum. Manag Audit J 7(3):25–29

38. Jackling B, De Lange P (2009) Do accounting graduates' skills meet the expectations of employers? A matter of convergence or divergence. Acc Educ: Int J 18(4–5):369–385
39. Brickner DR, Mahoney LS, Moore SJ (2010) Providing an applied-learning exercise in teaching fraud detection: a case of academic partnering with IRS criminal investigation. Issues Acc Educ 25(4):695–708
40. Lee C-C, Cefaratti M, Rose-Green E (2015) The incremental benefit of a forensic accounting course to creativity. J Forensic Investig Acc 8(1):157–167
41. Bierstaker JL, Brody RG, Pacini C (2006) Accountants' perceptions regarding fraud detection and prevention methods. Manag Audit J 21(5):520–535

Differences in Managers vs. Audience Tone: UK FTSE 350 Earnings Conference Calls

Salah Kayed, Rasmi Meqbel, Omar Arabiat, Huthaifa Al-Hazaima, and Ezzat Ghaidan

Abstract In this study, the linguistic tones of managers and audience during the UK earnings conference calls have been extracted and examined the differences between them. By using a sample of non-financial UK firms listed in the FTSE 350 index over the period 2010–2015, the authors found that manager tones convey much more optimism (less pessimism) than the audience counterparts in the UK earnings conference calls. This result supports academics and practitioners in understanding the language communication during the UK earnings conference calls. Overall, this study is beneficial for regulators, policymakers, and professionals, in evaluating the narrative information disclosed during the UK earnings conference calls.

Keywords Earnings conference call · Tone management · Agency theory · Corporate communication

1 Introduction

The disparity in information offered and the conflict in incentives between stakeholders and managers drive rising demand for corporate communications [31]. More particular, information asymmetry exists between corporate managers and outsiders. For the most part, outsiders don't have access to the same amount of information about

S. Kayed (✉) · R. Meqbel · O. Arabiat · H. Al-Hazaima · E. Ghaidan
Department of Accounting, Faculty of Economic and Administrative Sciences, The Hashemite University, Zarqa, Jordan
e-mail: salahk@hu.edu.jo

R. Meqbel
e-mail: rasmi.meqbel@hu.edu.jo

O. Arabiat
e-mail: omars@hu.edu.jo

H. Al-Hazaima
e-mail: huthaifa@hu.edu.jo

E. Ghaidan
e-mail: ezat@hu.edu.jo

A. M. A. Musleh Al-Sartawi et al. (eds.), *From the Internet of Things to the Internet of Ideas: The Role of Artificial Intelligence*, Lecture Notes in Networks and Systems 557,
https://doi.org/10.1007/978-3-031-17746-0_50

a company's investments as the company's managers do. Because of this, it is difficult for outsiders to assess the company's financial performance and investment potential. Thus, to prevent the significant implications of such disagreement between managers and stakeholders, early prior studies encourage managers to connect with outsiders' users and reveal extra financial information to them (e.g., [20, 24]. As a result, investors and capital providers can better assess management's performance and investment opportunities by using such corporate communication derived from the management environment to mitigate information asymmetry and agency conflicts between outsiders and insiders [5, 25]. Moreover, capital providers have the option to monitor the management's use of their resources through corporate transmission of financial information [5]. The importance of tone (i.e., sentiment as a means of communication has recently been recognized in academic circles. In the capital market, tone is crucial and fundamental information has incremental explanatory power,once disclosed, it immediately influences the market return. [15, 17, 18, 21, 23, 38]. Managers utilize it to communicate their performance to market players. As a result, managers can utilize tone as a tool to reinforce consumers' views of financial performance and eliminate information asymmetry [16–18, 21, 23, 34, 36, 38]. Sustainability reports, news media articles, corporate filings, earnings press releases, corporate environmental disclosure, and earnings conference calls are all examples of places where tone can be used. The tone in the earnings conference call is the focus of this research.

Today, earning conference calls are one of the expressive mediums of communication between firm's management and market participants [39]. Earning conference calls help management circulate and interactively discuss information with audience, particularly analysts, which helps produce more useful information to stakeholders. An earning conference call is important communication medium with unique features such as corporate fillings, that distinguishes it from other formal communication tools. Earning conference call can provide orthogonal information which effectively influence the decisions-making of stakeholders [10, 22, 30, 37, 38]. Participants of earnings conference call are usually analysts who are considered as sophisticated users and accounting and financial experts. Information provided in such conference call has a dynamic frame (Blau, 2015), compared to other formal communication tools that have a static frame. Scripts of earnings conference calls can be shown and presented differently according to the nature of discussion and interaction between participants including angers, analysts and other stakeholders. Despite its importance as a source of information in the capital market [12, 22], earnings conference calls can be more vulnerable to a managerial vague than other reports [11]. This managerial vague according to [32] can be due to firm's management deception in earnings conference calls. The conflict between researchers in evaluating this medium of management commination motivates the authors to gain deeper insights into this communication channel. The concentration in this study is on UK earnings conference calls, since most of previous studies consider US calls and very limited studies look at UK calls. By using a sample of non-financial UK firms listed in the FTSE 350 index over the period 2010–2015, this study attempted to conduct in-depth investigation of conference call transcripts to identify and compare managers and audience's linguistic tones. The study aimed to answer the question that says 'Do linguistic tones differ significantly between managers and audience during the UK earnings conference call?'.

This study is the first that examines the differences between mangers' and audience's tone in UK earnings conference calls. This study is closely related to [9] who study the differences between mangers' and analysts' tone in earnings conference calls. However, we differ from them as they use US market, while our study uses UK market.

The rest of the paper is structured as follows. A background of tone disclosure is provided in the next section. Then, the sample selection, data sources, tone's measure are described in the methodology section. The results' discussion, as well as the study's conclusion, are presented in the final two sections.

2 Background

Tone is considered as one of the narrative disclosures types. It has been studied in various aspects of capital market. Within the US context, prior studies have found that stock returns are positively linked to the tone in news media articles, according to [40]. Similarly, several researchers have found that tone influences market reaction positively, which indicates that when managers disclose an optimistic tone in earnings press releases, the stock price reacts instantly [15, 17, 18], in earnings conference calls [23, 38] or in MD&A section of corporate filling [21]. However, [7] found that the qualitative disclosures optimistic tone, given in conjunction with an earnings release, is only favourably connected to the market reaction when the earnings announcement does not include quantitative earnings guidance. In the UK, [41] show that the frequency of words in annual report narratives of FTSE 350 firms related with "activity" and "realism" groups in Diction software forecast later price increases, even after controlling for a variety of variables. The narratives in the yearly reports of UK FTSE 350 firms are also the topic of [42]. By dividing the entire number of positive terms in Henry's (2008) lexicon by the total number of words in annual reports, they discover a positive association between the level of positivity in annual reports and abnormal returns in short windows surrounding financial report disclosure dates [2]. [31] employ textual analysis to analyse narrative disclosures by source. They show that the market is affected by statements that are stemmed from both management and financial analysts, who are the main part of the audiences in the earnings conference call. Prior literature, however, has not addressed the differences between these two sources in the UK earnings conference call setting. Thus, this study is motivated to address the following question:

Are there significant differences between managers' and audience's tone in UK earnings conference calls? If so, then there will be different level of informativeness of tone disclosed in the call and this will affect the investors' decision when re-evaluating the firm's value.

Table 1 Sampling details

Number of firms classified under FTSE 350 in any year from 2010 to 2015	471
Less: financial firms	(155)
Less: firms with missing data during the sample period	(1)
Final Number of firms	315

Table 2 The collection process of earnings conference calls' transcripts

Earnings conference calls' transcripts available during the period 2010–2015 for 315 firms	955
Less: earnings conference calls' transcripts that have wrong format which lead to be wrongly divided	(54)
Earnings conference calls' transcripts used in the sample	901

3 Research Methodology

3.1 Sample Selection and Data Sources

Our study sample consists of UK firms listed on FTSE 350 during the period 2010–2015.

This index consists of largest 350 firms based on their market capitalisation. The reason for choosing it, is that not all firms have their earnings conference call, they majority are from large firms. Therefore, this index is suitable for the study purpose. To avoid selection bias, the sample comprises of firms that are listed in London stock exchange (LSE) over the whole sample period (2010 – 2015), and have been classified under FTSE 350 in any year over the sample period. We started our period from 2010 to avoid the effect of the financial crises that began in 2007 and had an influence on several company to get bankrupted, thus on the FTSE list changed dramatically from the period 2007 to 2009, and this will result in reducing the sample size. We also end our sample period in 2015 to avoid any potential effects of Brexit[1] on tone disclosure.

Data for tone disclosure was collected manually from firm's annual conference call, we rely on the Refinitiv eikon database to collect firms annual conference call.

Table 1 shows the results of the number of firms selected in the sample:

After identifying the firms, we then extracted the earnings conference calls' transcripts for these firms. Table 2 describes the collection process of earnings conference calls' transcripts.

[1] Brexit refers to the British exit from the European Union (EU), where on 23 June 2016, UK voters chose the UK to leave the EU (Los et al., 2017).

3.2 Variable Measurement

Tone in earnings conference call

This study employs [35] word list to derive the frequency of positive versus negative words shown in the annual earnings conference call. Following [6, 9] and [13], who recently study the tone in earnings conference call, the tone is constructed through calculating the difference between the frequency of positive and negative words spoken by the managers or audience in earnings conference calls scaled by the summation between them.

An earnings conference call comprises two parts (i.e., presentation, and questions and answers (Q&A)). We targeted manager's tone in both presentation and Q&A parts. To do so, PYTHON software is employed to distinct managers' words from audience's words in Q&A part in each earnings conference call transcript. In line with [26], 2008) and [42], the textual-analysis software, DICTION, is employed in this study to obtain the number of words frequencies for managers' words in presentation and Q&A parts, and audience's words in Q&A part based on [35] word list.

4 Results

The results of this study add to the literature on tone management in the context of UK earnings conference call. The results relate to an analysis to the differences between the managers' and audiences' tones in UK earnings conference call. The results show that the average of optimistic management tone is (0.340) while it is (−0.259) for the audience. This result means that managers use more optimistic words in their speech than the audience do. Figures 1 and 2 illustrate the difference between tones of managers and audiences. The Figures show that unlike the audience's tone which is less than zero in most cases, most of the managers' tone is above zero. This indicates that the audience speak pessimistically during the call compared to managers.

Fig. 1 Management tone

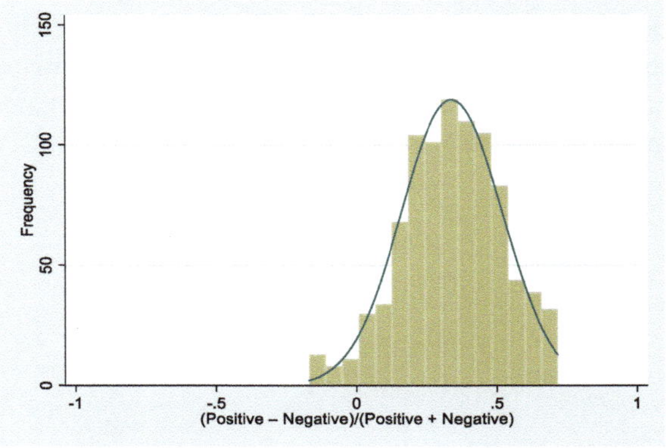

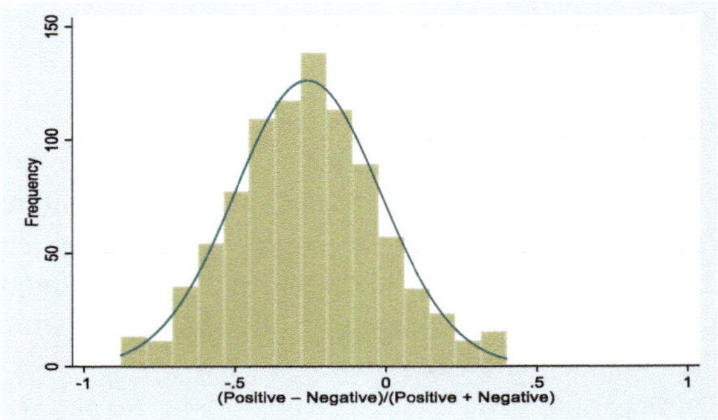

Fig. 2 Audience tone

This study also investigated the significant differences in the means of management and audience tone. According to Panel A in Table 3, significant differences between the mean of management tone and the mean of audience tone exist. Table 3 shows that the number of observations for the management tone in the sample is 901 whereas it is 896 for the audience tone. The difference in number is 5 observations which belong to 4 excluded transcripts due to absence of audience's questions or comments, and a transcript that does not show any positive or negative words from [35] list of words in the audience text.

In Panel B in Table 3 however, the study divided the management and audience tones into 3 sub-groups including low, medium, and high, and ordered them based on terciles as per the separated dimensions of management tone and audience tone. A difference-in-means test was then performed within each sub-group to determine whether the significant large result found in Panel A is fairly distributed over the entire sample, otherwise, evenly derived by a subset of the sample. The first order is based on the management tone while the second one is based on the audience tone. In each tercile, significant differences are found between management and audience tones. However, the difference is found to be always for the favour of management tone as it carries always greater values than the values of the audience tone (all p-values = 0.00). This result indicates that managers are able to use more optimistic tones in their speech in UK earnings conference call than the audience can do. The reason behind the audience inability to speak optimistically can be due to their attempt to restrict managers to push them disclose true information. This argument supports the argument of [9] who also found that managers' tone was much more positive than analysts' tone in US earnings conference call.

Table 3 Management and audience tone differences

	No. of Observations	Management Tone	Audience Tone	Tone Difference	p-Value
A. Tone differences	901 management, 896 audience	0.34	-0.259	0.599***	0
B. Tone differences by tercile					
Management Tone sort	No. of Observations	Management Tone	Audience Tone	Tone Difference	p-Value
Low	301 management, 299 audience	0.144	-0.333	0.477***	0
Medium	300 management, 299 audience	0.346	-0.254	0.600***	0
High	300 management, 298 audience	0.532	-0.189	0.721***	0
Audience Tone sort					
Low	301 management, 298 audience	0.388	-0.378	0.766***	0
Medium	300 management, 300 audience	0.366	-0.271	0.638***	0
High	300 management, 298 audience	0.266	-0.128	0.394***	0

p-values are in parentheses. *** denotes statistical significance at the 1% level.

5 Conclusion

The dynamic meeting between firm's managers (i.e., CEO and CFO), and the audience (i.e., financial analysts and other market participants), in UK earnings conference calls is the main focus of this paper. By using a sample of non-financial UK firms listed in the FTSE 350 index over the period 2010–2015, we compare the linguistic tones of managers and audience and show that the most optimistic tone during an earnings conference call is employed by managers, whereas the most pessimistic tone is stemmed from audience during a such call. More clearly, tones disclosed by managers are significantly more positive than audience tones. This finding contributes to the understanding of the communication roles in an earnings conference call played by managers and audience. Clearly, this finding suggests that investors and decision-makers should be more cautious and pay more attention to managerial incentives when evaluating the narrative content of managerial disclosures. The finding also highlights the important role of audience's questions during the call. Specifically, the pessimistic tone in audience's questions could help users in obtaining the truthful information from an earnings conference call. Future studies in this filed should consider the market reaction of tone of managers and audience in UK earnings conference calls.

References

1. Abernathy JL, Beyer B, Rapley ET (2014) Earnings management constraints and classification shifting. J Bus Financ Acc 41(5–6):600–626
2. Al-Sartawi A (2018) Institutional ownership, social responsibility, corporate governance and online financial disclosure. Int J Crit Acc 10(3/4):241–255
3. Barkemeyer R, Comyns B, Figge F, Napolitano G (2014) CEO statements in sustainability reports: Substantive information or background noise? Acc Forum 38(4):241–257
4. Beretta S, Bozzolan S (2008) Quality versus quantity: the case of forward-looking disclosure. J Acc Audit Financ 23(3):333–376
5. Beyer A, Cohen DA, Lys TZ, Walther BR (2010) The financial reporting environment: review of the recent literature. J Acc Econ 50(2–3):296–343
6. Blau BM, DeLisle JR, Price SM (2015) Do sophisticated investors interpret earnings conference call tone differently than investors at large? Evidence from short sales. Journal of Corporate Finance, vol. 31: 203–219
7. Bonsall SB, Bozanic Z, Fischer PE (2013) The informativeness of disclosure tone. Working paper. The Ohio State University
8. Boudt K, Thewissen J (2018) Jockeying for position in CEO letters: Impression management and sentiment analytics. Financ Manag 48(1):77–115
9. Brockman P, Li X, Price SM (2015) Differences in conference call tones: managers vs. analysts. Financ Anal J 71(4):24–42
10. Brown S, Hillegeist SA, Lo K (2004) Conference calls and information asymmetry. J Acc Econ 37(3):343–366
11. Bushee BJ, Gow ID, Taylor DJ (2018) Linguistic complexity in firm disclosures: obfuscation or information? J Acc Res 56(1):85–121
12. Bushee BJ, Matsumoto DA, Miller GS (2003) Open versus closed conference calls: the determinants and effects of broadening access to disclosure. J Acc Econ 34(1–3):149–180
13. Chen JV, Nagar V, Schoenfeld J (2018) Manager-analyst conversations in earnings conference calls. Rev Acc Stud 23(4):1315–1354
14. Cho CH, Roberts RW, Patten DM (2010) The language of US corporate environmental disclosure. Acc Organ Soc 35(4):431–443
15. Davis AK, Tama-Sweet I (2012) Managers' use of language across alternative disclosure outlets: earnings press releases versus MD&A. Contemp Acc Res 29(3):804–837
16. Davis AK, Ge W, Matsumoto D, Zhang JL (2015) The effect of manager-specific optimism on the tone of earnings conference calls. Rev Acc Stud 20(2):639–673
17. Davis AK, Piger JM, Sedor LM (2012) Beyond the numbers: measuring the information content of earnings press release language. Contemp Acc Res 29(3):845–868. https://doi.org/10.1111/j.1911-3846.2011.01130.x
18. Demers E, Vega C (2011) Soft information in earnings announcements: News or noise? Working paper. INSEAD
19. Dimson E, Nagel S, Quigley G (2003) Capturing the value premium in the United Kingdom. Financ Anal J 59(6):35–45

20. Dye RA (1998) Investor sophistication and voluntary disclosures. Rev Acc Stud 3(3):261–287
21. Feldman R, Govindaraj S, Livnat J, Segal B (2010) Management's tone change, post earnings announcement drift and accruals. Rev Acc Stud 15(4):915
22. Frankel R, Johnson M, Skinner DJ (1999) An empirical examination of conference calls as a voluntary disclosure medium. J Acc Res 37(1):133–150
23. Frankel R, Mayew WJ, Sun Y (2010) Do pennies matter? Investor relations consequences of small negative earnings surprises. Rev Acc Studies 15(1):220–242
24. Gjesdal F (1981) Accounting for stewardship. J Acc Res 19(1):208–231
25. Healy PM, Palepu KG (2001) Information asymmetry, corporate disclosure, and the capital markets: a review of the empirical disclosure literature. J Acc Econ 31(1–3):405–440
26. Henry E (2006) Market reaction to verbal components of earnings press releases: event study using a predictive algorithm. J Emerging Technologies in Acc 3:1–19
27. Henry E (1973) Are investors influenced by how earnings press releases are written?. J Bus Commun 45(4):363–407
28. Hsieh CC, Hui KW (2011) Analyst report readability in high-technology firms. Working paper, The Hong Kong University of Science and Technology
29. Huang X, Teoh SH, Zhang Y (2014) Tone management. Acc Rev 89(3):1083–1113
30. Kimbrough MD (2005) The effect of conference calls on analyst and market underreaction to earnings announcements. Acc Rev 80(1):189–219
31. Kothari SP, Li X, Short JE (2009) The effect of disclosures by management, analysts, and business press on cost of capital, return volatility, and analyst forecasts: a study using content analysis. The Acc Rev 84(5):1639–1670
32. Larcker DF, Zakolyukina AA (2012) Detecting deceptive discussions in conference calls. J Acc Res 50(2):495–540
33. Li F (2008) Annual report readability, current earnings, and earnings persistence. J Acc Econ 45(2–3):221–247
34. Li F (2010) The information content of forward-looking statements in corporate filings—a naïve Bayesian machine learning approach. J Acc Res 48(5):1049–1102
35. Loughran T, McDonald B (2011) When is a liability not a liability? Textual analysis, dictionaries, and 10-Ks. J Financ 66(1):35–65
36. Miller G, Piotroski J (2002) Forward-looking earnings statements: determinants and market response. Working paper. Harvard Business School and University of Chicago
37. NIRI (2004) Standards of practice for investor relations. Vienna: national investor relations institute
38. Price SM, Doran JS, Peterson DR, Bliss BA (2012) Earnings conference calls and stock returns: the incremental informativeness of textual tone. J Bank Financ 36(4):992–1011
39. Sakhnini Z, Elali W, Al-Sartawi A (2022) Behavioral and non-behavioral factors and the level of adapting and implementing Fintech and e-banking in Bahrain: suggested model. In: Musleh Al-Sartawi AMA (ed) Artificial Intelligence for Sustainable Finance and Sustainable Technology, vol 423. Lecture Notes in Networks and Systems. Springer, Cham, pp 188–193. https://doi.org/10.1007/978-3-030-93464-4_19
40. Tetlock PC (2007) Giving content to investor sentiment: The role of media in the stock market. J Financ 62(3):1139–1168
41. Wisniewski TP, Yekini LS (2015) Stock market returns and the content of annual report narratives. Acc Forum 39(4):281–294
42. Yekini LS, Wisniewski TP, Millo Y (2016) Market reaction to the positiveness of annual report narratives. British Acc Rev 48(4):415–430

Artificial Intelligence and Its Impact on Accounting Systems

Aysha Khaled AlKoheji and Abdalmuttaleb Al-Sartawi

Abstract Artificial intelligence is called as forth industrial revolution. The concept of artificial intelligence is to help humans in doing tasks and taking decisions by using machines. Nowadays technologies and artificial intelligence impacting our lives in many fields, and it keeps developing and changing in a rapid pace. People and companies are trying to cope with this advanced technology and apply it in ease their live in many areas. This literature review is important for business and organizations. It is essential for organization's business owner's and top managements to under- stand the power of artificial intelligence on business, how artificial intelligence improve their businesses sales, reduce costs, and enhance performance in many tasks in the firms the challenges that can face companies who is investing in artificial intelligence. Moreover, this literature review can be important to accountants also to understand the future of the profession and how AI can affect their career. Also, how to develop their skills. This literature review explained what artificial intelligence is and how it started. Examples of artificial intelligence applications in different fields like health, education, and business and the application in accounting in specific the big 4 accounting firms in the world. Moreover, this paper shows the tasks that can be replaced by AI, the impact of artificial intelligence on accounting field in specific in regard of how tasks can be replaced by artificial intelligence, the routine and time-consuming tasks that can replaced by AI. Also benefits and challenges of implementing AI was explained with an analysis of AI application in some organization in Malaysia. Further- more, it discussed the future of accounting position when applying AI and how this will impact the new accounting role, required skills and abilities. At last future research area was discussed.

Keywords Artificial intelligent systems · Accounting · AI applications · Industry 4.0. · Accounting systems

A. Khaled AlKoheji—Freelancer.

A. Khaled AlKoheji
Manama, Bahrain

A. Al-Sartawi (✉)
Ahlia University, Manama, Bahrain
e-mail: amasartawi@hotmail.com

A. M. A. Musleh Al-Sartawi et al. (eds.), *From the Internet of Things to the Internet of Ideas: The Role of Artificial Intelligence*, Lecture Notes in Networks and Systems 557,
https://doi.org/10.1007/978-3-031-17746-0_51

1 Introduction

In this literature review, some procedures were followed to guarantee a high standard of research. In the beginning, a wide research of peer reviewed journals was conducted using abroad key terms such as: Artificial intelligence, accounting, accounting systems, accounting software, and AI Application. Searching was conducted on databases for instance EBSCOHOST and google scholar.

Also, for the reference section for each article found have been used to find more articles as well. Moreover, key journals published on: Scientific Research Publishing Inc, International Scientific Publications, Advances in Social Science, Education and Humanities Research, and the CPA journal. This research process excludes all peer-reviewed articles before 2018.

2 Literature Review

2.1 What is AI

Is the creation of intelligence by machines that are programmed to take decisions and think like humans which is not real intelligence but created by humans through machines using programming that's where the name artificial intelligence come. The main part of artificial intelligence is machine learning which means specific program can adopt to any new data or information that entered the system and configure and learn from it without any interaction of humans.

As Lee and Farzana stated that artificial intelligence is the fourth industry revolution as programmers make machines learn from past experiences, adapt to new data, act and perform like human intelligence. (Cheah Saw & Farzana Parveen, 2020).

The main idea that comes to people's mind is robots but it's just machines that can imitate human intelligence such as decision making but in faster speed than human's brain.

2.2 Beginning of AI

The beginning of this concept was in 1956 at university of Dartmouth by John McCarthy and other experts. From that time to 1970s computers was developing to store and save more information and data. Moreover, computers become faster, cheaper, and available to more people. In 1980 the concept of deep learning, which is machines can learn from using data and experience [15].

The first application of artificial intelligence was on China in 2017 in more than 15 department and finance department was one of them. today many countries are devel- oping and trying to use artificial intelligence in different sectors and fields such as health, commerce, education, and agriculture [14].

2.3 *Application of AI in Daily Life*

Furthermore, everybody can notice how artificial intel ligence are affecting our lives from the smartphones that can understand our speech and transitioning it to typed words to using internet and browsing, the order of displaying the result is also IA as it shown based on habits of search and preferences for every person, until the self-driving cars which became a true and produced by many famous companies like Tesla and BMW, it is artificial intelligence.

AI in health field.
AI are affecting every aspect of our life for example in healthcare, as [19] stated in his article that at Children's National Medical Center in Washington a robot demonstrates a soft tissue surgery, the team was supervising the surgery and the stitching, a nd they conclude that the robot finishes this mission better even than human surgeon. Further- more, many positive impacts on healthcare field such as fast and accurate diagnostic and reduce medical errors [19].

AI in education field.
In education AI is growing rapidly, teachers are using AI to mark the tests and find gaps in students' performance with recommendations to change the way of teaching a specific lesson if a high percentage of the class did not answer it.

AI in business field.
For the time being in business sector, AI have a great impact in cutting cost and time, make faster decisions and increase productivity. As technology and AI is developing and growing every day and experts are using AI more and more in business, the application of AI will be high in the next years and most of the companies and businesses will invest in using those machines.

3 Application of AI in Accounting

3.1 *Application of Big 4 Companies*

In accounting sector, CPA firms can use AI to enhance productivity and efficiency, the big 4 companies (Deloitte, EY, KPMG, and PwC) had already started investing in AI. For example: EY started to use AI in analyzing lease of contracts as EY stated that it is easier to capture relevant information of the contract [3]. Additionally, EY used drones also for inventory stock count during the auditing process as drones can count the num- ber of stocks for example cars and report data directly to the global auditing platform "EY CANVAS". Furthermore, Deloitte company also developed a self-reviewed sys- tem that automates reviewing of documents and they claim that this technology reduced the time spent in reviewing document by 50% [12].

As big 4 companies are investing and working to develop AI technologies, small firms around the world should also consider starting to invest on AI immediately as they will be left behind and outdated due to the dramatically adopting of AI globally [4].

3.2 Tasks that Can Be Replaced by AI

Many tasks can be replaced by IA and organizations and accounting firms should consider investing in transferring them to AI technologies such as:

Book Keeping.
Bookkeeping is recording all business transactions such as revenue, sales, payroll, and expenses in day-to-day basis. It is a time consuming and routine task, that can be fully transferred to a system through machine learning technologies as the logic behind dou- ble entry is to set specific coding for the entries so the system itself can segregate the business transaction and sort them into the ledgers [2]. The main advantage from automa- tion bookkeeping is the accuracy of the transactions and information as there is a pos- sibility of error and mistakes which is human -being natural comparing with the AI, there is no errors or mistakes. As this task is taking a lot of time every day, automate it will save a lot of time for the accountant which he can concentrate in other complicated tasks [11].

Revenue forecasting.
revenue forecasting is a process to estimate revenue of the company for a certain period. As forecasting revenue is very critical task as the result of it will affect decision that managers use it to set the budget of the company, using machine learning will increase the accuracy of the forecast and that will lead to a better performance of the company, as accountants may spend a lot of time completing this task and reviewing the data provided with a chance of human errors [5].

Financial reporting.
Financial reporting is a good opportunity of automation as all information of the report are taken from the data base and the general ledger of the company so it will be easy to transfer to machine learning with less errors and reducing number of hours of accountants preparing all required reports [7, 9].

4 Benefits of AI on Accounting

The most important benefit of AI in all fields and in accounting as well is reducing time and cost. As machine learning is faster than human's brains, it completes the task in much less time comparing to humans doing the task, it will reduce the time of doing the task so the accountant will just need to review gathered information [17]. Time saved can be used in many other complicated tasks than routine tasks. Moreover, the cost also will be reduced as each company will need much less hours of working if they depend on AI.

Furthermore, the accuracy of information provided through machine learning will be very high comparing to traditional accounting when accountant have to enter all transactions, doing financial reports and reviewing all those information, it will take a lot of resources such as time an d money, and also there will be a very high possibility of mistakes but if the company depending on systems or AI the possibility of mistakes will reduce to minimal and even if there is an error the system can detect it easily and flag it to the accountant to take the required action instead of spending a lot of time searching for the error [6, 21].

Preventing and deducting fraud is also another important task as it is easy for ma- chine learning to trace and detect any unusual or up normal transaction such as tax avoidance or missing of cash and many other examples that may damage the company.

As these actions are hard for the managers or owners of the companies to track or detect, machine learning can easily identify these activities as it is not normal activities in the algorithm of the program [11].

"According to the Association of Certified Fraud Examiners, the average organization loses five percent of its annual revenue to internal fraud. Organizations and auditors can typically only audit 10 percent of expense reports manually, leaving most of the potential fraud to go undetected." [20]. However, AI can audit 100% of all reports and financial data by detecting irregularities in data. Also, AI can help auditor to identify fraudulent spending or receiving. In fact, the more data analyzed by AI, the smarter it will be to distinguish frauds [18].

It is essentials especially for small and medium companies where they do not segre- gate duties of accountants and managers due to the size of company and shortage of employees. So, applying AI will reduce the possibility of frauds [8].

Moreover, AI will save the environment because of reduction usage of paper and printing as most of the data will be in systems [16].

5 Challenges of AI on Accounting

AI is as any other technology facing some challenges at the beginning that may affect.

organizations and companies' decisions to invest in AI such as:

5.1 Lack of Experience

One of the most concerns in business and accounting fields in specific is lack of expe- rience in initial stage as it is a new technology that countries are trying to understand and apply. although China applied AI in 2016 into some departments such as Finance.

department but still China is still on its early stages of AI as they are using automation technology with clear algorithm which means it is processing large number of repeated scenarios without the ability of learning from new data and taking decisions which called weak [15] AI. Still there is a gap between AI and accounting work that needs to be developed so AI can take later most of the work steps such as financial analysis [14].

5.2 Algorithm Bias

It is a systematic repeatable error from the computer system that le d to unfair conclusions. Many reasons are causing this problem such as: design of algorithm and poor input of data. [1].

This bias can lead to a wrong figures and results and may be affect the decisions of the firm. Firms who are using algorithm must be sure of the design of algorithm and the data.

5.3 Data Privacy and Security

Data privacy and security is one of the main concerns globally as AI systems have the access to a massive amount of information and data. Organizations and accountants must ensure control and unauthorize access and usage of information as this can lead to many consequences [1].

6 AI Application in Malaysia

A study [13] was conducted to investigate the use and impact of artificial intelligence.

(AI)-based accounting software among organizations in Malaysia.

The study includes 9 organizations from different sectors and that are implementing AI based accounting software. Although these organizations were using different ac- counting software programs, but the usage was almost the same among them, the main purpose was to store the images of the invoices in the server to avoid loss or destruction of invoices.

Additionally, the second common usage was to auto capture the invoice information, moreover, some organizations was using this feature to pulls all the inform ation re- quired from the document and fill the entry automatically to Accounts Payable.

As a result, fraud cases possibilities will reduce because if any invoice is not match- ing with PO, it will be blocked from any action from the system. History log and users' activity is an important usage also as every amendment on any transaction processed is shown with the time, date, username.

The benefits of AI application were numerous, main improvement is in Account payable functions efficiency as it is automated, enhancement in customer service and productivity. Moreover, flexible working style was one of the benefits and reducing costs of manpower and save time. (Cheah Saw & Farzana Parveen, 2020).

This example can be used as a base ground for other companies who are eager to use.

AI in their business.

7 Future of the Accounting Profession

Many experts and accounting professionals are arguing whether applying AI will be a consequence to replace accountants completely, chaw lee and Farzana st ates that ac- counting field is one of the fields that will be enhanced by AI. But still the field needs human effort in taking decision in critical situations as AI can replace specific tasks not the full tasks of an accountant. [17].

Consequently, in the short term, the demand of entry -level accountant will decrease as most of the routine tasks will be done by AI, as a result some of entry-level account- ant will lose their jobs. However, application of AI should be considered as an oppor- tunity not a challenge for the people in accounting field. But in the long-term account- ants should take the advantage from AI, develop themselves, update their knowledge and cope with the AI technology [19].

From other hand, automation specific tasks give accountant the opportunity to allo- cate their resources in learning and developing their skills for example how to use AI data and reports. Moreover, they can spend more time to build a relationship with cli- ents or corporates.

8 New Role of Accountants

Moreover, with this transformation in the accounting industry because of implementing AI, accountants must develop and improve themselves in different aspects to be able to face the impact of AI on the field. First, computer skills are now at the top of require- ments for any accountant and in future also it will be a requirement to learn not only basic computer skills but some computer programming and coding techniques to de- velop their data capabilities. This will help accountants in transferring knowledge to other departments also.

Second, leadership skills are increasing in importance as accountants will participate more in company's management and collaboration with other departments rather than doing paper works which will be delegated to AI, as a result, leadership skills is an essential in this case. Accountants who take this opportunity to fill the gap in perform- mance and knowledge. This will give an advantage for skilled accountants ov er the rest [19].

Third, decision making ability is important also after this transformation as account- ant will have a huge amount of data, information and reports and accountant will have evaluate and take the decision for short term and long-term decisions according to the data given from AI. Also due to the change of accounting roles and responsibilities,

accountants will be involved in developing AI such as training or testing modeling in projects to help experts in AI to understand and manage the data.

AI is going to change a lot of things in accounting filed. It will decrease time, cost and increase efficiency of the work. Thus, from the examples of implementation of AI- based accounting software case study in Malay sia, companies show the positive impact of application of AI. Moreover, in future more companies will go to AI so the future of accounting position will change in some areas and number of accountants may be af- fected and decreased, but AI cannot replace accountants at all. The field still needs humans to continue but the new accounting role and the skills required will changed in consistent with AI adoption [11].

9 Future Research Area

As the whole AI concept is a new concept, most of the literature reviews discussing the concept itself and the possible impact and challenges but there is lack of case studies for companies that are using AI, and this is due to the low number of companies and organizations that are using AI. Nowadays organizations in developed countries are slowly turning to AI. More case studies and article should be conducted to document their experience, benefits, and challenges also. This can be as a guideline for a lot of organizations that have doubts entering this field [10].

As a conclusion, this literature review discussed the beginning of AI and how it started, the impact of artificial intelligence in different fields like health, education, and business. Moreover, the literature review explained accounting tasks that can be re- placed by AI, benefits and challenges of AI application, and impact of application AI in profession as well as the new accounting role and skills required in future as a result of the expand use of AI in accounting field supported with an example of AI application in organizations in Malaysia.

A good implication of AI can increase productivity, efficiency, and accuracy, and cut time and cost as the case study of organizations in Malaysia proved that also.

References

1. Alacron J, Fine T, Cory N (2019) Accounting AI and machine learning: applications and challenges
2. Alansari Y, Al-Sartawi A (2021) IT governance and E-banking in GCC listed banks. Procedia Comput Sci 183:844–848
3. Al-Okaily M, Al-Sartawi A, Hannoon A, Khalid AA (2022) Information technology governance and online banking in Bahrain. In: Musleh Al-Sartawi AMA (ed) Artificial Intelligence for Sustainable Finance and Sustainable Technology, vol 423. Lecture Notes in Networks and Systems. Springer, Cham, pp 611–618. https://doi.org/10.1007/978-3-030-93464-4_60
4. Al-Qassim M, Chen W, Al-Sartawi A (2022) Blockchain in environmental compliance and enforcement. In: Musleh Al-Sartawi AMA (ed) Artificial Intelligence for Sustainable Finance and Sustainable Technology, vol 423. Lecture Notes in Networks and Systems. Springer, Cham, pp 393–398. https://doi.org/10.1007/978-3-030-93464-4_39
5. Al-Sartawi A (2018) Ownership structure and intellectual capital: evidence from the GCC countries. Int J Learn Intellect Cap 15(3):277–291
6. Al-Sartawi A (2018) corporate governance and intellectual capital: evidence from gulf cooperation council countries. Acad Acc Financ Stud J 22(1):1–12
7. Al-Sartawi A (2018) Institutional ownership, social responsibility, corporate governance and online financial disclosure. Int J Crit Acc 10(3/4):241–255
8. Al-Sartawi A (2020) Does it pay to be socially responsible? Empirical evidence from the GCC countries. Int J Law Manag 62(5):381–394
9. Al-Sartawi A (2020) Shariah disclosure and the performance of Islamic financial institutions. Asian J Bus Acc 13(1):133–160
10. Al-Sartawi A (2020) Social media disclosure of intellectual capital and firm value. Int J Learn Intellect Cap 17(4):312–323
11. Eleonora P-T (2018) How artificial intelligence is challenging accounting profession
12. Hazelbaker T, Lin P (2019) Meeting the challenge of artif icial intelligence
13. Lee CS, Tajudeen F P (2020) Impact of artificial intelligence on accounting: evidence from Malaysian organizations. Asian J Bus Acc 13(1)
14. Luo J, Meng Q, Chai Y (2018) Analysis of the Impact of artif icial intelligence application on the development of accounting industry 850
15. Musleh Al-Sartawi AMA, Razzaque A, Kamal MM (eds) (2021) Artificial intelligence systems and the internet of things in the digital era, vol 239. Lecture Notes in Networks and Systems. Springer, Cham. https://doi.org/10.1007/978-3-030-77246-8
16. Musleh Al-Sartawi AMA (2020) E-learning improves accounting education: case of the higher education sector of Bahrain. In: Themistocleous M, Papadaki M, Kamal MM (eds) Information Systems, vol 402. Lecture Notes in Business Information Processing. Springer, Cham, pp 301–315. https://doi.org/10.1007/978-3-030-63396-7_20
17. Musleh Al-Sartawi AMA (ed) (2022) Artificial Intelligence for Sustainable Finance and Sustainable Technology, vol 423. Lecture Notes in Networks and Systems. Springer, Cham. https://doi.org/10.1007/978-3-030-93464-4
18. Musleh Al-Sartawi AM, Hussainey K, Razzaque A (2022) The role of artificial intelligence in sustainable finance. J Sustain Financ Invest 1–6
19. Tai M (2020) The impact of artif icial intelligence on human society 340
20. Vordenbaeumen H (2019) 3 Ways accountants can implement AI today
21. Zehong L, Li Z (2018) The impact of artif icial intelligence on accounting

How Gender Affected the Acceptance of Remote Work for Accountants During Coronavirus Lockdown in Jordan

Qutaiba Adeeb Odat and Merwiey Alaqrabawi

Abstract During the coronavirus outbreak, accountants in Jordan shifted their ordinary routine to remote working under the government and health ministry regulations. This paper investigates the moderating effect of gender on the actual use of technology and other determents believed to be critical factors influencing the usage. These factors are compiled under three theoretical frameworks, namely, technology acceptance, unified theory of acceptance and use of technology, and diffusion innovation theory. Technology acceptance compiled two factors: perceived ease of use and perceived usefulness. While unified theory of acceptance and use of technology consists of the moderate factor gender. Likewise, the diffusion innovation theory consists of the system's technological features, such as the complexity used in this paper. The overall results of this paper are based on previous scholars' findings and results from different technology acceptance and usage contexts. Researchers indicate that even with the mixed results presented by previous literature, research assumes that gender behavioral patterns toward remote working differ based on the time of risks and crises such as coronavirus and the time period.

Keywords Remote work · Gender · TAM · UTAUT · Diffusion innovation theory · Coronavirus

1 Introduction

The novel impact of the coronavirus pandemic has affected most of the world's careers and introduced new rules that state social interactions and distances. As a matter of fact, the pandemic has shifted most careers into remote work to flatten the curve of the disease [1]. Based on stay at home rules and orders, accountants worldwide were required to contribute to their jobs from home. These sudden procedures might shock

Q. A. Odat (✉) · M. Alaqrabawi
The Hashemite University, Zarqa, Jordan
e-mail: 2070014@std.hu.edu.jo

M. Alaqrabawi
e-mail: 2070005@std.hu.edu.jo

A. M. A. Musleh Al-Sartawi et al. (eds.), *From the Internet of Things to the Internet of Ideas: The Role of Artificial Intelligence*, Lecture Notes in Networks and Systems 557,
https://doi.org/10.1007/978-3-031-17746-0_52

many organizations with no risk or crisis plans to accept technology by employees, especially accountants. The tragic pandemic of covid-19 has impacted many areas in the world, especially the business and economic environment [2]. Therefore, most jobs shifted into remote work are highly increased to help companies sustain their operations, such as accounting.

Accountant jobs are a critical focus for many studies that underlined accounting role in the context of pandemics or risks that impact individuals, firms, and sectors [1, 3]. From the early years of 2008, studies showed how accounting has aided the extreme side effects of the financial crisis and how firms managed the overall decision making [4]. Consistently, recent studies have extended the focus rather than the role of accountants only but also on the responses and behaviors of accountants within the pandemic such as coronavirus [5]. In addition, the adoption of remote working contributes the providing a revolutionary era for the use of accounting systems, whether in public or private sectors [1]. According to [6], remote work for an accountant has shaped how pandemics are perceived, interpreted, and tackled to contribute to firms' best fit capabilities and tools to cope with the uncertainty of events.

However, studies showed that coronavirus exposed the impact of gender inequality in technology acceptance and usage. Several scholars showed that there are inequality forms in accounting roles and performance, especially in Gender. The coronavirus pandemic has exposed these forms of inequality in Gender and the intensified substantial ranging in different developed or developing countries. Therefore, the importance of this paper is to make awareness and examine the gender stereotypes' response and role during the pandemic and threats. Moreover, this paper will highlight the potential acceptance of male and female accountants in Jordan during the coronavirus pandemic. Andrew, Baker [7] noted that "if inequality is created and maintained through discourse and ideology, accounting has an essential role in producing more equitable futures." Therefore, accounting has a vital role in providing equable futures within risks [8].

This paper will draw together to identify the potential factors that affect the usage and acceptance of remote working for accountants based on Gender during coronavirus. The theme of this paper will provide a theoretical framework to extend the knowledge of accounting and accounting information system research. In addition, this paper will seek to point out how the acceptance and the usage of remote work practices differ based on Gender. The rest of this paper is planned to be divided into five sections. The first section complied with the introduction, importance of this paper, and contribution. The third section is the previous literature review. The fourth section is a theoretical framework and hypothesis development. The fifth section will draw the paper design and methodology. The fifth section is the discussion of the findings and conclusion.

2 Literature Review

2.1 *Accountants Remote Working During Coronavirus*

Based on Papadopoulou and Papadopoulou [9], the coronavirus pandemic has impacted the operation of many jobs, especially accounting, pushing to shift to remote jobs. Research has presented that the world economy was on edge during the outbreak; relatively, the accountant's role was much more pronounced at that time [10]. Coronavirus has raised the complexity of accounting in firms where there is an increase in demand for material procedures. The research discussed different aspects of accounting and their shifting into new patterns in various contexts [11].

According to JABIN [12], the situation has crucial implications for utilizing computers in firms and using digital technology within the rapid coronavirus growth. Therefore, accountants must add value to the existing issues and risks by using these new digital technologies to face such risks and obstacles. The study shows that remote jobs in the accounting profession must remain in business even after the outbreak vanished. In contrast, Ardiny [13] found a novel result stating that accounting careers face a significant and complex time disclosing information regarding the firms and their performance within this outbreak.

Furthermore, studies stated several changes raised through the use of new technology in adopting remote jobs [12]. In addition, literature shows different utilizations for accountants' remote jobs, such as training and meetings [10]. The high level of job insecurity has risen while employees face cybersecurity issues in adopting this type of job [14]. The literature review, in this case, has grown in discussing how these significant changes have advanced the accounting careers and that these changes will become the new normal [15, 16].

2.2 *Gender and Technology Acceptance*

As perceived, there is a lack of studies investigating gender as a significant factor influencing accountants' adoption of remote jobs. However, a study by Nash and Churchill [17] found that women were mainly the employees who participated in decisions and designed solutions for the problem faced during the outbreak. In addition, a field study conducted by Dunatchik, Gerson [18] examined the shifting from ordinary to remote jobs and how genders alter them. The study concludes that women have reported more significant stress than men in remote working shifting events. Literature approved those gender roles behavioral may adapt to the lived realities. The study also stated that in the long term, the consequences would depend on the level of experience for both women and men dealing with the shifting to remote jobs [19].

On the other hand, some studies found no significant gender or age group differences; however, remote working poses a challenge for employees due to workplace isolation, family disruption, peer absence, and working too much or not working at all [20]. These findings back up a recent study that found no gender differences in ICD device acceptability [21]. In addition, even though each study had an enormous sample size, the number of females in each study was much lower than the number of males. A more prominent female presence in future studies could be beneficial [18]

3 Theoretical Framework and Propositions Development

This paper designed and integrated research model that consists of three theories namely, the technology acceptance model (TAM), unified theory of acceptance and use of technology (UTAUT) and diffusion innovation theory. TAM theory illustrates how users accept the technology [22, 23]. In the designed model, as illustrated in Fig. 1 consisting of two main factors affecting using the adopted technology: perceived ease of use (PEOU) and perceived usefulness (PU) [24]. Likewise, the UTAUT theory one of the most comprehensive and most experimental models for assessing users' acceptance and tolerance of the technology [25]. In the specific context of this paper, the (PU) and (PEOU) refers to the extent to which he or she believes that using remote jobs is enhanced (i.e., PU). In addition to the extent to which he or she believes that using remote jobs will be easy and free from effort (i.e., PEOU). Meanwhile, gender diversity adopted factor from UTAUT refer to people's behavior that can be interpreted and depicted by the social psychology perspective of this model. Based on a social psychology framework, the studies show that men are more pragmatic and task-oriented than women [26]. In addition, the gender roles and behaviors literature stated that men have personal traits such as masculinity which produce higher confidence and are more likely to face risks. The study reported that females are more likely to prefer easy task routines and fewer risk-takers.

Moreover, other social psychology presented gender roles theories that the relationship between (PU) and technology usage is more vital for male users who have reported higher usefulness perceived from using a new system [27]. Hence, it is evident by Assaker [26], Ong and Lai [28] that women users have a more robust (PEOU) by which they value more the easiness of using a new system attributed to their fewer risks taking nature. Despite that, there is debate and mixed results across several studies on the moderating effect of Gender and the adoption of behavioral technology patterns. For example, Tan and Ooi [29] found that Gender is not a critical factor influencing the PEOU and the PU of users' intention to use technology. However, Wong, Teo [30] study failed to verify any moderating relationship between the gender differences and their effect on the PEOU and the PU of users' behavior. Therefore, in this paper, we have indicated that the impact of Gender on the relationship between the PEOU and the PU is more substantial for male users. In addition, studies revealed a difference in the use of technology based on genders [31].

According to Huang, Hood [32], this paper reported positive attitudes among women toward networking and social media. Subsequently, we propose the following proposition that literature findings will support to confirm in the context of accountants' remote working during coronavirus in Jordan.

P1 and P2: the impact of the determinants (PEOU and PU) on the actual use of accountants' remote working is moderated by Gender.

P3: the impact of the actual use of accountant's remote working is moderated by Gender.

Diffusion innovation theory is one theoretical framework that compiles technology features. The framework is designed by Rogers [33] to describe the user's choice to adopt or not an innovation according to their beliefs. Complexity is one of the technological features that appeared in diffusion innovation theory as a core diminution of the model. According to Rogers and Shoemaker [34], technology complexity is "the degree to which an innovation is perceived as relatively difficult to understand and use." Several studies showed that technological complexity induces a positive experience for men more than for women [31]. Studies revealed that men and women differ in the overall perception of their ability to use the technology but not in the tasks or goals achievement. In addition, women have reported more anxiety levels than men do within completing a task, influencing the usage of the systems [35]. For this reason, we indicate that women might undertake less complex technological systems. Therefore, the following proposition will be further supported by literature findings to confirm the accountant's remote working during coronavirus in Jordan.

P4: the impact of the technological complexity on the actual use of accountants' remote working is moderated by Gender.

4 Research Methodology

This paper adopts an interpretive approach designed to explore the practical differences between male and female accountants working remotely in the context of coronavirus. Researchers have concluded multiple literature reviews to examine the set of propositions that have conducted a similar study in the same area. Researchers design an integrated research model to assist in investigating the key factors found in previous papers under three theoretical lenses, technology acceptance unified theory of acceptance and use of technology and the diffusion innovation theory shown in Fig. 1. Therefore, researchers have analyzed the literature and presented their overall results accordingly. Figure 1 incorporates remote working perceived ease of use, usefulness, complexity, and actual use in relation to gender. Table 1 shows a summary for overall previous literature results that adopted to interpret this paper results.

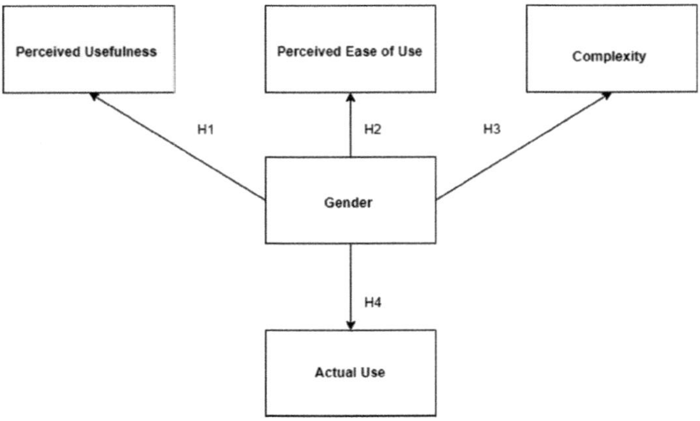

Fig. 1 Integrated research model

5 Findings and Discussion

The main of this study is to investigate the gender moderating impact on accountants' remote working during coronavirus in Jordan. The study has looked into two theoretical lenses, technology acceptance and the diffusion innovation theory, to depict the relationships in the research model. This study is based on previous scholars' findings and results from different technology acceptance and usage contexts. The set of propositions of this study is supported by previous literature; therefore, we have indicated the following expected results. As perceived, many studies did not find any gender impact in the overall technology users' usage. We found a diverse perception of technology among women and men based on several previous findings.

Due to the socialization's assumptions and gender roles congruent, women are more likely to comply with rules than men, which might indicate that women are higher in accepting new technology. One might argue that these facts, which some studies did; however, we assume that these findings depend on the geographical contexts (i.e., developed or developing country) or the time of conducting the study. In addition, a summary of literature review that used to interpret this paper designed to sum up for a conclusion, as shown in Table 1.

Many studies that used a similar approach as ours found a pattern of results. For example, Sobieraj and Krämer [31] reported that in technology, perceived usefulness and perceived ease of use there are depending on the technology feature. These findings support our assumption that the usage of behavioral patterns might change slightly or entirely in times of crisis and risks such as coronavirus outbreaks. Moreover, developed countries such as Jordan are one area under research for technology acceptance and utilization [36].

Table 1 Literature review summary

	TAM		UTAUT	Diffusion innovation theory
	Gender > PEUO	Gender > PU	Gender > Actual use	Gender > Complexity
Wong, Teo [30]	Gender (+)	Gender (+)		
Mahdi and Al-Dera [37]			Gender (Male) (−)	
Huang, Hood [32]			Gender (Female) (+)	
[35]				Gender (Female) (−)
Hungund and Kiran [27]			Gender (−)	
Hiran and Henten [38]			Gender (Male) (+)	
Assaker [26]	Gender (Female) (+)	Gender (Male) (+)	Non-significant	Gender (Female) (−)
Dunatchik, Gerson [18]			Gender (Male) (+)	
Cho, Cheon [39]	Gender (Female) (+)			
Sobieraj and Krämer [31]			Gender (Female) (−)	Gender (Female) (−)

However, it is recommended that further researchers use a quantitative method approach to test the propositions. In addition, it is suggested that future researchers implement a comparison study between two eras of time, such as before coronavirus and after coronavirus breakout. Moreover, a different area in the world might differ in results, such as a more developed country.

As much as any study, there is a limitation to this study which has been discussed based on previous literature review; it is suggested to be supported by a solid test to approve a theoretical point of view. However, this study will provide a more comprehensive view of remote working and its acceptance decision in private and public firms in Jordan. In addition, this study provides a critical determent that influences the adoption of remote working for firms that will make firms rethink and adjust their strategic plans before implementing the change. Lastly, this study could work as a risk plan and plan A to introduce managers where to start before implementing remote working in a culture that has a diverse gender.

The recommendations formulated according to the study results to improve the comparison between male and female accountants working remotely in the context of covid-19 are of high practical value. The practical importance is represented in applying the theoretical proposals and recommendations contained in the study by governmental and private institutions in the development and implementation of remote work in different contexts. The conclusions obtained during the study can serve as a methodological basis for the creation of comprehensive research by monitoring remote work and the impact and efficiency of using available resources in males and females.

References

1. Leoni G et al (2021) Accounting, management and accountability in times of crisis: lessons from the COVID-19 pandemic. Account Audit Account J 34(6):1305–1319
2. AlQudah AA, Salloum SA, Shaalan K. (2021) The role of technology acceptance in healthcare to mitigate COVID-19 outbreak. Emerging Technologies During the Era of COVID-19 Pandemic 348:223
3. Sargiacomo M et al (2021) Accounting for natural disasters from a historical perspective: a literature review and research agenda. Account Hist 26(2):179–204
4. Al Mahameed M, et al (2020) Social accounting in the context of profound political, social and economic crisis: the case of the Arab spring. Account Audit Accountability J
5. Rouleau L, Hällgren M, de Rond M (2020) Covid-19 and our understanding of risk, emergencies, and crises. J Manag Stud
6. Steccolini I (2018) Accounting and the post-new public management: re-considering publicness in accounting research. Account Audit Accountability J 32(1):255–279
7. Andrew J, Baker M, Guthrie J (2021) Accounting, inequality and COVID-19 in Australia. Account Audit Accountability J 34(6):1471–1483
8. Piketty T (2020) Capital and ideology. Harvard University Press, Cambridge
9. Papadopoulou S, Papadopoulou M (2020) The accounting profession amidst the COVID-19 pandemic. Int J Account Financ Report 10(2):39
10. Rinaldi L et al (2020) Accounting in times of the COVID-19 pandemic: a forum for academic research in accounting forum. Taylor & Francis
11. He P et al (2020) Accounting index of COVID-19 impact on Chinese industries: a case study using big data portrait analysis. Emerg Mark Financ Trade 56(10):2332–2349
12. Jabin S (2021) The Impact COVID-19 of on the accounting profession in Bangladesh. J Industr Distrib Bus 12(7):7–14
13. Musleh Al-Sartawi AMA (eds) (2022) Artificial intelligence for sustainable finance and sustainable technology. ICGER 2021. Lecture Notes in Networks and Systems, vol 423. Springer, Cham. https://doi.org/10.1007/978-3-030-93464-4
14. Delfino GF, van der Kolk B (2021) Remote working, management control changes and employee responses during the COVID-19 crisis. Account Audit Accountability J
15. Odat QA, Alshurafat H, Masadeh WM (2021) Factors affecting accountants transition to online working during COVID-19 pandemic: a proposed model. DR. DY Patil B-School, Pune, India : 528
16. Vyas L, Butakhieo N (2021) The impact of working from home during COVID-19 on work and life domains: an exploratory study on Hong Kong. Policy Design Pract 4(1):59–76
17. Nash M, Churchill B (2020) Caring during COVID-19: a gendered analysis of Australian university responses to managing remote working and caring responsibilities. Gend Work Organ 27(5):833–846
18. Dunatchik A et al (2021) Gender, parenting, and the rise of remote work during the pandemic: implications for domestic inequality in the United States. Gend Soc 35(2):194–205
19. Reichelt M, Makovi K, Sargsyan A (2021) The impact of COVID-19 on gender inequality in the labor market and gender-role attitudes. Eur Soc 23(1):S228–S245

20. Prasad K, Vaidya RW, Mangipudi MR (2020) Effect of occupational stress and remote working on psychological well-being of employees: an empirical analysis during COVID-19 pandemic concerning information technology industry in hyderabad. Indian J Comm Manag Stud 11(2):1–13
21. Ng J et al (2020) Age, sex, and remote monitoring differences in device acceptance for patients with implanted cardioverter defibrillators in Canada. CJC open 2(6):483–489
22. Davis FD (1989) Perceived usefulness, perceived ease of use, and user acceptance of information technology. MIS Q 13(3):319–340
23. Delfino GF, van der Kolk B (2021) Remote working, management control changes and employee responses during the COVID-19 crisis. Account Audit Accountability J 34(6):1376–1387
24. Liu I-F et al (2010) Extending the TAM model to explore the factors that affect intention to use an online learning community. Comput Educ 54(2):600–610
25. Mansour MMO (2020) Acceptance of mobile banking in Islamic banks: integration of DeLone and McLean IS model and unified theory of acceptance and use of technology. Int J Bus Excell 21(4):564–584
26. Hannoon A, Abdalla YA, Musleh Al-Sartawi AMA, Khalid AA (2021) Board of Directors Composition and social media Financial Disclosure: The Case of the United Arab Emirates. In: Musleh Al-Sartawi AMA (ed) The Big Data-Driven Digital Economy: Artificial and Computational Intelligence, vol 974. Studies in Computational Intelligence. Springer, Cham, pp 229–241. https://doi.org/10.1007/978-3-030-73057-4_18
27. Hungund S, Kiran KB (2019) Influence of firm-level factors on adoption of the innovation approach among software product SMEs: evidence from the Bangalore region of India. Int J Business Innov Res 20(2):221–232
28. Ong C-S, Lai J-Y (2006) Gender differences in perceptions and relationships among dominants of e-learning acceptance. Comput Hum Behav 22(5):816–829
29. Tan GW-H, Ooi K-B (2018) Gender and age: do they really moderate mobile tourism shopping behavior? Telemat Inf 35(6):1617–1642
30. Wong K-T, Teo T, Russo S (20112) Influence of gender and computer teaching efficacy on computer acceptance among Malaysian student teachers: an extended technology acceptance model. Australasian J Educ Technol 28(7)
31. Sobieraj S, Krämer NC (2020) Similarities and differences between genders in the usage of computer with different levels of technological complexity. Comput Hum Behav 104:106145
32. Musleh Al-Sartawi AMA (ed) (2021) The big data-driven digital economy: artificial and computational intelligence, vol 974. Studies in Computational Intelligence. Springer, Cham. https://doi.org/10.1007/978-3-030-73057-4
33. Rogers EM (1995) Diffusion of Innovations: modifications of a model for telecommunications. Die diffusion von innovationen in der telekommunikation, vol. 17, pp 25–38. Springer, Berlin. https://doi.org/10.1007/978-3-642-79868-9_2
34. Rogers EM, Shoemaker FF (1971) Communication of innovations; a cross-cultural approach
35. Scherer R, Siddiq F (2015) Revisiting teachers' computer self-efficacy: a differentiated view on gender differences. Comput Hum Behav 53:48–57
36. Jaradat M-IR, Twaissi NM (2010) Assessing the introduction of mobile banking in Jordan using technology acceptance model. Int J Interact Mobile Technol 4(1)
37. Mahdi HS, Al-Dera ASA (2013) The impact of teachers' age, gender and experience on the use of information and communication technology in EFL teaching. English Lang Teach 6(6):57–67
38. Hiran KK, Henten A (2020) An integrated TOE–DoIframework for cloud computing adoption in the higher education sector: case study of Sub-Saharan Africa, Ethiopia. Int J Syst Assur Eng Manag 11(2):441–449
39. Cho J et al (2021) Digital advertising policy acceptance by out-of-home advertising firms: a combination of TAM and TOE framework. Int J Adv :1–19

The Modified Effect of Governance on the Planning Internal Audit Tasks in Achieving the Added Value of Transactions in Jordanian Service Companies

Sanaa Nazami Maswadeh and Abdullah M. Alazzha

Abstract this study aimed to identify the modified effect of governance on the planning internal audit tasks in achieving the added value of transactions. The population of the study consists of all Jordanian services companies listed on the Amman stock exchange which is around 139 companies according to Amman stock statistics in 2020. The researcher distributed 200 questionnaires for internal auditors in these companies. The number of questionnaires included in the analysis was 195 questionnaires. Multiple regression analysis was used to test the study's hypotheses. The study reached many results; most notably that governance plays an important role in reinforcing the importance of the internal audit planning tasks with its dimensions consisting of (task objectives, task scope, and allocating task resources) in achieving added value of Jordanian service company's transactions. The study recommended the need for scholars and researchers to shed more light on the planning of internal audit tasks and link them to new variables such as market value of shares and other performance indicators.

Keywords Internal audit · Planning tasks · Task objectives · Task scope · Allocating task resources · Governance · Added value of transactions · Jordan · Services companies

1 Introduction

Companies have paid great attention to the internal audit tasks, believing in the importance of the services performed by these tasks. Therefore, developing the performance of internal auditors and improving the quality of internal audits has become extremely important to reach the desired goals of companies. The profession of internal audit has become concerned with the existence of an integrated set of standards and foundations that must be taken into account in companies to monitor their activities and ensure the achievement of an added value for their transactions

S. N. Maswadeh (✉) · A. M. Alazzha
Accounting Department, School of Business, Jadara University, Irbid, Jordan
e-mail: sanaam@jadara.edu.jo

A. M. A. Musleh Al-Sartawi et al. (eds.), *From the Internet of Things to the Internet of Ideas: The Role of Artificial Intelligence*, Lecture Notes in Networks and Systems 557,
https://doi.org/10.1007/978-3-031-17746-0_53

to achieve their success and objectives, and to ensure their real contribution to the growth of the national economy.

The Institute of Internal Auditors (IIA) believes that good governance depends on the information generated by the four parties that make up the governance system; i.e. the board of directors, the executive management, the internal auditors, and the external auditors. Hence, the Internal Audit Department is one of the main parties concerned with evaluating corporate governance. The principles and rules of corporate governance also emphasized the need to develop the internal audit tasks and activate their role within the framework of governance mechanisms, with a focus on its relationship with the internal audit committee and the external auditor [1].

The study of [2] stressed that internal audit leads to adding value to the company's transactions through the tasks performed within the framework of governance, which includes providing information to management at all levels, providing independence for each of the internal auditors, and creating an open dialogue between them. This reflects positively on the performance of the internal auditor, as [3] confirmed that there is a relationship between the profession of internal audit and maximizing the value of the parties benefiting from corporate governance.

Researchers discussed how the internal audit tasks add value and decided that the purpose of value addition for internal audit begins with defining who is the beneficiary of internal audit services; when the beneficiary of the internal audit services was the executive management, as the internal audit represented the eyes and ears of the management, it was contributing to adding value by ensuring the full protection of the company and ensuring compliance with laws and legislations through the recommendations it made to the executive management.

By paying attention to planning internal audit tasks, which aims at reducing the risks to which companies are exposed in all aspects and fields, and to achieve the company's goals and strategy. The lack of interest in planning the internal audit tasks is expected to negatively affect the efficiency of the company's performance, which is reflected in the confidence of investors in it and threatens the future of these companies in growth, survival, and continuity.

Thus, the study problem can be expressed by asking the following main question: What is the effect of governance on planning internal audit tasks with its various dimensions; i.e. task objectives, task scope, and allocating task resources in achieving the added value of the transactions of Jordanian service companies listed in Amman Stock Exchange?

The importance of the current study is that it deals with a topic of great importance, which seeks to highlight the importance of the modified effect of governance on the planning internal audit tasks in achieving the added value of the transactions carried out by Jordanian service companies, to direct their efforts towards taking care of governance processes and planning internal audit tasks, which leads to adding value to its transactions by improving the efficiency and effectiveness of its performance, ensuring that the internal audit task, with its tasks, assurance, and consultancy, contributes positively to improving and increasing the chances of achieving the organization's goals, improving procedures and transactions, and reducing risks to the lowest acceptable levels in companies and adding value to them [4]. It is expected that this study will contribute to enriching the accounting literature on the subject of the modified effect of governance on planning internal audit tasks in achieving the added value of transactions, hoping that this study will be one of the pillars in building studies for researchers in the future.

None of the previous studies in Jordan—according to the researcher's knowledge—dealt with the study of planning the internal audit tasks and the added value of the company under governance. The current study aims at analyzing and diagnosing the effect of planning for internal audit tasks on achieving an added value for the company's transactions in an environment characterized by applying governance, which seeks to protect the rights of shareholders and stakeholders in the company.

2 Literature Review

Governance expresses the rules, regulations, and standards that define the foundations of the relationship between the company's management on the one hand and the stakeholders associated with the company on the other hand. It is difficult to activate the principles of governance unless the effectiveness and independence of the audit committees and the activities of internal and external audits are strengthened [5]. By referring to Standard (IIA-2110) of the Institute of Internal Auditors about ***governance***, we find that it states that "the internal audit activity must evaluate the company's governance processes, and propose appropriate recommendations for improvement."

Regarding the Institute of Internal Auditors' Internal Audit Standard (IIA-2210) about ***Task Objectives***, internal auditors must define task objectives by making a preliminary assessment of the risks related to the activity being reviewed. Task objectives must reflect the results of that assessment, and internal auditors must take into account when developing task objectives the possibility of serious errors, fraud, non-compliance, or other risks. The objectives of the consulting task should address governance, risk management, and control processes, within the scope agreed upon with the concerned entity.

The Internal Audit Standard (IIA-2220) about ***Task Scope*** states that the planned task scope must include all relevant records, systems, and people to achieve its objectives. Written work programs should be drafted to achieve those goals so that these programs include the procedures necessary to identify, evaluate, analyze and validate the information during the audit process.

The Internal Audit Standard (IIA-2230) about ***Allocating Task Resources*** states that internal auditors must determine the appropriate and sufficient resources needed to achieve the objectives of the audit task. The personnel necessary to carry out the task are employed based on an assessment of its nature, complexity, time constraints, and the resources available for its implementation.

The added value of the company's transactions is represented in the positive effect achieved by the company's activities that will benefit it now or in the future through conducting internal audits to improve the effectiveness and efficiency of the company's operations, commitment to integrity, transparency, and accountability; in addition to making recommendations based on analysis and evaluation of data to reach confirmation of what is concerned with managing risks to an acceptable level and controlling them to achieve the proper application of governance [6]

The study, [7] is to provide evidence on the effectiveness of corporate governance practices and their effects on audit quality in Qatar. The results showed that the independence of the Board of Directors and the CEO duality and internal audit committees have a significant relationship with audit quality. Moreover, company size, business complexity, and financial leverage influence audit quality.

[8] That tried to identify the effect of internal audit procedures on achieving the qualitative characteristics of accounting information in the financial reports of Kuwaiti commercial banks. A questionnaire was distributed to members of the study population represented by the internal auditors of Kuwaiti commercial banks. One of the most important findings of the study is that the internal audit procedures, with all their control measures, affect the publication of financial statements whose data are relevant and reliable, and which can be relied upon in the process of making decisions by users of financial reports.

In their study [6] pinpointed the role of added value in achieving the governance pillars of independence and responsibility in the Al-Rafidain State Company for the implementation of dams in Iraq. The study reached many results; most notably that internal audit is an essential pillar of the economic entities because of its effective role in putting them on the right track, correcting deviations, and providing appropriate solutions. The added value is considered a standard for internal audit and a standard for judging the efficiency and effectiveness of the performance because of their positive effect on improving the efficiency and effectiveness of the organization's performance through commitment to integrity, transparency, accountability, risk management, and the proper application of governance principles.

The study [9] examines the effect of cognitive-behavioral skills and technical skills related to participation in information technology governance (ITG) for internal auditors of Egyptian commercial banks in assessing risks, specifically fraud risks and control risks. The results showed that the internal auditors who have an analytical cognitive skill and participated in the (ITG) were able to assess the fraud and control risks significantly higher than the internal auditors with the intuitive cognitive style and the internal auditors who did not participate in the ITG. The study emphasized the need for internal auditors to possess behavioral skills and technical skills to be able to participate in more strategic roles, and thus, add value to their organizations.

In the study of [10], which aimed to determine the extent to which internal audit contributes to creating added value in a sample of Algerian shareholding companies, this study was conducted based on a questionnaire designed and distributed to a random sample of internal auditors. The results of the study showed that internal audit contributes to creating the added value of the company's transactions by evaluating the internal control system, risk management procedures, and governance processes.

In their study, [11] identified the effect of corporate governance and auditing on economic growth in Islamic countries. The study concluded that corporate governance and auditing have a positive and significant effect on the economic growth of Islamic countries.

In their study, (Dinh, Pham, and Nguyen, 2021) aimed at researching the internal audit characteristics that affect the effectiveness of internal audits in steel enterprises in Vietnam. The study was based on a survey of 114 answers representing owners, managers, and internal auditors from 38 steel companies in Vietnam for the period from 2018 to 2019. The results showed that the effectiveness of internal audit in Vietnam steel companies is affected by various internal factors, including the volume, the scope, competence, independence, operational foundations, and the methods of internal audit. The study emphasized the importance of expanding the internal audit scope and not underestimating it and reforming the internal governance structure to increase the effectiveness of the internal audit.

In the study of [13] that studied the impact of the internal audit quality on the financial performance of Yemeni commercial banks, the primary data for the study was collected through a questionnaire distributed to internal auditors. The results revealed that adherence to internal audit standards, independence of internal auditors, and governance has a significant positive impact on the financial performance of Yemeni banks. The size and meetings of the internal audit committees had a little positive impact on the financial performance of the banks. Moreover, the results showed that the internal audit's reliance on electronic technologies has a positive impact on improving financial performance.

Relying on literature review and previous studies, this study seek to study the effect of planning the internal audit tasks according to the Internal Audit Standard (IIA-2210) "***Task Objectives***", (IIA-2220) "***Task Scope***", and (IIA-2230) "Allocating Task Resources", as independents variables, on added value of the company transaction as dependent variable with the modified effect of governance as a moderator variable.

The following main hypothesis will be tested:

There is significant effect at the level ($\alpha \leq 0.05$) of governance on **planning internal audit tasks** in achieving the added value of the transactions of Jordanian service companies listed in the Amman Stock Exchange.

The sub-hypotheses are derived from the main hypothesis related to study the significant effect at the level ($\alpha \leq 0.05$) of governance on **-task objectives, task scope, and allocating task resources-** in achieving the added value of the transactions of Jordanian service companies listed in the Amman Stock Exchange.

3 Study Methodology

This study relied on the descriptive analytical approach in analyzing the data of the study, collecting primary data through a questionnaire that was developed and distributed to internal auditors in Jordanian service companies at the beginning of the 2021 year.

3.1 Study Sample and Population

The researcher conducted a comprehensive search for all the study population, which consisted of 139 Jordanian service companies according to the statistics of the Amman Stock Exchange at the end of 2020. The questionnaires were distributed to the internal auditors in a simple random sampling method, and the number of distributed questionnaires reached 200 questionnaires, of which 198 were retrieved. The number of valid questionnaires for analysis was 195.

3.2 Study Variables

The study tool is a designed questionnaire that consisted of four parts. The first part included general information related to the demographic factors of the study sample members, including (educational qualification, experience years in the current field of work, job title, and professional certificates in the field of auditing).

As for the second part, it consisted of a set of 30 paragraphs, explaining the independent variable of the study represented by the dimensions of planning the internal audit tasks, and by 10 paragraphs for each of task objectives, task scope, and allocating task resources.

The third part consisted of 10 paragraphs to measure the dependent variable, which is the added value of transactions.

The fourth part consisted of 10 paragraphs explaining the modified variable represented in governance.

To analyze the questionnaire data, it was relied on the five-point Likert scale in answering the questions, and its scores were (5, 4, 3, 2, and 1) which express a very high degree of agreement, a high degree of agreement, a medium degree of agreement, a low degree of agreement, and a very low degree of agreement, respectively.

3.3 The Statistical Methods Used

To achieve the objectives of the study and analyze the collected data, the (SPSS) program was relied upon, and many suitable statistical methods were used; including frequencies and percent descriptions to measure demographic characteristics of the study sample members. Shapiro–Wilk to test the normal distribution of the study data, the inflation variance factor (VIF), the Durbin-Watson test, to test the validity of the study data for applying multiple regression. The mean and the standard deviation were used to measure the agreement of study members of the study variables; the multiple linear regressions were used to test the study hypothesis. The regression equation of the study appeared as follows:

$$\mathbf{AV} = \alpha_0 + \beta_1 OM + \beta_2 SM + \beta_3 RM + \beta_4 G + \beta_5 OM{*}G + \beta_6 SM{*}G + \beta_7 RM{*}G + \epsilon.$$

Where:

AV: Added Value to Transactions.
OM: Task Objectives.
SM: Task Scope.
RM: Allocating Task Resources.
G: Governance.
OM*G: Interaction between Task Objectives and Governance.
SM*G: Interaction between Task Scopes and Governance.
RM*G: Interaction between Allocating Task Resources and Governance.

4 Statistical Analysis Results

After analyzing the data of the study, the following important statistical results will discuss:

4.1 Testing Validity of the Study Data

By reviewing the demographic characteristics of the study sample, the sample members possess scientific and practical qualifications in addition to professional certificates that make them able to understand and answer the questionnaire items properly.

To determine the reliability of the Questionnaire, the study calculated the internal consistency using Cronbach's alpha, the values of Cronbach's alpha reliability coefficients for the study variables ranged between 0.70 and 0.79, all of which are statistically acceptable values. However, [14] indicated that the acceptance rate of Cronbach's alpha coefficient is 0.60 or more.

The Shapiro–Wilk test has used to test the normal distribution of the study data, the level of significance for all Shapiro–Wilk test values was greater than (0.05), indicating that all the variables in the current study follow the normal distribution [15].

By examining the values of the Variance Inflation Factor (VIF) test for all independent variables, all the independent variables related to the planning of internal audit tasks and the modified variable governance were less than (5), which means that there is no Multicollinearity among the independent variables of the study, and therefore all of them can be relied upon in the regression model.

To test that the data is free of autocorrelation, which is the existence of a relationship between regression random errors calculated by the least-squares method, the Durbin-Watson Test, which is symbolized by the symbol DW, was conducted and compared with two values extracted from the table for this test at the level of significance α, the number of observations n and the number of variables k, and these two values are symbolized by the symbols dl (minimum limit) and du (maximum limit), it was found that there is no autocorrelation for the main study models, as shown below in Table 1.

4.2 Testing the Study Hypotheses

The multiple regression equation has been applied and Table 1 shows the results that have been reached.

Regarding Table 1 and comparing the first regression model, which tested the effect of planning audit tasks on the added value of transactions, with the second regression model, which was introduced to the first model, governance, and interactions between independent variables and governance, to test the modified effect of governance on planning audit tasks to achieve the added value of transactions, the increased value of the (Adjusted R^2) was (2%) as noted in table (1), due to the difference of (Adjusted R^2) between the second model and the first model, which reached 0.642 and 0.622 respectively, that was statistically significant values.

This means that governance modifying the effect of planning internal audit tasks on the added value of transactions in Jordanian service companies. The value of the (F) test was (87.935) with a statistical significance of (0.00), which is a statistically significant value that indicates the acceptance of the regression model and the ability of governance as a modified variable with other independent variables in influencing the added value of transactions.

Thus, the main null hypothesis is rejected, and the alternative hypothesis is accepted, which states that: There is a significant effect at the level ($\alpha \leq 0.05$) of governance as a modified variable on the planning of internal audit tasks and achieving the added value of the transactions of Jordanian service companies listed in Amman Stock Exchange. This result is in agreement with the study by[16], which showed that the internal audit task is linked to adding value to the company through the tasks it performs within the framework of governance.

This also agreed with the study of [10], which showed that there is an effect of internal audit on creating added value by evaluating the internal audit system, risk management procedures, and governance processes in Algerian companies, this also agreed with [12], pinpointed the importance of reforming the governance structure to increase the effectiveness of the internal audit. Likewise, this agreement with [13], showed that governance is one of the main factors for the quality of internal audit, in addition to the fact that governance has a significant positive effect on improving the financial performance of Yemeni banks.

Based on Table 1, it was found that there is a significant effect at the level ($\alpha \leq 0.05$) of governance as a modified variable on the task objectives, in achieving the added value of the transactions of Jordanian service companies listed in Amman Stock Exchange. This result agreed with the study of [17], which showed governance

Table 1 Multiple Regression Equation to Study Governance as a Modified Variable on the Planning of Internal Audit Tasks in Achieving the Added Value of Transactions

Model (1) AV = $\alpha_0 + \beta_1 OM + \beta_2 SM + \beta_3 RM + \epsilon$					
AV = -0.932 + 0.389(OM) + 0.361 (SM) + 0.542(RM)					
	Beta	T	Sig	VIF	Tolerance
constant	-0.932	-3.365	0.001		
Task Objectives (**OM**)	0.389	7.964	0.000	1.234	0.81
Task Scope (**SM**)	0.361	7.384	0.000	1.228	0.814
Allocating Task Resources (**RM**)	0.542	12.222	0.000	1.186	0.843
R = 0.792	**(R^2) = 0.628**	**Adjusted (R^2) = 0.622**			
F = 107.486		**Sig F. = 0.000**			
Durbin-Watson	(D-W) = 1.846	DI (Lower bound) = 1.643		DU (Upper bound) = 1.704	
Modified Effect of Governance on the Planning Internal Audit Tasks					
Model (2): AV = $\alpha_0 + \beta_1 OM + \beta_2 SM + \beta_3 RM + \beta_4 G + \beta_5 OM{*}G + \beta_6 SM{*}G + \beta_7 RM{*}G + \epsilon$					
AV = -2.012 + 0.501(OM) + 0.541 + 0.541 (SM) + 0.522(RM) + 0.403(G) + 0.603(OM*G) + 0.356(SM*G) + 0.158(RM*G)					
	Beta	T	Sig	VIF	Tolerance
constant	-2.012	-7.982	0.000		
Task Objectives (**OM**)	0.501	8.049	0.000	1.234	0.81
Task Scope (**SM**)	0.541	10.815	0.000	1.228	0.814
Allocating Task Resources (**RM**)	0.522	10.422	0.000	1.186	0.843
Governance (**G**)	0.403	8.442	0.000	1.181	0.847
OM* G	0.603	12.891	0.000	2.151	0.465
SM*G	0.356	7.477	0.000	2.354	0.425
RM*G	0.158	3.394	0.001	2.879	0.347

(continued)

Table 1 (continued)

Model (1) AV $= \alpha_0 + \beta_1 OM + \beta_2 SM + \beta_3 RM + \epsilon$			
R = 0.806	**(R^2) = 0.649**	**Adjusted (R^2) = 0.642**	
F = 87.935		**Sig F. = 0.000**	
Durbin-Watson	(D-W) = 1.665	DI (Lower bound) = 1.615	DU (Upper bound) = 1.633
*Dependent Variable: **(AV): Added Value**			

Source Result output from SPPS statistical analysis

had a positive effect on raising the company's efficiency by achieving its goals in the Saudi Electricity Company. It also agreed with [18] showed a positive relationship between internal audit quality, governance, and goals set.

It was also found that there was a significant effect at the level ($\alpha \leq 0.05$) of governance as a modified variable in achieving the added value of transactions that increases the effectiveness of the internal audit. This result agreed with the study of [12], which showed that expanding the scope of internal audit and not underestimating it, increases the effectiveness of the internal audit.

As shown in Table 1, there is a significant effect at the level ($\alpha \leq 0.05$) of governance as a modified variable on allocating task resources in achieving the added value of the transactions of Jordanian service companies listed in the Amman Stock Exchange.

5 Study Conclusions and Recommendations

After reviewing the results of the statistical analysis, the study reached the following most important conclusions:

The implementation of the audit tasks needs to work regularly and properly effectively and efficiently, which increases the importance of planning for internal audit tasks. This is related to determining internal audit objectives to be achieved for internal audit tasks, proper selection of the scope of internal audit tasks, supervision, and resource allocation in an efficient manner to achieve the specified and agreed-upon objectives.

The interest in planning the internal audit tasks enhances the added value of the company's transactions through the ability of companies to employ audit planning in achieving their goals to increase their ability to face the risks they are exposed to, in addition to defining the agreed scope of the required task, which indicates that the audit planning includes the development of a comprehensive strategy for the audit tasks that determine the necessary resources, whether material or human, and all associated administrative arrangements, which are considered among the central and essential issues for the successful implementation of audit plans.

Governance is an important mechanism that enables companies to ensure the proper management practically and leads to the protection of shareholders' property and provides fair and transparent information for all parties related to the company, which will reflect positively on achieving the added value of the company's transactions [21].

Increasing the confidence of the external auditor in the internal audit planning may reduce the costs of external audit, by increasing the attention of external auditors to the planning of internal audit tasks, and in the same field selecting the qualified internal auditors, training and developing their performance, in a manner that leads to the development of the company's performance and increasing transparency and confidence of the financial information it discloses [19, 20].

The Internal Audit Department supports the executive management by developing and evaluating corporate governance practices, so that the board of directors' reviews, develops, and amends the Governance Manual from time to time and whenever necessary. It also supports the organizational structure and functional division to develop the foundations of governance and pays great attention to proper governance practices to clarify the regulations and instructions necessary to define the duties and responsibilities of each employee in the company.

The study recommended the need for internal auditors to use more electronic technologies in carrying out internal audit tasks through the use of the electronic application to analyze and understand the internal audit environment. In addition to the need for scholars and researchers to shed more light on the planning of internal audit tasks and link them to new variables such as the market value of shares and other performance indicators; as well as conducting similar studies to identify the effect of governance as a modified variable on new internal audit standards that were not addressed in this study.

6 Study Limitations

This study was applied to service companies listed on the Amman Stock Exchange at the beginning of 2021 under the circumstances of the Corona pandemic. Accordingly, the researchers faced mobility difficulties, partial bans, and the disruption of service companies or part-time during the Corona pandemic period. In addition, the subject of the study was limited to addressing the modified effect of governance on planning internal audit tasks in achieving the added value of their transactions.

References

1. Aldalayeen B (2017) Does corporate governance affect the financial performance? Analysis of findings from Jordanian banks. Int J Bus Manag 12(4):123. https://doi.org/10.5539/ijbm.v12n4p123
2. Piskar F (2006) Quality audits and their value added. Int J Serv Stand 2(1):69–83
3. Vadasi C, Bekiaris M, Andrikopoulos A (2020) Corporate governance and internal audit: an institutional theory perspective. Corp Gov 20(1):175–190. https://doi.org/10.1108/CG-07-2019-0215
4. Al-Okaily M, Alqudah H, Al-Qudah AA, Alkhwaldi AF (2022) examining the critical factors of computer-assisted audit tools and techniques adoption in the post-COVID-19 period: internal auditors perspective. VINE J Inf Knowl Manag Syst. https://doi.org/10.1108/VJIKMS-12-2021-0311
5. Meshack SN (2015) Influence of corporate governance practices on working capital efficiency of manufacturing firms in Nairobi County. South Eastern Kenya University, Nairobi
6. Hamad S, Muhammad, Awad (2020) the added value of internal auditing and its role in achieving governance pillars represented by independence and responsibility. J Account Financ Stud 15(52):152–170

7. Awadallah E (2020) Measuring the effectiveness of selected corporate governance practices and their implications for audit quality: evidence from Qatar. Afro-Asian J Financ Account 10(1):24–47. https://doi.org/10.1504/AAJFA.2020.104402
8. Al-Hardan M (2020) The impact of the internal audit procedures on the quality of accounting information in Kuwaiti commercial banks. Ramah J Res Stud 43(1):23–41
9. Gomaa II, Sabry SH (2021) Enhancing internal auditors' risk assessment through governance and cognition: an Egyptian perspective. Afro-Asian J Financ Account 11(3):447. https://doi.org/10.1504/aajfa.2021.10038581
10. Sarra B (2020) Do internal auditors in the Algerian public organizations contribute to create added value? Strateg Dev Rev 10(4):46–65
11. Samara WA, Talalweh MA (2020) The effect of strength of auditing and accounting standards and corporate governance on economic growth: a case study of the Islamic world. J Bus Financ Econ 04(02):311–323. https://doi.org/10.37170/1986-004-002-017
12. Dinh TH, Pham DC, Nguyen TT (2021) Factors affecting the internal audit effectiveness of steel enterprises in Vietnam. J Asian Financ Econ Bus 8(1):271–283. https://doi.org/10.13106/jafeb.2021.vol8.no1.271
13. Hazaea SA, Tabash MI, Hazaea SFAK, Zhu J, Kuhali AAAL (2020) The impact of internal audit quality on financial performance of Yemeni commercial banks: an empirical investigation. J Asian Financ Econ Bus 7(11):867–875. https://doi.org/10.13106/jafeb.2020.vol7.no11.867
14. Al-Okaily A, Al-Okaily M, Teoh AP (2021) Evaluating ERP systems success: evidence from Jordanian firms in the age of the digital business. VINE J Inf Knowl Manag Syst. https://doi.org/10.1108/VJIKMS-04-2021-0061
15. Al-Sartawi A (2020) Social media disclosure of intellectual capital and firm value. Int J Learn Intell Cap 17(4):312–323
16. Khalaf TM, Safiya (2020) The role of electronic auditing in improving the quality of the internal audit profession and its impact on maximizing value for the beneficiaries of corporate governance. Albashaer Econ J 6(2):763–779
17. Baabad A (2020) Corporate governance role in improving the financial information quality within the Saudi electricity company. J Econ Adm Leg Sci 4(1):81–107
18. Lakhdar O (2017) The nature of the relationship between internal audit quality and corporate governance: risk management and internal control under standard no. 2100 (nature of work). J Econ Manag Commer Sci 3(17):158–175
19. Musleh Al-Sartawi AMA (ed) (2022) Artificial intelligence for sustainable finance and sustainable technology, vol 423. Lecture Notes in Networks and Systems. Springer, Cham. https://doi.org/10.1007/978-3-030-93464-4
20. Al-Okaily M (2021) Assessing the effectiveness of accounting information systems in the era of COVID-19 pandemic. VINE J Inf Knowl Manag Syst. https://doi.org/10.1108/VJIKMS-08-2021-0148
21. Abdulrasool FE, Turnbull SI (2020) Exploring security, risk, and compliance driven IT governance model for universities: applied research based on the COBIT framework. Int J Electron Bank 2(3):237–265

The Mediating Role of Strategic Planning in the Impact of Project Management Strategy on Firm Performance

Ahmed Alnawafleh, Abdalmuttaleb Al-Sartawi, Firas Almarashdah, and Aws Al-Okaily

Abstract This study aims at the role that the project management strategy plays in improving the performance of firms and through the companies' adoption of the concept of strategic planning, which in turn helps in knowing the company's plans in the short and long term. In order to improve the company's performance and the continuity of the company's survival in the market. The purpose of this article is to study the relationship between project management strategy and the performance of SMEs in Jordan and the mediating role of strategic planning. The researcher used the descriptive and analytical approach to analyze and classify the data, and 291 valid questionnaires were used, and the descriptive analytical approach was used to analyze the data. The "probability sampling method" was applied to collect the data. The main results are that the project management strategy affects the performance of companies and strategic planning. Strategic planning as a mediator influences the performance of firms through this study.

Keywords Project management strategy · Firm performance · Strategic planning

1 Introduction

Strategic thinking has become one of the basic necessities for the success of organizations of all sizes and sectors in which they operate. And an important requirement for large companies not only, but it has become a requirement for all businesses that aim at excellence, success and achieving a sustainable competitive advantage. The prevailing influence of operations and their strategies on the fate of projects and companies has led to a number of early research and studies on the positive

A. Alnawafleh · F. Almarashdah
School of Business, Jadara University 733, Irbid, Jordan

A. Al-Sartawi (✉)
Chairperson of Accounting Finance and Banking Department, Ahlia University, Manama, Bahrain
e-mail: amasartawi@hotmail.com

A. Al-Okaily
Graduate School of Business, Universiti Sains Malaysia, Penang, Malaysia

A. M. A. Musleh Al-Sartawi et al. (eds.), *From the Internet of Things to the Internet of Ideas: The Role of Artificial Intelligence*, Lecture Notes in Networks and Systems 557,
https://doi.org/10.1007/978-3-031-17746-0_54

relationship between operations strategy and project management [1].The strategy begins with defining the organizational objectives and their relative importance as long-term primary goals and secondary objectives through an action plan that can develop a competitive advantage, then identifying and analyzing the environment in terms of market characteristics and working to meet the needs of customers with products of better value [2]. The need for project management has emerged as an important input to support the organizational performance towards achieving the required positive change, taking into account the impact of change on individuals and their reactions and concerns about the future, which work to curb and restrict the process of change and pay attention to the reality of the work environment [3]. Increased interest in the importance of project management skills due to the rapidly growing global market with high competition, the search for quality, cost reduction and continuous improvement of goods and services [4]. As every organization must accept continuous change and conduct an internal environmental survey to ensure that this change is managed in the external environment and meet its requirements. The change begins with introducing new necessary processes and searching for measures and best practices to improve business. Updating or abandoning the old inappropriate practices in order to create flexibility and effectiveness towards the forces [5]. Market Successful companies are a major component of developing countries, and many economists regard them as the same as a driver in determining their economic, social and political development. To survive in a competitive business environment, every company has to operate in conditions of performance. Nowadays, firm performance has become a relevant concept in strategic management research and is often used as a dependent variable. Although it is a very common idea in the academic literature, there is hardly any consensus about its definition and measurement. Today, there are many initiatives and standards to enhance the concept of sustainability in the performance of institutions and companies, as the project management strategy plays a major role in improving the performance of companies through the use and adoption of the concept of strategic planning that is commensurate with the capabilities available to these companies [6].

2 Literature Review

2.1 Project Management Strategy

Project management is a priority in nearly every organization today because of its effectiveness in setting appropriate expectations about what can be provided, when, and how much money. It unites teams and coordinates efforts to achieve results - and without them, projects can veer from deadlines and fall out of scope [7].

Due to the urgent need for good project management within organizations, detailed knowledge bodies exist to provide instructions on how to better manage projects, and people dedicate their entire career to building this expertise. When projects are managed successfully, teams are able to do more than just deliver on what they promised. Often the end result is the complete strategic alignment between teams, departments, and the entire organization [8].

The project manager has many different responsibilities that fall into the different phases or processes of the project life cycle: initiation, planning, implementation, monitoring, controlling, and closing. Through each of these stages, the project manager introduces the project through idea through to completion [9].

The essence of project management is to support the implementation of the organization's competitive strategy to achieve the desired results. The organizations that use effective strategic activities within a specific timetable require them to realize a great deal of the importance of resources [10]. Managing strategic projects with a series of practices, procedures, processes, tools, and behaviors that, when considered holistically, describe the extent to which the organization creates effective links between project management practices and distinct business practices with the aim of advancing the overall strategic objectives of the organization [11]. It is also the general direction of the project and the guidelines on what to do and how to do it, in order to achieve a higher competitive advantage and better project outcomes [12]. The success of the project is seen from the extent to which it achieves the strategic objectives of the organization, and when the organization makes the link between its projects and its business strategy, it achieves its goals and desired goals in the best way [13].

2.2 Firm Performance

In the 50 s of the last century, the performance of the company was considered equivalent to organizational efficiency, which represents the degree to which the organization, as a social system with some limited resources and means, achieves its objectives without exerting excessive effort from its members. The criteria used to evaluate performance are productivity, flexibility and tensions between organizations [14]. In the first decade of the twenty-first century, the definition of firm performance focused mainly on the ability of the organization and its ability to efficiently utilize available resources to achieve achievements consistent with the company's goals, as well as considering its relationship to its goals Users [15]. Evaluating or measuring the performance of companies is one of the basic elements of the management control system, and the most important and difficult management accounting tasks. Therefore, an effective system for measuring the performance of operational processes must include basic indicators of performance, which are measures that study the activity of the company as a whole from the viewpoint of customers, in order to provide Feedback to help company members identify deficiencies and opportunities for improvement. Good performance measures help to match the objectives of the sub-units with the strategic goals of the company, and therefore the performance measurement system must include both financial and non-financial performance measures [16].

Firm performance is A financial statement that reports on the company's financial performance over a specific accounting period. It summarizes the revenues, costs, and expenses incurred over a period of time, usually a fiscal quarter or more. These reports give information about the company's ability (or lack thereof) to generate profit, through increasing revenues, reducing costs, or both [17].

The performance measurement of companies is no longer limited to the financial or quantitative standard only to reflect the true picture of the success of a company, or its departments and divisions. Performance reports based on digital indicators to show the amount of work are not expressive to measure an entity because it gives a specific view of performance that leads to late arrival of information. Consequently, being caught in the downward spiral of corrective actions and their exorbitant costs, instead of the simpler and less expensive preventive measures. In order to evaluate the performance of a company or its branches, we need a more comprehensive vision of a balanced distribution of measures that shows the outcome of the overall performance of the company, and there are methods and techniques for performance management, the most important of which is the use of the Balanced Score Card, which is a method used to evaluate the performance of a facility or unit performance and find a balance for successful performance standards [18].

2.3 Strategic Planning

Strategic planning is the process of an organization determining its strategy or direction, and making decisions about allocating its resources to pursue that strategy. The concept can also expand into control mechanisms to guide the implementation of the strategy [19].

Once the project is approved and formally approved, the planning stage begins, which is a stage in which the project is divided into various tasks and activities, preparing the schedule for completing tasks, estimating the total cost, determining the roles of work teams, preparing the preliminary plans necessary for the project and determining its scope and appropriate methods for its completion [20].

Strategic planning is the process of determining the direction of your company by evaluating your current location and the location you want to reach. The strategic plan is a document that includes the company's mission, vision, values, and long-term goals, as well as its operational plans to reach these goals [21].

In addition to preparing the structure of business division, preparing the organizational structure of the project and determining the critical path to ensure that all requirements of the planning stage are fully met and within the timeframe and quality standards [22].

Strategic planning is used in order to develop an organized and clear work plan for the project by focusing on all aspects and details necessary to achieve the general objective. Taking into account the economic, social and environmental changes that may contribute to making changes to the plan or upon its completion and amendment.

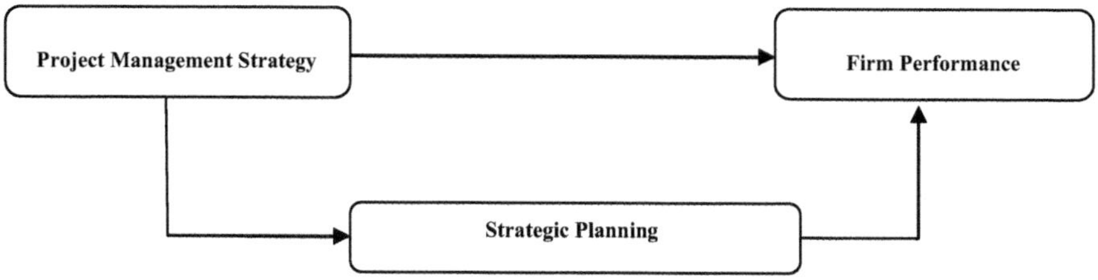

Fig. 1 Theoretical Framework

During the planning stage, the plans put in place must be evaluated to ensure their effectiveness, suitability for implementation, and their economic viability in terms of cost–benefit and expected financial returns based on analytical and statistical numbers.

The strategic planning process may take some time, but it is beneficial for everyone involved. As a small business owner, you will have a better idea of the goals and objectives you want to achieve and a path to do so. For your employees, the process can foster an increase in productivity—which contributes to business success [23].

2.4 Theoretical Framework

Based on the scholarly literature reviewed for this paper we develop a model shown in Fig. 1.

Therefore, in light of the previous literature mentioned and in pursuit of this study's goals, the following was hypothesized:

H1: *project management strategy has a positive effect on firm performance in SMES.*

H2: *project management strategy has a positive effect on strategic planning.*

H3: *strategic planning has a positive effect on firm performance in SMES.*

H4: *strategic planning mediates the relationship between project management strategy and firm performance in SMES.*

3 Methodology

3.1 Study Design and Data Collection

Data were collected from employees of Jordanian SMEs. 319 questionnaires were collected from 421 distributed questionnaires. In the list form, delete was used to clean up the dataset of missing values. Likewise, the non-interacting responses were omitted. The final data set used in the following analysis consists of 291 responses.

Constructs Measurement
The questionnaire, which is considered the main tool for the study, was divided into four sections, where the first section took the demographic information for the sample, while in the second section the independent variable was measured, which is the pioneering trend, and the third section dealt with the dependent variable, which is the performance of companies, and finally the fourth section dealt with the intermediate variable, which is the competitive advantage. A Likert scale with five points on all items in the current study to determine the respondent's answer, where the lowest value of 1 indicated "strongly disagree" and the highest value of 5 indicated "strongly agree" [e.g., 24–28].

3.2 Study Population and Sample

The study population consisted of managers of small and medium companies operating in Jordan at various administrative levels, as these companies operate in the fields of business and achieve profits and achieve the strategic goals of the company and the goals of the stakeholders, as there were about 48 companies, and the random sampling method was used to collect data from the study community. Where the sample size reached 319 distributed questionnaires, and the size of the 28 returned questionnaires was suitable for analysis, with a percentage of 96.21% of the total questionnaires.

3.3 Results

The distribution of the paragraphs of the questionnaire is displayed in Table 1, while showing which questions correspond to each variable.

After the researcher developed the questionnaire to include all dimensions of the independent and dependent variables so that they could test the hypotheses of the study through the questionnaire that was distributed. Prepared on members of the sample. The researcher used the Cronbach Alpha coefficient to measure the size of the internal consistency between the paragraphs of the questionnaire. In order to

Table 1 The distribution of the paragraphs of the questionnaire

Independent variables	Project management strategy	Q1–Q21
Moderating variables	Strategic planning	Q22–Q31
Dependent variables	Firm performance	Q32–Q45

Table 2 (Cronbach Alpha)

Variables	Reliability coefficient (R.c)
Project management strategy	0.810
Strategic planning	0.920
Firm performance	0.912
Total	0.**892**

verify the collected data and verify the distribution of the sample responses according to the five-point Likert scale, which indicates the extent of the respondent's approval of the paragraphs of the questionnaire. To obtain a high degree of reliability and reliability of the data.

The Internal consistency coefficients: Reliability coefficient (Cronbach Alpha) is displayed in Table 2.

By looking at Table 2, it became clear that is acceptable, as it is higher than 71%, meaning that the ratio of security transactions to the variables. In order to test the reliability of the questionnaire using SPSS to find the value of the reliability coefficient, this indicates the existence of an internal harmony between the corridors. 892 asserted that the theory is legitimate through reconnaissance.

In Table 3 the results of the regression analysis are presented about the relationship between project management strategy and firm performance.

The first hypothesis which means the effect of the project management strategy on the performance of companies was confirmed by looking at Table No. 3 where the value of F was a statistically significant function, and this means that there is a linear relationship between the independent variables and the dependent variables. R-Square equals the modifier 777. This indicates that the regression is estimated to be about 77% of the changes in the dependent variable. All t-values are meaningful when the level of significance is on a percentage basis.

The results of regression analysis on the relationship between project management strategy and strategic planning are displayed in Table 4.

The second hypothesis describing job changes has been confirmed, which is that the project management strategy greatly influences the strategic planning as the

Table 3 Project management strategy and firm performance (testing H1)

Model	B	Std. error	Beta	T	Sig
Constant	2.311	0.222		6.003	

F-value = 200.311(Sig. = 0.000) Adjusted R-Square = 0.777

Table 4 Project management strategy and strategic planning (testing H2)

Model	B	Std. error	Beta	T	Sig
Constant	1.999	0.889		13.351	0

.000 F-value = 01.999(Sig. = 0.000) Adjusted R-Square = 0.611

Table 5 Strategic planning and firm performance (testing H3)

	Standardized		Unstandardized coefficients		
Model	B	Std. error	Beta	T	Sig
Constant	1.910	0.370		4.001	000
Competitive advantage	0.891	0.070	0.070	13.001	000

Adjusted R-Square = 0.531

ratio of T-value was 88.999 and significant to the level of significance, modified R. The square equals 0.611 or 61% of the independent variables. Therefore, we found that all values of the entrepreneurial orientation T in the samples are statistically significant, and this proves the validity of the hypothesis by referring to Table 4. The results of regression analysis on the relationship between strategic planning and firm performance are displayed in Table 5.

That strategic planning affects the performance of the company. Through the results of Table 5, the modified R-Square equals 0.531 which means that the regression analysis may describe about 53 percent of the changes to the dependent variable. The T-value is statistically significant when the level of significance is one percent. This confirms the third hypothesis. The results of regression analysis on the relationship between project management strategy and firm performance, through strategic planning (mediator variable), using path analysis test (PLS), are displayed in Table 6.

Table 6 shows the results of the pathway analysis where the table shows the confidence interval value (LL, UL) (81.12%) which gives the minimum (LL) and maximum (UL) as well as the (VAF) values that were greater than (20%) and less than (81%), Which indicates a slight effect of the Strategic planning, that is, there is an effect on the (0.05) level of the project management strategy through the mediator variable [29–31].

Table 6 Project management strategy and firm performance, through strategic planning (testing H4)

Hypothesis	Relationship	Std. error	Beta	**t-value**	Bootstrapped confidence interval	Decision
					96% LL 96% UL	Accepted
FP	SP PPS CA	0.021	0.01212	4.444	0.042	0.188
Variable			VAF			
	CA-SP		81.12%			Partial mediation

4 Discussion

There was a positive effect of the project management strategy on the performance of companies. The first hypothesis proved its validity, with the existence of an important and positive relationship between the project management strategy and the corporate performance, which was statistically significant. It has also been emphasized that the project management strategy greatly affects strategic planning, as there is a statistically significant effect between the project management strategy and strategic planning, and this confirms the second hypothesis.

Based on Table 5, which showed that strategic planning affects the project management strategy, as the hypothesis has been proven correct and that there is a statistically significant effect between strategic planning and project management strategy. The last hypothesis (there is an indirect and statistical effect between project management strategy and corporate performance through strategic planning as an intermediate variable) showed that there is a partial effect of the mediating variable (strategic planning), on the relationship between project management strategy and corporate performance, based on the VAF technique (variance calculation) Which assumed that a VAF of less than 20% indicates no mediation, while a VAF of 20–81% indicates partial mediation, and that VAF > 81.12% indicates complete mediation.

5 Conclusion

In conclusion, project management is important in making sure that what needs to be delivered will be delivered and will add real value. Every organization has its strategic goals and every project must benefit from these goals. Risk Management: The importance of project management lies in the fact that it ensures that risks are managed. Well, it ensures that it does not become a problem. Where the researcher found that the project management strategy has an impact on the performance of companies and on strategic planning through the findings of the researcher, strategic planning has a role in contributing to setting a time frame for the implementation of the work by relying on developing an appropriate strategic plan that contributes to designing his own steps. Participation in transforming the planned actions into things that can be measured and applied until sufficient knowledge is achieved in obtaining acceptable results.

This study in this field contributed to bridging the literature gap in comparison with other studies. Because this study has its limits as it is limited to small and medium-sized companies in Jordan, where the results should not be measured and circulated to other countries. Therefore, the researcher advises small and medium companies urgently to follow the strategy of managing projects dynamically and actively in order to continue and survive and achieve the goals and results of the company, as the company has the ability to deliver its projects in a timely, appropriate and specific time without exposing them to losses and other costs. In the end, the researcher recommends conducting more studies on this topic in order to be comprehensive and broad.

References

1. Shekhar A (2004) Strategic project leadership: toward a strategic approach to project management. R&D Manag 34:569–578
2. Heerkens G (2007) Introducing the revolutionary strategic project management maturity model SPM 3 paper presented at the annual North American meeting of the Project Management Institute Atlanta, GA.P. I
3. Kenny C (2006) Linking corporate strategy to project management. Paper presented at the annual Latin American meeting of the project management institute, Santiago, Chile.
4. Green S (2005) Strategic project management: from maturity model to star project leadership. PMI World Congress
5. Alsafadi Y, Altahat S (2021) Human resource management practices and employee performance: the role of job satisfaction. J Asian Financ Econ Bus 8(1)
6. Sakhnini Z, Elali W, Al-Sartawi A (2022) Behavioral and non-behavioral factors and the level of adapting and implementing fintech and e-banking in Bahrain: suggested model. In: Musleh Al-Sartawi AMA (ed) Artificial Intelligence for Sustainable Finance and Sustainable Technology, vol 423. Lecture Notes in Networks and Systems. Springer, Cham, pp 188–193. https://doi.org/10.1007/978-3-030-93464-4_19
7. Asrilhant B, Meadows M, Dyson RG (2004) Exploring decisions support and strategic project management in the oil and gas sector. Eur Manag J 22(1):63–73
8. Anderson DK, Merna T (2003) Project management strategy– project management represented as a process based set of management domains and the consequences for project management strategy. Int J Project Manag 21(6):387–393
9. DyReyes J (2008) Strategic project management: aligning strategic business objectives with project management strategy. University of Oregon, University of Oregon, Applied Information Management
10. Milosevic D, Srivannaboon S (2006) A theoretical framework for aligning project management with business strategy. Proj Manag J 37(98):110
11. Nawafleh A, Alsafadi Y (2020) The effect of a high-commitment work system on improve organization citizenship behavior: an empirical study in Zain telecom company in Jordan Asian Journal of social science and management technology, ISSN: 2313–7410, Volume 2 Issue 4, July-August
12. Al-Qudah AA, Al-Okaily M, Alqudah H (2022) The relationship between social entrepreneurship and sustainable development from economic growth perspective: 15 'RCEP' countries. J Sustain Financ Invest 12(1):44–61
13. Dietrich P, Lehtonen P (2005) Successful management of strategic intentions through multiple projects- reflections from empirical study. Int J Project Manag 23:386–391
14. Georgopoulos BS, Tannenbaum AS (1957) A study of organizational effectiveness. Am Sociol Rev :534–540
15. Hansen GS, Wernerfelt B (1989) Determinants of firm performance: the relative importance of economic and organizational factors. In the Strateg Manag J :399–411
16. Alansari Y, Al-Sartawi A (2021) IT governance and E-banking in GCC listed banks. Procedia Comput Sci 183:844–848
17. Ivanov C, Avasilcăi S (2014) Performance measurement models: an analysis for measuring innovation processes performance. Procedia –Soc Behav Sci: 397–404

18. Man YS (2006) Performance measurement and management of third party logistics: an organizational theory approach. Hong Kong Baptist University
19. Almaaitah M, Alsafadi Y, Altahat S, Yousfi A (2020) The effect of talent management on organizational performance improvement: the mediating role of organizational commitment. Manag Sci Lett 10(12):2937–2944
20. Alsafadi Y, Aljawarneh N, Çağlar D, Bayram P, Zoubi K (2020) The mediating impact of entrepreneurs among administrative entrepreneurship, imitative entrepreneurship and acquisitive entrepreneurship on creativity. Manag Sci Lett 10(15):3571–3576
21. Safadi AL, Y. (2016) Case study, Cairo-Amman bank-Jordan: improving an organization by the use of smart phones and iPads which are considered the basics of the supply chain management (SCM). Invention J Res Technol Eng Manag 1(5):11–23
22. Musleh Al-Sartawi AMA (ed) (2022) Artificial Intelligence for Sustainable Finance and Sustainable Technology, vol 423. Lecture Notes in Networks and Systems. Springer, Cham. https://doi.org/10.1007/978-3-030-93464-4n
23. Hussein Y, Çağlar D (2019) The effect of high involvement work systems on organizational performance: the mediating role of knowledge-based capital. Manag Sci Lett 9(9):1361–1372
24. Al-Okaily M, Al-Okaily A (2022) An empirical assessment of enterprise information systems success in a developing country: the Jordanian experience. TQM J. https://doi.org/10.1108/TQM-09-2021-0267
25. Alsaad A, Al-Okaily M (2022) Acceptance of protection technology in a time of fear: the case of Covid-19 exposure detection apps. Inf Technol People. https://doi.org/10.1108/ITP-10-2020-0719
26. Al-Okaily M, Alqudah H, Al-Qudah AA, Alkhwaldi AF (2022) Examining the critical factors of computer-assisted audit tools and techniques adoption in the post-COVID-19 Period: internal auditors perspective. VINE J Inf Knowl Manag Syst. https://doi.org/10.1108/VJIKMS-12-2021-0311
27. Al-Okaily M, Alghazzawi R, Alkhwaldi AF, Al-Okaily A (2022) The effect of digital accounting systems on the decision-making quality in the banking industry sector: a mediated-moderated model. Global Knowledge, Memory and Communication. https://doi.org/10.1108/GKMC-01-2022-0015
28. Al-Okaily A, Al-Okaily M, Teoh AP, Al-Debei M (2022) An empirical study on data warehouse systems effectiveness: the case of Jordanian banks in the business intelligence era. EuroMed J Bus. https://doi.org/10.1108/EMJB-01-2022-0011
29. Al-Sartawi A (2018) Institutional ownership, social responsibility, corporate governance and online financial disclosure. Int J Critical Acc 10(3/4):241–255
30. Taamneh AM, Taamneh M, Alsaad A, Al-Okaily M (2021) Talent management and academic context: a comparative study of public and private universities. EuroMed J Bus. https://doi.org/10.1108/EMJB-08-2020-0088
31. Al-Okaily M (2021) Assessing the effectiveness of accounting information systems in the era of COVID-19 pandemic. Vine J Inf Knowl Manag Syst. https://doi.org/10.1108/VJIKMS-08-2021-0148

Diffusing IT Competence with Business Analytics Enables Data-Driven Culture for Achieving Evidence-Based Managerial Decisions for Assuring Sustained Competitive Advantage: A Literature Survey

Anjum Razzaque, Magdalena Karolak, Frank Lorne, and Sekhar Amba

Abstract Globalized economies have organizations seeking ways to implement Business Analytics (BA) though ample organizations fail in such initiatives. Hence, scholars embark on IT competence enabled BA to inspire a data-driven culture. Such is a crucial prerequisite for fostering long term organizational sustainable competitiveness. Even though the above rationale receives widespread attention, there is scant evidence comprehending whether IT competence enabled BA to empower an organizational data-driven culture to achieve long-term sustained competitive advantage. Current literature reviews are conducted predominantly through journals and academic conference papers. The literature review reveals a conceptual model composed of three propositions viable for future research's empirical confirmation. This study defuses IT competence, business analytics, data-driven culture, and organizational competition in one model, proving through critiqued literature that business analytics inspires mined data visualizations that encourage a data-driven organizational culture when enabled by IT competence. Such a culture, in turn, helps an organization achieve the evidence-based managerial decisions which help organizations sustain a long-term competitive advantage. Such a model bares theoretical and practical implications also portrayed in the article.

Keywords IT competence · Business analytics · Data-driven culture · Decision-making · Sustainable competition

A. Razzaque (✉) · M. Karolak
Ahlia University, Manama, Bahrain
e-mail: anjum.razzaque@gmail.com

F. Lorne
Zayed Univeristy, Dubai, UAE
e-mail: florne@nyit.edu

S. Amba
New York Institute of Technology, Vancouver, Canada
e-mail: smuni@nyit.edu

A. M. A. Musleh Al-Sartawi et al. (eds.), *From the Internet of Things to the Internet of Ideas: The Role of Artificial Intelligence*, Lecture Notes in Networks and Systems 557,
https://doi.org/10.1007/978-3-031-17746-0_55

1 Introduction

Modern firms adopt business analytics (BA) to achieve sustained performing organizations by accomplishing competitive amidst the globalized and constantly shifting corporate environments [14]. Firms acquire data from Supply chain (SC) partners for insights on possible risks and potential disruptions to avoid hampering SC performance [10]. Such information-sharing partnerships expect organizational implemented BA to harvest a data-driven culture [13, 34, 35]. Firms receive real-time data from their supply chain partners, where BA empowers insights to enhance firm competitiveness and improve long-term business performance [43]. However, several organizations fail in their BA implementations initiatives when they attempt to adopt BA. Hence, they continue hunting for paths that aid them in achieving a long-term sustainable competitive advantage [29]. When investigating why BA implementation initiatives fail, scholars constantly report a troublesome issue regarding the scant comprehension of the IT competence success factors that could enable BA. After all, it is common knowledge that IT infrastructures would underpin any BA initiative. Moreover, BA enriches when IT competence enables firms to comprehend internal processes vital for data-driven managerial decision-making [14, 36]. Therefore, as this study continues, the inquiry if IT competence enabled BA initiatives would succeed. However, there is another concern: scant literature reveals whether IT enables BA initiatives. Nevertheless, practitioners' insight reveals that IT competence influences BA. For instance, the effective utilization of IT tools like Hadoop facilitates BA to offer prescriptive and predictive analytics [15].

Do IT competence-enabled BA implementations inspire a data-driven culture for enhancing competitive advantage? Such an inquiry indicates two missing links that could help information systems research figure out. First, no study yet has confirmed, i.e. what added value IT competence contributes toward organizational goals [23]. In the cases of this study, organizational goals are a firm competitive advantage. One possible explanation is that past research investigates IT in isolation. In this case, substantial evidence proves that IT alone cannot enhance business processes to achieve performance and competitive advantage [12]. IT only adds value when deployed within business processes [49, 54. Since scholars pledge that IT competence enables BA and that for a firm to achieve competitiveness, they need to embrace the adoption of technologies [23]. Hence, this study proposes its literature-driven model (Fig. 1), which leverages IT competence to enable BA. Hence, BA usage inspires a data-driven culture to enhance firm competition [40]. Nevertheless, such a scenario is achievable by introducing BA with sophisticated statistical simulations, forecasting abilities, and machine learning capabilities [42]. This study significantly contributes to information systems research. Its literature-driven model (Fig. 1) first comprehends whether IT competence enabled BA to enhance its organizational data-driven culture to achieve a long-term sustainable competitive advantage.

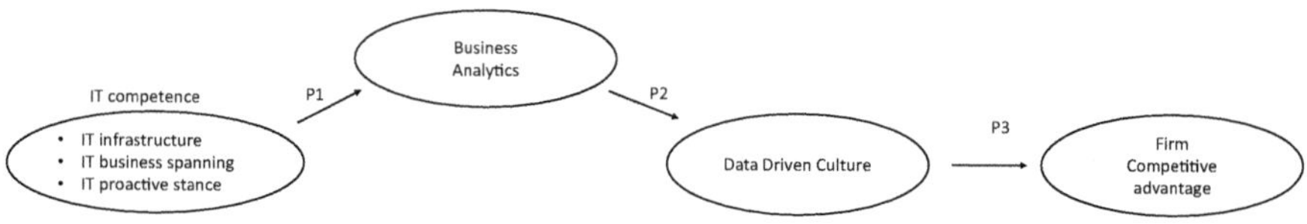

Fig. 1 Proposed research framework

2 Literature Review

An organization can construct, integrate, and reconfigure the internal and external capabilities to deal with dynamically shifting environments [45]. In the current globalized economies, organizations persistently reconfigure their resources or capabilities. Dynamic organizational environments emphasize the dynamic capabilities for integrating and reconstructing business resources [11]. Therefore, IT alone cannot add value and cannot directly influence competitive advantage. However, IT supports developing, optimizing, and integrating standard organizational resources for assuring IT competence, which plays a crucial role in achieving long-term competitive advantage [47, 48]. In other words, as per the case of this study, IT leveraged competency alone cannot directly influence organizational competitive advantage unless dynamic capability resources mediate [3]. Also, BA alone cannot practically facilitate a mediating body like data-driven culture to achieve competitive advantage. BA implementations constantly fail due to scant comprehension of BA's IT competence success factors. It is common knowledge that IT infrastructures underpin any BA implementation. Hence, BA initiatives require that they be enabled by IT competence, so organizations will achieve IT competence enabled BA implementations for comprehending their internal processes, which was deemed vital for data-driven managerial decision-making [14]. The challenge is that organizations are too busy calculating. IT invested return on investments. They expect a successfully implemented IT investment by pinpointing, building, customizing, and deploying databases, user interfaces, and analytical tactics. However, only IT competence enabled BA to form that dynamic capability resource to achieve a competitive advantage [2]. IT competence enabled BA combines to become a true dynamic capability resource, an organizational asset, for streamlining IT resources that manage the implemented BA capabilities [3, 39].

Suppose BA facilitates an intermediating capability resource bearing an absorptive capacity for data analytics to improve competitive advantage. Such an intermediating resource could prove crucial for strengthening a BA environment within an organization. When searching for an intermediating resource, BA and SC literature reveals "*big data analytics capabilities,*" which is deemed the appropriate intermediating resource that can get empowered by IT competence enabled BA implementation. [8] propose that BA implementations enable intangible big data analytics capabilities. Such a notion gets reaped when an organization conceives a culture where a blend of employees and managers start synergically perceiving an organizational challenge and deal with that challenge with a common organizational vision [44]. This study's case can utilize such a notion of organizational culture. Thus, an organization's data analytics capabilities are the organizational resource, referred to as organizational data-driven culture. Within such a culture, a blend of managers and employees synergically desire to gain skilful experience with and utilize big data and BA to make effective managerial decisions. Henceforth, top management gets committed to utilizing BA [36]. Thus, synergically, a blend of employees and managers begin to reap the fruit of their collegial collaboration for building an organization with; data-driven culture as they begin producing data-verified decisions. Considering that [8] warn organizations of intangible data analytics capabilities alone, they are ineffective mediators empowered by IT competence that enabled BA to achieve competitive advantage. Therefore, organizations require fostering a data-driven culture, and such a culture should get empowered by IT competence enabled BA. Hence, the role of IT competence enabled BA empowered data-driven culture adds value in the current globalized economies where scholars and practitioners seem keenly interested in the supportive ramifications of BA for enhancing performance [19, 37]. This concept is important to comprehend since BA's ability to optimize data analysis capabilities is crucial for enhancing managerial evidence-based decisions-. However, scholars report the need to be empirically justified to warrant assessing if IT competence enabled BA to be empowered by data-driven culture to enhance performance [9, 42]. Such an assessment is vital as scholars shed disunited consensus regarding the role of BA on performance, predominantly due to scant literature in this area.

3 A Conceptual Framework with Three Propositions

IT competence is an organization's capability for acquiring, deploying, and utilizing IT resources to strengthen corporate policies and business procedures [41]. IT competence alone cannot achieve organizational goals [46]; however, for an effective IT competence initiative, a well-supported production of and analysis of data for performing BA [4] is crucial. BA utilizes numerous sophisticated data analysis methods to solve problems and enhance organizational competitive advantage. BA is not referring to some technological machinery but is an integration of methodologies, administrative processes, and instruments for achieving the right information at the right time. BA examines such information and simulates probable consequences from the desired solutions to planning, sourcing, creating, and distributing [5]. IT competence integrates modules for conceiving new products/services [3]. According to [20], a flexible IT competence initiative led by experienced IT professionals enriches organizational performance. Complex business spanning, an IT competence component through appropriate IT infrastructure, enriches data sources' integration to analyze and visualize big data using BA for evidence-based managerial decision-making [18]. Organizations produce big data within and outside their boundaries. To ensure a well-developed IT proactive stance through IT business, IT infrastructure capacity is vital for implementing BA [7]. Organizations need such a resource for collecting, updating, and integrating big data [26]; to augment performance, hence should integrate IT competence with BA. Based on this argument:

Proposition 1 (P1): IT competency, enabled by IT infrastructure, IT business planning, and proactive stance, enables BA [38].

From the managers' interviewer from several case studies, [21] reveal that BA empowers managers to reap the benefits of data-driven decisions. However, BA can only be beneficial if employees' mindsets get incentivized to get driven by a data-driven culture. Employees get incentivized through workshops where literary data sessions get conducted, so employees comprehend the power of BA by reading, comprehending, and analyzing data encountered during daily tasks [51]. So, employees can relish the importance of making decisions driven by data, not opinions [30–34]. This way, organizations successfully implement BA to generate longer-term sustainable performance powered by a data-driven culture that aids in visualizing, predicting and cognitively perceiving big data [21]. According to [21], a data-driven culture is an organizational behavioural pattern, employees' synergic practice to prosper an organization. Organizations implement BA to apply data visualization techniques for customer insights and competitors' behaviours, thus deploying innovative strategies that allow BA to achieve insights into unpredictable situations and thus generate competitive advantage [50]. Such is possible when managers utilize BA through a data-driven organizational culture harness when decisions are evidence-based by big data [38]. Such organizational culture aids managers investigate buyers' perceptions to improve products/ services [25]. Hence, based on this argument:

P2: BA empowers organizational data-driven culture.

By harvesting a data-driven organizational culture, organizations innovate those business processes that aid in achieving a competitive advantage [6]. Effective management must collect and preserve customers' knowledge, analyze market needs, and study competitors. Such knowledge provides organizations with a competitive advantage. However, such knowledge is obtainable when employees harness a data-driven culture within their organization. The reason is that decisions get made not by opinions but when driven by BA, so BA drives a data-driven culture. When data-driven innovations drive an organization, it attracts more customers; thus, capitalizing on brand-new opportunities [28]. Scholars assert that decisions get data-driven insights to assure superior quality of managerial decision-making [17]. When employees and managers experience a robust data-driven culture [7], they achieve a competitive advantage by pinpointing potential markets for new products and services. It can redesign business processes to beat the competition [27]. Therefore, organizations should inspire and incentivize a BA empowered, data-driven culture so employees make evidence-based managerial decisions to sustainably compete for long-term [22]. Hence, based on this argument:

P3: Data-driven culture enhances organizational competitive advantage.

4 Discussion, Conclusion, Implications, and Future Research

This study conducts BA research. It embarked on research gaps that identified a need to include IT competence as a supportive variable for BA to facilitate organizations to harvest a data-driven culture for assuring competitive advantage through evidence-based managerial decision-making. Predominantly journal articles and conference papers were the main sources of knowledge for the write-up of this article. The critique of current literature theoretically validates that IT competence alone cannot aid any organization. Although IT competence enables BA through its supportive indicators, IT infrastructure, IT business planning, and IT proactive stance. In turn, BA empowers a data-driven culture. To retain competitive advantage, organizations must continually plan new products or services through BA empowered data-driven insights to constantly stay ahead of their competitors, thus, sustaining competitive performance over time [22]. In other words, organizations must align their customers' needs-driven perceived by managerial decisions that are constantly data-driven [49, 50] to achieve unique and quality products or services [22].

Such is the theoretical implication of Fig. 1 conceptual model. From the practical implications, organizations will better compete, especially during these turbulent times, if organizations build policies that encourage managers to bring onboard change management so a data-driven culture gets enabled by IT competence-driven BA. Future research should empirically assess this model within the case of proof and a public organization or sector to confirm the reliability and validity of the Fig. 1 model. From the point of view of managerial implications, IT competence diffused BA implementation projects empower a data-driven culture because BA supports organizations through its descriptive, predictive, and prescriptive analytics. Descriptive analytics mines past raw data in readable format through a post-mortem method to produce a managerial report and is applied to assess relationships between variables. Predictive analytics is a prominent big-data analysis using tools like Hadoop clusters to analyze large amounts of stored data using statistical algorithms blended with machine learning to predict the likelihood of future consequences via logical regressions, time series analysis, or decision trees. Mangers can apply predictive analytics to mitigate risks, detect fraud cases, and improve market campaigns, to name a few examples. Prescriptive analytics allow managers to make immediate data-driven probabilistic forecasted decisions based on pre-conditioned managerial set rules through three algorithms: fuzzy rule-established systems, decision trees, or switching neural networks (also referred to as artificial neural networks [16]. Such analysis can be analyzed using tools like R (via 3D models or multi-panel charts) or Tableau [1]. However, when organizations start depending on BA, automatically, a data-driven culture begins to conform.

Future research should extend the work of [24], who empirically confirm a significantly positive effect of IT competence on using BA for achieving a firm competitive advantage but this time applying such a model within the case of a public and private sector of developing countries. Furthermore, future research should critique the literature and empirically assess if there is a mediating role of BA between IT competence and data-driven culture. Hence, organizations will align their IT initiatives and streamline their BA capabilities for predicting and prescribing analysis, thus assuring a culture that yields data-driven, evidence-based managerial decisions over-opinionated decisions.

References

1. Ahmad N H (2020) TABLEAU for beginners: data analysis and visualization 101.
2. Akter S, Wamba SF, Gunasekaran A, Dubey R, Childe SJ (2016) How to improve firm performance using big data analytics capability and business strategy alignment? Int J Prod Econ 182:113–131
3. Musleh Al-Sartawi AMA, Razzaque A, Kamal MM (eds) (2021) Artificial intelligence systems and the internet of things in the digital era, vol 239. Lecture Notes in Networks and Systems. Springer, Cham. https://doi.org/10.1007/978-3-030-77246-8
4. Bharadwaj AS (2020) A Resource-based perspective on information technology capability and firm performance: an empirical investigation. MIS Q 24(1):169–196
5. Bose R (2009) Advanced analytics: opportunities and challenges. Ind Manag Data Syst 109(2):155–172
6. Büschgens T, Bausch A, Balkin DB (2013) Organizational culture and innovation: a meta-analytic review. J Prod Innov Manag 30(4):763–781
7. Ciampi F, Demi S, Magrini A, Marzi G, Papa A (2021) Exploring the impact of big data exploring the impact of big data orientation. J Bus Res 123:1–13
8. Dennehy D, Oredo J, Spanaki K, Despoudi S, Fitzgibbon M (2021) Supply chain resilience in mindful humanitarian aid organizations: the role of big data analytics. Int J Oper Prod Manag 41(9):1417–1441
9. Dubey R, Gunasekaran A, Childe SJ, Papadopoulos T, Blome C, Luo Z (2019) Antecedents of resilient supply chains: an empirical study. IEEE Trans Eng Manage 66(1):8–19
10. Dubey R, Gunasekaranb A, Childe SJ (2021) Empirical investigation of data analytics capability and organizational flexibility as complements to supply chain resilience. Int J Prod Res 59(1):110–128
11. Eisenhardt KM, Martin JA (2000) Dynamic capabilities: what are they? Strateg Manag J 21(10–11):1105–1121
12. El Sawy OA, Malhotra A, Park Y, Pavlou PA (2010) Research commentary—seeking the configurations of digital ecodynamics: it takes three to tango. Inf Syst Res 21(4):835–848
13. Galbraith JR (1974) Organization design: an information processing View. Interfaces 4(3):28–36
14. Goes PB (2014) Big data and IS research. MIS Q 38(3):3–8
15. Hindle GA, Vidgen R (2017) Developing a business analytics methodology: a case study in the foodbank sector. Eur J Oper Res 268(3):836–851
16. Jackson TW, Lockwood S (2018) Business analytics: a contemporary approach. Red Globe Press, London
17. Jaklic J, Grubljesi T, Popovic A (2018) The role of compatibility in predicting business the role of compatibility in predicting business. Int J Inf Manage 43:305–318
18. Jayachandran S, Sharma S, Kaufman P (2005) The role of relational information processes and technology use in customer relationship management. J Mark 69(4):177–192
19. Karolak M, Razzaque A, Al-Sartawi A (2021) E-services and M-services using iot: an assessment of the Kingdom of Bahrain. In: Musleh Al-Sartawi AMA, Razzaque A, Kamal MM (eds) Artificial Intelligence Systems and the Internet of Things in the Digital Era, vol 239. Lecture Notes in Networks and Systems. Springer, Cham, pp 523–533. https://doi.org/10.1007/978-3-030-77246-8_48

20. Kim G, Shin B, Kim KK, Lee HG (2011) IT capabilities, process-oriented dynamic capabilities, and firm financial performance. J Assoc Inf Syst 12(7):487–517
21. Kiron D, Ferguson R, Prentice PK (2013) From value to vision: reimagining the possible with data analytics. MIT Sloan Manag Rev 54(3):1–19
22. Kuncoro W, Suriani WO (2018) Achieving sustainable competitive advantage through product innovation and market driving. Asia Pac Manag Rev 23(3):186–192
23. Ling R, Yen DC (2001) Customer relationship management: An analysis framework and implementation strategies. JComputer Information Systems 41(3):82–97
24. Musleh Al-Sartawi AMA (ed) (2022) Artificial intelligence for sustainable finance and sustainable technology, vol 423. Lecture Notes in Networks and Systems. Springer, Cham. https://doi.org/10.1007/978-3-030-93464-4
25. Musleh Al-Sartawi AM, Hussainey K, Razzaque A (2022) The role of artificial intelligence in sustainable finance. J Sustain FinancInvest :1–6
26. Prahalad CK, Krishnan MS (2008) The new age of innovation driving co-created value through global networks. McGraw-Hill, New York
27. Ranjan J, Foropon C (2021) Big data analytics in building the competitive intelligence of organizations. Int J Inf Manage 56:102231
28. Ransbotham S, Kiron D (2017) Analytics as a source of business innovation. MIT Sloan Manag Rev
29. Ransbotham S, Kiron D, Kiron P (2016) Beyond the hype: The hard work behind analytics success. MIS Sloan Manag Rev 57(3):3–16
30. Razzaque A (2021) Artificial intelligence and IT governance: a literature review. In: Musleh Al-Sartawi AMA (ed) The Big Data-Driven Digital Economy: Artificial and Computational Intelligence, vol 974. Studies in Computational Intelligence. Springer, Cham, pp 85–97. https://doi.org/10.1007/978-3-030-73057-4_7
31. Razzaque, A., & Alalawi, M. (2015). knowledge sharing and positive emotions in virtual communities: the case of the healthcare sector. KS Global research international conference on emerging trends in multidisciplinary research (ETMR-2015). Istanbul, Turkey: KS Global Research
32. Razzaque A, Eldabi T, Jalal-Karim A (2013) physician virtual community and medical decision-making: mediating role of knowledge sharing. J Enterp Inf Manag 26(5):500–515
33. Razzaque A, Al-Sartawi AMM, Karolak M (2021) Rethinking tourism entrepreneurship amidst COVID-19 through the lens of artificial intelligence and big data analytics. In big data for entrepreneurship and sustainable development (pp. 175–188). CRC Press.
34. Razzaque A, Eldabi T (2018) Tackling obesity by diet quality, technology and social networks. Int J Food Eng 4(1):88–92
35. Razzaque A, Eldabi T (2020) Physicians social capital aids their medical decisions when they virtually share knowledge. Int J Knowl Manag Stud 11(3):229–257
36. Razzaque A, Eldabi T, Chen W (2020) Quality decisions from physicians' shared knowledge in virtual communities. Knowl Manag Res Pract 1–13. https://doi.org/10.1080/14778238.2020.1788428
37. Razzaque A, Eldabi T, Jalal-Karim A, Karolak M (2013) Culture sustained knowledge management infrastructure and architecture facilitate medical decision making. 2013 IIAS-IASIA Joint Congress. Manama, Bahrain: IIAS-IASIA.
38. Razzaque A, Eldabi T, Jalal-Karim A (2013) Physician virtual community and medical decision-making: mediating role of knowledge sharing. J Enterp Inf Manag 26(1):500–515
39. Sakhnini Z, Elali W, Al-Sartawi A (2022) Behavioral and non-behavioral factors and the level of adapting and implementing fintech and e-banking in bahrain: suggested model. In: Musleh Al-Sartawi AMA (ed) Artificial Intelligence for Sustainable Finance and Sustainable Technology, vol 423. Lecture Notes in Networks and Systems. Springer, Cham, pp 188–193. https://doi.org/10.1007/978-3-030-93464-4_19
40. Sambamurthy V, Zmud R (1997) At the heart of success: organizationwide management management. In: Sauer C, Yetton PW (eds) Steps to the Future: Fresh Thinking on the Management of IT-Based Organizational Transformation (Jossey-Bass Business & Management), 1st edn. Jossey-Bass, San Francisco, pp 143–163

41. Shanks G, Bekmamedova N (2012) Achieving benefits with business analytics systems: an evolutionary process perspective. J Decis Syst 21(3):231–244
42. Srinivasan R, Swink M (2018) An investigation of visibility and flexibility as complements to supply chain analytics: an organizational information processing theory perspective. Prod Oper Manag 27(10):1849–2186
43. Tanriverdi H, Venkatraman N (2005) knowledge relatedness and the performance of multibusiness firms. Strateg Manag J 26(2):97–119
44. Teece DJ, Pisano G, Shuen A (1997) Dynamic capabilities and strategic management. Strateg Manag J 18(7):509–533
45. Tippins MJ, Sohi RS (2003) IT competency and firm performance: is organizational learning it competency and firm performance: is organizational learning a missing link? Strateg Manag J 24(8):745–761
46. Vidgen R, Shaw S, Grant DB (2017) Management challenges in creating value from business analytics. Eur J Oper Res 261(2):626–639
47. Wade M, Hulland J (2004) Review: the resource-based view and information systems research: review, extension, and suggestions for future research. MIS Q 28(1):107–142
48. Wang L, Jin JL, Zhou KZ, Li CB, Yin E (2020) Does customer participation hurt new product development performance? Customer role, product newness, and conflict. JBusiness Research 109(C), 246–259
49. Yanqing D, Guangming C, Edwards JS (2020) Understanding the impact of business analytics on innovation. Eur J Oper Res 281(3):673–686
50. Yu BT, Ulukok E (2021) Google's use of artificial intelligence in a two-sided advertising market causing trust loss, and data monopolization. In: Musleh Al-Sartawi AMA, Razzaque A, Kamal MM (eds) Artificial Intelligence Systems and the Internet of Things in the Digital Era, vol 239. Lecture Notes in Networks and Systems. Springer, Cham, pp 383–397. https://doi.org/10.1007/978-3-030-77246-8_36
51. Zhu K, Kraemer KL (2005) Post-adoption variations in usage and value of e-business by organizations: cross-country evidence from the retail industry. Inf Syst Res 16(1):61–84

Accounting Information System and Islamic Banks' Performance: An Empirical Study in the Kingdom of Bahrain

Abdalmuttaleb Al-Sartawi, Zakeya Sanad, Munther Talal Momany, and Manaf Al-Okaily

Abstract Accounting information system has become an essential tool in any firm. In this study, the researchers investigated whether accounting information system has an impact on Islamic banks' organizational performance in Bahrain. The study focuses on testing three main elements of the accounting information system including human and physical resources and financial data. A questionnaire was designed to collect the study data. The population of the present research includes all the accountants, financial managers, and board members of the Islamic banks in Bahrain. The results revealed that there is a significant impact of human resources on Islamic banks organizational performance in Bahrain. The results of the research are significant for the banking sector, regulators, and other stakeholders in Bahrain as they reflect the extent to which the accounting information systems in the Islamic banks are effective.

Keywords Accounting information systems · Islamic banking · Performance · Bahrain

1 Introduction

Businesses today are operating in a highly competitive and extremely unpredictable economic environment. thus, there is a need for an information system that provides quick reports as a response to the complex business atmosphere. There has been

A. Al-Sartawi (✉) · Z. Sanad
Ahlia University, Manama, Bahrain
e-mail: amasartawi@hotmail.com

Z. Sanad
e-mail: zsanad@ahlia.edu.bh

M. T. Momany
Department of Accounting and Finance, American University in the Emirates, Dubai, United Arab Emirates

M. Al-Okaily
School of Business, Jadara University, Irbid 733, Jordan

A. M. A. Musleh Al-Sartawi et al. (eds.), *From the Internet of Things to the Internet of Ideas: The Role of Artificial Intelligence*, Lecture Notes in Networks and Systems 557,
https://doi.org/10.1007/978-3-031-17746-0_56

a strong argument in the accounting literature regarding that Accounting Information System (AIS) typically leads to a strategic success of any organization as it would assist the managers in collecting different data and generate useful reports [1–4]. Accounting generally represents a narrower concept compared to the AIS. Accounting is related to transactions recording procedure in organizations, whereas the AIS provides a broader concept of accounting which includes the techniques and controls within the information technology field [5–7]. AIS has provided broader facilities including retrieving financial statements' related data, providing capabilities of trend analysis, analyzing firms' various elements to measure the managerial performance [8]. Also, the system assists managers to understand their job more clearly and to reduce their uncertainty level before making any decisions [9, 10].

According to the literature, the AIS consists of three main aspects which are human resources, physical resources, and financial data [11, 12]. Human resources are professionals who can understand all types of firms' financial operations and able to handle them professionally. Human resources element in AIS deals with those who clearly understand how to deal with different types of accounting information and make sure to provide them in a timely manner [13, 14]. The second element, physical resources, represents all computer tools including software and hardware devices, storage media, and databases. It is essential for any effective AIS to have competent professionals as well as efficient devices that would help in achieving firms' objectives [15]. The last AIS aspect is the financial data. The inputs of any AIS are the financial statements and need to be accurate [16]. There are many factors that impact the performance of the banking sector all over the world. These factors probably related to the administrative or accounting work within banks. One of the factors that are still questionable through the writings of the accounting literature is the accounting information systems. Is it reliable and can it contribute in having a better firms' performance? the value of accounting information to the user depends on how reliable the information is, and information is said to be reliable when it has the attribute of relevance, concise and timeliness [17–19]. AIS can be maintained if a sound internal control system is applied in any firm [20]. Testing the usefulness of accounting information system (AIS) for better firms' financial performance has been an important research concern for many researchers [21]. For instance, prior studies were motivated to test the impact of AIS usage on firms' profitability including ROA and ROE and concluded that there is a significant association [22, 23]. On the other hand, some researchers argued that firms' performance might not be linked with the usage of AIS. Some authors argued that the firms' profitability is not associated with AIS and there are other factors that would affect profit [25–27]. Another justification could be that AIS may not be fully supported by the main aspects including human resources as mentioned earlier [28, 29]. However, overall, it is expected that AIS to be linked with the firms' outcomes because the reports generated from this system are used at different strategic, operational, and tactical levels. Besides, it is also expected that AIS would contribute to better exploitation of the available resources [30]. Reviewing the existing body literature related to AIS shows that there has been some research that investigated the definition and practices of AIS, but there have been a few studies that attempted to investigate the impact of AIS on enhancing

Organizational Performance. This is the literature gap that the current study attempts to fill. Hence, the current study attempts to investigate whether AIS has an impact on firms' performance of the Islamic banks in general since little research approached the relationship between the two variables in this critical sector in Bahrain. The problem can be stated in the following question" Is there an impact of accounting information systems on the performance of Islamic banks in Bahrain?".

2 Literature Review and Hypothesis Development

Due to globalization, companies worldwide compete in a very complex economic environment with extremely unpredictable circumstances. Therefore, companies started to change rapidly using technology and one of the most important changes that occur recently is the introduction of AIS. Many researchers tried to define AIS, for instance, AIS can be considered as a range of people and equipment used to collect data and deliver the needed financial information to the decision-makers at specific period [31–33]. [34–36] defined AIS as a system that assists managers in planning and controlling as well as providing related and reliable information for decision making purposes. We can also consider AIS is a computer-based system that guarantees a high level of accounting information quality that improves the control and boosts the support in a firm to provide information to the interested users [37, 38]. According to [39, 40], AIS is "a formal system for identifying, measuring, accumulating, analyzing, preparing, interpreting and communicating accounting information about a particular entity to a particular group". [51] said that AIS is a combination of persons, tools, strategies, and procedures that function together to gather data and convert it into useful reports that can be used for decision making. [52] mentioned that AIS is a computer-based electronic system applied by firms to gather, keep, analyze, and deliver the financial and accounting information mentioned in the financial statements with the aim of supporting the decision-making process.

According to the literature, AIS consists of three main aspects which are human resources, physical resources, and financial data [9, 53]. Human resources are professionals who can understand all types of firms' financial operations and able to handle them professionally. Human resources element in AIS deals with those who clearly understand how to deal with different types of accounting information and make sure to provide them in a timely manner [47, 54, 55]. The second element, physical resources, represents all computer tools including software and hardware devices, storage media, and databases. It is essential for any effective AIS to have competent professionals as well as efficient devices that would help in achieving firms' objective [17] The last AIS aspect is the financial data. The inputs of any AIS are the financial statements and need to be accurate [22, 56, 57]. The main purpose of AIS is to evaluate the past, present and future business occasions [17, 51, 58, 59]. This system mainly depends on data to generate relevant and reliable accounting information which could be used for taking different decisions such as pricing, outsourcing, capital investments, inventory policy, production levels, and customer service [18,

33, 60, 61]. Thus, it is expected that AIS should be interrelated with firms' performance and outcomes [62] In the literature, a number of researchers were motivated to test the association between AIS adoption and firms' outcomes [9, 63].

For example, [64] study results showed that AIS has a direct impact on decision-making and organizational effectiveness. [65] tested the impact of AIS on creating effective organizational performance and evaluated the impact of AIS on elements of organizational performance using Dubai Financial Market. The results indicated that AIS has a significant impact on enhancing the organizational performance to listed companies, but the study found no relationship between AIS and enhancing performance management.

[66] determined the benefits of the AIS integration in the operations of automobile companies as well as other firms in the same sector. The results showed that AIS is an important mechanism for organizations' effective management, controlling activities, and decision-making. In addition, the empirical results of this paper indicated that AIS has a significant impact on the organizational effectiveness of automobile companies in Kenya.

[67] investigated the impact of AIS system on organizational performance, especially the elements of making decisions and profitability level. The results revealed that the existence of quality and timely information using AIS does have a significant positive impact on enhancing the decision-making process and increasing the profitability level. [68] study was conducted in Ethiopia evaluated the impact of AIS on the organizational performance including customer welfare, inventory management, financial statements and operating cost. The results of this study confirm that AIS has a positive and significant impact on financial statement, customer welfare, inventory management, organizational performance, but has a negative relationship with operating cost.

[69] tested the impact of accounting managers' knowledge and top management support on AIS. The findings revealed that accounting managers' knowledge and top management support have a significant impact on the AIS in organizations. Moreover, this paper concludes that accounting information systems does have a significant impact on the performance management and organizational performance of an organization. The study of [22] examined the impact of AIS on organizational performance elements including quality improvement, effective decision making, and cost reduction of SMEs in Saudi Arabia. The findings of this study confirmed that AIS has a significant impact on all organizational performance aspects. [70] evaluated the sharia supervisory board experience and its role in adopting new systems to prepare the necessary information for the meetings and how it will affect the cost. Therefore, semi-structured interviews were conducted. The study concluded that a quality accounting application system such as security, accessibility, reliability use, data integration, and availability of correct and timely information were able to support company performance. In addition, the research study conducted by [3] evaluated the impact of AIS on enhancing the performance management of the petroleum industry in Iraq. To achieve the study aim, the author decided to collect data from the financial statements of ten petroleum companies in Iraq that were the most prominent contributor in GDP form year 2001 to 2018. The results of this study

showed that AIS has a significant positive impact on enhancing the performance of the organizational management. Also, [47] tested the association between AIS and organizational performance among commercial and Islamic banks in Jordan. The results revealed that information quality of AIS is a crucial element that would improve banks' organizational performance.

According to the previous discussion, the researcher formed one main hypothesis and three sub-hypotheses in order to investigate whether the AIS has an effective role in influencing Islamic banks in Bahrain in general and the three AIS elements in particular. The results of the research are significant for Islamic banks in Bahrain as they reflect the extent to which the accounting information systems in the Islamic banks in Bahrain are effective. The study hypotheses are as follow:

H1: *There is a significant impact of accounting information systems on performance in the Islamic banks in Bahrain.*

H1-a: *There is a significant impact of human resources as an element of the accounting information systems on performance in the Islamic banks in Bahrain.*

H1-b: *There is a significant impact of physical resources as an element of the accounting information systems on performance in the Islamic banks in Bahrain.*

H1-c: *There is a significant impact of financial data as an element of the accounting information systems on performance in the Islamic banks in Bahrain.*

Methodology:

A questionnaire was designed to collect the data. The population of the present research includes all the accountants, financial managers, and board members of the Islamic banks in Bahrain. The number of the population is not known for the researcher. The sample is selected based on the random accidental or available sample. This is one of the most commonly used methods of non-probability sampling techniques, especially in business research. This sampling technique is based on considering the available individuals or the available elements as the elements of the sample. The questionnaire collects data about the two variables of the research which are:

(1) Independent Variable: accounting information systems in the Islamic banks in Bahrain.

Based on the literature review, accounting information systems are usually made up of three elements:

- Human resources
- Physical resources
- Financial data

(2) Dependent Variable: Performance of the Islamic banks in Bahrain.

3 Research Reliability

The accepted level of alpha Cronbach is 0.7 (70%). In the following Table the values of alpha are presented. Overall, there is a high level of internal consistency between all the 20 items of the questionnaire as alpha Cronbach is 87.4% Table 1.

Table 1 Cronbach alpha values

Variables	No. of items	Cronbach Alpha
Independent: accounting information systems in the Islamic banks in Bahrain · Human resources · Physical resources · Financial data	15	0.889
Dependent: Performance of the Islamic banks in Bahrain	5	0.876
Total number of questionnaire items	20	0.874

Table 2 Respondents' ages

Gender	Frequency	Percentage %
Male	97	65
Female	53	35
Total	51	100

Table 3 Respondents' ages

Age	Frequency	Percentage %
25 or below	20	13.33
25 to 35	45	30
36 to 44	56	37.33
45 to 59	20	13.33
60 or more	9	6
Total	150	100

Table 4 Respondents' position

Position	Frequency	Percentage %
Accountant	90	60
Financial manager	57	38
Board member	3	2
Total	150	100

Table 5 Respondents' years of experience

Years of experience	Frequency	Percentage %
10 years or below	42	28
11 to 15 years	67	44.67
16 to 19 years	28	18.67
20 years or more	13	8.67
Total	150	100

Table 6 Summary for AIS element

Element of AIS	Number of items	Average mean score
human resources	5	3.97
physical resources	5	4.03
financial data	5	3.92
Average mean for all the 15 items	15	3.97

Table 7 Regression model between AIS and performance

R	Beta	R Square	Adjusted R Square	T-value	Sig
0.729[a]	0.729	0.616	0.614	15.92	0.000

Respondents' Demographic Information:
The total number of the collected questionnaires were 150. The responses were received from google forms where the questionnaire forms were uploaded and sent to the respondents. As shown in Table 2, the percentage of the male respondents weighs 65% from the total sample. The percentage of female respondents weighs35% of the total sample. Also, based on the data in Table 3, the highest percentage 37.3% is that for respondents who are 36 to 44 years. Those who are 25 to 35 represent 30% followed by the age group 25 or below and 45 to 59. The lowest percentage 6%, is for those who are 60 or more represent. Based on the data in the Table 4, the highest percentage 60% is that for respondents who are accountants. Those who are financial managers represent 38%. Those who are board members represent 2%. Besides, almost half of the respondents have an experience around 11 to 15 years Table 5.

4 Descriptive Statistics

In general, as shown in Table 6, the analysis of the three sub-variables that are elements of accounting information systems show that AIS is actually applied in an effective manner in the banking sector in Bahrain. This is attributed to the high

Table 8 Multiple regression between three elements of AIS and performance of the Islamic banks

Variables	Coefficients	t-value	Sig
	Beta		
Constant		3.475	0.001
human resources	0.170	2.794	0.006
physical resources	0.572	9.405	0.000
financial data	0.598	11.254	0.0002
R Square	0.476		

average scores that the sub-variables got (3.97) out of 5.00. The component that is most effective is physical resources (4.03); then human resources (3.97); and finally, financial data (3.92).

5 Regression Analysis

The regression analysis is done to test if there is an impact of the independent variable which is the AIS in the Islamic banks in Bahrain on the dependent variable which is performance of the Islamic banks in Bahrain. According to Table 7, it is found out that the relationship between AIS and performance of the Islamic banks is significant (sig = 0.000). Also, the AIS positively related to the performance of the Islamic banks as the t-value is 15.96. Further, Accordingly, the alternative hypothesis saying "There is a significant impact for accounting information systems on performance in the Islamic banks in Bahrain" is accepted. In addition, multiple regression between the three elements (human, physical and financial data) of AIS and performance of the Islamic banks is done to understand in depth this relationship. As shown in Table 8, it can be seen that there is a positive and significant impact of the three variables on firms' performance. Accordingly, the first sub hypothesis: "There is a significant impact for human resources as an element of the accounting information systems on performance in the Islamic banks in Bahrain" is accepted as the sig value is 0.006. Furthermore, the second sub hypothesis: "There is a significant impact for physical resources as an element of the accounting information systems on performance in the Islamic banks in Bahrain" is accepted as the sig value is 0.00. Finally, the third sub hypothesis saying "There is a significant impact for financial data as an element of the accounting information systems on performance in the Islamic banks in Bahrain" is accepted as the sig value is 0.0002.

6 Conclusion

The researchers investigated whether accounting information system has an impact on Islamic banks' organizational performance in Bahrain. The study focuses on testing three main elements of the accounting information system including human and physical resources and financial data. A questionnaire was designed to collect the study data. The population of the present research includes all the accountants, financial managers, and board members of the Islamic banks in Bahrain. The study has provided an understanding that AIS has become a system that is very important in Islamic banks because it generates important information for the banks for decision making and enhances the quality of accounting information generated, and most importantly, it has a significant impact on organizational performance in general. In particular, the human resources, physical resources, and financial data represent effective elements of AIS in Islamic banks in Bahrain. This shows that the performance of the Islamic banks in Bahrain has been witnessing outstanding improvements recently. Overall, based on the previous studies (i.e., Ali and Salem, 2020) and the current study's findings, it can be concluded that the AIS implementation especially in the emerging countries is still an interesting topic to be studied and there is a research gap that seems to be interesting subject for the future discussion. Accordingly, the study recommends developing AIS to contribute to future expectations about the liquidity situation and providing an adequate internal control system for each bank. In addition, it is important to use the computerized accounting system to raise the efficiency of accounting work. Finally, it is important to conduct future research to explore the effect of AI applications on AIS as well as audit efficiency.

References

1. Teru S, Idoku I, Ndeyati J (2017) A review of the impact of accounting information system for effective internal control on firm performance. Indian J Financ Bank 1(2):52–59
2. Borhan O, Bader O (2018) Investigating the impact of accounting information system on the profitability of Jordanian Banks. Res J Financ Acc 9(18):110–118
3. Al-Delawi AS Ramo WM (2020) The impact of accounting information system on performance management. Polish J Manag Stud 21
4. Horngren C, Harrison W, Bamber L, Willis B, Jones B (2005) Accounting, Pearson education. Financial Times/ Prentice Hall, Harlow
5. Musleh Al-Sartawi AMA (ed) (2022) Artificial intelligence for sustainable finance and sustainable technology, vol 423. Lecture Notes in Networks and Systems. Springer, Cham. https://doi.org/10.1007/978-3-030-93464-4
6. Ismail N (2009) Factors influencing the alignment of accounting information systems in small and medium-sized Malaysian manufacturing firms. J Info Syst Small Bus 1(1–2):1–20
7. Onaolapo A, Odetayo T (2012) Effect of accounting information system on organizational effectiveness: a case study of selected construction companies in Ibadan, Nigeria. Am J Bus Manag 1(4):183–189
8. Patel S, Bhavna P (2015) Effects of accounting information system on organizational profitability. Int J Res Anal Rev.

9. Al-Sartawi A (2020) Shariah disclosure and the performance of Islamic financial institutions. Asian J Bus Acc 13(1):133–160
10. Rehab U (2018) The impact of accounting information systems on organizational performance: the context of Saudi's SMEs. Int Rev Manag Mark 8(2):69–73
11. Rosa D Purfini AP (2019) Analysis effect quality of accounting information systems to support company performance. IOP Conf Ser: Mater Sci Eng
12. Al-Okaily M, Al-Sartawi A, Hannoon A, Khalid AA (2022) Information technology governance and online banking in Bahrain. In: Musleh Al-Sartawi AMA (ed) Artificial Intelligence for Sustainable Finance and Sustainable Technology, vol 423. Lecture Notes in Networks and Systems. Springer, Cham, pp 611–618. https://doi.org/10.1007/978-3-030-93464-4_60
13. Kashif B (2018) Impact of accounting information system on the financial performance of selected FMCG companies. Asian J Appl Sci Technol 2(3):8–17
14. Sanad Z, Al-Sartawi A (2021) Financial statements fraud and data mining: a review. In: Musleh Al-Sartawi AMA, Razzaque A, Kamal MM (eds) Artificial Intelligence Systems and the Internet of Things in the Digital Era, vol 239. Lecture Notes in Networks and Systems. Springer, Cham, pp 407–414. https://doi.org/10.1007/978-3-030-77246-8_38
15. Dandago K, Rufai A (2014) Information technology and accounting information system in the Nigerian banking industry. Asian Econ Financ Rev 4(5):655–670
16. Musleh Al-Sartawi AMA (ed) (2021) The big data-driven digital economy: artificial and computational intelligence, vol 974. Studies in Computational Intelligence. Springer, Cham. https://doi.org/10.1007/978-3-030-73057-4
17. Manchilot T (2019) Science Journal of Business and Management. Rev Determin Acc Inf Syst Adoption 7(1):17–22
18. Musleh Al-Sartawi AM, Hussainey K, Razzaque A (2022) The role of artificial intelligence in sustainable finance. J Sustain Financ Investment, 1–6
19. Ali BJ, AlSondos IAA (2020) Operational efficiency and the adoption of accounting information system (AIS): a comprehensive review of the banking sectors. Int J Manag 11(6)
20. Gupta M, Sikarwar TS (2020) Modelling credit risk management and bank's profitability. Int J Electron Bank 2(2):170–183
21. Fekade S, (2017) The impact of accounting information system on organizational performance : a case study on pharmaceutical fund and supply agency. AAU Institutional Repository
22. Budiarto DS, Prabowo MA, Djajanto L, Widodo K P, Herawan T (2018) Accounting information system (AIS) alignment and non-financial performance in small firm: a contingency perspective. In: International Conference on Computational Science and Its Applications, pp. 382–394
23. Gupta N (2019) Influence of demographic variables on synchronisation between customer satisfaction and retail banking channels for customers of public sector banks of India. Int J Electron Bank 1(3):206–219
24. Al-Sartawi A, Sanad Z (2019) Institutional ownership and corporate governance: evidence from Bahrain. Afro-Asian J Financ Acc 9(1):101–115
25. Abdulrasool FE, Turnbull SI (2020) Exploring security, risk, and compliance driven IT governance model for universities: applied research based on the COBIT framework. Int J Electron Bank 2(3):237–265
26. Hannoon A, Al-Sartawi AMAM, Khalid AA (2021) Relationship between financial technology and financial performance. In: Musleh Al-Sartawi AMA (ed) The Big Data-Driven Digital Economy: Artificial and Computational Intelligence, vol 974. Studies in Computational Intelligence. Springer, Cham, pp 337–344. https://doi.org/10.1007/978-3-030-73057-4_26
27. Al-Adaileh R (2008) Essentials of management information systems. Yazeed-Publications, Karak-Jordan
28. Chauhan N, Tekta P (2020) Fraud detection and verification system for online transactions: a brief overview. Int J Electron Bank 2(4):267–274
29. Bodnar G, Hopwood W (2010) Accounting information system, 10th edn. Pearson Education Inc., New York

30. Alansari Y, Al-Sartawi A (2021) IT governance and E-banking in GCC listed banks. Procedia Comput Sci 183:844–848
31. Shihadeh F (2020) Online payment services and individuals' behaviour: new evidence from the MENAP. Int J Electron Bank 2(4):275–282
32. Al-Qudah AA, Al-Okaily M, Alqudah G et al (2022) Mobile payment adoption in the time of the COVID-19 pandemic. Electron Commer Res. https://doi.org/10.1007/s10660-022-09577-1
33. Grande UE, Estebanez PR, Colomina MC (2010) The impact of accounting information systems (AIS) on performance measures: empirical evidence in Spanish SMEs. Int J Digit Account Res 11(2011):25–43
34. Chayjan MR, Bagheri T, Kianian A, Someh NG (2020) Using data mining for prediction of retail banking customer's churn behaviour. IntJ Electron Bank 2(4):303–320
35. Hla D, Teru SP (2015) Efficiency of accounting information system and performance measures. Int J Multi Curr Res 3:976–984
36. Ogah IJ (2012). An evaluation of the relevance of accounting system as a management decision tool in union bank of Nigeria PLC, Uyo branch of Akwa Ibom Greener. J Bus Manag Stud 3(1), 38–45
37. Al-Sartawi A, Reyad S, Madbouly A (2021) Shariah, presentation and content dimensions of Web 2.0 applications and the firm value of Islamic financial institutions in the GCC countries. J Islamic Market. Accepted article.
38. Memdani L (2020) Demonetisation: a move towards cashless economy in India. Int J Electron Bank 2(3):205–211
39. Samer M (2016) The impact of the effectiveness of accounting information systems on operational performance in public listed industrial companies in Jordan. J Soc Sci 5(3):263–276
40. Al-Qudah AA, Houcine A (2022) Stock markets' reaction to COVID-19: evidence from the six WHO regions. J Econ Stud 49(2):274–289. https://doi.org/10.1108/JES-09-2020-0477
41. Gogia J, Chakraborty D (2022) Open banking: a revolution in the Tech-Fin industry. Int J Electron Bank 3(2):100–120
42. Al-Sartawi A, Al-Okaily M, Hannoon A, Khalid AA (2022) Financial technology: literature review paper. In: Musleh Al-Sartawi AMA (ed) Artificial Intelligence for Sustainable Finance and Sustainable Technology, vol 423. Lecture Notes in Networks and Systems. Springer, Cham, pp 194–200. https://doi.org/10.1007/978-3-030-93464-4_20
43. Prakash A, Mahajan Y, Gadekar A (2022) Adoption of mobile money among internal migrant workers during the corona pandemic in India: a study focused on moderation by mode of payments. Int J Electron Bank 3(2):144–162
44. Alnajjar M (2017) Impact of accounting information system on organizational performance: a study of SMEs in the UAE. Retrieved from research gates. https://www.researchgate.net/publication/324062452
45. Sanad Z, Al-Sartawi A (2016) Investigating the relationship between corporate governance and internet financial reporting (IFR): evidence from Bahrain bourse. Jordan J Bus Adm 12(1):239–269
46. Al-Okaily M, Alqudah H, Al-Qudah AA, Al-Qadi NS, Elrehail H, Al-Okaily A (2022) Does financial awareness increase the acceptance rate for financial inclusion? An empirical examination in the era of digital transformation, Kybernetes. https://doi.org/10.1108/K-08-2021-0710
47. Ali BJ, Salem M (2020) Information quality and data quality in accounting information system: implications on the organization performance. Int J Psychosoc Rehabil 24(5):3258–3269
48. Al-Sartawi A (2020) Does it pay to be socially responsible? Empirical evidence from the GCC countries. Int J Law Manag 62(5):381–394
49. Al-Qudah AA (2022) Artificial intelligence in practice: implications for information systems research, case study UAE Companies. In: Musleh Al-Sartawi AMA (ed) Artificial Intelligence for Sustainable Finance and Sustainable Technology, vol 423. Lecture Notes in Networks and Systems. Springer, Cham, pp 225–234. https://doi.org/10.1007/978-3-030-93464-4_23

50. Bhambra RK (2022) Adoption of online banking in Goa amidst the pandemic. Int J Electron Bank 3(2):163–175
51. Karolak M, Razzaque A, Al-Sartawi A (2021) E-services and M-services using IoT: an assessment of the Kingdom of Bahrain. In: Musleh Al-Sartawi AMA, Razzaque A, Kamal MM (eds) Artificial Intelligence Systems and the Internet of Things in the Digital Era, vol 239. Lecture Notes in Networks and Systems. Springer, Cham, pp 523–533. https://doi.org/10.1007/978-3-030-77246-8_48
52. Al-Okaily M, Alqudah HM, Al-Qudah AA, Alkhwaldi AF (2022) Examining the critical factors of computer-assisted audit tools and techniques adoption in the post-COVID-19 period: internal auditors perspective. VINE J Inf Knowl Manag Syst. https://doi.org/10.1108/VJIKMS-12-2021-0311
53. Audenaert M, Decramer A, George B, Verschuere B, Van Waeyenberg T (2019) When employee performance management affects individual innovation in public organizations: the role of consistency and LMX. Int J Human Resour Manag 30(5):815–834
54. Al-Qudah AA, Al-Okaily M, Alqudah A, Ghazlat A (2022) Mobile payment adoption in the time of the COVID-19 pandemic. Electron Commer Res. https://doi.org/10.1007/s10660-022-09577-1
55. Almaqableh L, Reddy K, Pereira V, Ramiah V, Wallace D, Veron JF (2022) An investigative study of links between terrorist attacks and cryptocurrency markets. J Bus Res 147:177–188
56. Hall M (2010) Accounting information and managerial work. Acc Organ Soc 35(3):301–315
57. Al-Sartawi A (2020) Information technology governance and cybersecurity at the board level. Int J Crit Infrastruct 16(2):150–161
58. Harash E, Al-Timimi S, Alsaadi J (2014) Effects of financing on performance of small and medium enterprises (SMEs). Int J Manag 2(10):1–9
59. Almaqableh L et al (2022) Is it possible to establish the link between drug busts and the cryptocurrency market? Yes, we can. Int J Inf Manag 102488
60. Sanad Z, Shiwakoti R, Al-Sartawi A (2022) Gender diversity on corporate boards and earnings management: a review. In: Musleh Al-Sartawi AMA (ed) Artificial Intelligence for Sustainable Finance and Sustainable Technology, vol 423. Lecture Notes in Networks and Systems. Springer, Cham, pp 340–351. https://doi.org/10.1007/978-3-030-93464-4_34
61. Agarwal N (2021) Redefining banking: exchange traded savings and loans using cryptocurrencies. Int J Electron Bank 3(1):1–11
62. Samuel N (2013) Impact of accounting information systems on organizational effectiveness of automobile companies in Kenya. University of Nairobi Research Archieve
63. Al-Sartawi A (2020) Social media disclosure of intellectual capital and firm value. Int J Learn Intellect Cap 17(4):312–323
64. Soudani S (2012) The usefulness of an accounting information system for effective organizational performance. Int J Econ Financ 4(5):136–140
65. Musleh Al-Sartawi AMA, Razzaque A, Kamal MM (eds) (2021) Artificial intelligence systems and the internet of things in the digital era, vol 239. Lecture Notes in Networks and Systems. Springer, Cham. https://doi.org/10.1007/978-3-030-77246-8
66. Susanto A, Meiryani M (2019) The impact of environmental accounting information system alignment on firm performance and environmental performance: a case of small and medium enterprises s of Indonesia. Int J Ener Econ Policy 9(2):229
67. Hannoon A, Abdalla YA, Musleh Al-Sartawi AMA, Khalid AA (2021) Board of directors composition and social media financial disclosure: the case of the United Arab Emirates. In: Musleh Al-Sartawi AMA (ed) The Big Data-Driven Digital Economy: Artificial and Computational Intelligence, vol 974. Studies in Computational Intelligence. Springer, Cham, pp 229–241. https://doi.org/10.1007/978-3-030-73057-4_18
68. Trabulsi RU (2018) The impact of accounting information systems on organizational performance: the context of Saudi's SMEs. Int Rev Manag Mark 8(2):69–73

69. Al-Sartawi A (2015) The Effect of Corporate Governance on the Performance of the listed companies in the gulf cooperation council countries. Jordan J Bus Adm 11(3):705–725
70. Al-Sartawi A (2019) Performance of Islamic banks do the frequency of Shari'ah supervisory board meetings and independence matter? ISRA Int J Islamic Financ 11(2):303–321

Transformation of Managerial Accounting Trends in the Era of Digitalization

Maryam Ali Mansoor, Ebtisam Mohammed Salmanand, and Abdalmuttaleb Al-Sartawi

Abstract Digitalization can be termed as the use of different digital technologies within the business model which allow a business to have different types of value producing opportunities within the firm. Further, with the help of digitalization, the domain of management accounting has seen significant amount of change with which it has been able to improve to significant extent. Further, it is evident that the business models within an organization can be also affected by the digitalization and with the help of digitalization, management accounting within a business environment can be improved significantly. The finance function as well as the management accounting function can be greatly improved with the help of robotization and automation. Management accounting process can be denoted as the branch of accounting that delivers information as well as assists in understanding business transactions. With the help of management accounting function, the internal as well as external users of the company can understand the current business situation as well as improve decision making within the organization. It is essential for managers to execute different financial activities related to organization. Without help of management accounting, it is not possible for a business to improve significantly. Despite of different research on the management accounting and digitalization, there is a lack of analysis on the transformation of managerial accounting trends in the era of digitalization. Hence, the paper evaluates the topic and understand how digitalization has improved the overall management accounting process within a business environment.

M.A. Mansoor and E.M. Salmanand—Freelancer.

M. A. Mansoor
Manama, Bahrain

E. M. Salmanand
Manama, Bahrain

A. Al-Sartawi (✉)
Ahlia University, Manama, Bahrain
e-mail: amasartawi@hotmail.com

A. M. A. Musleh Al-Sartawi et al. (eds.), *From the Internet of Things to the Internet of Ideas: The Role of Artificial Intelligence*, Lecture Notes in Networks and Systems 557,
https://doi.org/10.1007/978-3-031-17746-0_57

1 Introduction

In the modernized business environment, researchers have found that digitalization can allow in improving different business functions and can improve the competitive position of the business. It is evident that the management accounting with digitalized software can allow more efficiency in recording the conventional financial entries as well as can support the decision-making activities within the competitive business environment [3]. While it is still not clear how management accounting process can reduce its limitations in response to the adherence to digitalization, it is evident that digitalization can significantly improve the efficiency of transactions as well as can play a supplementary role in the external reporting framework of the organization [14]. Within the external reporting framework of an organization, the management accounting can play a pivotal role and can allow in understanding of inward look function within the business environment.

Digitalization can further allow a firm to overcome the limitations related to the traditional management accounting functions. In recent times, the increase in the business competition and globalization has created opportunities as well as threats to business environment. Without analyzing the business transaction's digitally, a business can overcome the difficulties created by the traditional management accounting methods. The unit and functional level play an important role in the success and failure of a business [12]. Digitalization and the use of automation software can allow a business to analyze different variables as well as can also allow it to analyze the profound effects of business functions with the business. Strategic management accounting has responded to the business's strategic move in at least three ways: first, it demonstrates a clear link between organizational strategies and managerial functions; second, it employs innovative managerial accounting method to enable strategic cost estimates that inevitably aids in competitiveness; and third, it takes into account a competitive business environment characterized by increased use of data [4]. Although top business schools, experts, and academic works have championed this strategic trend of management accounting, it will fail if it overlooks networks and the complicated corporate environment.

2 Literature Review

In recent times, accounting industry officials were predicting the end of an age for traditional accounting, fearing that evolving technology, particularly automation, will bring about the end of bookkeeping as we recognize it. Indeed in 2019, accountants asked by Robert Half about the impact of technology on their job expressed concerns about being replaced, having less opportunities for creative problem-solving, including being unduly reliant on technology to execute commercial activities [10, 28]. There's no denying that the digital revolution has drastically changed the game. In order to become more competitive, big data has matured into a vital asset that must be leveraged. This transition, on the other hand, digitalization is not a threat, for companies who will leverage from it [1].

As per the recent research, it is evident that most of the companies are already introduced to the automated forecasting techniques as well as data analytics which will allow it to combine time series techniques as well as assist in the machine learning. With the help of management accounting, a business can ensure proper identification and control of the existing business challenges related to the management accounting. Without having access to proper data and having secure database, it is not possible for a business to prosper and improve the existing data storage [21, 24]. Without data storage, it will not be possible for the management accountant to evaluate different data sources and make proper decision based on the data research. Hence it is evident how digitalization in management accounting practices can allow in accounting process automation and better decision-making process within the organization.

Further, it is still remains unclear that how can be digitalization be applied to the business model and improve the system implementation and control. In this field, there are several factors that comes into consideration. In order to prosper, a business will require relevant as well as reliable data which will allow them in better financial decision-making process. It is core responsibility of the management to ensure that the increasing challenges of management accounting are reduced [9]. Companies which do not adhere to the process of digitalization are found to be affected by poor data quality and the same has been found to impact the internal control function of an organization. Self-service reporting with the help of digitalization can lead to better self-service-based reporting as well as better automation process in the business process. However, critiques have noted that digitalization can creation new information routines which many create more decentralized situation within the existing business framework and can create confusions across the users of automation. With that being said, there will be various challenges of digitalization in regard to the implementation of management accounting functions however the business must ensure that the finance function with the help of digitalization are sound and is under governance. Employees needs to be trained and further analytics may be needed which may further need analytical thinking [2, 25].

With the help of increased globalization across the world and globalization, it has been evident that the financial function of the business may get changed. Most of the finance function in large firms will not require further function-based approach however it is essential to emote that the small and medium sized organization may require its employees to learn about the changes associated with the digitalization in order to prosper further [5, 27]. Further, business will need to develop expertise in the field of big data and analytics which will allow it to aid in the transformation. The higher management of businesses will have to understand the ways in which it can bring changes in the corporate functions and business units in the light of digitalization. CEOs of a firm always tend to focus on the value creation and always tend to prioritize in the finance function of a business [17, 23]. The enterprise function of a business is found to have profound impact on the production as well as analysis of big data in the part of its control system [15]. Therefore, it will be required for the managers to understand how the increase in accounting research will create value for the firm. Without the implementation of big data automation, artificial intelligence and critical thinking, businesses will have hard time in generating depth and sound knowledge based on the field of management accounting. In recent times, the era of digitalization that have been connected with the context of big data. There are various factors that add value to the use of management accounting in business which includes the internet of things (IOT), digitalized supply chains as well as cloud computing. With the help of digitalization, different financial data within the business will improve significantly thereby adding value to the existing process and finance functions [6, 19, 21]. Digitalization within firms have allowed in creation of different quantities of information which are anticipated to grow at a rapid pace. Due to the evolving nature of the big data, it is possible within the business to understand and investigate different possibilities in relation to the current accounting atmosphere. Organizations, with the help of big data can understand the structured as well as unstructured financial information which is essential to comprehend for improvement in business function. Further it has been also evident that information systems can display important message and information which is required for firms to execute its strategic operations.

Organizational arrangements, corporate strategy creation as well as structuring of the information system can make it massive difference in the big data analytics and insights. Digitalization can improve better transformation of the information outputs within the business environment which can allow in further improvement of the accounting prospects within the work environment of an atmosphere. Despite of the fact that security issue remains one of the key implications affecting the framework of digitalization and it should be the key responsibility for an organization to understand the key issues related to the implementation and therefore most improve the existing business framework which will allow them better decision making within the management accounting functions [20]. As the organizational risks and interconnections have increased, there has been increase in the implementation of new applications within the accounting framework. Parameters related to accounting framework needs to be evaluated and maintained in order to ensure transparency within the management accounting process [6, 8, 26]. Digitalized data can ensure that the process is streamlined with the business operations and further the data can be traced however it cannot always ensure the efficiency within the framework in which management accounting process is adhered. Research can gain significant number of opportunities with the help of the digitalization process. It is evident as per the recent research that digitalization can greatly improve the speed and efficiency of the existing business transactions as well as has allowed firms to process maximum number of transactions at a time [11, 22]. The derivation of the analysis and prediction from the technological data and can be also interpret with the use of the digitalization trends. In line with the digital transformation, it has been evident that the management accounting system can allow a system to process more efficient decision-making system. For managing multi-platform-based networks and transactions, the use of AI and automation is required. Without digitalization, it is not possible for a business to process such transactions [18, 29].

However, the implementation of the digitalization process can be only possible if the digitalized environment is transformed, and the rapid pace can be only achieved with deeper understanding of the digitalization and associated factors. If accountant fails to acknowledge and understand the traditional technical skills, the implementation of digitalization within the organizational environment might get delayed [7, 30]. The process of the digitalization can also ensure improvement in customer service as well as will ensure increase in revenue for the business [19, 21]. With deeper understanding of the business functions and processes, firms can be able to make sure that the business has significant number of advantages within the new digitalization framework. If can get problematic for business to store different systems and access to them gets difficult with the increase of business size and area. Digitalization within the management accounting framework can remove the need of manual accounting processes and thereby allow the business to streamline the business transactions. Digitalization will also ensure proper storage of accounting information and therefore provide the business owners access to the real time information. In current business framework, it is very easy to gain insights from the financial data as well as understand how the business is performing within the industry. Once implemented, the digitalization process is very easy to use which will allow business to provide users of accounting information with access to the data related to the financial function of the business [13, 31].

3 Conclusion

On a concluding note, it is evident that the paper has been able to analyze the transformation of managerial accounting within the era of digitalization. The paper has been successfully able to identify how management accounting can be greatly improved in respect to the speed of information and accuracy of the data. Further, it has been also identified from the research that digitalization can allow business to use their current financial information more efficiently and provide the third-party users to improve in the current environment [16]. The multiplatform based financial transactions and networks can be managed with the help of the digitalization process. After globalization, there has been an evolution of the use of financial information and data within the business framework and without the understanding of the business function, it will not be possible for the business to improve the current accounting atmosphere within the organization. With use of automation and big data can allow a firm to comprehend its structured as well as non-structured financial information and understand how it can improve its financial functions significantly. Management accounting with the help of digitalization can ensure that the errors related to financial entries are reduced and the existing process is improved with great efficiency. The inward-looking function of the business can be also understood with the implementation of digitalization therefore the implementation of different variables related to digitalization are needed to be understood and interpreted.

References

1. Bhimani A (2020) Digital data and management accounting: why we need to rethink research methods. J Manag Control 31(1):9–23
2. Agarwal N (2021) Redefining banking: exchange traded savings and loans using cryptocurrencies. Int J Electron Bank 3(1):1–11
3. Oncioiu I, Bîlcan FR, Stoica DA, Stanciu A (2019) Digital transformation of managerial accounting-trends in the new economic environment. EIRP Proc 14(1)
4. Sanad Z, Shiwakoti R, Al-Sartawi A (2022) Gender diversity on corporate boards and earnings management: a review. In: Musleh Al-Sartawi AMA (eds) Artificial intelligence for sustainable finance and sustainable technology, ICGER 2021. Lecture notes in networks and systems, vol 423. Springer, Cham
5. Rikhardsson P, Yigitbasioglu O (2018) Business intelligence & analytics in management accounting research: status and future focus. Int J Account Inf Syst 29:37–58
6. Shihadeh F (2020) Online payment services and individuals' behaviour: new evidence from the MENAP. Int J Electron Bank 2(4):275–282
7. Musleh Al-Sartawi AM, Hussainey K, Razzaque A (2022) The role of artificial intelligence in sustainable finance. J Sustain Finance Invest 1–6
8. Bogasiu I, Ardeleanu N (2021) Advantages and disadvantages of digitalisation in accounting. EIRP Proc 16(1)
9. Al-Sartawi A (2016) Measuring the level of online financial disclosure in the Gulf Cooperation Council Countries. Corp Ownersh Control 14(1):547–558
10. Cagle MN, Yılmaz K, Doğru H (2020) Digitalization of business functions under Industry 4.0. In: Digital business strategies in blockchain ecosystems, pp 105–132. Springer, Cham
11. Al-Sartawi A (2018) Ownership structure and intellectual capital: evidence from the GCC countries. Int J Learn Intellect Cap 15(3):277–291
12. Wadan R, Teuteberg F, Bensberg F, Buscher G (2019) Understanding the changing role of the management accountant in the age of industry 4.0 in Germany. In: Proceedings of the 52nd Hawaii international conference on system sciences
13. Al-Sartawi A (2018) Corporate governance and intellectual capital: evidence from gulf cooperation council countries. Acad Account Financ Stud J 22(1):1–12
14. Evstafyeva E, Kislaya I, Kruchanova I (2019) Advanced accounting perspectives in the light of leading economic digitalisation processes. Adv Soc Sci Educ Humanit Res 359
15. Gulin D, Hladika M, Valenta I (2019) Digitalization and the challenges for the accounting profession. ENTRENOVA-ENTerprise REsearch InNOVAtion 5(1):428–437
16. Al-Sartawi A (2018) Institutional ownership, social responsibility, corporate governance and online financial disclosure. Int J Crit Account 10(3/4):241–255
17. Al-Okaily A, Al-Okaily M, Teoh AP, Al-Debei M (2022) An empirical study on data warehouse systems effectiveness: the case of Jordanian banks in the business intelligence era. EuroMed J Bus Vol. ahead-of-print No. ahead-of-print. https://doi.org/10.1108/EMJB-01-2022-0011
18. Bredmar K (2017) Digitalisation of enterprises brings new opportunities to traditional management control. Bus Syst Res Int J Soc Adv Innov Res Econ 8(2):115–125
19. Al-Sartawi A (2020) Does it pay to be socially responsible? Empirical evidence from the GCC countries. Int J Law Manage 62(5):381–394
20. Al-Sartawi A (2020) Social media disclosure of intellectual capital and firm value. Int J Learn Intellect Cap 17(4):312–323
21. Al-Sartawi A (2019) Assessing the relationship between information transparency through social media disclosure and firm value. Manage Account Rev 18(2):1–20
22. Karolak M, Razzaque A, Al-Sartawi A (2021) E-services and M-services using IoT: an assessment of the Kingdom of Bahrain. In: Musleh Al-Sartawi AM, Razzaque A, Kamal MM (eds) Artificial intelligence systems and the internet of things in the digital era, EAMMIS 2021. Lecture notes in networks and systems, vol 239. Springer, Cham

23. Almajali D, Al-Okaily M, Barakat S, Al-Zegaier H, Dahalin ZM (2022) Students' perceptions of the sustainability of distance learning systems in the post-COVID-19: a qualitative perspective. Sustainability 14(12):7353. https://doi.org/10.3390/su14127353
24. Al-Sartawi A (2020) Information technology governance and cybersecurity at the board level. Int J Crit Infrastruct 16(2):150–161
25. Sanad Z, Al-Sartawi A (2021) Financial statements fraud and data mining: a review. In: Musleh Al-Sartawi AM, Razzaque A, Kamal MM (eds) Artificial intelligence systems and the internet of things in the digital era. EAMMIS 2021. Lecture notes in networks and systems, vol 239. Springer, Cham
26. Al-Sartawi A (ed) (2021) The big data-driven digital economy: artificial and computational intelligence. In: Kacprzyk J (ed) Studies in computational intelligence, vol 974. Springer
27. Baidybekova SK, Shomshekova BK, Kydyrbayeva EO, Sharipov AK, Kasseinova MI (2021) Accounting and financial reporting in the context of globalisation and digitalisation: problems and improvements. Public Policy Admin 20(2):248–259
28. Musleh Al-Sartawi AM, Razzaque A, Kamal MM (eds) (2021) Artificial intelligence systems and the internet of things in the digital era. In: EAMMIS 2021. Lecture notes in networks and systems, vol 239. Springer, Cham
29. Gogia J, Chakraborty D (2022) Open banking: a revolution in the Tech-Fin industry. Int J Electron Bank 3(2):100–120
30. Prakash A, Mahajan Y, Gadekar A (2022) Adoption of mobile money among internal migrant workers during the corona pandemic in India: a study focused on moderation by mode of payments. Int J Electron Bank 3(2):144–162
31. Musleh Al-Sartawi AMA (eds) (2022) Artificial intelligence for sustainable finance and sustainable technology. In: ICGER 2021. Lecture notes in networks and systems, vol 423. Springer, Cham

An Exploration into People's Perception About the Future of Cryptocurrencies and its Impact on the Financial World: Evidence from an Emerging Market

Husain Ghuloom and Zakeya Sanad

Abstract Cryptocurrencies are now a popular subject of discussion. Companies and investors worldwide have already started to accept the usage of cryptocurrencies; however, others are still rejecting it. Until now, it is not easy to ascertain the opinion of the mass population specially in the emerging markets. Accordingly, the purpose of this study is to learn how cryptocurrencies are perceived by the mass population to better understand the possible success and usage in the emerging market. In particular, this study digs more to discover on a personal level how people perceive the new technology, cryptocurrencies, which had yet to be understood as this market is still at an early stage specially in the emerging markets region. Another important takeaway from this research is that it examined and expounded on people's future trends concerning the adoption of Cryptocurrencies. For the analysis, the research adopts a quantitative research methodology using a questionnaire. A 5-points Likert-scale questionnaire has been designed including questions for each variable. The data was collected from 300 respondents. The hypotheses have been tested and it is found that cryptocurrency is perceived to have a positive and significant impact on the financial world in the emerging market.

Keywords Cryptocurrency · Bitcoin · Financial world · Emerging market

1 Introduction

Globalization and digitization have been more intertwined in recent years. We can communicate with anybody, anywhere, at any time, thanks to our cell phones. Over the Internet, we share information and services in real-time. Trade is taking place all around the globe. In a world that is increasingly linked, a global reserve currency that is tied to a single country feels awkward [34]. With the rise of Bitcoin, this study

H. Ghuloom
Manama, Bahrain

Z. Sanad (✉)
Ahlia University, Manama, Bahrain
e-mail: zsanad@ahlia.edu.bh

A. M. A. Musleh Al-Sartawi et al. (eds.), *From the Internet of Things to the Internet of Ideas: The Role of Artificial Intelligence*, Lecture Notes in Networks and Systems 557,
https://doi.org/10.1007/978-3-031-17746-0_58

explores the benefits and potential hazards of using it as a worldwide currency in the future.

Several countries have asked for changes to the international monetary system in the wake of the 2008 financial crisis. For example, People's Bank of China Governor Zhou Xiaochuan thinks that "the ideal goal of reorganizing the international financial system is to establish a world's reserve currency that is not tied to any particular country's currency". Money has undergone a dramatic shift in the course of history. When money was invented, one of the earliest innovations was the use of precious metals in coinage [13]. The uniformity and marking of coinage allowed consumers to instantly understand the worth of the precious metals, avoiding the friction of weighing the precious metals for each transaction. In a short period of time, numerous countries all around the world began using this new idea. Carrying coined precious metals became more expensive when commerce moved away from town squares and onto ships and ports throughout the world. Because of this, money changed from coins to paper money, which was easier to carry about. When coins could not be carried for great distances, trust was necessary to use paper money.

As a result, trust became a crucial component of the currency. In the end, it was the state that stood as the security for that trust. New payment options like iDEAL in the Netherlands or specialist providers like PayPal or WeChat, which allow us to make near-instant payments, are available to us over the Internet. In today's globalized environment, these services are a reaction to what individuals want and what the economy requires. Inventing Bitcoin was Satoshi Nakamoto's solution to 2008's global financial crisis, which required faith in central banks and banks [3]. It is Nakamoto's belief that central banks and banks cannot be trusted since they have before and would continue to do so in the future [4]. He predicted that Bitcoin would evolve into an electronic cash system that allowed users to conduct transactions without the need for an intermediary, hence removing the necessity for an intermediary. More than a thousand other cryptocurrencies are already being exchanged, despite the fact that Bitcoin was first introduced in 2009 [18].

It is important to note that these cryptocurrencies are not perfect replicas of the Bitcoin system. Different cryptocurrencies have different consensus mechanisms, fees, and speeds, as well as different goals in terms of reducing friction in the economy. One thing that all cryptocurrencies have, however, is the absence of a third-party mediator. Intermediaries play an important role in the cross-border payments sector, which is part of the global financial system [19]. Transnational market transactions were worth $156 trillion in 2014, and this market will continue to expand as the world becomes increasingly interconnected. Cross-border payments are dominated by banks, which account for up to 95% of the market's income. Banks have done nothing to update the outdated systems in the back-office of cross-border payments, which remain an expensive and frustrating experience for clients, particularly small businesses, and individuals [20]. Transactions using cryptocurrencies are less costly. Even on the most crowded blockchains, median transaction prices remain lower than wire transfer rates. Besides, bitcoin transactions may be completed in a matter of minutes. Once the network has validated the block containing the transaction, it is complete, and the money are accessible for usage [28].

While Bitcoin offers several benefits, it also has significant drawbacks, including prices of cryptocurrencies may shoot through the roof, but they may also plummet to dizzying lows in the blink of an eye. In other words, if an investor wants a steady income, this isn't the ideal investment. Bitcoin and other cryptocurrencies are built on speculation, which makes them more susceptible to price changes. One of the primary drawbacks of cryptocurrencies is that it may have a disastrous effect on the value of individual coins. Also, the Reserve Banks have control over the whole financial system, this does not include digital money; centralized authority does not extend to cryptocurrencies [23]. Companies and Investors worldwide have already started to accept the usage of cryptocurrencies; however, others are still rejecting it. Until now, it is not easy to ascertain the opinion of the mass population specially in the emerging markets. Accordingly, the purpose of this study is to learn how cryptocurrencies are perceived by the mass population to better understand the possible success and usage in the emerging market.

Cryptocurrencies are a hot topic right now. So far, limited studies have explored the opinion of people regarding the future of crypto money and how it would affect the present financial system. In addition, other researchers have examined whether cryptocurrencies were the future of money [31]. However, this study digs more to discover on a personal level how people perceive the new technology, cryptocurrencies, which had yet to be understood as this market is still at an early stage specially in the emerging markets region. Another important takeaway from this research is that it examined and expounded on people's future trends concerning the adoption of Cryptocurrencies, as well as how

Cryptocurrencies have changed the existing financial environment.

Accordingly, the main aim of this study is to explore the public opinion about:

1- The future of Cryptocurrencies in the emerging market.
2- The impact of Cryptocurrencies on the current financial world.

2 Literature Review and Hypotheses Development

Global financial institutions have lost the trust of consumers during the 2008 crisis, according to the Organization for Economic Co-operation and Development [4]. A few days after the economy crashed on October 31 of 2008, Bitcoin and its Blockchain technology were presented as a peer- to-peer digital currency system. As stated by [9], Blockchain technology has continually offered notable uses, notably in the financial industry. "The core problematic with conformist currencies is all the trust that is necessary to make it function," said Bitcoin's creator, Satoshi Nakamoto, in a forum post [4]. Because it eliminates digital currency's double-spending issue, blockchain technology will serve as a trusted guarantee for all transactions [36].

The Cryptocurrencies market has been active since 2018, despite the bursting of several booms. Cryptocurrencies infrastructure continues to be strengthened by major financial firms like Fidelity Investments, ICE, and Nasdaq. There is a strong indication that institutional investors are embracing cryptocurrencies and getting more comfortable with this sort of asset class [21].

There are many different forms of digital money, including Bitcoin, Ethereum, Litecoin, Monero, and a variety of others. According to the findings of [35], the most attractive and useful cryptocurrency is the EOS cryptocurrency, it has the lowest amount of complexity and cost among the digital currencies under consideration, and which enables you to incorporate third-party apps into the system.

Apart from its efficiency and convenience, every technology has its own set of advantages and disadvantages. To give an example, cryptocurrencies provide the same benefits as private electronic money providers [23]. Central banks and banks adopting crypto imply that the unbanked throughout the world now have access to the same blockchain and currency that banks use because crypto is an open system [29].

Another important point is that throughout the history of fiat money, governments have had a strong relationship with the currency. We have seen in the past that when a currency is attached to a country or a government, it is inevitable that the currency loses its worldwide dominance. Indeed, the devaluation of a nation's currency is certain in the event of a war or government failure. Only gold has ever been acknowledged and had value across the world, no matter what state it was in at any given time in history. Cryptocurrencies might be used to supply liquidity and bridge value over the world if central institutions and banks started using them. One of the potential uses of cryptocurrencies is to facilitate frictionless cross-border transactions. Because there would be no central authority to issue the global currency, the system would not be as vulnerable to the failure of a single country as it is now [24]. In addition, many people believe that the price of Bitcoin and other cryptocurrencies is immune to inflation caused by limits or modifications imposed by governments throughout the world [20]. Inflation does not devalue this asset, which makes it a "haven" for investors to save their riches. Moreover, a key benefit that Cryptocurrencies offer over traditional currencies is the ability to conduct peer-to-peer transactions quickly, particularly in international business-to-business circumstances.

Cryptocurrency, on the other hand, has the disadvantage of not having a centralized administration that is responsible for resolving any issues that arise in all transactions; in addition, money laundering crimes are frequently committed, which poses a challenge for all those attempting to use a cryptocurrency and blockchain technology in the globalized world [11]. Also, the lack of cryptocurrency confidence could stem from the fact that digital files may be duplicated infinitely. This lack of trust is exacerbated by the fact that digital money can be spent several times, a condition known as "double- spending" [12]. Besides, [10] stressed that with the increasing usage of cryptocurrency, the security of these systems has become a significant source of worry. The security dangers and hazards that blockchain systems face, as well as the tools and procedures that can be used to avoid and deal with them in a more sophisticated way, are considered a major concern when dealing with cryptocurrencies [9]. For instance, digital currency, Etherium, was recently hacked for 50 million dollars and experienced a similar sort of loss [40].

Despite Bitcoin's immense potential, the Cryptocurrencies market is very volatile, and many purchases of the currency have been viewed as speculative at best. Cryptocurrencies' daily price amplitude is up to 10 times more than that of the money market daily. Bitcoin's value plummeted by 65 percent in a single month following the sell-off of major cryptocurrencies that began in January 2018 [42]. Also, currently, the legal status of Bitcoin exchanges in the United States is unclear [41]. Because they are digital assets, cryptocurrencies are taxed as capital gains by the Internal Revenue Service. However, more and more businesses in the United States begin to accept cryptocurrencies as a means of payment for products and services. eBay, Expedia, Uber, PayPal, Microsoft, and Expedia are just a few well-known companies [5].

Economics has traditionally highlighted the role of money as a medium of trade, in addition to the inherent features that increase the scalability and credibility of money, as well as the fact that money is the most liquid store of value [32]. However, the social institution of money changes in tandem with technological advancement and the introduction of cryptocurrency has the potential to have a significant influence on economic policy [22]. The wide range of distinct crypto coins has shown the technological viability of competing for private currency in a competitive environment. A growing number of central banks are concerned about preserving public sovereignty over the common unit of account, and some are proposing the creation of their own digital fiat money. It is expected that central bank digital currencies (CBDCs) will replace private cryptocurrency after the present phase of vigorous creative experimentation, the design of which will be influenced by purposeful public decisions and policies [8, 40].

Many people believe that crypto money is an illusion and a passing phase unless it becomes a fiat currency, which is unlikely to happen. Fiat currencies are chosen at the national or regional level, and there is no desire for a worldwide currency to be established. Using e-money is almost certain to become more popular if people feel that it is backed by reputable, long-lasting, and strong organizations [14]. While the debate regarding the usage, value, and long-term viability of crypto currencies continues, it is necessary to evaluate how they are being used in today's society, which has an impact on their worth in the long run. Even though they are not directly regulated, there is a compelling argument that wallet and exchange institutions should be subject to extensive regulation, the specifics of which are outside the scope of this study. If, for example, a payment services provider (PSP) desires to conduct business in the United Arab Emirates (UAE), it must submit an application to the Central Bank of the UAE for the necessary license. This is not the case in most countries since payment service providers (PSPs) are not considered money remitters within the meaning of the law. In the United States, for example, a situation arose that warranted an investigation into the operations of money transmitters and payment service providers to bring them under regulatory control. In response, the Financial Crimes Enforcement Network (FinCEN) defined money transmission and identified exemptions therein: Financial Crimes Enforcement Network (FinCEN) announced a final judgment on July 21, 2011, modifying terminology relevant to the money services business (MSB) regulation. "A person wherever located who is engaged in business, whether or not regularly or as an organized or licensed commercial concern, within the United States," according to the definition in the statute.

Cryptocurrency exchangers convert actual money from customers into virtual money, which they then transfer to a retailer who accepts the cryptocurrency. So far, limited studies have explored the opinion of people regarding the future of crypto money [25, 42]. In addition, other researchers have examined whether cryptocurrencies are the future of money [31]. However, this study digs more to discover on a personal level how people perceive the new technology, cryptocurrencies, which had yet to be understood as this market is still at an early stage specially in the emerging markets region [30]. Another important takeaway from this research is that it examined and expounded on people's future trends concerning the adoption of Cryptocurrencies. Therefore, the first hypothesis of the current study is:

Ha1: Cryptocurrency will significantly affect Future Trends.

Another important point is how cryptocurrencies would affect the present financial system and how it changed the existing financial environment. Bonds, equities, and international currencies are all examples of financial products that may be traded in financial markets. Financial markets facilitate the flow of information between people who need capital and those who have the capital to invest.

In today's world, such technology may easily be deemed mainstream, and its widespread use has had a significant influence on the broader financial markets [7]. As a result of technological advances, it has been predicted that money will disassociate itself from finance while remaining firmly rooted in the actual economy and continuing to serve as a method of payment. The handmaiden of finance will, however, most certainly continue to be money for the foreseeable future, since financial advances (also known as "fintech") continue to outnumber those targeted at funding the actual economy. limited studies have explored the opinion of people regarding the future of crypto money and how it would affect the present financial system [6]. Accordingly, the second hypothesis is:

Ha2: Cryptocurrency has a significant impact on Financial World.

3 Methodology

Throughout this study, the cryptocurrency serves as the independent variable with the future trends and financial world represent the dependent variable. The questionnaire was employed as a research instrument in this study to gather data. The questionnaire utilized in this study was a five-point Likert scale as respondents have a variety of options without becoming overwhelmed [37].

This study main aim is to focus on the perception of the future of cryptocurrency and its impact on the financial market. Therefore, according to the prior studies, it is unnecessary to choose a target sample according to their gender or age. Hence, this current study targeted all adults (18+ years old) because they have the capacity to invest or use cryptocurrencies in Bahrain. Following [25], a convenience sampling was the best choice for the current study as the chosen sample did not need to be

Table 1 Cronbach's Alpha (reliability statistics)

Variable	Cronbach's Alpha	No of items
Cryptocurrency (CC)	0.764	5
Financial world (FW)	0.788	5
Future trends (FT)	0.871	5

controlled and have certain representatives. Also, as stated by [15], convenience sampling reduces pressure from time constraints. The sample size should be at least 100 participants before doing the data analysis to increase statistical power [17]. The responses from 300 participants have been arranged and counted for the analysis.

4 Data Analyses

Cronbach's alpha value as shown in the below table are greater than 0.7 meaning that the data is acceptable and can be used for further analysis [2]. The following table indicates the results of reliability analysis by applying Cronbach's alpha test (Table 1).

5 Frequency Analysis

The research questionnaire includes four demographic characteristics such as age, gender, qualification, and the number of years in crypto. As shown in Table 2, the total number of respondents counted for the analysis of the study was 300. The table shows that 56% of the participants were male and 44%. The attribute age of the participants shows that most of the respondents' age is between 26 to 35. Moreover, the qualification information of the respondents shows that the majority has a bachelor's degree followed by master's degree, 66% and 30% respectively. According to the number of years in using crypto, most of the respondents had an experience from 1 to 2 years which means that the adoption of is relatively new in Bahrain.

6 Descriptive Statistics and Correlation

The following Table 3 shows the descriptive statistics of the study variables. The minimum of cryptocurrency is 1.20, the maximum is 5.00, the mean value is 3.4680, and the standard deviation is .78.

Table 2 Frequency analysis of the demographic information

	Demographic characteristic	Frequency	Percentage
Gender	Male	167	56
	Female	133	44
Age	25 or less	29	10
	26 to 35	140	47
	36 to 45	57	19
	46 to 55	39	13
	56 or above	35	11
Qualification	High school	33	11
	Bachelors	198	66
	Masters	59	20
	Doctorate	10	3
Years in crypto	1 to 2 years	135	45
	3 to 4 years	98	33
	5 to 6 years	56	18
	7 or above years	11	4

Table 3 Descriptive statistics

	N	Minimum	Maximum	Mean	Std. deviation
CC	300	1.20	5	3.468	0.787
FW	300	1.20	5	3.404	0.858
FT	300	1.00	5	3.744	0.966

The results of the financial world show a mean value of 3.404, a minimum value of 1.20, a maximum value of 5, and a standard deviation of .858. Furthermore, the results show that the minimum, maximum, mean, and standard deviation of the variable future trends are 1, 5, 3.744, and .966 respectively. Concluding, future trends show the highest value of mean among all other variables.

Next, correlation analysis is a statistical approach that is used to determine the strength of the linear connection between two variables [33]. Table 4 shows that the correlation coefficient of cryptocurrency with the financial world is .581 indicating a positive correlation between them. Moreover, the correlation coefficient between cryptocurrency and future trends is -.410 indicating a negative relationship between them. Furthermore, the correlation coefficient value of .045 between the financial world and future trends indicates that the correlation between them is weakly positive.

Table 4 Pearson correlation analysis

		CC	FW	FT
CC	Pearson correlation	1	0.581**	−0.041**
	Sig. (2-tailed)		0.000	0.689
	N	100	100	100
FW	Pearson correlation	0.581**	1	0.045**
	Sig. (2-tailed)	0.000		0.656
	N	100	100	100
FT	Pearson correlation	−0.041**	0.045**	1
	Sig. (2-tailed)	0.689	0.656	
	N	300	300	300

7 Regression Analysis

In order to determine which factors have an influence on a subject of interest, regression analysis is a trustworthy strategy to use [1]. Table 5 shows the results of regression analysis of cryptocurrency in relationship with future trends. The coefficient value of regression between cryptocurrency and future trends is −0.050 indicating a negative relationship between them, however, according to the p-value (0.689), it is insignificant. This indicates that that cryptocurrency is not significantly affecting future trends. Concluding from the above results and discussion, it is identified that the Ha1 hypothesis is rejected.

Next, the results of regression analysis of the independent variable (cryptocurrency) and the dependent variable (financial world) can be seen in Table 6. The coefficient value of the regression between cryptocurrency and the financial world is 0.633 indicating a strong positive relationship between them. The p-value of 0.000 shows that it is a significant relationship indicating that cryptocurrency has a significant impact on the financial world. The results concluded that the Ha2 hypothesis has been accepted which was assumed that "Cryptocurrency has a significant impact on Financial World".

Table 5 Regression analysis (CC in relationship with FT)

		Unstandardized coefficients		Standardized coefficients		
Model 1		B	Std. error	Beta	T	Sig
	(Constant)	3.917	0.440		8.897	0.000
	CC	−0.050	0.124	−0.041	−0.402	0.689
	R-square: 0.002			P-value: 0.689		

Dependent Variable: FT
Predictors (Constant): CC

Table 6 Regression analysis (CC in relationship with FW)

		Unstandardized coefficients		Standardized coefficients		
Model 1		B	Std. error	Beta	T	Sig
	(Constant)	1.209	0.319		3.795	0.000
	CC	0.633	0.090	0.581	7.063	0.000
	R-square: 0.337			P-value: 0.000		

Dependent Variable: FW
Predictors (Constant): CC

8 Conclusion

In recent years, globalization and technology have become more linked than ever before [38]. With the globe becoming more interconnected, having a global reserve currency that is tied to a single nation seems out of place. Taking into consideration the emergence of Bitcoin, this study investigates the advantages and possible dangers of adopting it as a global currency in the near future specially that no intermediate or government oversight is required for cryptocurrency transactions to be completed safely on the internet. The main goal of creating cryptocurrency was to create an environment in which the value of a currency could not be influenced by the actions of the government or other institutions. Their foundation is built on blockchain technology, which provides a secure and uncheckable method of conducting financial transactions for the first time [40, 43].

Because of the global financial crisis that began in 2008, a number of nations have called for reforms to the international monetary system. For example, the Governor of the People's Bank of China, Zhou Xiaochuan, believes that "the ultimate aim of rebuilding the international monetary system is to establish a global reserve currency that is separated from the sovereign governments" [16]. Therefore, this research was carried out in order to explore the public opinion regarding the influence of cryptocurrency and its adoption on the financial markets as well as what the future holds for cryptocurrency as a currency in Bahrain as an emerging market. The study was conducted to better understand the possible success and usage of cryptocurrency in the emerging market. For the analysis, the research adopts a quantitative research methodology using a questionnaire. A 5-points Likert-scale questionnaire has been designed including questions for each variable. The data was collected from 300 respondents. An online survey has been created through Google and the link has been sent to different people who use cryptocurrency platforms in Bahrain.

Overall, the study results concluded that the adoption of cryptocurrency can improve and develop a great change in the financial world significantly. This supports what several economic professionals believed which is that the financial market will undergo a significant shift after using cryptocurrency [39].

Businesspersons, industrialists, financers all throughout the world benefit from the many incentives provided by cryptocurrency [27]. It has made it simpler for entrepreneurs to expand their businesses into worldwide markets rather than limiting themselves to domestic markets. This has enabled vendors to establish connections and build confidence with customers in marketplaces that were previously unavailable, which has been very beneficial for developing countries [26]. Cryptocurrency has ushered in a new era of economic transactions that are based on technological innovation.

Furthermore, it has been concluded that the effect of cryptocurrency on future trends is insignificant and the relationship is negative. This means that it can negatively but insignificantly affect future trends. The market has attracted a large number of new customers as a consequence of its growth and has improved the efficiency of international commerce. However, even if the market has seen significant growth in recent years, it still has a long way to go before becoming a more commonly accepted form of money and it is perceived to have an insignificant impact on the future trends as shown in the current study results.

Accordingly, cryptocurrencies should be closely observed since they are likely to influence the financial market. Also, many issues are being raised about the long-term viability of this decentralised digital money, as well as the long-term viability of the regularly utilised traditional currency system. Can anything be done to make it even better? It is clear that the systems are far from ideal. A perfect system would be faultless and, at the same time, would support the growth of the economy to have a better impact on the future trends.

References

1. Abdi Y, Garavand AT, Sahamieh RZ (2018) Prediction of strength parameters of sedimentary rocks using artificial neural networks and regression analysis. Arab J Geosci 11(19):1–11
2. Adesta EY, Prabowo HA, Agusman D (2018) Evaluating 8 pillars of Total Productive Maintenance (TPM) implementation and their contribution to manufacturing performance. In: IOP conference series: materials science and engineering, vol 290, no 1. IOP Publishing, p 012024
3. Alansari Y, Al-Sartawi A (2021) IT governance and E-banking in GCC listed banks. Procedia Comput Sci 183:844–848
4. Albrecht C, Duffin KM, Hawkins S, Rocha VMM (2019) The use of cryptocurrencies in the money laundering process. J Money Laund Control 22:210–216
5. Al-Sartawi A (2020) Information technology governance and cybersecurity at the board level. Int J Crit Infrastruct 16(2):150–161
6. Al-Sartawi A, Sanad Z (2019) Institutional ownership and corporate governance: evidence from Bahrain. Afro-Asian J Finan Account 9(1):101–115
7. Al-Sartawi A, Reyad S, Madbouly A (2021) Shariah, presentation and content dimensions of Web 2.0 applications and the firm value of Islamic financial institutions in the GCC countries. J Islamic Mark
8. Auer R, Cornelli G, Frost J (2020) Rise of the central bank digital currencies: drivers, approaches and technologies
9. Bhadoria RS, Agasti V (2019) The paradigms of blockchain technology: myths, facts
10. Chang V, Baudier P, Zhang H, Xu Q, Zhang J, Arami M (2020) How Blockchain can impact financial services–the overview, challenges and recommendations from expert interviewees. Technol Forecast Soc Chang 158:120166

11. Dorofeyev M, Kosov M, Ponkratov V, Masterov A, Karaev A, Vasyunina M (2018) Trends and prospects for the development of blockchain and cryptocurrencies in the digital economy. Eur Res Stud J 21(3):429–445
12. Dörr J, Kowalski O, Nevskiy S (2019) Digitalization and monetary order: problems and prospects of cryptocurrency market regulation. Terra Economicus 17(4):6–22
13. Gipp B, Meuschke N, Gernandt A (2015) Decentralized trusted timestamping using the cryptocurrency bitcoin. arXiv preprint arXiv:1502.04015
14. Goodhart CAE (2000) Can central banking survive the IT revolution?, pp 1–34. London School of Economics, Financial Markets Group
15. Gravetter F, Forzano L (2012) Research methods for the behavioral sciences, 5th edn. Cengage
16. Haihong G (2017) The legacy of bretton woods and reconstruction of the international financial
17. Hair J (2010) Multivariate data analysis: a global perspective, 7th edn. Pearson
18. Hannoon A, Abdalla Y, Al-Sartawi A, Khalid A (2021) Board of directors composition and social media financial disclosure: the case of the United Arab Emirates. In: Al-Sartawi AMAM (eds) The Big data-driven digital economy: artificial and computational intelligence. Studies in computational intelligence, vol 974. Springer, Cham
19. Hannoon A, Al-Sartawi A, Khalid A (2021) Relationship between financial technology and financial performance. In: Al-Sartawi AMAM (eds) The big data-driven digital economy: artificial and computational intelligence. Studies in Computational Intelligence, vol 974. Springer, Cham
20. Howden E (2014) The crypto-currency conundrum: regulating an uncertain future. Emory Int'l L Rev 29:741
21. Kaseke F, Ramroop S, Mhwambi H (2021) A comparison of the stylised facts of bitcoin, ethereum and the JSE stock returns. Afr Finan J 23(2):50–64
22. Knell M (2021) The digital revolution and digitalized network society. Rev Evol Polit Econ 2(1):9–25
23. Lee D, Deng RH (2018) Handbook of blockchain, digital finance, and inclusion: Cryptocurrency, FinTech, InsurTech, and regulation
24. Matthews W, Driver R (2018) The Euro common currency: ties that bind. Bus Manag Rev 9(3):144–151
25. McMorrow J, Esfahani MS (2021) An Exploration into people's perception and intention on using cryptocurrencies. HOLISTICA–J Bus Public Adm 12(2), 109–144
26. Meaning J, Dyson B, Barker J, Clayton E (2018) Broadening narrow money: monetary policy with a central bank digital currency
27. Mikhaylov A (2020) Cryptocurrency market analysis from the open innovation perspective. J Open Innov Technol Mark Complexity 6(4):197
28. Al-Sartawi AMM, Razzaque A, Kamal MM (eds) (2021) Artificial intelligence systems and the internet of things in the digital era. EAMMIS 2021. Lecture Notes in Networks and Systems, vol 239. Springer, Cham
29. Al-Sartawi AMAM (eds) (2022) Artificial intelligence for sustainable finance and sustainable technology. ICGER 2021. Lecture Notes in Networks and Systems, vol 423. Springer, Cham
30. Al-Sartawi AMM, Hussainey K, Razzaque A (2022) The role of artificial intelligence in sustainable finance. J Sustain Finan Invest 1–6
31. Nandal N, Jora MN (2020) Impact of household income on investors attitude towards cryptocurrency. Ann Romanian Soc Cell Biol 449–458
32. Peneder M (2022) Digitization and the evolution of money as a social technology of account. J Evol Econ 32(1):175–203
33. Senthilnathan, S. (2019). Usefulness of correlation analysis. Available at SSRN 3416918.
34. Sigler K (2018) Crypto-jacking: how cyber-criminals are exploiting the crypto-currency boom. Comput Fraud Secur 2018(9):12–14
35. Sanad Z, Al-Sartawi A (2016) Investigating the relationship between corporate governance and internet financial reporting (IFR): evidence from Bahrain Bourse. Jordan J Bus Adm 12(1):239–269

36. Sokolenko L, Ostapenko T, Kubetska O, Portna O, Tran T (2019) Cryptocurrency: economic essence and features of accounting system. In: Reconstructing China's participation in the global order. Brill, pp 163–185
37. Sanad Z, Al-Sartawi A (2021) Financial statements fraud and data mining: a review. In: Al-Sartawi AMM, Razzaque A, Kamal MM (eds) Artificial intelligence systems and the internet of things in the digital era. EAMMIS 2021. Lecture Notes in Networks and Systems, vol 239. Springer, Cham
38. Turner- DM, Lamoureux RE, Stokes J, Litcher-Kelly L, Galipeau N, Yaworsky A, Shields AL (2018) Informing a priori sample size estimation in qualitative concept elicitation interview studies for clinical outcome assessment instrument development. Value Health 21(7):839–842
39. Victor F, Weintraud AM (2021) Detecting and quantifying wash trading on decentralized cryptocurrency exchanges. In: Proceedings of the web conference 2021, pp 23–32
40. Yarovaya L, Matkovskyy R, Jalan A (2021) The effects of a "black swan" event (COVID-19) on herding behavior in cryptocurrency markets. J Int Finan Markets Inst Money 75:101321
41. Yu Y, Li Y, Tian J, Liu J (2018) Blockchain-based solutions to security and privacy issues in the internet of things. IEEE Wirel Commun 25(6):12–18
42. Zhang W, Wang P, Li X, Shen D (2018) Some stylized facts of the cryptocurrency market. Appl Econ 50(55):5950–5965
43. Zhang Y, Chan S, Chu J, Nadarajah S (2019) Stylised facts for high frequency cryptocurrency data. Physica A 513:598–661

Knowledge Sharing Augments Healthcare Sector Performance in Bahrain

Mahmood Saeed Mustafa Alalawi, Shaimaa Mishrif, and Mukhtar Al-Hashimi

Abstract Knowledge sharing consists of several factors that contribute to the development of the performance of an organization. This research aimed to analyze the relationship between the six factors of knowledge sharing leadership, management practices, system, culture, motivation/commitment, and organization structure with the organizational performance. The literature review section describes the concept of organizational performance in general, explains the relationship between knowledge sharing and the factors which affect organizational performance and explores the relationship between knowledge sharing and Organizational Performance. The study used the questionnaire survey as the tool of quantitative methodology out of 150 questionnaires, only 115 were returned. A random sampling strategy was used for the distribution of the questionnaire. The hypotheses got tested through the application of simple linear regression. SPSS was used for testing the hypotheses.

Keywords Leadership · Management practices · System · Culture · Motivation commitment · Organization structure

1 Introduction

Knowledge sharing is defined transferring of skills and abilities among the employees as it determines the success and profitability of the organization through the passing of the knowledge throughout the employees working within the organization [42]. According to [42] Knowledge sharing can generally explain be defined as the process of making knowledge available to others within an organization [23, 24, 26, 27, 43]. The process of knowledge sharing assist managers in keeping the flow of knowledge

S. Mishrif
Ministry of Health, Sanabis, Bahrain

M. S. M. Alalawi (✉) · M. Al-Hashimi
College of Business and Finance, Ahlia University, Manama, Bahrain
e-mail: mahmood.alawi@gmail.com

A. M. A. Musleh Al-Sartawi et al. (eds.), *From the Internet of Things to the Internet of Ideas: The Role of Artificial Intelligence*, Lecture Notes in Networks and Systems 557,
https://doi.org/10.1007/978-3-031-17746-0_59

throughout an organization and assimilating it into practical applications. Knowledge is one of the most vital elements in an organization, as it has a vigorous relationship with performance and success [31, 42]. This exchange may be focused or unfocused, but it usually does not have a clear a priori objective. This definition provides a clear view of knowledge sharing within an organization [29, 42, 43]. Practically, Knowledge sharing can be described as the transfer of skills and abilities between employees. It is one of the most effective tools for increasing the performance of an organization and attaining competitive advantage [11, 43, 45]. It is widely believed that encouraging employees to work collaboratively and to share their different perspectives and knowledge is statuary to creating a knowledge sharing culture within an organization. It is necessary to encourage employees to work collaboratively to share the knowledge gained from different perspectives [54]. Implementing a knowledge-sharing culture among employees will help an organization meet its goals and objectives [12, 30, 31, 59]. Recently, many organizations have made immense investments in developing knowledge-sharing initiatives, specifically post the assessment of the numerous benefits that can be derived from the encouragement of the knowledge sharing concept within an organization [46–51, 57]. The capitalization and exploitation of resources based on knowledge can also be made possible by sharing knowledge among employees in all departments and throughout the organization [13, 14, 53, 57, 59]. Knowledge exchange occurs when employees provide knowledge to others yet search for knowledge from others [6, 8, 14]. Knowledge transfer comprises sharing the knowledge obtained from the source and applying and acquiring knowledge from the recipient [3, 27]. This description of the various aspects of knowledge sharing shows that knowledge sharing carries immense importance for the path of improvement in organizational performance, along with its development and the improvement of other operational activities and related functions [14, 23, 25].

According to [11], knowledge sharing can get defined as the transmission or distribution of individual knowledge. In addition, the employees with different ideas and experiences of the organization create new ideas and knowledge through the sharing and communicating that knowledge. Their study also noted that there are two different means of transferring knowledge within an organization transferring knowledge between the employees and transferring knowledge with the help of written documents. [8, 24] found that knowledge sharing provides the opportunity to address organizational needs and for competitive advantage. [7, 28] described knowledge sharing as a culture of social interaction involving exchanging employee knowledge, experiences, and skills through a whole department or organization. Relatively, in Bahraini companies, it was found that the knowledge sharing system was introduced in coherence with the reward system, leadership, communication, and trust among the organization. All these factors contribute to organizational culture, which is also one of the major determinants of the introduction of knowledge sharing [2, 54, 58].

It has been earlier studied that the system of knowledge sharing among organizations has become one of the major pre-requisite. In this context, it is also necessary to encourage knowledge sharing among the employees of the healthcare sectors of

Bahrain as they need to utilize the maximum of the knowledge sharing practices to improve the quality of healthcare for the patients. The major problem addressed in this particular study is thus to analyze the effect of knowledge sharing on organizational performance in the health sector of Bahrain by encouraging knowledge sharing among employees working within an organization so that they can improve their healthcare services to be given to the patients [34–37]. This study also tries to find out the relationship between the various factors of knowledge sharing, namely leadership, management practices, organization structure, system, communication or motivation, and the organization's culture upon the performance of the health care sectors of Bahrain [26, 32]. Until now, it has been understood that knowledge sharing makes a huge contribution towards improving organizational performance, but specific approaches to the implementation of knowledge sharing, especially among the health workers or within the environment of the health sector, are a problem. Due to an inability to identify accurate knowledge sharing methods, the health sector has been unable to formulate strategies for the stimulation of knowledge sharing for the specific purpose of organizational performance [29, 30].

2 Literature Review

The performance of an organization can be determined by factors such as the productivity of employees, increased volumes of sales and the accomplishment of organizational goals and objectives before a stipulated period [16]. As per Forehand & Madison (2000), health care centres are one of the most critical places where the services are to be rendered with adequate effectiveness and efficiency. There should be high performance in the provision of healthcare because it should focus more on the specialized performance through which the patients can heal themselves in the most attractive manner. The healthcare sectors can increase some of their strategies for ensuring better performance within the organization such as bringing innovations in the health care service, introducing new and advanced technologies in the services, implementing new rules for the encouragement of knowledge sharing and management system and arranging for advanced training and development programs for all the medical staff within the hospital [22–25, 52].

According to [16], major findings were that if an organization is involved in a struggle to make their employees highly productive and motivated in their work, they can bring in a high level of performance and profitability situation for a similar organization. If employees are provided with effective training and development programmes, or if an organization fulfils and understands all their needs and demands, there might be a feeling that the organization is taking care of them and respecting their needs and demands. In return, employees will also try to make their best efforts for the organization, which will again be useful in meeting organizational goals and objectives and achieving the highest point of success and performance [38]. The increasing productivity of employees can ensure an increase in the productivity of an organization [34]. Organizational performance can be measured with the use of indicators such as effectiveness, efficiency, relevance for stakeholders and financial stability. Forehand & Madison (2000) According to health care centres are one of the most critical places where the services are to be rendered with adequate effectiveness and efficiency. There should be high performance in the provision of healthcare because it should focus more on the specialized performance through which the patients can heal themselves in the most attractive manner. The healthcare sectors can increase some of their strategies for ensuring better performance within the organization such as bringing innovations in the health care service, introducing new and advanced technologies in the services, implementing new rules for the encouragement of knowledge sharing and management system and arranging for advanced training and development programs for all the medical staff within the hospital [23, 25].

These four indicators can be explained. First effectiveness can be determined in terms of its potential to meet or achieve its mission, vision, and strategy. Second, efficiency: The ability of an organization to manage its resources, such as information, physical, human, and financial resources. Third, relevance to stakeholders: The degree to which the stakeholders of an organization feel that their organization is relevant to their needs. Customers test the relevance of products or services by purchasing them. Fourth, financial stability: The factor which determines how stable an organization will be in terms of finance, not only for the short term but also for long-term tenure. First, organizational motivation: This factor consists of various elements of an and its employees. Those elements are understanding the employees and the incorporation of the mission of the organization, the connection between the historical background of a company and its present strategy. Second, external environment: The performance level of an organization can be increased by analyzing the external environment completely and making use of similar knowledge along with the set strategy which is followed by the company. Third, organizational capacity: the capacity of sales, operation, marketing, finance, HR, and IT gives rise to organizational capacity to deliver performance [33, 34]. The most important capacity is that of the leadership, which motivates the company to move towards strategic success [49]. [48–50] posit that knowledge sharing is one of the major elements that should be mandatory to be included in the healthcare sector and its services. It has been studied that there must be a proper collaboration between the experts to share knowledge and inform practise so that they can ensure contemporary work within the healthcare sector [5, 49, 51]. Apart from these, it has also been significant that if there is a proper trend of sharing knowledge among the staff, the healthcare can ensure for the high-quality performance along with the superior quality of attention and care to the patients, which can again help them to recover as soon as they can. Thus, in this section of the study, more emphasis will be given to the study of knowledge sharing in relation to the key factors that can raise the performance of the healthcare sectors as well [21, 57]. Knowledge sharing should be encouraged within the majority of organizations due to its benefits and large contribution towards the increase in organizational, as well as individual levels of performance. However, organizations must consider a few major points or key factors that can motivate the implementation or affect knowledge sharing for organizational performance [17, 18].

There is a significantly positive relationship between knowledge sharing practices and organizational performance [56, 57]. This was claimed because organizations tend to increase their performance and efficiency through the implementation of practices that encourage the sharing of knowledge among employees and help them to bring out systems or ideas that can develop productivity, level of performance and profitability as per [14, 17, 19]. It has also been observed that organizations tend to seek the best practices for their employees so that they can come up with new ideas and innovative skill, knowledge and technologies which would be beneficial for organizational performance [10]. The innovative skills of employees could result in an increase in organizational profitability. The potential benefits that the health care sector can achieve through the implementation of knowledge sharing are the facilitation of collaboration, protection of an organizations's intellectual capital, avoidance of a high rate of attrition and increase in the operational efficiencies [10]. Knowledge sharing is a strong tool that enhances organizations so that they improve their performance, along with an improvement in their employees' skills and attitudes, through the exchange and sharing of knowledge among employees of the organization [15, 16, 18].

3 Research Methodology

Various kinds of methodologies have been adopted for the study. The quantitative methodology has been adopted for the study. The primary data, or the quantitative data, has been collected through the distribution of questionnaires. Also presents the other required information, such as the sampling techniques used in the study along with the data collection, data analysis and the other related aspects of the study. The dependent and independent variables that assisted the researcher in framing the hypotheses can be better demonstrated by representing a conceptual framework diagram of the study of knowledge sharing and organizational performance [1]. If employees are convinced by following a knowledge sharing policy, it will also ensure their professional growth and development in their performance [50, 51] (Fig. 1 and Tables 1, 2, 3 and 4).

The results of the statistical analysis and the hypotheses testing results shall be discussed. This chapter discusses the key factors that influence the introduction of knowledge sharing for the purpose of increasing the efficiency of organizational performance in the health sectors of Bahrain. This study has also tried to explain the current research findings considering the previous studies done by the other researchers.

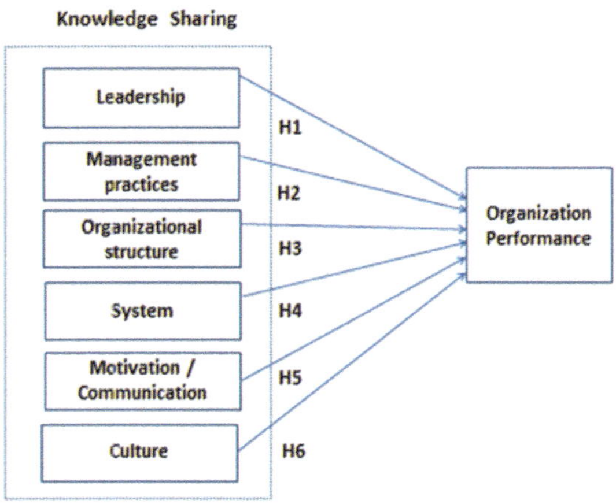

Fig. 1 Research model

Table 1 Demographic analysis

Gender	Frequency	Percentage (%)
Female	82	71.30
Male	33	28.70
Total	**115**	100
Age	Frequency	Percentage (%)
21–30 years	47	40.87
31–40 years	39	33.91
41–50 years	14	12.17
50 years and above	12	10.43
18–20 years	3	2.61
Total	**115**	**100**

Table 2 Minimum, maximum and standard deviation of responses

Construct	N	Min	Max	Mean	SD
Leadership	115	1.00	5.00	3.78	0.85
Management practices	115	1.00	5.00	3.65	0.93
Organization structure	115	1.00	5.00	3.62	0.91
Systems	115	1.00	5.00	3.61	0.89
Communication/Motivation	115	1.00	5.00	3.52	0.89
Culture prevalent	115	1.00	5.00	3.53	0.95
Organizational performance	115	1.00	5.00	3.57	0.89

Table 3 Simple linear regression

Model F	P-value (F)	Adjusted R^2	β	P-value (β)
69.47	0.00	0.381	0.647	0.00
86.90	0.00	0.435	0.633	0.00
86.90	0.00	0.435	0.633	0.00
113.27	0.00	0.501	0.713	0.00
69.22	0.00	0.380	0.617	0.00
69.22	0.00	0.380	0.617	0.00

Table 4 Summary (test of hypotheses)

Hypothesis (H)	Hypotheses	Test results
H1	There is an association between organizational performance and leadership	Accepted
H2	There is an association between organizational performance and its management practices	Accepted
H3	There is an association between organizational performance and organizational structure	Accepted
H4	There is an association between organizational performance and the several systems within an organization	Accepted
H5	There is an association between the communication or motivation of employees and organizational performance	Accepted
H6	There is an association between organizational performance and the culture prevalent within an organization	Accepted

4 Conclusion

Knowledge is one of the essential elements of an organization as it possesses a strong relationship with success and improved performance. Knowledge could be seen as the transfer of skills and abilities between employees, as it determines the success and profitability of an organization through passing the knowledge to the employees who work within the organization. It is also one of the most effective tools for increasing the performance of an organization and attaining competitive advantage. Knowledge has always been considered one of the most critical organizational resources, which aids organizations in achieving sustainable competitive advantage in a dynamic environment with immense competition [9, 15]. However, a common issue has also been noticed where organizations have been focusing more on the provision of staffing and training practices that emphasize recruiting employees with adequate knowledge, experience, skills, and competencies [55]. In this study, it was discovered that the major factors of knowledge sharing that affect the performance of organizations were leadership, management practices, organization structure, culture, motivation or commitment and system. All the hypotheses' statements were found to be positive.

Thus, it denotes that knowledge sharing can be successfully implemented or encouraged within organizations to raise their performance level by considering the above bespoke factors. It was found in this study that the major factors of knowledge sharing affecting the performance of organizations were leadership, management practices, organization structure, culture, motivation or commitment and system. These factors encourage the developed performance of the information and identify areas that need improvement for the smooth implementation of knowledge sharing within the health sector. Recommendations that can be provided for employees, managers and the owners of organizations are as follows: First, organizations must ensure formal knowledge capture and acquisition practices among all the healthcare sectors, such as expert systems, brainstorming, knowledge bases and post-action reviews. Second, they must also encourage communities of practice, succession planning, mentorship, coaching and orientation to ensure the identification of operational knowledge and sharing of available knowledge within the healthcare sectors of Bahrain. Third, organizations must introduce new knowledge-sharing policies, which should improve effective knowledge management to raise their performance level. Fourth, all the healthcare sectors must also ensure that their policies and practices are accessible to all staff members. Finally, management must introduce funds for knowledge sharing within organizations so that knowledge management initiatives can be supported and encouraged to improve organizational performance.

References

1. Akram F, Bokhari R (2011) The role of knowledge sharing on individual performance, considering the factor of motivation-the conceptual framework. Int J Multidisc Sci Eng 2(9)
2. Al-Shammari M (2014) The relationship between organizational culture and knowledge sharing in a GCC company. Int J Bus Manage Rev 2(6)
3. Azadehdel MR (2013) The relationship between knowledge sharing, innovation, and performance of manufacturing and trading companies in Guilan Province. IJCRB 5(5)
4. Becker B, Gerhart B (1996) The impact of human resource management on organizational performance: progress and prospects. Acad Manage J 39(4)
5. Bordoloi P, Islam N (2012) Knowledge management practices and healthcare delivery: a contingency framework. Electron J Knowl Manage 10(1)
6. Bradshaw R, Chebbi M, Oztel H (2015) Leadership and knowledge sharing. Asian J Bus Res
7. Cabrera EF, Cabrera A (2012) Fostering knowledge sharing through people management practices. Arizona State University
8. Carmeli A, et al (2013) Leadership, creative problem-solving capacity, and creative performance: the importance of knowledge sharing. Wiley Periodicals, Inc. 52(1)
9. Carton RB, Hofer CW (2006) Measuring organizational performance: metrics for entrepreneurship and strategic management research. Edward Elgar Publishing
10. Chen ET (2013) Knowledge management implementation in the healthcare industry. Role Knowl Manage
11. Chiu YW, Chien YC (2012) The effect of knowledge sharing on organizational performance: organizational citizenship behaviour and knowledge absorption as mediators. Int J Inf Technol Bus Manage 36(1)
12. Creswell JW (2014) Research design. Sage Publication, United States of America

13. D'Andrad RF et al (2016) Contributions of junior companies in an emerging market as collaborative partners for sharing knowledge and innovation 5(12)
14. Danish RQ et al (2013) Effect of knowledge sharing, participative decision making and transformational leadership on organizational performance. World Appl Sci J 24(10)
15. Darwish S, Ahmed U, Bunagan VD, Sison SM (2021) Women Entrepreneurs (WE): reshaping and sustaining the Gulf Cooperation Council (GCC) economy. In: Musleh Al-Sartawi A, Razzaque A, Kamal MM (eds) Artificial intelligence systems and the internet of things in the digital era, EAMMIS 2021. Lecture notes in networks and systems, vol 239, pp 455–466. Springer, Cham. https://doi.org/10.1007/978-3-030-77246-8_42
16. Dobre OI (2013) Employee motivation and organizational performance. Rev Appl Socio-Econ Res 5(1)
17. Eih Y, Shih (2011) Why individual share knowledge in the workplace? - the effects of motivation and context
18. Gavrea C et al (2011) Determinants of organizational performance: the case of Romania. Manage Market Challenges Knowl Soc 6(2)
19. Hatala JP, Lutta JG (2011) Managing information sharing within an organizational setting: a social network perspective. Perform Improv Q 21(4)
20. Hendriks PHJ (2004) Assessing the role of culture in knowledge sharing. University of Nijmegen the Netherlands
21. Islam Z et al (2011) Organizational culture and knowledge sharing: empirical evidence from service organizations. Afr J Bus Manage 5(14)
22. Karamat AU (2013) Impact of leadership on organizational performance. University of Applied Sciences
23. Karimi MJ (2011) The impact of communication on employee motivation in the Kenyan public service: a case study of postal corporation of Kenya. http://erepository.uonbi.ac.ke/xmlui/bitstream/handle/11295/Miriti_The%20impact%20of%20communication%20on%20employee%20motivation.pdf?sequence=3&isAllowed=y. Accessed 03 Nov 2016
24. Karolak M, Razzaque A, Al-Sartawi A (2021) E-services and M-services using IoT: an assessment of the Kingdom of Bahrain. In: Musleh Al-Sartawi AM, Razzaque A, Kamal MM (eds) Artificial intelligence systems and the internet of things in the digital era, EAMMIS 2021. Lecture notes in networks and systems, vol 239, pp 523–533. Springer, Cham
25. Kathiravelu SR et al (2014) Why organisational culture drives knowledge sharing? Elsevier
26. Khalid A, Ahmed M (2015) Impact of knowledge sharing on organizational learning: Moderating effect of organizational leadership. Asian J Manage Res 5(3)
27. KordValouei H et al (2014) A survey on the relationship between knowledge sharing and organizational performance (case study: Bandarabas Municipality). ISSN, vol 3, no 6
28. Lin HF (2007) Knowledge sharing and firm innovation capability: an empirical study. Int J Manpower 28(3/4)
29. Maduenyi S et al (2015) Impact of organizational structure on organizational performance. http://eprints.covenantuniversity.edu.ng/5296/1/Paper%20140.pdf. Accessed 03 Nov 2016
30. McKeen JD (2006) Knowledge management and organizational performance: an exploratory survey. IEEE
31. Mehrabi S et al (2013) Studying the relationship between the process of knowledge sharing and organizational agility among personnel of agriculture- Jahad Organization in ShareKord. ISSN, vol 3, no 5
32. Melo MFD et al (2013) Leadership and knowledge sharing: a case study. J Econ Bus Manage 1(2)
33. Mitchell H (2002) Strategic worth of human resources: driving organizational performance. Universalia
34. Momeni S et al (2013) Relationship between organizational structure and knowledge sharing. Interdisc J Contemp Res Bus 5(2)
35. Mustafa YMBY (2012) Factors influencing organizational performance in metro specialist hospital, Sungai Petani. Universiti Utara Malaysia, Kedah Darul Aman

36. Ngah R, Ibrahim AR (2012) The effect of knowledge sharing on organizational performance in small and medium enterprises. University of Kuala Lampur
37. Ngah R, Ibrahim AR (2010) The effect of knowledge sharing on organizational performance in small and medium enterprises. In: Knowledge management international conference
38. Oborn E, Barett M, Racko G (2010) Knowledge translation in healthcare: a review of the literature. Cambridge Judge Business School
39. Oye ND et al (2011) Knowledge sharing in workplace: motivators and demotivators. Int J Manag Inf Technol 3(4)
40. Ozlen MK (2015) Knowledge sharing behaviour of Bosnian enterprises. Account Manage Inf Syst 14(3)
41. Pai FY, Chang HF (2013) The effects of knowledge sharing and absorption on organizational innovation performance – a dynamic capabilities perspective. Interdisc J Inf Knowl Manage 8
42. Paulin D, Suneson K (2012) Knowledge transfer, knowledge sharing and knowledge barriers – three blurry terms in KM. Electron J Knowl Manage 10(1)
43. Potocki KA, Brocato RC (2011) A system of management for organizational improvement. Johns Hopkins Apl Tech Digest 16(4)
44. Lorne FT (2021) Digital social contracts with ai robots: some implications for Amazon.Com. In: Al-Sartawi A, Razzaque A, Kamal MM (eds) Artificial intelligence systems and the internet of things in the digital era -proceedings of EAMMIS 2021, pp 78–89. Springer, Istanbul
45. Razzaque A (2020) M-Learning improves knowledge sharing over e-learning platforms to build higher education students' social capital. SAGE Open 10(2):1–9. https://journals.sagepub.com/doi/10.1177/2158244020926575
46. Razzaque A, Jalal-Karim A (2010) The influence of knowledge management on EHR to improve the quality of health care services. In: 7th European, mediterranean and middle eastern conference on information systems, EMCIS 2010, Abu-Dhabi, 12 April 2010–13 April 2010. Abu-Dhabi: proceedings of the European, mediterranean and middle eastern conference on information systems: global information systems challenges in management, EMCIS
47. Razzaque A, Al-Sartawi AMM, Karolak M (2021) Rethinking tourism entrepreneurship amidst COVID-19 through the lens of artificial intelligence and big data analytics. In: Big data for entrepreneurship and sustainable development, pp 175–188. CRC Press
48. Razzaque A, Karolak M (2011) Building a knowledge management system for the e-health knowledge society. J Econ Develop Manage IT Finance Market 2(2):23–40
49. Razzaque A, Eldabi T, Jalal-Karim A (2013) Physician virtual community and medical decision-making: mediating role of knowledge sharing. J Enterp Inf Manag 26(5):500–515
50. Rasula J et al (2012) The impact of knowledge management on organizational performance. Econ Bus Rev 14(2)
51. Shaheen A et al (2013) Employees training and organizational performance: mediation by employees performance. Interdisc J Contemp Res Bus 5(4)
52. Sakhnini Z, Elali W, Al-Sartawi A (2022) Behavioral and non-behavioral factors and the level of adapting and implementing fintech and E-banking in Bahrain: suggested model. In: Musleh Al-Sartawi AMA (ed) Artificial intelligence for sustainable finance and sustainable technology, ICGER 2021. Lecture notes in networks and systems, vol 423. Springer, Cham
53. Tan CL, Nasurdin AM (2011) Human resource management practices and organizational innovation: assessing the mediating role of knowledge management effectiveness. Electron J Knowl Manage 9(2)
54. Waheed H et al (2013) Mediating role of knowledge sharing: organizational Performance for competitive advantage and innovation. Afr J Bus Manage 7(7)
55. Wang S, Noe RA (2010) Knowledge sharing: a review and directions for future research. Hum Resour Manage Rev 20
56. Yu BT, Ulukok E (2021) Google's use of artificial intelligence in a two-sided advertising market causing trust loss, and data monopolization. In: Al-Sartawi A, Razzaque A, Kamal MM (eds) Artificial intelligence systems and the internet of things in the digital era-proceedings of EAMMIS 2021, pp 383–397. Springer, Istanbul

57. Zahari ASM et al (2014) The influence of knowledge sharing on organizational performance among insurance companies in Malaysia. J Appl Environ Biol Sci 4(55)
58. Zin IN (2013) Knowledge sharing approaches in Malaysian construction organizations for improved performance. University of Salford

Printed by Printforce, the Netherlands